ENVIRONMENTAL
GEOLOGY
Today

Robert L. McConnell
University of Mary Washington

Daniel C. Abel
Coastal Carolina University

JONES & BARTLETT
LEARNING

World Headquarters
Jones & Bartlett Learning
5 Wall Street
Burlington, MA 01803
978-443-5000
info@jblearning.com
www.jblearning.com

Jones & Bartlett Learning books and products are available through most bookstores and online booksellers. To contact Jones & Bartlett Learning directly, call 800-832-0034, fax 978-443-8000, or visit our website, www.jblearning.com.

Substantial discounts on bulk quantities of Jones & Bartlett Learning publications are available to corporations, professional associations, and other qualified organizations. For details and specific discount information, contact the special sales department at Jones & Bartlett Learning via the above contact information or send an email to specialsales@jblearning.com.

Production Credits

Chief Executive Officer: Ty Field
President: James Homer
Chief Product Officer: Eduardo Moura
Executive Publisher: William Brottmiller
Publisher: Cathy L. Esperti
Senior Acquisitions Editor: Erin O'Connor
Editorial Assistant: Rachel Isaacs
Production Editor: Leah Corrigan
Marketing Manager: Lindsay White

Manufacturing and Inventory Control Supervisor: Amy Bacus
Composition: Circle Graphics, Inc.
Cover Design: Kristin E. Parker
Photo Research and Permissions Coordinator: Lauren Miller
Rights and Photo Research Assistant: Ashley Dos Santos
Cover Image: (top) © Ingram Publishing/Thinkstock; (bottom) Courtesy of Mark Reid/USGS
Printing and Binding: Courier Companies
Cover Printing: Courier Companies

To order this product, use ISBN: 978-1-4496-8487-7

Library of Congress Cataloging-in-Publication Data
McConnell, Robert L.
    Environmental geology today / Robert L. McConnell, Daniel C. Abel. — 1st ed.
        p. cm.
    ISBN 978-0-7637-6445-6
    1. Environmental geology. I. Abel, Daniel C. II. Title.
    QE38.M39 2014
    550—dc23
                                                2013015261

6048

Printed in the United States of America
17 16 15 14 13   10 9 8 7 6 5 4 3 2 1

BRIEF CONTENTS

TABLE OF CONTENTS

PREFACE

■ TO THE INSTRUCTOR

We are immersed in predictions and forecasts. New York Yankee great Yogi Berra is credited with saying, "It's tough to make predictions, especially about the future," and this is especially true about **energy and natural resources**, one of the topics covered in our new book, *Environmental Geology Today*. For example, ten years ago, academics and government forecasters confidently predicted, based on a set of given **assumptions**, that the U.S. would remain dependent on coal for over half of our electricity generation well into the future. Moreover, the same forecasts were asserting that the country would become ever more dependent on foreign sources of oil, mostly from countries unfriendly to us, as **recoverable oil reserves** declined. However, rapid development of new technologies involving **hydrofracturing and directional drilling**, coupled with increasing energy efficiency, have put the U.S. on the road to self-sufficiency in oil and the development of an export capability in natural gas.

Gas is rapidly supplanting coal as a source of electricity, with complex but likely positive impacts on the U.S. economy and environment. Other new sources of energy have had equally strong impacts on a global scale, and not all have been positive; prior to 2011, and despite the 1986 Chernobyl nuclear disaster in the Ukraine, many countries of the European Union were set to obtain a major portion of their electricity from nuclear power. The 2011 Fukushima disaster in Japan altered that trend, such that the German government announced plans to close all of its nuclear power plants before mid-century. Thus, disasters or "accidents" continue to affect human history, not to mention national economies. Another example: the 1989 *Exxon Valdez* tanker disaster prompted the U.S. to require all oil tankers to be double-hulled.

Geohazards represent another important segment of this book. Geohazards include earthquakes and associated tsunami, volcanic eruptions, floods, and mass movement events such as landslides and tropical storms, to name a few. Geohazards have been impacting human history since at least the 14th century BCE, when the Minoan civilization, centered on Crete, was destroyed, probably by cataclysmic eruptions on Santorini, a volcano in the eastern Mediterranean. How likely are such events to recur, and what should be our response to the perceived threat? How likely is the U.S. to experience another disastrous earthquake like the 1994 Northridge quake in Southern California? What impacts do forecasting, planning, and zoning have on the damage done by natural hazards? What about tsunami? In 2011 Fukushima, Japan's nuclear reactors were destroyed by a gigantic series of tsunami. They had prepared for a 5-meter high event, but the tsunami crested at over *15 meters*. How can we predict and plan for such events?

As for **climate change**, in 2012, the U.S. federal government proposed regulating the carbon emissions from major emitters like power plants and began to implement strict new efficiency measures for the transport sector. Moreover, global installation of wind energy "farms," both onshore and offshore, could significantly alter how climate change impacts humanity and the biosphere.

Considering the relentless—though slowing—**human population growth**, many assert that this is the ultimate environmental issue. Berra is also credited with saying of a favorite restaurant, "Nobody goes there anymore. It's too crowded." Crowding notwithstanding, everybody is going "there": coastal cities already contain more than half the global human population, and these numbers are forecasted to increase significantly in the decades ahead. How will this **coastal** population growth be affected by geohazards? What challenges and opportunities does coastal urban growth present?

Another aspect of environmental geology is human interaction with the **water cycle**. Today, the greatest demand for water is from **agriculture**. As developing countries expand economically and industrial production accelerates, demand for clean water will increase. This, coupled with the growth in global demand from a richer human population, could, and probably will, put increasing stress on our remaining sources of fresh water. At the same time, in countries like China, for example, many freshwater rivers are too polluted for industries, much less municipal use.

Finally, basic to agriculture are **soils**. Soils are critical to a healthy planet, but what are soils? They store **carbon**, are the basis for agriculture, provide nutrients to plants, and harbor a teeming biota that scientists are only beginning to understand.

Pedagogical Approach

In this book, we have striven to inculcate an appreciation for, and the ability to use, many aspects and principles of **critical thinking**. Critical thinking, in turn, depends in part on the ability to manipulate units of the metric system, a familiarity with basic math, and understanding of the **scientific method**.

Indeed, the importance of critical thinking led us to write our first textbook when we were colleagues at the University of Mary Washington. We grew impatient with a teaching style centered on the faculty member as "lecturer" and expert and the student as "scribe" and novice. We felt, and feel, that such an approach encourages students to be passive rather than active learners and leads to an unhealthy dependency on the faculty person as "expert."

We have also found that many students are afraid of math, are rusty in its use, or have been superficially trained in arcane fields of calculus. This lack of math skills often leaves students unprepared to deal with the complexity of environmental geology issues. Moreover, we are continually surprised to discover how many bright students can't do three things: understand and USE the units of the metric system, use scientific notation, and critically evaluate complex environmental issues.

Most of a student's discomfort with math is generally founded in frustration. For example, making one error in a series of calculations can render the whole effort useless. We believe that, in the absence of a learning disability, solving most math problems requires no special aptitude, only clear sequential instructions and attention to detail.

One of our major objectives is thus to help develop math literacy (**numeracy**) among today's students. We understand that many students have some "math anxiety," so we have included a section entitled "Population Math Made Easy," in which we use a step-by-step method to take students through examples of some calculations. We even show sample keystrokes involved in using common calculators. We believe this method will gradually build the student's confidence enough to trust his or her own efforts. Familiarity with basic math is one of the important skills necessary for fully understanding environmental geology, and without these skills, the student's only option is to make choices on the basis of which "expert" is most "believable." Such skills involve the ability to manipulate large numbers using scientific notation and exponents and occasionally the ability to use compound growth equations containing natural logs and so on. We show you how.

It is our goal that this book be provocative, factually accurate, and up-to-date. *Environmental Geology Today* is meant to be the basis for an issues-oriented, introductory, seminar, or upper-level course. In addition, our text introduces concepts such as sustainability and development and considers the impact of global trade where appropriate.

Becoming educated is much more than simply acquiring skills, however. We therefore have two additional objectives: to provide students with the knowledge and intellectual standards necessary to apply critical thinking to environmental studies and sustainability, and to foster their ability to critically evaluate issues. You can use projects to develop students' critical thinking skills in a deliberate and structured way. By their nature, they require students to integrate topics from across sub-disciplines to measure, analyze, and evaluate each issue using the discipline and method of a scientist. As such, *Environmental Geology Today* is as much an *interactive workbook* as a traditional textbook. We expect students to have access to standard references in environmental, physical, and natural sciences and to have access to and know how to use the Internet. Indeed, chapters contain URLs to websites for up-to-the-minute information.

Another key feature that we have included in the text is the **Case Studies** incorporated throughout each chapter. These case studies are drawn from current events and hot topics, and are designed to further engage the students in the material and to encourage the application of students' critical thinking skills.

We also trust that you, as an expert in your field and with your own perspectives, will supplement the information in this book with your own comments, introductions, and critical comments on the questions we ask your students. The chapters are user-friendly and based on solid science. We define key terms and keep jargon to a minimum. When important terms are introduced they are bolded and defined where necessary. We introduce and explain a few key mathematical formulas using a step-by-step, nonthreatening approach that we hope you will appreciate. There is also a comprehensive glossary and an index to topics in addition to sample test questions. Each chapter has a detailed **Summary**.

To encourage rigorous critical thinking, there are assessment questions integrated into each issue. Critical thinking implies using a set of criteria and standards by which the reasoner constantly assesses his or her thinking. Self-assessment and reflection are at the core of critical thinking.

Additional Suggestions for the Classroom

We devote a detailed section to "critical thinking" in the first section of this book. We believe *it is vitally important that you encourage your students to read this material, as they will be asked to apply these standards and criteria throughout the book.*

Many questions have been designed to be provocative. This may lead you or your students to perceive a "bias" in the wording of some of the questions. Although we have made the content as factual as possible, we do have strong convictions about these issues. *Convictions, however, are not biases.* Based on the scientific method, our views as scientists are subject to change as evidence supporting our convictions changes. As such, we are constantly testing the *assumptions* that we use when approaching complex environmental issues. Indeed, this aspect of our approach can be turned to a major advantage. Ask your students to look for examples of bias in the questions, and then discuss with them the difference in science between "bias" and "conviction." No doubt it will prove a fruitful activity and may lead students into research (perhaps to "prove us wrong"), which is the essence of progress in the search for scientific truth.

Additionally, we have found that many students who would be reluctant to participate in a classroom discussion will willingly contribute in the relative anonymity of an electronic discussion group. Students can debate the issues and grade each other, and we encourage you to try to have students debate the questions we raise, time permitting. While debating and the expression of opinion are important, any opinions on science must be informed by a thorough background in the subject being debated.

Work done on these questions can take the place of one or more exams or quizzes, freeing you for other activities and providing students with a less threatening way to earn class credit than an all-exam format. We believe that students retain more information from work on projects and reports than from cramming for tests. At the end of the term, you might pass out a study sheet detailing specifically what material or questions from the Issues will be covered on the final exam, if you choose to test them on their work in this manner.

Please contact us at rmcconne@umw.edu and dabel@coastal.edu with questions and comments.

■ TO THE STUDENT

As environmental scientists, we care deeply about environmental issues, and we feel that you, as a responsible citizen who will have to make increasingly difficult choices in the years ahead, need to be concerned about them as well. Here are a few of our reasons:

- A STRONG scientific consensus exists that our growing human population is having a measurable and harmful effect on the composition of the planet's atmosphere. The world's leaders continue a dialogue on how to respond to human-induced global climate destabilization, a result of global warming. Even though we can't yet be certain what the scale of these changes will be, evidence suggests that the impact will, on the whole, be negative, and could be catastrophic for hundreds of millions of people crowded into many of the planet's coastal cities.

- Marine scientists are concerned about the very survival of many marine organisms, including some that form the basis of major world fisheries. In fact, entire ecosystems such as coral reefs may be at risk. The ocean's ability to absorb our waste, toxic and otherwise, is certainly limited, and these limits have likely been exceeded.

- Although we in the United States and a few other areas of the world have made impressive strides in improving or at least slowing the degradation of our air, water, and soil, our relentlessly growing human numbers threaten this progress.

- Growing levels of material consumption in developed countries coupled with less-regulated international commerce (free trade) are placing increased stress on critical ecosystems such as tropical and boreal forests, in turn gravely threatening the planet's species diversity.

- Environmental issues such as water conflicts in the Middle East and the transnational impact of air and water pollution could destabilize international relations and lead to regional conflicts.

- Fossil fuels continue to be the basis of industrial and postindustrial society. Their extraction, transportation, and use impose significant costs on the planet. Much of this cost is externalized (or dumped) onto the environment. As the 2010 Deepwater Horizon Platform incident in the Gulf of Mexico illustrated, addressing environmental issues should involve an inclusion of these costs.

- Geohazards—earthquakes, volcanic eruptions, floods, landslides, etc., cause untold billions of dollars in damage and sometimes kill hundreds of thousands of people, as did the 2004 Sumatran earthquake and tsunami. Although such horrific loss of life hasn't happened in the United States to date, it theoretically could. We need to understand geohazards and how to respond to them.

- Agriculture, and the soils that are essential to it, is the basis of human society. However, rates of soil degradation continue at appalling rates in much of the developing world. How will this affect a human population that is projected to grow to nearly 9 billion by 2040?

We hope you will find this book to be a provocative introduction to a number of these issues and to many others that you may have never even thought about. These are real-life issues, not hypothetical ones, and you need certain basic skills to fully understand them.

- You must be familiar with the units of the metric system.
- You must be able to use a few mathematical formulas to understand some of the topics we cover. We show you how to do this.

- You should develop the habit of rigorously assessing your thinking, and you should apply certain critical thinking skills and techniques when discussing the implications of your calculations.
- You should realize that critical thinking requires practice, dedication, time, and an open mind.

Becoming educated is much more than acquiring skills. Therefore, we have two further objectives: to provide you with the knowledge and intellectual standards necessary to apply critical thinking to environmental studies, and to foster your ability to critically evaluate issues, a skill without which mechanical skills are of limited value. To that end, we have included a section called "Basic Concepts and Tools: Using Math and Critical Thinking," in which we illustrate and describe each criterion for critical thinking, as well as the framework within which these criteria are applied. It is very important that you read this section carefully and do all of the exercises in it *before* you begin the analyses of the issues.

Let's consider an example: *growth*. The word is used in many societal, economic, demographic, and environmental contexts, including growth of the economy, growth of the population, growth of impervious surfaces, growth of food production, and growth of energy use. Assessing the impact of growth requires an understanding of a few simple equations, such as the compound interest equation. You should be able to use it accurately and understand its implications. We will show you how.

As our national and global population grows and changes and our relationships with other nations and peoples evolve, environmental issues will become even more complicated. Domestically, demographic and ethnic changes are becoming more important, which in turn requires an enhanced ability to critically evaluate issues.

We hope you will be challenged by the issues discussed in this text and that you will research them and become an "expert" on the topics yourself. In fact, if we may be allowed a hidden agenda, this is it.

Finally, while you are testing your own thinking and reasoning, test ours as well. Look for examples of "bias" in the "Value-added" questions, and be prepared to support your conclusions and discuss them with your fellow students and with your instructor.

A Note on Conventions Used in This Book

When we express rates, concentrations, and so on—for example, milligrams per liter, people per hectare, tonnes per year—we use all acceptable formats, including "per" (as above), the slash (mg/L, people/hectare, tonnes/year), and at times negative exponents (mg L^{-1}, people hectare^{-1}, tonnes year^{-1}).

A Word about Calculators

A handheld calculator can make analyzing environmental issues easier, or it can be a source of great frustration. It all depends on how carefully you put the information into the calculator and how familiar you are with how to use it. We recommend buying the simplest calculator you can find that will carry out the tasks you need performed. These are typically sold as "scientific calculators." *We also encourage you to take the time to learn how to use this calculator properly.* The calculator should perform the basic data manipulation functions—such as adding, subtracting, multiplying, dividing, determining squares and square roots—and have a single, rather than multiple, memory that is easy to use. Some additional features you should look for include the following:

- Parentheses () keys
- A yx key
- A reciprocal (1/x) key
- Ability to do simple statistics, including means (x) and standard deviations (S)
- An **ex** key
- A **LN** (natural log) key
- An **exp** or **EE** key

■ ANCILLARIES

For Instructors

An *Instructor's Media CD*, compatible with Windows and Macintosh platforms, provides instructors with the following resources:

- The PowerPoint **Image Bank** contains all of the illustrations, photographs, and tables (to which Jones & Bartlett Learning holds the copyright or has permission to reproduce electronically). These images are inserted into PowerPoint slides so that instructors can quickly and easily copy individual images into existing lecture slides.
- The PowerPoint **Lecture Outlines** provides lecture notes and images for each chapter of *Environmental Geology Today*. Instructors with the Microsoft PowerPoint software can customize the outlines, art, and order of presentation.

To receive a copy of the Instructor's Media CD, please contact your sales representative.

The **Test Bank** consists of over 500 exam questions and is available online. Qualified instructors can download these text files from our catalog page; please contact your sales representative for instructions.

For Students

Jones & Bartlett Learning hosts a **Navigate Companion Website** that provides content designed exclusively for *Environmental Geology Today* free with purchase of every new print textbook. The site features an array of study tools, including study quizzes, weblinks for further exploration, an interactive glossary, animated flashcards, and crossword puzzles. To access the site, please visit **go.jblearning.com/ EnvironmentalGeologyCW**.

■ ACKNOWLEDGMENTS

This book could not have succeeded without the aid of numerous friends and colleagues. We would also like to thank our team at Jones & Bartlett Learning for their hard work and understanding as we brought this book to bear—in particular, Erin O'Connor, Rachel Isaacs, Leah Corrigan, Lauren Miller, and Ashley Dos Santos have been patient and helpful at every stage of the process.

Finally, we appreciate the invaluable feedback from the following reviewers, whose comments and suggestions materially improved the manuscript:

Zsuzsanna Balogh-Brunstad, Hartwick College
Mahmood M. Barbooti, Montclair State University
Nathalie Brandes, Montgomery College
Jeff Chaumba, University of Georgia
Marta Clepper, University of Kentucky
Kenneth Galli, Boston College
Thomas Gardner, Trinity University
Jonathan Gilligan, Vanderbilt University
Katherine Grote, University of Wisconsin, Eau Claire
Mohammad Hassan Rezaie-Boroon, California State University, Los Angeles
Ezat Heydari, Jackson State University
Leslie Kanat, Johnson State College
Jonathan Levy, Miami University
Jamie MacDonald, Florida Gulf Coast University
Johnson Olanrewaju, Gannon University
Emma C. Rainforth, Ramapo College
Usha Rao, St. Joseph's University
John B. Ritter, Wittenberg University
John Rockaway, Northern Kentucky University
Dennis Ruez, University of Illinois, Springfield
Cassandra Runyon, College of Charleston
Conrad Shiba, Centre College
Kevin Svitana, Otterbein College
Gina Szablewski, University of Wisconsin, Milwaukee
Dean Whitman, Florida International University
Shelley Whitmeyer, James Madison University

Robert L. McConnell
Daniel C. Abel

JONES & BARTLETT LEARNING TITLES IN PHYSICAL SCIENCE

Astronomy Activity and Laboratory Manual
Alan W. Hirshfeld

Climatology, Third Edition
Robert V. Rohli & Anthony J. Vega

Dynamic Earth: An Introduction to Physical Geology
Eric H. Christiansen

Earth's Evolving Systems: The History of Planet Earth
Ronald Martin

Earth's Natural Resources
John V. Walther

Environmental Oceanography: Topics and Analysis
Daniel C. Abel & Robert L. McConnell

Environmental Science, Ninth Edition
Daniel D. Chiras

Environmental Science: Systems and Solutions, Fifth Edition
Michael L. McKinney, Robert M. Schoch, &
   Logan Yonavjak

Essential Invitation to Oceanography
Paul R. Pinet

Essentials of Geochemistry, Second Edition
John V. Walther

Igneous Petrology, Third Edition
Alexander McBirney

In Quest of the Universe, Seventh Edition
Theo Koupelis

Invitation to Oceanography, Sixth Edition
Paul R. Pinet

Laboratory Manual to accompany Invitation to Oceanography
Karl M. Chauffe & Mark G. Jefferies

Meteorology: Understanding the Atmosphere, Fourth Edition
Steven A. Ackerman & John A. Knox

Principles of Atmospheric Science
John E. Frederick

Principles of Environmental Chemistry, Third Edition
James E. Girard

Restoration Ecology
Sigurdur Greipsson

How to Approach the Issues in This Book

The Scientific Method, Critical Thinking, and Logical Fallacies

© Kheng Guan Toh/ShutterStock, Inc.

CHAPTER OUTLINE

■ SCIENCE AS A WAY OF KNOWING

Science is one way of explaining our universe. Besides science, other ways of knowing include the humanities, belief systems, myths, and math. They are similar in that each has both a body of lore, knowledge, or information, as well as its own processes for discovering truths, answering questions, and solving problems. Scientific thought arose as an alternative to or—more accurately—a rejection of ideas that were accepted because a ruler or religious authority declared their truth without empirical (based on observation or experiment) **evidence**. This shift marked a revolution in our understanding of the natural world. It explained how the world works on the basis of observation and experimentation. Science thus differs from other ways of knowing in one essential way: It is the only form of discovery in which the process is the scientific method.

■ THE SCIENTIFIC METHOD

Students sometimes see the **scientific method** as an intimidating formal series of jargon-filled steps that scientists use to uncover truths or find answers. We have tried to abandon an arcane, philosophical approach and instead identify the scientific method's essence: the formation and rigorous, objective testing of hypotheses.

Steps of the Scientific Method

Typically, the first step of the scientific method is observation, either some phenomenon of nature or observation of an experiment. In this context, observation means more than merely seeing. It uses all of the human senses as well as the vast array of measurement techniques, some of which detect signals that human senses are incapable of receiving. Our knowledge of the Earth's core, mantle, and crust, for example, is based in part on the behavior of seismic waves, only some of which human senses can perceive, as discussed elsewhere in this text. Another example: Accurate measurements of global temperature date back to the mid to late 1800s, but scientists use proxies (substitutes) such as tree rings, corals, ice cores, and lake sediments to reconstruct high-resolution paleoclimate temperatures from at least 500,000 years ago.

Here's an important point: If an event or process cannot be observed, either directly or indirectly, it cannot be explained by science. In the hands of a scientist, or even a curious lay person, observations and experiments lead to hypotheses, which can be educated guesses, carefully crafted explanations, or even questions. Consider the observation made first in the 1970s that Caribbean coral reefs were (and still are) suffering increased frequencies of disease and overgrowth by algae (FIGURE 1-1).

It has long been known that the coral polyps, the minute living animals that form the calcareous foundation of the reef, are very sensitive to environmental factors such as temperature, salinity, light, nutrients, and sediment in the water. Researchers noticed that the onset of the coral decline coincided with increasing aridity and **desertification** in northern Africa (FIGURE 1-2). Now, of course, there could have been some other factor causing the coral decline, because corals are sensitive to a variety of environmental insults ranging from temperature extremes

FIGURE 1-1 Dead coral reef. Coral disease and overgrowth by algae may be due to a variety of human impacts, including:
- Sediment and nutrient pollution from airborne dust and terrestrial runoff
- Overexploitation and damaging fishing practices
- Engineering modification of shorelines
- Global climate change causing coral bleaching, rising sea levels, and potential acidification of shallow marine water threatening the ability of corals to form skeletons

Courtesy of David Burdick/NOAA.

to overfishing. This observation, however, led to a **hypothesis**. Scientists proposed that dust from Africa resulting from drought and a loss of vegetative cover (desertification) was blown across the Atlantic Ocean by prevailing winds (coming from the northeast at those latitudes) and caused the coral problem, either through smothering the coral or by transportation of pathogens that harmed the coral. To formulate their hypothesis, they had to be familiar with the nature of global wind belts.

This story shows a key feature of the scientific method: Explanations can (and frequently must) be changed as new evidence becomes available. A *hypothesis* is thus a proposed explanation of some phenomenon. The word phenomenon is from Greek *phainomenon*, meaning "to appear." It refers to any fact or experience that can be sensed and scientifically described.

After you have devised a hypothesis, the next step is to test it. Testing implies that your hypothesis can be falsified, that is, your hypothesis can be shown to be incorrect. Guesses, explanations, and questions that cannot be tested and falsified—like things that cannot be observed—are not science.

This *testability* requirement also means that science cannot make value judgments because qualities such as goodness or beauty cannot be subjected to scientific tests.

Similarly, a scientist, acting solely as a scientist, does not conclude that something is good or bad, beautiful or ugly, and so forth.

For the coral reef example, tests of the dust hypothesis were, in some cases, consistent with the hypothesis. Satellite photos showing dust clouds emanating from sub-Saharan Africa led to the calculation that several hundred million **tonnes**[1] of dust are transported over the Atlantic Ocean annually.

FIGURE 1-2 The scientific method in practice. Scientists make observations and then formulate hypotheses to explain what they observed. Next, these hypotheses are tested and retested and, if they withstand the rigor of scientific scrutiny, they are provisionally accepted. If data contradict the working hypothesis, then the hypothesis is either reformulated or rejected and replaced with one consistent with the new data.

Evidence suggests a strong link between African dust and coral health, but more research is needed

Here are some important points about hypotheses. If the data[2] you collect support your hypothesis, then your hypothesis can be tentatively accepted. A hypothesis is never proven, as science does not deal in proofs, despite claims to the contrary in advertisements for consumer products. Moreover, acceptance of a hypothesis may be only temporary because scientists suspend judgment on making final determinations, except in very unusual situations. The original hypothesis may need to be revised or even abandoned completely as new information becomes available, new approaches to testing are undertaken, and/or as new technology becomes available. This means that scientists do not "jump to conclusions."

Science also demands that tests of hypotheses be repeatable. Other researchers must be able to duplicate your findings, for example. Thus, science advances very cautiously. If the data do not support a hypothesis, then the hypothesis is modified or rejected, no matter how good it may have seemed or how much a scientist wanted to accept it. In 1989, to much fanfare, Drs. Stanley Pons and Martin Fleischmann shook the scientific community with the major announcement that they had achieved so-called cold fusion, nuclear fusion at room temperature, in a beaker of water. If true, their discovery would have led to the large-scale production of a cheap and infinite energy source. Other scientists were unable to replicate their experiment, however, thus invalidating the original finding as unscientific.

■ SCIENTIFIC THEORIES AND UNIFYING PRINCIPLES

Some hypotheses considered central to the understanding of a discipline—that is, some branch of science—have been subjected to an enormous amount of testing and, having withstood this level of scrutiny, have come to be regarded as

scientific theories. In science, a theory is a broadly accepted explanation for an important phenomenon; in other words, a scientific theory denotes a truth. This definition is very different from the nonscientific dictionary connotation of the word *theory*, which is simply conjecture, and implies considerable doubt. *This is a very important distinction that you should carefully note and remember.* A scientific theory could never be referred to as "only" a theory because extensive testing and retesting leave virtually no room for doubt.

Plate tectonics and evolution are theories in geology and biology, respectively. Plate tectonics and evolution also represent **unifying principles** in their disciplines. A unifying principle is one that offers an overarching, or unifying, explanation for seemingly diverse phenomena and assembles them into a coherent whole.

■ CAUSE–EFFECT VERSUS CORRELATION

Let us revisit the African dust hypothesis that we used to explain the decline of some Caribbean corals. The evidence obtained through scientific research suggests three things: There were no coral health problems when dust concentration was low; the first appearance of African dust was associated with coral disease and death, and higher dust levels resulted in increased levels of disease and mortality.

This research shows a connection between dust levels and coral health. Such a comparison that demonstrates a relationship between two variables—in the previously mentioned case, dust concentration and coral health—is known as a **correlation**. Correlations are important to the advancement of science and in many cases may be the only data available from which we can draw conclusions. Unfortunately, erroneous conclusions are also made from spurious correlations, especially in the media.

A much stronger case for accepting a hypothesis can be made if there is a better link, preferably verified experimentally, between a cause and an effect. Currently, a **cause–effect** relationship between coral health and African dust has not been firmly established, although the hypothesis remains a possible explanation for the observations. The possibility also remains that dust is one contributing factor in the decline of reefs, along with pollution, higher ocean temperatures associated with climate change, or some other factor or factors.

To summarize: science offers a way to understand our natural world that incorporates numerous "firewalls"—that is, phenomena must be observable, and hypotheses must be testable and falsifiable. Scientists also suspend judgment, and try to refrain from assessments of value.

In the following section, we examine how policy makers use scientific information.

■ SCIENCE AND PUBLIC POLICY: THE PRECAUTIONARY PRINCIPLE AND SCIENTIFIC UNCERTAINTY

At the start of the third millennium, there is an overwhelming consensus among Earth scientists that our growing human population is changing the composition of the planet's atmosphere. Even though scientists cannot yet be certain what the effect of these changes will be, the preponderance of evidence suggests that the impact will be, on the whole, negative and could be catastrophic for hundreds of millions of people crowded into the planet's coastal cities as well as for entire ecosystems like coral reefs, mangroves, and tropical rainforests. Although the present level of agreement among scientists has grown over the past decades to a consensus, there is still vigorous debate on the magnitude, timing, and nature of specific impacts.

Science, as you have learned, advances cautiously, in accordance with the principles of the scientific method. In the case of global warming and climate change, the scale of the phenomenon is so large and the subject so complex that achieving scientific certainty of the impacts is likely not possible. In this section, we introduce you to two approaches to applying science to policies such as climate change. They are the **precautionary principle** and the **principle of scientific uncertainty**.

Scientific certainty, as the preceding pages should have impressed on you, is very difficult to achieve. Even things we consider to be "laws" can be modified with new observations. Because the 100% level of certainty is thus not a practical threshold for accepting hypotheses, scientists typically use the 95% standard, which basically means that a hypothesis is accepted if 95% of the observations of a test are in line with the hypothesis, the other 5% varying because of random chance, experimental error, or some other factor.

When it comes to artificial chemicals like the category called **persistent organic pollutants (POPs)**, scientists can rarely be 95% certain as to their impacts on individuals or ecosystems, as there are so many variables. In those cases, according to the "Rio Declaration" from the 1992 United Nations Conference on Environment and Development,

> "In order to protect the environment, the precautionary approach shall be widely applied.... Where there are threats of serious or irreversible damage, lack of full scientific certainty shall not be used as a reason for postponing cost-effective measures to prevent environmental degradation."

This standard is known as the *precautionary principle.*

Embedded in this principle is the notion that "there should be a reversal of the burden of proof, whereby the onus should now be on the operator or polluter to prove that an action *will not* cause harm, rather than on science to prove that harm (is occurring or) will occur."[3]

Another way to express this principle is "better safe than sorry." The products of science and technology are often brought to the marketplace without adequate investigation into any possible long-term effects on human health and the global environment. Some examples are the uses of POPs, lead, and mercury detailed elsewhere in this text.

In most industrialized nations, the so-called burden of proof falls not on the producers of goods, but rather on those who allege that they have suffered harm. Known as the **risk paradigm**, this is the basis of our tort system of civil law. As a result of the proliferation of new products, government agencies like the Food and Drug Administration, the Environmental Protection Agency, and the Federal Trade Commission, to name but a few, are sometimes unable to keep pace. For example, in deciding whether to approve pharmaceuticals for the market, the Food and Drug Administration uses the precautionary principle and requires drug companies to submit evidence that their products are safe and effective *before* they are sold. Even this approach, however, does not prevent many potentially harmful products to be marketed (**FIGURE 1-3**).

Although adult individuals have recourse to law if they believe they have been injured, fetuses, children, wildlife, and ecosystems have no such means of redress. Strict adherence to the precautionary principle in the view of many could facilitate democratic oversight.

Similarly, under the precautionary principle, a potentially serious threat such as global warming or the proliferation and buildup of organochlorines (a form of POP) in the ocean would trigger action to address the threat even if the science is not yet conclusive but is supported by the preponderance of available evidence.

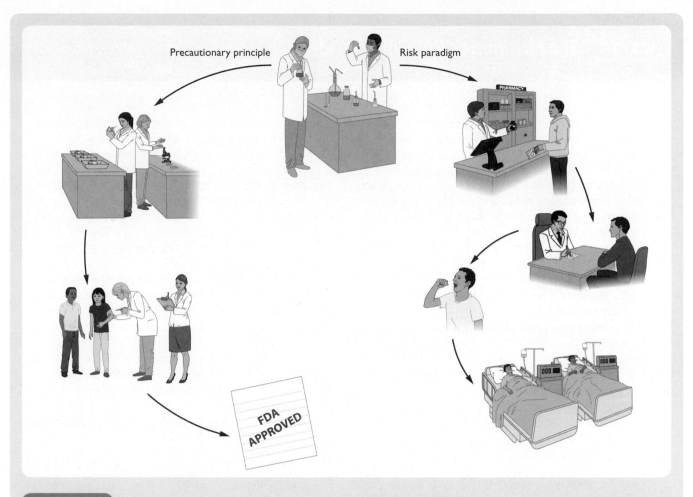

FIGURE 1-3 The precautionary principle versus risk paradigm in determining the safety and efficacy of pharmaceuticals. Consider these different approaches in the context of environmental geology, for example, would you build a subdivision on the site of a landfill or hazardous waste dump? What are the advantages and disadvantages of using each approach?

■ CRITICAL THINKING

It behooves us all to use the mind's power effectively, which involves **critical thinking**. Much of our thinking is spontaneous, is often emotional, and is rarely analytical and reflective. As such, it contains prejudice, bias, truth and error, inspiration, and distortions—in short, good and bad reasoning, all mixed together. Critical thinking essentially requires that we apply analysis, assessment, and the rules of logic to our thought processes.[4]

Some scientists equate critical thinking with the application of the scientific method, but we think critical thinking is a far broader and more complex process. Critical thinking involves developing skills that enable you to dissect an issue (analyze it) and put it together (synthesize it) so that interrelationships become apparent. It involves identifying *assumptions*, the basic ideas and concepts that guide our thoughts. Critical thinking also encourages an appreciation for our own and others' points of view, which is important when approaching complex environmental issues.

Too often, analyzing complex issues leads some to a belief that everyone is "entitled" to an **opinion** that should be respected (**BOX 1-1**). We do not necessarily concur;

In Other Words

John Milton said in *Paradise Lost*:

The mind is its own place
And in itself can make
A hell of heav'n
And a heav'n of hell.

The McDonald's Hot Coffee Incident

When you hear *McDonald's Hot Coffee Incident*, do you immediately know and react to it? Surprisingly, even though the incident occurred in 1992, and was not of any historical importance (and of no direct importance to environmental geology, but please bear with us), a surprisingly large number of students are aware of the event. More significantly, many people have an opinion on the matter. Are you one of them?

We have discovered that a majority of people aware of the case view it the quintessential frivolous lawsuit. They believe a woman driving a car spilled hot coffee on herself while driving and collected several million dollars from suing McDonald's.

Let's quickly examine the facts that emerged from the case. First, according to McDonald's operations and training manual, the brewing temperature for their coffee should be between 195° and 205°F and the coffee was supposed to be held between 180° and 190°F.

These details should stimulate a few questions, for example, at what temperature is coffee typically served? (The answer is between 135° and 160°F, although some establishments serve it hotter). Why would McDonald's serve their coffee at a higher temperature? (1. To satisfy customer demand, because many people do not drink their coffee for 20 minutes or longer after purchasing it and expect it to be hot when they do. 2. Perhaps doing so also extracts more flavor from inexpensive coffee beans.) At what temperature does coffee cause serious injury? (At about 180°F for a full skin-thickness burn). Was McDonald's aware that consumers had complained their

coffee being too hot? (Yes, as approximately 1,000 complaints had been lodged.)

The case involved a 79-year-old woman passenger in a stopped car (a former department store clerk who had never before filed suit against anyone) who had placed the cup of hot coffee between her knees and, while attempting to remove the plastic lid, spilled the entire contents into her lap. She required hospitalization. A jury awarded her $2.7 million in punitive damages, which a judge later reduced to $480,000.

If you had one, was your opinion based on adequate facts and information? If not, does knowing the facts now change your opinion? If your answer to both questions was no, do you believe that your current opinion is more well-informed?

We use this example to illustrate the importance of knowing facts before forming an opinion, and to stress that it is acceptable to have no opinion on an issue until you are well-informed on the issue.

You will have many opportunities to have opinions on issues involving environmental geology (e.g., Should deepwater or, for that matter, any drilling be allowed in the Gulf of Mexico? Is building your home in a floodplain a good idea? Is radon gas a problem in your house?). You'll be a lot happier, and feel a great deal smarter, if you have the facts in hand before forming opinions and making decisions. Critical thinking requires an awareness and understanding of the facts surrounding an issue, especially when it results in an opinion.

however, problem-solving demands a willingness to listen for content to what others are saying. Talking is easy, but listening is not.

To develop critical thinking skills, all of us must learn to use a set of intellectual standards as an "inner voice" by which we constantly test and hone our reasoning, but the standards must be set in an appropriate framework in order for true critical assessment to take place.

The following describes the intellectual standards we should apply when assessing the quality of our reasoning. This is the basis for critical thinking, which in turn is the approach that we try to apply throughout this book.[5]

Intellectual Standards: The Criteria of Solid Reasoning

Clarity. Clarity is the most important standard of critical thinking. If a statement is not clear, its accuracy or relevance cannot be assessed. For example, consider the following two questions:

1. What can we do about marine pollution?
2. What can citizens, regulators, and policy makers do to ensure that toxic emissions from industry, transportation, and power generation do not cause irreversible ecological damage to the marine environment, or harm human health?

Accuracy. How can we find out if a statement is true? A statement can be clear but not accurate. For example, the assertion "Clean coal is not a major contributor to climate change" is inaccurate.

Precision. A statement can be clear and accurate, but not precise. For example, we could say that, "there are more floods in the United States than ever before." That statement is clear and accurate; however, how many more floods are there? 1? 1,000? 1,000,000? (There is a difference between the way many scientists use the word *precision* and the more general way it is used here.) A lack of precision is the basis of much advertising.

Relevance. How is the statement or evidence related to the issue we are discussing? A statement can be clear, accurate, and precise, but not relevant. Assume that we are given the responsibility to eliminate the harmful marine environmental impact of mercury emitted from coal-burning power plants, and we invite public comment on our proposals. Someone might say, "Coal-burning power plants provide 100,000 jobs in this state alone." That statement may be clear, accurate, and precise, but it is not relevant to our specific responsibility of removing mercury.

Breadth. Are we considering all lines of evidence that could provide us with some insight in addressing an issue? Is there another way to look at this question? For example, in assessing the impact of African dust on coral reefs that we discussed previously, we also must consider other causes, such as global warming, increased sedimentation due to deforestation of slopes on Caribbean Islands, and so forth.

Depth. Is a proposed solution realistic? How does it address the real complexities of an issue? This question is one of the most difficult to tackle because here is where reasoning, "instinct," and moral values may interact. The points of view of all who take part in the debate must be carefully considered. For example, politicians have offered the statement "just don't do it" as a solution to the problem of teenage drug use, including smoking. Is that a realistic solution to the problem, or is it a superficial approach? How would you defend your answer? Is your defense grounded in critical thinking?

Logic. Does one's conclusion clearly follow from the evidence? Why or why not? When a series of statements or thoughts are mutually reinforcing and when they exhibit the intellectual standards described above, we say they are logical. When the conclusion does not make logical sense, is internally contradictory, or not mutually reinforcing, it is not "logical." We give examples of logical fallacies later in the chapter.

■ APPLYING INTELLECTUAL STANDARDS IN A CRITICAL THINKING FRAMEWORK

The intellectual standards described previously are essential to critical evaluation of issues, but there are more factors to be considered. The following criteria constitute the framework in which these standards should be applied.

Point of view. What viewpoint does each contributor bring to the debate? Is it likely that someone who has a job on an ocean fishing fleet would have the same view on marine sanctuaries as someone who does not? Why or why not?

Identifying a point of view does not mean that the point of view should automatically be accepted or discounted. We should strive to identify our own point of view and the bases for this, we should seek other viewpoints and evaluate their relevance, and we should strive to be fair-minded in our assessment. Few people are won over by having their ideas ridiculed. Furthermore, our points of view are often informed by our assumptions, which we address later here.

Evidence. All problem-solving is, or should be, based on evidence and factual information. Our conclusions or claims must be based on sufficient relevant evidence.

The information must be laid out clearly. The evidence against our position must be evaluated, and we must be open to new evidence that challenges our conclusions.

Purpose. All thinking to solve problems has an obvious purpose. It is important to have a clear understanding of that purpose and to ensure that all participants understand what that purpose is. Because it is easy to wander off the subject, it is advisable to check periodically to make sure that the discussion is still on target. For example, students working on a project occasionally stray into subjects that are irrelevant and unrelated, although they may be interesting or even seductive. It is vitally important, therefore, that the issue being addressed must be defined and understood as precisely as possible.

Assumptions. The Intergovernmental Panel on Climate Change's (IPCC) Fourth Assessment Report (2007) projected a rise in sea level by 2100 of between 18 to 59 cm, *assuming no contribution by melting ice sheets* (emphasis ours). When melting ice sheets and other factors are taken into account, the estimated sea level rise by 2100 is conservatively projected to be a meter or more. Reports of the IPCC projected sea level rise frequently omitted this seminal assumption.

All reasoning and problem-solving depends on assumptions, which are *statements accepted as true without proof*. For example, students show up in class because they assume that their professor/teacher will be there. We should identify our assumptions, and always be ready to examine and evaluate them. They often need to be revised in the light of new evidence.

Now, before we analyze our own assumptions, let us summarize some characteristics of sound reasoning. Critical thinking requires that we do the following:

- Continually exercise our thinking skills
- Eliminate irrelevant topics and explain why they are irrelevant
- Come to well-reasoned conclusions and solutions

Moreover, we should strive to understand concepts, key terms, and phrases essential to our discussion (such as greenhouse gas, ocean circulation, etc.).

Additionally, the effective reasoner continually assesses and reassesses the quality of his or her thinking in light of new evidence. Finally, one must be able to communicate effectively with others.

Next, let's do an exercise to allow you to test your assumptions about an issue.

Assumptions About Government's Role in Protecting the Environment

To assess the importance and power of assumptions in guiding your reasoning, take the following self-directed quiz. In it, you will identify your assumptions in defining government's responsibility to protect the environment, determine your assumptions as to the proper level of government that may act, determine your position on the precautionary principle and, finally, identify your assumptions concerning the extent to which individuals, governments, or institutions may impose costs on others with or without their knowledge or consent.

Thomas Jefferson and Government. Most politicians and many Americans probably consider themselves to have "Jeffersonian" principles. Read the following quotation taken from Thomas Jefferson's First Inaugural Address delivered on March 4, 1801.

> *What more is necessary to make us a happy and prosperous people? Still one thing more, fellow citizens—a wise and frugal Government, which shall restrain men from injuring one another, shall leave them otherwise free to regulate their own pursuits of industry and improvements, and shall not take from the mouth of labor the bread it has earned.*

Concept Check 1-1. In a clear sentence or two, explain what you think Jefferson meant by the phrase "which shall restrain men from injuring one another". Now assess the breadth of your response by considering this and the next three questions.

Concept Check 1-2. Do you think he was referring solely to thugs who physically brutalize their fellow citizens? Explain.

Concept Check 1-3. Could he logically also have been referring to citizens who sought to poison others? In other words, is restraining "poisoners" a legitimate role of government? Explain your answer.

Concept Check 1-4. Now, what if a citizen or organization dumps a toxin into water or air that all citizens depend on or if a citizen or organization fills in a wetland that performed valuable ecological functions on which local residents depend? May government under Jefferson's principle restrain that person or organization?

Your answer to these questions will define your assumptions as to the proper role of government. On what did you base your assumptions?

The Proper Level of Government That May Act

Next, evaluate your assumptions about the level of government, if any, which may properly intervene in environmental issues.

One of the major discoveries of the past two decades has been the extent to which much marine pollution is **transboundary** in nature. For example, one third of the air pollution affecting the Oregon coast comes from marine vessels outside the 5-km (3-mi) territorial limit controlled by the state, and from power plants in Asia thousands of kilometers away. This situation is repeated over and over across the country and around the world.

Concept Check 1-5. In the light of the transboundary nature of pollution, is it appropriate that local government by itself (e.g., the coastal city of Newport, Oregon) bear the responsibility for protecting its own environment? May the states and federal government have a legitimate role? Should international agencies be involved? Explain and justify your answers.

Your answer will help evaluate your assumptions about the extent to which state, federal, or global agencies have responsibilities to intervene to protect local environments.

The Question of Externalities

Economists define *externalities* as any cost of production not included in the price of the good. An example would be environmental pollution or health costs resulting from burning diesel fuel, not included in the price of the fuel. Another example is the cleanup costs paid by governments resulting from animal waste pollution of water bodies from large-scale meat-processing operations. In this example, the price of chicken or pork at your local supermarket is lower than it would be if all environmental cleanup costs were included in the price of the meat.

Assumptions About Corporations

Here is another choice quotation from Thomas Jefferson on the impact of those new organizations called corporations. Read Jefferson's words, and then respond to the following question:

> I hope we shall take warning from, and example of, England, and crush in its birth the aristocracy of our moneyed corporations, which dare already to challenge our Government to trial, and bid defiance to the laws of our country.

Concept Check 1-6. Do you share or reject Jefferson's opinions concerning corporations? Justify your conclusion.

Concept Check 1-7. Prepare a list of positive and negative contributions corporations make to our environment and economy. Do you conclude corporations have "too much power" in contemporary life? Why or why not?

If you are interested in corporate power, research the 1886 U.S. Supreme Court ruling: *Santa Clara County v. Southern Pacific.* The Court ruled that Southern Pacific was a "natural person" entitled to the protections of the U.S. Constitutions Bill of Rights and 14th Amendment.[6] In 2010, the U.S. Supreme Court affirmed the principle of corporate personhood. After researching these cases, answer Concept Check 1-8.

Concept Check 1-8. Do you believe the Court acted correctly in deciding that a corporation was a person? Is the 1886 ruling relevant to the 21st century? Why or why not? If you did not research the case, do you feel you are still entitled to an opinion? Should your opinion be given as much credence as someone's who did research the case? Why or why not?

After having thoughtfully responded to the previous scenarios, you should now have a better awareness of the assumptions that you bring to the analysis of issues in environmental geology that you are about to undertake in *Environmental Geology Today*.

Summary of Critical Thinking

Intellectual standards by which critical thinking is carried out are clarity, accuracy, precision, relevance, breadth, depth, and logic. These standards are applied in a framework delineated by points of view, assumptions, evidence or information, and purpose. We encourage you to return to this section whenever you need to refresh and polish your critical thinking skills.

Knowledge and Opinion

The philosopher Mortimer Adler pointed out[7] that there is no contradiction in the phrase "true opinion," nor is it redundant to speak of "false opinion." Thinking about the subject matter of this book requires that we separate truth and knowledge from opinion, and false opinion from true opinion.

There are few things that are both incorrigible and immutable: One example is the statement that the sum of a finite whole is greater than any of its parts. Sometimes definitions are self-evident truths: For example, a triangle has three sides.

Few scientific concepts meet these rigid standards. Does that mean that everything else is opinion? That depends on how we define opinions. Some opinions deserve the status of knowledge, even if they are not immutable and incorrigible. Examples are theories that have an overwhelming body of supporting evidence. We can say with confidence that such ideas are true at a particular time. Should new evidence come to light, we may have to evaluate our theory.

Contrast these kinds of opinions with personal prejudices, which we assert often without any evidence or force of reason to support them. We may, for example, believe that "the government has no business telling people what to do with their land." This is what philosophers would call a "mere" opinion, as opposed to a "true" opinion, which we illustrated earlier. There is nothing "wrong" with having mere opinions, but we all should recognize them for what they are.

Common Logical Fallacies

Much of your information concerning scientific issues will come from the media: television, magazines, radio talk shows, newspapers, and blogs and other websites. These sources often exhibit evidences of poor reasoning, such as logical fallacies (as well as mere opinions). Learn to identify them to ensure that you are getting the best information possible. Here, in no particular order, are some of the more common examples of **logical fallacies**:

- The fallacy of composition: assuming that what is good for an individual is good for a group. An example is standing at sports events—this is an advantage to one person but not when everybody does it.
- The fallacy of starting with the answer: including your conclusion in your premises or assumptions. For example, "America runs on high levels of energy consumption. We can't have the American lifestyle without energy. Thus, America can't afford to cut energy use." Here, the arguer is simply defining his way out of the problem. By this reasoning, we would have to continue to increase energy consumption forever, an obvious impossibility.
- The fallacy of hasty generalization: "Senegal and Mali have very low levels of energy consumption. They are very poor countries. Low levels of energy consumption lead to poverty." How is poverty measured? Are there any "wealthy" countries that have relatively low levels of energy consumption? Are there any relatively poor countries that have high levels of energy consumption?
- The fallacy of false choice: stating an issue as a simplistic "either–or" choice when there are other more logical possibilities. "Those who don't support fossil fuel use want to go back to living in caves."
- Fallacy of an appeal to deference: accepting an argument because someone famous supports it.

- Fallacy of *ad hominem* argument (literally, "at the person"): attacking a person (or his or her motives) who advocates a position without discussing the merits of the position.
- The fallacy of repetition: the basis of most advertising: repeating a statement without offering any evidence. "Population growth is good. People contribute to society. We need population growth to survive."
- The fallacy of appealing to tradition: "Coal built this country. Eliminating coal use would threaten our society."
- The fallacy of appealing to pity: "The commercial fishing industry supports millions of American families. We have to support them."
- The fallacy of an appeal to popularity: "Seventy five percent of Americans support this position." Perhaps the poll asked the wrong question. Perhaps the respondents did not have enough information to properly respond, or perhaps they were uninformed, and so forth.
- The fallacy of confusing coincidence with causality: "After passage of the Endangered Species Act, jobs in sawmills fell 70%; therefore, the Endangered Species Act was bad for the economy." Were there other possible explanations for the drop in jobs?
- The fallacy of the rigid rule: "Hard-working people are good for the economy. Immigrants are hard-working people; therefore, the more immigrants we have, the better for our economy." "Large numbers of immigrants commit crimes. Crimes are bad for the economy; therefore immigration should be curtailed."
- The fallacy of irrelevant conclusion: using unrelated evidence or premises to support a conclusion. "Development raises the value of land and provides jobs. Developed land pays more taxes than undeveloped land; therefore, any available land should be developed."

CHAPTER SUMMARY

1. Science is the discovery of truths about the universe using the scientific method.
2. Science differs from other ways of knowing in that it relies on empirical evidence.
3. The scientific method begins with observation; if an event or process cannot be observed, either directly or indirectly, it cannot be explained by science.
4. Observations lead to hypotheses, or educated explanations of some phenomenon.
5. Hypotheses are tested, with one possibility being that the hypothesis can be falsified, or shown to be incorrect.
6. Science makes no value judgments.
7. Hypotheses are never proven; they are simply either accepted or rejected.
8. Tests of hypotheses must be repeatable, or the hypothesis is not valid.
9. Broad hypotheses that have withstood enormous scrutiny may become scientific theories, which are not conjectures, but denote truths.

10. Unifying scientific principles, like evolution or plate tectonics, offer an over-arching, or unifying explanation for seemingly diverse phenomena and assembling them into a coherent whole.

11. A comparison that shows a simple relationship between two variables is known as a correlation.

12. Correlations based on cause and effect make a more powerful case for accepting hypotheses.

13. Scientific certainty is very difficult to achieve; scientists thus typically use the 95% standard (a hypothesis is accepted if 95% of the observations of a test are in line with the hypothesis).

14. The precautionary principle and risk paradigm are alternate approaches to regulate the introduction of chemicals into the environment.

15. Critical thinking is higher order thinking that essentially requires that we apply analysis, assessment, and the rules of logic to our thought processes.

16. Intellectual standards of critical thinking include clarity, accuracy, precision, relevance, breadth, depth, and logic.

17. The framework of critical thinking includes point of view, evidence, purpose, and assumptions.

18. Having an opinion on an issue comes with the responsibility to know the underlying facts of the issue.

19. You should be aware of common logical fallacies, which are evidence of poor reasoning.

KEY TERMS

accuracy
assumptions
breadth
cause–effect
clarity
correlation
critical thinking
depth
desertification
evidence
hypothesis

logic
logical fallacy
opinion
persistent organic
   pollutant (POP)
point of view
precautionary
   principle
precision
principle of scientific
   uncertainty

purpose
relevance
risk paradigm
science
scientific method
scientific theory
tonne
transboundary
unifying principle

REVIEW QUESTIONS

1. Science differs from other ways of knowing in that
   a. science has both a body of knowledge and a process.
   b. science answers questions about the natural world.
   c. science relies upon empirical observation.
   d. only science and the humanities use the scientific method.

2. Which of the following is a part of the scientific method?
   a. Making an observation.
   b. Formulating a hypothesis.
   c. Testing the hypothesis.
   d. All of the above.

3. Because science relies on observation, the Earth's temperature 500,000 years ago is unknown because humans had not yet appeared on Earth and thus could not measure it.
   a. True
   b. False

4. The hypothesis that dust from Africa was killing some coral in the Caribbean was
   a. tested and not rejected.
   b. conclusively demonstrated.
   c. proven.
   d. rejected.

5. The Pons and Fleischman experiment was
   a. scientific because it tested a valid hypothesis.
   b. scientific because it could solve the energy crisis.
   c. unscientific because nuclear fusion could never occur in a beaker.
   d. unscientific because it could not be repeated.

6. A scientific theory
   a. is an explanation for a phenomenon.
   b. is not conjecture.
   c. has withstood the scrutiny of repeated testing.
   d. all of the above.

7. The idea that dust from the Sahara has killed Caribbean corals is
   a. a unifying principle.
   b. a scientific theory.
   c. both a and b.
   d. none of the above.

8. The precautionary principle
   a. is used to regulate the introduction of pharmaceuticals.
   b. is a unifying principle in science.
   c. puts the burden of proof on the polluter.
   d. is none of the above.

9. In assessing the cause of coral die-offs, if we consider only dust and not other causes such as global warming or increased sedimentation, we violate the critical thinking criterion of
   a. clarity.
   b. accuracy.
   c. depth.
   d. relevance.

10. The statement, "Those who don't support fossil fuel use want to go back to living in caves" is an example of which logical fallacy?
    a. False choice.
    b. Starting with the answer.
    c. Straw man.
    d. None of the above.

FOOTNOTES

[1] A *tonne*, as you will see throughout this book, is also known as a *metric tonne* and equals 1,000 kg, or 2,200 pounds. A *ton*, or *short ton*, equals 2,000 pounds.

[2] The word *data* is plural and thus takes the plural form of a verb. The singular form of *data*, that is, one piece of information, is known as a *datum*. In practice, the word *data* is sometimes used to denote the singular, but technically this is incorrect.

[3] Glegg, G., and P. Johnston. 1994. The policy implications of effluent complexity. In *Proceedings of the Second International Conference on Environmental Pollution*, Vol. 1, p. 126. London: European Centre for Pollution Research.

[4] Paul, R., and L. Elder. 2000. *Critical Thinking—Tools for Taking Charge of Your Learning and Your Life.* Upper Saddle River, NJ: Prentice Hall.

[5] We obtained the basis for much of the information on critical thinking from the International Conferences on Critical Thinking and Educational Reform, sponsored by the Foundation and Center for Critical Thinking (http://www.criticalthinking.org). This section is adapted from *Environmental Issues: An Introduction to Sustainability* by Robert L. McConnell and Daniel C. Abel (2008, Pearson Prentice Hall, Upper Saddle River, NJ).

[6] This interpretation is disputed by many legal scholars who argue that in fact the court did not reach that conclusion.

[7] Available at http://www2.franciscan.edu/plee/adler.htm.

The Language of Environmental Geology

Basic Assumptions and Principles

CHAPTER OUTLINE

Although the field of environmental geology is vast and encompasses a wide range of concepts, issues, and facts, you should remember that underlying all of this information are a few basic principles and themes. You should not accept these principles without evaluation. In other words, do they pass the standards we lay out in our section on critical thinking? Are they clear? Precise? Accurate? Sufficiently broad? What point of view, if any, can you identify in each?

You should apply the standards of critical thinking to everything you read in this book.

Here, greatly simplified and summarized, are what we believe to be the central themes of the book.

■ THE CHALLENGE OF HUMAN POPULATION GROWTH AND RESOURCE CONSUMPTION

Human population growth and accompanying resource consumption are the ultimate challenges facing humanity.

There are three components to this principle: 1) **exponential growth**; 2) increased consumption of natural resources; and 3) Hardin's Third Law of Human Ecology.

Exponential Growth

The growth of human numbers is *exponential* and, during the twentieth century, was unparalleled in human history. Exponential growth simply means growth by a percentage of the starting number (say, 1.3% per year) rather than by the addition of a fixed quantity (e.g., 80,000,000 people added each and every year). Although we'll cover this topic more rigorously elsewhere in this text, you should realize that growth by a percentage means that each year an increasingly larger number is added to the starting population, and thus exponential growth can be thought of as *growth upon growth*.

At the beginning of the twentieth century, global population was approximately 1.65 billion, and at its end over 6.1 billion, an overall growth of 270% (FIGURE 2-1). Over the same interval, life expectancy more than doubled, from 30 to over 60 years. And, as tragic as they have been, AIDS and other epidemics, wars, and famines barely put a dent in this growth on a global scale (although regionally these affect populations drastically). Using simple math, one can demonstrate that this growth is not sustainable. If the present growth rate of 1.3% per year continues, the human population would expand to cover the planet's dry land surface at a density of one person per square meter in a little over 500 years! Imagine the lines at the Department of Motor Vehicles in the year 2500!

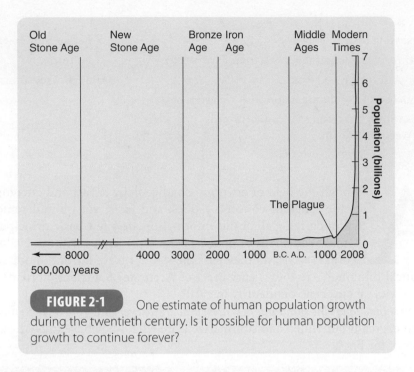

FIGURE 2-1 One estimate of human population growth during the twentieth century. Is it possible for human population growth to continue forever?

Increased Consumption of Natural Resources

Just as global population has grown, so too has there been an increase in the consumption of natural resources, as well as in the alteration and degradation of natural landscapes. Moreover, the allocation of these resources has been mainly to those who could best afford them, while all of society is saddled with the costs. The United States, for example, comprises less than 5% of the world's population, yet uses nearly a quarter of the world's resources.

We believe that there are long-term, quantifiable risks and costs associated with an unequal distribution of resources. For example, lead and mercury poisoning from sources such as lead paint, air pollution, and consumption of mercury-tainted fish are environmental health problems that disproportionately impact disadvantaged people, mainly children. These poisoned children frequently suffer loss of cognitive (reasoning) abilities. Research indicates that environmental toxins such as lead can induce violent behavior in children as well. Some of the costs of this loss are borne by society at large. Thus, they represent costs that could be avoided by curtailing release of these toxins into the environment.

Hardin's Third Law of Human Ecology

Lastly, the impact of a population on its environment, be it an individual ecosystem or the entire planet, is a function not only of the absolute number of people but also the impact per person. This relationship is known as *Hardin's Third Law of Human Ecology* (after environmental provocateur Garrett Hardin; **FIGURE 2-2**). Although population growth is highest in developing countries, the impact per person thus peaks in the developed world.

So the ultimate question is: will we learn to control our population, or will natural forces do the work for us? And, as we will show, human population growth has implications for almost all other topics covered in this book. For example,

FIGURE 2-2 Garrett Hardin's "Third Law of Human Ecology." How does one go about measuring impact per person? Suggest some ways and justify your choices. Courtesy of Lalo Alcaraz.

we must understand the impact that humans are having on the planet's climate, because many of these impacts may not be reversible over intervals of decades to centuries. We will return to this principle frequently throughout the book, so remember it well.

■ IS WATER THE ULTIMATE NATURAL RESOURCE?

Basic economics teaches that as a commodity becomes more and more expensive, people increasingly turn to substitutes. During the 1970s, American electric utilities turned away from oil and began to build more and more coal-fired power plants, due to the high price and increasing unreliability of oil supplies. Water is an absolute requirement for life, and there is no substitute.

As global population increases and population becomes more urbanized, and we face the uncertain consequences of global warming and climate change, a continuous and ever-growing supply of clean drinking water is not guaranteed. According to the World Resources Institute, water stress already affects more than 40 percent of the world's population, and this number is projected to grow to almost 50 percent by 2025. Water suitable for agriculture and industry is likewise threatened. Many rivers in China are already unusable for industrial processes, much less for agriculture or municipal supplies.

A corollary of this principle is that we must understand the nature and implications of the **hydrologic cycle**, which we will explain elsewhere in this text (FIGURE 2-3). One way that humans have responded to the need for water is to build dams and reservoirs, but this practice is becoming increasingly controversial globally (FIGURE 2-4).

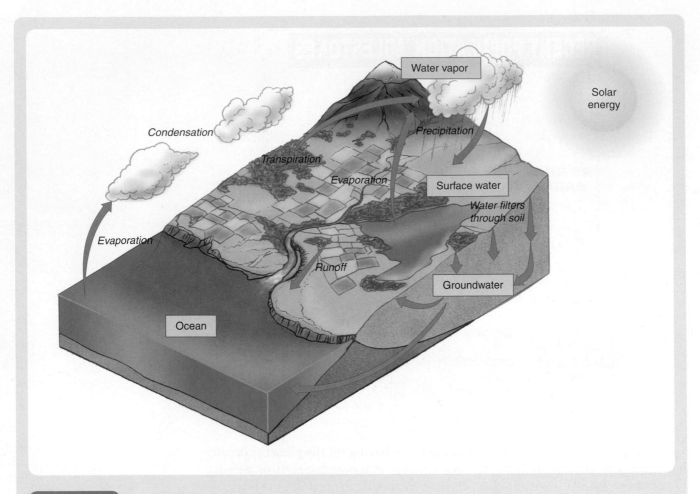

FIGURE 2-3   A diagram of the hydrologic cycle. Precipitation, infiltration, transpiration, evaporation, and runoff are the labeled components of the cycle. How could climate change affect the components of the hydrologic cycle?

FIGURE 2-4   The Nile River from space. The Sudan and Egypt are in conflict over uses of Nile water. Note the Aswan Dam (arrow) whose construction resulted in major changes in the northern Nile Valley, in the fisheries of the eastern Mediterranean, and in the rate of coastal erosion along the Mediterranean coast of Egypt. Dam construction is increasingly controversial globally. Courtesy of JSC PAO Web Team/NASA.

◼ SOILS: NEXT TO WATER, THE MOST IMPORTANT PLANETARY RESOURCE?

Before the words "global warming" entered our vocabulary, if you were to ask an environmental scientist to name the world's most pressing environmental problem, the answer might well have been **topsoil loss**. Although this issue has been displaced from the headlines, it still remains vitally important. Many soil scientists fear that soils are being degraded by nutrient loss, erosion, and **salinization** (salt buildup), at rates that far exceed their regeneration or purification rates. Thus, they may not support agricultural production sufficient to meet the nutritional needs of a growing human population

◼ ACTIONS HAVE CONSEQUENCES

This can also be expressed as ***Hardin's First Law of Human Ecology***, "We can never do merely one thing." This expression connotes the interconnectedness of all things on the planet.

Part of our problem is a failure to ***make connections between connected events and processes***. For example, clearing of forests in Nepal causes increased severity of floods in downstream Bangladesh.

Here is another example: An act as seemingly simple as drinking a can of carbonated soda has attached to it a lengthy range of social, economic, and environmental impacts that one rarely considers. Let's ignore the product itself and consider only the container. The can, for example, is made of aluminum, which is refined from **bauxite** ore. Bauxite is found in extensive deposits in Australia, Surinam, Jamaica, and India (**FIGURE 2-5**). During processing, four metric tons (a metric ton,

FIGURE 2-5   A bauxite mine. In this case, tropical topsoil concentrate (bauxite) is the ore being mined. © James L. Amos/Corbis.

abbreviated *MT* and referred to as a "tonne," equals 1000 kg or 2200 lb) of bauxite ore yields two MT of alumina and two MT of waste. The two MT of alumina further reduce to one ton of aluminum metal, which still must be processed further. The whole process consumes a great deal of electricity; up to 3% of the electricity used in the United States is used to refine aluminum.

The electricity to extract the aluminum is only part of the cost of producing the aluminum can. Other costs include the extraction and shipping of the ore, the cost to restore the surface environment after mining is completed (or, if restoration is not carried out, this cost is simply "dumped" onto the local environment and residents), the energy to ship the ore (or concentrate) to the refinery, and the cost to transport and distribute the can. A few of the environmental costs include air and water pollution from the manufacturing process; threats to the survival of salmon because of building dams for cheap electricity; the introduction into harbors and bays of exotic species that traveled in the ballast water of ore tankers; destruction of the surface during strip-mining (bauxite mining is among the most destructive mining practices); dredging channels to maintain navigable waters for bauxite tankers; and space in landfills for discarded cans (FIGURE 2-6). Of course aluminum cans are recycled at a fairly high (but declining) rate, which can significantly diminish these costs. But remember, we didn't even consider plastic and glass bottles, which have their own impacts, and we ignored the contents of the bottles as well.

This analysis of the impacts of aluminum cans is an example of **product life cycle assessment (LCA)**, in which the environmental costs of a product are calculated over the product's entire life cycle, from extraction of raw materials to handling of wastes. Many of these costs are not included in the price one pays for the product. The price tag of efforts to restore salmon populations, for example, isn't included in the price you pay for a six-pack of cola, even though the principal cause of the salmon's demise is blockage of their migratory routes by

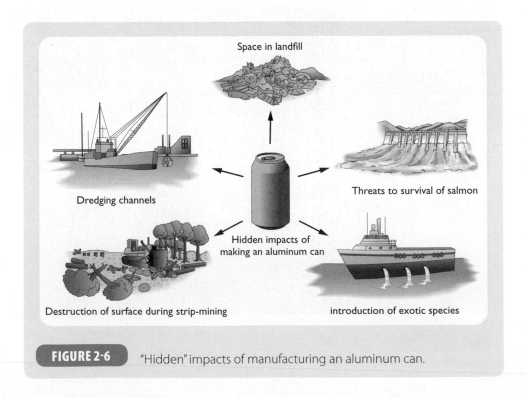

FIGURE 2-6   "Hidden" impacts of manufacturing an aluminum can.

dams built to provide cheap and abundant electricity for aluminum production. Such *externalities* (a term economists use to describe the costs to society and the environment that are not reflected in the price of a product) are paid for, sooner or later, by the rest of society or by the environment. The price of your cola and other products also is kept low by **subsidies**, an **externality** in the form of financial assistance given to industries that typically reduces the price of their product. Also, maintenance of waterways by the U.S. Coast Guard and Army Corps of Engineers subsidizes the cargo ship industry. Externalities in general and subsidies in particular encourage use of a commodity in excess of the level that the commodity would be used if all the costs, including externalities, were included in the price.

■ NATURE'S TRANSBOUNDARY IMPACTS

Many environmental impacts are **transboundary**; that is, they are not contained within arbitrary political boundaries. Governments thus must work together to control such impacts. For example, one-third or more of certain types of air pollution experienced by Northeastern states like New York are "imported" from as far away as the Midwest and Canada. Dust from Africa (**FIGURE 2-7**) may be an important trigger of asthma in the Caribbean and the southeastern United States and, as you have already seen, also may adversely affect the health of coral reefs. Pollution from Asian coal-fired power plants is already affecting the west coast of North America, and by 2012 the radiation plume from Japan's Fukushima nuclear power plant, heavily damaged in the Tōhoku earthquake and tsunami in 2011, had crossed the Pacific Ocean and reached California.

FIGURE 2-7 Photo of dust from the Sahara taken on November 11, 2006. Land degradation in the Sahel (southern border of the Sahara) may contribute dust to such storms, which have been implicated in increased rates of asthma attacks among children in the Western hemisphere. Courtesy of NASA.

■ SUSTAINABILITY

Sustainability, or sustainable development, was defined in 1987 by the World Commission on Environment and Development as development that meets the needs of the present without compromising the ability of future generations to meet their own needs.

Sustainability transcends and in many ways supersedes environmentalism. It involves a transformation from a wasteful linear model of resource use in which natural resources (be they living or nonliving) are extracted, used, then thrown "away" to a cyclical model built around reduction, recycling, and reuse.

Quantifying Sustainability

In 1999, the Environmental Performance Measurement Project at Yale University issued its first Environmental Sustainability Index (ESI; changed to Environmental Performance Index or EPI in 2006), ranking nations on the extent to which their societies approached sustainability (FIGURE 2-8). "No country is on a sustainable trajectory—and the ESI demonstrates this," said Gus Speth, Dean of the Yale School of Forestry and Environmental Studies. In most cases, insufficient data exist to determine each nation's ESI accurately. Their rankings are, therefore, only crude comparative measurements of societal sustainability, but they represent a starting point to quantify sustainability.

Concept Check 2-1. Examine Figure 2-8. Which regions are most sustainable, according to the ESI? The least? Discuss why you think this is so.

Concept Check 2-2. What is the EPI ranking of the United States? Does this seem high or low to you? Explain your answer.

Concept Check 2-3. Explain how sustainability indices like the EPI might be used by policy makers.

Targets for a Sustainable Planet

The Sustainable Communities Network sets general targets for a Sustainable Planet Earth. They are as follows:

- Creating community
- Growing a sustainable economy
- Protecting natural resources
- Governing sustainably
- Living sustainably

Achieving sustainability requires addressing a number of seemingly unrelated issues. These include population, energy, climate change, biodiversity, fisheries, forests, manufacturing and industry, and justice and equity (FIGURE 2-9).

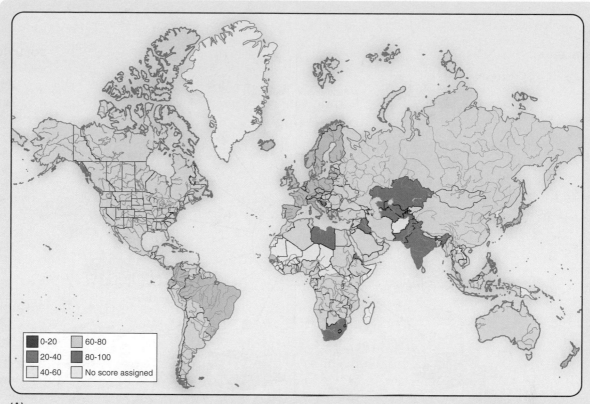

(A)

EPI Rank strongest performers			EPI Rank strong performers			EPI Rank modest performers			EPI Rank weaker performers			EPI Rank weakest performers		
1	Switzerland	76.69	11	Germany	66.91	47	Georgia	56.84	85	Togo	48.66	121	Tajikistan	38.78
2	Latvia	70.37	12	Slovakia	66.62	48	Australia	56.61	86	Algeria	48.56	122	Eritrea	38.39
3	Norway	69.92	13	Iceland	66.28	49	United States of	56.59	87	Malta	48.51	123	Libyan Arab	37.68
4	Luxembourg	69.2	14	New Zealand	66.05		America		88	Romania	48.34		Jamahiriya	
5	Costa Rica	69.03	15	Albania	65.85	50	Argentina	56.48	89	Mozambique	47.82	124	Bosnia and	36.76
6	France	69	16	Netherlands	65.65	51	Cuba	56.48	90	Angola	47.57		Herzegovina	
7	Austria	68.92	17	Lithuania	65.5	52	Singapore	56.36	91	Ghana	47.5	125	India	36.23
8	Italy	68.9	18	Czech Republic	64.79	53	Bulgaria	56.28	92	Dem. Rep. Congo	47.49	126	Kuwait	35.54
9	United Kingdom	68.82	19	Finland	64.44	54	Estonia	56.09	93	Armenia	47.48	127	Yemen	35.49
10	Sweden	68.82	20	Croatia	64.16	55	Sri Lanka	55.72	94	Lebanon	47.35	128	South Africa	34.55
			21	Denmark	63.61	56	Venezuela	55.62	95	Congo	47.18	129	Kazakhstan	32.94
			22	Poland	63.47	57	Zambia	55.56	96	Trinidad and	47.04	130	Uzbekistan	32.24
			23	Japan	63.36	58	Chile	55.34		Tobago		131	Turkmenistan	31.75
			24	Belgium	63.02	59	Cambodia	55.29	97	Macedonia	46.96	132	Iraq	25.32
			25	Malaysia	62.51	60	Egypt	55.18	98	Senegal	46.73			
			26	Brunei Darussalam	62.49	61	Israel	54.64	99	Tunisia	46.66			
			27	Colombia	62.33	62	Bolivia	54.57	100	Qatar	46.59			
			28	Slovenia	62.25	63	Jamaica	54.36	101	Kyrgyzstan	46.33			
			29	Taiwan	62.23	64	Tanzania	54.26	102	Ukraine	46.31			
			30	Brazil	60.9	65	Belarus	53.88	103	Serbia	46.14			
			31	Ecuador	60.55	66	Botswana	53.74	104	Sudan	46			
			32	Spain	60.31	67	Côte d'Ivoire	53.55	105	Morocco	45.76			
			33	Greece	60.04	68	Zimbabwe	52.76	106	Russia	45.43			
			34	Thailand	59.98	69	Myanmar	52.72	107	Mongolia	45.37			
			35	Nicaragua	59.23	70	Ethiopia	52.71	108	Moldova	45.21			
			36	Ireland	58.69	71	Honduras	52.54	109	Turkey	44.8			
			37	Canada	58.41	72	Dominican Republic	52.44	110	Oman	44			
			38	Nepal	57.97	73	Paraguay	52.4	111	Azerbaijan	43.11			
			39	Panama	57.94	74	Indonesia	52.29	112	Cameroon	42.97			
			40	Gabon	57.91	75	El Salvador	52.08	113	Syria	42.75			
			41	Portugal	57.64	76	Guatemala	51.88	114	Iran	42.73			
			42	Philippines	57.4	77	United Arab Emirates	50.91	115	Bangladesh	42.55			
			43	South Korea	57.2	78	Namibia	50.68	116	China	42.24			
			44	Cyprus	57.15	79	Viet Nam	50.64	117	Jordan	42.16			
			45	Hungary	57.12	80	Benin	50.38	118	Haiti	41.15			
			46	Uruguay	57.06	81	Peru	50.29	119	Nigeria	40.14			
						82	Saudi Arabia	49.97	120	Pakistan	39.56			
						83	Kenya	49.28						
						84	Mexico	49.11						

(B)

FIGURE 2-8 (A) Map and (B) list of Environmental Performance Index (EPI) scores by country. The EPI is based on 25 environmental indicators. Countries with higher EPI scores are considered more sustainable than those with lower scores. Which countries and regions are most sustainable? Least? Discuss your answers.

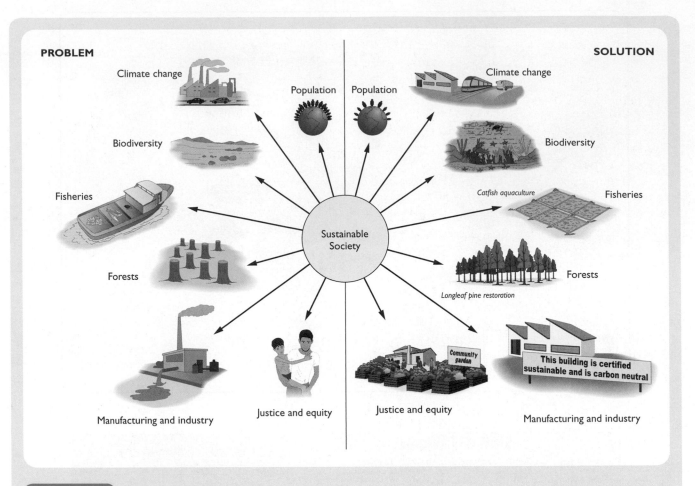

PROBLEM

Climate change

Population

SOLUTION

Climate change

Population

Biodiversity

Biodiversity

Fisheries

Catfish aquaculture

Fisheries

Sustainable Society

Forests

Longleaf pine restoration

Forests

Manufacturing and industry

Justice and equity

Justice and equity

Community garden

This building is certified sustainable and is carbon neutral

Manufacturing and industry

FIGURE 2-9 Achieving a sustainable society requires addressing the following areas: population, energy, climate change, biodiversity, fisheries, forests, manufacturing and industry, and justice and equity.

Human Population

As mentioned above, human numbers must eventually become stabilized, as it is physically impossible for population growth to continue forever. The only questions are at what level will growth end and whether growth will end as a result of human actions or by natural processes like famine, disease, and war. Ageing societies are typical of developed nations. Populations dominated by the young are typical of developing ones. Large numbers of young people provide great promise for societies, but also impose great costs. The readers of this book are mainly young, and it is they who will address—or not address—these challenges. The next century should prove to be one of the most interesting and potentially rewarding centuries in the entire span of human history. A deeper examination at human population growth is found elsewhere in this text.

Energy

Sustainable societies cannot be built on nonrenewable energy resources. Humans use almost unimaginable amounts of energy and generate vast amounts of pollution from it. Fossil fuel burning emits pollutants like SOx and NOx (oxides of sulfur and nitrogen), particulates, and heavy metals such as mercury. These have now been distributed throughout the globe from airborne fossil fuel emissions. Moreover, the transport of petroleum by tanker inevitably results in oil spills that degrade and may destroy sensitive marine habitats.

Energy conservation and the use of renewable fuels provide cost-effective and sustainable alternatives that generate little if any air and water pollution. Subsidizing the production of coal and oil-based nonrenewable energy makes little sense in a world threatened with rapid climate change and accelerating species loss. Here, too, change is coming. Wind energy is the fastest growing energy source in Europe and North America, supported by government subsidies, which partly offset subsidies for fossil fuels. Five of the top ten large-scale development of offshore "wind farms" under construction or planned are in the United Kingdom; the others are located elsewhere in Europe and more smaller scale operations are being planned. Such developments, while offering virtually pollution-free energy, must be carefully constructed and monitored so as to avoid harming the local marine environment. Likewise, offshore oil and gas development, a fixture in the oceans since the 1950s, must be restricted and, where necessary, more carefully undertaken in marine environments already stressed by overfishing, climate change, sediment pollution, and so forth. Energy is considered further elsewhere in the text.

Climate Change

Although climate change is a well-documented fact of planetary history—the Earth has gone through several megacycles of "greenhouse" and "icehouse" conditions—the *speed* with which human-induced climate change is occurring is virtually unprecedented. Too-rapid change overwhelms the ability of natural ecosystems to adapt. For example, rising sea levels may flood and destroy coastal salt marshes if sea level rise is too fast for these communities to migrate inland, and rising shallow-water ocean temperatures can lower oxygen content and threaten communities like coral reefs already operating throughout the oceans at temperatures near their thermal maxima. The impacts of climate change are imperfectly understood, but on the whole will test the capability of a human species that in the past has "subdued" a seemingly limitless Earth. Moving to sustainable societies may be essential to address the impacts of climate change. We consider climate change more fully elsewhere in this text.

Biodiversity

Our very survival ultimately relies on maintaining the integrity of marine and terrestrial ecosystems that we barely understand. An ecosystem is a geographic area that includes all living organisms with their physical surroundings and the natural cycles that sustain them (such as the hydrologic cycle). All these elements are connected.

Altering any one component affects the others in a particular ecosystem. Biodiversity, the mix of organisms within an ecosystem, is particularly critical for sustainability because of the specialized and often little-understood roles each species plays in maintaining the dynamic state of ecological balance. Moreover, little is known about key ecosystems like soils and the deep ocean. Esthetics and ethics also must play a part, as humans may be able to survive, at least for the short term, on an Earth with drastically reduced species diversity. The question then becomes this: Do we wish to make a decision to eradicate species and ecosystems without the input of our descendants? That our ancestors did so in ignorance is no excuse for our perpetuating such behavior.

Fisheries

Eventually, the impact of our growing demand for fish protein must be addressed. Contemporary industrial-style fishing methods impose significant environmental costs. For example, deep-ocean trawling has been compared with clear cutting a forest, only on a much larger scale. Moreover, per capita yields from global fisheries are declining, and expansion of aquaculture often occurs at the expense of coastal ecosystems and wild fish stocks. Many fish species, such as bluefin tuna, are in precipitous decline. Thriving and diverse aquatic wildlife are necessary for healthy marine and freshwater ecosystems. It is therefore critical that those dependent on fisheries and aquatic ecosystems use these resources responsibly. This can mean creating marine-protected zones that are large enough to maintain species diversity, away from human interference; however, protected zones alone may not be enough. Large fish and marine mammals characteristically have dangerously high levels of toxic artificial chemicals in their tissue from pollution sources perhaps thousands of kilometers away. Protecting aquatic wildlife could be aided through sustainable aquaculture. For example, growing plant-eating (herbivorous) fish such as carp, catfish, and tilapia puts less strain on resources compared with growing carnivorous species like salmon, which usually must be fed feed made with wild-caught fish. Shrimp farming puts great stress on coastal ecosystems, as mangrove communities are often destroyed to make room for shrimp ponds, especially in developing countries such as Honduras and Vietnam, which in turn export the farmed shrimp to the United States. Populations of many oceanic species are at critical levels because of industrial style fishing practices as well as massive national subsidies for fishing fleets. Ironically, the destruction of ocean fisheries coincides with an increased demand for fish resulting from its recognition as a health food.

Forests

Although trees have economic value as a raw material, in many cases the environmental services provided by forests far transcend the economic value of trees. Mature trees maintain desirable microclimates and retard sediment loss that can poison near-shore marine ecosystems such as coral reefs. Forests store carbon, mediating climate change and thereby reducing excess carbon dioxide dissolved in seawater, which can lead to acidification of the oceans.

The United States leads the world in percentage of forest cover lost. The world's largest wood and paper producing region is not tropical rainforests, but rather in the U.S. south, the most heavily industrially logged region in the world.

Globally, forests are clear-cut and converted to tree plantations (e.g., loblolly pine in the U.S., palm in Indonesia), which do not provide ecosystem services comparable to those of the intact forests preceding them. In many cases the new trees are genetically engineered. Supplanting natural genetic diversity with monoculture increases the risks of rapid propagation of disease and loss of forest products. Finally, deforestation and other types of resource extraction may lead to negative economic impacts in rural communities.

Manufacturing and Industry

The Industrial Revolution generated wealth beyond humanity's dreams but also generated waste in unprecedented quantities, some of it artificial chemicals never before seen on Earth. Processing this waste is rapidly exceeding the capacity of natural systems, which did not evolve in the presence of many of these chemicals. In nature, waste eventually becomes something else's food. In human societies, waste is everywhere, and it indicates inefficiency. Pollution is one form of waste. Approaching "wasteless" production must become the norm in human activity, as it is in the marine realm, and progress is being made: The European Union has set a goal of ending landfill disposal by 2025. Many businesses have found that waste reduction and even elimination can enhance profitability, and progress has been made here as well, as some nations have agreed to phase out or eliminate the most harmful kinds of persistent organic pollutants (POPs). However, the growth of human populations and the universal association between increasing wealth and increasing waste pose critical problems for a world, five sixths of whose population is trying to develop along Western-style free-market lines.

Justice and Equity

The pursuit of justice and equal opportunity are key ingredients in building a sustainable society. Examples of injustice are a lack of adequate housing, a lack of access to education, poor sanitation, an inadequate supply of pure water, exposure to environmental toxins, and environmental degradation related to industrial pollution. Rich societies ignore these issues at their peril. Furthermore, injustice often drives rapacious and unsustainable use of natural resources such as bush meat "harvesting" in the African rain forest, coral reef fisheries, and mangrove wetlands.

Sustainable Consumption: An Oxymoron?

We are in the twilight of the era of the "myth of unlimited resources." Sustainability involves adopting responsible patterns of buying, consumption, and reproduction, thereby consuming minimal energy and fewer resources. Responsible consumption is based on education, not coercion, in a democratic society. Unfortunately, industrial and postindustrial societies are philosophically based on the myth of ever-increasing consumption, in turn driving ever-increasing production: the "growth" concept.

Development

Here we use *development* to refer to a complex set of changes that convert the economy of a society based on subsistence agriculture to one in which most of the employed inhabitants work in manufacturing or services.

Two hypotheses, much simplified, purport to explain such relationships as may exist between development and the environment.

1. Development harms the environment. Many environmentalists assert that development leads to some or all of the following: destructive land-use practices, mining of marine resources, injurious levels of air emissions and fossil-fuel use, and water pollution. Moreover, high levels of population growth in many developing countries exacerbate environmental degradation, leading to misery, child prostitution, civil wars, and the like and encouraging large-scale migration.

2. Development eventually improves the environment. Many economists and some environmentalists, while acknowledging harmful levels of environmental pollution in countries in early stages of development, cite considerable empirical evidence that (1) population growth rates decline as development proceeds and (2) rates of some forms of environmental pollution decline as per capita income increases. Newer forms of technology may be less polluting than older forms, but also tend to require high capital expenditures. Countries with higher per capita incomes tend to have cleaner environments along with increased consumption of goods and services. As nations become richer and middle classes expand, demands for scarce goods such as wild fish stocks expand, and tougher environmental standards may also; however, in many cases, passage of comprehensive legislation to preserve the coastal environment in a rich country simply leads to mining of marine resources for export from poorer countries. Most recently, the decline in fish stocks from the Atlantic off Northwest Africa has been due to increased exports to Europe and a concomitant decline in traditional fisheries.

■ THE STUDY OF EARTH'S HAZARDOUS PROCESSES

Earth's hazardous processes include earthquakes, slope failures, volcanic eruptions, floods, erosion, hurricanes, tsunami, brush fires, desertification, and sinkhole collapse. Added to these are hazardous human activities and impacts, including mining, damming of rivers, fossil fuel extraction, sea level change, environmental health hazards (e.g., lead, mercury, radon), groundwater pollution, waste disposal, and coastal pollution (FIGURE 2-10). Our objective should be to minimize their effects on human societies, and to sharply reduce or eliminate anthropogenic contributions to them.

Major advances in the understanding of the causes of natural events, like earthquakes, have allowed scientists to estimate probabilities for the occurrence of such events. The U.S. Geological Survey (USGS), for example, estimates that the probability of a major earthquake along portions of the San Andreas fault in California to be between 60% and 70% during the next 30 years. Governments can use such information to minimize damage and plan for major earthquakes, and individuals can make informed decisions about where to live.

Both natural and human-caused **geohazards** may be low-probability events. It is worthwhile noting, however, that low-probability events—given enough time— *ultimately will occur.*

■ ADDRESSING ENVIRONMENTAL GEOLOGY ISSUES

We are all "entitled to an opinion," it is said. We do not necessarily agree. We believe that accompanying this alleged entitlement is an obligation: the responsibility to know and understand the facts, issues, and concepts that allow us to make

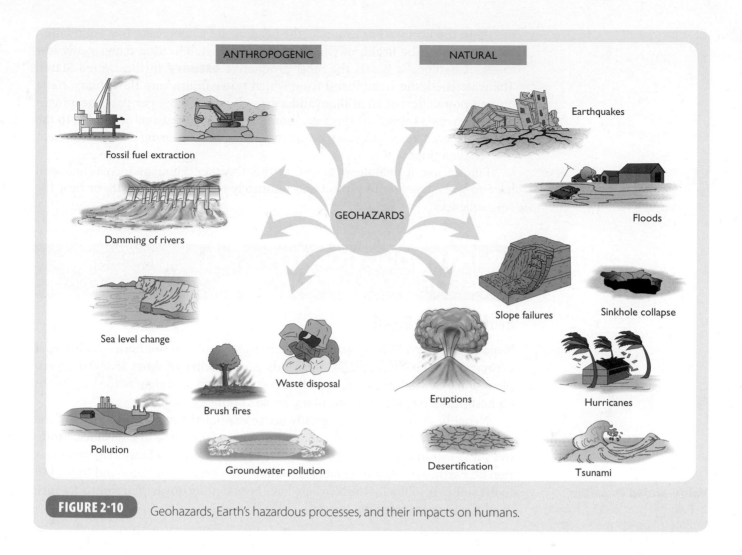

ANTHROPOGENIC   NATURAL

Fossil fuel extraction

Damming of rivers

Sea level change

Pollution

Brush fires

Waste disposal

Groundwater pollution

GEOHAZARDS

Eruptions

Desertification

Earthquakes

Floods

Slope failures

Sinkhole collapse

Hurricanes

Tsunami

FIGURE 2-10   Geohazards, Earth's hazardous processes, and their impacts on humans.

informed decisions and judgments. Therefore, although many expressions of opinion deserve a respectful hearing, not all opinions have equal standing, and certainly not in science, because positions and opinions are not always well thought-out. All citizens living under representative governments should learn to think logically and critically.

And we also have a responsibility to be numerate (that is, the ability to handle numbers well) and ecological (to understand the basic principles by which our environment works). One of the consequences of our innumeracy is a failure to comprehend **cumulative impact**. For example, many of us routinely consume soft drinks in aluminum cans or plastic bottles, and some of us "throw them away," an act of inconsequential environmental impact on an individual level. Multiply this by the tens of thousands of times this is repeated daily, however, and the impact becomes meaningful, considering that containers are recyclable, manufacturing the product consumes considerable energy, and landfill space is becoming limited.

Reason also leads us to appreciate and take pride in the great strides that have already been made to protect and restore our environment. Having that said, it is also true that great challenges remain that will test our inventiveness and mettle to the utmost. For example, the head of the Chesapeake Bay Foundation recently said, "The health of Chesapeake Bay has stabilized at dangerously low levels." Chesapeake Bay is the most productive **estuary** in the United States. The watershed, the area of land from which water drains into the estuary, has a human population of 16 million, and a growth rate of 1.3% per year, meaning it doubles every 54 years. At this rate, more than 200,000 persons are added to the watershed each year. Can you see why protecting and restoring this environment is so challenging?

Finally, like it or not, we are all in this together. Humans increasingly will have to work together to protect and ultimately restore planet Earth, or bear the consequences.

■ INTRODUCTION TO GEOHAZARDS AND RISK

Geohazards in Nepal

Nepal, a country a little bigger than Arkansas, is located in southern Asia between China and India (**FIGURE 2-11**). This landlocked country of about 28,000,000 people is home to Mt. Everest, the world's tallest mountain, and eight of the world's ten highest peaks. Nepal is also one of the world's poorest countries.

Nepal is situated in a geologically active earthquake region, and many of the steep slopes associated with its mountains are a prime cause of landslides. Moreover, it is within one of the most extensive belts of monsoon (seasonal) precipitation on Earth. Many of our planet's rainfall records were set in Nepal and India, just to the south. And finally, the Nepalese have been cutting down their forests for fuel, export, and space for agriculture.

Value-added Question
2-1. In the year 2000, 1 million Nepalese rupees were worth approximately 14,552 U.S. dollars. Use the table shown in Figure 2-12 to calculate the combined average annual loss from earthquakes, floods, landslides, and avalanches.

Value-added Question
2-2. In 2000, the total value of goods and services exchanged in Nepal, a quantity known as a country's *gross domestic product, or GDP,* was 5,560 million (or 5.56 billion) U.S. dollars. (A) What percent of the GDP was the combined average annual loss from earthquakes, floods, landslides, and avalanches? (B) Is the loss as large from a financial perspective? (C) A human perspective?

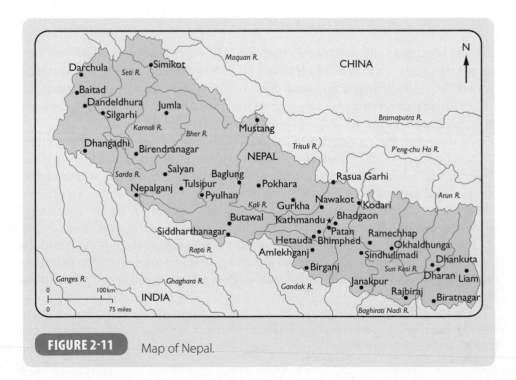

FIGURE 2-11   Map of Nepal.

Type of disaster	Number of death	Number of families affected	Number of livestock killed	Number of houses damaged	Loss of agricultural land (ha)	Loss of property (million Rupees)
Fire	61	4010	771	3323	0.0	205.0
Epidemics	656	1166	16	0	0.0	0.0
Earthquake	40	3799	784	5898	0.0	634.7
Flood, landslide and avalanche	330	32828	2371	7143	5712.6	748.9
Hailstorm, windstorm and thunderbolt	40	6012	127	980	0.0	157.9
Total	1127	47814	4069	17344	5712.6	1746.4

Source: DPTC, 1998 and Ministry of Home

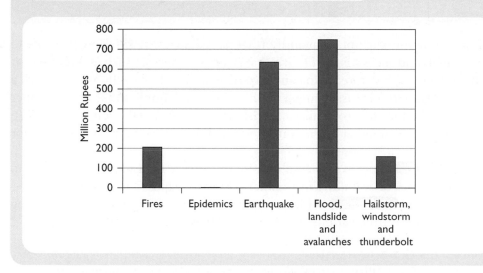

FIGURE 2-12   Average annual loss of lives and property in Nepal. The two greatest hazards are _____ and _____.

Losses from various hazards in Nepal are shown in **FIGURE 2-12**.

What conclusion could we draw from this brief introduction? That *geology is a fundamental criterion to use in the assessment of hazards and risk.*

■ DEFINING HAZARDS AND RISK

It bears repeating that *geology is a fundamental criterion to use in the assessment of hazards and risk.*

How would you define a "geohazard?" Is *risk* the same as *hazard?*

A geological hazard, or *geohazard,* arises out of humanity's interaction with the natural environment. Geohazard risk assessment has three aspects: assessing the risk of personal harm, assessing the risk of financial loss, and assessing the extent to which people are willing to accept risk from the hazard, assuming they are free to do so.

We will limit our consideration in this book to earthquakes, volcanic eruptions, floods, and slope failures (mass movement). We will not include impact from wildfires and violent weather, which, of course, may be considerable.

Global Impact of Geohazards

Over the decade of the 90s, the World Bank estimated yearly losses from floods, earthquakes, volcanoes, and associated landslides for developing countries at $35–$40 billion U.S. According to the reinsurer Munich Re, global disaster losses were $378 billion in 2011.

As much as half of humanity lives in regions that are significantly at risk from geohazards. In many countries, poverty is a contributing factor in risk. For example, most of the residents of Peru live in adobe structures that can be severely damaged or destroyed in an earthquake, and cannot afford safer housing. Similarly, the January 10, 2010 Haiti earthquake left 3 million Haitians in need of emergency aid and 2 million homeless in the Western Hemisphere's poorest country.

One of the objectives of this book is to educate you about hazards, how they affect us, and what we can do about them.

■ ASSESSING RISK

How would you attempt to quantify (that is, measure) "risk"? How valuable to individuals and society is *qualitative*, that is, non-numerical, such as "elevated" or "above average," risk forecasting and hazard analysis? How would you use such a warning? *Quantitative* assessment, on the other hand, implies that some number can be applied to the probability of a risk or a hazard. Would you know how to compare such a number to risks in everyday life that we more or less take for granted?

TABLE 2-1 shows examples of risk. In it, activities are listed which purport to increase the risk of death by 1 in 1 million to anyone exposed to the risk.

Now study **TABLE 2-2**, which shows one estimate of the probability of dying in any given year from various causes. Some of these are described as "accidents."

Concept Check 2-4. (A) How would you define an accident? Using that definition, identify the items in Table 2-2 that you think are truly accidents, and identify those that you think are not accidents. (B) Select those for which you need other information to decide, and state what kind of information you would need. (C) To hone your analysis skills, look at the data in the table for deaths from earthquakes in Iran and California. Why do you think the risk is so much greater in Iran? List some reasons (hypotheses) and then tell what kind of information you would need to test your hypotheses.

- -

Concept Check 2-5. (A) *Assuming* (see section on critical thinking elsewhere in this text) the numbers in Table 2-2 are accurate, assess the geohazard risks compared to others on the list in daily life. If you were allocating government resources to attempt to manage risk, where would you direct those resources? (B) What other information would you like to have to more comprehensively answer this question? For example, many of us carry out several of these activities simultaneously: we may smoke, drive a car, own a firearm, *and* live in an earthquake-prone region.

- -

Now let's select three kinds of hazards and discuss some aspect of the relation between hazard and risk for each. They are volcanic processes, earthquakes, and floods.

TABLE 2-1	Everyday Risk Activities That Increase Risk of Death by 1 in 1 Million

Activity	Cause of Death
Smoking 1.4 cigarettes	cancer, heart disease
Living 2 months with a smoker	cancer, heart disease
Spending 1 hour in a coal mine	black lung
Spending 3 hours in a coal mine	"accident"
Traveling 6 minutes by canoe	"accident"
Traveling 10 miles by bicycle	"accident"
Traveling 150 miles by car	"accident"
Flying 1,000 miles by jet	"accident"
Flying 6,000 miles by jet	cancer, radiation
Having 1 chest x-ray	cancer, radiation
Eating 100 charcoal-broiled steaks	cancer from benzopyrene

(a unit of risk known as a *micromort*)

TABLE 2-2	Probability of Dying in Any Year From Various Causes

Smoking 10 cigarettes a day	1 in 200
Working in Chinese coal mines	1 in 800
All natural causes, age 40, United States	1 in 850
Firearm, United States	1 in 7,000
Homicide, Europe	1 in 100,000
Road crash, United States	1 in 7,000
Earthquake, Iran	1 in 23,000
Earthquake, California	1 in 2,000,000
Home accident	1 in 25,000
Flood, Bangladesh	1 in 50,000
Flood, United States	1 in 2,000,000
Lightning strike	1 in 10,000,000
Volcanic eruption, United States	1 in 400,000,000

Assessing Financial Losses from Earthquakes

It is clear that many hazards in life are more "dangerous" than geohazards, if by dangerous we mean the "average" risk of loss of life, but part of risk assessment is also assessing financial losses. For example, you will see later that the Kobe, Japan earthquake in 1995 cost that country the equivalent of at least $200 billion (U.S.) as well as thousands of tragic deaths. In fact, financial impacts may outweigh impacts from injuries and loss of life from geohazards throughout the world. How do we assess financial impact of geohazards?

CASE STUDY   Assessing Volcanic Hazards at Weed, California

In this case study, you will use maps to assess the hazard from volcanic activity to Weed, California. The several types of volcanic flows and deposits mentioned in this chapter are defined and described fully elsewhere in this text.

Look at **Figure CS 1-1A**, a view of Weed, California with Mt. Shasta in the background. Weed and vicinity (pop. ~ 8000) is actually built on an enormous volcanic mudflow that formed during an eruption of Mt. Shasta 300,000 years ago. **Figure CS 1-1B**, shows a mudflow hazard zone map of the Shasta area with the site of Weed indicated.

The eruption of Mt. St Helens in 1980, which killed 57 people, produced great clouds of volcanic ash that contaminated water supplies, killed crops, and disrupted transportation, among other things. One of the key kinds of data we need to assess risk from ash deposits is the prevailing wind direction near the site of an eruption. At Shasta, winds blow persistently (more than 90% of the time) from the northwest, west, and southwest, toward the east, northeast, and southeast. Examine **Figure CS 1-1C**, a map showing the hazard from volcanic ash at Shasta that does not take into consideration wind direction.

Using what you know about wind direction, where is ash most likely to accumulate?

During an eruption, you might assess the risk to Weed from lava flows as very great. **Figure CS 1-1D** is a lava flow hazard map. Based on this, assess the potential for lava flow damage to Weed. **Figure CS 1-1E** assesses the risk to the region from destructive pyroclastic flows and lateral blasts, which are like hurricanes but loaded with searing hot ash. Assess the risk to Weed from these events.

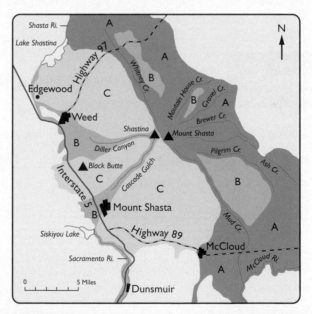

FIGURE CS 1-1B   A mudflow hazard zone map of Mt. Shasta. Note the scale—length of the bar is 5 miles (8 km). How far is Weed from Shasta's crest? Based on Weed's location and the map, what is the risk the city faces from another volcanic mudflow? Modified from Crandell, D. R. & D. R. Nichols. 1987. Volcanic Hazards at Mount Shasta, California. Reston, VA: U.S. Department of the Interior, U.S. Geological Survey.

FIGURE CS 1-1A   Weed, California with Mt. Shasta in the background. Courtesy of USGS.

Continued ▶

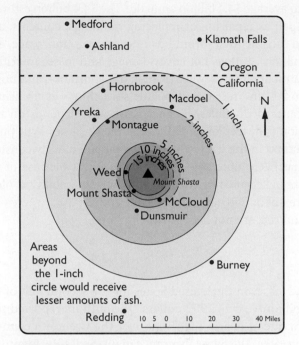

FIGURE CS 1-1C Volcanic ash hazard map for Mt. Shasta. Modified from Crandell, D. R. & D. R. Nichols. 1987. Volcanic Hazards at Mount Shasta, California. Reston, VA: U.S. Department of the Interior, U.S. Geological Survey.

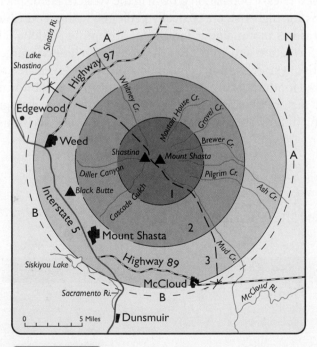

FIGURE CS 1-1D A Lava flow hazard map for Mt. Shasta and vicinity. Modified from Crandell, D. R. & D. R. Nichols. 1987. Volcanic Hazards at Mount Shasta, California. Reston, VA: U.S. Department of the Interior, U.S. Geological Survey.

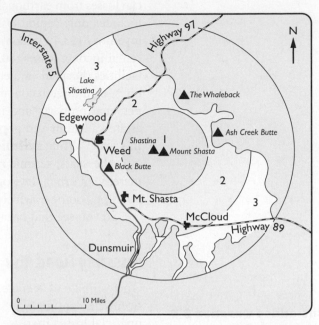

FIGURE CS 1-1E Zones of relative risk from pyroclastic flows and lateral blasts. Modified from Crandell, D. R. & D. R. Nichols. 1987. Volcanic Hazards at Mount Shasta, California. Reston, VA: U.S. Department of the Interior, U.S. Geological Survey.

The essence of risk assessment is trying to determine how likely a destructive event is to occur or recur, and what its effects will be, based on geological data.

Write out the kinds of information you think a geologist would need to know to assess the hazard posed to Weed by Mt. Shasta. Two factors you would likely list would be the age and nature of all lava flows and ash deposits from Shasta during historic time, for example.

Next, assuming the hazard from another eruption could be determined, what use should be made of that information? Should, for example, authorities have the right to forcefully evacuate residents for their own protection, or should they simply publicize the information and trust citizens to use their best judgment? What about people with young children (children can't choose), who decide to stay and risk an eruption? How certain should geologists have to be in order for authorities to order an evacuation? What form should their warning have to take? Suppose they were to issue a report concluding that Shasta had a 90% chance of erupting during the next year? During the next month? During the next week? A 55% chance? Under what conditions would you be inclined to evacuate? The answers to these questions go to the heart of hazard analysis and risk assessment, and how people respond to it.

The U.S. Federal Emergency Management Agency (FEMA) estimates U.S. financial losses from earthquakes to average $4.4 billion annually. The $4.4 billion estimate includes only capital losses such as repairing or replacing buildings, contents, and inventory ($3.49 billion), and income losses from business interruption, wage, and rental income losses ($0.93 billion). It does not cover damage and losses to critical facilities (hospitals, water treatment plants, sewage plants), transportation and utility lines, or indirect economic losses. Indirect losses include such things as uninsured personal property and insurance deductibles. Financial risk also depends on real estate valuation and population growth. Former FEMA Director James Witt put it this way: "While **seismic hazard** in the United States has remained fairly constant over the years, seismic *risk* has been increasing substantially. That's because of the increase of urban development in seismic hazard areas and the vulnerability of older buildings, some of which were not built to adequate seismic code."

So you see that hazard and risk are not necessarily the same.

Assessing Flood Risk

Value-added Question 2-3.
How much greater would the U.S. flood losses have to be to reach 1% of the 2012 GDP of $15,290 billion?

One estimate of average flood losses in the United States for the second half of the twentieth century was $3.9 billion/year in 1999 dollars. The authors reported that financial losses increased at a rate of nearly 1.2%/year over the twentieth century. Globally, flood risk is much greater. The Swiss insurance giant, Swiss Re, identified a number of countries in which flood losses could reach or exceed 1% of gross domestic product (GDP) over the next decade. They are Argentina, Ecuador, Brazil, China, Honduras, and Nicaragua.

TABLE 2-3 shows Swiss Re's assessment of potential flood losses for Honduras, a country with one of the highest rates of population growth in the Western Hemisphere.

Here again, the reasons for high risk are basically geologic, with a contributing human factor. Honduras has mountainous terrain (contributing to slope failure risk), due to crustal deformation and volcanism. Its native vegetation is largely rainforest that is very effective at storing rainfall, but much of it has been cleared for fuel and agriculture. Finally, Honduras is in a belt of high risk from tropical storms.

In summary, the basic factors causing flood *hazard* are natural and geologic; flood *risk* results from living in flood-prone areas, and unwise land-use decisions.

TABLE 2-3	Flood Risk in Honduras (Swiss Re)	
Event (RI)*	**Loss of Capital Stock***	**As % of Total Capital Stock**
10 yr	$100 million	1%
50 yr	$650 million	5%
100 yr	$1,000 million	12%
500 yr	$4,000 million	31%

*RI = Recurrence Interval
**Examples are buildings, roads, railroads, and port facilities.

■ CONCLUSION: GEOHAZARDS AND RISK ASSESSMENT

We define geohazards to include earthquakes, volcanic activity, floods, and mass movement.

Hazard assessment is similar to, but not the same as, **risk assessment**. Hazard assessment involves understanding and quantifying the geological processes that can cause disasters in a region. Risk assessment involves determining what the impact on human populations in the affected area will be. In most cases, *hazard mitigation is not as successful as risk mitigation.* Basic to both are an understanding of the geology of an area, and the impact humans have on their natural environment, especially land-use changes. Education of a population at risk from geohazards is an essential process in risk mitigation.

© NinaMalyna/ShutterStock, Inc.

CHAPTER SUMMARY

1. Human population growth and resource consumption are the ultimate challenges.
2. Water may be the ultimate natural resource.
3. We must protect our soils, which, next to water, may be the most important planetary resource.
4. Actions have consequences and many of them are unforeseen and harmful.
5. Many impacts are transboundary in nature.
6. Hazardous Earth processes and activities should be intensely studied.
7. We must use reason to address environmental geology issues.
8. Reason dictates that sustainability must be the aim of all successful societies.
9. Geology is a fundamental criterion to use in the assessment of hazards and risk. An objective of this book is to educate you about hazards, how they affect us, and what we can do about them. We define geohazards to include earthquakes, volcanic activity, floods, and mass movement.
10. Half of humanity lives in regions that are significantly at risk from geohazards.
11. Geohazard risk assessment has three aspects: assessing the risk of personal harm, assessing the risk of financial loss, and assessing the extent to which people are willing to accept risk from the hazard.
12. Over the decade of the 1990s, the World Bank estimated yearly losses from floods, earthquakes, volcanoes, slope failure, and tropical storms for developing countries at $35–$40 billion U.S. The reinsurer Munich Re reported that global disaster losses were $378 billion in 2011.
13. The essence of risk assessment is trying to determine how likely a destructive event is to occur or recur, and what its effects will be, based on geological data.
14. The U.S. Federal Emergency Management Agency (FEMA) estimates U.S. financial losses from earthquakes to average $4.4 billion annually.
15. Average flood losses in the United States for the second half of the twentieth century were $3.9 billion/year (in 1999 dollars).
16. In most cases, *hazard mitigation is not as successful as risk mitigation.*

KEY TERMS

bauxite

cumulative impact

estuary

exponential growth

externality

geohazard

hazard assessment

hydrologic cycle

product life-cycle
  assessment

risk assessment

salinization

seismic hazard

subsidy

sustainability

topsoil loss

transboundary

REVIEW QUESTIONS

1. Which phrase best describes the changes to human population during the twentieth century?
   a. Human population increased to 1.6 billion and life expectancy halved.
   b. Human population increased to over 6 billion and life expectancy halved.
   c. Human population increased to more than 6 billion and life expectancy doubled.
   d. Human population increased to 1.6 billion and life expectancy doubled.

2. The present human population growth rate per year is closest to which number?
   a. 1.3%/year
   b. 2.7%/year
   c. 0.1%/year
   d. 7%/year

3. Which is the correct relationship between U.S. population and global resource consumption?
   a. The United States comprises 1% of global population and consumes 10% of global resources.
   b. The United States comprises less than 5% of global population and consumes a quarter of global resources.
   c. The United States comprises 10% of global population and consumes 10% of global resources.
   d. The United States comprises less than 5% of global population and consumes 5% of global resources.

4. According to the World Resources Institute, what percentage of the world's population is facing "water stress"?
   a. None
   b. All
   c. About 5%
   d. About 40%
   e. About 75%

5. Which best describes the state of the planet's soils?
   a. Soils are being degraded at rates that exceed their regeneration or purification rates.
   b. Soils are being degraded but not at faster rates than their regeneration or purification.
   c. Soils are being regenerated at rates that far exceed their degradation rates.
   d. Because farmers put fertilizer on soils, they are protected from degeneration.

6. What percent of U.S. electricity production is used to refine aluminum, up to half of which is discarded?
   a. 50%
   b. 10%
   c. less than 0.1%
   d. up to 3%
   e. 25%

7. Which is an example of transboundary pollution?
   a. dust from African storms polluting Caribbean coral reefs
   b. lead pollution next to a lead smelter
   c. air pollution from Asian power plants affecting the U.S. West Coast
   d. all of these (a–c)
   e. a and b only
   f. a and c only

8. The Environmental Performance Measurement Project's Environmental Sustainability Index (ESI)
   a. ranks nations on the basis of sustainability.
   b. is an accurate index based on available data.
   c. ranks the United States as the world's most sustainable nation
   d. ranks Africa as the world's most sustainable region

9. The Sustainable Communities Network targets for a Sustainable Planet Earth include all but the following:
   a. Creating community
   b. Continually expanding economics
   c. Protecting natural resources
   d. Governing sustainably

10. Achieving sustainability requires addressing which of the following issues:
    a. population
    b. energy and climate change
    c. biodiversity, fisheries, and forests
    d. manufacturing and industry, and justice and equity
    e. all of the above

11. Use the Doubling Time equation, T = 70/R (see Appendix 2, Basic Population Math) to determine how many years it will take the population of the Chesapeake Bay watershed, presently 16 million, to double. Rate of growth is 1.3% per year. T = doubling time, R = % increase per year (1.3% would be entered as 1.3).
    a. 5.4 years
    b. 540 years
    c. 54 years
    d. 10 years

12. What portion of humanity lives in regions susceptible to geohazards?
    a. all
    b. half
    c. a quarter
    d. 10%

13. Which activity has the highest probability of causing death in any year?
    a. Working in Chinese coal mines.
    b. Owning a firearm in the United States.
    c. Home accident in the United States.
    d. Road crash in the United States.

14. The siting of towns can give an idea of the risk you accept in living there. The town of Weed, California is built on
    a. an active volcano.
    b. the San Andreas Fault.
    c. swelling clays.
    d. a volcanic mudflow.
    e. an abandoned landfill

15. The reason for high hazard risk to residents of Honduras, as documented by Swiss Insurer Swiss Re, is
    a. the country's long civil war.
    b. land mines.
    c. its geologic setting.
    d. the policies of the country's Marxist government.

Human Population Growth

The essence of the population debate is not the absolute number of people on the planet *per se*; it is whether or not the Earth has exceeded its human **carrying capacity**, the maximum number of people that the planet's resources can sustain in the long term without irreversibly harming the planetary ecosystem. What then is the Earth's human carrying capacity and have we exceeded it? If we have, what are the consequences now and into the future?

Point

One view of human population is that the carrying capacity of the planet is essentially limitless. The late business economist Julian Simon argued that the Earth could support 20 billion or more people. A chapter in his book, *The Ultimate Resource II: People, Materials, and Environment,* is entitled *Can The Supply Of Natural Resources—Especially Energy—Really Be Infinite? Yes!*

He wrote:

> Adding more people causes problems, but people are also the means to solve these problems. The main fuel to speed the world's progress is our stock of knowledge, and the brakes are (a) our lack of imagination, and (b) unsound social regulation of these activities.
>
> The ultimate resource is people—especially skilled, spirited, and hopeful young people endowed with liberty—who will exert their wills and imaginations for their own benefit, and so inevitably benefit not only themselves, but the rest of us as well.

Counterpoint

Simon's view is disputed by many population and environmental scientists. They cite evidence suggesting we have already exceeded the planet's carrying capacity. For example, biologist Garrett Hardin writes:

> On the American scale of living the carrying capacity of the Earth is only about one-hundredth as great as it would be if people would be content with the barest minimum of goods . . .
>
> Moreover, some [amenities] impose costs that cannot be stated . . . the solitude of lonely beaches, access to wilderness and areas rich in flowers, birds, and butterflies, together with time to enjoy these amenities as well as music and the visual arts . . .
>
> . . . Thus does the problem of the optimum human population become inextricably woven with problems of value.

The late oceanographer Jacques Cousteau was more pessimistic:

> The road to the future leads us smack into the wall. We simply ricochet off the alternatives that destiny offers: a demographic explosion that triggers social chaos and spreads death, nuclear delirium and the quasi-annihilation of the species . . . Our survival is no more than a question of 25, 50 or perhaps 100 years.

Which view is correct? After you have read the following chapter, and indeed the entire book, judge for yourself. Accurately assessing the Earth's carrying capacity is one of humankind's most important tasks, and environmental geology plays a major role in this endeavor.

◼ INTRODUCTION

Why is human population growth included as one of the first chapters in this and other environmental geology texts?

Although humans exert a substantial positive impact with our enormous ingenuity, as Julian Simon asserted above, we also have a profound physical impact on our local, regional, and global geological environment, and *our geological environment also impacts us.*

First, we take up space—space that at one time was forest, marsh, prairie, or hillside. The space an individual occupies may be minimal, as in cities in India,

FIGURE 3-1 (A) A shack in Chennai, India, one of the world's fastest developing economies. (B) A favela in Salvador, Brazil. (C) Corrugated metal huts outside of Johannesburg, South Africa. A–C © Pluff Mud Photography.

Brazil, or South Africa (**FIGURE 3-1**) or it may be large, as in newer American single-family houses, which averaged over 2,400 ft² by 2013 (**FIGURE 3-2**).

Because humans must eat, we require land for agriculture, and much of that land is irrigated. In most instances this land is also fertilized, usually with industrially produced fertilizers. Fertilizers contribute to **nutrient pollution**, which sounds like an oxymoron (How can one have *too many* nutrients?) but is in fact one of the most pervasive forms of water pollution. Excess fertilizer runs off from farmlands and turfgrass into waterways and ultimately to the ocean. There, the fertilizers promote the growth of minute *phytoplankton*, which die and decompose, in the process depleting the water of oxygen. Globally, this phenomenon leads to the production of huge expanses of ocean where life cannot survive. These areas are known as **dead zones**. **FIGURE 3-3** shows the Gulf of Mexico dead zone in 2010. Nutrient pollution is largely responsible for the environmental crisis of the largest U.S. estuary, the Chesapeake Bay.

An additional pollution load is generated by animal waste during meat production, using so-called animal factories or factory farms (also known as **Concentrated Animal Feeding Operations, or CAFOs**). Agricultural pollution frequently causes large-scale fish kills in streams draining agricultural areas and CAFOs. Perhaps most significant, soil erosion and loss of fertility threaten the

TABLE 3-1

TABLE 3-1	Characteristics of Housing in the U.S., India, China, and Ghana		
Country	**Average Size Sq. ft. (year)**	**Avr. Household Size (year)**	**Area per Person Sq. ft.**
United States	2,438[1] (2009)	2.6[5] (2009)	
China	753[2] (2001)	3.0[6] (2008)	
India	499[3] (2008)	4.3[3] (2008)	
Ghana	323[4] (2005/06)	4[7] (2005/06)	

[1] http://www.census.gov/const/C25Ann/sftotalmedavgsqft.pdf
[2] http://english.peopledaily.com.cn/200109/25/eng20010925_80981.html
[3] 63rd round survey of the National Sample Survey Organisation, cited in The Times of India (http://timesofindia.indiatimes.com/33_of_Indians_live_in_less_space_than_US_prisoners/articleshow/3753189.cms)
[4] http://www.statsghana.gov.gh/docfiles/glss5_report.pdf
[5] http://factfinder.census.gov/servlet/SAFFFacts
[6] http://www.womenofchina.cn/Data_Research/Latest_Statistics/18158.jsp
[7] http://www.statsghana.gov.gh/docfiles/glss5_report.pdf

Value-added Question 3-1.

TABLE 3-1 shows characteristics of housing in the United States, India, China, and Ghana. Use the data in the table to calculate the average floor space per person in square feet for each nation. Discuss your findings.

ability of agriculture to feed the growing human population. FIGURE 3-4 depicts some adverse environmental impacts of industrial agriculture.

People also require transportation. Roads generate polluted runoff (which even includes the toxic dust from tires wearing down!), cover permeable land with impervious paving, and take formerly productive land off the tax base. Roads also take an enormous toll on animals (including humans), divide habitats (which can accelerate species loss), and provide human access to wilderness and forestland, frequently

FIGURE 3-2 Typical new U.S. house, this one in Pawley's Island, South Carolina. What would happen to demand for commodities (lumber, cement, metals, etc.) and what would the environmental impact be if the Indian, Brazilian, and South African homes shown in Figure 3-1 would be built to resemble newer American ones? © Pluff Mud Photography.

Gulf of Mexico dead zone shown in a false-color satellite image. This dead zone is caused largely by nutrient pollution from agriculture in the Mississippi River watershed. Reds and oranges represent high concentrations of phytoplankton that, when decomposed, will decrease the water's oxygen content. Courtesy of NASA.

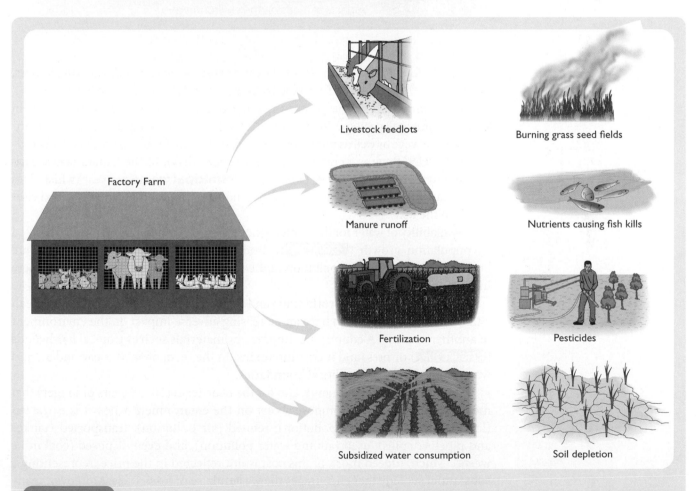

Livestock feedlots

Burning grass seed fields

Factory Farm

Manure runoff

Nutrients causing fish kills

Fertilization

Pesticides

Subsidized water consumption

Soil depletion

FIGURE 3-4 Summary of some adverse environmental impacts of industrial agriculture, as practiced in the United States, parts of Western Europe, Australia, and elsewhere. Impacts include: greenhouse gas emissions; fertilizer, pesticides, and manure in runoff to nearby streams; air pollution, fossil fuel use; subsidized water consumption to grow surplus crops; and soil depletion.

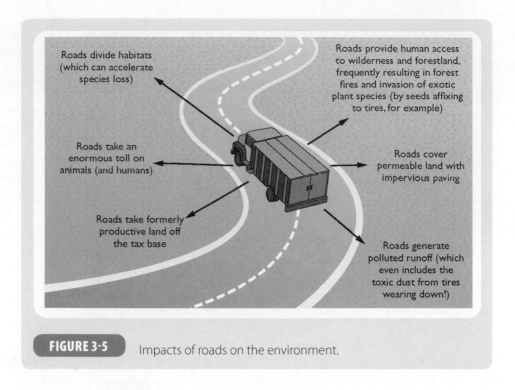

FIGURE 3-5 Impacts of roads on the environment.

Roads divide habitats (which can accelerate species loss)

Roads provide human access to wilderness and forestland, frequently resulting in forest fires and invasion of exotic plant species (by seeds affixing to tires, for example)

Roads take an enormous toll on animals (and humans)

Roads cover permeable land with impervious paving

Roads take formerly productive land off the tax base

Roads generate polluted runoff (which even includes the toxic dust from tires wearing down!)

resulting in forest fires and invasion of exotic plant species (by seeds affixing to tires, for example; **FIGURE 3-5**). This issue is discussed in detail elsewhere in the text.

Waste produced by the growing number of humans is another major problem in environmental geology. In developed countries, a growing amount of human-generated sewage as well as municipal and industrial waste is being produced. According to the U.S. Environmental Protection Agency (EPA), in the United States, each person produces around 730 kg (1,606 lb) of **municipal waste** per year (which does not include sewage, mining waste, or other industrial waste; these are discussed elsewhere in the text).

Availability of water for domestic, agricultural, and industrial uses also is affected by population growth (**FIGURE 3-6**). Both surface and groundwater sources are becoming endangered by pollution and excessive withdrawals (detailed elsewhere in the text).

Freer (less-regulated) world trade and a more integrated global economy mean citizens of one country can have an increasing adverse impact on the environment in another country. A country can import raw materials such as tropical hardwoods (**FIGURE 3-7**) or ores, and it can ship toxins, in the form of solid waste and air and water pollution, across political boundaries.

Humans also require energy, and for the near-term (10–20 years or longer) that means fossil fuels, which impose a cost on the environment when it is extracted (land degradation, water pollution), refined (air pollution), transported (tanker and pipeline spills), used (air and water pollution), and even disposed (coal mine waste). Much, indeed most, of this cost is not reflected in the price, representing a substantial subsidy that encourages the wasteful use of fossil fuels. Moreover, many countries still actively subsidize fossil fuel use by keeping prices artificially low.

While human needs are taxing precious natural resources, our ever-increasing population is concentrating in areas at risk from potent **geohazards**. Here are a few examples:

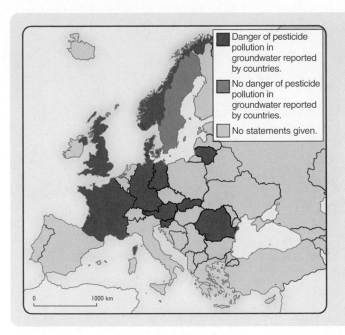

Danger of pesticide pollution in groundwater reported by countries.

No danger of pesticide pollution in groundwater reported by countries.

No statements given.

0   1000 km

FIGURE 3-6 Water use affected by population growth. Danger of groundwater pesticide pollution in European countries is shown in red; green indicates no data.

- At least 60% of the planet's human population lives within 100 km of the coast (**FIGURE 3-8**), and coastal cities are reaching sizes unprecedented in history. The population of São Paolo, Brazil, for example, is approaching 20 million. Thus ever-increasing numbers of people are exposed to flooding from tropical storms as well as to coastal erosion and sea-level rise accompanying global climate change.
- Las Vegas, Nevada is the fastest growing metropolitan region in the United States. It has a voracious appetite for energy and fresh water, and is expanding into areas at serious risk from catastrophic flash floods.
- California is one of the fastest growing states in the United States. Major population centers in California are located close enough to the San Andreas Fault to suffer billions of dollars of damage from even moderate earthquakes.

(A)

(B)

FIGURE 3-7 (A) Malaysian teak plantation. © Louise Murray/Visuals Unlimited/Getty Images. (B) Teak from Malaysia sits in a shipyard in Chennai, India. © Pluff Mud Photography.

(A)   (B)

FIGURE 3-8 (A) Stilt houses at Tai O fishing village near Hong Kong. (B) Shacks in the shadow of Ho Chi Minh City (Saigon), Vietnam. A and B © Pluff Mud Photography.

Value-added Question 3-2.
The United States and world populations in 2012, were about 314 million (written in scientific notation as 314×10^6) and 7 billion (7×10^9), respectively. What percentage of the world's population is that of the United States?

Value-added Question 3-3.
The United States and world, respectively, use about 100 and 462 quads of energy annually (this may sound foreign to you, but a quad equals 10^{15} BTUs, in other words, a LOT of energy). What percentage of the world's energy use is that of the United States?

Value-added Question 3-4.
The United States and world, respectively, produce about 5.9 and 29.2 billion tonnes (recall that one *tonne* equals 1,000 kg, or about 2,200 pounds) of the main greenhouse gas, carbon dioxide. What percentage of the world's carbon dioxide emissions use is that of the United States?

- China, with more than 1.3 billion people, is desperately trying to curtail its population growth, often using methods that many in more democratic countries find offensive. Rampant industrialization in China is producing quantities of pollution and greenhouse gases that literally are beginning to threaten the entire planet.
- Population growth in Central America is very high compared to the United States. This puts increasing numbers of people at risk from destructive earthquakes, tropical storms, and slope failure. It also helps drive land degradation, and migration to the United States.
- The Middle East (Jordan, Egypt, the Palestinian Authority, Israel, Iran, and Saudi Arabia, for example) has some of the highest rates of population growth on Earth. This growth combines with age-old ethnic and religious animosity and disputes over water rights to produce perhaps the planet's most unstable political region. One wonders what the region will be like with twice the present population competing for scarce resources within two decades.

So, although each additional human being contributes something unique to the planet, and may be the source of problem-solving ingenuity, each of us places physical demands on Earth as well: the more humans, the greater the potential demand.

Finally, people in developed countries like the United States, the European Union, Canada, and Australia consume disproportionately more resources than residents of developing countries.

Concept Check 3-1. Summarize the impacts of human population growth.

Concept Check 3-2. As stated above, people in developed countries consume disproportionately more resources than residents of developing countries. (A) Explain whether your calculations in the last two questions support this claim. (B) List additional information that would enable you to more fully support our assertion.

Some consider these to be issues of **environmental justice**, and seek ways to provide greater access to resources to all residents of Planet Earth.

Biologist Garrett Hardin's **Third Law of Human Ecology** states that the "*total impact of a population on its environment is determined by the absolute population size multiplied by the impact per person.*"

Concept Check 3-3. In the above discussion we have ignored the impact per person. (A) In what ways does the impact on the planet of a U.S. resident differ from your perception of the impact of a resident of a developing country? (B) Discuss whether you think the use of resources by Americans is a fair allocation of the planet's resources.

Next, we are going to examine world population growth patterns.

■ MEASURING POPULATION GROWTH

Statistics used to measure population growth include *fertility rate, rate of natural increase, birth rate, death rate,* and *net migration rate.*

Fertility Rate

One widely used definition of **fertility rate** is the *average number of children born to a woman during her lifetime.* This is also referred to as the **total fertility rate (TFR)**. TFRs in 2009 range from a high of 7.75 in the African country Niger to 1.09 for the Asian island nation Singapore. TFR of the United States and the world are 2.05 and 2.56, respectively.

The value of TFR at which global population growth would be zero is about 2.33 children per woman. In industrialized countries, this **replacement fertility rate** is about 2.1, whereas in developing countries it ranges from 2.5 to 3.3.

Concept Check 3-4. Why do you think developed countries have a much lower replacement fertility rate than developing countries? In other words, why must women in developing countries give birth to more children than in developed countries, if population is to remain stable in both groups?

Concept Check 3-5. TFR is not the best indicator of a country's population growth. Suggest why this is so.

Other commonly used statistics to determine rate of population growth are **birth rate** (typically, the number of live births per year per thousand of population) and **death rate** (often expressed as the number of deaths per 1,000 of the population per year) and net migration.

Natural Increase and Birth Rate

The **rate of natural increase (or decrease)** in a country's population can be calculated by subtracting the death rate from the birth rate. This does not include emigration or immigration (hence the term *natural*), which can materially change a country's population in many ways. For example, immigrants from developing countries may have children at higher rates than indigenous residents. Thus, over time, the population growth rate would increase with immigration. The rate of population growth that includes migration is the *overall increase* (or decrease).

Death Rate

During the twentieth century, death rates came down in most countries, due to better medical care and better hygiene. Birth rates were often slower to change, especially in developing countries, which had historically experienced high child mortality rates. Death rates by country range from a high of near 30 per 1000 in war-torn countries like Mozambique, to far lower rates in developed countries like the United States (8.4) the United Kingdom (10.2), and Canada (7.6).

Concept Check 3-6. Suggest hypotheses to explain the disparities in death rates among countries.

To calculate a country's overall population growth rate (as opposed to its *natural rate of growth*), we must consider net migration rate. In practice, simply subtract the death rate from birth rate, and add the net migration rate (or subtract if it is negative). To present this as a percent, simply divide the result by 10. Population math, growth rate calculations, and step-by-step examples of other population growth calculations can be found elsewhere in this text. (You may need a scientific calculator for some of the computations.)

Value-added Question 3-5. Examine **TABLE 3-2**, which shows birth rates, death rates, and net migration rates for China, India, and the United States, the world's three most populous countries, in 2012. Use the data to calculate the population growth rate per thousand population in each country.

Concept Check 3-7. As we constructed Table 3-2, two numbers from different countries stood apart from the others to us. Use your crystal ball to guess which numbers these were, and then explain why you think they stood out to us (i.e., explain how each contributed to population growth in the respective country).

Concept Check 3-8. **FIGURE 3-9** shows global population growth rates. (A) Describe any geographic pattern you see. (B) Where are high birth rate countries? (C) Where are birth rates low? Remember these values give you only a part of the data you need to evaluate global population growth. (D) What other information do you need to know?

TABLE 3-2	Mid-year 2012 Birth, Death, and Net Immigration Rates for China, India, and the U.S. (CIA World Factbook)		
	Birth Rate (per thousand population)	**Death Rate (per thousand)**	**Net Immigration (Immigration— Emigration, per thousand)**
China	12.31	7.17	−0.33
United States	13.68	8.39	3.62
India	20.60	7.43	−0.05

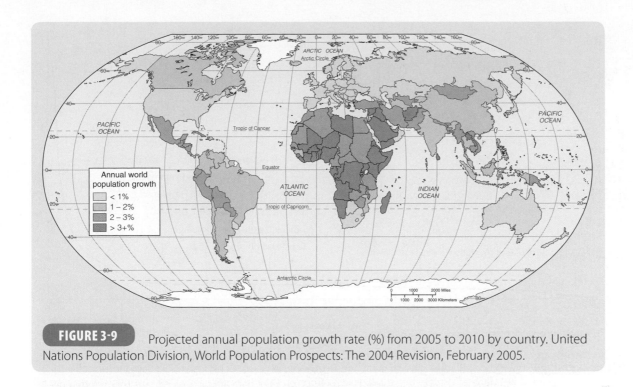

FIGURE 3-9 Projected annual population growth rate (%) from 2005 to 2010 by country. United Nations Population Division, World Population Prospects: The 2004 Revision, February 2005.

■ EXPONENTIAL GROWTH AND ITS IMPACTS

On October 12, 1999, the United Nations "celebrated" its "**Day of 6 Billion**," the day at which the Earth reached an estimated human population of 6 billion. In 2012 the world population surpassed 7 billion.

According to the U.S. Census Bureau[1], the Earth's population was 5 billion in 1987, 4 billion in 1974, 3 billion in 1959, and 1 billion in about 1825 (**FIGURE 3-10**).

FIGURE 3-11 shows United Nation (UN) population projections to 2050, based on differing *assumptions* of growth rates.

Before we can analyze the effect of the impact of population growth, it is necessary to understand the simple math that describes such growth.

When any quantity, such as the rate of oil consumption, the human population, or a bank account, grows at a fixed rate, or percentage, per year, say 2% or 12% (as contrasted with growing by a fixed *quantity* every year, say 500,000 barrels, or 80 million people, or $1,000 every year), that growth is said to be *exponential*. **Exponential growth**, including population growth, is calculated using the compound interest formula, also known as the **compound growth equation**, the same one used to calculate interest on funds in bank accounts. Using it is not as intimidating as it may at first seem, and is demonstrated below. If you consider yourself "math phobic," first read the following and answer the questions using your best effort. You may surprise yourself by discovering that patience and repetition may allow you to use the compound growth equation and its variations. If you still have difficulty with the math, however but are comfortable with basic addition, subtraction, multiplication, and division, an alternative *but less accurate* approach is demonstrated at the end of this text.

The compound growth equation is:

$$\text{Future value} = \text{Present value} \times e^{rt}$$

In words, we'd say *the future value of some entity* (e.g., the human population of a country, the brown bat population of the United States, or the amount of recoverable

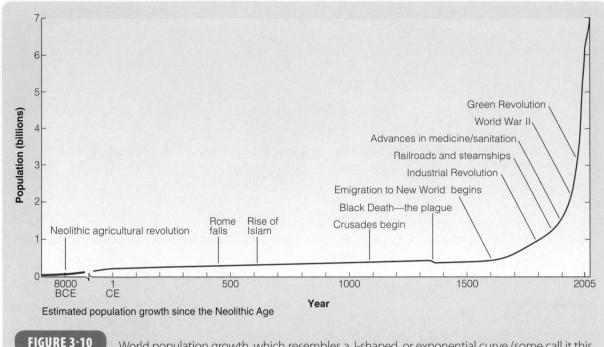

Estimated population growth since the Neolithic Age

FIGURE 3-10   World population growth, which resembles a J-shaped, or exponential curve (some call it this the "hockey stick" graph). How reliable do you think the first 9,000 years of data are? Would it make much difference in the graph if the estimates for early human populations were off by a factor of two? Why or why not? When did human population growth begin to accelerate? Why? U.S. Census Bureau, International Programs Center, 2001.

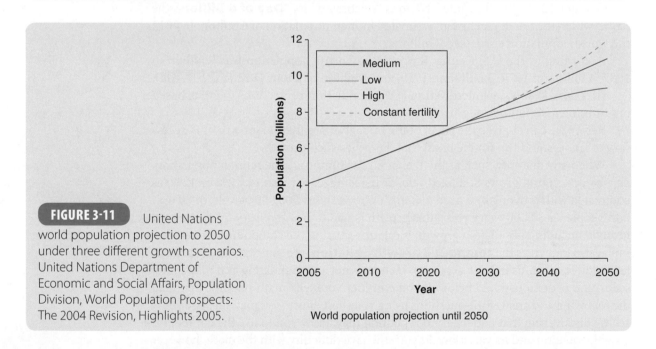

FIGURE 3-11   United Nations world population projection to 2050 under three different growth scenarios. United Nations Department of Economic and Social Affairs, Population Division, World Population Prospects: The 2004 Revision, Highlights 2005.

World population projection until 2050

oil in the Gulf of Mexico) *equals the present value times e to the are-tee.* The unit *e* equals the constant 2.71828 . . . , r equals the rate of increase, and t is the number of years (or other units of time) over which the growth is to be measured. Replacing the words and symbols entirely with symbols (which makes it look unnecessarily intimidating), the equation reads:

$$N = N_0 \times e^{rt}$$

The variable N_0 (pronounced N-sub-zero) is the quantity at time zero, the starting point.

If necessary, take a deep breath and reread the previous section.

The compound growth equation allows demographers (people who study human population characteristics) to *project* future population size, assuming that we know the size of the current population as well as the population growth rate. This equation is central to understanding exponential growth and is one of the few worth memorizing. The compound growth equation is summarized and explained at the end of the text.

In using the compound growth equation to project population growth (that is, to estimate it at some point in the future) we make two assumptions. First, we assume that our data are accurate. Second, we assume that the population growth rate we are using (in this case, birth rate minus death rate only) remains constant over the period we are projecting. Neither assumption may be absolutely correct. Here are some reasons why.

Population Estimates

In the early twentieth century, Sir Josiah Stamp, of Great Britain's Inland Revenue, remarked on collecting data (which he called *statistics*):

> *The government is very keen on amassing statistics. They collect them, add them, raise them to the Nth power, take the cube root and prepare wonderful diagrams. But you must never forget that every one of these figures comes in the first instance from the village watchman, who just puts down what he damn well pleases.*

Population estimates are compiled from censuses all over the world, including the U.S. Census, the latest of which was conducted in 2010. As you might imagine, making accurate counts or even estimates in developing countries, and in large nations like China and India, is difficult.

And growth rates can vary. Still, the compound growth equation is a particularly useful tool for examining growth scenarios and planning for the future.

You also can calculate how long it would take a population of a given size to grow (or decrease) to a different size at a specified growth rate using $t = (1/k) \ln(N/N_0)$.

The doubling time equation calculates how long it takes for a population or, for that matter, any entity, that is growing exponentially, to double in size. The *doubling time* equation is $T = 70/r$, where T is time and r is the rate of change. In the doubling time formula, in contrast to the other growth formulas, r is entered as the decimal growth rate × 100. For example, a growth rate of 7% (which is equivalent to 0.07) is entered as 7. The doubling time formula is explained and demonstrated at the end of the text.

To solve the problem using the doubling time equation, substitute 1.5 for r and divide. The answer is ~47 years.

Using the simple modification of the compound interest formula introduced above, $t = (1/k)\ln(N/N_0)$, we could determine when, at a growth rate of 1.13% per year, the world's 2010 population of 6.8 billion would grow to occupy the Earth at a density of one person per square meter (1 person/m^2) of dry land. The Earth has approximately 1.31×10^{14} m^2 of dry land. Thus, this number also represents the population size that would fill the Earth if one person occupied an area of one meter. Try this calculation, and then check the end of this text to see if your answer is correct.

BOX 3-1 demonstrates some of the calculations you just learned in the context of understanding the concept of ecological footprints.

Value-added Question 3-6.
The mid-year 2012 world population was 7.0 billion and was growing at a rate of 1.11%. Project what the world population would be in 2025, 2050, and 2100 assuming this constant growth rate. For this and following VAQs, consult the population math equations and samples at the end of the text.

Value-added Question 3-7.
Given an annual growth rate of 1.5%, how long would it take a population of 5 billion to grow to 10 billion?

BOX 3-1   *Party of 314 Million? I Suppose We Could Seat You in Texas*

Some critics of the idea of a limited **human carrying capacity** on Earth assert that everyone in the United States could fit comfortably inside the state of Texas, thereby refuting that there are too many people on the planet. Let's evaluate that claim. The mid-2012 U.S. population was 314 million (or, in scientific notation 314×10^6). The area of Texas is 261,914 sq. mi. We will first calculate how many square feet each person would occupy if all U.S. residents lived in Texas (1 mi^2 = 27,800,000 square feet).

$$314 \times 10^6 \text{ people}/261,914 \text{ mi}^2 = 1,199 \text{ people}/\text{mi}^2$$

Expressed per square foot:

$$1,199 \text{ people}/\text{mi}^2 \times 1 \text{mi}^2/27,900,000 \text{ square feet}$$
$$= 4.30 \times 10^{-5} \text{people}/\text{square foot}$$

or 0.0000430 people per square foot

That is, there is space In Texas for 0.0000430 people per square foot, or about 23,255 square feet per person. So, the claim is technically correct. Everyone in the United States could fit comfortably in Texas. In fact, the entire 2012 global population of 7.0 billion could fit inside Texas, and each would have a cozy 1043.9 sq. ft.

Of course, people need more than just space to live. In fact, by some estimates each U.S. resident uses 1.1 million square feet of *biologically productive land and sea* to live, a number known as our **ecological footprint**. This includes land to raise food, produce other natural resources, and absorb waste. Thus, if 310 million Americans each use 1.1 million square feet (ignoring in this case that it's not just *area* that we need, but *biologically productive* area), then collectively we would require 3.4×10^{14} square feet (340 trillion ft^2). But there is only 7.3×10^{12} sq. ft in Texas (261,914 mi̶^{2} $\times$ 27,900,000 ft^2/mi̶^{2}), meaning that we would need about 46 Texas's. So, we could *fit* comfortably in Texas; we just couldn't eat, etc., *unless we reduced our ecological footprint.*

Ecological footprints vary by country (**TABLE 3-3**). When these are added together, we currently need at least 1 to 2 more Earths to supply the demand. Because we have only one planet, this means we are using resources *unsustainably*, that is, we are taking resources from future generations.

TABLE 3-3   **Estimates of Ecological Footprints for Selected Countries***

Country/Region	Population (millions, 2007)	Ecological Footprint (global hectares per person, 2007)
Low Income	1,303.6	1.19
Afghanistan	26.29	0.62
Bangladesh	157.75	0.62
Ghana	22.87	1.75
Rwanda	9.45	1.02
Nigeria	147.72	1.44
Lower Middle Income (e.g., China, India, Colombia)	3,505.9	1.64
Upper Middle Income (e.g., Brazil, Mexico, S. Africa)	817.4	3.31
High Income	1031.6	6.09
Australia	20.85	6.84
Canada	32.95	7.01
France	61.71	5.01
Qatar	1.14	10.51
U.S.	308.67	8.00
World	6,670.8	2.70

*Data from: Ewing B., D. Moore, S. Goldfinger, A. Oursler, A. Reed, and M. Wackernagel. 2010. *The Ecological Footprint Atlas 2010.* Oakland: Global Footprint Network.

Concept Check 3-9. Read Box 3-1 and then answer this and the following questions. Discuss ways in which the ecological footprint of the planet's inhabitants could be reduced so that we live sustainably on Earth, that is, we do not require 1.4 Earths.

Concept Check 3-10. Increasingly, the American standard of living is used as a goal for developing countries. How might it change the global ecological footprint if every one of the 7.0 billion inhabitants of the Earth in 2012 were to consume resources like Americans do?

Concept Check 3-11. Examine Table 3-3 Does a country's ecological footprint necessarily correlate with standard of living or quality of life? Cite examples and explain your answer.

Concept Check 3-12. List ways in which you might reduce your ecological footprint.

CASE STUDY — Population Growth, Flooding, and Sustainability in Bangladesh

Background © Homero Technologies/AbleStock.com/Thinkstock; Title © iStockphoto/Thinkstock

Bangladesh (**Figure CS 1-1**), a country about the size of Illinois, Wisconsin, or Florida, lies on the northern shore of the Indian Ocean. Bangladesh was originally part of Pakistan after the Indian subcontinent gained independence from Britain in 1947. In 1970, Bangladesh seceded from Pakistan. Its population in mid-2012 was nearly 161 million on a total area of 144,000 km^2 (55,000 mi^2). The mid-2012 population of the United States was about 314 million on a total area of 9.8 million km^2 (3.8 million mi^2).

Floods in Bangladesh

Bangladesh is vulnerable to catastrophic flooding from rivers as well as from tropical storms. More than half the country lies at an elevation of less than 8 meters (26 feet) above sea level. Moreover, more than 17 million people live on land that is less than 1 meter (3.3 feet) above sea level. As a result, about 25% to 30% of the country is flooded each year. This can increase to 60% to 80% during major floods.

River flooding in Bangladesh results from some combination of the following: (1) excess rainfall and snow melt in the catchment area, especially in the foothills of the Himalayas (an area where many of the Earth's rainfall records were set); (2) simultaneous peak flooding in all three main rivers; and (3) high tides in the Bay of Bengal, which can dam runoff from rivers and cause the water to "pond up" and overtop its banks.

In addition, changes in land use in the catchment area can result in profound changes in the flood potential of Bangladesh's rivers. For example, deforestation of the Himalayan foothills for agriculture, fuel, or habitation makes the hills more prone to flash flooding (**Figure CS 1-2**).

Then, during the torrential monsoon rains, mudslides can dump huge volumes of sediment into the rivers. This sediment can literally pile up in the river channels, reducing channel depth and thereby the channel's capacity to transport water. The result is more water sloshing over the rivers' banks during floods and more disastrous floods in Bangladesh. The silt can smother offshore reefs when it finally reaches the ocean, aided by land-use practices that clear coastal mangrove forests (**Figure CS 1-3**). Geologists estimate that the drainage system dumps at least 635 million tonnes (1.4×10^{12} pounds) of sediment a year into the Indian Ocean, which is four times the amount of the present Mississippi River.

Western-style development may be contributing to Bangladesh's increased vulnerability to natural disasters. Recently built roads blocked floodwaters, thereby increasing flooding from cyclones and the damage and loss of life. Truly catastrophic floods can result when high tides coincide with tropical storms. In the early 1990s, such a combination took over 125,000 lives. Floods in 1998 caused 1000 deaths and left 20 million homeless. Major floods also occurred in 2004 and 2010. About one tenth of all Earth's tropical cyclones occur in the Bay of Bengal (Figure CS 1-3), and approximately 40% of the global deaths from storm surges occur in Bangladesh.

Storm surges result when cyclones move onshore and the winds push a massive wall of seawater onshore with them. Surges in excess of 5 meters (16.4 feet) above high tide are not uncommon, and thus, you can imagine the impact such a storm can have on a country in which half the land is within 8 meters (26.2 ft) of sea level!

Continued ▸

Exponential Growth and Its Impacts **59**

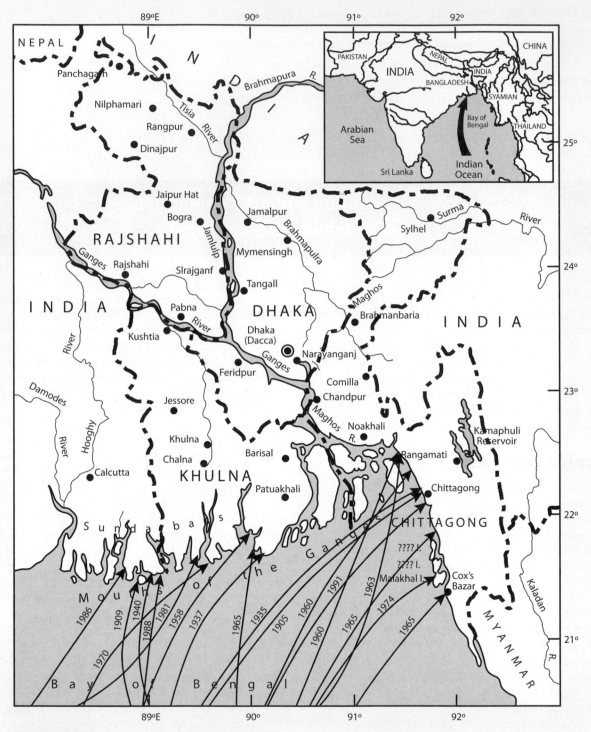

FIGURE CS 1-1 Location of Bangladesh, showing tracks of major historical typhoons.

Continued ▶

FIGURE CS 1-2 Deforestation on the slopes of the Himalayas, which leads to flash floods. © Galen Rowell/Corbis.

(A)

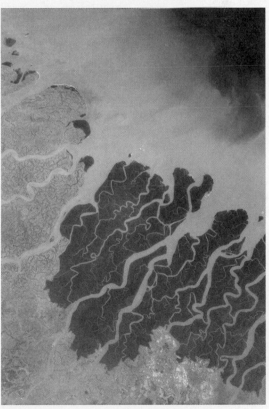

(B)

FIGURE CS 1-3 (A) Clearing of tropical mangroves for a golf and gambling resort on the island of North Bimini, Bahamas. The fate of the remaining mangroves has not been determined. Photo courtesy of Dr. Steve Kessel. (B) Satellite photo showing deforestation of mangroves in the Ganges delta in Bangladesh. The light areas (center-left and lower-left) represent deforested mangrove areas now inhabited by Bangladeshis. Courtesy of NASA.

Value-added Question 3-8.
The population of Dhaka, Bangladesh was estimated to be 8 million in 1998. First, calculate the doubling time (i.e., when the population would increase to 20.6 million) at a growth rate of 6% per year, using the doubling time formula, $t = 70/r$ (refer to the end of the text for an explanation of population math). The population was estimated at 10.8 million in 2010. How did this agree with the value you calculated from your use of 1998's growth rate?

Value-added Question 3-9.
Using the 2010 growth rate of 3.9%, estimate when the population of Dhaka will reach 21.6 million (i.e., double the 2010 population).

Concept Check 3-13. Bangladesh is a trivial producer of CO_2 compared with nations like the United States and China, and it faces potential disaster from sea level rise because of global climate change and tropical storms. What responsibility, if any, do you believe the main carbon-emitting nations have to reduce the potential for environmental disaster in Bangladesh?

Concept Check 3-14. Given that impacts of geohazards like flooding are already major problems in Bangladesh and are expected to worsen as global climate change progresses, do you see any way in which Bangladesh can become a more sustainable society? Explain your answer.

■ ISSUES OF AN AGEING GLOBAL POPULATION

Author Julian Simon called young people the "ultimate resource." Yet in many countries of Western Europe and the former Soviet Union, a decline in the birth rate has led to a decline in the numbers of young people. Study **FIGURE 3-12**, which shows **population pyramids** of France and India.

Member states of the European Union (EU) have identified the ageing of the population as a significant sociopolitical challenge. But it could also have environmental impacts as well. For example, elderly residents drive less than younger citizens and depend more on more efficient public transportation. Moreover, seniors generally prefer smaller houses than citizens with small children, which can cost less to heat.

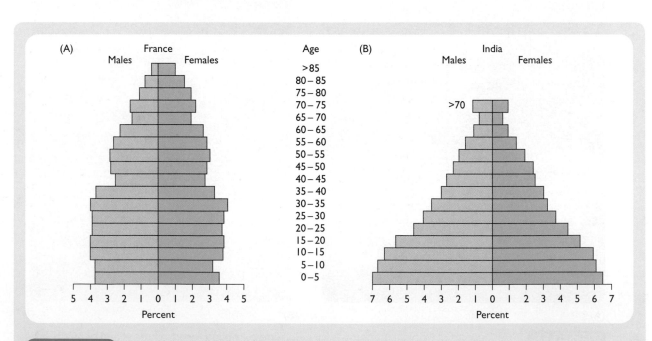

FIGURE 3-12 Population bar graphs for France and India. Which would better satisfy Simon's requirement for an abundance of young people to solve population-related problems? Do you think that young people need to be educated to be able to solve problems? If so, which country would have the highest potential education bill? Predict what would happen if that country chose not to spend the money necessary to educate its young people. Data from the US Census Bureau.

The EU, and indeed any entity that chooses to control its human population, will need to find ways to use and keep fresh the expertise and skills of an ageing workforce—especially in an increasingly knowledge-based economy. Social services including healthcare systems will need to adapt to the demands of an older U.S. population. For example, the number of people 85 or above is expected to triple from 7 million in 2000 to 19 million in 2050.

In the field of environmental health, an ageing population may have to depend more on preventative care than reactive treatment. Moreover, an older population may be more susceptible to environmental toxins than a traditional one. This will require precautionary elimination of carcinogens and mutagens in the environment as well as policies to reduce the incidence of atmospheric and water pollution. Global climate change, to the extent that it increases climatic extremes, could severely affect an older population.

On the other hand, older people with more free time may become more involved in environmental issues, and thus may contribute to environmental protection to a greater extent than at present.

At the other end of the spectrum are countries with young populations like Vietnam and India, whose median ages are 27.4 and 25.9 years, respectively (compared to 36.8 and 44.3 for the United States and Germany, respectively.)

Concept Check 3-15. In what ways might the environmental impact of a more youthful population structure like Vietnam's differ from that of a country with an ageing population, like Germany?

■ CARRYING CAPACITY REVISITED

Recall that the Earth's human carrying capacity is defined as the maximum number of people that the planet can support in the long term without (1) degrading our cultural and social environments and (2) harming our physical environment in ways that would adversely affect future generations. Living within the Earth's carrying capacity according to these specifications means living *sustainably*.

Unfortunately, experts disagree on the theory and/or calculation of the Earth's human carrying capacity, and thus estimates are subject to dispute. Why is this so?

First, many, including Julian Simon, have argued that the term *human carrying capacity* is meaningless. In their 1984 book entitled, *The Resourceful Earth,* Simon and Herman Kahn assert, "Because of increases in knowledge, the earth's 'carrying capacity' has been increasing . . . to such an extent that the term . . . has by now no useful meaning."

Second, while carrying capacities for nonhumans are calculated by ecologists and are based on numerical data and mathematical models, assessing human carrying capacities involves, in addition to ecologists, a group that includes economists, policymakers, sociologists, theologians, citizen activists, and so forth. These groups rarely talk with each other and, when they do, often disagree with each other when it comes to defining what constitutes degradation.

Finally, in their own ecosystems, other organisms are typically limited by their food supply, but humans can import food from outside of their region (if one does not consider the Earth as a single region). Humans also can export waste and pollution by air and water to areas outside their immediate surroundings. Also, by buying raw materials or manufactured goods from outside their region, some humans can avoid the environmental impact of producing the materials themselves. Perceptions of the planet's carrying capacity therefore vary with where an individual lives.

■ POPULATION IMPACT

Complicated as it may be, the human carrying capacity of the Earth can be estimated, assuming that different populations have different impacts based on their technology, consumption, and ethics as well as the simple number of individuals.

For example, the impact of 500 million vegetarians would have significantly different environmental impact compared with 500 million meat-eaters. Consider, for example, water supplies. To produce 1 ton of grain, about 1000 tons of water are required. Globally, an estimated 40% of all grain goes to meat and poultry production. High levels of meat consumption thus put additional stresses on global water supplies.

In terms of energy production and consumption, the impact of industrialized countries varies widely. France, for example, generates over 70% of its electricity by nuclear power while China generates electricity largely with coal using ageing power plants with few environmental controls. France therefore has a significantly different regional energy/environmental impact than China, although China is increasing its use of renewable energy.

When estimating global human carrying capacity, scientists study changes in the environment as well as the rate that these changes occur. Some useful indicators of environmental change include the loss of biodiversity (e.g., species extinctions), declines in fisheries, topsoil loss, degradation of water and air quality, increase in populations of invasive species, and increased greenhouse gas levels in the atmosphere.

And finally, carrying capacity often may have less to do with **population density** and ultimately more to do with the impact of a society's technology, resource demand, and waste generated.

■ CARRYING CAPACITY AND ECOLOGICAL FOOTPRINTS

In 2012 the population of the United States exceeded 314 million. Are our natural resources successfully supporting 314 million people? How do our imports of electronics and other consumer goods that fuel economy and standard of living affect the global environment? How fast is our population growing? Will our natural resources be able to support a growing and aging population? If so, for how long? Which segment of our population is growing the fastest? How much of our growth is due to legal immigration? Illegal immigration?

By now, you should have concluded that the exact human carrying capacity of the planet or of a smaller area is very difficult to determine, especially for a country as large and diverse as the United States. Despite the difficulty, it is essential that estimates of human carrying capacity be made so that policymakers can act to ensure that our environment is able to support human life and natural biodiversity into the future.

Concept Check 3-16. List several indicators, environmental or otherwise, that may support the conclusion that the planet has exceeded its carrying capacity. Be as specific as you can.

- -

To live sustainably in an area, human populations must remain at or below that area's carrying capacity. **Sustainability**, or *sustainable development* as it is sometimes called, refers to development that meets the needs of the present without compromising the ability of future generations to meet their own needs or, more specifically, not using resources faster than they can be replenished. However, as we have already discussed, humans in one area are able to import food, water, and manufactured products from other areas and export waste to other areas. These are typically not factored into carrying capacity estimates.

Another way of examining the impact of humans is to estimate the overall impact each of us has, a measure known as our ecological footprint (Table 3-3). Every individual has an ecological footprint that extends well beyond the geographical area in which that person lives. The most comprehensive analysis of ecological footprints of nations has been conducted by the Global Footprint Networks and is published in their Ecological Footprint Atlas 2010[2] presents estimates for how much of the Earth's area we appropriate for our needs. The average U.S. resident, for example, uses 8.0 **global hectares** (19.8 ac) (the term *global hectare* is used to denote one hectare of biologically productive land or water of average global productivity. Using the term *hectares* leads to an overestimate of productive space, since much of the planet's surface is marginally productive or unproductive) to support his or her lifestyle. This includes farmland, forests, mines, dumps, schools, hospitals, roads, playgrounds, shopping malls, etc.

The biosphere contains approximately 11.4 billion hectares (28.2 billion ac) of biologically productive area (2.0 billion ha [4.9 billion ac] of ocean and 9.4 billion ha [23/2 billion ac] of land), that is 11.4 billion *global hectares*. Table 3-3 shows global hectares per capita for continents and selected countries. Refer to the table as necessary to answer the following questions.

Concept Check 3-17. Table 3-3 and your calculations show that U.S. residents use a disproportionate portion of the Earth's resources compared to other countries. Discuss reasons why this is so.

- -

Concept Check 3-18. At the beginning of this chapter, we asked why human population growth is a major topic in an environmental geology textbook. Do you agree that human population growth is central to the field of environmental geology? Defend your answer.

- -

Concept Check 3-19. Using evidence from this chapter and beyond, discuss whether you think we have exceeded the Earth's human carrying capacity.

- -

Value-added Question 3-10.
The United States and world populations at mid 2012, respectively, were about 314 million (314×10^6) and 7 billion (7×10^9). If the average U.S. resident uses 8 global ha (A), what percentage of the planet's biologically productive area does that represent? (B) If every person on the planet had the same ecological footprint as a U.S. resident, how many global hectares would that require? (C). What percentage of the Earth's biologically productive area does that represent?

Value-added Question 3-11.
Select a country in Table 3-3 whose ecological footprint is less than 1. Name the country, its population, and its ecological footprint. What percentage of the planet's biologically productive area does that represent?

■ CONCLUSION

Human numbers must eventually become stabilized, as it is physically impossible for population growth to continue forever. The only questions are at what level will growth end and whether growth will end as a result of human actions or by natural processes like famine, disease, and war. Ageing societies are typical of developed nations. Populations dominated by the young are typical of developing ones. Large numbers of young people provide great promise for societies but also impose great costs. Developing countries are experiencing the most rapid population growth. Is this a problem? Do developed countries have a responsibility to help them control this growth? Specifically, what measures, if any, should be taken? To what extent is population control a moral rather than a scientific or environmental issue?

The readers of this book are mainly young, and it is you who will or will not address these challenges. The next century should prove to be one of the most interesting centuries in the entire span of human history.

CHAPTER SUMMARY

1. The essence of the population debate is whether or not Earth has exceeded its human carrying capacity, the maximum number of people that the planet's resources can sustain in the long term without irreversibly harming the planetary ecosystem.

2. Humans have a profound physical impact on our local, regional, and global geological environment, and our geological environment also impacts us.

3. Our ever-increasing population is concentrating in areas at risk from potent geohazards.

4. Each human being contributes something unique to the planet, and may be the source of problem-solving ingenuity, but each of us places physical demands on Earth.

5. People in developed countries consume disproportionately more resources than residents of developing countries. Some consider these to be issues of environmental justice.

6. The total impact of a population on its environment is the product of the absolute population size and the impact per person.

7. Statistics used to measure population growth are fertility rate, rate of natural increase, birth rate, and death rate. Fertility rate is the average number of children born to a woman during her lifetime. The rate of natural increase in a country's population can be calculated by subtracting the death rate from the birth rate.

8. Emigration and immigration can materially change a country's population in many ways.

9. The population of the Earth surpassed 6 billion in 1999, and now exceeds 7 billion.

10. When any quantity, such as human population, grows at a fixed rate, or percentage, per year, that growth is said to be exponential.

11. Exponential growth, including population growth, is calculated using the compound growth equation, $N = N_0 \times e^{rt}$

12. The compound growth equation allows demographers to project future population size, assuming a current population, and the population growth rate.

13. Developing countries like Bangladesh are experiencing the most rapid population growth.

14. Member states of the European Union have identified the ageing of their population as a significant sociopolitical challenge, but it could also have environmental impacts as well. In the field of environmental health, an ageing population may have to depend more on preventative care than reactive treatment. Moreover, an ageing population could be more susceptible to environmental toxins than a traditional one.

15. Countries like Vietnam and India have more youthful populations, and hence a young workforce.

16. The Earth's human carrying capacity is defined as the maximum number of people that the planet can support in the long term without (1) degrading our cultural and social environments and (2) harming our physical environment in ways that would adversely affect future generations. Living within the Earth's carrying capacity according to these specifications means living sustainably.

17. Calculating the Earth's human carrying capacity is difficult in practice.

KEY TERMS

birth rate

carrying capacity

compound growth equation

Concentrated Animal Feeding Operations (CAFOs)

Day of 6 Billion

dead zones

death rate

ecological footprint

environmental justice

exponential growth

fertility rate

geohazard

global hectare

Hardin's Third Law of Human Ecology

human carrying capacity

municipal waste

nutrient pollution

population density

population pyramid

rate of natural increase/decrease

replacement fertility rate

sustainability

total fertility rate (TFR)

REVIEW QUESTIONS

1. The essence of the population debate is
   a. the absolute number of people on a continent.
   b. the age distribution of a country's population.
   c. whether we have exceeded Earth's human carrying capacity.
   d. the fact that the planet's population can live comfortably inside Texas.

2. A major source of nutrient pollution is
   a. volcanoes.
   b. plastics manufacture.
   c. oil refining.
   d. fertilizers.

3. Runoff from CAFOs can cause
   a. fish kills in streams.
   b. mercury buildup in invertebrates.
   c. increased sediment retention in streams.
   d. all of the above.
   e. none of the above because runoff from CAFOs is subject to regulation.

4. Vehicular access to wilderness can cause
   a. the spread of rap music.
   b. the spread of viral diseases.
   c. the spread of invasive species.
   d. an increase in arctic snowmelt.

5. The "average" American produces about what amount of municipal waste each year?
   a. less than 70 kg.
   b. about 10,000 kg.
   c. about 100 kg.
   d. about 700 kg.

6. In the United Kingdom, average municipal waste production per person is about 25 kg/week. How does this compare to the average for Americans?
   a. it is much less.
   b. it is much greater.
   c. it is about the same.

7. The percent of humans who live within 100 km (62 miles) of the ocean is
   a. about 25%.
   b. about 10%.
   c. about 90%.
   d. about 60%.
   e. 100%.

8. The fastest growing metropolitan region in the United States is
   a. Las Vegas, Nevada.
   b. Las Vegas, New Mexico.
   c. Los Angeles, CA.
   d. Washington, D.C.

9. The Middle East (Israel, Palestine, Jordan, Egypt, etc.)
   a. has some of the highest rates of population growth in the world.
   b. has some of the lowest population growth rates in the world, due to terrorism.
   c. has some of the lowest population growth rates, due to government family-planning activities.
   d. has ample water for twice its present population

10. The United States and world populations at the beginning of 2011, respectively, were about
    a. 314 million and 7 billion.
    b. 314 billion and 7 trillion.
    c. 3.14 million and 700 million.
    d. 3.14×10^6 and 7×10^9
    e. a and d only.

11. Biologist Garrett Hardin's Third Law of Human Ecology states
    a. humans have dominion over the planet.
    b. a population's impact is determined by the product of the population size and the impact per person.
    c. a population's impact is determined exclusively by the absolute number of people.
    d. the human and the fish can peacefully coexist.

12. Fertility rate may be defined as
    a. birth rate minus the death rate.
    b. live births less aborted births.
    c. the average number of children born to women in the country.
    d. the total number of children born to women in a country per year.

13. Subtracting the death rate from the birth rate yields the
    a. rate of natural increase.
    b. rate of emigration.
    c. rate of immigration.
    d. total population change per year.

14. Exponential growth measures growth that occurs
    a. at a given number per year.
    b. at a given rate per year.
    c. in developed countries without immigration.
    d. in developing countries without reliable censuses.

15. The *replacement fertility rate* for developing countries is
    a. greater than that for developed countries.
    b. less than that for developing countries.
    c. between 2.5 and 3.3.
    d. a and c only.

16. Use the doubling time equation, $t = 70/k$, to calculate how many years it will take for a population growing at 3% per year to double.
    a. 33 years
    b. 11 years
    c. 23 years
    d. 210 years

17. The population of a country was 100 million in 2000. When will it be 200 million, at a growth rate of 2%?
    a. 2035
    b. 2050
    c. 2100
    d. 2010

18. Earth's human population is growing at 1.2% per year, and was approximately 6 billion in 2000. How long would it take for it to reach 12 billion, assuming no change of rate?
    a. 100 years
    b. 5.8 years
    c. 58 years
    d. 580 years

19. Bangladesh is vulnerable to
    a. flooding because half the country lies at an elevation less than 8 m.
    b. flooding because of heavy rainfall in the Himalayas.
    c. flooding because of deforestation of the Himalayan foothills.
    d. all of the above.

20. The human carrying capacity of the planet
    a. is set each year by the United Nations.
    b. is a concept about which there is much disagreement.
    c. is exactly 7 billion.
    d. depends on how many acres are in a hectare.

FOOTNOTES

[1] U.S. Census Bureau. Total Midyear Population for the World: 1950–2050. Retrieved August 16, 2012 from http://www.census.gov/ipc/www/worldpop.html

[2] Retrieved August 16, 2012 from http://www.footprintnetwork.org/en/index.php/GFN/

CHAPTER 4

Global Tectonics and Geohazards

CHAPTER OUTLINE

■ INTRODUCTION

In this chapter, we introduce material that you must learn if you are to understand, among other important concepts, why and where earthquakes and volcanoes occur, and, therefore, the origin of the many geological hazards they generate. First, we will discuss the concept of the geological *model*. Then we will review the *scientific method* and continue with a brief look at the history of recent scientific thought on the nature of the Earth's interior. Finally, we present a *model* of the Earth's interior to set the stage for our discussion of global tectonics, the great unifying theory of geology.

■ THE CONCEPT OF THE MODEL

If scientists want to know the Earth's composition and structure at any depth, the easiest way to find out would be to drill a hole and collect a sample (**FIGURE 4-1**).

In the United States alone, over 250,000 holes have been drilled exploring for oil and gas!

How deep into the interior do you think we have succeeded in drilling (the radius of the Earth is about 6600 km, or 4100 mi)? Science-fiction writers to the contrary, scientists have succeeded in drilling less that ¼ of *one percent* of the way to the Earth's center, or only about 14 km (8.7 mi)!

The reason we have not drilled deeply into the Earth is that temperatures and pressures increase as depth increases, and the conditions eventually become too extreme for drilling material to hold its strength. Solid steel drill pipe starts to act like angel-hair pasta. The cost becomes prohibitive as well. This illustrates a very important point about our knowledge of the Earth's interior: *almost all our knowledge is indirect.*

Concept Check 4-1. Recall that knowledge is not scientific if it is not observable. Discuss whether our knowledge of the Earth's interior is scientific, given that we have drilled only ¼ of one percent into the planet.

- -

Because we don't have direct samples from boreholes into the Earth's center, we must content ourselves with making **models** of the interior based on indirect evidence. And we must constantly test the validity of these models, which is the scientific standard of "suspended judgment."

An Analogy: A Model of an Aircraft

Models in geology are *abstract representations of a general concept*. As kids, many of us put together model airplanes. To make those models, we bought kits

(A) THE *GLOMAR CHALLENGER*

(B) THE *JOIDES RESOLUTION*

(C) THE *CHIKYU*

FIGURE 4-1 Ships designed for deep-sea drilling. (A) The *Glomar Challenger,* a unique drilling vessel 122 meters long, could manage about 7.6 kilometers of drill pipe. Courtesy of Integrated Drilling Program. (B) The *Joides Resolution,* about 300 meters long, can handle over 9.1 kilometers of drill pipe and operate safely in heavier seas and winds than the Glomar Challenger could. Courtesy of Ocean Drilling Program, Texas A&M University. (C)The 210-meter-long *Chikyu* can drill in water depths up to 2.5 kilometers and carries enough drill pipe to continue 7.5 kilometers below the sea floor. © Kyodo/Landov.

consisting of the exterior parts of the plane all reduced by the same amount to the original airplane: in other words, a *scale model.* But what if you had to design a *general model* for an aircraft—something that immediately would be recognized by everybody as an aircraft, without being an actual model of any real aircraft? Obviously you would first have to decide *what characteristics are common to all aircraft.* Perhaps you'd conclude that all aircraft have wings, a body (fuselage), one or more engines, and a tail of some sort. So, you might make a general model of an aircraft with these four characteristics. However, they may be of different sizes and shapes. For example, some planes have straight wings and some have wings that are swept back at an angle. Some have propellers for engines and others have jets. But these specific variants do not mean that you couldn't make a general model of a plane that everyone could recognize *as a plane*; they would only have to know something about the variations in plane types.

This is the basis for the geological use of the model: we try to boil down the *system* we are trying to model to the basic and essential characteristics that all variants share. Models are essential to a geologist: in sedimentary geology we make models of sedimentary environments, like rivers, in order to be able to recognize river sediment from ancient sedimentary rocks.

Concept Check 4-2. Find photographs of at least three different deltas. Design a model of a delta (in other words, determine common features of these deltas that would characterize all deltas). Describe your model.

A VERY BRIEF REVIEW OF THE SCIENTIFIC METHOD

Let's review the scientific method, because we use it to make our model of the Earth's interior. Being scientists, geologists use **hypotheses** to explain some natural phenomenon, and then gather evidence to test their hypotheses. They also may design experiments to test them. After this, the hypothesis has either been supported or rejected and, if supported, geologists may elevate it to a **theory** (a widely accepted explanation for a phenomenon); if rejected, it's back to the drawing board.

MODELING THE EARTH'S INTERIOR

A model of the Earth's interior shows the general structure such as: interior layers drawn to *scale,* where interior boundaries are, what the **states of matter** are (whether solid, liquid or gas, for example), and the composition of the interior, in as much detail as necessary. We may not know the exact composition of each zone, or the layers may vary in composition, so our model might try to show an "average" composition; or our model might try to show how compositions vary with depth, or laterally.

To make our model, we use all available evidence that will allow us to put *constraints* on what our model looks like. For example, we know (from its gravitational interactions with other bodies in the Solar System) the Earth's overall **density** (an indicator of the mass of a substance relative to its volume, in other words, how heavy it is) is 5.52 g/cm³, so our interior material must *average* 5.52 in density, even though the density of each zone or layer may be higher or lower.

The Importance of Earthquake Waves

The energy stored in a stick when bent is released quickly when the stick breaks. Some of the energy is released as *sound*: a sharp crack is heard. Sound is a form of energy and breaking the stick generates sound waves. In the same way, breaking rocks along *faults,* which are fractures in the Earth along which movement takes place, also releases energy. Several types of energy waves are generated (discussed elsewhere in this text; **FIGURE 4-2**).

Seismic *wave paths* and velocities inside the Earth depend on the property of the material through which the waves pass. If we can track the wave path, we can tell a lot about the nature of the material along its path (**FIGURE 4-3**).

Summary of Evidence Used to Model the Earth's Interior

We use evidence from the following sources to make our model of the Earth's interior.

- The composition and distribution of surface rocks. There are three categories of rocks: **igneous**, which form from a molten mass called **magma**; **sedimentary**, which form by erosion of pre-existing rocks or by organisms, like corals; and **metamorphic**, a "hybrid" rock formed when a pre-existing rock is subjected to heat and pressure, causing the original minerals to align themselves, recrystallize, or both. Besides outcrops, we have hundreds of

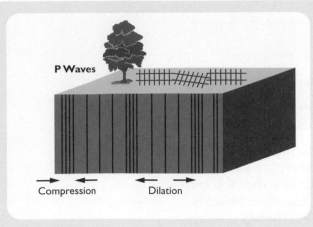

P Waves

Compression   Dilation

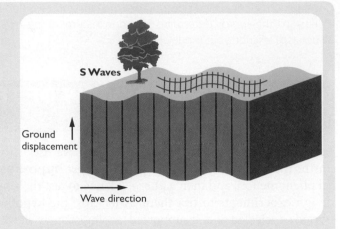

S Waves

Ground displacement

Wave direction

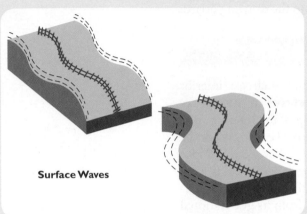

Surface Waves

FIGURE 4-2   Types of earthquake waves, how they move, and their relative velocities. P waves are the fastest waves, and are known as compressional waves because they exert a push and pull (think of a wave moving in a Slinky toy). S waves arrive after P waves, and they displace the ground perpendicularly to the direction of propagation. Surface waves are the slowest but may be the most destructive, since they displace the ground like rolling ocean waves and/or side-side. A more complete description of these waves is found elsewhere in this book.

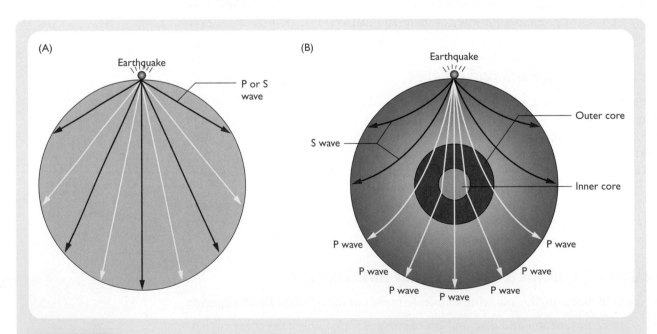

FIGURE 4-3   Changes in seismic wave velocities with depth. P waves penetrate through liquid, while S waves are reflected. Internal zones where waves abruptly change velocity are known as discontinuities. They are important internal boundaries and indicate a change of state, of composition, or both. (A) How waves would travel through an Earth with a homogeneous interior. (B) Actual wave paths, which demonstrate the increase in density with depth and the existence of the core.

Fig. 1. — La météorite de Peary sur camion en face du muséum d'Histoire Naturelle à New-York.

FIGURE 4-4 A meteorite recovered from Greenland. The snow cover helped researchers spot the dark meteorites. © Mary Evans Picture Library/Alamy Images.

thousands of samples of the shallow continental crust from boreholes, so we can estimate the average composition of the Earth's continents.

- Meteorites, the "raw material" of the original Earth. Meteorites, objects from space that have survived impact on Earth, commonly are either an iron-nickel alloy (about 6%), or composed of silicate minerals containing the elements silicon and oxygen (about 93%; **FIGURE 4-4**).

- Thousands of drill holes and dredge samples of oceanic sediment and underlying bedrock. We also have seismic data that show where boundaries exist in the oceanic crust. We can confidently infer the composition of these layers.

- We know the Earth's average density is 5.52 g/cm³, which puts strong limits on the kinds of substances that can possibly make up the interior. **TABLE 4-1** shows densities of some common substances, which should give you an idea of what a density of 5.52 means.

TABLE 4-1	*Some Common Substances and Their Densities*
Diamond	3.5
Pyrite	5.2
Gold (22 carat)	17.5
Cement	1.6
Concrete	2.4
Cast iron	7.7
Marble	2.7

- The physics of the Earth's rotation dictates that the densest material must be at the Earth's center and densities must decrease outward (and of course we know the average density of the outer layer from direct samples).
- When an earthquake occurs, **seismic waves** are generated, which travel through the Earth, making it "ring like a bell." Like light off a mirror, seismic waves can bounce off surfaces inside the Earth where densities abruptly change. We therefore know from the behavior of seismic waves as they travel through the Earth where the major boundaries are located between subsurface materials, and we know something about their properties, because rock properties determine seismic wave velocity. For example, other things being equal, the higher the density of rock, the faster seismic waves travel through it. Figure 4-3 shows how waves, in this case sound waves, are reflected and bent (refracted) by the various layers inside the Earth. Internal zones where waves abruptly change velocity are known as **discontinuities**. They are important internal boundaries and indicate a change of state, composition, or both.
- The Earth's powerful magnetic field puts limits on the possible composition and state of the planet's interior (**FIGURE 4-5**).

Using such information, we could hypothesize that the Earth is made of material that *gradually* increases in density from that at the surface to a denser mass at the center, so that the overall average density is 5.52. Crustal rocks, that is, rocks near the Earth's surface, have densities in the range of 2.5 to 3.1, but no common natural rock has a density approaching 5.52.

Applying data from seismic waves, we find that this model won't work, because seismic wave velocities (which tell us the density of Earth materials) change abruptly at several *discontinuities* within the Earth. So we have to modify our model by hypothesizing separate *layers* inside the Earth, rather than material that gradually increases from surface to center.

Concept Check 4-3. Explain why not altering our hypothesis given the existence of the seismic data would have violated the "rules" of science.

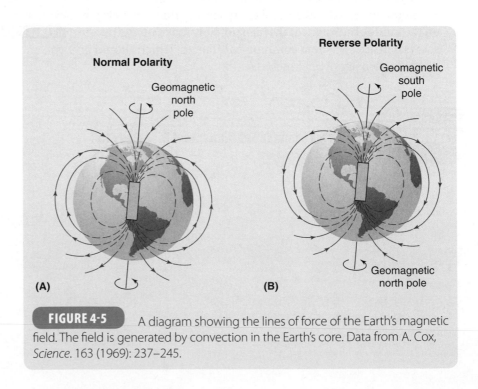

Normal Polarity

Reverse Polarity

Geomagnetic north pole

Geomagnetic south pole

Geomagnetic north pole

(A)   (B)

FIGURE 4-5   A diagram showing the lines of force of the Earth's magnetic field. The field is generated by convection in the Earth's core. Data from A. Cox, *Science*. 163 (1969): 237–245.

Concept Check 4-4. Explain the concept of a geological model. Based on the information in the preceding sections, explain the model of the Earth's interior.

■ THE INTERIOR: LITHOSPHERE, ASTHENOSPHERE, CRUST, MANTLE, AND CORE

One of the most fundamental discoveries about the Earth was that the interior was composed of *layers* with recognizable boundaries. Geologists at first divided the Earth into three layers: the **crust** (outermost material), an underlying **mantle**, and a **core** at the center, based on hypothesized differences in rock *composition*. **FIGURE 4-6A** shows the major zones within the Earth. Material may vary in composition, and properties such as strength (which measures how it responds to deforming force), or state (whether solid, liquid, or gas). Strength and state in turn depend on the temperature and pressure to which the material is subjected.

Later study of seismic waves and the Earth's magnetic field, coupled with laboratory research, allowed detail to be added to this model. The *core* was hypothesized to consist of an inner solid portion and an outer liquid portion. Two types of crust were found: **oceanic**, comprising **basaltic** (*basalt* is a dark-colored, fine-grained, igneous rock) rocks below sediment, and **continental**, composed of all three rock types, but on average having a composition near that of granite. The lower mantle was denser than the upper mantle. The upper mantle was found to contain a zone near its top where seismic wave velocities were lower than above and below: This zone became known as the *low-velocity zone,* or **asthenosphere**. The

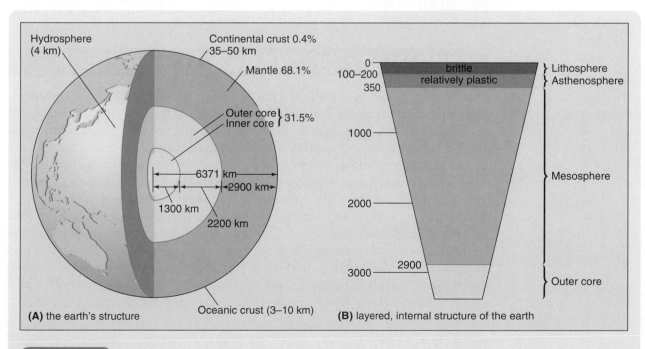

(A) the earth's structure

(B) layered, internal structure of the earth

FIGURE 4-6 Earth's internal structure. Percentages indicate the relative masses of the core, mantle, and continental crust. Where is most of the Earth's mass concentrated? (A) Layering based on chemical composition and physical state. (B) Layering based on response to a deforming force. The lithosphere was defined to include the crust (both continental and oceanic: see arrows) and the rigid uppermost portion of the mantle. The asthenosphere lies entirely within the upper mantle.

FIGURE 4-7 A peridotite sample, brought up by lava known to have originated in the mantle. Peridotite takes its name from the gem-quality form of the mineral olivine—peridot—which is a magnesium-rich silicate mineral. Silicon is the second most abundant element in the Earth after oxygen. It thus should be no surprise then that minerals made mainly of silicon and oxygen should be so common. © Slim Sepp/ShutterStock, Inc.

term is derived from the Greek *asthenos* (meaning weakness or loss of strength), which means a weak, plastic (that is, deformable), partially molten zone, to distinguish it from the **lithosphere** (Greek, *lithos* = rock), a strong, rigid zone, above it (FIGURE 4-6B). The lithosphere sits atop the asthenosphere and includes the crust (both continental and oceanic) and the rigid uppermost portion of the mantle.

The upper mantle's composition is best described as **peridotite**, a magnesium-rich rock containing abundant olivine, based on several lines of evidence (FIGURE 4-7). First, peridotite is a rock, which on being partially melted, yields a *basaltic* magma, and basalt is the universal bedrock of the ocean basins (below a veneer of sediment), so it is likely that the ocean basin bedrock formed from partially melted peridotite. Furthermore, the density and properties of peridotite are consistent with the behavior of seismic waves traveling through the upper mantle. Finally, it is possible to derive peridotite from the composition of stony meteorites, the probable raw material from which the planet was built more than 4 billion years ago.

■ SUMMARY OF THE EARTH'S INTERIOR

Our model at present is like this: The Earth is a layered body that has discrete boundaries separating the layers. These layers, based on compositional differences from the surface inward are, crust, mantle, and core. The uppermost few hundred km of the Earth are divided into a lithosphere (containing the crust and uppermost mantle) and an asthenosphere, or low-velocity zone; a partially molten zone entirely within the mantle. The crust is of two types: oceanic, consisting mainly of basaltic rocks below a variable layer of sediment, and continental, composed of all three rock types, but on average having a composition near that of granite. The outer core is molten and the inner core solid. The composition of the core is mainly iron, with some nickel and minor elements such as sulfur, silicon, and/or potassium. The upper mantle is of peridotitic composition (a material which, if partially melted, yields a basaltic magma) and the lower mantle is probably made up of densely packed oxide minerals, with a similar composition to the upper mantle but made up of denser crystal structures.

■ RADIOACTIVE DECAY AND INTERIOR HEAT

One additional factor affects the nature and behavior of the Earth's interior and surface: radioactive decay (FIGURE 4-8). **Radioactivity** refers to the "decay" of unstable atomic nuclei. (Decay here means the radioactive atom changes to a non-radioactive element by emitting particles and energy.) We discuss radioactive decay further elsewhere.

The reason we introduce radioactivity here is that it produces a great deal of *heat* in the Earth's interior, and radioactive heat powers the global process we call plate tectonics, which is introduced below. The fact that radioactive decay is irreversible that is, once an atom has decayed it no longer produces heat, helps us understand how the planet's crust has evolved over the past 4+ billion years. Thus, heat available to power plate tectonics, including volcanic activity, has declined over geologic time.

■ HEAT TRANSFER

Heat produced by radioactive decay moves from high concentrations to areas where there is less heat, forming a heat *gradient*. The same thing happens when you put a pot on a hot burner on the stove. The pot warms by heat that migrates from high heat (the burner) toward low heat regions (the pot's contents). In general the Earth's

Uranium 238 (U238) Radioactive Decay

Nuclide	Half-life
uranium-238	4.47 billion years
thorium-234	24.1 days
protactinium-234	1.17 minutes
uranium-234	245000 years
thorium-230	8000 years
radium-226	1600 years
radon-222	3.823 days
polonium-218	3.05 minutes
lead-214	26.8 minutes
bismuth-214	19.7 minutes
polonium-214	0.000164 seconds
lead-210	22.3 years
bismuth-210	5.01 days
polonium-210	138.4 days
lead-206	stable

FIGURE 4-8

Radioactive decay. The U238/Pb206 decay series. Heat produced by radioactive decay has decreased over the Earth's history. If the Earth's heat engine is based on radioactive decay, and the rate at which plates move is related to heat produced, how do you think plate motion rates have changed over planetary history? How do you think plate motion rates will change going into the future? Will they stay the same, increase, or decrease? On what do you base your answer?

interior is hot and the surface is cold, so heat tends to move from the interior toward the surface. Three ways that heat can move are: by *conduction*, by *convection*, and by **radiation** (FIGURE 4-9).

Conduction

In the solid parts of the Earth's interior, **heat transfer** is accomplished by molecular motion, like sticking a cold iron bar in a hot fire (Figure 4-9). This is also called **conduction**, and it is a relatively slow process.

Convection

If there is any liquid or gas present in a system, heating can make it behave like a pot of soup on the stove: *convection cells* or plumes will form, carrying hot soup at the pot's bottom to the surface, and displacing colder soup at the top of the pot to the bottom, where it is heated. **Convection** will continue as long as the pot is being heated and at least part of it is liquid or gas, and this is apparently the process by which much of the heat in the mantle is transferred towards the surface. Computer simulations of models of the mantle show that concentrations of heat will form such convection cells, as shown in Figure 4-9.

Convection cells and plumes powered by radioactive decay are the means by which new oceanic bedrock is made, and new oceans may be formed in the process. The oceanic bedrock is composed of basalt and basaltic rock.

The great *mass* of material being moved by convection, as well as the material's high viscosity (resistance to flow) limits the speed of the convection currents to a few

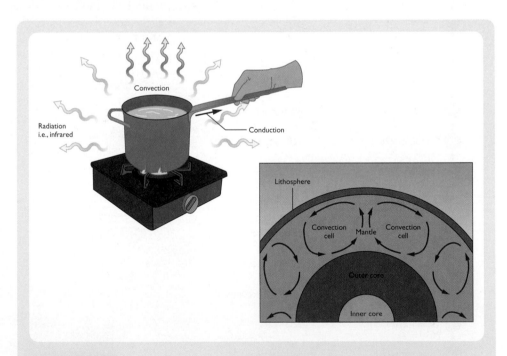

FIGURE 4-9 (A) Three modes of heat transfer: conduction, convection, and radiation. Radiation here does not mean radioactivity; radiation means transferring heat by electromagnetic radiation (e.g. by microwaves). (B) Formation of a convection cell in the Earth's interior. Heat may be transferred rapidly by convection, and the material need not be entirely liquid for convection to occur.

centimeters a year and this is not coincidentally the rate at which the lithospheric plates, which we'll discuss below, are moving at the present time. Because continents are imbedded in these plates, they are passively carried over the Earth's surface like rafts.

GLOBAL TECTONICS AND THE DYNAMIC EARTH MODEL

Plate Tectonics as a Unifying Theory of Geology

Now, we're ready to introduce the concept of **plate tectonics**. The word *tectonics* has Latin/Greek origins and means *of or relating to building,* in this case, of the structures on the Earth's surface. Plate tectonic theory states that the surface of the Earth is covered by seven major rigid lithospheric plates and a number of smaller plates (**FIGURE 4-10**), all of which are in motion, and not all in the same direction or at the same speed. Plate tectonics explains the majority of the Earth's large-scale geologic structures and processes, from island arcs like the Aleutians to events like earthquakes and, as one might expect for a concept of such overarching significance, has been called a **unifying principle** of geology.

Continental Drift and Seafloor Spreading

The theory of plate tectonics evolved from two older ideas, **continental drift** and **seafloor spreading**. The concept of continental drift is credited to German meteorologist Alfred Wegener who, during the early twentieth century, postulated that the location of the Earth's continents in the distant past was different from their current location, in other words, the continents had "drifted" over geological time. Based in part on the jigsaw puzzle fit of the current continents that was evident on maps at the time, for example, the east coast of South America meshing almost seamlessly with the west coast of Africa, and using evidence from earlier researchers, Wegener proposed the existence of a supercontinent known as

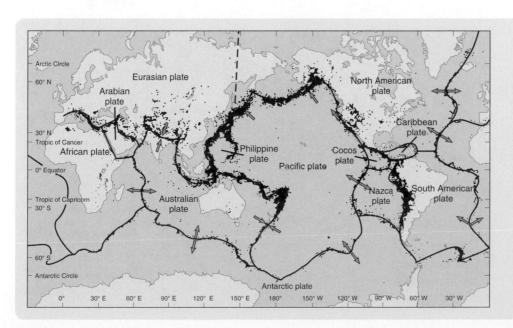

FIGURE 4-10 Earth's lithospheric plates. Clusters of black dots indicate earthquake activity. Arrows show relative plate motion. Plates are generally very thin (tens of kilometers is typical) relative to their other dimensions. Therefore, they are rather like the fractured shell of an egg in scale. Data from Stowe, K.S. *Ocean Science.* John Wiley & Sons, Ltd., 1983.

Pangaea (from the Greek meaning "all land") surrounded by single large ocean, **Panthalassa** ("all sea"), about 200 million years ago. Wegener marshaled additional supporting evidence, including the presence of the late Paleozoic fossil plant *Glossopteris* on India, Australia, South Africa, and South America, areas dissimilar in their current climate and thus unlikely to support these plants unless under a similar climate regime in the past when these areas were all connected. Other evidence included the contiguous distributions of fossil reptiles across ocean basins, an event unlikely unless the basins were closed and the continents had been connected. Geological structures like coal veins and mountains of the same age showed similar distributions across ocean basins. There was evidence of ancient glaciation in tropical parts of South America, India, Africa, and Australia, a phenomenon best explained by these locations once having existed as part of an Antarctic supercontinent (FIGURE 4-11). Despite a seeming abundance of geologic, paleontologic, and climatologic substantiation for his idea, Wegener's concept of continental drift was not widely accepted by the scientific community because a sound explanation for the mechanism by which the continents moved could not be made. How could granitic continents plow through the denser basaltic oceanic crust, scientists wondered?

In the late 1950s–early 1960s, the advent of post–World War II technology, namely SONAR (SOund Navigation And Ranging), allowed for more extensive mapping of the sea floor. The most extraordinary finding was the existence of 65,000 km (40,400 mi) ridge system—an underwater mountain chain—on the sea floor and deep-ocean trenches (FIGURE 4-12).

Other evidence came from new discoveries of magnetism of fossil rocks, **paleomagnetism**. This new arena of study revealed that information on the direction and intensity of the Earth's magnetic fields at a particular time was locked into the iron in magma when it solidified into rock at the ridges. It was

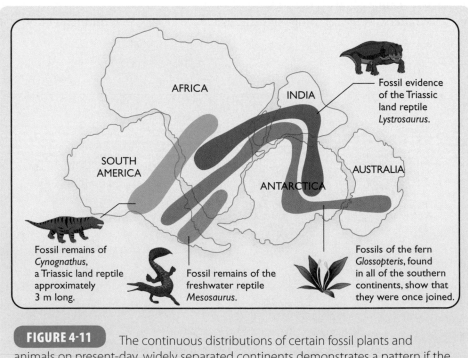

FIGURE 4-11 The continuous distributions of certain fossil plants and animals on present-day, widely separated continents demonstrates a pattern if the continents are rejoined. Modified from This Dynamic Earth, USGS.

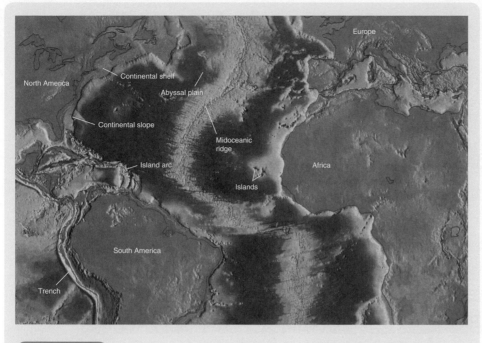

thus learned that the current Earth's magnetic field has undergone complete reversals in the past, 170 during last 76 million years (**FIGURE 4-13**). When a **magnetometer**, a device developed at the Scripps Institution of Oceanography (**FIGURE 4-14**) to measure magnetic properties of the sea floor, was deployed, it revealed a pattern of magnetic "stripes" parallel to the ridge and symmetrically distributed around it (**FIGURE 4-15**).

This information was synthesized by Harry Hess of Princeton University in 1962. Hess proposed the theory of seafloor spreading. According to this theory, new seafloor is produced at the ocean ridges from magma, which solidifies and moves very slowly bilaterally away from the ridge, preserving information on the current direction and intensity of the Earth's magnetic field. When the field reversed, evidence of this reversal was similarly locked into the new sea floor. The striped pattern of alternating magnetic orientations symmetrical to the ridge revealed by the magnetometer thus was explained (Figure 4-15).

In addition to the paleomagnetic evidence, further validation of seafloor spreading came from radiometric dating of the ocean crust, which revealed (1) that oceanic crust is geologically young, much younger than continental crust (180 million years for the oldest oceanic crust compared to nearly 4 *billion* years for the oldest continental crust) and (2) the farther away from the ridge, the older the oceanic crust was, and there was a symmetrical distribution of ages around the ridge. New seafloor therefore was produced at the ridges and subducted ("destroyed") at the trenches. Moreover, this idea did not rely on the continents plowing through the ocean (recall that this was a major objection to Wegener's theory of continental drift), but rather both continents and oceans move.

Concept Check 4-5. Explain the concept of paleomagnetism and discuss its role in understanding seafloor spreading.

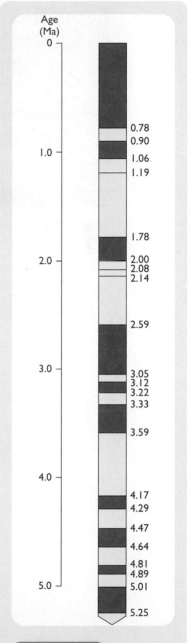

FIGURE 4-13 Reversals of the Earth's magnetic fields over the last 5 million years. Red represents periods of normal polarity. Yellow represents reversals.

FIGURE 4-14 The survey ship *Pioneer*, which first deployed the magnetometer, steaming under the Golden Gate Bridge in the late 1950's. The discovery of magnetic striping on the seafloor contributed to our understanding of the theory of seafloor spreading. Renowned marine geologist, H. W. Menard called the discovery of the magnetic stripes, "among most significant geophysical surveys ever made." Courtesy of NOAA.

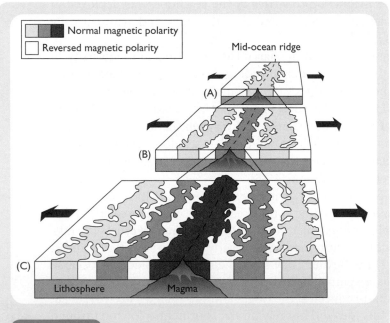

FIGURE 4-15 The pattern of magnetic stripes parallel to the ridge and symmetrically distributed around it. The oldest are shown by the pale brown color, the youngest are shown by the darkest brown color. What was the significance of these stripes? Modified from The Dynamic Earth, USGS.

Plates and Plate Boundaries

Plates are segments of the lithosphere (see Figure 4-10). The geological "action"—mountain-building, earthquakes, volcanoes—is mainly concentrated at **plate boundaries**, because here is where plates are jostling against one another, relieving great stresses.

The three types of plate boundaries are: *divergent, convergent, and transform.*

Divergent Boundaries

If the plates move away from each other, a **divergent boundary** is formed. Rift zones, like the East African Rift, form where continents are ripped apart. Here, ascending convection cells reach the surface and spread out at the base of the lithosphere, tugging at it, heating it, and thereby weakening it. With enough time, the spreading convection cell may fracture the thinned and weakened overlying lithosphere, allowing molten basaltic magma to rise from the asthenosphere and fill the cracks between the spreading plates (Figure 4-12).

Oceans grow and contract as ocean floor rock forms and is in turn consumed, and thus modern oceans exhibit various stages of development. **FIGURE 4-16** illustrates the

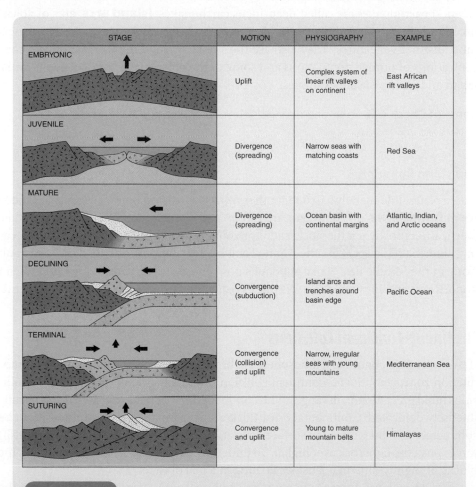

STAGE	MOTION	PHYSIOGRAPHY	EXAMPLE
EMBRYONIC	Uplift	Complex system of linear rift valleys on continent	East African rift valleys
JUVENILE	Divergence (spreading)	Narrow seas with matching coasts	Red Sea
MATURE	Divergence (spreading)	Ocean basin with continental margins	Atlantic, Indian, and Arctic oceans
DECLINING	Convergence (subduction)	Island arcs and trenches around basin edge	Pacific Ocean
TERMINAL	Convergence (collision) and uplift	Narrow, irregular seas with young mountains	Mediterranean Sea
SUTURING	Convergence and uplift	Young to mature mountain belts	Himalayas

FIGURE 4-16 The Wilson Cycle, which shows the evolution of ocean basin development. Data from Wilson, J.T. *American Philosophical Society Proceedings* 112 (1968): 309–320; Jacobs, J.A., et al. *Physics and Geology.* McGraw-Hill, 1974.

Wilson Cycle, which shows the evolution of ocean basin development. Representative examples could be the Rift Valley of East Africa, the Gulf of California, the Gulf of Aden, and the North Atlantic.

Convergent Boundaries and Subduction Zones

Where plates move together a **convergent boundary** results. At convergent boundaries old oceanic lithosphere is recycled back into the mantle. A place where this happens is called a **subduction zone** (FIGURE 4-17).

Convergent boundaries may be of three types: if two masses of continental crust are brought together at a subduction zone, a *continent–continent convergent boundary* is formed. If oceanic lithosphere is subducted beneath continental lithosphere, an *ocean–continent convergent boundary* is formed. And if two oceanic lithosphere segments converge, an *ocean–ocean convergent boundary* is formed (FIGURE 4-18).

Modern subduction zones are usually found with **trenches**. Trenches are arcuate (that is, curved) topographic lows that form by the incessant downward tugging on the oceanic lithosphere. Earthquakes are distributed along subduction zones from near the surface (shallow-focus), through intermediate depths (intermediate-focus), to depths where the subducting material gets too weak to store stress (about 700 km). Deepest earthquakes are called deep-focus earthquakes. **Island arcs**, arcs of volcanic islands formed by subduction at ocean–ocean boundaries, are one of the planets most distinctive features. They also are the site of some of the world's highest concentrations of human populations and some of the most severe zones of geohazards. Japan and Indonesia are examples.

Concept Check 4-6. Name and distinguish between the types of plate boundaries.

Transform Boundaries

When plates slide laterally past one another, *transform or shear* boundaries result. **Transform boundaries** on continents are recognized by the presence of strike-slip faults (vertical fractures where the blocks slide horizontally past each other) like the San Andreas in California (FIGURE 4-19), the Alpine Fault in New Zealand, and the Dead Sea Fault Zone in the Middle East Strike-slip faults pose enormous hazards to growing human populations and infrastructure. We discuss earthquake hazards elsewhere in this text.

Continent-Continent Collisions

If two masses of continental lithosphere are brought together by plate motion at a subduction zone, a continent–continent collision eventually occurs (Figure 4-18). Collisions build great mountain ranges, with continental crust greatly thickened and usually intensely deformed. Continental crust thickens because it is too low in density to be subducted, so it piles up at the surface. Isostatic (Greek *isos* = equal, *stasis* = standstill) forces push the light rock ever higher. To understand **isostasy** (FIGURE 4-20), imagine blocks of slightly differing densities floating in a pail of water.

The Himalayas formed when India collided with Asia, nearly doubling the thickness of the continental crust there. The resulting uplift and subsequent erosion of the light crustal rocks generates huge volumes of sediment, which are presently being deposited in the floodplains of the great rivers draining the Himalayas. Some

Text continues on page 91.

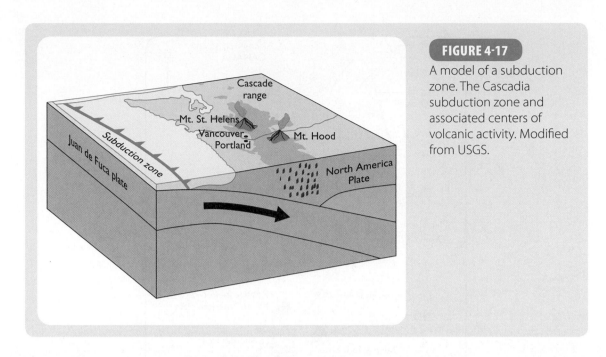

FIGURE 4-17
A model of a subduction zone. The Cascadia subduction zone and associated centers of volcanic activity. Modified from USGS.

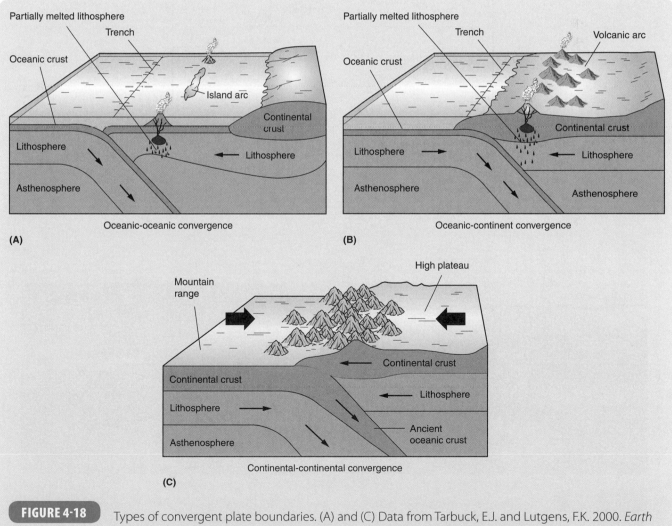

FIGURE 4-18 Types of convergent plate boundaries. (A) and (C) Data from Tarbuck, E.J. and Lutgens, F.K. 2000. *Earth Science* 9th ed, Upper Saddle River, NJ: Prentice-Hall. Figure 7.13 (p. 194). (B) Courtesy of USGS/NPS..

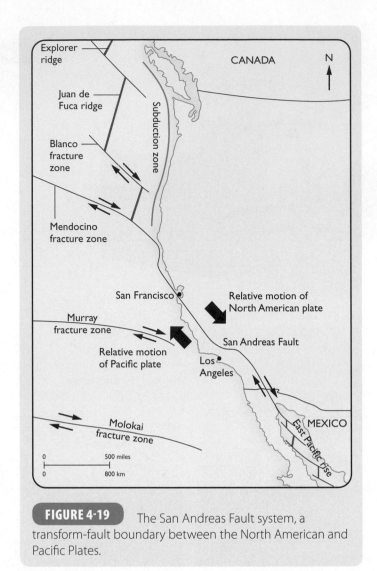

FIGURE 4-19 The San Andreas Fault system, a transform-fault boundary between the North American and Pacific Plates.

FIGURE 4-20 Two barges on the Saigon River in Vietnam, one with cargo, one without, float at different levels due isostatic equilibrium. © Pluff Mud Photography.

An online travel guide[1] describes Sumatra thusly:

> "Anchored tenuously in the deep Indian Ocean, this giant island is still as wild and unpredictable as the Victorian-era jungle-seekers dreamed. Millennia of chaos erupting from the earth's toxic core or from the fierce ocean waves create and destroy in equal measure. When the earth and sea remain still, the past's death and destruction fertilise a verdant future."

Located in western Indonesia, Sumatra is the sixth largest island in the world. Despite the "millennia of chaos erupting from the Earth's toxic core or from the fierce ocean waves," 50 million people call Sumatra home.

From a geological perspective, Sumatra lies near a convergent plate boundary. The Indian (also known as the Indo-Australian) Plate lies to the south and east. To the north and west is the Eurasian Plate (**Figure CS 1-1**). The Indian Plate moves northerly at a rate of about 40–50 mm/ year (1.6–2 in/year). It subducts beneath a portion of the Eurasian plate (technically, the Burma microplate, on which also sit the Andaman and Nicobar Islands in the Bay of Bengal) creating the **Sunda Trench** as well as a large fault called an *interplate thrust*. This Sunda fault runs for approximately 5,500 km (3,300 mi).

These plates creep past each other until large portions of the plates lock or "get stuck" in some locations, building up tremendous forces. When these plates rupture, the stored force is released, resulting in an earthquake (detailed elsewhere in this text).

Although earthquakes along the Sunda fault and within both the subducting Indian and overriding Eurasian Plates are not uncommon, the massive earthquake that struck 60 km off the west coast of northern Sumatra on December 26, 2004 surprised geologists. The event, known as the **Sumatra-Andaman Earthquake**, and the resulting **tsunami**, enormous water waves generated when the seismic energy

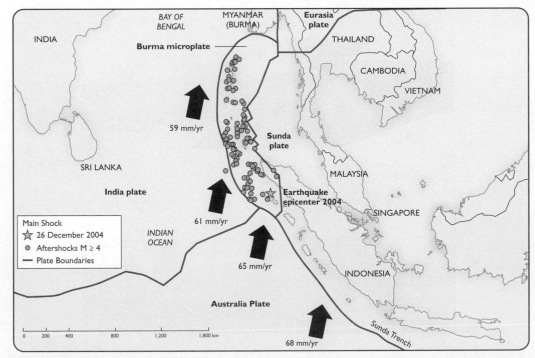

FIGURE CS 1-1 Tectonic setting for the 2004 Sumatra-Andaman Earthquake.

Continued ▶

released by the earthquake entered the ocean, killed 250,000 people in Indonesia, Thailand, India, Sri Lanka, and elsewhere around the Indian Ocean. It was one of the largest earthquakes ever recorded as well as one of the greatest natural disasters in human history.

The earthquake registered > 9.0 on the Richter scale, releasing energy said to be equivalent to that of 10,000 atomic bombs. The overriding Burma microplate was vertically displaced, elevating the seafloor by several meters and setting in motion the monstrous tsunami. Two "N-waves" traveling in opposite directions were generated, a local tsunami and a distant tsunami (**Figure CS 1-2**). The local wave traveled toward Indonesia, Thailand, and nearby islands, while the distant tsunami moved across the Bay of Bengal in the direction of India, and Sri Lanka, and was even detected by tide gauges along the Atlantic coast of North America from Nova Scotia to Florida. The height of the wave varied from 2 to 3 m along the African coast to 10 to 15 m at Sumatra.

Concept Check 4-7. If you could use one adjective to describe the geological setting of Sumatra, in light of this case study, what word would you choose? Explain your choice.

- -

At locations of the tsunami's maximum impact, the force of the waves was so powerful that structures and people stood little chance (**Figure CS 1-3**). Along Sumatra's coastline, waves may have reached heights of 15 to 30 m over a distance of 100 km.

Further away, however, according to a report in *Science* magazine,[2] "areas with coastal [forest] were markedly less damaged than areas without." The trees referred to were mangroves, and they absorbed much of the energy of the waves, protecting the area landward of the mangrove forest. According to the report, mangrove area had been reduced before the tsunami (between 1980 and 2000) by 26% in the five countries most affected by the tsunami, from 5.7 to 4.2 million hectares (14.1–10.4 million acres) because of human activities that included aquaculture and development of resorts. Neil Burgess, a co-author of the study, stated,

"Just as the degradation of wetlands in Louisiana almost certainly increased Hurricane Katrina's destructive powers, the degradation of mangroves in India magnified the tsunami's destruction. Mangroves provide a valuable ecological service to the communities they protect."

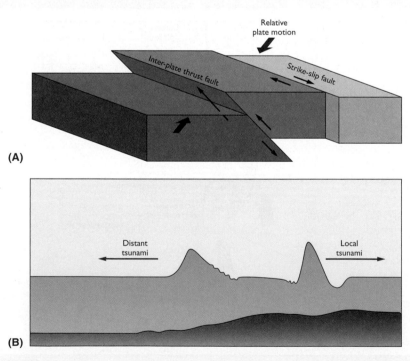

(A)

(B)

FIGURE CS 1-2 (A) Diagram of the type of vertical displacement of the Burma microplate that elevated the seafloor by several meters and set in motion the disastrous tsunami. (B) The resulting N-waves of the tsunami.

FIGURE CS 1-3 A view of the destruction caused by the 2004 Indonesian earthquake and tsunami. © Claudio Gallone/agefotostock.

Concept Check 4-8. Summarize the major points of this case study.

Value-added Question 4-1. The displacement of the fault occurred along a 1,200-km long segment. If the earthquake moved at a rate of 2.5 km/sec, how long (in minutes and seconds) did it take for the fault to propagate from its hypocenter (origin) to its extreme 1,200 km away?

Value-added Question 4-2. The initial velocity of the tsunami was 700 km/hr. Assuming that the tsunami maintained that speed (even though it slows in shallow water), how long would it take for it to reach: (A) India (approximate distance 2,000 km); (B) Sri Lanka (1,570 km); (C) Thailand (500 km); (D) Sumatra (60 km); (E) Myanmar (1,800 km); (F) Somalia (5,000 km); (G) Maldives (2,500 km)?

of the sediment is finding its way into the Indian Ocean, where it is building massive *submarine fans* (**FIGURE 4-21**).

The deformation of the continental crust north of the Himalaya extends at least 2,000 kilometers from the collision zone. Some of the planet's most destructive earthquakes have occurred along faults here.

Mantle Plumes and Hot Spots

FIGURE 4-22 shows the Hawaiian **island chain**. Island *chains* like Hawaii form as a result of movement of oceanic plates over **mantle plumes**. Plumes are believed to form due to abnormally high concentrations of heat that produce vertical, candle-flame shaped masses containing magma of basaltic composition. As they reach the surface they form *hot spots*. Magma rises to the surface mainly by convection, forms hot spots, and may punch holes in the overlying lithosphere, forming volcanic islands if the overlying lithosphere is oceanic.

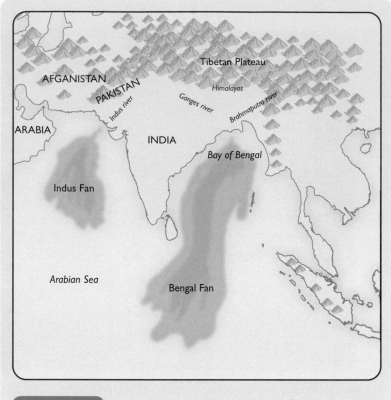

FIGURE 4-21 The Indus and Bengal submarine fans, formed of thousands of meters of sediment eroded at the Himalayan collision zone. The weight of the sediment depresses the oceanic crust in another example of isostatic adjustment.

Hot spots may also form along divergent boundaries. Iceland is formed by hot spot volcanism at the **Mid-Atlantic Ridge**. Hot spots under oceanic lithosphere produce volcanoes emitting mainly basalt lava and tephra (**tephra**, from the Greek for ash, describes volcanic rock fragments and lava of varying sizes that are explosively blasted into the air or lifted by hot gases). A good example of such a volcano is Hawaii's Mauna Loa, the world's largest volcano.

In an ocean, a *line* of volcanic islands formed as the oceanic plate passes over the hot spot, as contrasted with the *arc* of islands at a subduction zone (Figure 4-18). The orientation of the island chain traces the direction of plate movement. The Hawaiian Plume is presently situated immediately southeast of the Big Island of Hawaii, and, in addition to the active volcanoes on Hawaii, the plume is forming a new island called Loihi, presently entirely underwater (Figure 4-22).

Plumes may persist for tens of millions of years, and apparently stay near the same place in the mantle.

Volcanoes and Plate Tectonics

Volcanoes will be discussed in more detail elsewhere in this text. Geologists recognize *plate boundary volcanoes* and *hot-spot volcanoes*. The **plate tectonic setting** strongly influences magma type produced, which in turn determines viscosity (*viscosity* refers to internal resistance to flow and can thought of as a liquid's thickness) and ultimately risk, as we discuss elsewhere in this text.

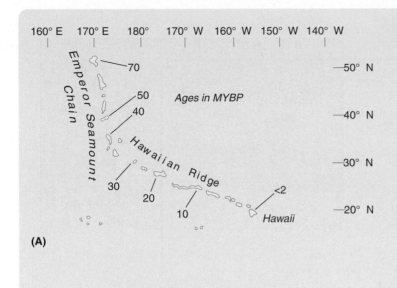

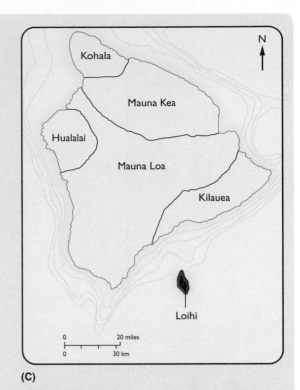

(A)

(B)

(C)

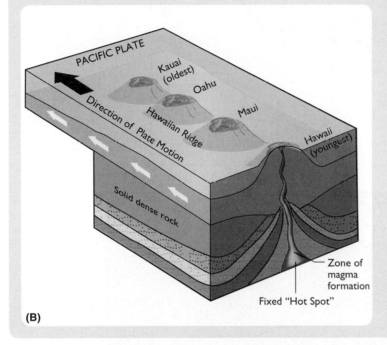

FIGURE 4-22 (A) The Hawaiian Islands are made up of basaltic islands above seal level, and seamounts, below. Data from Dalrymple, G.B., et al., *American Scientist* 6 (1973): 294–308; Claque, D.A. and R.D. Jarrard, *Geol. Soc. Am. Bull.* 84 (1975): 1135–1154. (B) Plumes form as heat is transferred from the core/mantle boundary. The magma rises to the surface mainly by convection, forming hot spots, and may punch holes in the overlying lithosphere, forming volcanic islands if the overlying lithosphere is oceanic. Courtesy of USGS. (C) The underwater island of Loihi, the newest island in the Hawaiian chain. Courtesy of USGS.

Plate Boundary Volcanoes

Plate boundary volcanoes can form at divergent and convergent boundaries.

Much volcanic activity occurs at divergent boundaries and at rift zones. Recall that partial melting of mantle peridotite produces basaltic magma. Because there is no continental (granitic) crust to contaminate this magma at mid-ocean ridges, basalt lava is overwhelmingly produced. Volcanoes of Iceland (**FIGURE 4-23**) are the most visible examples of these types, and they are mainly basalt.

Volcanoes of convergent boundaries form either at *ocean–ocean boundaries* or *ocean–continent boundaries*.

FIGURE 4-24 shows the Indonesian arc. These islands are made up largely of volcanic rocks. Island arc volcanoes derive their name from the broad curve of their volcanic islands. Indonesia, Japan, and the Aleutians are examples.

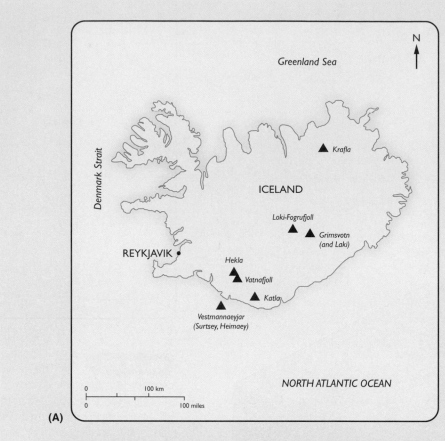

(A)

(B)

FIGURE 4-23   (A) Location of volcanoes of Iceland. Courtesy of USGS. (B) Photo of the volcano Krafla. © iStockphoto/Thinkstock.

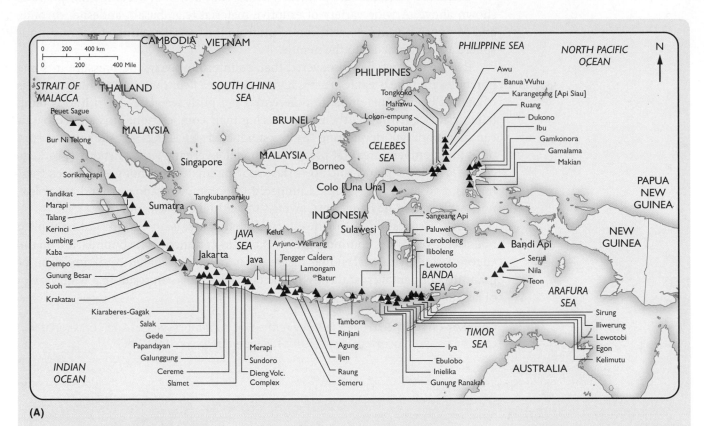

(A)

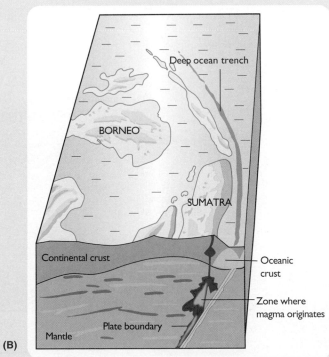

(B)

FIGURE 4-24 (A) The Indonesian Arc, showing some of the sites of active volcanoes. (B) Diagram depicting how these islands arc formed.

Island arc volcanoes are produced by the subduction of oceanic lithosphere as shown above. A wide variety of eruptive material (e.g., basalt, andesite, rhyolite, and compositions in between) can result from the partial melting of oceanic lithosphere as it is carried into the mantle.

Recent research has discovered that, counterintuitively, the pressure in some boreholes into the oceanic crust is higher at the sea *floor* than in the hole hundreds of meters below. This pressure differential can push seawater down into cooling basaltic rocks, hydrating them and altering their composition. This altered basalt can then undergo "dewatering" as it is subducted (**FIGURE 4-25**). The water driven off at high temperatures and pressures can melt overlying rocks. Eruptive products can be lava flows, pyroclastic flows (fast-moving mixtures of hot gas and rock fragments), and tephra eruptions. Compared to oceanic volcanoes, island arc volcanoes are much more dangerous, owing to their higher silica content.

The Cascades of the Northwestern United States, is an ocean–continent plate boundary. Volcanoes commonly form at ocean–continent plate boundaries, on the edges of continents. Ascending magma may be contaminated by assimilation of granitic continental crust through which it passes. This can result in a wide variety of lava and pyroclastic eruptions.

Concept Check 4-9. Write a few key terms that contrast divergent and convergent volcano types and processes. Which are most dangerous to humans? Why?

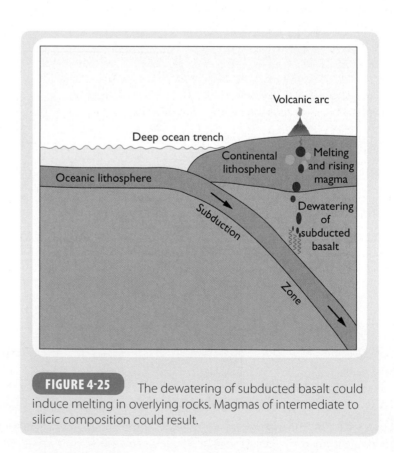

FIGURE 4-25   The dewatering of subducted basalt could induce melting in overlying rocks. Magmas of intermediate to silicic composition could result.

CHAPTER SUMMARY

1. Because we don't have direct samples from boreholes into the Earth's center, we must content ourselves with making models of the interior based on our indirect evidence.

2. Geologists use hypotheses to explain some natural phenomenon, and then gather all evidence to test their hypotheses.

3. If the hypothesis is supported by evidence, geologists may elevate it to a theory.

4. A model of the Earth's interior shows the general internal structure, where interior boundaries are, what the states of matter are, and the composition of the interior, in as much detail as necessary.

5. We use evidence from the following sources to make a model of the Earth's interior. Seismic wave paths and velocities inside the Earth depend on the property of the material through which the waves pass. If we can track the wave path, we can put constraints on the nature of the material along its path. We have excellent knowledge of the composition and distribution of surface rocks. We also use evidence from meteorites. We have hundreds of drill holes and dredge samples of oceanic bedrock, and so we know its near-surface composition very well. We know the earth's average density is 5.52 g/cm³. The physics of the Earth's rotation dictates that the densest material must be at the Earth's center. The existence and nature of the magnetic field puts limits on the possible composition and state of the planet's interior.

6. One of the most fundamental discoveries about the Earth was that the interior was composed of layers with recognizable boundaries.

7. Geologists divide the Earth into three layers: the crust (outermost material), an underlying mantle, and a core at the center. The core was found to consist of an inner solid portion and an outer liquid portion. The upper mantle was found to contain a zone near its top where seismic wave velocities were lower than above and below; this zone became known as the low-velocity zone, or asthenosphere, and is partially molten. The upper mantle is of peridotitic composition and the lower mantle is probably made up of densely packed oxide minerals.

8. Radioactivity refers to the decay of unstable atomic nuclei. Radioactive heat powers the global process we call plate tectonics. Three ways that heat can be transferred are: by conduction, by convection, and by radiation. Convection cells and plumes powered by radioactive decay are the means by which new oceanic bedrock is made, and new oceans may be formed in the process.

9. Plate tectonic theory holds that the Earth is covered by seven major rigid lithospheric plates and a number of smaller plates.

10. Plate tectonics explains the majority of the Earth's large-scale geologic structures and processes and has been called a Unifying Principle of Geology.

11. The theory of plate tectonics evolved from two older ideas, continental drift and seafloor spreading.

12. Plates move at a few centimeters a year.

13. The three types of plate boundaries are: divergent, convergent, and transform.

14. At convergent boundaries old oceanic lithosphere is recycled back into the mantle. A place where this happens is called a subduction zone.

15. In 2004 one of the Earth's most powerful earthquakes occurred at the subduction zone of the convergent Indian and Eurasian Plates. The quake and resulting tsunami killed 250,000 people in Indonesia, Thailand, India, Sri Lanka, and elsewhere around the Indian Ocean.

16. Transform boundaries on continents are recognized by the presence of strike-slip faults like the San Andreas in California.
17. If two masses of continental lithosphere are brought together by plate motion at a subduction zone, a continent–continent collision eventually occurs.
18. Island chains like Hawaii are hypothesized to form as a result of movement of oceanic plates over mantle plumes.
19. Volcanoes can be classified as plate boundary volcanoes and hot-spot volcanoes. Island arc volcanoes are produced by the subduction of oceanic lithosphere. Continental volcanoes, like island arc volcanoes, are more dangerous than oceanic volcanoes, and may occasionally erupt with explosive and even cataclysmic force.

KEY TERMS

asthenosphere
basaltic
conduction
continental crust
continental drift
convection
convergent boundary
core
crust
density
discontinuity
divergent boundary
fault
heat transfer
hypothesis
igneous
island arc
island chain

isostasy
lithosphere
magma
magnetometer
mantle
mantle plume
metamorphic
Mid-Atlantic ridge
model
ocean trench
oceanic crust
paleomagnetism
Pangaea
Panthalassa
peridotite
plate
plate boundaries
plate tectonics

plate-tectonic setting of volcanoes
pyroclastic
radiation
radioactivity
seafloor spreading
sedimentary
seismic wave
states of matter
subduction zone
Sumatra-Andaman Earthquake
Sunda Trench
tephra
theory
transform boundary
trench
tsunami

REVIEW QUESTIONS

1. A hypothesis supported by experiments and observation can be elevated to a
   a. paradigm.
   b. fact.
   c. proof.
   d. theory.

2. The Earth's average density is closest to
   a. 5.5.
   b. 1.
   c. 55.
   d. 10.

3. The two commonest kinds of meteorites are made of
   a. uranium-lead and iron-nickel.
   b. uranium-lead and silicate minerals.
   c. silicate minerals and potassium-sulfur.
   d. silicate minerals and iron-nickel.

4. Which is true about our knowledge of the Earth's interior?
   a. We have direct drill samples through the mantle.
   b. We have direct drill samples of the outer core.
   c. We have direct drill samples of the outer few km only.
   d. We have no direct drill samples of oceanic sediments, only the continents.

5. Zones within the Earth where seismic waves abruptly change velocity are called
   a. discontinuities.
   b. attenuated zones.
   c. continuities.
   d. batholiths.

6. Which is the most accepted model of the Earth's core?
   a. The inner core is liquid and the outer core solid.
   b. The inner core is liquid and the outer core gaseous.
   c. The inner core is solid and the outer core is liquid.
   d. The entire core is solid, according to magnetic field data.

7. Another name for the low-velocity zone in the upper mantle is
   a. the discontinuity.
   b. the lithosphere.
   c. the ophiolite sequence.
   d. the asthenosphere.

8. The composition of the upper mantle is most nearly which of the following?
   a. peridotite
   b. basalt
   c. granite
   d. iron-nickel alloy

9. Heat transfer occurs in three ways:
   a. conduction, induction, and radioactivity.
   b. conduction, convection, and radiation.
   c. subduction, attenuation, and shear.
   d. shear, transform, and divergence.

10. Plate motion, powered by radioactive decay, occurs at an average rate of
    a. a few km/year.
    b. a few meters/year.
    c. a few centimeters/year.
    d. a few nanometers/year.

11. Which of the following is considered a unifying principle of geology?
    a. evolution
    b. continental drift
    c. seafloor spreading
    d. basalt
    e. the principle of radiometric dating

12. Evidence for continental drift includes all of the following except
    a. the jigsaw puzzle nature of continents.
    b. the presence of the remains of early humans on Africa and South America.
    c. the presence of the late Paleozoic fossil plant *Glossopteris* on India, Australia, South Africa, and South America.
    d. the contiguous distributions of fossil reptiles across ocean basins.

13. In the late 1950s-early 1970s, paleomagnetism showed
    a. that the Earth's poles periodically reversed.
    b. that all rocks are magnetic.
    c. that seafloor is produced at deep sea trenches and consumed at mid-ocean ridges.
    d. none of the above.

14. The theory of seafloor spreading was confirmed in the 1960s by
    a. magnetic striping parallel to and symmetrical around the ocean ridges.
    b. current weather patterns across the planet.
    c. advances in molecular biology.
    d. all of the above.

15. Plates are
    a. segments of the asthenosphere.
    b. segments of the lithosphere.
    c. segments of the biosphere.
    d. formed at subduction zones.

16. New oceans form at a _____ type of plate boundary.
    a. divergent
    b. subduction/convergent
    c. aseismic
    d. discontinuous

17. At convergent boundaries, older lithosphere is
    a. recycled back into the mantle.
    b. converted to core.
    c. converted to peridotite.
    d. all of the above.

18. When plates slide laterally past each other,
    a. transform or shear boundaries result.
    b. no earthquakes occur.
    c. earthquakes are all deep (>500 km) focus.
    d. earthquakes occur, but none has to date caused damage.

19. Island *chains* like Hawaii are hypothesized to form as a result of movement of oceanic plates
    a. over mid-ocean ridges.
    b. over old continents.
    c. over mantle plumes.
    d. away from each other at a divergent boundary.

20. Island arc volcanoes are produced by
    a. the subduction of oceanic lithosphere.
    b. movement of plates over hot spots.
    c. continent–continent collisions.
    d. any of the above.

21. Which is/are accurate contrasts between continental and oceanic volcanoes?
    a. Continental volcanoes are more explosive and dangerous than oceanic volcanoes.
    b. Oceanic volcanoes tend to be basalt, while continental volcanoes can vary in composition.
    c. Ascending basaltic magma under continents can melt continental crust and change composition.
    d. All of the above are correct.
    e. None of the above is correct.

22. Doubling the thickness of the continental crust by formation of the Himalayas causes them to rise, and piling sediment derived from weathering of the Himalayas on the north Indian Ocean causes it to sink. These are examples of the Principle of
    a. Uniformity.
    b. Plate Activity.
    c. Isostasy.
    d. Discontinuity.

FOOTNOTES

[1] Introducing Sumatra. Retrieved August 19, 2012 from http://www.lonelyplanet.com/indonesia/sumatra

[2] Danielsen, F., et al. 2005. The Asian tsunami: A protective role for coastal vegetation. *Science* 310: 643.

5

CHAPTER OUTLINE

■ INTRODUCTION: GOING TO EUROPE? MAYBE NOT THAT YEAR

In April, 2010, up to 4,000 flights to and from Europe were suddenly cancelled, with airspace closed throughout Scandinavia and in the United Kingdom. The cause was ash from a volcanic eruption in Iceland. According to BBC News, the U.K.'s airspace restriction was the worst in living memory. As many as 600,000 people may have been impacted. Particles of ash in the vast eruptive cloud, which had been ejected many kilometers into the atmosphere, could have clogged jet engines and caused a disastrous loss of power.

■ ERUPTION OF LAKE NYOS, CAMEROON

On the night of August 21, 1986, a dreadful catastrophe overtook nearly 1,500 sleeping villagers in valleys draining Cameroon's Lake Nyos (FIGURE 5-1). Thousands of livestock and untold wildlife also perished. Geologists believe that the lake belched a vast cloud of carbon dioxide (CO_2) gas that, being heavier than air, flowed close to the ground as a lethal invisible cloud down valleys and away from the lake. Victims were asphyxiated as far as 27 km from the lake. The area around the lake was immediately evacuated.

Volcanologists believed that CO_2 gas seeped into the base of the 200-meter deep crater-lake from hot springs emanating from magma below. The solubility of CO_2 under pressure in water was very high—think of the fizz in a bottle of pop or soda—such that the lake could hold up to *five times its volume* in CO_2. This meant that the amount of CO_2 in the lake if it were to come out of solution would fill a basin five times the size of the present lake. And the CO_2 buildup continues. Researchers estimated in 2000 that the lake contained at least 300 *million* cubic meters of CO_2. This CO_2 remains trapped in the bottom water of the lake until the water is disturbed, which causes the gas-laden bottom water to rise. At Lake Nyos, the CO_2 came out of solution violently, as a gas–water mixture, shooting up about 55 m (260 feet), and then the gas moved laterally at a speed of over 70 km/h (45 mi/hr)!

FIGURE 5-1 (A) Location map of Cameroon. (B) Lake Nyos. Courtesy of Jack Lockwood/USGS.

Scientists set up a system of pipes to allow CO_2 to escape from the lake's depths little by little, so as to avoid another catastrophe (**FIGURE 5-2**). They also installed an early warning system of sirens and strobes to alert residents of nearby villages.

But is it the end of the nightmare? There are still dangerously high CO_2 levels in Lake Nyos, and perhaps even worse, scientists fear that the volcanic rock that forms a natural dam to keep water in the lake could collapse. If the dam should fail,

FIGURE 5-2 Pipes to drain carbon dioxide gas from magma below Lake Nyos. Courtesy of Bill Evans/USGS.

the upper 40 meters of water would spill out, leading to an immediate geyserlike eruption of dissolved gas as well as a flood that could reach all the way into Nigeria (see map) and affect as many as 10,000 people.

One thing is certain: it was not the first such toxic gas eruption from Lake Nyos and it will not be the last.

■ POPULATIONS AT RISK: "EDUCATION IS ESSENTIAL"

Science, you will recall, is a way of discovering knowledge about our surroundings using the scientific method. Scientists, however, are sometimes put into positions of making important, even life-and-death decisions based on their findings, or providing advice to others who make the decisions.

Imagine that you are a volcanologist studying a volcano near a city. Suddenly you see the first signs of volcanic activity. These signs are collectively known as **precursors**, and can include small earthquakes, changes in the position of the volcano's surface, and changes in the composition of gases emanating from the volcano. Although these precursors point to an eruption, they don't specify _when_ it will occur (precursor events can last for months) or the _magnitude_ of the future eruption.

A major explosive eruption could kill thousands, while evacuating an area prematurely could lead to significant economic and social disruptions whose effects could be felt for many years.

What would you recommend to authorities?

This hypothetical scenario illustrates the importance of understanding the geological processes of our planet. Indeed, the Federal Emergency Management Agency (FEMA) says in reference to volcanic hazards: **Education is Essential**. We hope you will study this information carefully. Your life or the life of someone you love might someday depend on it.

■ HISTORICAL IMPACT OF VOLCANOES

Volcanic eruptions since 1700 have killed more than 260,000 people, destroyed whole cities, and disrupted local economies for months to years. **TABLE 5-1** shows the major eruptions since 1500 and their effects.

Italy's Mt. Vesuvius (see below) has erupted over 25 times since the eruption that buried the Roman cities Herculaneum and Pompeii without a trace for 1500 years. Since then, Naples and its suburbs have grown into a city of over two million, all of whom are at risk when—not _if_—Vesuvius next erupts. **FIGURE 5-3** shows the proximity of Naples to Vesuvius.

Other regions with high and growing populations at risk from volcanic eruptions include Indonesia, the Philippines, Central America, Japan, and southwestern Alaska. All these areas have experienced eruptions in the past decade, and more are certain. In the United States, the Aleutian Island volcanoes are very active, and volcanoes in the Cascades of British Columbia, Washington, Oregon, and Northern California pose a threat to communities throughout the Pacific Northwest. **FIGURE 5-4** shows a map of active volcanoes of the West Coast. While no volcanic event in the past 500 years has come close to the truly cataclysmic eruptions in recent prehistory, the United States Geological Survey (USGS) estimates that at least 500 million people face potential risk from a volcanic eruption.

TABLE 5-1	Deadliest Volcanic Eruptions Since 1500 A.D.		
Eruption	**Year**	**Casualties**	**Major Cause**
Nevado del Ruiz, Colombia	1985	25,000[1,3]	Mudflows[3]
Mont Pelée, Martinique	1902	30,000[1] (29,025)[2]	Pyroclastic flows[2]
Krakatau, Indonesia	1883	36,000[1] (36,417)[2]	Tsunami[2]
Tambora, Indonesia	1815	92,000[1,2]	Starvation[2]
Unzen, Japan	1792	15,000[1] (14,030)[2]	Volcano collapse, Tsunami[2]
Lakagigar (Laki), Iceland	1783	9,000[1] (9,350)[2]	Starvation[2]
Kelut, Indonesia	1586	10,000[1]	

[1] Tilling, Topinka, and Swanson, 1990, Eruptions of Mount St. Helens: Past, Present, and Future: U.S. Geological Survey General Interest Publication, 56p.
[2] Blong, R.J., 1984, Volcanic Hazards: A Sourcebook on the Effects of Eruptions: Orlando, Florida, Academic Press, 424p.
[3] Wright and Pierson, 1992, Living With Volcanoes: The U.S. Geological Survey's Volcano Hazards Program: U. S. Geological Survey Circular 1073, 57p.
[4] Spall, H. (ed.), 1980, Earthquake Information Bulletin: July-August, 1980, v.12, no.4, 167p.

Though most studies focus on their destructiveness, volcanoes do provide significant economic benefits. Ash and **lava** enrich soils, and volcanic products like pumice and lava are used for construction and abrasives.

A Sample of Twentieth Century Volcanic Disasters

Here are three examples, compiled by the United Nations Educational, Scientific and Cultural Organization (UNESCO), illustrating the destructive power of volcanoes.

 Naples, Italy with Vesuvius in the background. Naples and its environs are home to more than two million people.
© iStockphoto/Thinkstock.

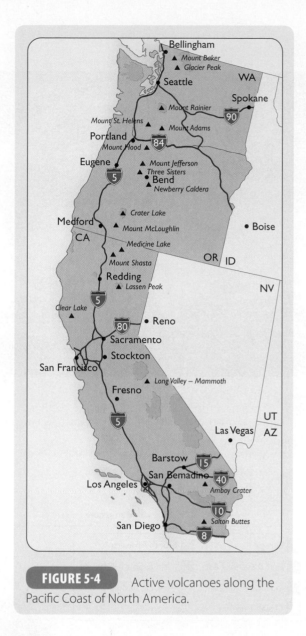

FIGURE 5-4 Active volcanoes along the Pacific Coast of North America.

Mt. Pinatubo, Philippines

Inactive for over *600 years,* the eruption of Pinatubo on June 15–16, 1991 was the biggest eruption of the twentieth century, slightly edging out Alaska's Mt. Katmai, in 1912 (**FIGURE 5-5**). More than 200,000 people were evacuated, and at least 320 were killed. The region was economically impacted in part due to the closing of the U.S.'s Clark Air Force base as a result of the eruption.

The eruption of Pinatubo had *global effects.* Slightly cooler than usual temperatures worldwide, and brilliant sunsets and sunrises have been attributed to the fine ash and gases blasted high into the stratosphere, producing a cloud of debris that drifted around the world. The sulfur dioxide (SO_2) in this cloud, at least 20 million tons in mass, combined with water to form **acid aerosols** (aerosols are fine particles or droplets suspended in a gas, in this case, air). These, along with ash, act like tiny mirrors, reflecting sunlight and thereby cooling temperatures in some regions by as much as 0.5°C. The cooling effect of eruptions depends in part on the sulfur oxide (SOx, a mixture of sulfur oxides that includes SO_2) emissions as shown in **FIGURE 5-6** .

FIGURE 5-5 The main eruption of Pinatubo lasted 15 hours, sending ash at least 30 km into the sky, and setting off pyroclastic flows and lahars (volcanic mudflows: see Table 5-2). Courtesy of USGS.

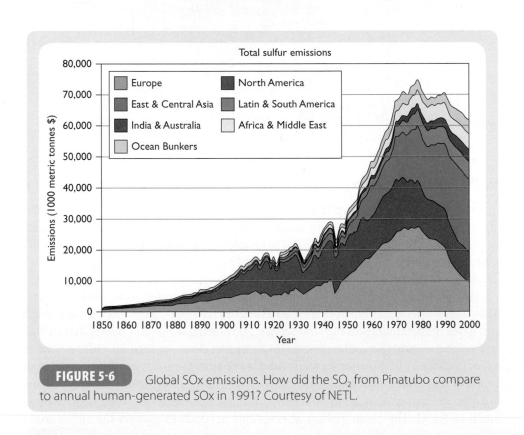

FIGURE 5-6 Global SOx emissions. How did the SO_2 from Pinatubo compare to annual human-generated SOx in 1991? Courtesy of NETL.

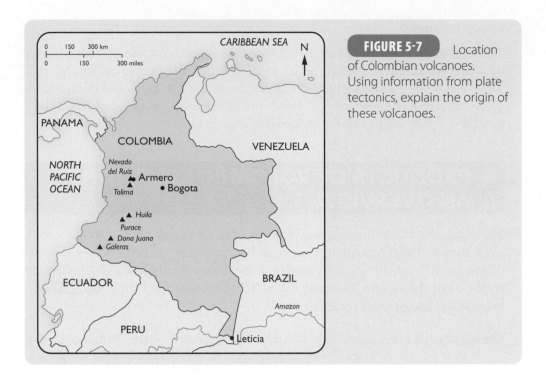

FIGURE 5-7 Location of Colombian volcanoes. Using information from plate tectonics, explain the origin of these volcanoes.

Nevado del Ruiz, Colombia

FIGURE 5-7 shows the location and geologic setting of Nevado del Ruiz and other Colombian volcanoes. After being dormant for 140 years, the volcano erupted incandescent *pyroclastic* material (that is, rocks, debris, etc. resulting from explosive volcanic eruptions) on November 13, 1985, which melted the mountain's ice cap and set off volcanic mudflows. Following four riverbeds at speeds reaching 30 km/hr (19 mi/hr), they buried the towns of Armero and Chinchina late at night killing 24,740 people, injuring 5,455, and destroying 5,650 homes. **FIGURE 5-8** shows Armero, Colombia buried beneath these volcanic mudflows. We further discuss Nevado del Ruiz in our section on lahars later in the chapter.

FIGURE 5-8 Armero, Colombia buried beneath volcanic mudflows. © Steven L. Raymer/National Geographic Stock.

Agung, Indonesia

Two explosive eruptions in March and May 1963 generated a hot ash cloud and set off *lahars* (volcanic mud flows), which killed at least 1,145 people and injured 296. These eruptions also destroyed over 50,000 hectares (124,000 acres) of farmland, 11,000 hectares (27,000 acres) of forest and swept away bridges and roads. Like Pinatubo, it may have caused a measurable drop in global temperatures.

■ PRODUCTS OF VOLCANIC ERUPTIONS: AN OVERVIEW

Many of the world's 500 most active volcanoes erupt infrequently but, as you have seen, when they do the effects can be disastrous. Nearby towns can be buried by deposits of **ash** (which can also threaten aircraft in the area and pollute water supplies), by **pyroclastic flows**, *lava flows,* and *lahars* (these terms are defined in **TABLE 5-2**). We further discuss these materials below.

Concept Check 5-1. Look at Table 5-2 and review the definitions of the italicized terms above. Which of these do you think the general public believes would pose the greatest threat? Do you agree?

TABLE 5-2	**Quick Reference Definitions of Volcanic Terms** (definitions from http://volcanoes.usgs.gov)

Photo courtesy of USGS.

Ash rock, mineral, and volcanic glass fragments smaller than 2 mm (0.1 inch) in diameter, which is slightly larger than the size of a pinhead.

Photo courtesy of USGS.

Lahars Lahar is an Indonesian word for a rapidly flowing mixture of rock debris and water that originates on the slopes of a volcano. Lahars are also referred to as volcanic mudflows or debris flows.

TABLE 5-2

Quick Reference Definitions of Volcanic Terms (Continued)
(definitions from http://volcanoes.usgs.gov)

Lateral Blasts . . . almost instantaneous expansion (explosion) of high temperature-high pressure steam present in cracks and voids in the volcano and of gases dissolved in the magma . . . can be compared in some ways to the sudden removal of the cap or a thumb from a vigorously shaken bottle of soda pop, or to punching a hole in a boiler tank under high pressure.

Photo courtesy of Richard P. Hoblitt/USGS.

Lava Flow masses of molten rock that pour onto the Earth's surface during an effusive eruption.

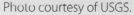
Photo courtesy of USGS.

Photo courtesy of USGS.

Pyroclastic Flow a ground-hugging avalanche of hot ash, pumice, rock fragments, and volcanic gas that rushes down the side of a volcano as fast as 100 km/hour.

Tephra a general term for fragments of volcanic rock and lava, regardless of size, that are blasted into the air by explosions, or carried upward by hot gases in eruption columns or lava fountains.

Photo courtesy of USGS.

■ CLASSIFYING VOLCANOES

A Rationale

Volcanoes may be classified in several ways. Unfortunately, the existence of several different schemes and their specific jargon can be bewildering. The reasons we include several systems is that they often have special uses. For example, hazard specialists classify volcanoes according to their explosiveness. Finally, you may run across these specialized terms both in and out of class. As you may have already learned, studying a science is not unlike learning a foreign language, in that in both you must master a vocabulary before you can apply the science, or understand it in context.

Classifying Volcanoes by Shape and Eruptive Product

Geologists often classify volcanoes by shape and eruptive product. We discuss each type briefly below.

Shield Volcanoes

Shield volcanoes are composed of basalt (a dark-colored, fine-grained, igneous rock), have broad, nearly flat-to gently sloping flanks, and are shaped like an ancient warrior's shield (**FIGURE 5-9**). They are built up from numerous freely-flowing (that is, low **viscosity**) basalt flows. Because the lavas are not highly viscous, each lava flow can flow relatively far from the vent, and the lava's abundant volcanic gases, mainly H_2O with some CO_2, bubble off easily. *Lava tubes,* natural underground conduits through which the lava spreads, allow the lava to reach great distances from the source.

Shields do not build up a cone, as do *stratovolcanoes* (see below) and *cinder cones.* Shield volcanoes are typical of oceanic *hot spots* and mid-ocean ridges, places where basaltic magmas form by partial melting of the Earth's mantle (discussed elsewhere in this text). The largest volcanoes on the Earth are shield volcanoes, and include Mauna Loa and Medicine Lake Volcano in northern California. Olympus Mons on Mars (**FIGURE 5-10**), one of the largest known volcanoes in the Solar System, is also a shield volcano.

Stratovolcanoes

Also called *composite* volcanoes because they are *composites* of lava and ash deposits, stratovolcanoes erupt both lava flows and ash of intermediate-to-felsic chemical composition (**FIGURE 5-11**). *Intermediate* composition refers to having a *silica* (SiO_2) concentration of between 52% and 63%. *Felsic* refers to light-colored magma and rocks enriched with higher concentrations of silica. Lava of **stratovolcanoes** include *andesite, dacite,* and *ryolite.* Andesite lava (first applied to lavas of the Andes) is intermediate in composition. *Dacite* lava is considered intermediate-felsic, and *rhyolite* lava is felsic. Stratovolcanoes are the signature volcanoes of ocean–continent convergent boundaries and island arcs. They are conical and steep and are built over periods of hundreds of thousands of years.

The higher **silica content** (55%–>70%) of the magma of stratovolcanoes results in higher viscosities (resistance to flowing, or "thickness"), and causes the lava's abundant volcanic gas to be more easily retained. The trapped gas in turn can build up enormous pressures, which may lead to eruptions of awesome power. These eruptions are very dangerous and result in most of the loss of life and property

(A)

(B)

FIGURE 5-9 (A) Belknap Shield Volcano, Oregon. Courtesy of USGS. (B) Mauna Loa, a shield volcano and also the world's largest volcano, covering half of the island of Hawaii. Courtesy of USGS.

damage sustained during eruptions. According to the USGS, stratovolcanoes comprise 699 of Earth's 1,511 volcanoes known to have erupted in the past 10,000 years. FIGURE 5-12 shows the structure of a representative stratovolcano.

Cinder Cones

The earliest known depiction of volcanic activity is a wall painting in what is now Turkey, of a **cinder cone** erupting, around 6200 BCE (FIGURE 5-13). Cinder cones are relatively small but steep conical volcanoes, emitting much pyroclastic matter of relatively coarse size called *cinder*. They are predominantly of intermediate composition. They are relatively benign, but can cause damage close to the crater. Cinder cones commonly occur on the flanks of shield volcanoes, stratovolcanoes, and calderas (see below).

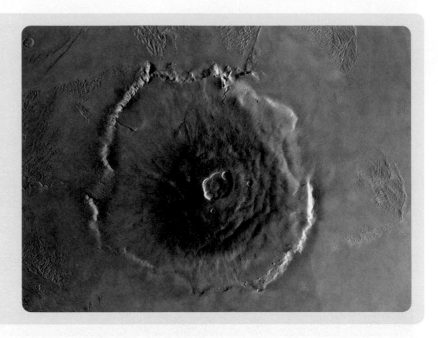

FIGURE 5-10 Olympus Mons, the largest Martian volcano, is a shield volcano. What type lava would it have likely therefore erupted? © Stocktrek Images/Thinkstock.

FIGURE 5-11 Mt. Rainier, WA, a stratovolcano, with portions of Seattle in the foreground. © Hemera/Thinkstock.

Lava Domes

Lava domes are usually small and dome-shaped, and extrude highly viscous, silicic magma. Because of this high viscosity, dome-building lava does not move far from the vent before solidifying. Mt. Lassen in California is an example (**FIGURE 5-14**). Lava domes can be very dangerous, because very silicic lava can trap gas, leading to very explosive eruptions. Deadly pyroclastic flows can also emanate from lava domes.

Calderas

Calderas are the remains of cataclysmic eruptions that literally have blown an existing volcano to bits, leaving behind a crater, which may fill with water. While

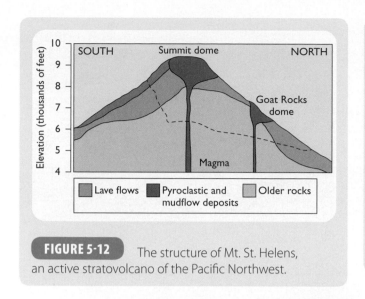

FIGURE 5-12 The structure of Mt. St. Helens, an active stratovolcano of the Pacific Northwest.

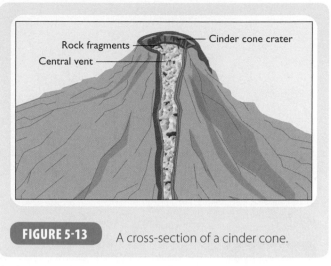

FIGURE 5-13 A cross-section of a cinder cone.

(A)

(B)

FIGURE 5-14 (A) Lassen Peak, California, a lava dome. Courtesy of USGS. (B) Structure of a lava dome.

the origin of calderas is complex, they sometimes form over hot spots (discussed elsewhere in this text), which underlie continental crust, such as Yellowstone. Caldera-forming eruptions are the largest eruptions on Earth. For example, the Fish Canyon eruption in southwestern Colorado about 25 million years ago erupted more than 5,000 km³ (1200 mi³) of magma from La Garita Caldera, enough to bury the entire state of Colorado to a depth of nearly 12 m! Santorini, in the eastern Mediterranean, which has been implicated in the destruction of Minoan civilization in ~1400 BCE, is a caldera, as is Crater Lake, Oregon (**FIGURE 5-15**).

The 23 km-wide (14 mi) Valles Caldera (**FIGURE 5-16**) is one of the largest geological features in New Mexico. It is a product of the Rio Grande Rift System (discussed elsewhere in this text), and formed between 1.0 and 1.4 million years ago. Collapse of the caldera generated vast explosions, forming clouds of super-hot gas, ash, and other fragments of lava. When these deposits, blown high into the atmosphere, collapsed, they were so hot the particles fused to form extensive deposits called **welded tuffs**. Tuffs are glass-rich mixtures of volcanic rock, fragments, and ash.

(A)

(B)

FIGURE 5-15 Two calderas: (A) Santorini in the Aegean, and (B) Crater Lake, southern Oregon. Both were the sites of cataclysmic eruptions. (A) © Andrei Nekrassov/ShutterStock, Inc. (B) Courtesy of Mike Doukas/USGS.

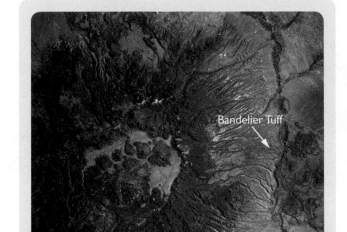

FIGURE 5-16 Aerial photo of the Valles Caldera, New Mexico and the Bandelier Tuff, near Los Alamos, New Mexico. Courtesy of Image Science and Analysis Laboratory, NASA-Johnson Space Center. "The Gateway to Astronaut Photography of Earth."

Yellowstone National Park lies over a mantle hot spot, and consists of three enormous calderas that erupted about 2, 1.2, and 0.6 million years ago (**FIGURE 5-17**). The most recent caldera is 45 km (28 mi) across and 75 km (47 mi) long!

Summary: Comparison of Volcanoes

TABLE 5-3 is our attempt to organize each type volcano in one place for you.

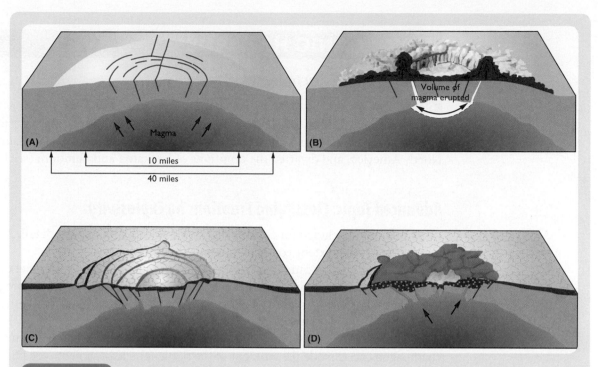

FIGURE 5-17 Formation of the Yellowstone Caldera. The area is still very active with geysers and hot springs. (A) Magma uplift causes fracturing in overlying crust. (B) Explosive volcanic activity evacuates magma chamber. (C) Collapse of surface. (D) Late stage volcanism. Modified from Geological Survey Bulletin 1347, National Park Service.

TABLE 5-3

Volcano Types by Shape, Viscosity, Product and Intensity of Eruption

Type	Shape	Viscosity (%SiO$_2$)	Product	Intensity/ Violence (See below)	Plate Tectonic Setting
Shield		Low (~50%)	Mainly basaltic lava	Hawaiian (low)	mainly oceanic hot spots and mid-ocean ridges
Cinder Cone		Low (~50%)	Mainly basaltic lava and tephra	Hawaiian and Strombolian	continental and oceanic
Stratovolcano/ Composite		Medium—High	Tephra and lava, basalt to rhyolite	Strombolian and Plinian	island arc and ocean–continental convergent
Dome		High (60–70%)	Rhyolite lava and tephra	Pelean and Plinian	mainly ocean–continent and continental
Caldera		High (60–70%)	Tephra	Plinian (high)	mainly ocean–continent and continental hot-spots

◼ VOLCANIC HAZARDS

Volcanic Explosivity Index

FIGURE 5-18 shows the **Volcanic Explosivity Index (VEI)** as well as volumes of eroded materials at different levels on the scale. Note that caldera eruptions (Yellowstone and Long Valley) produced the highest VEIs ever measured for North America, and dwarfed the eruptions of Krakatoa and Tambora (see below).

Advanced Topic: Classifying Eruptions by Explosivity

[This section is included for courses that cover volcanic hazards in detail. Ask your instructor to what extent you need to know these terms.]

Even though a volcano may exhibit different intensities of eruption at different times, and eruptions may change from one type to another as an eruption progresses, five general levels of explosivity are recognized. We list and describe them here in order of increasing violence.

Eruptions are powered by gas pressure, somewhat like a violently shaken bottle of champagne can explode the cork out of the bottle. The least violent type of eruption is termed *Hawaiian* and is characterized by far-spreading, runny (low-viscosity) flows from central **vents** or fissures (cracks), occasionally accompanied by lava fountains, which are jets of lava sprayed into the air by the expansion of gas bubbles in the molten rock. Hawaiian volcanoes, though not explosive, may erupt dangerous lava flows and tephra near the vent.

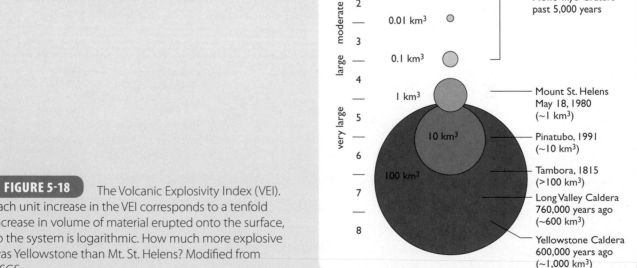

FIGURE 5-18 The Volcanic Explosivity Index (VEI). Each unit increase in the VEI corresponds to a tenfold increase in volume of material erupted onto the surface, so the system is logarithmic. How much more explosive was Yellowstone than Mt. St. Helens? Modified from USGS.

Strombolian eruptions are characterized by lava flows of medium viscosities, usually accompanied by a lava-fountain that produces volcanic bombs and cinders (large pyroclastics; **FIGURE 5-19**).

Vulcanian eruptions are characterized by viscous, intermediate to felsic magmas that form short, thick flows around vents.

Concept Check 5-2. We introduced the terms *intermediate* and *felsic* in this section. Define them.

Very viscous or solid fragments of lava are violently ejected from these vents during Vulcanian eruptions (**FIGURE 5-20**).

Peléan eruptions are similar to Vulcanian eruptions but have even more viscous lava. Domes form around the vents, and ash flows commonly accompany the dome formations. In a Peléan eruption, such as occurred on the Mayan Volcano in the Philippines in 1968, gas, dust, ash, and incandescent lava fragments are blown out of a central crater, collapse back, and form lobe-shaped *glowing avalanches* that move

FIGURE 5-20 Paricutin Volcano, which erupted in 1947, was a Vulcanian-type eruption. Dense clouds of ash-laden gas exploded from the crater and rose high into the atmosphere. © INTERFOTO/Alamy.

Mt. St. Helens erupts in 1980. Courtesy of USGS.

down-slope at velocities as great as 160 km/h (100 mi/h). Such eruptive activity can cause catastrophic destruction and loss of life if it occurs in populated areas, as demonstrated by the devastation of St. Pierre during the 1902 eruption of Mount Pelèe on Martinique, West Indies, when nearly 30,000 residents perished.

The most violent eruptions, such as that of Mt. St. Helens in 1980, are termed *Plinian* after Pliny the Elder, who was killed in the Vesuvius eruption of CE 79. They include the violent ejection of large volumes of volcanic ash, followed by collapse of the central part of the volcano (FIGURE 5-21).

Hazardous Volcanic Products or Processes

Volcanoes may erupt **tephra**, *lava*, and *gas*. We describe each of these next.

Tephra

Collectively, pyroclastic matter is called *tephra*. Pyroclastic material of increasing size is called *ash, lapilli, blocks,* and *bombs* (FIGURE 5-22).

Concept Check 5-3. We defined the term *pyroclastic* and used it several times. What does pyroclastic mean?

Ash is gritty and abrasive. Moreover, it can, if saturated with water, collapse structures like roofs on which it comes to rest. A one-inch deep layer of ash weighs about ten pounds per square foot (479 Pa), about as much as a cement slab used as a paving

FIGURE 5-22 Photo of volcanic tephra emanating from Mt. Cleveland, Alaska. Ash is extremely abrasive, composed as it is of much volcanic glass. Many cows starved during the aftermath of a 1783 eruption on Iceland when they tried to eat grass covered with ash. © NASA/Science Photo Library.

stone. Tephra from the 1991 eruption of Mt. Pinatubo in the Philippines caused the collapse of many poorly constructed houses near the volcano.

Tephra can pollute water supplies and coat crops and other vegetation, making it inedible for livestock or wild animals. It can blot out the sun, bringing darkness to midday. Ash in the atmosphere can affect the world's climate where it may circle the globe and take years to settle out (**FIGURE 5-23**). Tambura volcano erupted in Indonesia in 1813 and produced the so-called "year without summer" (see Indonesia section later in the chapter for a discussion).

Value-added Question 5-1.
Assuming that a 1-inch deep layer of ash weighs 10 pounds per square foot, calculate the total weight of this level of ash on (A) a car (60 square feet); and (B) a typical American house (2,500 square feet).

FIGURE 5-23 The ash cloud from space produced by the 1996 eruption of Soufriere Hills Volcano, Montserrat, West Indies. Which way are prevailing winds blowing? Courtesy of Jesse Allen and Robert Simmon/NASA, based on data from the MODIS Science Team.

When an ash cloud combines with rain, dilute acids may form that can burn the skin, eyes, and respiratory system. Even more potent *hydrochloric acid rains* also have been reported. Acid rains can degrade water supplies, harm vegetation, corrode exposed structures, and dissolve fabric.

Sudden heavy demand for lighting and air conditioning during an ash fall may cause a drain on power supplies, leading to a partial or full power failure. Ash clogs machinery of all kinds (and lungs as well) and poses a threat to aviation because particles can damage aircraft systems and clog jet engines (see Introduction). Ash can be carried hundreds of km downwind and thus can pose a threat to regions far from the eruptive center. Because it is easily carried by winds, ash may pose a hazard for weeks after the eruption.

Lava Flows

Lava flows are coherent masses of hot, partially or completely molten rock that flow down-slope, generally following valleys. They may extrude from the main volcanic cone or from nearby sources. Because lava flows are slow moving and take predictable paths, they generally pose little direct threat to human life; however, they can damage or destroy towns and villages. Heat from the lava burns vegetation and can cause forest or grass fires. Lava flows can dam rivers, forming lakes that can flood.

As you saw above, lava characteristics vary with the silica (SiO_2) content of the magma; those with low silica content (basalts) tend to have low viscosities and thus flow readily. Moreover, in basalts the low viscosity of the lava means that gas ordinarily bubbles relatively gently from the surface, often producing *vesicles,* fluid-filled cavities in the lava.

Owing in part to sensationalized movies, the public thinks lava flows are a major geohazard. In fact, lava flows are less-important threats than other volcanic products, and can even be controlled to some extent by the construction of levees and dams to channel or block the flow of lava. During the 1973 eruption of Eldfell volcano, on the island of Heimaey off the coast of Iceland, residents sprayed seawater onto the advancing lava front, eventually freezing it and blocking the lava's further advance.

As silica content of magma increases, pyroclastic deposits increase in abundance. Thus, almost all eruptions of basaltic lava are flows, but many eruptions of intermediate to felsic lavas consist of pyroclastic material.

Toxic Gases

Toxic gases may emanate from both active and inactive volcanoes. We began this chapter with the disaster caused by the eruption of CO_2 gas from Lake Nyos. Carbon dioxide is heavier than air and therefore will collect in low-lying areas. It is difficult to detect because it is both odorless and colorless. Other gas eruptions from Lake Nyos and other sites in the vicinity have occurred.

Examples of other volcanic gases that are toxic are sulfur dioxide (SO_2), hydrogen sulfide (H_2S), and hydrogen chloride (HCl).

Although CO_2 is a significant component of gas from many eruptions, total carbon dioxide (CO_2) emissions from terrestrial and submarine volcanoes are imprecisely known. One scientist estimated that annual anthropogenic (human-caused) CO_2 emissions are at least 150 times greater than those from volcanoes.

Hazardous Flows and Falls

Tephra is a component of many types of flows and falls, depending on the nature of the magma and the explosivity of the eruption. We discuss these hazards next.

A pyroclastic flow (**FIGURE 5-24A**), also called a *nuée ardente,* is a ground-hugging incandescent avalanche of tephra and gas that can move down the side of a volcano at speeds that can reach 160 km/h (100 mi/h; **FIGURE 5-24B**). The temperature within the flow can reach 700°C (1,300°F), high enough to incinerate anything combustible in its path. Once the flow has stopped, the extremely hot fragments may flatten and weld together, powered by the intense heat and the weight of the overlying material,

(A)

(B)

Ash cloud

Pyroclastic flow

1 cm

(C)

FIGURE 5-24   (A) A pyroclastic flow on the island of Montserrat, West Indies. © Photovolcanica.com/ ShutterStock, Inc. (B) Elements of a pyroclastic flow. Refer back to Table 5-2's photo of a pyroclastic flow. (C) A welded tuff. Courtesy of U.S. Geological Survey Professional paper 658-A/USGS.

forming a welded tuff (FIGURE 5-24C). Most pyroclastic flows consist of two parts: a *basal flow* of coarser fragments that hugs the ground, and a turbulent *ash cloud* that can rise 15 km above the basal flow. Ash may fall from this cloud over a wide area downwind from the flow. In the 1984 Philippine eruption of Mt. Mayon, ash from pyroclastic flows fell 50 km downwind of the volcano. A pyroclastic flow, ranging in composition from ash to boulders, can destroy virtually everything in its path. It can knock down, shatter, bury, or carry away nearly anything it meets. The extreme temperatures can ignite vegetation and wooden structures including houses. Pyroclastic *surges* are more gas-charged and energetic, and thus less restricted by topography. They can even surmount ridge tops.

Pyroclastic flows and surges are extremely dangerous. The high mobility of these flows threatens anyone nearby, such that even ridge tops and valley slopes may be unsafe.

The May 18, 1980 eruption of Mt. St. Helens generated a series of horizontal explosive eruptions called **lateral blasts**, when the sudden collapse of the volcano's north side in an avalanche released pressure on the superheated steam- and magma system within the volcano, like removing the cap from a vigorously shaken bottle of soda pop.

The blast's initial velocity of ~ 400 km/h (250 mi/h) rapidly increased to nearly 1100 km/h (680 mi/h) and destroyed an area of 600 km^2 (230 mi^2). Trees 2 m (6 ft) in diameter were mowed down like matchsticks as far as 25 km (16 mi) from the volcano (FIGURE 5-25).

The eruption of Mt. St. Helens killed at least 54 people and did over $1 billion in damage.

Lahars

Many geohazards are complex and not easily classified. **Lahars** are a good example. Lahars are also referred to as *volcanic mudflows*. Lahar is an Indonesian word applied to a hot to cold mixture of water and rock fragments flowing down the slopes of a volcano, often following river valleys. In motion a lahar looks like wet concrete that carries rock debris ranging in size from clay, to boulders more than 10 m in diameter (FIGURE 5-26). Lahars hundreds of meters wide and tens of meters deep can flow several tens of meters per second (typically 15–60 mi/h). For comparison, a track star can run short distances at about 7 m/sec (15–20 mi/h).

A lahar can incorporate water from melting snow and ice and any river it may overrun. By eroding and incorporating rock debris, and additional water, lahars can easily increase their initial size by a factor of ten.

They form in a variety of ways: as a result of rapid melting of snow and ice by pyroclastic flows, from intense rainfall on loose volcanic rock deposits, by spillout of a lake dammed by volcanic deposits, or from debris avalanches. Lahars move faster on steep slopes nearest their source and attain speeds greater than ordinary floodwaters downstream. The highest speed measured on the slopes of Mt. St. Helens during the 1980 eruption was 150 km/h (90 mi/h) and the lowest, in the lower valleys, was about 4 km/h (2 mi/h). They may attain depths of hundreds of meters in the canyons near their point of origin but spread out over valleys and low ridges downstream. They can reach sites tens of km from their origin.

Lahars can erode the sides of river channels causing bank collapse. Anything built along those banks may then be swallowed by and incorporated into the lahar. A very large volume lahar may even overtop or destroy a dam. According to USGS scientists, the mudflows that accompanied the Mt. St. Helens eruptions "damaged or destroyed more than 200 buildings, ruined 44 bridges, buried 27 km of railway and more than 200 km of roadway, badly damaged three logging camps, disabled

(A)

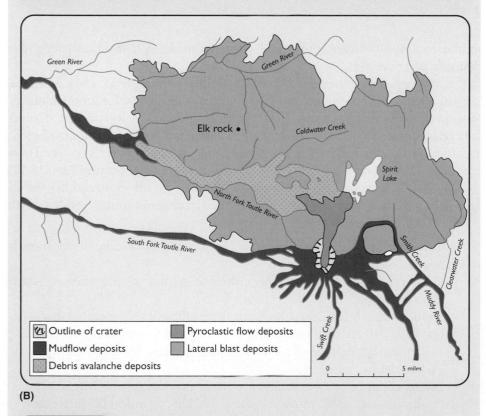

Green River

Green River

Elk rock •

Coldwater Creek

Spirit Lake

North Fork Toutle River

South Fork Toutle River

Smith Creek

Clearwater Creek

Swift Creek

Muddy River

 Outline of crater

Mudflow deposits

Debris avalanche deposits

Pyroclastic flow deposits

Lateral blast deposits

0   5 miles

(B)

FIGURE 5-25   Effects of the Mt. St. Helens eruption of 1980. (A) Blown down trees from the eruption. Courtesy of Lyn Topinka/USGS. (B) Distribution of deposits and impact.

several community water supply and sewage disposal systems, and partly filled channels and reservoirs."

A prime example of lahars occurred the night of November 13, 1985, when they formed after pyroclastic flows and surges mixed with glacial ice atop Columbia's Nevado del Ruiz (**FIGURE 5-27**). These lahars poured down river valleys, draining the crater. Ironically, one month before, geologists had produced and distributed a hazard zone map, identifying the nearby town of Armero as gravely threatened by an eruption. Armero was buried by lahars, killing at least 23,000 people (Figure 5-27).

The deadly lahars accompanying the 1985 eruption offers several key lessons for scientists, emergency-response professionals, and communities located downstream of such volcanoes:

- Catastrophic lahars can be generated on ice- and snow-capped volcanoes by relatively small eruptions.
- The surface area of snow is more critical than total volume when evaluating lahar potential.
- Dumping hot rock debris on snow is insufficient to generate lahars—the two materials must be *mechanically mixed* for sufficient heat transfer.
- Lahars can experience significant volume increase by engulfing water and surface debris.
- Lahars confined to river valleys can maintain relatively high velocities, and can have catastrophic impacts as far as 100 km (60 miles) downstream.
- Unless authorities take warnings of volcanologists seriously and evacuate threatened populations, tens of thousands of people can die needlessly.

Some areas have had a long frequency of lahar activity. Geologists have mapped at least 55 lahar deposits in the valleys draining Mount Rainier, Washington State that formed during the past 10,000 years. In the past 5,600 years, at least 6 and possibly as many as 13 lahars formed. These lahars have an approximate recurrence

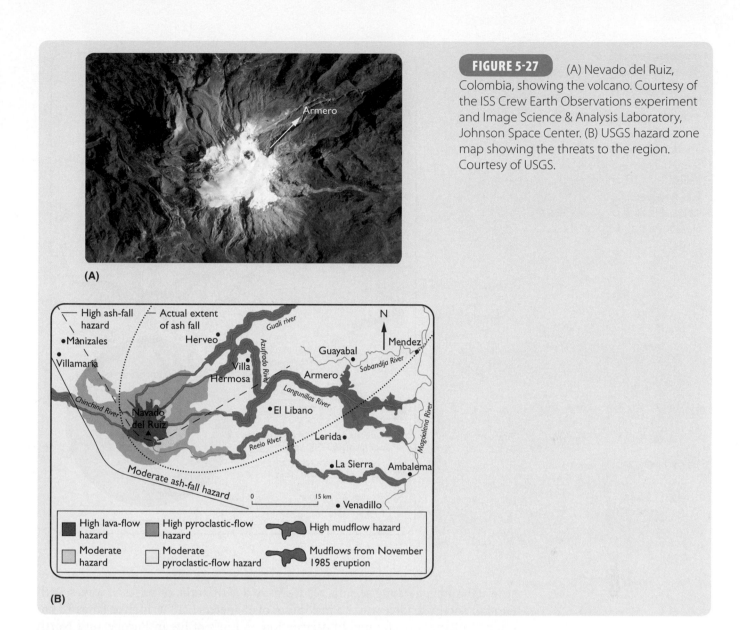

FIGURE 5-27 (A) Nevado del Ruiz, Colombia, showing the volcano. Courtesy of the ISS Crew Earth Observations experiment and Image Science & Analysis Laboratory, Johnson Space Center. (B) USGS hazard zone map showing the threats to the region. Courtesy of USGS.

(A)

High ash-fall hazard — Actual extent of ash fall

Manizales
Villamaria
Herveo
Villa Hermosa
Guayabal
Armero
Mendez
El Libano
Navado del Ruiz
Lerida
La Sierra
Ambalema
Venadillo

Guali river
Azufrado River
Sabandija River
Langunillas River
Chinchiná River
Recio River
Magdalena River

N

Moderate ash-fall hazard

0   15 km

High lava-flow hazard
High pyroclastic-flow hazard
High mudflow hazard

Moderate hazard
Moderate pyroclastic-flow hazard
Mudflows from November 1985 eruption

(B)

interval of 500 to 1,000 years, which means they can pose a threat to population centers like Tacoma (**FIGURE 5-28**).

Post-Eruption Tsunami

Danger from volcanic eruptions is not limited to land, however. The September 2001 edition of the scientific journal *Geophysical Research Letters* (GRL) contains an article by geoscientists Simon Day (University College, London) and Stephen Ward (University of California). Their report focuses on the potential for disastrous seismic sea waves, or *tsunami,* from an eruption of Cumbre Vieja volcano, located in the Canary Islands, off Africa's west coast (**FIGURE 5-29**).

As you will see later, most tsunami are generated by earthquakes. Few earthquakes in the Atlantic Basin are big enough to generate tsunami with enough energy to damage coastal communities around the Atlantic. However, an eruption on Cumbre Vieja could cause a massive landslide of up to 700 km³ (170 mi³) of material, which is twice the volume of the Isle of Man in the Irish Sea. Such an event would set off a tremendous submarine landslide, displacing enormous volumes of

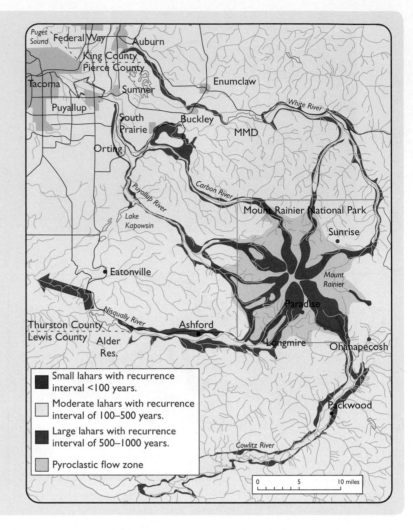

FIGURE 5-28 A hazard map of Mt. Rainier showing population centers threatened by lahars. The youngest of the lahars inundated most of the Puyallup (Pyew-ALLIP) River valley about 500 years ago, reaching depths of at least 50 m. Below what is now the town of Electron, the lahar spread across the 2 to 3 km-wide valley floor, reaching depths as great as 10 m. Today, several communities are built atop this deposit, called by geologists the "Electron Mudflow." Courtesy of USGS.

Map legend:
- Small lahars with recurrence interval <100 years.
- Moderate lahars with recurrence interval of 100–500 years.
- Large lahars with recurrence interval of 500–1000 years.
- Pyroclastic flow zone

seawater and generating seismic sea waves in a *wave train,* or series of waves, with enough energy to force waves a maximum of 65 meters (213 ft) high as far as 16 km (10 mi) inland, causing untold destruction and loss of life in Europe, and North and South America. Other geologists call this a "worst-case" scenario, however, as it is probably imprecise to say, as many do, that tsunami generate massive "walls of water." Rather, a tsunami usually appears as a rapidly rising or falling tide, a series of breaking waves, or even as a tidal bore when it approaches shore.

The last eruption of Cumbre Vieja was in 1971. It remains active.

■ VOLCANIC REGIONS OF THE WORLD

One of the most important regions of volcanic activity on Earth is the Pacific Rim (**FIGURE 5-30**). We will illustrate Pacific Rim activity and hazards from the regions indicated on the map. We begin with Indonesia.

Indonesia

Indonesia leads the world in many volcano statistics. It has the largest number of historically active volcanoes (76), and its 1172 dated eruptions are only exceeded

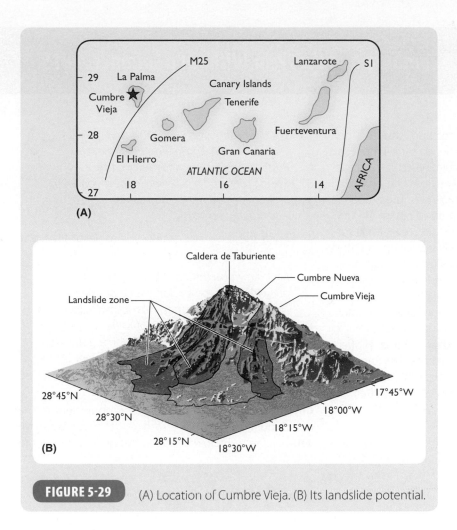

FIGURE 5-29 (A) Location of Cumbre Vieja. (B) Its landslide potential.

by Japan's 1274 (**FIGURE 5-31**). These two regions have produced one-third of the known explosive eruptions, but Indonesia has suffered the highest numbers of fatal eruptions, 105.

Tambora (see Case Study: Tambora and the "Year Without a Summer," 1815) and Krakatau are the sites of two of the Earth's largest eruptions in the past two hundred years.

Demographics and Volcanic Hazards in Indonesia

Indonesia is an equatorial archipelago of about 17,000 islands sitting between the Indian and Pacific Oceans. It is the world's fourth most populous country, with a population of 248 million as of 2012. The growth rate was 1.1% per year. To quantify the hazards posed by volcanoes to Indonesia, review this chapter as well as the plate tectonics setting of the country from elsewhere in this text.

Concept Check 5-4. What types of volcanoes do you expect here? Explain.

Concept Check 5-5. What volcanic products and processes pose the greatest threat to humans? How likely are such products and processes here, based on your answer to Concept Check 5-4?

Value-added Question 5-2.
Based on the growth rate of 1.1% per year, and using the doubling time relationship $T = 70/r$ (where T is doubling time in years, and r is rate, in this case, 1.1), in how many years will Indonesia's population double?

Value-added Question 5-3.
How many people would have been added in 2013, based on a 2012 population of 248 million?

Tambora is an immense stratovolcano situated on Sumbawa Island along the east Sunda Arc. Eruptions began in early April of 1815, which culminated in the great eruption of April 10–11. Earthquakes from the eruption were felt more than 500 km (310 mi) away. The eruption spewed 150 km³ (36 mi³) of tephra up to 43 km (27 mi) height from its summit. It plunged a region as far as 60 km (37 mi) west of the volcano into darkness for up to three days. Ash settled as far as 1300 km (800 mi) from the volcano, while fine particulates, gas and aerosols drifted around the world for at least three years.

In the words of geographer Gary Sutter, "summer 1816 in the Northern Hemisphere brought abnormally low daily temperatures, especially daily minima. Higher than average snowfall occurred in northern New England and Canada and heavy rains fell in Western Europe. The 1816 summer average temperatures on the east coast of North America, between 1.5°C and 2.5°C (34.7°–36.5°F) below average, were the coldest on record.

There were widespread crop failures, delayed harvests, and famine. The poet George Gordon, Lord Byron was inspired by the event to write *Darkness*:

The bright Sun was extinguish'd, and the stars
Did wander darkling in the eternal space
Rayless and pathless, and the icy earth
Swung blind and blackening in the moonless air;
Morn came and went—and came,
And brought no day . . ."

From the diary of Chauncey Jerome of Plymouth, Connecticut:

I well remember the 7th of June . . . dressed throughout with thick woolen clothes and an overcoat on. My hands got so cold that I was obliged to lay down my tools and put on a pair of mittens . . . On the 10th of June, my wife brought in some clothes that had been spread on the ground the night before, which were frozen stiff as in winter.[1]

Concept Check 5-6. If an eruption equivalent to that of Tambora in 1815 were to occur today, how might you be affected? How about someone in a developing country?

Value-added Question 5-4. The area of Indonesia is 1,826,440 km² (705,192 mi²), slightly less than three times that of Texas. What was the population density (in number of people per km² and mi²) as of mid-2012? What will the density be when the population doubles?

Concept Check 5-7. Assess the pressure such a population growth rate will place on Indonesia's forests and cropland.

Japan

Japan has a history of recorded volcanic activity going back to the eruption in the 20-km-wide caldera of Aso in 553 CE. Japan has 118 volcanoes with eruptions that have been dated.

Concept Check 5-8. Examine Figure 5-30 and assess the potential risk to Japan from volcanoes and earthquakes based on its plate tectonic setting.

Active Volcanoes of the United States and Canada

American Cascades

The Cascade Range extends from southern British Columbia to northern California (**FIGURE 5-32**). **TABLE 5-4** summarizes the eruptive history of Cascade volcanoes. Eruptions have occurred at an average rate of one to two per century during the past 4,000 years.

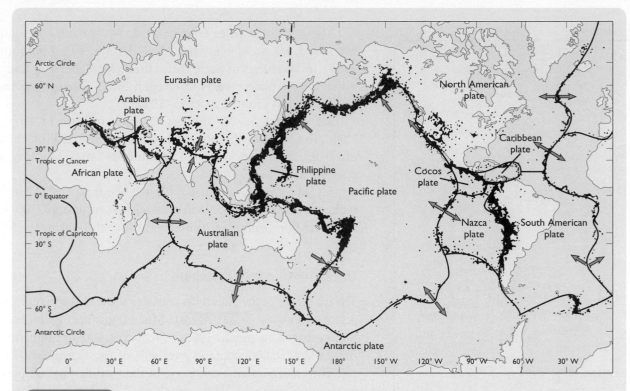

FIGURE 5-30 A map of the Pacific Rim, showing plate tectonic features. Data from Stowe, K. S. *Ocean Science* (New York: John Wiley & Sons, Ltd., 1983.

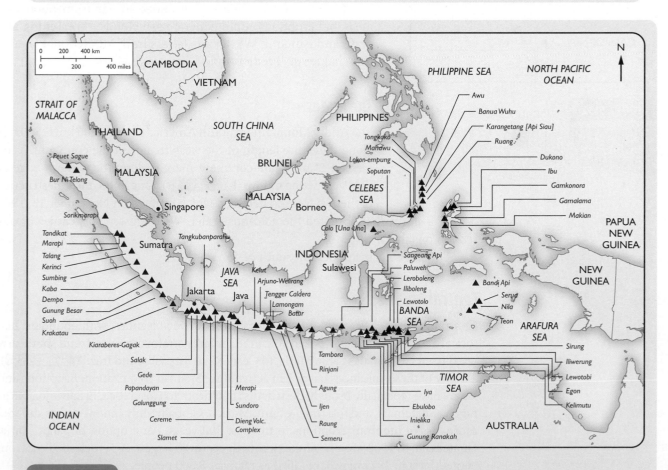

FIGURE 5-31 Indonesia's active volcanoes, with eruptions since 1900. Assess the hazard faced by Jakarta, Indonesia's largest city.

FIGURE 5-32 Volcanoes of the 1,600 km-long Cascade Range produced by the subduction of the Juan de Fuca plate under the North American Plate.

The USGS classifies Glacier Peak, Mt. Adams, Mt. Baker, Mt. Hood, Mt. St. Helens, and Mt. Rainier as being potentially active in Washington State. Mt. Hood, Mt. Jefferson, and the Three Sisters pose threats to Oregon. Mt. Lassen in northern California last erupted in 1915–1917.

Mt. St. Helens was the last Cascade volcano to erupt, in 1980. The VEI for the eruption was classified as Plinean, and ash from the eruption settled certainly as far as Montana and possibly as far as Colorado.

Mt. Mazama (Crater Lake) blasted about 50 km³ (12 mi³) of tephra into the atmosphere in a cataclysmic eruption 7700 years ago. Heat flow from magma below the lake still warms the water.

Threat from Aleutian Volcanoes

The island arc of the Aleutians contains some of the planet's most active volcanoes (FIGURE 5-33). Mts. Augustine, Redoubt, and Spurr erupted during the 1990s. Anchorage, with its population of more than 250,000, contains over 40% of Alaska's 600,000+ people. The volcanoes sit astride one of the world's most frequented air routes. During the 1990s, ash clouds brought down at least one commercial jetliner (which was able to land safely), dumped ash on Anchorage, and closed numerous schools.

The rest of the Aleutian volcanoes are far from populated areas, but eruptions could generate ash clouds, threatening aircraft, and tsunami. We discuss lahars from Mt. Redoubt later in this chapter in "Predicting and Tracking Lahars."

The Andes

The Andes Mountains of South America extend 4100 km (2500 mi) along the entire length of the western side of the continent. They formed at an ocean–continent margin along the western edge of the Pacific Ring of Fire (FIGURE 5-34). One of the most destructive of these is Nevado del Ruiz, which erupted on November 13, 1985 during a storm that obscured the volcano's crater. We discussed this earlier in our section on lahars.

Andean volcanism dies out south of 45° south.

Central America

Central America has some of the world's most active volcanoes, with abundant historical activity. Nearly two-dozen volcanoes, mainly stratovolcanoes, trend southeastward from Guatemala's Santa Maria, to Costa Rica's Turrialba and Irazu (FIGURE 5-35). Twenty of these volcanoes have been active since Spanish colonization. Eruptions and lahars have repeatedly brought destruction and death to Costa Rica, El Salvador, Nicaragua, Guatemala, and to a lesser extent, Mexico. Globally, only Indonesia has suffered more loss of life from eruptions in the past 500 years. An eruption of Santa Maria, Guatemala, in 1902 was one of the twentieth century's largest. A pyroclastic flow from Santa Maria in 1929 may have killed as many as 5,000 people.

| TABLE 5-4 | Summary of Cascade Volcano Eruptions |

Volcano and Location	Activity
BRITISH COLUMBIA	
Meager Mtn.	Most recent activity 2500 y BP.
WASHINGTON	
Mount Adams	3 in the last 10,000 years, most recent between 1,000 and 2,000 years ago
Mount Baker	5 eruptions in past 10,000 years; mudflows more common (8 in same time period); pyroclastic flows, mudflows, ashfall in 1843.
Glacier Peak	8 eruptions in last 13,000 years; pyroclastic flows and lahars
Mount Rainier	14 eruptions in last 9,000 years; also 4 large mudflows pyroclastic flows and lahars
Mount St. Helens	19 eruptions in last 13,000 years; pyroclastic flows, mudflows, lava, ashfall. Last activity 1980.
OREGON	
Mount Hood	repeated eruption 1,500–200 y BP. Potential for significant ash eruptions.
Mount Jefferson	Most recent activity 15,000 y BP. Potential for tephra and lahars.
Sisters Complex	Contemporary ground uplift of up to 10 cm near South Sister. Most recent activity 2,000 y BP. Pyroclastic flows and tephra.
CALIFORNIA	
Lassen	12 eruptive episodes during past 50,000 y. Last events 1915–1917. Lateral blasts, debris flows affected areas up to 5 km away.
Mono Lake/Long Valley	Tephra, flows mostly early Holocene; some may have occurred as recently as 500 y BP.
Shasta	minor tephra and flows, debris flows may be as young as 200 y BP.

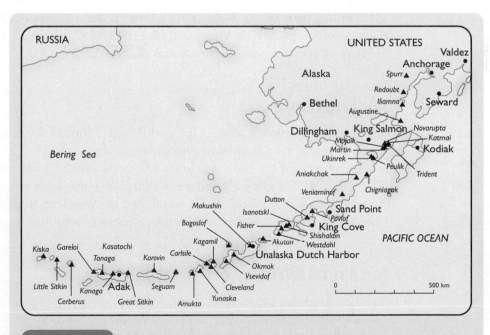

FIGURE 5-33   The Aleutian Arc's volcanoes. Mts. Augustine, Redoubt, and Spurr, which erupted during the 1990s, are highlighted. All are stratovolcanoes in the Aleutian Arc's northeast end.

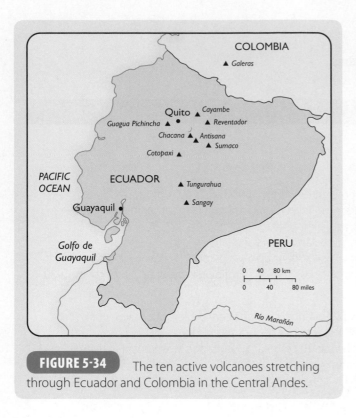

FIGURE 5-34 The ten active volcanoes stretching through Ecuador and Colombia in the Central Andes.

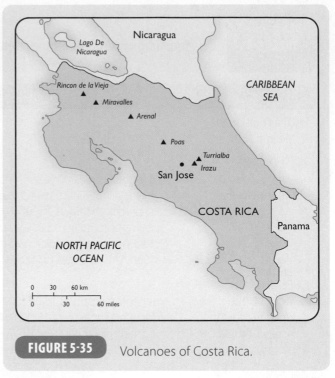

FIGURE 5-35 Volcanoes of Costa Rica.

West Indies

One of the great volcanic disasters in history occurred when Mt. Pelèe erupted in 1902, killing as many as 30,000 people from *nuées ardentes* (**FIGURE 5-36A, B**). An eruption of Soufriere Hills Volcano during 1995–1997 destroyed much of the town of Plymouth on the island of Montserrat (**FIGURE 5-36C**).

Southern Europe

Italy and Greece contain some of the world's most famous volcanoes (**FIGURE 5-37A**). North of Crete lies Santorini, a popular resort area, which ~3,380 years ago was the site of one of the most fearsome eruptions ever to have affected humanity (**FIGURE 5-37B**). Etna has erupted many times since the dawn of the Common Era. An eruption during World War II was filmed by American troops. Vulcano itself, from which the process derives its name, is located in southern Italy.

East African Rift Volcanoes

Nyiragongo poses the greatest threat to human lives among the many East African active volcanoes. In 1977, about 70 people were killed when one side of the Nyiragongo volcano collapsed, flooding the town of Goma with lava flowing at 40 mph, one of the fastest lava flows on record. Goma was further affected by eruptions during 2000, forcing thousands to flee as refugees.

FIGURE 5-36 (A) Eruption of Mt. Pelèe in Martinique, West Indies, in 1902. Courtesy of Library of Congress Prints and Photographs Division. (B) Impact of the eruption. Courtesy of the Library of Congress. (C) Impact of the Soufriere Hills Volcano on Montserrat during 1995–1997. © Michael Utech/iStockphoto.

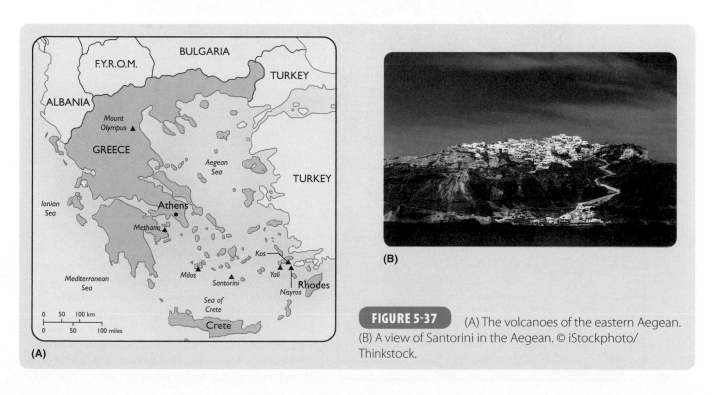

FIGURE 5-37 (A) The volcanoes of the eastern Aegean. (B) A view of Santorini in the Aegean. © iStockphoto/Thinkstock.

CASE STUDY *Vesuvius and Classicism*

Background © Hemera Technologies/AbleStock.com/Thinkstock. Title © iStockphoto/Thinkstock.

A volcanic eruption may have led to the revitalization of Western culture during the seventeenth century. Workmen excavating the foundation of a building near Naples, Italy tapped the interior of a vast subterranean amphitheater, part of the city of Pompeii that had been buried by the eruption of Mt. Vesuvius in the year 69 CE. The excavations revealed a treasure-trove of Roman mosaic and statuary, which gave rise to the "classical revival" in art. The move-

ment known as classicism dominated art until the end of the eighteenth century.

Even today, tourists by the thousands flock to Pompeii, and excavations continue in the resort and retirement community of Herculaneum (**Figure CS 1-1A**). Today, Vesuvius remains a dangerous volcano, and threatens the more than 2 million people who live in and around Naples (**Figure CS 1-1B**).

(A)

(B)

FIGURE CS 1-1 (A) During the eruption of 69 CE, Vesuvius buried Pompeii and Herculaneum, two wealthy Roman cities whose rediscovery and excavation in the seventeenth century gave rise to the Classical Period of art. © deepblue-photographer/ShutterStock, Inc. (B) Vesuvius threatens the more than 2 million people who live in and around Naples.

■ FORECASTING ERUPTIONS

Threat to Bend, Oregon

Volcanologists are actively researching the history of volcanic eruptions to try to determine their **recurrence intervals**, that is, some estimate of how regularly volcanoes erupt. Information that might signal *when* a volcano was about to erupt would be invaluable. The fast-growing resort-and-retirement city of Bend, Oregon, for example, and its 80,000+ residents, would greatly benefit if eruption of the nearby Sisters volcanic complex, which is showing signs of renewed activity after 1,000+ years of dormancy, could be anticipated (FIGURE 5-38).

Recognizing Precursor Events

Volcanologists have long suspected that certain *precursor* events precede eruptions. They have sought to use patterns of precursor events to predict eruptions, and thus save lives. Research to identify such precursors has focused on (1) analysis of seismic data from eruptions, (2) measurements of gas composition and quantity, and (3) land-level changes of volcanoes. The most useful warnings are those that indicate an eruption is likely to be imminent.

In the case of lahars, scientists have known that events like lava dome collapse often generate lahars, and have attempted to set up early warning systems to identify such precursors and to track the movements of resulting lahars.

We here describe two apparently successful attempts to predict eruptions and track erupted material, both of them devised and carried out by USGS personnel. First we describe the successful prediction and tracking of lahars at Mt. Redoubt, Alaska. Next we describe the analysis of "long period seismic events" associated with some volcanic eruptions.

Predicting and Tracking Lahars

A new automated detection system developed by a USGS scientist relies on a series of *acoustic-flow monitor* (AFM) stations installed downstream from a volcano. It is

FIGURE 5-38   Bend, Oregon showing its proximity to the Sisters Volcanoes. © ABCRounds/iStockphoto.

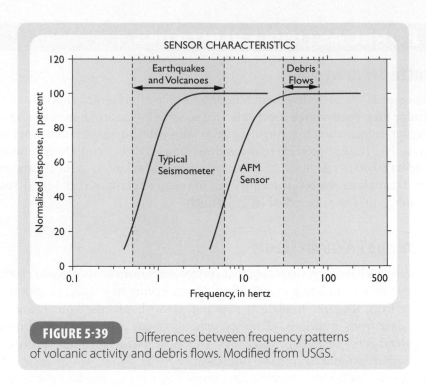

FIGURE 5-39 Differences between frequency patterns of volcanic activity and debris flows. Modified from USGS.

now installed downstream of volcanoes in the United States, Indonesia, Philippines, Ecuador, Mexico, and Japan.

Each station consists of a seismometer that picks up ground vibrations from an approaching and passing lahar, and a sensor that analyzes the signal. A radio at the station sends the data to a base station, usually a volcano observatory. The system is solar-powered.

Each second, the sensor measures the amplitude of the waves detected by the seismometer. At regular intervals (usually every 30 minutes) data are sent to the base station. An emergency message is sent any time the vibrations exceed a programmed threshold for longer than 40 seconds. The microprocessor continues to send alerts every minute for as long as the amplitude stays above the threshold level.

An AFM is a seismometer sensitive to higher frequency ground shaking than a typical seismometer used for recording earthquakes and volcano activity. An AFM has a frequency response of 10 to 250 Hz (hertz, or cycles per second; 10 Hz would thus be 10 vibrations per second) whereas a typical seismometer has a frequency response of < 2 Hz (**FIGURE 5-39**).

Ground vibration generated by lahars is predominantly in the frequency range of 30 to 80 Hz, whereas vibrations generated by earthquakes, volcano tremors, and explosive eruptions are predominantly < 6 Hz. This difference in frequency response allows scientists to distinguish between lahars and other natural ground-vibrating events.

A system installed at Redoubt volcano, Alaska successfully identified and tracked lahars generated by lava-dome collapse of the volcano in 1990.

Detecting and Predicting Volcanic Eruptions

Volcanologists have had considerable success in differentiating seismic signals produced by volcanic activity from those produced by other kinds of events, and using these differences to recognize when volcanoes are likely to become active. For

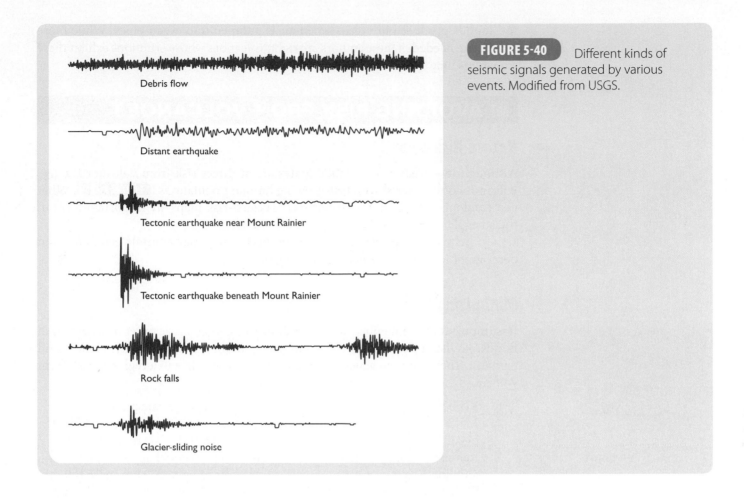

Debris flow

Distant earthquake

Tectonic earthquake near Mount Rainier

Tectonic earthquake beneath Mount Rainier

Rock falls

Glacier-sliding noise

example, scientists can identify "shocks" from different sources: wind, elk herds, a hovering helicopter, volcanic explosions, avalanches, lahars, nuclear explosions, and earthquakes (**FIGURE 5-40**).

One recent discovery has the potential to give scientists the ability to forecast precisely when a volcano will erupt. Made by USGS geologist Bernard Chouet, it involves recognizing the significance of a distinctive kind of seismic wave pattern generated by volcanoes: **long period events** (**FIGURE 5-41**). These seismic "signatures" differ from the typical shock waves generated by eruptions. Eruption shock waves begin abruptly, while long period events build up slowly and taper off. Chouet and colleagues have concluded that long period events are generated by pulses of magma and gas moving into the volcano just before an eruption. They liken these pulses to

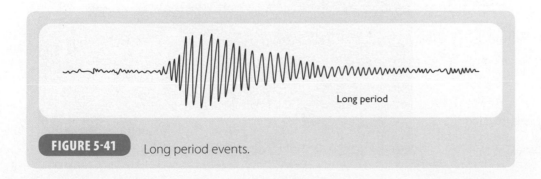

Long period

FIGURE 5-41 Long period events.

the sounds produced in organs when air is pumped into organ pipes. While more studies are needed, if these authors are right, volcanoes whose eruptions exhibit these signatures may become predictable.

■ AVOIDING RISK FROM VOLCANOES

Hazard-Risk Maps

Although few areas in the United States are at direct risk from volcanoes, active volcanoes often coexist with fast-growing human populations (**FIGURE 5-42**). What can residents do to minimize their risk? Hazard-risk maps are important tools in the struggle to mitigate the impacts of eruptions.

To help scientists evaluate risk to property, they may compile economic and demographic data for the endangered region.

Minimizing Risk

It is important to remember that volcanic eruptions can cause other disasters: flash floods, landslides and mudflows, acid precipitation, destructive ash clouds, and tsunami. They can even generate violent storms! If you live in a region at risk from volcanoes, here are some tips from USGS and FEMA.

1. Keep a pair of goggles and some breathing masks in a safe place.
2. Make evacuation plans. These can be especially important since roads often become clogged with evacuees who try to leave at the last minute.
3. Remember that volcanoes can hurl incandescent material up to 35 km (20 mi), and floods, airborne ash and toxic gas can spread up to 200 km (over 100 mi). So if you live close enough to the eruption, evacuate and move to higher ground as soon as you can. Follow instructions from authorities. Many of the deaths from Mt. St. Helens occurred because residents refused to evacuate.
4. Have the types of disaster supplies on hand that are described on FEMA's website (www.fema.gov). Check on neighbors, especially those who are elderly or handicapped.
5. Volcanic ash can contaminate water supply. Store emergency water as soon as a hazard is announced.
6. Lateral blasts can be unimaginably deadly. Stay away from active volcanoes!

FIGURE 5-42 Santa Maria volcano looms in the background of Quetzaltenango, Guatemala. © Bert de Ruiter/Alamy Images.

CHAPTER SUMMARY

1. On the night of August 21, 1986, Cameroon's Lake Nyos belched a vast cloud of carbon dioxide (CO_2) gas that killed nearly 1800 sleeping villagers and untold animals and wildlife.

2. The Federal Emergency Management Agency (FEMA) says in reference to volcanic hazards: Education is Essential.

3. Volcanic eruptions since 1700 have killed more than 260,000 people, destroyed cities, and disrupted local economies for months to years. While no volcanic event in the past 500 years has come close to the truly cataclysmic eruptions in recent prehistory, the U.S. Geological Survey estimates that at least 500 million people face potential risk from a volcanic eruption.

4. Inactive for over 600 years, the eruption of Pinatubo in the Philippines on June 15–16, 1991 was the biggest eruption of the twentieth century. The eruption had short-term global effects on climate.

5. After being dormant for 140 years, Nevado del Ruiz in Colombia erupted incandescent pyroclastic material on November 13, 1985 that melted the mountain's ice cap and set off volcanic mudflows, killing nearly 25,000 people. Volcanologists had warned that an eruption was imminent.

6. Two explosive eruptions in March and May of 1963 of Agung volcano, Indonesia, generated a hot ash cloud and set off lahars (volcanic mud flows), which killed at least 1,148 people.

7. Volcanoes may erupt tephra, lava, and gas. There are several ways to classify volcanoes, depending on the purpose of the classification.

8. Shield volcanoes are composed of basalt, have broad, nearly flat-to gently sloping flanks, and are shaped like an ancient warrior's shield.

9. Also called composite volcanoes, stratovolcanoes erupt both lava flows and ash of intermediate to felsic composition.

10. Cinder cones are relatively small but steep conical volcanoes, emitting much pyroclastic matter of relatively coarse size called cinder.

11. Lava domes are usually small and dome-shaped, and extrude highly viscous, silicic (up to and exceeding 70% SiO_2) magma.

12. Calderas are the remains of cataclysmic eruptions that have blown an existing volcano to bits, leaving behind a crater, which may fill with water.

13. The Volcanic Explosivity Index (VEI) measures the volume of material erupted by a particular eruption.

14. Five general levels of explosivity are recognized in order of increasing violence. They are Hawaiian, Strombolian, Vulcanian, Pelean, and Plinian.

15. Pyroclastic matter is collectively called tephra. Tephra is a major volcanic hazard. It is a component of many types of flows and falls, depending on the nature of the magma and the explosivity of the eruption.

16. Lava flows are coherent masses of hot, mainly molten rock that flow down slope, generally following valleys.

17. Toxic gases may emanate from both active and inactive volcanoes. One scientist estimated that annual anthropogenic CO_2 emissions are at least 150 times greater than those from volcanoes.

18. Most pyroclastic flows consist of two parts: a basal flow of coarser fragments that hugs the ground, and a turbulent ash cloud that can rise 15 km above the basal flow.

19. Lateral blasts are horizontal, explosive eruptions.

20. Lahar is an Indonesian word applied to a hot to cold mixture of water and rock fragments flowing down the slopes of a volcano, often following river valleys. They are also called volcanic mudflows.

21. On the night of November 13, 1985, lahars formed when pyroclastic flows and surges mixed with glacial ice atop Columbia's Nevado del Ruiz. One month before, geologists had produced and distributed a hazard zone map, identifying the town of Amero as gravely threatened by an eruption. Amero was buried by lahars, killing at least 23,000 people.

22. Geologists have mapped at least 55 lahar deposits in the valleys draining Mount Rainier, Washington State that formed during the past 10,000 years.

23. An eruption of Cumbre Vieja volcano, located in the Canary Islands could generate disastrous scismic sea waves.

24. One of the most important regions of volcanic activity on Earth is the Pacific Rim. Indonesia leads the world in many volcano statistics. Japan has a history of recorded volcanic activity going back to 553 AD. The island arc of the Aleutians contains some of the planet's most active volcanoes.

25. The U.S. Geological Survey (USGS) classifies Glacier Peak, Mt. Adams, Mt. Baker, Mt. Hood, Mt. St. Helens, and Mt. Rainier as being potentially active in Washington State. Mt. Hood, Mt. Jefferson, and the Three Sisters pose threats to Oregon. Mt. Lassen in northern California last erupted in 1915–1917.

26. There are ten active volcanoes in the Andes.

27. Central America has some of the world's most active volcanoes, with abundant historical activity.

28. One of the great volcanic disasters in history occurred when Mt. Pelèe erupted in 1902, killing as many as 30,000 people in the West Indies.

29. Italy and Greece contain some of the world's most famous volcanoes. South of Crete lies Santorini, a popular resort area, which ~3380 years ago was the site of one of the most fearsome eruptions ever to have affected humanity. The eruption of Vesuvius in 69 CE may have led to the revitalization of western culture during the seventeenth century.

30. The East African Rift Valley contains many sites of active volcanism.

31. Volcanologists are actively researching the history of volcanic eruptions to try to determine their recurrence intervals, that is, some estimate of how regularly volcanoes erupt. Patterns of precursor events are being identified to predict eruptions. Considerable success has been achieved in differentiating seismic signals produced by volcanic activity from those produced by other kinds of events, and using these differences to recognize when volcanoes are likely to become active. For lahars, an automated detection system is now installed downstream of volcanoes in the United States, Indonesia, Philippines, Ecuador, Mexico, and Japan.

32. Hazard–risk maps are important tools in the struggle to mitigate the impacts of eruptions.

33. It is important to remember that volcanic eruptions can cause other disasters: flash floods, landslides and mudflows, acid precipitation, destructive ash clouds, and tsunami.

KEY TERMS

acid aerosols	lava domes	stratovolcanoes/composite
ash	long period events	volcanoes
caldera	precursors	tephra
cinder cone	pyroclastic flows	vents
Dust Veil Index	pyroclastic material	viscosity of flows
lahars	recurrence interval	Volcanic Explosivity Index
lateral blasts	shield volcanoes	welded tuff
lava	silica content of lava	

REVIEW QUESTIONS

1. The disaster at Lake Nyos, Cameroon was due to
   a. formation of a large lahar.
   b. a pyroclastic flow.
   c. a toxic gas cloud.
   d. all of the above.

2. The Roman cities Herculaneum and Pompeii were destroyed by an eruption of
   a. Vesuvius.
   b. Santorini.
   c. a volcano.
   d. Etna.

3. Active volcanoes of the United States are located in
   a. Southern California.
   b. Utah.
   c. Alaska.
   d. all of the above.

4. The biggest eruption of the twentieth century was
   a. Santorini.
   b. Lassen.
   c. Pinatubo.
   d. Nevado del Ruiz.

5. Which eruption was known to have had global effects?
   a. Timbura
   b. Pinatubo
   c. Krakatau
   d. all of the above
   e. none of the above.

6. Lahars (volcanic mudflows) caused great loss of life during an eruption of
   a. Lassen.
   b. Nevado del Ruiz.
   c. Rainier.
   d. all of the above.

7. Which volcano is likely to pose the greatest hazard during an eruption?
   a. shield
   b. cinder cone
   c. stratovolcano
   d. dormant volcano

8. Cataclysmic eruptions that literally blow an existing volcano to bits, leaving behind a crater, form
   a. calderas.
   b. lahars.
   c. cinder cones.
   d. shield volcanoes.

9. Which U.S. National Park is situated on a caldera?
   a. Crater Lake
   b. Yellowstone
   c. both of the above
   d. neither of the above

10. The name for eruptions with the greatest Explosivity Index is
    a. Hawaiian.
    b. Plinian.
    c. Calderian.
    d. Vesuvan.

11. Volcanic eruptions are powered directly by
    a. radioactive decay
    b. gravity
    c. gas pressure
    d. density

12. Pyroclastic material is collectively called
    a. ash
    b. tephra
    c. siliciclasts
    d. pyrons

13. During which eruption did workers actually stop a lava flow?
    a. St. Helens, Washington
    b. Vesuvius, Naples
    c. Eldfell, Iceland
    d. Mauna Kea, Hawaii

14. The eruption of Mt. St. Helens generated a series of
    a. lateral blasts.
    b. toxic gas eruptions.
    c. tsunami.
    d. all of the above.

15. The Nevado del Ruiz eruption in Colombia was noteworthy in that
    a. there was no loss of life.
    b. the eruption was predicted but the predictions were ignored.
    c. the toxic gases accompanying the eruption killed many thousands of people.
    d. Colombia's capitol city was destroyed.

16. Eruption of Cumbre Vieja in the Canary Islands could generate a series of massive tsunami through what process?
    a. a gigantic underwater landslide
    b. an enormous gas pressure release
    c. a catastrophic caldera eruption
    d. eruption of radioactive decay products

17. The country with the largest number of historically active volcanoes is
    a. Japan.
    b. Italy.
    c. United States.
    d. Indonesia.

18. The "year without a summer" in 1816 resulted from the eruption of
    a. Vesuvius.
    b. Rainier.
    c. Tambora.
    d. Krakatau.

19. Which was the last volcano to erupt in the lower 48 United States?
    a. St. Helens
    b. Lassen
    c. Mauna Loa
    d. Nevado del Ruiz

20. Nevado del Ruiz is in what volcanic region?
    a. the West Indies
    b. Central America
    c. the Andes
    d. the Pacific Plate

21. The great disaster of Mt. Pelèe in 1902 occurred in which volcanic region?
    a. the Andes
    b. Central America
    c. the West Indies
    d. none of the above

22. The eruption that is generally credited with destroying Minoan civilization in around 1400 BC was
    a. Vesuvius.
    b. Santorini.
    c. Pelèe.
    d. Cumbre Vieja.

23. Lahars can be tracked arising out of their particular
    a. ground vibration frequency.
    b. surface temperature.
    c. seismic wave structure.
    d. infrared pattern.

24. "Long period events" may be useful in predicting
    a. the onset of a volcanic eruption.
    b. the damage done by an eruption.
    c. earthquake damage from an eruption.
    d. all of the above.

25. The eruption that may have stimulated the art revival called Classicism was
    a. Vesuvius.
    b. Santorini.
    c. Volcano.
    d. Etna.

FOOTNOTE

[1] From Stommel, H. and Stommel, E. 1979: The year without a summer. *Scientific American* 240, 176–80, 182, 184, 186, as quoted in Oppenheimer C. (2003). Climatic, environmental and human consequences of the largest known historic eruption: Tambora volcano (Indonesia) 1815. *Progress in Physical Geography,* Vol. 27, pp. 230–259.

CHAPTER OUTLINE

CASE STUDY *A Tale of Two Earthquakes*

The Costliest Earthquake: Kobe, 1995

The January 17, 1995 Kobe earthquake (**Figure CS 1-1**) was the most damaging and costly to hit Japan since the Kanto earthquake in 1923, which destroyed large areas of Tokyo and Yokohama and killed 143,000 people.

In the 1995 earthquake, Kobe and adjacent cities suffered at least 5,500 deaths and 27,000 injuries. *One-fifth of Kobe's 1.5 million people were left homeless and more than 103,000 buildings were destroyed.* Kobe resident Shigemitsu Okino, who witnessed the bombing of Kobe at the end of World War II, compared the earthquake to that experience: "It looked a lot like this. The difference is, we could hear the planes coming, but the earthquake was silent."[1]

Before the earthquake, the Port of Kobe was one of the largest in the world. The port was severely damaged by the 1995 earthquake. Parts of wharfs and warehouses sank and were covered with water. As a result of earthquake damage, Kobe permanently lost business to the ports of Yokohama, Osaka, and even to some ports in South Korea, in spite of the billions spent at Kobe on rebuilding. Kobe is currently the fourth busiest port in Japan.

The local government estimated the cost of restoring basic functions at about $100 billion US. Total losses, including losses of privately owned property and reduction in business activity, approached *$200 billion,* or *10 times*

FIGURE CS 1-1 Collapsed elevated freeway, Kobe earthquake. © Dario Mitidieri/Edit/Getty Images.

greater than losses from the devastating 1994 Northridge, California earthquake, which we will discuss later. Such a loss would represent about 5% of Japanese Gross Domestic Product (GDP, a measure of the annual total value of goods and services). For comparison, estimated direct monetary cost to the U.S. economy of the World Trade Center terrorist attacks was about $30 billion, or 0.3% of the U.S. 2001 GDP.

Continued ▸

The Deadliest Earthquake: Tangshan, 1976

As of 2011, 24 earthquakes in recorded history had killed at least 50,000 people each. The greatest disaster in recent history—and perhaps the greatest of all time—occurred in 1976 when at least 240,000, and possibly as many as 600,000 people, were killed in Tangshan, China, a city of 1,000,000+ about 160 km (100 mi) east-southeast of Beijing (**Figure CS 1-2**).

Meng Jiahua was working in a coal mine the evening of the Tangshan earthquake. He hurriedly evacuated the mine when he felt the first tremors. He was horrified by what he saw: "All the houses and buildings were flattened. Every-thing was in ruins. The living were lined up along one side of the roads and the corpses on the other. The strange thing is that almost none of the miners working underground were killed. Only those sleeping in their beds at home died."[2]

The Tangshan earthquake may even have had political repercussions. Because Chairman Mao Tse-tung died a few months afterwards, some observers took the earthquake as an omen of his demise.

Earthquakes pose extraordinary threats to the populations and economies of many nations. Some of the nations at greatest risk, in addition to Japan and China, are Italy, Turkey, Iran, and the United States (see seismic risk maps, below).

FIGURE CS 1-2 Tangshan after the earthquake. Photo from Hebei Provincial Seismological Bureau, 1976. Page 103, Earthquake Information Bulletin, v.15, no. 3. Courtesy of USGS.

■ UNDERSTANDING EARTHQUAKES

Stress, Strain, Rock Strength, and Faults

The U.S. Geological Survey (USGS) defines an **earthquake** as a shaking of the ground caused by movement of rocks along *faults* within the crust. *Faults* are simply breaks in rocks where one side moves relative to the other. A crack in the crust is only a fault if there is movement of the rocks on either side. Faults exist *between* lithospheric plates, that is, at plate boundaries (discussed elsewhere in this text) as well as *within* plates. **FIGURE 6-1** illustrates faults. The overwhelming majority of, but not all, earthquakes result from the release of energy that builds up slowly at plate boundaries.

Earthquakes happen because stress builds up in rocks. **Stress** is simply *force per unit area* (for example, tons per square inch, or dynes per square cm), like pushing on a wall. If the rocks on either side of a fault slide more or less smoothly by each other, no stress or minimal stress builds up, like when you slide your hands loosely by each other. This smooth, gradual movement of rocks is called **creep**, or **aseismic slip** and does not produce significant earthquakes. As you might imagine, however, the stress along faults can increase when the rocks are not smooth and they periodically lock. You can demonstrate this by pressing your hands tightly together and sliding them by each other. Notice that your hands will "catch" if you press hard enough. If rocks are strong, they can store much stress, a form of potential energy. You stored this stress when your hands briefly stuck. You could store stress in a stick by bending it.

Rocks eventually respond to stress by **strain**. Strain occurs as a change in volume, shape, or both. **FIGURE 6-2** shows what a deformed rock looks like, and examples of stress and strain.

As stress increases, strain may build up, and rocks may break or rupture when the stress exceeds the strength of the rock. The more brittle a rock is, the more likely it is to fracture. The more plastic (ductile), a rock is, the more it resists fracturing. To visualize one kind of strain, again slide your hands past each other while pressing tightly. Strain occurs when your hands slide again after catching, or when the stick breaks when you bent it too far. The energy stored in your hands or the stick was quickly released. So, the potential energy when your hands stuck or the stick bent was stress. The kinetic energy of your hands sliding or the stick breaking was strain. When this happens to rocks sliding by each other, the result is an earthquake, a phenomenon also called *seismic slip*. Some of the strain energy is released as sound: a sharp crack is heard when the stick broke. This is sound energy. In the same way, rocks that break along faults release energy as we explain below.

Seismologists (scientists who study earthquakes) call this sequence of events explaining how earthquakes occur the **elastic-rebound theory**. To summarize, forces cause plates on opposite sides of a fault to shift. These plates may move smoothly (aseismic slip) or they may store energy and slowly deform. Once their rupture strength is exceeded, a concept we'll tell you about next, the energy is released as a sudden movement, and an earthquake (seismic slip) occurs. Because earthquakes reoccur along some faults on a periodic basis, this sequence of events is also known as the *earthquake cycle*. **FIGURE 6-3** illustrates the main points of the elastic-rebound theory and the earthquake cycle.

Concept Check 6-1. Explain the elastic-rebound theory.

(A)

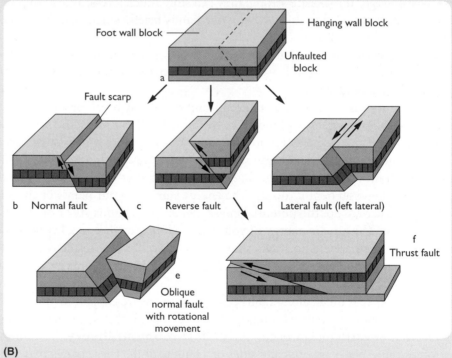

Foot wall block — Hanging wall block

Unfaulted block

a

Fault scarp

b   Normal fault

c   Reverse fault

d   Lateral fault (left lateral)

e   Oblique normal fault with rotational movement

f   Thrust fault

(B)

FIGURE 6-1   (A) Fault in Mesozoic rocks, Somerset, U.K. Jurassic (200–145 my) age on left, Triassic (200–250 my) age on right. Courtesy of Ashley Dace. (B) Illustration of faults, showing (a) area before faulting showing fault plane; (b) normal fault; (c) reverse fault; (d) lateral fault; (e) oblique fault, showing complex offset; (f) thrust fault. A fault scarp may be seen where the fault breaks the surface.

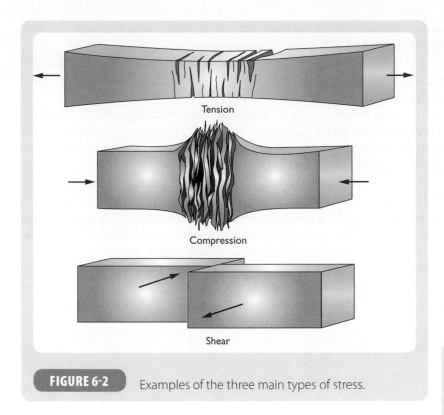

FIGURE 6-2 Examples of the three main types of stress.

The amount of stress that rocks can store is known as a **rock's strength**. The higher a rock's strength, the higher its strain (that is, the amount of energy released) can be. The strength of rocks varies with four factors: temperature, composition, water content, and **confining pressure**. Rocks store more stress—that is, they are stronger—at lower temperatures, where they are brittle and tend to fracture. At higher temperatures, rocks are weaker and deform more easily.

Rocks whose composition is dense, like the igneous rock gabbro (classification of rocks is found elsewhere in this text), are typically stronger than loosely-cemented, lighter (in weight) sedimentary rocks like sandstone.

Water content is one of the most important factors affecting rock strength. The amount of water in a rock varies widely and, as you might expect, rocks with higher water content are weaker than those with low water content, although the explanation is too complex to present here. Basalts, dense igneous rocks which make up most of oceanic crust, have low water content and are typically stronger than upper-mantle rock with higher water content.

Finally, rocks with high confining pressure, the pressure exerted on the rock from all directions by enclosing and overlying rocks, are weaker; they are less brittle and more plastic. All rocks beneath the surface have confining pressure. The deeper the rock, the higher the confining pressure.

If rocks are very strong, they can store lots of stress (that is, energy), which will likely result in a few large earthquakes. If rocks

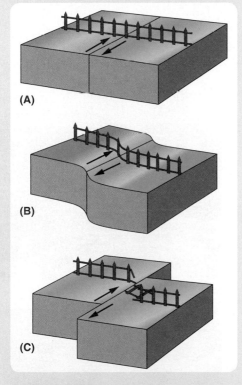

(A)

(B)

(C)

FIGURE 6-3 The elastic rebound theory of earthquakes. (A) The surface pictured before stress. (B) Stress build-up and the surface deform (strain) relieving stress. The two surfaces are offset along the fault. (C) The new equilibrium after deformation. This is called the earthquake cycle, because it tends to occur periodically at somewhat regular intervals.

are weak, however, many small earthquakes are likely, because the rocks will not be able to store enough stress for large earthquakes to occur.

Movement along faults can occur along *planes* or across broad *zones*. **FIGURE 6-4A** shows the surface expression of the San Andreas Fault. **FIGURE 6-4B** is a diagram of elements of the complex fault zone.

The San Andreas often juxtaposes rocks of different compositions, resulting in varying rock strengths along and across the fault. This *heterogeneity* causes the response to stress to vary. Segments of the San Andreas where weak rocks are found break frequently, yielding many small earthquakes.

Segments consisting of stronger rocks can allow great stresses to build up, and may yield relatively fewer, much stronger earthquakes. In general, many small earthquakes are preferable to a fewer big ones. Stress that has slowly accumulated within the rock, in some instances over centuries, can be released quickly when the strength of the rock is exceeded, commonly with devastating results.

The "Good Friday" Alaska earthquake of 1964 was one of the most powerful of the twentieth century, and killed 125 people. Most of its built-up stress was released in about *180 seconds*. The total time of shaking was about 3 minutes 40 seconds.

Concept Check 6-2. Do you agree with the authors that many small earthquakes are preferable to a single big one? Explain your answer.

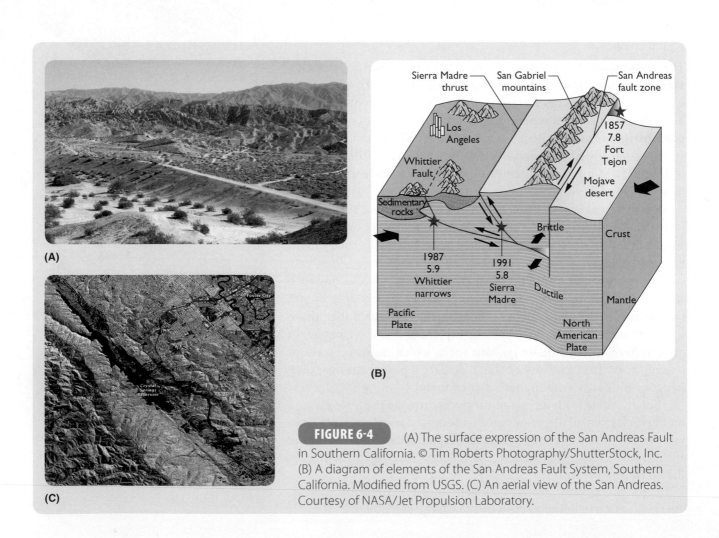

(A)

(C)

(B)

FIGURE 6-4 (A) The surface expression of the San Andreas Fault in Southern California. © Tim Roberts Photography/ShutterStock, Inc. (B) A diagram of elements of the San Andreas Fault System, Southern California. Modified from USGS. (C) An aerial view of the San Andreas. Courtesy of NASA/Jet Propulsion Laboratory.

Locating Earthquakes

The point inside the earth where an earthquake orig-inates is called the **focus** or *hypocenter*. The point on the Earth's surface directly above the focus is called the **epicenter** (FIGURE 6-5).

Enormous amounts of energy are released when earthquakes occur. A moderate earthquake (say, 4 to 5 on the Richter Scale, which we describe shortly) releases the equivalent of a small atomic bomb, where a large earthquake (8.5 on the Richter Scale), releases over 85 megatons of energy, nearly twice as much as the largest thermonuclear bomb tested. Some of this energy can be released as heat, and some may be used to crush rocks along the fault zone. However, much of it is released as **seismic waves** (*seismic* refers to vibrations associated with earthquakes).

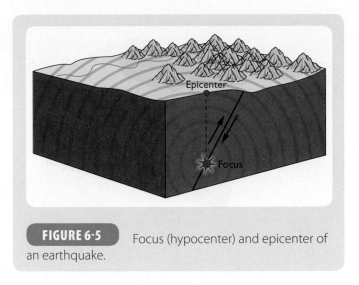

FIGURE 6-5 Focus (hypocenter) and epicenter of an earthquake.

There are three common types of seismic waves: Primary (P) waves, Secondary (S) or Shear waves, and Surface waves. FIGURE 6-6 shows the three types of seismic waves, how they move, and how they cause damage (discussed in our treatment of global tectonics).

Seismographs are instruments that record the passing of seismic waves. A record of an earthquake recorded by a seismograph is a **seismogram**. Using

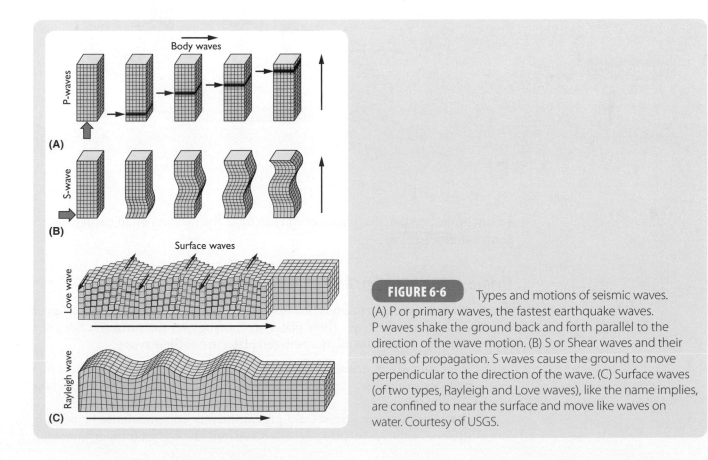

FIGURE 6-6 Types and motions of seismic waves. (A) P or primary waves, the fastest earthquake waves. P waves shake the ground back and forth parallel to the direction of the wave motion. (B) S or Shear waves and their means of propagation. S waves cause the ground to move perpendicular to the direction of the wave. (C) Surface waves (of two types, Rayleigh and Love waves), like the name implies, are confined to near the surface and move like waves on water. Courtesy of USGS.

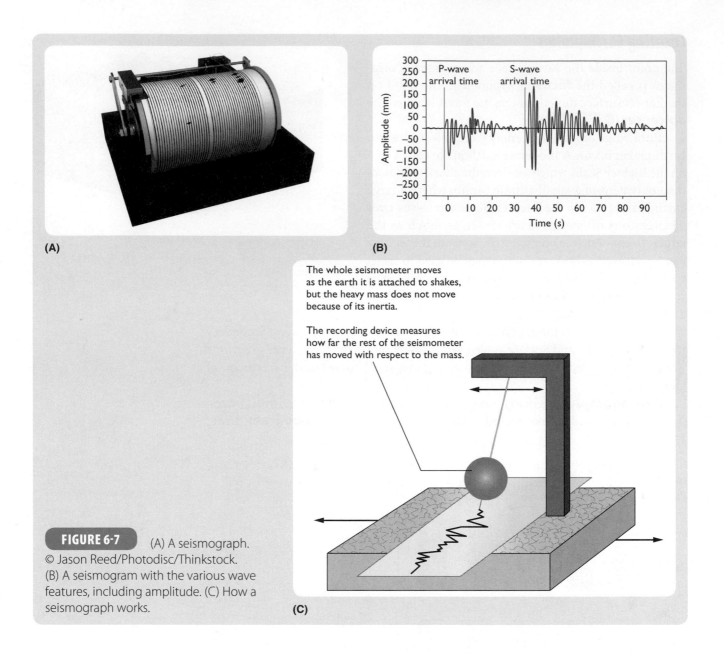

FIGURE 6-7 (A) A seismograph.
© Jason Reed/Photodisc/Thinkstock.
(B) A seismogram with the various wave
features, including amplitude. (C) How a
seismograph works.

The whole seismometer moves
as the earth it is attached to shakes,
but the heavy mass does not move
because of its inertia.

The recording device measures
how far the rest of the seismometer
has moved with respect to the mass.

(A)

(B)

(C)

seismograms, earthquake scientists can calculate the location of an earthquake as shown in **FIGURE 6-7**.

Origin of Earthquakes: Role of Global Tectonics

As you read elsewhere in this text, the *lithosphere,* composed of the crust and upper mantle, is broken into *plates.* These plates are moving over the surface at rates of a few centimeters a year. *Boundaries* between plates are of three types:

1. *Divergent,* where stresses are mainly *tensional* (which refers to the stress that pulls something apart), and depths to earthquakes (focal depths) are shallow (70 km or less);

2. *Convergent,* where stresses are predominantly compressional (squeezing stresses), and focal depths may reach 700 km. These great focal depths result

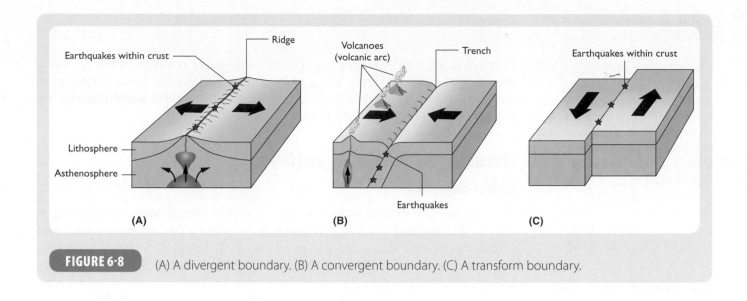

FIGURE 6-8 (A) A divergent boundary. (B) A convergent boundary. (C) A transform boundary.

from the presence of huge rigid slabs of colder oceanic lithosphere in the process of being "recycled" back into the mantle at subduction zones (which, you will recall, are convergent boundaries where dense oceanic lithospheric plates subduct, or move under, less dense continental plate).

3. *Transform*, where stresses are parallel to the fault plane. They are typified by **strike-slip** faults like the San Andreas, and generate shallow-focus earthquakes.

FIGURE 6-8 shows the typcs of plate boundaries and the locations of earthquakes that characterize them. **FIGURE 6-9** shows the ten largest earthquakes measured in the lower 48 states, along with years of occurrence.

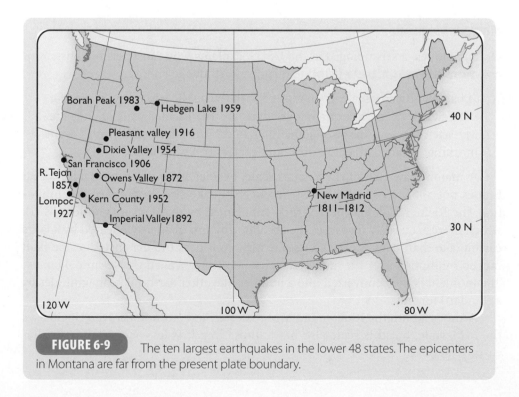

FIGURE 6-9 The ten largest earthquakes in the lower 48 states. The epicenters in Montana are far from the present plate boundary.

Value-added Question 6-1.
If the projected frequency of major earthquakes along the Cascadia subduction zone is approximately one every 550 years, what is the probability (expressed as a percentage) of such an earthquake occurring any single year? Explain whether you think this probability should justify actions such as stricter building codes or restricting development from areas at higher risks for earthquakes.

Value-added Question 6-2.
Given that there is a 32-fold increase in the amount of energy released for every increase on 1 on the Richter scale, how much more energy is released in the M7 earthquake compared to the M1 earthquake?

Recall that the overwhelming majority of earthquakes result from the release of stresses that build up slowly along faults at plate boundaries. Some earthquakes, however, occur well *within* a plate. They are called **intraplate earthquakes**. We will explain and illustrate them below.

Measuring Earthquake Size and Impact

Richter Magnitude

The well-known Richter scale is named after the seismologist who devised it, Charles Richter, and is a rough measure of the relative size, or magnitude, of an earthquake based on the amount of ground shaking during the event. When a seismogram records an earthquake, the recorded spikes (Figure 6-7) vary enormously in amplitude with earthquakes of different sizes. The Richter scale is a logarithmic scale, meaning that there is a tenfold difference in the amplitude of the seismogram (that is, the shaking or ground motion) for every one unit difference in the scale. What's more, the difference in energy level between say, an earthquake of **Richter magnitude** 5 and 6, is 32 times. For example, the ground movement for a quake of Richter Magnitude (which we abbreviate as M) at 100 km from the epicenter, the standard distance, is only 10 μm (10 micrometers, or 10 millionths of a meter). For an M7 earthquake, the recorded ground movement on the seismogram is 1 meter, or 10,000,000 μm. Thus, there is a million times greater movement in an M7 than a M1 earthquake.

Magnitude is only one of the factors that determine earthquake damage. We discuss other factors below. **FIGURE 6-10** shows a number of important earthquakes, the Richter magnitudes, and the estimated energy release.

Moment Magnitude

Dr. Lucy Jones, a seismologist at the USGS, explains why most seismologists now use **moment magnitude (MM)** to quantify earthquake energy:

> "I hate the Richter scale. Even more, I wish the public had never heard of it. I am sick of trying to explain logarithms . . . The Richter scale offends me. My scientific soul despises something so arbitrary and empirical. I want meaning; I want connection to a physical reality. I want units."

A major drawback of the Richter scale, besides its logarithmic scale, is that it does not allow an accurate measure of earthquake size. Because of this limitation, the Richter scale has been largely supplanted by moment magnitude (MM), which more accurately measures the absolute amount of energy released by a large earthquake than does the Richter scale. MM considers the distance that the fault moved multiplied over the fault plane area, a quantity called the seismic *moment*. The moment is then converted into a number like other earthquake magnitudes by a standard formula.

Although both the Richter and MM scales are in common use, the USGS uses the MM scale, but this scale also has a limitation: It is less accurate for smaller earthquakes.

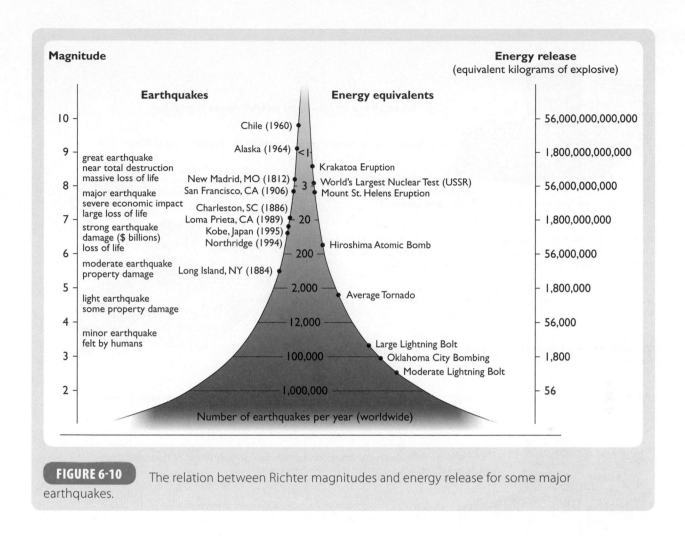

FIGURE 6-10 The relation between Richter magnitudes and energy release for some major earthquakes.

Intensity

Another measure of the strength of an earthquake is its **intensity**, which expresses the *relative damage caused* to people and structures. Intensity is measured using a Modified Mercalli scale (**FIGURE 6-11**).

Concept Check 6-4. Which of the indices, Richter, moment magnitude, or Mercalli, is the most useful for disaster planning and emergency relief? For geologists? For people choosing where to live? Explain your answers.

Isoseismal Maps

An **isoseismal** (eye-so-SIZE-mal) **map** shows the intensity of a particular earthquake. Such a map uses *isoseismal* (iso = same) contours, that is, lines of equal intensity to represent the damage done by an earthquake. Surveys are conducted that ask residents in the affected area to describe the effect of the earthquake by picking one or more terms on the Mercalli scale that best describe their

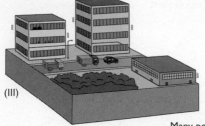
(III)

I. Not felt except by a very few under especially favorable conditions.

II. Felt only by a few persons at rest, especially on upper floors of buildings.

III. Felt quite noticeably by persons indoors, especially on upper floors of buildings. Many people do not recognize it as an earthquake. Standing motor cars may rock slightly. Vibrations similar to the passing of a truck. Duration estimated.

IV. Felt indoors by many, outdoors by few during the day. At night, some awakened. Dishes, windows, doors disturbed; walls make cracking sound. Sensation like heavy truck striking building. Standing motor cars rocked noticeably.

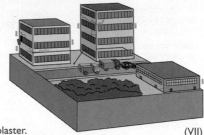

V. Felt by nearly everyone; many awakened. Some dishes, windows broken. Unstable objects overturned. Pendulum clocks may stop.

VI. Felt by all, many frightened. Some heavy furniture moved; a few instances of fallen plaster. Damage slight.

(VII)

VII. Damage negligible in buildings of good design and construction; slight to moderate in well-built ordinary structures; considerable damage in poorly built or badly designed structures; some chimneys broken.

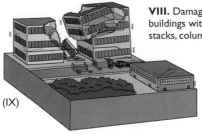

VIII. Damage slight in specially designed structures; considerable damage in ordinary substantial buildings with partial collapse. Damage great in poorly built structures. Fall of chimneys, factory stacks, columns, monuments, walls. Heavy furniture overturned.

IX. Damage considerable in specially designed structures; well-designed frame structures thrown out of plumb. Damage great in substantial buildings, with partial collapse. Buildings shifted off foundations.

(IX)

X. Some well-built wooden structures destroyed; most masonry and frame structures destroyed with foundations. Rails bent.

XI. Few, if any (masonry) structures remain standing. Bridges destroyed. Rails bent greatly.

XII. Damage total. Lines of sight and level are distorted. Objects thrown into the air.

(XII)

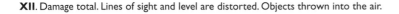

FIGURE 6-11 The intensity categories on the Mercalli scale. Earthquake damage will vary depending on geological setting, construction materials and quality, and distance from the epicenter, among other things.

perceptions of the damage the earthquake caused. Some isoseismic maps are less precise than others, especially if the earthquake happened before the twentieth century. **FIGURE 6-12** shows such a map for the 1811 New Madrid earthquake. This was one of the largest to have affected the interior of North America since European settlement.

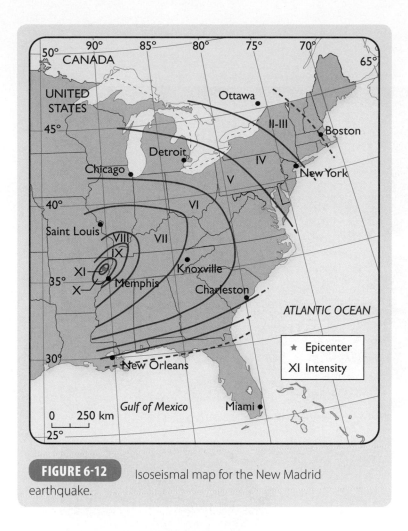

FIGURE 6-12 Isoseismal map for the New Madrid earthquake.

Earthquakes and Blind Thrusts

Between 1769 and 2011, approximately 20 large earthquakes (13 with a Richter magnitude of 6–6.9, and 7 with magnitudes of 7 or above) were recorded in Southern California. The largest earthquake in the Los Angeles Basin during recent times was the January 1994 magnitude 6.7 Northridge Earthquake (**FIGURE 6-13A**). The Northridge earthquake cost $40 billion, killed 57, injured more than 9,000, and displaced more than 20,000 people. **FIGURE 6-13B** shows the Sylmar Interchange on Interstate 5 in Southern California. This interchange was damaged in both the 1971 and 1994 earthquakes. During the 1994 quake a California Highway Motorcycle Patrolman was killed when one of the ramps collapsed.

As a result of the Northridge earthquake, hospitals across the state spent over $4.6 billion to upgrade their facilities, so as to be able to function after a magnitude 6.8 earthquake in response to state law.

The Northridge earthquake occurred on a **thrust fault**, now called the Northridge Thrust, not previously known to exist because it never broke the surface. Such faults are now called **blind thrusts**. Seismologists recently determined that the 1995 Kobe earthquake may also have occurred along a blind thrust as well.

Since the Northridge quake, geologists have discovered more blind thrusts under the Los Angeles Basin. Seismologists are intensely investigating these faults to try to

(A)

(B)

(A) A car at a dealership that was crushed in the Northridge earthquake. (B) Damage to the Sylmar Interchange on Interstate 5 in Southern California as a result of the 1994 Northridge earthquake. Both photos © spirit of america/ShutterStock, Inc.

understand the previously unappreciated threat they may pose to that region's large and growing population.

Intraplate Earthquakes

As we noted above, some earthquakes occur *within* lithospheric plates far from plate boundaries, and are called *intraplate* earthquakes. Very large earthquakes have occurred in Asia many hundreds of km north of the continent–continent collision zone under the Himalayas. In addition, seismic activity in the Rocky Mountains of North America occurs more than a thousand km from the boundary between the Pacific and American plates.

Intraplate earthquakes can cause extensive damage because of their magnitude, and because the stronger crust of plate interiors transmits seismic energy more efficiently than the fractured crust of plate boundaries. For example, the December 1811 New Madrid intraplate earthquake (see below) rang church bells in Boston and damaged buildings as far as 1500 km (900 miles) away on the American East Coast (see Figure 6-13).

Three regions in North America are most prone to intraplate earthquakes. They are the South Atlantic Coastal Plain (Charleston SC, 1886 earthquake), the **New Madrid Fault Zone (NMFZ)** of the continental Interior (New Madrid earthquakes, 1811–1812), and the St. Lawrence Seaway (FIGURE 6-14).

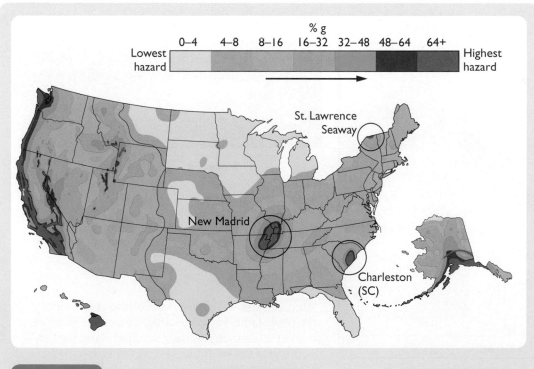

FIGURE 6-14   Most intraplate earthquakes in eastern North America occur in the three circled regions, as shown on this Earthquake Hazard map of the United States, which shows the threat of ground-shaking.

New Madrid

Figure CS 2-1A shows epicenters of earthquakes along the New Madrid Fault Zone (NMFZ). It is interpreted to lie above an ancient *failed rift* or gap (**Figure CS 2-1B**), the Reelfoot Rift. Geologists believe that as much as 750 million years ago a rift began to form under the present site of New Madrid. The rift stopped spreading and eventually filled with sediment. After millions of years, the rift zone had become buried under a blanket of younger sedimentary rocks. The failed rift is bounded by inactive fault systems, but they remain sites of relative crustal weakness. With initiation of westerly movement of the North American Plate during the Triassic Period 250–200 million years ago, as the present Atlantic began to open, these old zones of crustal weakness were stressed.

In addition, geologists have measured the *heat flow* under the rift zone and have found it to be unusually high. While the impact of heat flow is complex, high heat flow could weaken the crust and make it more likely to fail under tensional stresses, leading to earthquakes.

Today, studies at the University of Memphis (TN) and elsewhere seek to determine how often earthquakes occur along the NMFZ and associated regions, that is, the **recurrence interval** (RI: see below). These studies suggest that earthquakes comparable to the New Madrid group are most likely to recur every few hundred to one thousand years.

Now do the following calculation to evaluate the limitations of this concept.

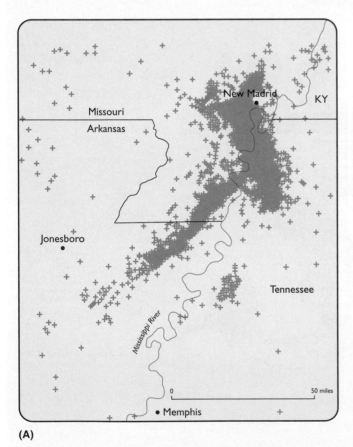

(A)

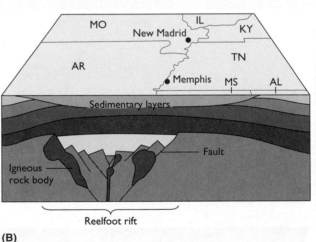

(B)

FIGURE CS 2-1 (A) Epicenters of earthquakes along the New Madrid Fault Zone (NMFZ). Since 1974, more than 4,000 earthquakes have occurred along the NMFZ. Courtesy of USGS. (B) The Reelfoot Rift below the New Madrid Fault Zone.

Value-added Question 6-3. Here is a hypothetical list of the years in which damaging earthquakes occurred along a fault: 50, 450, 1100, 1400, 1811, 2010, and 2030. Determine the average time (in years) between earthquakes. What confidence would you place in such a calculation if you were a local civil defense official?

Charleston, South Carolina, 1886

One of the largest earthquakes ever to have affected continental North America hit an area around Charleston, South Carolina on August 31, 1886 at 9:50 PM and lasted less than one minute (**Figure CS 2-2**). It had an estimated Richter magnitude of 6.6–7.3, roughly comparable to the 1994 Northridge earthquake, and reached an Intensity of X on the Modified Mercalli scale. It killed about sixty people and caused extensive damage in Charleston, and in an oval area within about 50 km of the city, trending northeast, parallel to the coast. Its effects were felt hundreds of km away. It knocked down chimneys as far away as Richmond, Virginia. At least three hundred **aftershocks** occurred during the 30 years following the earthquake.

The 1886 Charleston earthquake remains the "**design earthquake**" that all major structures (dams, nuclear power plants, etc.) erected within the region since passage of the National Environmental Policy Act in 1969 must be able to withstand. Although the New Madrid earthquakes can be explained as movement along a failed rift with some degree of confidence, and numerous small earthquakes continue to occur there, a satisfactory explanation for the Charleston earthquake so far has eluded researchers. No fault broke the ground, but there is evidence that the region was seismically active before European settlement. Some geologists believe that the Charleston earthquake could have occurred above a buried rift zone in a manner similar to the New Madrid quakes.

Concept Check 6-5. One of this book's authors resides in Pawley's Island, South Carolina, about 75 miles northeast of Charleston. Study the isoseismal map of the Charleston Earthquake shown in Figure CS 2-2. If you were he, would you buy earthquake insurance? What additional information would you want before making this decision?

(A)

(B)

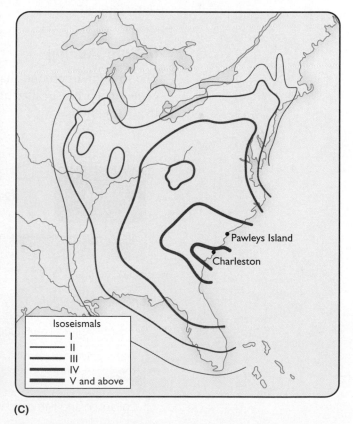

Pawleys Island

Charleston

Isoseismals
——— I
——— II
——— III
——— IV
▬▬▬ V and above

(C)

FIGURE CS 2-2 The 1886 Charleston, South Carolina earthquake. (A) Aftermath of the earthquake. (B) A damaged building. (C) Isoseismal map of the earthquake. A–C courtesy of USGS.

■ HOW EARTHQUAKES CAUSE DAMAGE

The damage *resulting from* an earthquake is related to many factors, only one of which is magnitude, which we mentioned earlier. Probably the most important three factors are:

- Proximity to the epicenter
- Type of rock or sediment underlying a site
- Types of structures built on the site

Other important factors are:

- Slope of the land (steep slopes are more prone to slope failure, i.e., landslides)
- Proximity to the shoreline (seismic sea waves called **tsunami** can damage coastal cities)
- Focal depth: Shallow-focus earthquakes can be more damaging than deeper-focus earthquakes, because the seismic energy of deeper focus earthquakes is dissipated as it moves away from the focus.

Naturally, structures built on a fault that ruptures will be damaged. We consider how earthquakes cause damage in detail below.

The main ways by which earthquakes *cause* damage are: (1) strong ground shaking; (2) the effects of surface failures (surface failure, ground cracking, liquefaction, subsidence); and (3) water waves. Most damage to buildings from earthquakes results from ground shaking. **TABLE 6-1** lists direct and indirect ways earthquakes cause damage.

Ground Shaking

The strength of ground shaking decreases with distance from the earthquake's epicenter in a process called *attenuation*, although not necessarily by the same amount in all directions.

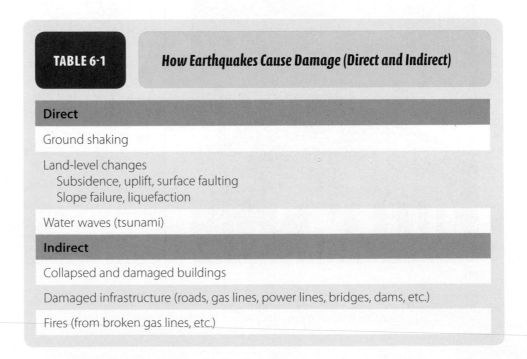

TABLE 6-1	How Earthquakes Cause Damage (Direct and Indirect)
Direct	
Ground shaking	

Land-level changes
    Subsidence, uplift, surface faulting
    Slope failure, liquefaction	

Water waves (tsunami)	
Indirect	
Collapsed and damaged buildings	
Damaged infrastructure (roads, gas lines, power lines, bridges, dams, etc.)	
Fires (from broken gas lines, etc.)	

Ground shaking can increase by **amplification** of shock waves. Amplification occurs where earthquake waves of certain frequencies pass from stronger into weaker geologic materials.

High wave amplitudes can cause great damage to structures. During the 1989 Loma Prieta Earthquake, the part of San Francisco (about 42 km, or 70 mi, to the north of the epicenter) that suffered the greatest damage was built on debris from the 1906 earthquake that had been piled on "reclaimed" San Francisco Bay mud (FIGURE 6-15).

Elevated highways and overpasses are especially prone to damage by earthquakes. FIGURE 6-16 shows collapsed overpasses and elevated freeways.

Strong shaking of long duration (2+ minutes) is one of the most damaging effects of great earthquakes.

Water Waves (Tsunami)

Tsunami, often mistakenly called tidal waves, are seismic sea waves caused by submarine landslides, or sea-floor displacement, related to volcanic or earthquake activity or, rarely, by meteorite impact.

Because water cannot be compressed, a rapid displacement of the sea bottom or a massive landslide will instantaneously transmit enormous energy into the water. This will take the form of a series of high-energy waves, radiating outward like pond ripples from the focus of the generating event. This *train* of tsunami waves arrives at the shore over an extended interval of time. The initial tsunami will be followed by others, a few minutes to a few hours later. The tsunami caused by the 1883 eruption of Krakatau killed more than 30,000 people and was recorded in the English Channel the next day. The time interval between tsunami may lure the curious into the coastal zone, sometimes to be swept away with the next wave.

Tsunami can be extremely destructive to life and property. Tsunami wave "trains" are commonly 100 km or more from crest to crest and may travel at speeds exceeding 900 km/h in the open ocean. They can traverse the entire Pacific Ocean in 20 to 25 hours! In 1960, a magnitude 8.5 earthquake off Chile triggered a tsunami that arrived 15 hours later in Hilo, Hawai'i, destroying most of the downtown and killing 61 residents (FIGURE 6-17).

Between 1895 and 1995, 454 tsunami were recorded in the Pacific, killing over 51,000 coastal residents. An 1886 tsunami on the coast of Japan alone killed about 26,000 people. As a result of this activity, one might have suspected that the Japanese would have been prepared for a devastating tsunami. But, on March 11, 2011 that proved not to be the case, as an enormous earthquake measuring 8.9 on the Richter Scale produced a series of tsunami that devastated extensive sections of the coast, and destroyed nuclear reactors situated at Fukushima (FIGURE 6-18) immediately on the coast. The nuclear plant was protected by a seawall designed to withstand a 6-meter tsunami; unfortunately, the maximum height of the waves reached nearly 15 meters (46 feet). Authorities admitted that the Japanese utility TEPCO had produced a mere one-page safety plan for the reactor site.

Vulnerability of the United States to Tsunami

The King County (Washington) Office of Emergency Preparedness reported that stresses off the coast of Washington "could generate an earthquake measuring 9+ on the Richter scale. This would cause coastal areas to drop up to six feet in minutes

(A)

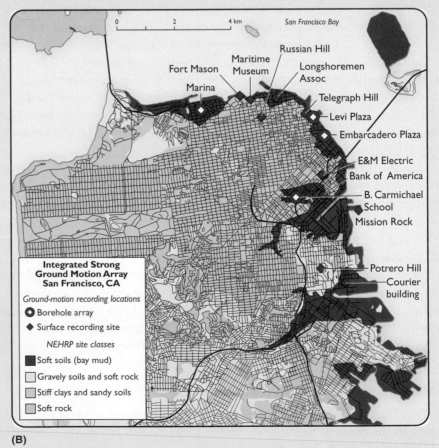

San Francisco Bay

Russian Hill

Fort Mason   Maritime   Longshoremen
              Museum   Assoc

Marina   Telegraph Hill

                           Levi Plaza

                           Embarcadero Plaza

                           E&M Electric
                           Bank of America

                           B. Carmichael
                           School
                           Mission Rock

**Integrated Strong
Ground Motion Array
San Francisco, CA**

Ground-motion recording locations

⬡ Borehole array
◆ Surface recording site

NEHRP site classes

◼ Soft soils (bay mud)
☐ Gravely soils and soft rock
☐ Stiff clays and sandy soils
☐ Soft rock

                           Potrero Hill
                           Courier
                           building

(B)

FIGURE 6-15 (A) A house destroyed by ground shaking during the Loma Prieta earthquake of 1989. Slide X-4, U.S. Geological Survey Open-File Report 90-547. Courtesy of USGS. (B) A geologic map of San Francisco Peninsula. The weakest rocks, bay mud, are shown in brown.

Collapsed overpass on Golden State Freeway in Southern California resulting from1971 San Fernando earthquake. Highways on stronger sediment were little damaged. California rebuilt the Nimitz freeway in the San Francisco Bay area with earthquake-resistant features. Courtesy of R.E. Wallace/USGS.

and would produce a tsunami from British Columbia to Mendocino, California. Such an earthquake would last several minutes and produce catastrophic damage. These earthquakes are believed to occur approximately every 550 years." The last one occurred approximately 300 years ago.

In 1964, tsunami generated by the "Good Friday" Alaska Earthquake killed 11 people and caused $7.4 million of damage to Crescent City, California. The city remains at risk. Floods from tsunami destroyed several coastal communities in Alaska during the 1964 Good Friday earthquake.

Most at risk are the people and property on beaches, at low-lying areas of coastal towns and cities, and near tidal flats or river mouths.

On December 26, 2004, a massive earthquake struck 60 km off the west coast of northern Sumatra. The energy released by the displacement of the seafloor caused a tsunami that killed 250,000 people in Indonesia, Thailand, India, Sri Lanka, and elsewhere around the Indian Ocean.

Largely as a result of the death and destruction of the Sumatran tsunami, the National Oceanic and Atmospheric Administration (NOAA) expanded its

(A)

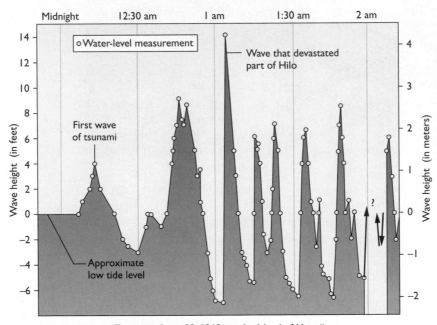

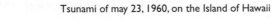

First wave of tsunami

Approximate low tide level

Wave that devastated part of Hilo

○ Water-level measurement

Tsunami of may 23, 1960, on the Island of Hawaii

(B)

(C)

FIGURE 6-17 (A) Damage from the 1960 tsunami in Hilo, Hawai'i, which destroyed most of the downtown and killed 61 residents. The tsunami arrived 15 hours after a magnitude 8.5 earthquake off Chile. Warning sirens were sounded more than three hours before the first tsunami waves arrived, but residents apparently were confused about their meaning. Courtesy of NOAA. (B) Locals were fooled by the relatively low height of the first tsunami waves and, as a result, many thought the danger had passed, only to be overwhelmed by a wave three-times as high a short time later. Modified from Atwater, Brian F. et al. Surviving a Tsunami - Lessons from Chile, Hawaii, and Japan. Circular 1187, Volume 1. USGS. (C) Tsunami warning sign in downtown Hilo, Hawai'i. © Pluff Mud Photography.

Deep-ocean Assessment and Reporting of Tsunami (DART®) program from 6 to 39 buoys (**FIGURE 6-19**). The program's goal was to improve the ability to detect tsunami early and thus give sufficient warning to areas in the path of a developing tsunami.

Other Effects

Fires

Fires caused most of the damage in the 1906 San Francisco earthquake. The earthquake ruptured water lines so that firefighters had no water pressure. Widespread use of gas for lighting (electricity had not yet saturated the lighting market) meant that gas lines ruptured and the wooden buildings readily caught fire.

During the 1989 Loma Prieta earthquake, fires from ruptured natural gas lines in San Francisco's Marina District again resulted in serious damage. Fires also can occur in modern homes during an earthquake if gas hot water heaters topple, rupturing the gas line.

See our discussion below on protecting yourself against earthquakes.

Floods

Floods resulting from earthquakes usually are caused by tsunami impacting the coast. However, earthquakes can damage or destroy dams, resulting in potentially catastrophic floods. During the 1971 San Fernando earthquake, the Van Norman Dam near North Hollywood was damaged, necessitating the evacuation of tens of thousands of residents who lived below the dam (**FIGURE 6-20**).

Earthquakes can induce land-level changes that can cause rivers to leave their channels and flood adjacent areas. Land-level changes can also affect the coast. Tens

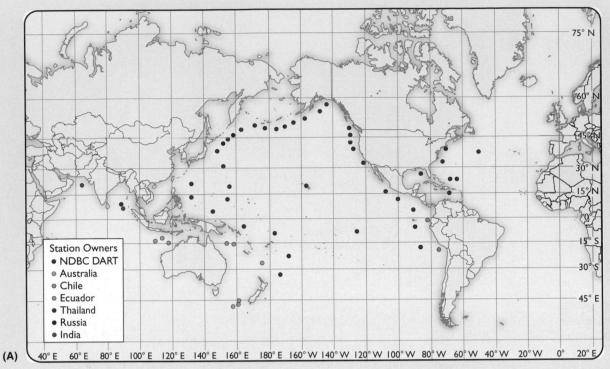

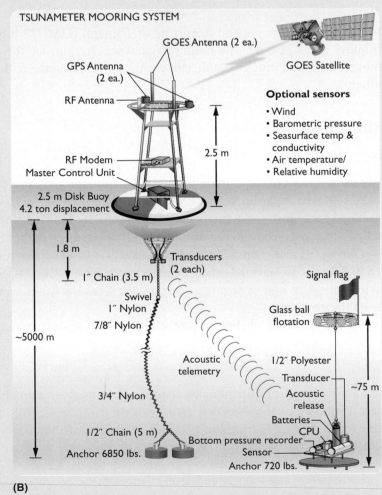

FIGURE 6-19
(A) Current deployment of NOAA DART® buoys. (B) Diagram of a DART® buoy. Modified from Overview of first operational DART® Mooring System, NOAA.

(A)

Station Owners
- NDBC DART
- Australia
- Chile
- Ecuador
- Thailand
- Russia
- India

75° N
60° N
45° N
30° N
15° N
0°
15° S
30° S
45° E

40° E 60° E 80° E 100° E 120° E 140° E 160° E 180° E 160° W 140° W 120° W 100° W 80° W 60° W 40° W 20° W 0° 20° E

(B)

TSUNAMETER MOORING SYSTEM

GOES Antenna (2 ea.)
GOES Satellite
GPS Antenna (2 ea.)
RF Antenna

Optional sensors
- Wind
- Barometric pressure
- Seasurface temp & conductivity
- Air temperature/
- Relative humidity

2.5 m

RF Modem
Master Control Unit

2.5 m Disk Buoy
4.2 ton displacement

1.8 m

Transducers (2 each)

1″ Chain (3.5 m)
Swivel
1″ Nylon
7/8″ Nylon

~5000 m

Acoustic telemetry

Signal flag

Glass ball flotation

1/2″ Polyester
Transducer
Acoustic release
Batteries
CPU

~75 m

3/4″ Nylon

1/2″ Chain (5 m)
Anchor 6850 lbs.

Bottom pressure recorder
Sensor
Anchor 720 lbs.

The Van Norman Dam, showing liquefaction damage from the 1971 earthquake. Courtesy of R.E. Wallace/USGS.

of thousands of square kilometers of the Alaskan Coast experienced permanent elevation changes during the 1964 earthquake.

Slope Failure

Slope failures (e.g., "landslides" or slope collapse) are a near-universal result of earthquakes. In the 1964 Alaska earthquake, most slope failures occurred along bedding plane failures in bedrock, probably triggered by violent ground shaking. **FIGURE 6-21** shows an example of slope failure that occurred during an earthquake.

Even low-magnitude earthquakes can produce dangerous rockfalls. One person was killed in the Magnitude 6 earthquake in Klamath Falls, Oregon, in 1973 when the car he was driving was crushed by a boulder from the rockslide caused by the earthquake, which caused numerous rockfalls and rockslides in road-cuts and on steep slopes.

Earthquakes can also cause **liquefaction** of water-saturated fine sediments, which can strongly amplify shock waves. During

FIGURE 6-21 Example of slope failure that occurred during an earthquake. Courtesy of NOAA/NGDC, Jose C. Borrero, University of Southern California.

liquefaction, violent shaking reduces the internal cohesiveness of sediment and can cause it to collapse and flow like a fluid.

In soils, liquefaction usually occurs when they are saturated, that is, in soils in which the space between individual particles partially is filled with water. When undisturbed (that is, before the shaking of an earthquake), the friction between the particles as well as the bonds between water molecules hold the soil particles together. When agitated by the shaking of an earthquake (or other event), the friction disappears and the bonds between the water molecules are temporarily broken (the cohesive forces disappear) and the soil then acts like a liquid (**FIGURE 6-22**). During liquefaction, the ground above may sink or even pull apart.

When liquefaction occurs, the ability of a substrate to support foundations for buildings and bridges is reduced. Liquefied soil also exerts higher pressure sideways on retaining walls, which can cause them to tilt or slide. Ground failures commonly occur above liquefaction zones.

Although many areas of the United States are prone to liquefaction, we do not include a map because liquefaction effects are generally site-specific, that is, in localized areas rather than across wide swaths of landscape.

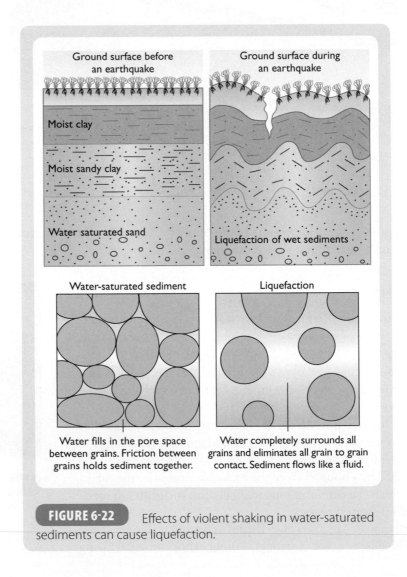

FIGURE 6-22   Effects of violent shaking in water-saturated sediments can cause liquefaction.

Examples of liquefaction that occurred during earthquakes include:

- San Fernando, California, 1971
  During the 1971 San Fernando, CA earthquake, an underwater slide occurred in the reservoir behind the lower Van Norman Dam. Figure 6-20 shows the dam after the slide damage, caused by liquefaction. Officials immediately ordered the evacuation of thousands of residents (including relatives of one of your book's authors) in North Hollywood and other cities below the dam. The dam barely avoided collapse.

- San Francisco, California, 1906
  Liquefaction can be indirectly blamed for 85% of the total damage to San Francisco during the earthquake of 1906, by inducing lateral spreading that broke water-supply pipelines. This prevented containment of the fire in San Francisco that destroyed about 500 city blocks.

- Loma Prieta, California, 1989
  Approximately $100 million out of a total of $5.9 billion in property damage resulted from liquefaction. Liquefaction of substrate below the road caused extensive damage to highways. Oakland International Airport was damaged 40+ miles (60 km) from the epicenter of the earthquake. The main 10,000-foot runway of the airport, built on compacted fill over "reclaimed" bay mud, was severely damaged by liquefaction.

- Niigata, Japan, 1964
  This earthquake killed 26 and destroyed or damaged over 12,000 houses (**FIGURE 6-23**). About ⅓ of the city sank as much as 2 m (6.5 ft) as a result of soil liquefaction.

FIGURE 6-23　Soil liquefaction combined with poorly constructed foundations caused the collapse of these apartment buildings during the 1964 magnitude 7.4 Niigata, Japan earthquake. Photo by T. L. Youd. Japan, 1964. USGS.

Figure CS 3-1 shows regions in the Bay Area that are most vulnerable to liquefaction, which can be divided into two categories: (1) areas covered by fill poured into San Francisco Bay since 1845 to develop 77 miles^2 (200 km^2) of tidal and submerged areas; and (2) areas along existing and filled stream channels and flood plains, particularly those areas with deposits less than 10,000 years old.

Protecting against liquefaction is expensive. Beginning in 1985, California utility Pacific Gas and Electric (PG&E) undertook a 25-year, $2.5 billion program, known as the Gas Pipeline Replacement Program (GPRP). As a result of the GPRP, many pipeline upgrades were installed before and after the Loma Prieta earthquake. The newer pipelines are significantly less vulnerable to earthquake effects, including liquefaction, differential settlement, violent shaking, and ground strain, than the older types of pipe installed 50 to 100 years ago.

Experts argue that it is rarely cost-effective to retrofit roads or airport runways. Depending on the epicenter of a future earthquake, road closures and airport damage can be expected similar to that that occurred as a result of the Loma Prieta earthquake. Oakland and San Francisco International Airports are built on liquefiable fill on bay mud. Runways at San Jose International Airport cross a number of ancient stream channels.

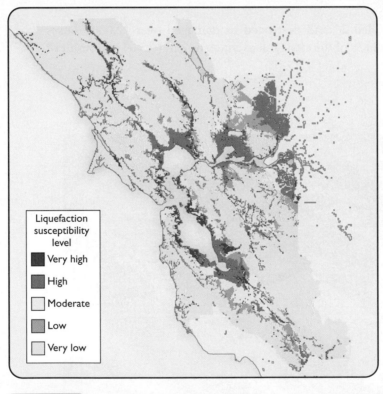

Liquefaction susceptibility level

■ Very high
■ High
□ Moderate
■ Low
□ Very low

FIGURE CS 3-1 Land vulnerable to liquefaction in the Bay Area.

Two terms have been used interchangeably in the popular press in discussions of whether or not scientists can predict earthquakes. **Prediction** implies some ability to state with a degree of confidence where and when an earthquake is likely to occur. **Forecasting** implies an ability to predict the likelihood of an earthquake of a given magnitude occurring within *a specified time interval*. Much confusion is introduced when these terms are used incorrectly.

The ability to forecast earthquakes would be greatly aided by identifying any periodicity of earthquake activity in an area; that is, determining the *recurrence interval*. Predicting earthquakes would be facilitated by the ability to recognize events that immediately precede earthquakes, called **precursors**.

Recurrence Intervals of San Andreas, California Earthquakes

The San Andreas Fault system in the San Francisco Bay Area contains faults with varying characteristics, e.g. different lengths and rates of movement. According to the U.S. Geological Survey, this heterogeneity results in a broad range of earthquake magnitudes and frequency of occurrence. A large fault break requires the build-up of an enormous amount of stress and hundreds of years may pass between surface ruptures on a single fault. The average time between such ruptures is known as the *recurrence interval* (RI) of that fault. You should not confuse the use of RI here with the slightly different use of the concept as applied to floods (covered elsewhere in this text for a comparison). **FIGURE 6-24** shows earthquakes that have occurred along the San Andreas system in the Bay Area since 1850. Recurrence interval studies are used in earthquake probability forecasting (**TABLE 6-2**).

Concept Check 6-6. Stop here and state what use you, as a citizen or public official, would make of Table 6-2. Can it be used to state when an earthquake will happen? Would it help you decide where to buy a house? Why or why not? Could it help decide where to put major infrastructure (highways, hospitals, etc.)? Explain.

- -

Studies of prehistoric fault movements, called **paleoseismicity**, have provided data like in **TABLE 6-3**, which shows the approximate dates of historic and prehistoric earthquakes.

Earthquake Precursors

Many seismologists believe that if they study earthquakes carefully enough they should be able to discover some event or sequence of events that precedes earthquakes. Such events are called *precursors*. There is evidence that animals may sense changes in a region preceding an earthquake, perhaps "hearing" sounds generated by stress release beyond the range of human hearing. Earthquake waves have frequencies that are usually out of the range of human hearing but may well be within the range of animals.

Value-added Question 6-4.
The range in years (difference between shortest and longest interval) between earthquakes according to Table 6-3 is 4 to 332 years. Calculate the average interval between earthquakes here, ignoring the oldest and youngest earthquakes since their intervals cannot be discerned.

- -

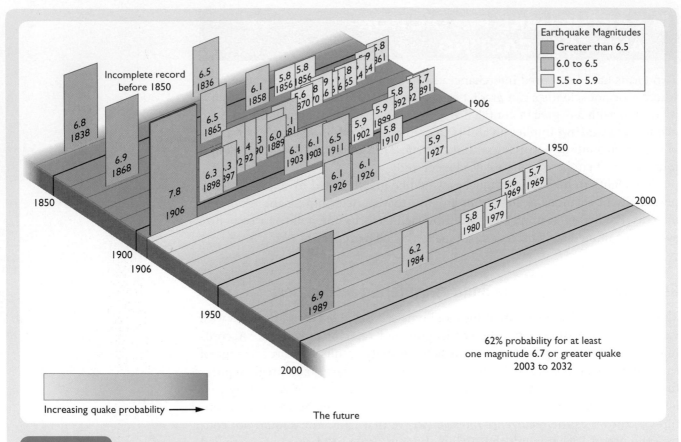

Major earthquakes along the San Francisco Bay portion of the San Andreas system, along with probabilities for future earthquakes. Modified from Working Group on California Earthquake Probabilities. Bay Area Probabilities. USGS.

TABLE 6-2	Probability of at Least One M ≥ 6.7 Earthquake Before 2030 (USGS)	
Fault system		**Probability (%)**
San Gregorio		10
San Andreas		21
Hayward-Rodgers Creek		32
Calaveras		18
Concord-Green Valley		6
Greenville		6
Mount Diablo		4
Background		9
Regional aggregate		70*

* This means that a 70% chance of at least a Magnitude 6.7 earthquake exists, before 2030.
Data from USGS.

TABLE 6-3	Previous San Andreas Fault Ruptures at Pallett Creek, CA	
Est. event date	**Possible date range**	**Years until next event**
January 9, 1857	January 9, 1857	greater than 141*
December 8, 1812	December 8, 1812	4.08
1480	1465–1495	332
1346	1329–1363	134
1100	1035–1165	246
1048	1015–1081	52
997	981–1013	52
797	775–819	200
734	721–747	63
671	658–684	63
before 529	???–529	greater than 142

* The number of years between the last earthquake and 1998, when the table was compiled. To obtain "years until next event", add the number in the right column to the next lowest date in the far left column.
Data from USGS.

Many seismologists hypothesize that recognizable land-level and/or groundwater-level changes could precede earthquakes. Others point to changes in rock density, emission of electromagnetic radiation in distinctive frequencies, or radon emission as potential precursors. However promising, empirical research on most precursors is complicated by the scarcity of data, and evidence remains largely at the anecdotal level.

Precursors and the Dilatancy-Diffusion Model of Earthquakes

Two models used to explain aspects of earthquakes are the *elastic-rebound* (Figure 6-3), which we described above, and the **dilatancy-diffusion** models. The dilatancy-diffusion model of earthquakes in general holds that, just before an earthquake, stress causes the rocks about to fail to expand (to *dilate*), so that their volume increases and their density decreases, an example of strain. Numerous tiny cracks appear along the fault plane along which rupture eventually is concentrated.

If this is true, geophysical measuring devices ought to be able to detect the subtle changes in density, electrical resistivity (an index of resistance to flow of electricity), radon emission, and even water content, because if rocks begin to crack, water could infiltrate the cracks. Research in this area is ongoing, with mixed results.

Palestine, considered a holy place by three religions, has been a dangerous place to live since human settlement. Earthquakes are especially frequent. The destruction of Sodom and Gomorrah and the collapse of the walls of Jericho could have been the result of earthquakes.

Figure CS 4-1 shows a map of seismicity in the Middle East.

Concept Check 6-7. (A) In which countries are earthquake epicenters concentrated? (B) At which depth zone are earthquake foci concentrated? Shallower focus earthquakes are more dangerous than others. (C) Which countries have the highest risk? Locate Jericho and Gomorrah on the risk map. (D) Next, think about and write down what sort of *substrate* is likely to slow down, and thus increase the amplitude of, seismic waves. If you included lake muds, you'd be right.

Note that Gomorrah is located beside the Dead Sea, and was likely built on lake sediment. Finally, recall that fires for cooking, etc. would likely have been kept burning nearly constantly in dwellings, because they could not have been easily restarted. Collapse of buildings could have triggered numerous fires.

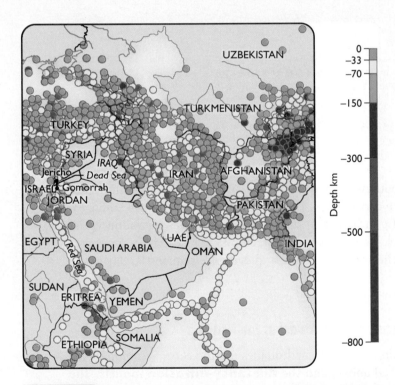

FIGURE CS 4-1 Seismicity of the Middle East, along with location of Jericho and Gomorrah.

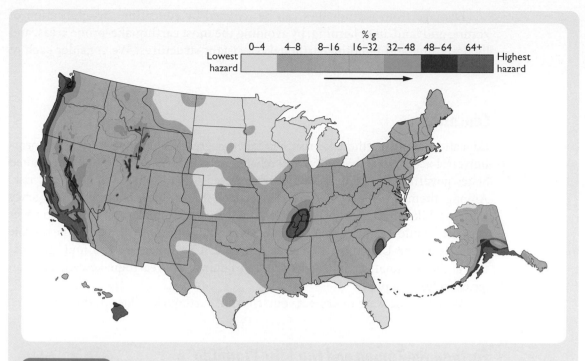

FIGURE 6-25 A shaking risk map of the United States published by the USGS based on ground acceleration. Units are percents of g, the acceleration of a falling object due to gravity. Colors depict the levels of horizontal shaking that have a 2% chance of being exceeded in a 50-year period. Courtesy of USGS.

■ EVALUATING EARTHQUAKE RISK

Earthquake Risk for the United States: Seismic Risk Maps

The USGS compiles data on earthquakes by state. For your state's earthquake history, go to http://earthquake.usgs.gov/earthquakes/states/.

Evaluating earthquake hazard begins with **seismic risk maps**. These maps show the probability of *ground acceleration* for a given locality. Ground acceleration is the major cause of earthquake damage.

The USGS publishes seismic risk maps for the United States. Seismic risk maps usually show the expected strength of ground acceleration, based on historic earthquake magnitudes, and subsurface material. An example is shown in **FIGURE 6-25**.

Concept Check 6-8. Examine Figure 6-25. Where is seismic risk the highest? Does it coincide with any highly populated areas? Which ones? You can find state population information from http://www.census.gov.

■ REDUCING RISK FROM EARTHQUAKES

Economic impacts of earthquakes can be devastating. For example, a comprehensive report on earthquake risk in Canada concluded, "a moderate earthquake close to Vancouver would result in disastrously large insured losses, bankrupting many companies."

Damage from earthquakes can be reduced by education, by application of zoning and land-use planning, by avoiding the most earthquake-prone areas, and by designing and building earthquake-resistant structures. We consider each of these next.

Education

Education about earthquake risk has taken a great leap forward with the near-universal access to authoritative and reliable websites. Many agencies of the United States government maintain web pages devoted to earthquakes and seismic risk. Among them are the USGS and the Federal Emergency Management Agency (FEMA). Likewise, states and localities at risk from earthquakes maintain web pages with useful information. Interested citizens should begin with a study of the seismic risk map of the United States maintained by USGS. Canadians can find information at the National Earthquakes Hazard Program (http://earthquakescanada.nrcan .gc.ca/index-eng.php).

We list some ways to protect yourself from earthquakes later in this chapter.

Earthquake Zoning and Land-Use Planning

Seismic risk zoning is applied with varying degrees of success in the United States, Canada, Japan, and a few other countries. It takes into consideration estimates of earthquake probability, ground shaking to be expected, distances to faults, and the nature of substrate. Such studies are primarily designed for geologists and engineers as well as for insurance companies who use these studies to set rates. **Microzoning** is an especially effective technique that we discuss below.

Seismic Risk Zoning in California

The California Department of Conservation (CDC) is required by the Seismic Hazards Act of 1990 to identify and map the most prominent earthquake hazards in the state, to help minimize earthquake damage. The Seismic Hazard Zone Mapping Program of the CDC ultimately will chart areas prone to liquefaction and earthquake-induced landslides throughout California's principal urban and major growth areas.

Concept Check 6-9. Reexamine Figure CS 3-1, a hazard map for the North San Francisco Peninsula. Which of the following areas are most prone to liquefaction?

1. areas distributed around the periphery of the peninsula
2. areas distributed at random throughout the peninsula
3. areas located in belts, trending across the peninsula.

- -

Concept Check 6-10. Describe how maps such as that shown in Figure CS 3-1 could help local planning officials. Is this a legitimate role of government? In other words, should potential property buyers be held responsible for evaluating the risk of a particular piece of property? Should the real estate industry be entrusted with that responsibility? Where should the taxes come from to pay for such maps, assuming it is a legitimate role of government? Should the federal government require all localities to provide such data? Should state governments? Why or why not? Explain the reasons for your answer.

- -

Designing Earthquake-Resistant Buildings

How Earthquakes Damage Buildings

Contrary to popular belief, buildings are rarely damaged by faults moving under the building. Rather, they are normally damaged during an earthquake in two ways, (1) failure of substrate (soil, sediment, bedrock) beneath buildings, and (2) earthquake waves shaking the building. The longer the shaking, the more building damage generally results. Horizontal shaking is more damaging than vertical shaking.

Resonant Frequencies

Recall that the ground motion during earthquakes is caused by the passage of seismic waves. Seismic waves vary in duration, amplitude, and frequency (in units of cycles/sec). Buildings, like all objects, have a "natural frequency" of vibration, the frequency at which it vibrates when disturbed. To visualize this, imagine a ringing bell. The tone of any bell depends on its size and composition. It always rings at its own natural frequency. Likewise, each building has its own frequency that depends on its mass, how "flexible" it is, etc.

When seismic wave frequencies are close to a building's natural frequency, it is said that the building and the ground shaking are in **resonance** with one another. Resonance tends to increase or *amplify* the shaking of the building. Therefore, buildings suffer the greatest damage from ground motion at a seismic frequency close or equal to the natural frequency of the structure.

Various techniques can be used to "dampen" a building's response to shaking by seismic waves. Some buildings are built on giant "shock absorbers" to dampen or absorb energy. The St. John's Health Center in Santa Monica, California spent $20 million on such a system in a reconstruction of one of its wings severely damaged during the 1994 Northridge earthquake.

Additional construction practices include strong internal bracing and support from steel rods bundled inside concrete.

Building Design and Earthquake Damage

Well-designed buildings in earthquake country have the following attributes:

- Simple square or rectangular ground plan
- Solid foundation, with load-bearing walls solidly attached to it
- Floors solidly attached to the load-bearing walls
- Foundations built on strong bedrock (such as crystalline igneous rocks and well-cemented sandstones), and not on weak material like mudrocks or landfills

The building housing the headquarters of AutoZone Corp. in Memphis, Tennessee is built on the Mississippi Embayment, an area of ancient river sediments covering parts of several southern states and prone to rare but significant intraplate earthquakes. It was designed to be "earthquake resistant" by using "base isolation," an engineering technique that absorbs shock from earthquakes.

In the San Francisco earthquake of 1906, *ornate facades* of buildings were shaken off, sometimes onto heads of residents taking cover under the building's walls.

In Turkey, California, Japan, and elsewhere, *buildings without sufficient support between floors* often collapse and pancake (**FIGURE 6-26**). Support can be retrofitted to existing buildings, but it is expensive and often unsightly. It is almost always cheaper to design the structure correctly the first time.

In Mexico City during the 1985 earthquake many unreinforced buildings collapsed, but reinforced buildings were much less damaged. Mexico City's situation is

Earthquake-resistant buildings are not a recent development. Istanbul's Hagia Sophia (**Figure CS 5-1**) was built by the order of the Byzantine Emperor Justinian in the sixth century. It lies on the North Anatolian Fault, one of the Middle East's most active faults. While it has been damaged several times by earthquakes, it has survived for 1400 years in one of the most earthquake-prone regions of the world.

Elsewhere in this book we asked you to suggest reasons why earthquakes in Iran are so much more deadly than those in California. If you suggested that building type could be responsible, you were right. An old maxim among geoengineers is, "earthquakes don't kill people, buildings do."

Many developing countries do not appropriate sufficient resources to build earthquake-resistant structures. Typical structures damaged or destroyed in Iran were one-story to several-story **unreinforced buildings** (that is, no metal rods inside the walls or additional bracing to support it). These are made of bricks and covered with mud (adobe) and built on the ground without foundations. Adobe can collapse during an earthquake, because shaking can separate walls at corners and pull them apart (**Figure CS 5-2**).

FIGURE CS 5-2   Damage to unreinforced houses during the Bam, 2003 earthquake, in southern Iran. Image courtesy of Space Imaging.

FIGURE CS 5-1   Hagia Sophia in Istanbul, Turkey.
© VLADJ55/ShutterStock, Inc.

complicated by the fact that the city's substrate consists of fine-grained lake deposits, compliments of Spanish colonists who drained most of the lake for building sites. Building on weak-substrate causes most building damage and up to 80% of deaths in earthquakes.

Because substrate is so important, it is critical to know as much as possible about substrate under building sites before construction. This has led to a mapping procedure called *microzoning* in earthquake-prone countries like Mexico, Japan, Italy, and the United States.

Microzoning is *the characterization and mapping of earthquake-relevant ground conditions* in a community. The most important factors of such an assessment are:

- Detailed soil mapping
- Mapping areas of land instability, including areas prone to liquefaction
- Mapping the location of all known faults

Microzoning should include subsurface mapping to identify and trace buried faults like blind thrusts.

FIGURE 6-26 A new structure that partially collapsed during the 1971 Sylmar (San Fernando) earthquake in Southern California. This building, the Medical Treatment and Care Building in the Olive View Hospital complex, was completed in 1970 at a cost of $25 million. The four towers contained stairs and day-room areas and were separated four inches from the main building. The towers were supported by concrete pillars that collapsed, causing the towers to overturn. A collapsed stairwell is visible just above the photo's center. The first story nearly collapsed, and the building was ultimately demolished. Courtesy of USGS.

Protecting Yourself From Earthquakes

You can take a number of actions to protect yourself if you live in, or visit, areas prone to earthquakes.

First,

- Find out if you live in an earthquake hazard zone. Consult the national risk map above. The four states with the greatest earthquake risk are California, Nevada, Washington, and Alaska, but many U.S. states have at least some earthquake risk. In Canada, the St. Lawrence Seaway has been the site of some moderately damaging earthquakes, as has the West Coast. If you live in an earthquake-risk zone, contact your local emergency preparedness office or planning office.
- Remember that earthquakes can strike at any time, can be over in seconds, and also can be followed by damaging aftershocks. Most people are killed by falling debris or crushed in buildings, not by ground motion.
- Consult the USGS website for more information.

Earthquake "Proofing" Your House

Many websites offer detailed information about making your house more earthquake proof. FEMA offers instructions for securing your water heater, furnace, heat pump, etc., and strengthening your house, foundation and roof. Go to http://www.fema.gov for details.

Background © Hemera Technologies/AbleStock.com/Thinkstock, Title ©iStockphoto/Thinkstock

Here is a summary of the efforts being undertaken by public and private agencies in the San Francisco Bay Area in response to the USGS published estimate of a 67% chance of another great earthquake in the region by 2020. These expensive measures contribute to the State's chronic budget problems.

- The California Department of Transportation (CALTRANS) is strengthening area bridges and elevated freeways.
- BART, the Bay Area Rapid Transit system, has an earthquake safety program that includes a vulnerability assessment and making the system "seismically safe."
- The East Bay Municipal Utility District (EBMUD) has proposed a seismic improvement program and has held public meetings to educate ratepayers and residents.
- The Pacific Gas and Electric Company (PG&E) replaces 100 miles of vulnerable natural gas pipelines each year, reinforces and replaces at-risk components of the distribution system (transformers, circuit-breakers, etc.), is retrofitting substations and other key buildings, and has implemented a company-wide seismic safety plan.
- Local governments including Berkeley, San Francisco, San Leandro, Sunnyvale, and Oakland, are using funds from general-obligation municipal bonds to strengthen public facilities and educate employees and citizens.
- San Francisco is spending $350 million in bond proceeds to finance upgrades to private unreinforced masonry buildings throughout the city.
- Corporations including Chevron, Hewlett-Packard, IBM, and Genentech are inspecting and reinforcing buildings.

Earthquake protection is expensive!

CONCLUSION

Americans are among the most mobile people on Earth. Chances are most of us will spend some time in a region at some risk from earthquakes. Nearly one out of every eight Americans presently lives in California. Knowing about earthquakes and what to do during earthquakes could be some of the most important information you take from this course.

POSTSCRIPT

Earthquakes from Injection or Withdrawal of Underground Fluids

Recent developments in oil and gas production involve use of the practice called "hydrofracturing" or "fracking." This process injects fluids into oil and/or gas-bearing deposits, often thousands of feet underground, to prop open fractures in the rock and make extraction of the oil and gas more effective. By using this method, the U.S. has increased its oil and gas reserves, and could become self-sufficient in oil and a net exporter of gas for the first time since 1970. Fracking, however, by injecting fluids at high pressure deep underground, could cause earthquakes. While no major quakes have so far been attributed to fracking, the widespread use of fracking should, under the precautionary principle, be accompanied by strict measures to monitor ground stability.

CHAPTER SUMMARY

1. Total losses to Japan from the 1995 Kobe earthquake were estimated at the equivalent of $200 billion. Such a loss would have represented about 5% of Japanese Gross Domestic Product.

2. The greatest earthquake in recent history, and perhaps the greatest of all time, occurred in 1976 when at least 240,000 and possibly as many as 600,000 people were killed in Tangshan, China.

3. An earthquake is a shaking of the ground, caused by movement of rock masses along faults within the crust.

4. Stress is force per unit area acting on rock. Rocks respond to stress by strain: that is by a change in volume, shape, or both.

5. The strengths of rocks (how much stress they can store) depend upon composition, water content, temperature, and confining pressure.

6. Faults are breaks in the crust along which movement has occurred.

7. The point of rupture inside the Earth where an earthquake originates is called the focus or hypocenter. Projecting the focus onto the surface defines the earthquake's epicenter.

8. Seismic waves result from earthquakes. Three common types of seismic waves are: Primary (P) waves, Secondary (S) or Shear waves, and Surface waves.

9. The epicenter of an earthquake can be accurately located by determining the distance to the epicenter from at least three seismograms.

10. Most earthquakes occur at plate boundaries. Earthquakes occurring within a plate are called intraplate earthquakes. Three regions in North America are prone to intraplate earthquakes: the South Atlantic Coastal Plain (Charleston, SC, 1886); the New Madrid Fault Zone (NMFZ) of the continental Interior (New Madrid earthquakes, 1811–1812); and the St. Laurence Seaway.

11. The Richter magnitude is an estimate of the amount of energy released by a given earthquake.

12. Moment magnitude is based mainly on the area of the earthquake rupture and the average fault displacement.

13. Intensity expresses the relative damage an earthquake causes to people and structures. An isoseismal map plots the intensity of a particular earthquake.

14. A thrust fault not previously known to exist because it never broke the surface is called a blind thrust.

15. The damage caused by an earthquake is related to many factors, only one of which is magnitude. Others include the proximity to the epicenter, the kind of rock or sediment underlying a site, and types of structures built on the site.

16. The main ways by which earthquakes cause damage are (1) by strong ground shaking, (2) by the effects of surface failures (surface failure, ground cracking, liquefaction, subsidence), and (3) by water waves.

17. Tsunami, often mistakenly called tidal waves, are seismic sea waves caused by displacement of the sea floor. The West Coast of the Americas is vulnerable to tsunami.

18. Fires can cause extensive damage during earthquakes. Fires caused most of the damage in the 1906 San Francisco earthquake, for example.

19. Slope failures are a near-universal result of earthquakes.

20. Liquefaction occurs when water-saturated material is shaken so violently that the sediment loses strength and begins to flow. Liquefaction effects are generally site-specific.

21. Earthquake prediction implies an ability to state with some degree of confidence when an earthquake is likely to occur.

22. Earthquake forecasting implies an ability to predict the likelihood of an earthquake of a given magnitude occurring within a specified time interval.

23. The study of prehistoric fault movements is called paleoseismicity.

24. Precursors are events or sequences of events that precede earthquakes.

25. Two theories used to explain earthquakes are the elastic-rebound theory and the dilatancy theory.

26. Evaluating earthquake hazard begins with seismic risk maps. These maps show the probability of ground acceleration for a given locality.

27. Damage from earthquakes can be reduced by education, by application of zoning and land-use planning, by avoiding the most earthquake-prone areas, and by designing and building earthquake-resistant structures.

28. Buildings are normally damaged during an earthquake in two ways, (1) failure of substrate beneath buildings, and (2) earthquake waves vibrating the building. All buildings have natural periods, and frequencies, of vibration.

29. When seismic wave frequencies are close to a building's natural frequency, it is said that the building and the ground motion are in resonance with one another. Resonance tends to amplify the shaking of a building.

30. Engineering practices and techniques can be used to "dampen" a building's response to shaking from seismic waves.

31. Building on weak substrate causes most building damage and up to 80% of deaths during earthquakes. To address this problem, microzoning is often employed. Microzoning is the mapping of earthquake-relevant ground characteristics in a municipality.

32. Earthquake protection is expensive, but citizens can protect themselves and their property with proper education.

KEY TERMS

aftershock

amplification

aseismic slip

blind thrusts

confining pressure

creep

design earthquake

dilatancy-diffusion theory of earthquakes

earthquake

earthquake forecasting

earthquake precursors

earthquake prediction

earthquake recurrence interval

elastic rebound theory of earthquakes

epicenter

focus

fracking

intensity

intraplate earthquake

isoseismal map

liquefaction

microzoning

moment magnitude

New Madrid Fault Zone (NMFZ)

normal faults

paleoseismicity

resonant frequency of buildings

reverse faults

Richter magnitude

rock strength

seismic risk map

seismic waves

seismogram

seismograph

strain

stress

strike-slip faults

thrust faults

tsunami

unreinforced buildings

wave amplification

REVIEW QUESTIONS

1. The response to stress is
   a. P waves.
   b. S waves.
   c. strain.
   d. amplitude.

2. Which is true concerning rock strength?
   a. Rocks are stronger at higher water content.
   b. Rocks are weaker at higher temperature.
   c. Poorly cemented sedimentary rocks are stronger than crystalline rocks.
   d. Rocks are weaker at higher confining pressure.

3. Breaks in the crust along which movement occurs are
   a. joints.
   b. bedding planes.
   c. dikes.
   d. faults.

4. The point of rupture inside Earth where an earthquake originates is
   a. a joint.
   b. the intensity.
   c. the magnitude.
   d. the focus.
   e. none of the above.

5. Instruments that record the passage of seismic waves are
   a. seismographs.
   b. seismograms.
   c. spectrometers.
   d. geodolites.

6. Most earthquakes occur
   a. at plate boundaries.
   b. within plates.
   c. at aseismic ridges.
   d. in the Mediterranean Sea.
   e. in the Atlantic Ocean.

7. An estimate of *the amount of energy released* by a given earthquake is
   a. a seismogram.
   b. hypocentral distance.
   c. Richter magnitude.
   d. Richter Intensity.

8. The *relative damage an earthquake causes* to people and structures is measured by
   a. intensity.
   b. Moment magnitude.
   c. Richter magnitude.
   d. an intraplate diagram.

9. The 1994 Northridge, CA and 1995 Kobe, Japan earthquakes both probably occurred along
   a. the same plate boundary.
   b. a subduction zone.
   c. a strike-slip fault.
   d. a blind thrust fault.

10. Intraplate earthquakes are capable of causing great damage because
   a. the stronger continental crust of plate interiors transmits seismic energy more efficiently than the fractured crust of plate boundaries.
   b. most people live in regions prone to intraplate earthquakes.
   c. they are much more common than earthquakes at plate boundaries.
   d. all of the above are true.

11. Which of these was an intraplate earthquake?
   a. Kobe, Japan, 1995
   b. Charleston, South Carolina, 1886
   c. all of the above
   d. a and b only
   e. b and c only.

12. Most damage to buildings from earthquakes results from
   a. tsunamis.
   b. being situated on a fault.
   c. ground shaking.
   d. fire.

13. The strength of ground shaking decreases with distance from an earthquake. This is called
   a. intensity.
   b. amplification.
   c. attenuation.
   d. interference.

14. Which rock type, other things being equal, would amplify seismic waves the most, resulting in greater damage?
   a. unconsolidated wet mud
   b. shale
   c. cemented sandstone
   d. unfractured granite.

15. The Marina District of San Francisco, heavily damaged during the 1989 Loma Prieta earthquake, is built on
   a. massive granite.
   b. the San Andreas fault.
   c. "reclaimed" Bay mud and debris from the 1906 earthquake.
   d. old salt ponds, paved over and landscaped.

16. Seismic sea waves caused by displacement of the sea floor are
   a. tidal waves.
   b. tidal bores.
   c. sturzstroms.
   d. tsunami.

17. Most of the damage from the 1906 San Francisco earthquake resulted from
    a. fires.
    b. falling debris from ornate buildings.
    c. ruptures along faults.
    d. land subsidence.

18. A type of mass-movement in which the cohesiveness of a soil is reduced by intense shaking is
    a. liquefaction.
    b. solifluction.
    c. creep.
    d. attenuation.

19. Which is true concerning earthquake precursors?
    a. All earthquakes have precursor events.
    b. Some but not all earthquakes have precursor events.
    c. Evidence on the validity of earthquake precursors remains largely anecdotal.
    d. Earthquake precursors can be used in earthquake prediction, but not in earthquake forecasting.

20. Seismic risk maps show
    a. the probability of ground acceleration for a given locality.
    b. how dangerous each dwelling will be during an earthquake.
    c. the likelihood of an earthquake of a given magnitude occurring in a jurisdiction.
    d. the intensity variation of an earthquake of a given magnitude in a locality.

21. When seismic wave frequencies are close to the natural frequency of a building,
    a. the resultant seismic waves do no damage to the building.
    b. the shaking of the building is reduced.
    c. the shaking of the building is amplified.
    d. the building is said to be "out of resonance."

22. Which buildings would most likely be most damaged during an earthquake, other things being equal?
    a. unreinforced buildings without foundations, on mudrock substrate
    b. reinforced buildings with foundations
    c. reinforced buildings with foundations on crystalline bedrock
    d. unreinforced buildings with foundations, on sandstone substrate

23. Much of Mexico City was damaged during the 1985 earthquake. The situation was made worse by the city's being largely situated on
    a. granitic substrate.
    b. substrate made of reclaimed construction material.
    c. sandstone substrate with many fractures.
    d. substrate consisting largely of fine-grained lake deposits.

FOOTNOTES

[1] http://www.seismo.unr.edu/ftp/pub/updates/louie/kobe/kobe-time.html
[2] http://special.scmp.com/chinaat50/Article/Fulltext_asp_ArticleID-19990928191 042300.html

CHAPTER OUTLINE

■ INTRODUCTION: ACCESS TO SAFE WATER

In 2013, about 1.1 billion of the planet's 7 billion people lacked access to safe water. And 2.6 billion did not have access to adequate sanitation. According to the United Nations (UN), 80% of all illnesses in developing countries are related to contaminated water. Global water demand is projected to grow by 40% over the next 15 years. If this issue is to be addressed, hundreds of billions of dollars must be invested in water infrastructure globally over the next five decades.

■ THE HYDROLOGIC CYCLE

Water at the Earth's surface moves through a series of *states* (liquid, solid, vapor) and *sites*. This relationship is called the water or **hydrologic cycle**. In all but the driest climates, water flow (runoff) at the surface collects into waterways, called (in order of increasing flow) *rills, creeks,* and *rivers.* Waterways may be dammed naturally or by humans (or beavers!) to form lakes. Water that sinks into the ground may adhere to soil particles or flow through interconnected pore spaces in soil, sediment, and rock to form groundwater. **FIGURE 7-1** shows the proportion of surface water in each of the various *reservoirs.*

From Figure 7-1 you can see that the overwhelming proportion is in ice and groundwater, with the atmosphere and surface streams holding only trivial amounts.

This should illustrate the vital significance of groundwater and lakes in water supply scenarios. However, at the same time biologists increasingly view river water and lake water as necessary for the health and even survival of critical terrestrial

In Other Words

The moment one starts using freshwater beyond the rate at which it can be replenished, the hydrologic cycle is endangered.

Dinyar Godrej
Water, the facts. In
New Internationalist,
2003, p. 18

Value-added Question 7-1.
From the data in Figure 7-1, determine by what factor the volume of water in freshwater lakes exceeds that of rivers. By what factor is fresh groundwater greater than water in rivers? Explain whether this surprises you.

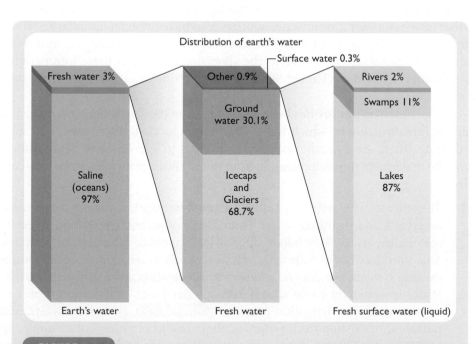

FIGURE 7-1 The global distribution of water. Where is the water? Modified from The USGS Water Science School. USGS.

While one might think diamonds are more valuable than water, it really depends on how thirsty we are.

Idea attributed to
Adam Smith
An Inquiry into the Nature and Causes of the Wealth of Nations

ecosystems. More and more of these sources thus may be unavailable to humans if an increasing value is placed on environmental protection.

WHAT IS WATER WORTH?

Most things that have economic value have substitutes to which users can turn if the price becomes too high: natural gas for oil, mass transit for autos, spinach for broccoli, and the like. But there is no substitute for water. Moreover, the demand for water is growing with human population and affluence. Until fairly recently, most people in wealthy countries had little inkling of the ecological, economic, and social consequences of too little clean water.

SUPPLY: AVAILABILITY OF WATER

Due mainly to population increase, global per capita water supplies by 2013 were one third lower than they were in 1970, and water quality was declining in many areas. Water scarcity has become a major obstacle to sustainable development. By United Nations estimates, two thirds of humanity will face shortages of clean freshwater by the year 2025. Even in wealthy (developed) countries, water problems may become serious. Some toxic organisms like *Cryptosporidium* are already resistant to chlorination, the most widespread technique used to purify drinking water. In the view of many hydrologists, better management of planetary water resources is humanity's most urgent global need.

The Water Sector

Worldwide, 260 rivers—the Danube, the Volga, the Nile, the Euphrates, the Ganges, and the Mekong, to name a few—flow through two or more countries. Likewise, neither lake basins nor groundwater aquifers recognize national boundaries. Three fifths of the world's population live in the watersheds of international freshwater systems. No global doctrine governs the allocation and use of these international water bodies. Even in the United States, water doctrines and laws vary from state to state.

The term **water sector** has been used to refer to the scientific, economic, and political *infrastructure* by which water supplies are located, distributed, and used worldwide. There are a number of points you need to appreciate to better understand the material in this section, aside from the scientific principles we will discuss below.

1. The water sector is extremely **capital-intensive**: huge investments are required to extract and distribute water. Should economic and population forecasts materialize, at least $180 billion U.S. will be required to expand global supplies. This sum does not include what will be required to rehabilitate or modernize existing systems, nor does it include routine operation and maintenance costs.

2. Water is wasted on a vast scale. Leaking water pipes, evaporation from reservoirs, inefficient irrigation practices (**FIGURE 7-2**), household waste, and pollution of existing fresh water supplies are all examples. For example, up to 20% of New York City's water supply disappears unmetered, presumably through leaking underground pipes. A project to fix some of the leaks was expected to cost more than $240 million.

3. The benefits of abundant clean, fresh water extend beyond those that are easily measured by economics. For example, should lack of water or polluted supplies generate large numbers of refugees, the costs will be borne by other segments of society or other nations.

4. Water distribution systems are natural monopolies, because it is inefficient or impractical to have numerous, competing irrigation systems or public water supply systems.

5. Competitive markets for water either don't exist or are inefficient.

6. The pricing of water is often if not usually inefficient. Governments are usually pressured to build and operate water distribution systems, and the prices charged are often not market-based. Charging a price for water that includes all costs is often politically difficult, for example the federal irrigation systems in the U.S. West, where politicians are fearful of offending large-scale water users. Riots may occur in poor countries if water prices are raised.

7. Although water circulates globally, its use by humans is local. Thus, water problems are best dealt with on the regional to local scale. National and international agreements are an essential framework for local and regional solutions, however. It has been said that "all of us live downstream," which means that, because water circulates globally, we are all vulnerable to the effects of water pollution.

■ DEMAND: FACTORS THAT AFFECT WATER USE

The amount of water used anywhere depends on supply and demand. The following factors determine water use.

1. Population numbers, distribution, and wealth. Demand grows with population and income, and urban dwellers have different patterns of water use than rural people.

2. Technology. Industrialization usually increases water demand and may lower water quality, thus reducing effective supply. Technological innovation has drastically lowered water use by industry in some countries, which use is often quite sensitive to price. Similarly, agriculture is the planet's greatest water user, and that use is commonly inefficient because water use is generally subsidized. Where prices or policies require efficiency, agriculture can drastically lower demand, as Israel has demonstrated. Water demand for agriculture in Israel dropped from 1400 million cubic meters (mcm) to 1180 mcm in 2007, even as demand for municipal supplies increased.

3. Economic conditions. A drop in the price of an agricultural commodity may lower demand for water to irrigate that crop. Or, overuse of a groundwater reservoir may destroy the supply, forcing abandonment of the land, or conversion to dryland farming.

4. Environmental conditions. Global climate change will likely have profound impact on regional water use. Pollution of water supplies lowers available water as effectively as drought.

Concept Check 7-1. CheckPoint: List and briefly discuss factors that determine water use.

■ INSTREAM VERSUS WITHDRAWAL USES OF WATER

For surface water supplies, a fundamental contrast in water use is that between instream use, and withdrawal or offstream use of water. **Instream uses** include recreational uses, protection of aquatic habitats, navigation, and generation of hydropower, although the latter may have profoundly negative impacts on the dammed water body. Many instream uses do not impair water's availability for future or downstream use.

Withdrawal removes water from the water body where it may be consumed and not returned, or ultimately returned either in a degraded form, or relatively unchanged. Irrigation water is the biggest **consumptive use**, and while municipal users may eventually return the water after use, it is usually degraded from its former state, which may render it less fit for reuse by others. **FIGURE 7-3** illustrates water use in the United States.

■ IMPACT OF CONTAMINATED WATER

According to the UN, as many as 3 million children die each year from issues related to polluted water or water scarcity. Moreover, it is one of the major factors driving mass migration, creating increasing numbers of human refugees.

Water pollution impacts human health in three main ways.

First, humans need access to at least 50 cubic meters (50,000 L, 13,200 gal) a year for survival, both for drinking and to meet minimal hygiene demands.

Second, contaminated water increases exposure to contagious diseases. Drinking or bathing in water containing animal or human waste facilitates transmission and proliferation of disease-bearing organisms. The most common water-borne infectious and parasitic diseases include hepatitis A, diarrheal diseases like cholera (**FIGURE 7-4**), typhoid, roundworm, guinea worm, leptospirosis, and schistosomiasis.

Value-added Question 7-2.
How much water does a human need *per day*? Report your answer in liters and gallons. Discuss whether you think this number represents a lot of water or a little.

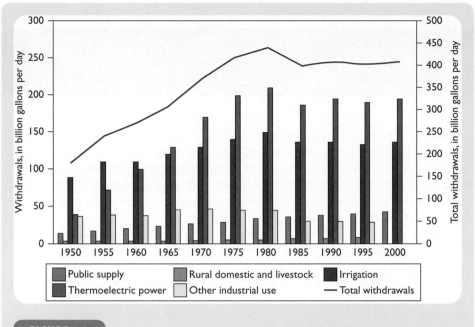

FIGURE 7-3 Water use in the United States to 2000. List the four largest water users in the United States by sector, in order. How has water use changed over the period measured? Modified from Water Use in the United States. *The National Atlas of the United States.*

In developing countries, according to the World Bank, diarrheal diseases alone cause an estimated 3 million deaths and 900 million illnesses annually, mainly affecting children.

Third, surface and groundwater can dissolve and/or transport inorganic and organic chemicals, heavy metals, algal toxins, and other unsafe substances. These produce chronic illnesses, cancers, birth defects and other mutations, and impaired immune system function from direct (ingesting contaminated water), or indirect (eating plants or animals harvested from contaminated waters) exposures. Trace levels of antibiotics, antidepressants, and sex hormones have been found in drinking water.

In Other Words

The biggest challenge to battling the cholera epidemic is sanitation.

Henry Gray
Water and Sanitation
Emergency Coordinator,
Médecins Sans Frontière

FIGURE 7-4 Open sewers in Accra, Ghana. Unsanitary conditions like this are major sources of waterborne infectious and parasitic diseases. © Pluff Mud Photography.

In the United States, water plays a critical role in the "politics, economy, and life of seventeen Western states," in the words of former Senator Bill Bradley. For example, Southern and eastern California is essentially a desert. Virtually no rain falls from April through October, yet the state is home to 38 million people and adding upwards of 400,000 new legal residents a year. About ⅘ of all the water used in California by humans is in agriculture, even though agriculture provides only about 4% of the state's gross domestic product (GDP) and only 1% of its jobs. Water has been heavily subsidized to agribusinesses for decades. Politically powerful agribusinesses historically paid cents on the dollar for federally subsidized water. In parts of California's 450-mi long Central Valley, 325,000-gallon units (1 acre-foot, or a volume equivalent to an area about the size of a football field one foot deep) could be had for about $7.50. At the same time, farmers in San Diego County were paying $400 for the same amount, and residents of Santa Barbara were charged $1,900.

Value-added Question 7-3. Recalculate Santa Barbara's cost ($7.50/325,000 gal) in dollars/cubic meter (one cubic meter = 264.2 gal). Oceanside, California, just north of San Diego, operates a desalination plant using brackish water and produces high-quality water by reverse osmosis for about $0.50 per m³. Compare the cost of Santa Barbara water to that of Oceanside.

But even municipal users got what amounted to a free ride. In many California cities, water use was, until recently, too cheap to meter. Even during droughts, water users paid a flat fee, meaning conservation couldn't be easily encouraged by raising water's price. This low price led to massive waste: most irrigation ditches were made of unlined dirt (**Figure CS 1-1**), which meant that up to a quarter of the water shipped from California's rivers to farm fields trickled into the soil before reaching its destination. Still more evaporated from ditches in 100-degree heat, and up to 25% yearly evaporated from vast reservoirs behind dams.

Low, heavily subsidized prices for water imposed heavy environmental costs on wildlife as well. Rivers like the San Joaquin routinely ran dry, and seasonal runs of salmon and steelhead disappeared, forcing commercial fishers out of business. Before passage of the Omnibus Water Bill in 1992, much of the water supplied by the federal government at subsidies of up to 90% was used to grow surplus crops; in other words, crops for which there was an insufficient market.

As if California didn't have enough water troubles, California's neighbors are growing in population and water use rapidly. In fact, Nevada and Arizona are two of the fastest growing states in the United States. Fueling California's growth throughout the twentieth century was cheap water—much of which came from the Colorado River. But California will have to reduce its use of that river's water as Nevada and Arizona claim their full share. A recent extraordinary agreement to restore the great Sacramento-San Joaquin Delta-San Francisco Bay ecosystem will require substantial water to be left in the water body, rather than exported to Southern California. The next twenty years could see precedence-shattering innovation to reduce water waste and restore ecosystems throughout the West. But even those measures might not be sufficient. Computer forecasts suggest the West, and especially California, could become more drought-prone over the next century, as the climate reverts to the climate of the pre-historic past as deduced from tree rings.

FIGURE CS 1-1 An unlined dirt irrigation ditch. © Tom Grundy/ShutterStock, Inc.

WATER AND DEVELOPMENT: THE PEOPLE'S REPUBLIC OF CHINA

While development is generally offered as a solution to traditional maladies such as infections and parasitic diseases, development could simply exchange one set of toxins for another, especially in its early stages, when insufficient investment is placed in maintaining environmental quality. As a result, "modern" diseases such as cancer and diabetes, and "epidemics" such as obesity and tobacco use generally increase alongside the decline of traditional maladies. This is the **risk transition concept**.

According to the World Resources Institute, this risk transition interval is illustrated by the recent history of the Peoples Republic of China (PRC). The double burden of declining (but still substantial) traditional maladies, and increasing modern ailments, poses a major challenge to China's healthcare system. The relationship between water and health in China may help shape China's present and future patterns of disease, and is probably a major factor in migration patterns now.

China ranks relatively low in water availability per capita (**FIGURE 7-5**). Approximately 700 million Chinese—about half the population—must use water that contains levels of animal and human waste that fail to meet minimum drinking water quality standards. Three reasons account for China's degraded water:

1. Rapid and unregulated expansion of industry;
2. Failure to invest in infrastructure to meet growing urban and suburban water needs; and
3. Expansion of agriculture, with reliance on sewage effluent to irrigate crops.

Industries scattered throughout rural China are gradually replacing obsolete state-owned concerns. These new concerns generally are unregulated and have few if any pollution-control facilities. As a result, contaminants such as organic matter,

FIGURE 7-5   A man collects dead fish in Donghu Lake in central China. According to local officials, the combination of pollution and high temperatures killed the fish. © AP Photo.

acids and bases, the nutrients nitrogen and phosphate, organic chemicals such as phenols, inorganic chemicals such as cyanide, and heavy metals such as lead, cadmium, mercury, and chromium, are among the major water pollutants detected in water bodies near rural residential areas. The impacts of these toxins fall heaviest on the very young and very old. Specific health effects can be found elsewhere in this text.

Urban areas likewise experience universally chronic, and typically acute, pollution loads. Each year, more than 30 billion tons of urban sewage is discharged into rivers, lakes, or seas in China with virtually none receiving even rudimentary treatment. Chinese researchers report that untreated sewage usually contains dangerous levels of pathogenic (disease-causing) microorganisms.

Most urban rivers are also polluted to varying degrees by inorganic and organic chemicals, many of them known to be carcinogenic (cancer-causing), **mutagenic** (changing the genetic material), or both. A cause of stomach and liver cancers is believed to be exposure to infectious agents such as hepatitis B. Some researchers found associations between gastric diseases and elevated cancer rates in areas irrigated with sewage as well as industrial wastewater. Some bacteria that are found in contaminated drinking water are known to cause ulcers. High levels of organic matter in drinking water means that when the water is chlorinated, toxic chemicals such as chloroform and trihalomethane are formed when chlorine reacts with the organic matter. These substances in elevated quantities are thought to pose a seriously increased cancer risk (up to one in one hundred), but chlorinating does protect against many microbial pathogens.

Agriculture is a major source of water pollution in China. Since the 1980s, government has allowed farmers to sell surplus produce, prompting many farmers to adopt so-called "**Green Revolution**" technologies common to Western industrial agriculture: intensive fertilizer and pesticide application together with use of hybrid seeds. As a result, the use of pesticides and fertilizers has been increasing in China. Between 1990 and 1996 alone China's production of active ingredients used to make pesticides increased by ten percent. After Japan, China is Asia's largest user of pesticides. Fertilizer use increased by at least one third during the 1990s, but inefficient application meant that as much as 70% of the fertilizer was wasted, leading to high levels of nitrate in groundwater. Exposure to nitrate in water can cause "blue baby syndrome" (methanoglobulinemia), a potentially fatal disorder so named because the diminished ability of the blood to carry oxygen results in blue skin coloration.

Concept Check 7-2. Apples imported from China are becoming a major global source, displacing locally produced fruit. Suggest reasons why this is happening.

Yet while water pollution is a persistent threat, more than half of Chinese cities are experiencing severe water shortages, which also could be called demand *longages*.

Concept Check 7-3. Explain whether you agree that the above may be a cause for the increases in Chinese nationals trying to gain entry into the United States and other countries.

Concept Check 7-4. Explain whether you agree or disagree with the idea that cheap imports of consumer goods from China could be subsidized by an insufficient investment in environmental protection in that country. What information would you like to have in order to answer the question more fully?

■ WATER AND GEOPOLITICS

Although there have been no major international conflicts strictly over water, the lack of clean water has been described by some scientists as potentially politically destabilizing. The diversion by India of a portion of the Ganges led to **salination** of the Ganges downstream, and to resultant "waves of environmental refugees" from Bangladesh into India, in the words of geographer A. T. Wolf of the University of Alabama. A dam planned in Laos on the Mekong River may lead to political friction in Cambodia, Thailand, and Vietnam, all downstream of the project. And, as we will see below, failure by Israelis and Palestinians to agree over distribution of groundwater could forestall hopes for Middle East peace.

Middle East Peace and Access to Water

Since the advent of Jewish repatriation in Palestine around the turn of the twentieth century, conflicts over the region's scarce water have been growing. They now pose one of the thorniest problems stalling a comprehensive Middle East peace. At the center of this conflict is the disposition of groundwater underlying the West Bank region, called by the Israelis Judea and Samaria, access to which is claimed by both Israelis and Palestinians. As if this were not severe enough, further complicating the issue is the rapid population growth in the region on both sides, much of which on the Israeli side is coming from immigration.

Groundwater is the most important source of fresh water in the region and comes mainly from **aquifers** (underground zones of water beneath the water table) located and recharged in the West Bank. Components of the *hydrologic cycle* that are relevant here are as follows: total annual precipitation is estimated at 2600 mm (102 in) of which only about 600 mm/year (24 in/yr) infiltrates into the ground. The **safe yields** of the three main aquifers shown in TABLE 7-1 .

Water Supplies in Gaza

Water supplies in the Gaza Strip (FIGURE 7-6) are even more restricted than supplies elsewhere. Groundwater is the only source of fresh water, and while the

TABLE 7-1	Safe Yields of 3 Main Aquifers in the Middle East	
Aquifer system	**Number of aquifers**	**Annual safe yield (mcm/y)**
Western Basin	2	350
Northeastern Basin	2	140
Eastern Basin	6	125

mcm = million cubic meters
The amount of groundwater in mcm/y that was, as of 1998, not being used was estimated at 78 mcm/y. This is the amount of additional supply available.

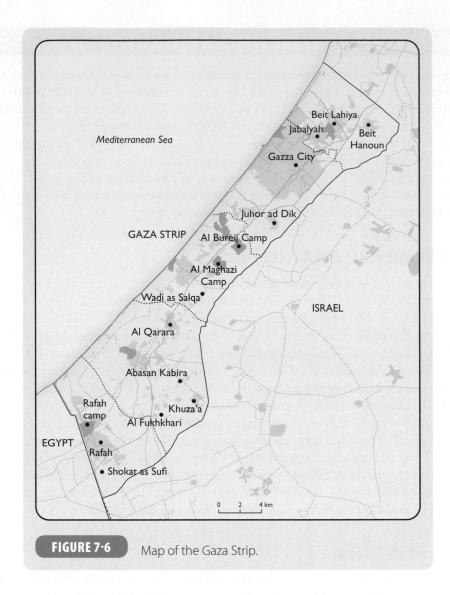

FIGURE 7-6 Map of the Gaza Strip.

present safe yield of the aquifers is estimated at 65 mcm/y, at least 100 mcm/y is being withdrawn. Thus, each of these systems is being "mined" at the rate of at least 35 mcm/y. Because the systems are in contact with seawater, the aquifers are being poisoned by **saltwater intrusion** (**FIGURE 7-7**), and will eventually be unusable, should present trends continue. The population of Gaza is increasing, so these trends will likely worsen. Near half of groundwater has been used to irrigate crops; if irrigation efficiencies increase, more supply would be available to protect the system from saltwater incursion.

Conclusion: Water and the Middle East

Water was one of the major areas of disagreement between Israelis and Palestinians during the negotiations leading to the interim "Oslo B" agreement, signed in Washington D.C. in 1995. While Israel for the first time recognized Palestinian water rights in the West Bank, the interim agreement left details to be negotiated in future deliberations.

In the meantime, total Palestinian population in West Bank and Gaza exceeded 2.8 million. While an average Israeli uses 60 gallons of water daily, based on pres-

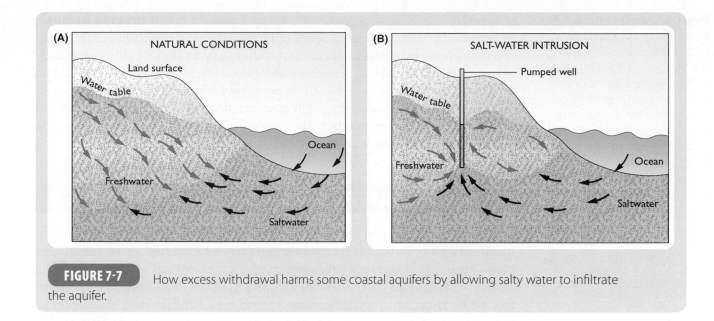

FIGURE 7-7 How excess withdrawal harms some coastal aquifers by allowing salty water to infiltrate the aquifer.

ent water allocations, a Palestinian has, on average, access to less than 15 gallons per day, or fewer than 25 cubic meters (25,000 L, 6600 gal) of water per year for domestic and industrial uses. Should a goal be set to allocate to each Palestinian the minimum of 50 cubic meters per year recognized to meet basic domestic needs, at least 70 mcm/y of supply will have to be found in a region which is already near capacity in terms of water use.

Approximately 25% of Israel's water supply comes from West Bank aquifers. The present allocation of groundwater is described as, "the abilities of the strong to impose their wills on the weak." Furthermore, Israeli surface water use has resulted in pollution and salination of the Jordan River downstream from Israeli diversions, rendering the water unsuitable for West Bank farmers.

Future water needs cannot be met by increasing supply from presently existing sources. Thus, conservation, new supplies, or a combination of both will be needed along with measures to address population growth, if the issue of Middle East water allocations is to be fairly settled. And if the issue is not fairly settled, it is unlikely that the region will experience a long-term decline in hostilities.

Concept Check 7-5. Population growth in Israeli-occupied Palestine and among Arab Israelis is among the highest on Earth. Suggest the long-term implication for this in terms of water supplies and political stability in the Middle East.

■ AUGMENTING SUPPLIES FROM BRACKISH OR MARINE SOURCES

Figure 7-1 showed that most of the planet's fresh water exists, for the present at least, as ice. Can we tap the seemingly limitless supplies in the world's oceans? It depends on two things: how much users are willing to pay, and how much environmental impact we will tolerate. Saudi entrepreneurs have already proposed towing icebergs from the Antarctic to the Middle East, and because some icebergs are as large as Rhode Island, even if they partially melt, they would provide appreciable

Value-added Question 7-4.
Israeli farmers used on average 5,000 cubic meters of water per hectare (10,000 m², or 2.47 ac) of irrigated crops in 1990. At a charge of $1.00 per cubic meter for reverse osmosis water, how much would they pay per hectare per year? How much would they spend per m²? How much water would they apply per m² per year?

water. More realistic sources involve using a process called **reverse osmosis** to pump salty water through a series of filters, removing dissolved matter. Many reverse osmosis plants are in operation, producing up to 40,000 cubic meters daily at costs of $0.60–$4.00 U.S. per cubic meter (1,000 liters). These costs are based on very low electricity prices of $0.04 U.S. per kilowatt-hour (kwh). Reverse osmosis produces wastewater containing high concentrations of dissolved solids, some of which, such as selenium, may be toxic. Disposal of such fluids could be problematic at high volumes, although disposal in saline aquifers is a likely alternative.

■ WATER SUPPLIES AND CLIMATE CHANGE

Studies of ancient trees in the Sierra Nevada have yielded valuable insight into climatic variability in California over the past several thousand years. Bristlecone pines, the oldest living trees, have rings by which the trees can be accurately dated, and some of them are 4,000 years old. Each year's growth yields a pair of rings: a narrow winter ring and a broader summer growth ring. Similarly, lake sediments can indicate variations in precipitation by study of fossil spores and seeds. Such studies confirm that California's climate over the past millennium has been extremely variable, with severe drought conditions persisting for intervals of time ranging from decades to centuries.

We can now supplement this historical record with computer forecasts of future climates based on changing levels of greenhouse gases: California's climate is forecast to become more extreme than in the recent past. Should such forecasts, or something even more severe, become the norm, such climatic extremes will critically stress the state's water storage and delivery system, and may force agonizing choices between protecting aqueous ecosystems and providing the massive amounts of water consumed by industrial agriculture, commerce, and municipalities.

California has two advantages: First, its residents are already familiar with drought conditions and are used to conserving water, and many of its residents are foreigners who are used to much less water consumption than native-born Americans. Second, agriculture uses 80% of the state's water, and much of that use is either wasted or is used to grow crops for which there is little demand.

■ CONCLUSION: WATER IN THE TWENTY-FIRST CENTURY

Water demand promises to be one of this century's most challenging issues. Human populations are projected to grow to 9 billion by the middle of the century, and as of 2013 many of the planet's citizens lack access to safe, abundant supplies. Growing food to feed the new arrivals will place additional strains on ecosystems already stressed by pollution, dams, and **mining of aquifers**. Finally, climate change induced in significant part by human activity will no doubt change precipitation and evaporation patterns regionally.

One reason for some optimism is ironically the waste inherent in most contemporary water use. Efficient pricing must replace water subsidies, at the very least, for the planet and its human passengers to have a chance at a safe, adequate water supply.

CHAPTER SUMMARY

1. In 2013, 1.1 billion of the planet's 7 billion people lacked access to safe water, and 2.6 billion did not have access to adequate sanitation.
2. Global water demand is projected to grow by 40% over the next 20 years.
3. Water at the Earth's surface moves through a series of states and sites. This relationship is called the water, or hydrologic, cycle.
4. The overwhelming proportion of global fresh water is in ice and groundwater, with the atmosphere and surface streams holding only trivial amounts.
5. Due mainly to population increase, global per capita water supplies by 2013 were one third lower than they were in 1970, and water quality was declining in many areas.
6. Three fifths of the world's population lives in the watersheds of international freshwater systems. No global doctrine governs the allocation and use of these international water bodies.
7. The term water sector has been used to refer to the scientific, economic, and political infrastructure by which water supplies are located, distributed, and used worldwide.
8. The amount of water used anywhere naturally depends on supply and demand.
9. For surface water supplies, a fundamental contrast in water use is that between instream, and withdrawal, or offstream, uses of water.
10. According to the UN, up to 3 million children die each year from issues related to polluted water or water scarcity. It is one of the major factors driving mass migration, creating increasing numbers of human refugees.
11. About ⅘ of all the water used in California by humans is used in agriculture, even though agriculture provides only about 4% of the state's GDP and only 1% of its jobs.
12. Low, heavily subsidized prices for water encourage waste, and impose heavy environmental costs.
13. The next 20 years could see precedence-shattering innovation to reduce water waste and restore ecosystems throughout the West.
14. Computer forecasts suggest the West, and especially California, could become more drought-prone over the next century, as the climate reverts to the climate of the pre-historic past as deduced from tree rings.
15. The relationship between water and health in China may help shape China's present and future patterns of disease, and is probably a major factor in migration patterns now. Approximately 700 million Chinese—over half the population—must use water that contains levels of animal-and-human waste that fail to meet minimum drinking water quality standards.
16. The lack of clean water has been described by some scientists as potentially politically destabilizing.
17. At the center of the Israeli-Palestinian conflict is the disposition of groundwater underlying the West Bank region. Water supplies in the Gaza Strip are even more restricted than supplies elsewhere. Future water needs cannot be met by increasing supply from presently existing sources.
18. Many reverse osmosis plants are in operation producing fresh water from brackish or salt water, producing up to 40,000 cubic meters daily at costs of $0.60–$4.00 U.S. per cubic meter (1,000 liters).
19. Water demand promises to be one of this century's most challenging issues.

KEY TERMS

acre-foot

aquifer

capital-intensive

consumptive
  water use

Green Revolution

hydrologic cycle

instream water use

mining groundwater

mutagenic

reverse osmosis

risk-transition
  concept

safe yield (of aquifer)

salination

saltwater intrusion

water sector

withdrawal use
  of water

REVIEW QUESTIONS

1. Global water demand is projected to grow by what percent over the next 20 years?
   a. 100%
   b. 80%
   c. 5%
   d. 40%

2. According to the United Nations, what percentage of all illness in developing countries is related to contaminated water?
   a. all
   b. 80%
   c. half
   d. 10%

3. Most fresh water is present
   a. in glaciers and groundwater.
   b. in the atmosphere.
   c. in soil moisture.
   d. in rivers.

4. The most widespread technique used to purify water is
   a. filtration.
   b. ozonation.
   c. chlorination.
   d. heating.

5. The statement that the water sector is capital-intensive means
   a. most infrastructure is around capital cities of developing countries.
   b. most infrastructure costs are related to the high cost of labor.
   c. huge investments are required to extract and distribute water.
   d. water is wasted on a grand scale.

6. The statement that water distribution systems are natural monopolies means
   a. to have many competing systems would be inefficient or impractical.
   b. water distribution systems should all be privately owned.
   c. water distribution systems should all be publicly owned.
   d. each individual should ideally be responsible for his/her own water quality.

7. Industrialization usually has what effect on water demand?
   a. It decreases water demand because most people live in cities.
   b. It decreases water demand because substitutes are readily available.
   c. It increases demand.
   d. It has no effect on demand.

8. The planet's greatest water user is
   a. hydroelectric plants.
   b. agriculture.
   c. municipal supplies.
   d. electric utilities.

9. Recreation, hydropower, and navigation are
   a. examples of offstream water use.
   b. examples of groundwater use.
   c. examples of instream use.
   d. examples of water waste.

10. About ⅘ of the water used in California by humans is used in
   a. bathing.
   b. toilet flushing.
   c. agriculture.
   d. lawn-watering.

11. One acre-foot of water is equal to about
   a. 325,000 gallons.
   b. 325 gallons.
   c. one billion gallons.
   d. 1,000 gallons.

12. The superposition of traditional maladies with maladies indicative of early development is being experienced by China. This concept is called
   a. the protodevelopment concept.
   b. the nascent malady concept.
   c. the risk-transition concept.
   d. the incipient development concept.

13. The two largest users of pesticides, a major source of water pollution, in Asia are
   a. China and Japan.
   b. China and North Korea.
   c. Japan and North Korea.
   d. Japan and South Korea.

14. Most of the water supply for the West Bank, Palestinian Authority, is
   a. the Jordan River.
   b. the Dead Sea.
   c. the Sea of Galilee.
   d. groundwater.

15. Water supply in Gaza is entirely from
   a. desalination plants in Egypt.
   b. bottled water provided by the United Nations.
   c. groundwater.
   d. the Nile River.

16. Aquifers in Gaza in contact with seawater are being mined. This leads to
    a. influx of seawater—saltwater incursion.
    b. influx of heavy metals.
    c. magnesium poisoning.
    d. an increasing risk of major earthquakes.

17. What percent of Israel's water is reported to come from West bank aquifers?
    a. all
    b. 25%
    c. 50%
    d. 75%

18. Reverse osmosis plants
    a. are plants that take up toxic metals like selenium.
    b. are plants that emit toxic metals when bruised.
    c. are facilities that may be used to purify oil in refineries.
    d. are facilities that may be used to purify water.

19. Computer forecasts suggest that the climate of California will become
    _____ as a result of global climate change.
    a. colder
    b. warmer
    c. wetter
    d. more extreme

8

CHAPTER OUTLINE

POINT-COUNTERPOINT

Groundwater Dispute in Arkansas

A groundwater dispute is playing itself out in Arkansas, and the implications are national. Rice farmers in Arkansas are overusing one of the state's biggest *aquifers* (underground zones of water beneath the water table), the Alluvial Aquifer (**Figure PC 1-1**). The dispute is over a $200 million federal plan to provide $300,000 per rice farmer to divert water from the White River to irrigate rice fields. Rice farmer John Kersieck says, "We really don't have a water problem. There's plenty of water in the river. They've just got to divert it."

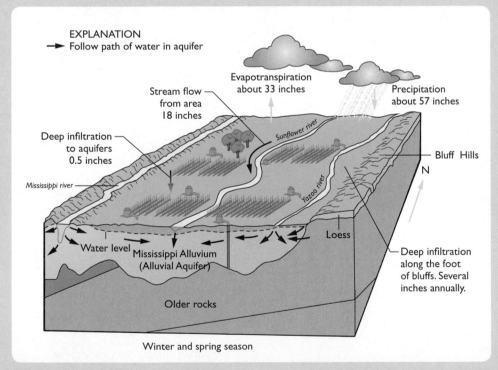

FIGURE PC 1-1 Recharge of the Alluvial Aquifer in Arkansas (black arrows) may take place from surface streams in winter. The aquifer is used for irrigation in the dry summer months. Rice cultivation uses a great deal of water. Modified from: Sumner, D.M., and Wasson, B.E., 1984, Geohydrology and simulated effects of large ground-water withdrawals on the Mississippi River Valley alluvial aquifer in northwestern Mississippi: U.S. Geological Survey Open-File Report 84-822, 82.

Those opposing the program cite environmental and financial reasons. G. Alan Perkins, an Arkansas attorney, talks of an "imminent aquifer failure." Attorney David Carruth favors a program that supports teaching farmers to use water more efficiently and expressed concerns over protecting the White River.

The Alluvial Aquifer underlies 32,000 square miles (83,000 km²) of Arkansas, Tennessee, Missouri, Mississippi, and Louisiana. The problem arose in Arkansas because even with abundant rainfall, farmers have relied heavily upon irrigation to increase rice yields. In one Arkansas county alone withdrawals from the Alluvial Aquifer increased from 115 million gallons/day (mgd) to 560 mgd between 1965 and 2000.

Water in the aquifer is used mainly for rice and fish farming. It is not a priority for municipal supplies because of its "hardness" (see later in the chapter), a high content of elements like iron and calcium. But even without municipal use, by 2015 the aquifer is projected to be functionally drained and unable to supply the 1,000 farms presently cultivating 250,000 acres.

Who are the stakeholders—groups of individuals with an interest in the outcome—here? Should the farmers be allowed to divert river water? What are the implications—economic, environmental, and otherwise—of diverting the water? Of doing nothing? Which choice would you make?

■ INTRODUCTION

During the past 50 years, groundwater depletion has occurred in many countries around the world. Withdrawing water faster than it can be replenished is called *mining*. Prolonged mining can irreversibly drain aquifers. Examples include the Ogallala Aquifer, located in the High Plains of the central United States (see later in the chapter), where more than half the groundwater in storage has been withdrawn in some areas; and the North China Plain, where overuse of shallow aquifers is depleting deep aquifers with wells more than 1,000 m (3,280.8 ft) deep. Overuse of groundwater has also birthed grandiose schemes for interbasin transfer of river water.

In arid regions, much of the groundwater being used today was emplaced during wetter conditions thousands of years ago and is not presently being replaced at a rate equal to consumption, raising serious concerns about present withdrawal rates.

The Middle East is one of the Earth's most unstable political regions, with some of the world's highest rates of population growth as well as groundwater use. Israel and the Palestinian Authority are equally dependent on groundwater for irrigation and public water supplies (discussed elsewhere in this text). A comprehensive peace in that region is unlikely without a comprehensive groundwater use treaty.

Global groundwater *depletion* (use in excess of natural recharge rate) has been significant enough to contribute to sea-level rise during the past century, as a result of water pumped from wells returned to the sea, either by runoff or by evapotranspiration followed by precipitation (see below).

Groundwater *contamination* poses some of science's greatest challenges. Aquifers underlying nuclear-weapons production sites are some of the most contaminated in the world. One of these sites, the Hanford atomic-weapons reservation adjacent to the Pacific Northwest's Hanford Lab, is the most contaminated subsurface site in the United States (FIGURE 8-1). Cleanup of the site—if possible at all— could take decades and cost billions of dollars. Such sites represent the "avoided costs" of nuclear weapons programs over the past half-century.

Our misuse of groundwater represents, in the words of University of Arizona professor Robert Glennon, an example of "the unlimited human capacity to ignore reality."

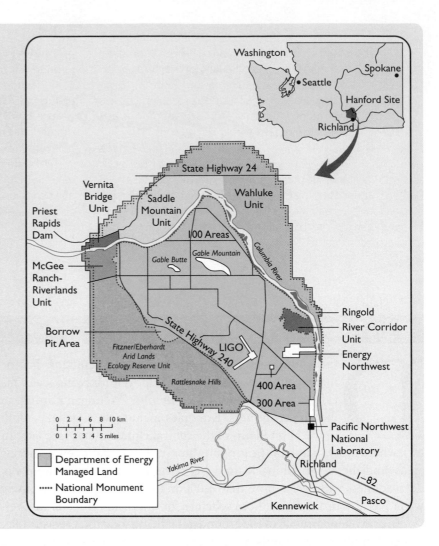

FIGURE 8-1 Location of the Hanford weapons site, Washington State.

■ GEOLOGIC PRINCIPLES

Groundwater and the Hydrologic Cycle

Recall that *infiltration* is one component of the **hydrologic cycle**. Precipitation that does not immediately run off, or that is not evaporated or transpired by vegetation, may enter pores in soil and sediment. Water that seeps below the root zone of plants becomes groundwater; this process is called **recharge**. Layers of rock or sediment that store and transmit water are called **aquifers** (FIGURE 8-2). What we call groundwater, or subsurface water, is a part of an interrelated *system* that includes the geologic **substrate** (rock, sediment) containing that water, the *boundaries* of flowing underground water masses, the *sources* of the water (such as recharge), and *sinks* (places where groundwater is lost or flows out of the system, such as springs, flow between aquifers, or wells; FIGURE 8-3).

Water may be stored within an aquifer for geological time intervals (thousands to millions of years). Under natural conditions, the travel time of water from areas of recharge (sources) to areas of discharge (sinks) can range from less than a day to more than a million years. Water stored within the system can range in age from recent precipitation, to water trapped in the sediments when they were deposited during the geologic past.

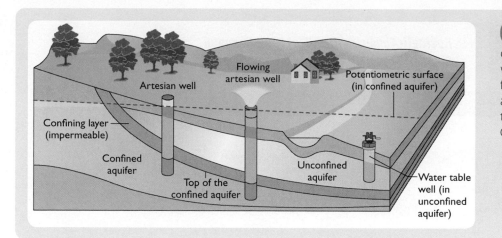

FIGURE 8-2 Model of an aquifer and associated wells. A potentiometric surface forms when water in a confined aquifer is under pressure from the weight of overlying rock. Courtesy of USGS.

Describing Groundwater Flow

Porosity and Permeability

Porosity is a measure of the amount of pore space in material. **Hydraulic conductivity** (K) is a measure of the ability of a substance to transmit a fluid. See **TABLE 8-1**. Permeability is similar to hydraulic conductivity but takes into account the properties of the fluid being transported: For example, natural gas is transported more freely through a substance of a given K than is relatively more viscous water. **Permeability** depends on the size, shape, orientation, and abundance of pore spaces, and on the *extent to which these pores are connected*. Materials that freely conduct fluids are said to be permeable: those that don't are impermeable (**FIGURE 8-4**).

Concept Check 8-1. CheckPoint: Distinguish between porosity and permeability of a material.

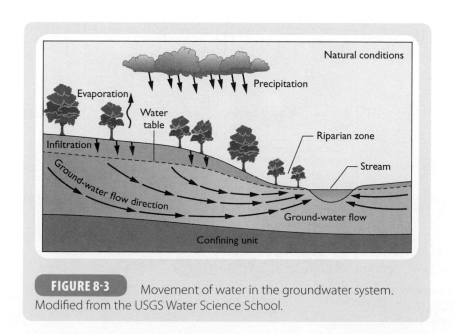

FIGURE 8-3 Movement of water in the groundwater system. Modified from the USGS Water Science School.

TABLE 8-1

Hydraulic Conductivities (K) of Some Common Geologic Materials

Substance	Hydraulic Conductivity (m/sec)*
Sediment	
Gravel	3×10^{-4} to 3×10^{-2}
Coarse sand	9×10^{-7} to 6×10^{-3}
Medium sand	9×10^{-7} to 5×10^{-4}
Fine sand	2×10^{-7} to 2×10^{-4}
Clay	1×10^{-11} to 4.7×10^{-9}
Sedimentary rock	
Karst limestone	1×10^{-6} to 2×10^{-2}
Sandstone	3×10^{-10} to 6×10^{-6}
Shale	1×10^{-13} to 2×10^{-9}
Crystalline rock	
Permeable basalt	4×10^{-7} to 2×10^{-2}
"Impermeable" basalt	2×10^{-11} to 4.2×10^{-7}
Fractured ig/mm rock	8×10^{-9} to 3×10^{-4}
Weathered granite	3.3×10^{-6} to 5.2×10^{-5}
Unfractured ig/mm rock	3×10^{-14} to 2×10^{-10}

*To put this value in perspective, there are 31,536,000 seconds in a 365-day year. We can also express this as 3.1×10^{7}. (So for a value of K = 3.1×10^{7}, water will move 1 meter in a year.)

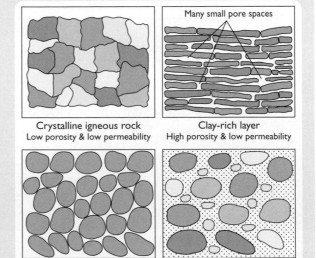

FIGURE 8-4 Illustrations of porosity and permeability.

Aquifers, Aquitards, and Aquicludes

As we stated above, layers of rock or sediment that store and transmit water are called *aquifers*. **Aquitards** are impermeable units that may be able to hold water but not readily transmit it. Clay-rich shale and pumice are good examples of aquitards. (Some hydrologists use the term **aquiclude** for a substrate that can neither hold nor transmit water. Unfractured granite is an example.) Shale may be over 50% porous, but the pore spaces are so small that permeability is almost nonexistent.

Concept Check 8-2. In shale gas deposits, the shale is porous and the pore spaces contain natural gas, but the rock is not permeable. How do you think scientists could 'unlock' these rocks and extract the gas?

Although some faults can retard groundwater flow, rock fractures, especially intersecting sets of fractures, often form pathways through which groundwater can travel, regardless of the nature of the original rock. Wells drilled into such rocks can yield a significant supply of groundwater under favorable conditions (**FIGURE 8-5**).

Once in the groundwater zone, water will flow under the influence of gravity at rates that can vary by orders of magnitude (powers of ten). As you will see, flow rates are usually slow except in areas underlain by carbonate rocks, where caves, sinkholes, and fractures may allow very rapid movement. Factors that control flow in groundwater systems will be discussed in detail below.

There are two types of aquifers: *unconfined* and *confined*.

In an **unconfined aquifer**, the water is not confined by a low-permeability overlying bed and is recharged from the surface. Attributes of near-surface groundwater systems are shown in **FIGURE 8-6**. The **unsaturated zone** is the surface zone where pores are empty or where water clings to the sides of pores by *capillary attraction,* that is, adhesion of the water molecules to the solid surface of the pores. Pore spaces in the underlying **saturated zone** are filled with water. Separating these two zones is the **water table**: thus, *the water table marks the top of the zone of saturation.* Near-surface aquifers are especially prone to contamination by human activity, because any contaminant dumped on the surface could enter the zone of saturation.

The water table roughly parallels the topography. The *elevation* in feet or meters above sea level of the water table fluctuates with precipitation in humid regions like the eastern United States, Western Europe, and southern China (**FIGURE 8-7**).

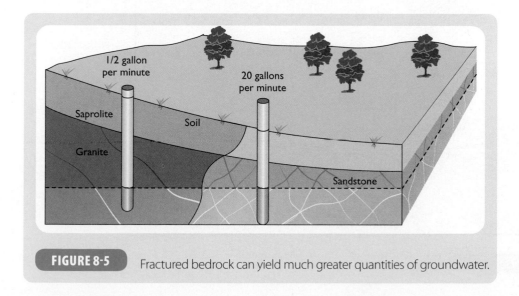

FIGURE 8-5   Fractured bedrock can yield much greater quantities of groundwater.

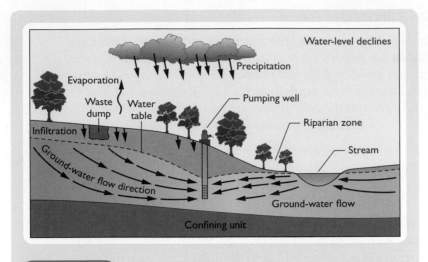

FIGURE 8-6 A near surface groundwater system. The aquifer is recharged directly by surface precipitation. Note that such systems can be easily polluted by surface sources, like waste dumps. Modified from USGS.

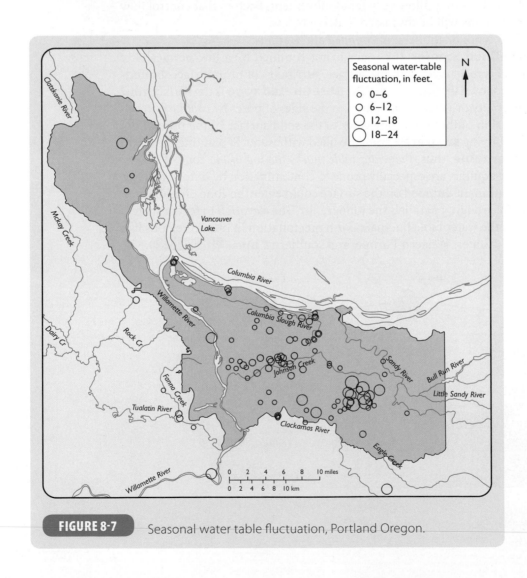

FIGURE 8-7 Seasonal water table fluctuation, Portland Oregon.

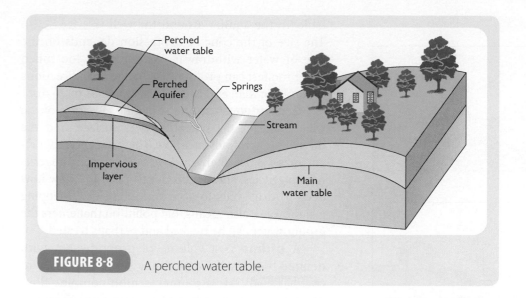

FIGURE 8-8 A perched water table.

Perched water tables, that is, a water table positioned over the normal water table, can develop locally (**FIGURE 8-8**). Note the development of springs where the perched water table intersects the surface in Figure 8-8.

Aquifers that are sandwiched between impermeable layers are called **confined aquifers**. While surface aquifers are characterized by a water table, confined aquifers are characterized by a hypothetical level known as a **potentiometric surface**. A potentiometric surface is *the elevation to which water in the confined aquifer would rise if a well were sunk into it*. This surface exists because pressures on confined aquifers are much higher than atmospheric pressure, because water anywhere in the aquifer must bear the weight of all the water in it that is higher in elevation as well as the overlying material. A well allows water in the aquifer to flow towards the zone of lower pressure. A well drilled into a confined aquifer allowing water to rise above the top of the aquifer is called *artesian,* from the region in the south of France where Roman engineers first developed such systems as public water supplies.

Some artesian systems can be sources of high-quality water. Such water has no toxic substances and low concentrations of Total Dissolved Solids (TDS), but many deep artesian aquifers have TDS values too high for use as drinking water, though they may be acceptable in industrial applications or for irrigation.

Water Movement in Aquifers

Whether in confined or unconfined aquifers, we can view groundwater flow as a balance of forces, much like our view of the causes of slope failure, discussed elsewhere in the text. The driving force for groundwater flow is *gravity*. The resisting force is *friction* between water and substrate pore surfaces. Groundwater flow is very different from flow in rivers. In most aquifers, water flows along torturous pathways among tiny sediment grains and along minute cracks, dissipating much of the potential energy provided by gravity. This means that groundwater flow is usually very slow: orders of magnitude slower than streamflow. One estimate of "average" groundwater flow rate is 30 meters (98 ft) per year, while the Willamette River in Oregon can flow at a rate of 30 centimeters (12 in) per second.

Ideally, water in aquifers flows at right angles to water table contours in unconfined aquifers. Because water will flow from high pressure towards lower pressure, wells drilled into aquifers can create **cones of depression**, a valley in the water

Value-added Question 8-1.
How many times faster does the Willamette flow than "typical" groundwater? There are about 3.1×10^7 seconds in a year.

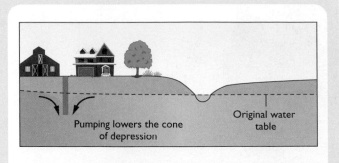

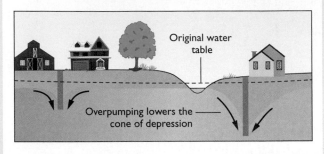

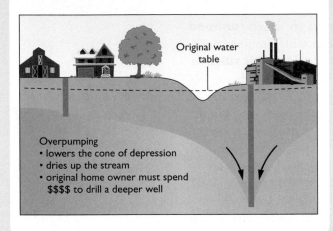

FIGURE 8-9 Direction of groundwater flow is shown here, along with a cone of depression, resulting from too rapid withdrawal of water from the aquifer.

table, in the aquifer around the well (FIGURE 8-9). The size of the cone of depression depends on the rate of water withdrawal. Withdrawing too much water too fast can profoundly change flow directions around the well, and cause adjacent wells to go dry.

Groundwater maps show the elevation of the water table or the potentiometric surface, and the direction of groundwater flow. These are very important maps, since many individuals, farmers, government agencies, and organizations need to know how deep to drill wells to access groundwater. Mapping groundwater also means that pollution that enters the groundwater can be tracked and perhaps treated.

A **plume** is a volume of groundwater, with defined boundaries, containing some contaminant (FIGURE 8-10). Most plumes exhibit complex flow. They pose a hazard to water users sometimes far from the source of the contamination.

Recharge of unconfined aquifers may be *diffuse* or *localized*. Diffuse recharge occurs over a wide area, and occurs via infiltration of *precipitation* through the unsaturated zone. *Localized* recharge is recharge from surface-water bodies (rivers, lakes, etc.; FIGURE 8-11). In general, the importance of diffuse recharge decreases as the *aridity* (dryness) of a region increases. This makes groundwater in arid regions depend on through-flowing rivers, or even leaking irrigation canals, for replenishment. For this reason, as water demand increases, greater stress is placed on both surface water and groundwater in arid regions (See our discussion of Albuquerque, NM later in the chapter.)

Recharge *rates* for surface aquifers depend on precipitation, bedrock geology, soil type, and surface topography. For example, aquifers in flat regions generally receive more recharge than aquifers underlying surfaces with high relief, because water is more likely to run off in the latter case.

Precipitation in humid regions that infiltrates does not necessarily become recharge, because it may be intercepted in the soil zone and eventually returned to the atmosphere by evaporation or (plant) transpiration. The percentage of precipitation that becomes recharge varies widely. It is determined by factors like weather, permeability, and composition of surface soils, nature or absence of vegetation, local topography, presence and style of development, and depth to the water table. For example, over a six-year period United States Geological Survey (USGS) scientists estimated recharge in the Great Bend area of central Kansas to be only 10% of the yearly precipitation of 58.5 cm. In some years, no recharge occurred at all.

Groundwater flows through aquifers from *recharge areas* to *discharge areas*. During rainy seasons, water is added to the aquifer and water pressure in the aquifer increases, increasing discharge rate. As precipitation slows, recharge declines, pressure drops, and discharge drops. In humid regions like the eastern United States,

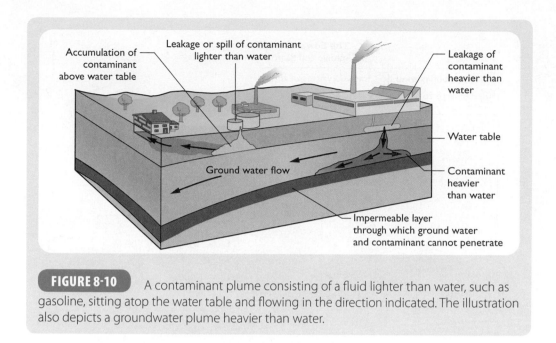

A contaminant plume consisting of a fluid lighter than water, such as gasoline, sitting atop the water table and flowing in the direction indicated. The illustration also depicts a groundwater plume heavier than water.

natural recharge usually occurs during winter and spring, and drops off during summer. Droughts can lower surface groundwater levels precipitously. Wells that have been drilled to insufficient depths can, and often do, run dry.

Confined aquifers are more or less sealed at the top, but they may receive some recharge from leaks, that is, water that enters the aquifer from adjacent strata. Other than from leaks, they are recharged via precipitation in discrete *recharge areas.* The Edwards Aquifer is the sole source of water for over 1.3 million people in and near San Antonio, Texas, as well as for ranchers and farmers throughout south-central Texas (**FIGURE 8-12**). Recharge areas are much like stream watersheds in that *recharge areas can be far removed from where the water is used.* Land-use practices in recharge areas can degrade water quality and can impact users far away. For examples of how humans affect groundwater recharge, see the following section.

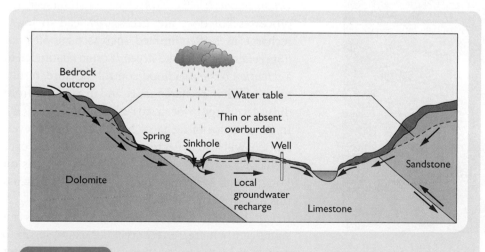

FIGURE 8-11 Localized groundwater recharge. Modified from Hampson, P.S., Treece, M.W. Jr., Johnson, G.C., Ahlstedt, S.A., and Connell, J.F., 2000, Water Quality in the Upper Tennessee River Basin, Tennessee, North Carolina, Virginia, and Georgia 1994–98: *U.S. Geological Survey Circular* 1205, p. 32.

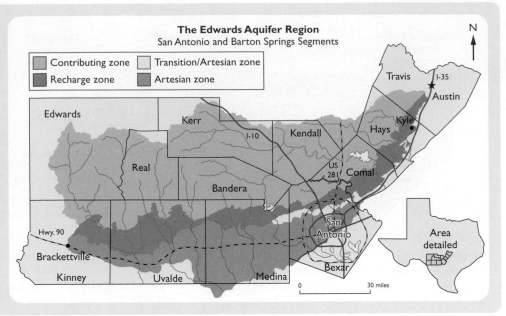

FIGURE 8-12 The Edwards and Trinity Aquifers, with recharge area near San Antonio. Suggest some uses government officials, charged with responsibility for planning and regulating development, might make of such a map. In the contributing zone, surface water flows towards the recharge area. The Transition/Artesian and Artesian zones are buried by younger rocks. Modified with permission from Gregg Eckhardt.

The Edwards Aquifer Region
San Antonio and Barton Springs Segments

Contributing zone   Transition/Artesian zone
Recharge zone   Artesian zone

Human Practices That Affect Recharge

It is easier to identify human practices that influence recharge than to *quantify* these effects. Here are some examples.

- *Clearing of vegetation* may lead to an increase in recharge rates (e.g., the Niger Basin in Africa. To appreciate why, think of the hydrologic cycle, and recall that plants transpire water though their leaves).

- *Irrigation* can increase recharge rates but also may cause **salination** of soils and aquifers when evaporation of water leaves salts and fertilizer behind in the soil, as in parts of California's San Joaquin Valley and southern Iraq.

- *Urbanization* (**FIGURE 8-13**) changes natural recharge processes but does not necessarily lead to lower recharge, as is often intuited. Increased runoff from *impervious* surfaces like streets is often channeled to a *retention basin* as a flood control measure, resulting in a shift in recharge areas, and in a change from diffuse recharge to localized recharge (**FIGURE 8-14**). In the absence of recharge basins, however, development will lead to decreased recharge. Paving over recharge areas is a serious concern for officials on Long Island, New York and elsewhere. We discuss this in more detail below.

- *Irrigation canals*, *leaky water mains,* and *sewers* can significantly increase recharge. For example, up to a quarter of the water transmitted through water mains in Göteborg (Gothenburg), Sweden's second largest city, leaks out of pipes. Old water distribution systems in the United States likewise commonly lose over 10% of municipal water to leaks.

FIGURE 8-13 Urbanization in Southern California. Estimate the percentage of impervious surface (roofs, streets, and driveways) in the housing tract. © Ron Chapple Studios/Thinkstock.

Groundwater Budgets

The amount of water "stored" in an aquifer can be expressed using this simple (yet frequently overlooked) relationship:

Change in storage = Amount added (recharge)
− Amount withdrawn (depletion)

The *recharge rate* over time thus must be determined in order to calculate a **safe withdrawal rate**; that is, one that will not deplete the aquifer or lead to undesirable results such as formation of a cone of depression, subsidence of the surface, or **saltwater incursion**. Although surface water supplies are renewable, groundwater supplies may be renewable or not, depending on the nature of the aquifer, the rate at which water is withdrawn, and on the extent to which water enters the aquifer. Some confined aquifers are recharged at such slow rates that withdrawal of large volumes of water is irreversible. *This is one of the most important facts that you should remember about groundwater.*

FIGURE 8-14 A retention basin, constructed to intercept and trap runoff. © EugeneF/ShutterStock, Inc.

We mentioned that in many regions of the world groundwater is *mined;* that is, *withdrawn at rates that exceed recharge.* Mining will eventually deplete aquifers, rendering them useless, so determining the recharge rate is very important.

However, recharge rates are unknown for many aquifers. For example, the aquifer underlying El Paso, Texas and adjacent areas in Mexico has been tapped at unsustainable rates for decades, judged by the rate at which water levels are dropping (**FIGURE 8-15**). Officials in the United States and Mexico differ by an order of magnitude in their estimates of recharge rate, however. The Mexican National Water Commission estimates a recharge rate of 235,000 acre-feet/year (AF/Y), while the Texas Water Development Board estimates a recharge rate of 24,000 AF/Y!

But even if recharge rates *were* known, depleted aquifers are often still mined. This situation is of life or death importance in the Middle East. In the Gaza Strip for example, groundwater pumping exceeds recharge by up to 18 million cubic meters a year, causing water levels to fall steadily in wells. And in Jordan, an important aquifer was mined to such an extent during 1985–1994 that the water level fell 10 meters.

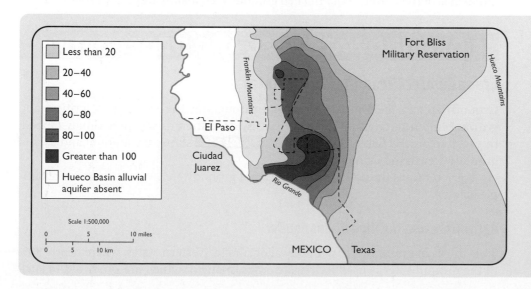

FIGURE 8-15 Decline in El Paso alluvial aquifer 1903 to 1989, in feet. Modified from Ryder, Paul D. *Ground Water Atlas of the United States*, HA 730-E. USGS.

Falling water levels in aquifers impose additional costs on users in that, as natural pressure is lost, the water must be pumped out using electricity or diesel fuel.

Surface Water–Groundwater Interaction

Groundwater and Stream Flow

Groundwater and surface water may be complexly interconnected. **Springs** may form where aquifers break the ground surface. This water may then run into streams, becoming runoff. Groundwater may recharge stream flow during summer months when stream discharges are low, and streams can recharge groundwater during times of high discharge.

Stream flow has two components: **base flow**, the amount of a stream's discharge provided by groundwater, and **runoff**, from precipitation. Without precipitation, the stream's flow is dependent solely upon groundwater. Streams with a base flow component flow all year and are called *perennial* streams (see flood information elsewhere in this text); those without base flow are called *ephemeral* streams. Many smaller streams in the more arid Southwest are ephemeral. Because stream flow and groundwater are connected, excessive groundwater withdrawal could cause a stream to run dry during the dry season.

Once thought to be of little consequence and thus ignored, the interactions of groundwater with lakes, wetlands, estuaries, and oceans now are recognized as important processes. For example, coastal groundwater discharge is equivalent to as much as 40% of total stream discharge to the ocean in summer along the coast of South Carolina, USA.

Groundwater and Coastal Ecosystems

Many varieties of coastal environments and ecosystems are found along the coasts of North America. The importance of groundwater as a source of freshwater and nutrients to these ecosystems has only recently begun to be quantified. Fresh groundwater discharge to coastal wetlands and estuaries is an important process because of the *habitats* these areas provide for plants, microorganisms, fish, and other wildlife, and because of their extremely high productivity. Examples of areas where inputs from groundwater are a concern are: Chesapeake Bay; Florida Bay; Cape Cod, Massachusetts; and Great South Bay, Long Island. In Chesapeake Bay, as much as one third of the Bay's excess nitrogen is derived from contaminated groundwater, mainly from farm runoff.

■ GROUNDWATER USE AND DEGRADATION

Groundwater Use in the United States

Groundwater in the United States, and throughout the world, is of critical and growing importance as a water supply, as surface streams are increasingly tapped, reserved to protect habitats, or polluted. **FIGURE 8-16** shows the amount and uses of groundwater in the United States.

Agriculture is the nation's biggest user of groundwater. We illustrate one example below.

Agriculture and the High Plains Aquifer

Two-thirds of groundwater withdrawal occurs in Western states, where nearly 80% of the groundwater is used to irrigate crops, many of which are produced in such

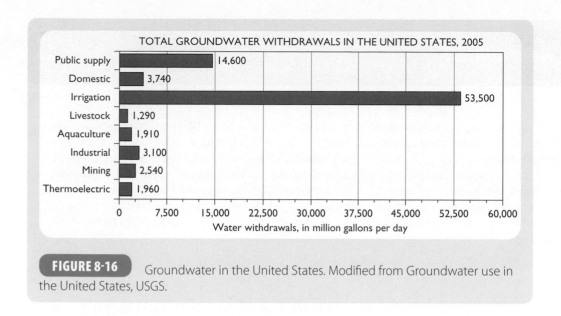

TOTAL GROUNDWATER WITHDRAWALS IN THE UNITED STATES, 2005

Public supply	14,600
Domestic	3,740
Irrigation	53,500
Livestock	1,290
Aquaculture	1,910
Industrial	3,100
Mining	2,540
Thermoelectric	1,960

Water withdrawals, in million gallons per day

FIGURE 8-16 Groundwater in the United States. Modified from Groundwater use in the United States, USGS.

surplus that their growers are basically wards of the U.S. Department of Agriculture.

One area where groundwater withdrawal threatens the future of agriculture is in the region of the United States known as the "High Plains." One of the most extensive aquifers in North America underlies this region: the High Plains, or Ogallala, Aquifer (HPA; **FIGURE 8-17**). The HPA underlies parts of eight states. Water levels in the HPA fell steadily between 1980 and 1998 (**FIGURE 8-18**). Groundwater withdrawals for irrigation were 18.0 million acre-feet in 1980 and 15.7 million acre-feet in 1990, according to the USGS.

Municipal Demand for Groundwater

More than half of the U.S. population depends on groundwater for their drinking water, including coastal cities without access to adequate surface water supplies. Miami, Florida; San Antonio, Texas; and Memphis, Tennessee are prime examples of large cities that depend on groundwater, but there are many others. Some states are even more dependent: New Mexico and Florida tap groundwater for more than 90% of public supplies (see Albuquerque, New Mexico later in the chapter).

One of the best-studied groundwater systems used for municipal supplies is the groundwater of Nassau and Suffolk Counties, Long Island, New York (**FIGURE 8-19**).

Groundwater Contamination

Contamination of groundwater is usually the result of human activities on the land surface. For example,

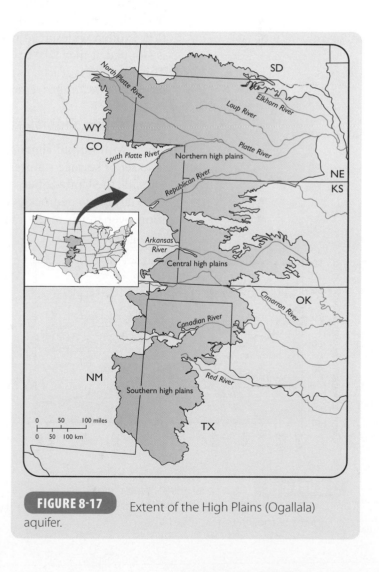

FIGURE 8-17 Extent of the High Plains (Ogallala) aquifer.

Most people have heard of "hard" water, but do you know what it means? Hardness is a prime consideration in selecting a municipal water supply.

Water that contains abundant calcium (Ca) and magnesium (Mg) is said to be **hard**. The hardness of water is expressed in terms of the amount of calcium or magnesium carbonate that would be formed if the water were evaporated. Water is soft if it contains 0 to 60 mg/L of "hardness," moderately hard from 61 to 120 mg/L, hard between 121 and 180 mg/L, and very hard if more than 180 mg/L. Very hard water is not desirable for domestic uses because it will deposit minerals on the inside of pipes, boilers, and tanks. Hard water can be softened, but it is not always desirable to do so. One way to soften water is to use filters that exchange the Ca and Mg for sodium (Na), but intake of high-sodium water is not desirable for people with high blood pressure.

agricultural activities can add soluble nitrogen (as nitrate, for example), pesticides, herbicides, and fungicides to groundwater. Because groundwater usually moves slowly, except in regions underlain by carbonate rocks, years may pass before a pollutant released on the land surface above an aquifer is detected in water taken from the aquifer some distance away. Unfortunately, this means that contamination is often widespread before being detected. Even if release of the contaminant is stopped, it may take many years for an aquifer to be cleaned of the contaminant. This is why *it is crucial to prevent groundwater contamination rather than to have to attempt remediation* (removal of hazardous substances).

Karst Terrains and Groundwater Contamination

Areas with humid climates underlain by carbonate bedrock develop a characteristic set of surface features and attributes, called **karst** topography (FIGURE 8-20). Outside the United States karst is also abundant.

Carbonate bedrock is very soluble in the slightly acidic waters (pH 5–6) common over much of the Earth's surface. In areas prone to acid precipitation (also called acid rain), acidity may increase by an order of magnitude or even more. Natural joints in carbonate bedrock are readily widened and extended by solution, and may intersect to form complex channels and underground caverns. Joints open to the surface widen, and underground caverns can collapse to form surface sinkholes. Surface streams may even disappear into underground channel ways.

In karst areas, groundwater discharge rates, measured in meters per year in other types of strata, may increase substantially, often by orders of magnitude. Moreover, surface pollutants can readily enter groundwater and contaminate much greater volumes of aquifers much more quickly, because no adequate filtering mechanism exists. Thus, areas with carbonate bedrock can experience severe groundwater contamination, necessitating expensive, and often ineffective, cleanup procedures.

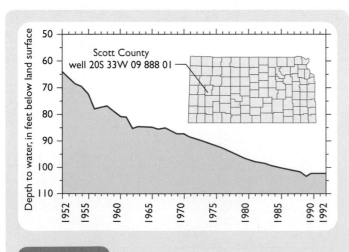

FIGURE 8-18 One example of water level decline in the High Plains Aquifer, Nebraska. Modified from Water Resources of the United States, USGS.

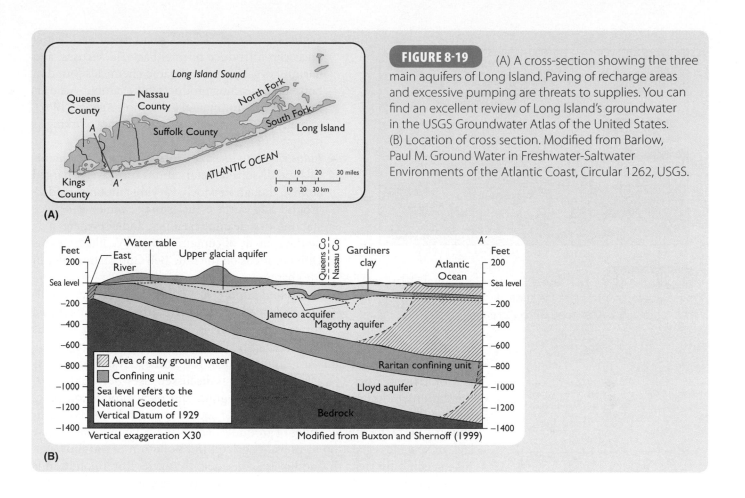

FIGURE 8-19 (A) A cross-section showing the three main aquifers of Long Island. Paving of recharge areas and excessive pumping are threats to supplies. You can find an excellent review of Long Island's groundwater in the USGS Groundwater Atlas of the United States. (B) Location of cross section. Modified from Barlow, Paul M. Ground Water in Freshwater-Saltwater Environments of the Atlantic Coast, Circular 1262, USGS.

Sources of Groundwater Contamination

Human activities that degrade groundwater include mining, agricultural and industrial activities, sewage from septic tanks, and leaking landfills, underground storage tanks, and waste dumps. Indeed, any contaminant dumped onto the surface potentially can contaminate groundwater.

Except where contaminated water is injected directly into an aquifer, essentially all groundwater pollutants enter the aquifer from the land surface. Sinkholes are an especially troublesome case as they can provide a direct connection between surface contamination and an aquifer, thus allowing contaminants in surface runoff to move rapidly into groundwater with no filtering by surface material (**FIGURE 8-21**). Any area underlain by carbonate rock is susceptible, but Florida, with its high population growth and intensive industrial agriculture, is a notable example. Others are Texas and Kentucky. We discuss Florida groundwater in more detail later in the chapter; other geohazards of carbonate terrains are covered elsewhere in this text.

Here are some sources of groundwater contamination (**FIGURE 8-22**).

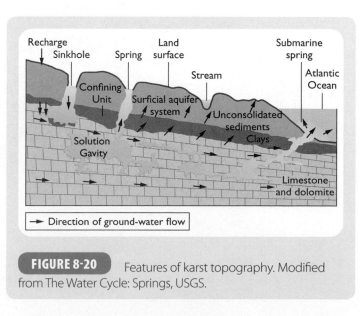

FIGURE 8-20 Features of karst topography. Modified from The Water Cycle: Springs, USGS.

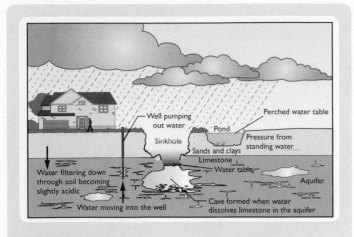

FIGURE 8-21 Diagram of karst in Florida. Note that the water table intersects the surface in the sinkhole, making groundwater contamination easy.

- *Septic systems* (FIGURE 8-23) in rural residential areas can introduce nitrate, bacteria, viruses, and harmful organic and inorganic chemicals found in household cleaning products, for example. According to the Environmental Protection Agency (EPA), as many as 25% of septic systems are not working as designed (FIGURE 8-24).
- *Industrial activities* may introduce organic chemicals and dissolved metal cations (positively charged atoms), though in widely varying amounts. Underground and surface fuel storage tanks at service stations, dry cleaning facilities, and elsewhere, may be the sources of contamination. We illustrate this with our focus on the Blue Bell Gulf Station later in the chapter.
- *Roads and parking lots* contribute petroleum pollutants leaked from vehicles, metals from exhaust fumes, and salt from use during winter months (FIGURE 8-25).
- *Older sanitary landfill and waste dumps,* whose **leachate** may contain a mixture of toxic chemicals (FIGURE 8-26). This is discussed elsewhere in the text.
- *Agriculture,* especially the industrial-style grain and meat production practiced in the developed world, can be a major source of groundwater contamination. One of the shortcomings of groundwater law in the United States is the lack of specific legislation to prevent agricultural pollution of groundwater. EPA specifically lacks this authority: consequently, agricultural pollution is one of the most widespread sources of groundwater contamination. And, because groundwater can contribute to surface water bodies like rivers and estuaries, agricultural pollution is a significant contaminant there as well.

Sources like runoff from farm fields are called *non-point sources* because they don't come from a pipe or other discrete source.

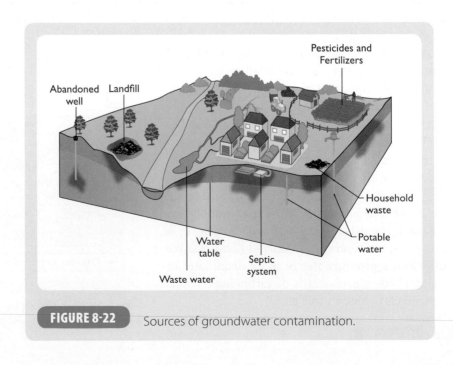

FIGURE 8-22 Sources of groundwater contamination.

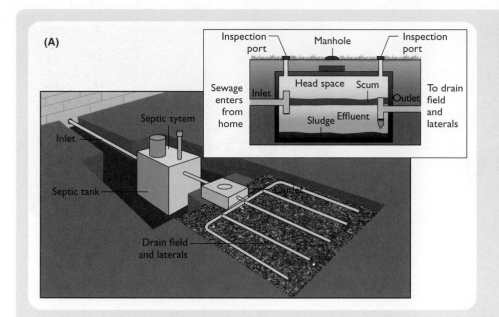

(A)

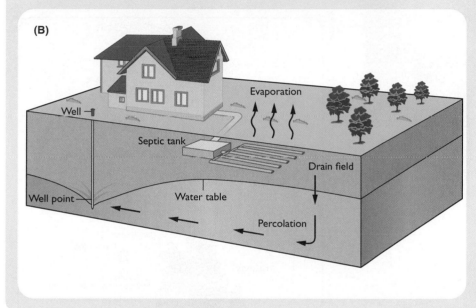

(B)

FIGURE 8-23 (A) Diagram of a septic system. After the sewage enters the tank, solids are partially degraded and settle and oil and grease float to the surface. Wastewater is discharged into the soil through the drainage field, where it is further "treated" by soil microorganisms before entering groundwater. (B) Any toxins put into the wastewater can easily contaminate the groundwater.

An example of agricultural pollution can be found in the Mississippi Alluvial Plain. Study **FIGURE 8-27**. Do you live in the "Cotton Belt?" Do you wear cotton T-shirts? Application of pesticides to cotton (*Gossypium hirsutum*) may be contaminating shallow groundwater over much of the southern United States. Cotton growers use as much as 7 kilograms (15.4 lb) per hectare (2.47 ac, or 100 × 100 meters—a little larger than two football fields) of herbicide and 5 kilograms (11 lb) per hectare of insecticide, to kill weeds and the boll weevil, *Anthonomus grandis* (**FIGURE 8-28**). These applications of pesticides are three to five times greater per hectare than applications of pesticides to corn, yet there have been no regional studies of pesticide in groundwater in the cotton belt even though many different pesticides have been used.

Herbicides, pesticides, and fertilizer from industrial agriculture are among the most widespread contaminants in shallow groundwater of rural areas in the United States. They are also major contributors to contaminated runoff resulting from regional floods and tropical storms.

FIGURE 8-25 Polluted runoff from motor vehicles drains from a parking lot. Is this an unpaid cost of using fossil fuels? Why or why not? © fstop123/iStockphoto.

FIGURE 8-26 Leachate from a landfill. © Environmental Images/Universal Images Group/age fotostock.

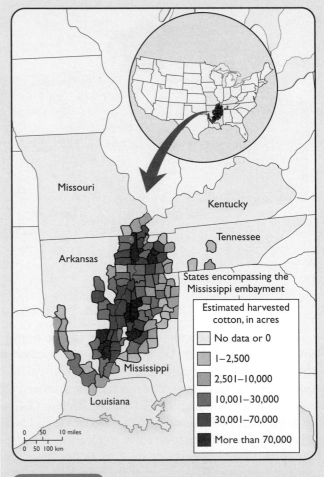

FIGURE 8-27 Cotton growing areas of the United States. Modified from Thurman, E. Michael, et al. Occurrence of Cotton Pesticides in Surface Water of the Mississippi Embayment, USGS.

FIGURE 8-28 (A) The boll weevil, *Anthonomus grandis*, a native of Central America and a major cotton pest in the southeastern U.S. © Ariel Bravy/ShutterStock, Inc. (B) Pesticide application to cotton, southeastern United States. Two techniques reduce pesticide use in cotton: mixed cropping, in which other crops grown near cotton house insects which prey upon cotton pests, and use of cotton plants genetically modified to produce a toxin against some cotton pests. Despite these, pesticides are still widely used. Modified from Thurman, E. Michael, et al. Occurrence of Cotton Pesticides in Surface Water of the Mississippi Embayment, USGS.

(A)

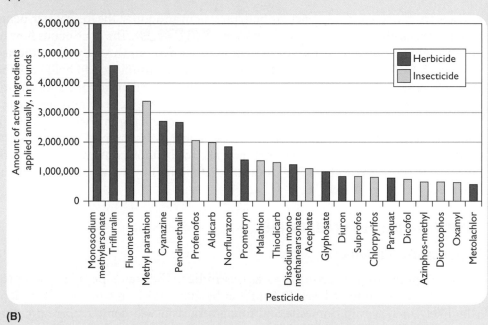

(B)

Groundwater Mining and Subsidence

Excessive pumping of groundwater from aquifers can remove that part of the mechanical support for the overlying strata once provided by pore-fluid pressure, because water is incompressible. That job is now transferred to the "granular skeleton" or framework of the aquifer itself. If too much water is withdrawn, the supporting pressure exerted by water can be lowered such as to irreversibly compress the framework of the aquifer (**FIGURE 8-29**). Surface *subsidence* often ensues, which can damage surface infrastructure. It can damage and disrupt facilities built for drainage, flood protection, and water conveyance, such as artificial levees and irrigation canals. Examples of areas with large subsidence due to groundwater withdrawal include the California Central Valley; Venice, Italy; Houston, Texas; and Mexico City. Issues of subsidence are discussed elsewhere in the text.

FIGURE 8-29 Space photo taken of Venice, Italy in 2001. The city, at sea level, has been subsiding due to groundwater withdrawal, causing the tides and rising sea level to have an increasingly devastating effect on Venice's buildings. Courtesy of Earth Sciences and Image Analysis Laboratory at Johnson Space Center.

Saltwater Incursion

Where fresh water and salt water are both present in an aquifer, the fresh water, being lighter, floats on a wedge or lens of salt water (FIGURE 8-30). The continuous flow of fresh water towards the sea keeps the salt water at bay, but should wells withdraw too much fresh water, the fresh water flow rate may be insufficient to hold back the lens of denser salt water, allowing the latter to infiltrate farther into the aquifer. *Salt water incursion (SWI)* often leads to abandonment of productive municipal wells.

Many coastal areas of the United States are experiencing salt-water incursion. Sites as far apart as San Francisco Bay; Pascagoula, Mississippi; Orange County, California; and Long Island, New York have had to abandon contaminated wells, or have taken measures to fight SWI. Research in coastal Egypt and China has documented SWI there as well.

Human Impact on Groundwater

Groundwater in Florida

Florida's population is growing as fast as India's. The state's population is projected to increase to at least 20.7 million by 2025. With this increase as well as the likely increases in domestic and foreign tourism will come a corresponding increase in water demand and greater potential for groundwater contamination.

In Florida, rainfall is the only surface source of fresh water south of the Suwannee River (FIGURE 8-31). As a result *ninety percent* of this state's population relies on groundwater for drinking water supplies. Over half of all other water needs including agriculture, industry, mining operations, and electric power generation are supplied by groundwater. Florida has several important aquifers: the two major ones are the Biscayne Aquifer and the Floridan Aquifer. We describe each of them briefly below.

The Floridan Aquifer

The Floridan aquifer is one of the largest and most productive aquifer systems in the world. It underlies about 100,000 square miles (259,000 sq km) of Florida and adjacent states, and reaches a total thickness exceeding 1,000 m (3,280 ft) in central Florida (FIGURE 8-32). The aquifer system consists mainly of thick carbonate units (limestone and dolomite) of Tertiary age (from 65 million to 2.6 million years

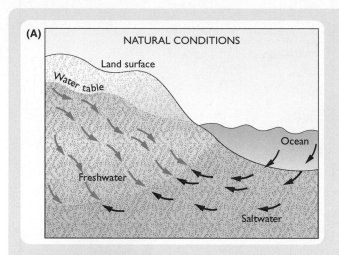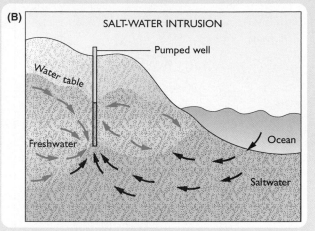

FIGURE 8-30 (A) The salt water/fresh water interface in an aquifer. Courtesy of USGS. (B) Salt water moves into an aquifer due to excessive withdrawal of fresh water.

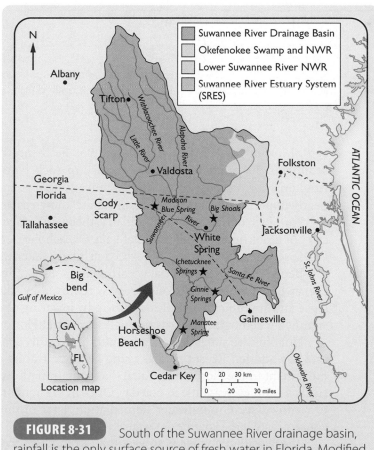

FIGURE 8-31 South of the Suwannee River drainage basin, rainfall is the only surface source of fresh water in Florida. Modified from Suwannee River Basin and Estuary Initiative, USGS.

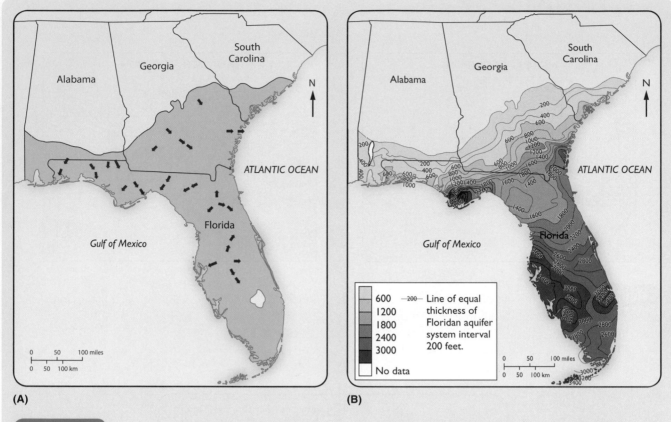

FIGURE 8-32 (A) Extent of Floridan aquifer complex, southeastern United States, showing pre-development groundwater movement (arrows). (B) Contour lines show thickness of Floridan Aquifer, southeastern United States. How does the thickness change over the extent of the aquifer? Parts (A) and (B) modified from Miller, James A. *Ground Water Atlas of the United States: Alabama, Florida, Georgia, and South Carolina.* HA 730-G. USGS.

ago). Most of the Floridan aquifer complex lies beneath at least 100 m (328 ft) of sediments. Its multiple components vary in rock type and thus in permeability. Aquitards forming the base of this aquifer system in much of Florida consist of a nearly impermeable form of calcium sulfate known as bedded anhydrite.

For much of the state, the Floridan is the main source of fresh water, but is generally not usable in the region south of Lake Okeechobee, because total dissolved solids (TDS; see earlier in the chapter) exceed drinking water standards.

Because the aquifer's base is carbonate, solution by acid waters occurs nearly everywhere. Sometimes enough rock is removed by solution to cause surface collapse and sinkholes. Sinkholes can form by lowering of water table as well, and are significant geological hazards throughout central Florida.

Not only are sinkholes a hazard, but they readily conduct surface contaminants into the aquifer. Recent studies using geologic tracers (dyes, etc) to follow water movement in the Floridan revealed that contaminations can enter the aquifer from surface sources in as little as one day. Once in the aquifer, contaminants are difficult to remove (see Groundwater Remediation, later in the chapter) and can quickly contaminate large volumes of water.

The Floridan aquifer is so vast that, even given withdrawals of over 4 billion gallons (11 billion liters) a day, there are only a few sites where large *cones of depression* (**FIGURE 8-33**) persist. In coastal regions where the aquifer is in contact with seawater, saltwater intrusion may displace freshwater if the aquifer is mined.

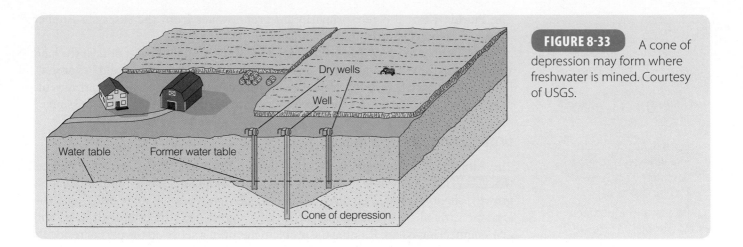

FIGURE 8-33 A cone of depression may form where freshwater is mined. Courtesy of USGS.

The Biscayne Aquifer

The Biscayne aquifer underlies an area of about 4,000 square miles (10,000 square km) and is the principal source of water for Dade and Broward Counties and the southeastern part of Palm Beach County in southern Florida (**FIGURE 8-34**). Cities that depend on the Biscayne aquifer for water supply include Boca Raton, Pompano Beach, Fort Lauderdale, Hollywood, Hialeah, Miami, Miami Beach, and Homestead.

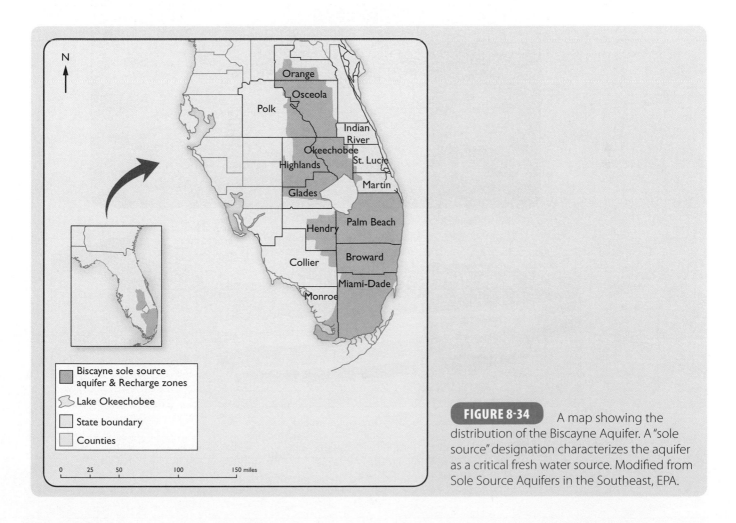

FIGURE 8-34 A map showing the distribution of the Biscayne Aquifer. A "sole source" designation characterizes the aquifer as a critical fresh water source. Modified from Sole Source Aquifers in the Southeast, EPA.

The Florida Keys also rely primarily on groundwater from the Biscayne Aquifer transported by pipeline from the mainland.

Throughout much of its area, the top of the Biscayne aquifer is at or near the land surface, with thin or absent soil cover. Because it is highly permeable and lies at shallow depths everywhere, the Biscayne is easily contaminated, especially from agriculture, septic tanks and many other types of surface runoff, including auto emissions.

Groundwater in New Mexico

The Albuquerque Metropolitan Area is one on the West's fastest growing regions (FIGURE 8-35A). According to the New Mexico Geological Survey, rapid population growth, shallow alluvial aquifers (FIGURE 8-35B), great topographic relief, and the extreme variability of precipitation are raising a growing number of hydrologic and geo-engineering issues. Feeding the problem has been a sense on the part of residents that they had an inexhaustible supply of water.

Withdrawals from the aquifer peaked around 1993 at 160,000 acre-feet per year (52 billion gal).

(A)

(B)

FIGURE 8-35 (A) The Albuquerque metropolitan area. © J. Bicking/ShutterStock, Inc. (B) Cross-section showing geology of the area around Albuquerque. The Santa Fe Group is a massive aquifer that ranges from 3,000–14,000 feet thick, and is predominantly coarse sediments.

Residential development is rapid along the east side of the Rio Grande Rift (Sandia and Manzano Mountains), where many houses use septic tanks and private water wells. Rapid development could deplete the aquifers in this region.

Moving Toward a Sustainable Supply for Albuquerque

In 1997 Albuquerque adopted a plan to attempt a sustainable water supply over the long term. In the future, the city plans to use only the amount of water from the deep aquifer that can be replenished by natural recharge—about 65,000 acre-feet/year (21 billion gal). The remainder of the supply is proposed to come from river water, recycled water, and shallow groundwater. Treaty obligations between the United States and Mexico governing use of Rio Grande water (called the Rio Bravo in Mexico) may complicate this strategy, however.

Critics have questioned the City's commitment to a sustainable water-use plan at the same time it seeks to attract water-intensive industries like computer chip fabricators. To assess industrial water use, consider Intel. The company's Rio Rancho plant makes silicon wafers that must be rinsed ~ 20 times in extremely pure water. Industry analysts estimate an average of 2,840 gallons are needed to produce one six-inch wafer and perhaps twice that to produce an eight-inch wafer. Intel's new chip factory proposed to make about 30,000 eight-inch wafers a month.

Most of the water would be returned to the Rio Grande through sewage treatment plants, but the water would not recharge the aquifer.

Value-added Question 8-2. Compare Intel's monthly water use to make 8-inch chips alone to that of an inefficient golf course, that uses about 1 million gallons a day.

Conservation and Per Capita Water Use

Albuquerque also proposes to aggressively promote conservation, with a goal to reduce water use by 30%. Reaching this goal could mean saving billions of gallons per year, and help to bring depletion and recharge into balance. The city's 2009 per capita water use of 159 gallons per day was down from a high of 250 gallons per day, but still ranks among the highest in the Southwest, while water rates are among the lowest.

Concept Check 8-3. Explain whether you think per capita use of 159 gallons a day includes more than personal water use at home? What else might be included in this volume?

Per-household consumption varies considerably. Half of Albuquerque's households use more than three fourths of municipal water: the top one-half of users by volume are consuming at three times the rate of the bottom one-half.

To enhance efficiency and conservation, the city, like many others, offers rebates for water-saving washing machines and low-flush toilets.

Rate increases are planned to average 4.5% per year, equivalent to raising the average monthly water and wastewater bill by about $9/month. But water rate increases are inevitable, since continuing to rely exclusively on groundwater would require spending an additional $43 million on new wells alone, plus about $4 million/year in Operation and Maintenance costs. Other major expenses would be triggered by declining water levels in the aquifer, costs for arsenic treatment (see below), and to remedy damage caused by land subsidence. The costs of land subsidence in particular could be enormous.

Arsenic in Albuquerque Water

Arsenic (As; also discussed elsewhere in this text) is a toxic substance and a multimillion-dollar issue for water utilities in New Mexico and other areas of the Southwest, where arsenic occurs naturally in rocks and sediment, and thus in drinking water. For the Albuquerque Water Utility alone, meeting EPA's mandated 10 part per billion (ppb) As limit could cost from $190 million to $380 million. Costs of $250 million would double water rates (which are still low by national standards).

Concept Check 8-4. Do you think increasing water rates would be more likely to promote additional demand, or more conservation?

Concept Check 8-5. *Tiering* means charging users for water based on how much they use, with high-volume users paying significantly more per unit of water. Should water rates be tiered? Why or why not?

■ GROUNDWATER REMEDIATION

Long-Term Contamination May Be Irreversible

Cleaning contaminated groundwater (*remediation*) is difficult and costly, according to the National Research Council (NRC), the nation's foremost scientific advisory body. The NRC concluded that remediation of EPA Superfund sites (areas contaminated with hazardous substances listed for federally-mandated clean-up) was largely limited by technical problems. In their words, ". . . at the bottom of the problem are technical difficulties that would be hard to counter even with near-perfect planning procedures in an ideal institutional setting." Indeed the NRC was pessimistic about the future potential for remediating contaminated groundwater in ways that are cost-effective. Instead, the NRC recommended setting as the higher priority protecting as-yet unpolluted groundwater from contamination, because, ". . . *long-term contamination may be virtually irreversible*"(our emphasis).

Remediation Techniques and Strategies

Even though the NRC was pessimistic about the cost-effectiveness of many groundwater remediation strategies, a body of federal and state law requires cleaning of polluted groundwater. Therefore, we list and describe the options for treating polluted groundwater and soil here.

In-situ Versus Ex-situ Methods

Technologies for remediating polluted groundwater may involve **ex-situ** methods, in which the groundwater is pumped out (or soil is removed) and treated, and **in-situ** (in its existing place) methods, in which pollutants are treated within the aquifer. *Ex-situ* methods ("pump-and-treat") are most common. We first list and briefly describe *in-situ* methods.

In-Situ Methods

- Air sparging: This involves injecting gas (usually air or oxygen) under pressure into well(s) installed within the zone of saturation to volatilize (vaporize) dissolved contaminants if they are present as immiscible liquids (those that don't dissolve in water, like gasoline), or contaminants attached to clays within soils. Ideally, the *volatilized* contaminants migrate upward and are removed upon reaching the zone of aeration. This method is most applicable for volatile organic contaminants (benzene, xylene, gasoline, etc.), in substrate with moderate to high permeability.

- <u>Fracturing</u>: This process uses explosives or injection of high pressure water into wells drilled into contaminated fracture systems to expand the fractures and provide a channel by which pollutants can be removed.
- <u>Groundwater Recirculation Wells</u>: This method works best for removal of volatile contaminants like gasoline. It basically uses a well field to inject water or an inert gas under pressure to "push" the polluted water towards a collection site. This process has been used successfully to extract oil and gas from reservoirs, where it is called *enhanced recovery* (discussed elsewhere in this text).
- *In Situ* <u>Flushing</u>: Another method used in oil extraction, *in-situ* flushing injects detergent-like fluids through wells that "scrub" the pollutants out of the groundwater. This method is typically used with recirculation wells.
- *In Situ* <u>Stabilization/Solidification</u>: This method injects materials into the polluted zone to isolate, solidify, or contain the contaminant, or to convert the pollutant to a nontoxic form.
- <u>Permeable Reactive Barriers (PRBs)</u>: Used to remediate shallow plumes with a lower bounding aquitard. A trench may be dug to intercept the plume, into which chemicals or microorganisms are placed to neutralize contaminants. PRBs are best for systems with relatively high permeability and adequate knowledge of aquifer parameters.
- <u>Thermal Enhancement</u>: Various techniques (steam injection, for example) may be used to heat the groundwater plume to volatilize pollutants or make them more easily extracted. Can be expensive if natural gas is used to heat the water.
- <u>Electrokinetics</u>: A low-intensity direct current is set up across pairs of electrodes stuck in the ground on each side of a contaminated area of soil. This causes ions to migrate towards electrodes. Electrokinetics is effective against heavy metals, radionuclides, and organic contaminants from saturated or unsaturated soils, sludges, or sediments, and is applicable to materials of variable permeability.

Some *In-Situ* methods are biological:

- <u>Bioslurping</u>: Easily the drollest sounding remediation method, bioslurping uses vacuum pumps and wells drilled into polluted soils or shallow groundwater. The vacuum slowly draws volatile contaminants upwards, and sucks in air to replace withdrawn polluted material. The air in turn enhances aerobic activity, which can break down some toxic materials.
- <u>Intrinsic Bioremediation</u>: Basically a "do nothing" approach. Uses natural organic activity in soils or shallow groundwater to degrade contaminants, or to convert toxics to less-toxic states, which can then be remediated at less cost. The effectiveness of the approach must usually be documented prior to approval.
- <u>Monitored Natural Attenuation</u> (MNA): Similar to but more comprehensive than intrinsic bioremediation, MNA uses natural physical, chemical, and biologic processes in the polluted zone to eliminate pollution or convert it to a less toxic state. To be effective it should reduce the mass, toxicity, mobility, volume, or concentration of the pollutant to levels that do not threaten human health. Obtaining consensus in establishing such levels, however, can be controversial.
- <u>Phytoremediation</u>: Also called "plant remediation," this is using plants to eliminate pollution in soils and groundwater. Plant roots can absorb or remove heavy metals, and incorporate toxins into plant tissue that is then harvested. Because it relies on living plants, phytoremediation is most effective where groundwater is within 3 m (10 feet) and soil contamination is within 1 m (3 feet) of the ground surface.

Ex-Situ Methods

Ex-situ remediation requires the polluted groundwater to be removed and treated. A common treatment is carbon filtration, which will remove many forms of toxins, especially heavy metals and organic compounds. This is the method approved for remediating the Blue Bell Gulf site described below. Carbon filtration is a known technology and relatively inexpensive. On the other hand, it may take years to decades to fully remediate a contaminated site.

Remediation Example: The Blue Bell Gulf Station, Pennsylvania

Hundreds of thousands of underground storage tanks have been used to store petroleum products and other materials in the United States for nearly 100 years. A large portion of older stainless steel tanks leak. Remediation of contaminated sites has been required since passage of the Resource Conservation and Recovery Act (RCRA, pronounced "Rik-Ra") in 1976.

One of numerous sites contaminated with petroleum fluids is the Blue Bell Gulf site in Whitpain Township, Montgomery Co., Pennsylvania. Whitpain Township is one of the oldest townships in Montgomery County, founded in 1701, which has been called one of the best places in the Philadelphia suburbs to live.

A leak of more than 12,500 gallons (48,000 L) of gasoline from the Blue Bell Gulf station was discovered in May 1998, after fumes exploded in a well house across the street. State Department of Environmental Protection (DEP) officials said the station's leak-detection equipment was malfunctioning, and that the owner failed to submit required inspection records for 1996 and 1997. The station was ordered closed in January 1999 because the station owner failed to demonstrate to the DEP "the ability to assume responsibility for the remaining corrective action in this case."

Two nearby homes were declared unsafe and evacuated for several years because of toxic fumes, and 14 other nearby homes with their own wells had to be given emergency supplies of city water.

The site contains a service station building and four dispenser islands. It is surrounded by commercial and residential properties, and is situated on a local topographic high. Site geology consists of a thin veneer of soils and weathered bedrock overlying fractured sandstones and mudstones of the Triassic Stockton Formation. (Classification of Rocks is found elsewhere in this text.) The Blue Bell Gulf Station site sits on, or next to, a major NE-SW trending fracture zone, which readily transmits groundwater. Streams in the area complexly interact with shallow groundwater.

The DEP prepared a map of the contaminant plume (**FIGURE 8-36**). Although the aquifer does not appear to be influenced by pumping from a municipal well field located approximately one mile north of the plume boundary, groundwater extraction from a nearby golf course does impact the aquifer and could alter the migration path of the northern portion of the plume according to the DEP.

Hydrology of the Stockton Formation

The contaminant plume is within the Stockton Formation. The Stockton is a complex, heterogeneous, multi-aquifer *system*. The upper part is unconfined. The lower layers behave as a leaky confined aquifer system.

Groundwater flow within these rocks occurs mainly along *fractures* within, and cutting across, rock units. The direction of groundwater flow is predominantly toward the northeast. Flow rates along major fracture systems may be even higher than the 4.3 feet/day rate estimated from aquifer test data, based on the relatively rapid expansion of the groundwater contaminant plume.

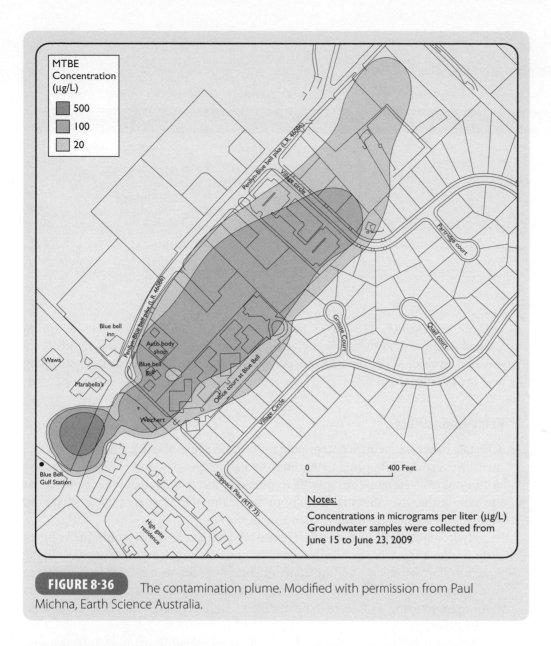

MTBE
Concentration
(μg/L)

- 500
- 100
- 20

Penllyn-Blue ball pike (L.R. 46086)

Village circle

Partridge court

Quail court

Grouse Court

Blue bell inn

Penllyn-Blue bell pike (L.R. 46086)

Auto body shop

Blue bell gulf

Wawa

Marabella's

Office court at Blue Bell

Village Circle

Weichert

Blue Bell Gulf Station

Skippack Pike (RTE 73)

High gate residence

0   400 Feet

Notes:

Concentrations in micrograms per liter (μg/L)
Groundwater samples were collected from
June 15 to June 23, 2009

FIGURE 8-36   The contamination plume. Modified with permission from Paul Michna, Earth Science Australia.

Groundwater Contamination

A small area immediately south of the spill site contains soil contaminants that exceed State Health Standards. Most of the groundwater contamination rests in the shallow portion (<75 feet below ground surface) of the aquifer. **TABLE 8-2** lists contaminants, their maximum concentration, and their health effects.

An 850 foot-wide groundwater contaminant plume, consisting of gasoline-derived constituents MTBE and BTEX (chemical compounds containing benzene, toluene, ethylbenzene, and xylene) issues from the station and extends to depths exceeding 200 ft (65 m). The underground plume of the Whitpain leak is a quarter- to a half-mile long, and is slowly creeping northeasterly. It is within 2,000 ft (600 m) of Wissahickon Creek, a tributary of the Schuylkill River.

Although several feet of pure contaminant was discovered "floating" on ground-water below the site shortly after the leak was discovered, recent measurements

TABLE 8-2 **Contaminants in the Blue Bell Plume**

Contaminant	Concentration (max)	MCLG*	Hazard
Benzene	4,300 ppm	5 ppb	carcinogen, mutagen
Toluene	33,000 ppm	1.0 ppm	liver, kidney, nerv. sys.
Ethylbenzene	2,900 ppm	0.7 ppm	poss. mutagen
Xylenes	20,000 ppm	10 ppm	liver, kidney, nerv. sys.
MTBE (methyl tertiary-butyl ether)	not available	–	poss. mutagen, carcinogen

*MCLG = Maximum Contaminant Level Goal set by EPA

indicate less than 0.5′ (15 cm) of measurable pure contaminant remains. A spring located within the northeast portion of the groundwater plume contains significantly elevated levels of MTBE/BTEX. The spring discharge is being treated via air stripping and carbon filtration.

Remediation Plan

The state began extracting contaminant by wells on September 23, 1998 as an interim measure, recovering approximately 50 to 60 gallons per day, and commissioned the drilling of over 50 monitoring wells to precisely locate the contaminant plume. At the same time, they initiated a study to determine the most cost-effective way to remove soil and groundwater contaminants.

Hydrologists first needed to determine which fracture zones within the bedrock system were impacted. They drilled cores to locate fracture systems in the underlying Stockton Formation because the contaminants were following the fractures. Water from each fracture system was analyzed to detect any contaminants. Using this information, monitoring wells were to be installed to intercept the zones where contamination exists.

Accordingly, the State plans a $5.5 million remediation plant, to treat approximately 200 gallons of groundwater per minute. The system will consist of up to 9 extraction wells and a 70′ × 100′ treatment plant. It will extract groundwater and remove volatile organic compounds (VOCs) with air stripping, and will incorporate carbon absorption units that will ensure zero air discharge of VOCs. The purified water will then be reinjected or used.

Concept Check 8-6. Is protecting groundwater a legitimate role of government, even if only a few citizens might be impacted, and then only economically and not medically? Who should bear the cost of remediation?

Concept Check 8-7. Should the precautionary principle be applied to the siting of all underground storage tanks, because most will leak eventually? Use critical thinking principles to support your answer.

■ GROUNDWATER LAW AND DOCTRINES: WHO OWNS GROUNDWATER?

Here we discuss groundwater laws and doctrines.

According to the National Research Council, groundwater law in the United States is the result of "a bewildering mix of state court decisions and state statutes" (laws) that may be holdovers from pioneer days. There is no national groundwater law or doctrine.

Concept Check 8-8. Groundwater basins and aquifers inconveniently do not stop at state boundaries. Suppose users in two adjacent states have a disagreement on who gets to use water in an aquifer. How would they settle their dispute?

- -

There are several doctrines, or traditional rules, which states follow when dealing with conflicting claims on groundwater. We list and describe them below. First we deal with the states that base their doctrines on *common law*. Under common law, a person is generally allowed to conduct an activity unless an aggrieved party sues to stop the activity, alleging harm.

Common Law Doctrines

Common law doctrines hold that the landowner has the right to groundwater under his/her land. Details follow.

The Absolute Ownership Doctrine: British common law decreed that the landowner is owner of the groundwater beneath his/her land. The landowner is not liable for excessive groundwater use because, under nineteenth century British law, the flow of groundwater was "unknowable." Only Texas retains a vestige of this rule, but several entities in Texas have modified the rule to allow regulation of groundwater use.

The Reasonable Use Rule: Most of the central and eastern U.S. states, along with Arizona, allow a landowner to use groundwater so long as the use is "beneficial" (that is, useful) and not "unreasonable" or wasteful. If a plaintiff sues a user, the defendant may be held liable under common law for damages arising out of any "unreasonable use" of the resource.

Concept Check 8-9. Discuss whether application of this rule discriminates against persons based on their inability to hire a lawyer, go to court, and win a lawsuit against an unreasonable user of groundwater.

- -

Western Correlative Rights. California and Nebraska hold that a user may be "restrained" if groundwater supply is "inadequate," in that they may be required to proportionally reduce their groundwater use.

Eastern Correlative Rights. This doctrine holds that users are unrestricted unless conflicts occur, in which case water will be allocated on the basis of "most beneficial use" as decided by a court or court appointee.

Deciding which use is most beneficial is based on a number of factors such as the purpose of a use, the economic value, and the amount of harm the use causes.

Statutory States

Some states *regulate* groundwater use. They fall into two categories.

Permit States. Florida, Iowa, Wisconsin, and Minnesota require a permit to be obtained before groundwater wells are drilled and water is extracted. Supplies are rationed at times of shortage to protect public water supplies.

Appropriation States. Other than the major water-using states (Texas, Nebraska, Arizona, California), western states (**TABLE 8-3**) apply the "prior appropriation" doctrine, meaning the right to groundwater is based on a state permit, and the earlier permits have priority over later permits. In other words, "first in time, first in right."

Disputes Over Groundwater Allocation

In many jurisdictions, groundwater use is based on ownership of the land above the aquifer. Under the absolute ownership doctrine, pumping may not be restricted to avoid harm to others, although in Texas, "malicious" use may be restricted.

In statutory states, groundwater use is based on a permit, but most users were "grandfathered in" when the system was started.

In "appropriation" states, those with the most senior rights have priority during shortages without regard to use. However, a "junior" user may sue any senior user alleging "unreasonable" use.

Well owners also have water rights issues. It is common for a well to experience reduction in yield due to excessive pumping by an adjacent well. Figures 8-9 and 8-33 show a cone of depression. Both wells might create a cone of depression, but the greater user's cone could overlap the lesser user's, causing the latter's yield to fall or run dry. In common law states, the aggrieved user has little recourse except to drill a deeper well. This is called "the race to the pump house."

Concept Check 8-10. Under the common-law scenario, what incentive is there for users to conserve water in a drought? What advantage would a wealthy user have over a less-wealthy user?

TABLE 8-3	States Whose Rights to Groundwater Are Based on a State Permit*	
New Mexico	Wyoming
Idaho	Colorado
Montana	Washington
Oregon	Kansas
Nevada	Utah
Oklahoma	New Jersey
South Dakota	Florida

*Earlier permits have priority over later permits. This doctrine arose during the California gold rush.

In "reasonable use" states like Arizona, the aggrieved party can sue, but a court would likely award damages only if the defendant's use was "wasteful" or "excessive," and these terms would have to be interpreted by the court. Only Nebraska enforces a well-spacing rule to reduce conflicts.

Summary of Groundwater Depletion Issues

To summarize, *there is no national groundwater law or doctrine.* Groundwater doctrines and applications vary by states and may even vary by county! Many disputes are settled by state court rulings or by arbitration.

Groundwater depletion is a chronic problem in heavy-use Western states: California, Nebraska, Arizona, Colorado, Texas, Kansas, Nevada and New Mexico, as well as in Long Island and Florida. States deal with overdraft issues based on their groundwater doctrine. In common law states, users can generally pump as much as they want unless their use is deemed wasteful or they do not own the land under their well. States that use a permit system can ration users, but give priority to "senior" permit holders. In California the state can charge users a fee for withdrawing water above a state-determined **safe yield**.

Transferring Water Rights

Western states allow water rights to be sold; in fact, many fast-growing cities like San Diego, California are beginning to depend on buying water rights from prior appropriators like irrigation districts. If an owner wishes to sell rights to his/her water, the **consumptive use** must first be determined; that is, how much the present user is taking from the aquifer that is not being replaced. Then, that amount can be transferred to the new user, generally by use of a pipeline. Although up to the present most water transfers involve surface water, groundwater transfers are common in Arizona.

■ PROTECTING GROUNDWATER QUALITY

Federal Groundwater Protection Legislation

Laws and doctrines that seek to protect groundwater or remediate polluted groundwater are even more diverse than allocation doctrines. First, *there is no single overriding federal law.* Rather, authority to protect groundwater is divided and fragmented between federal agencies. Under the 1996 Amendments to the Safe Drinking Water Act, EPA can designate an aquifer as a "sole source" aquifer; that is, an aquifer that is important as a community's main drinking water source. Individual states like Virginia can do the same thing.

Concept Check 8-11. Discuss whether you think it is a proper role of the federal government to protect groundwater, or it is an example of "big government" trampling on individual or state's rights.

Resource Conservation and Recovery Act

The Resource Conservation and Recovery Act (RCRA) seeks, among other things, to protect groundwater supplies. RCRA has three relevant subtitles: Subtitle C regulates hazardous waste; Subtitle D regulates nonhazardous solid waste; and Subtitle I regulates underground storage tanks. Because the RCRA was passed in 1976, it excludes hazardous wastes generated before that date (but see CERCLA section). Implementation of RCRA was primarily responsible for the closure of almost 70% of landfills nationwide, as operators chose not to spend the sums required to comply with the law.

Here is how RCRA attempts to protect groundwater:

- The Law mandates a *permit and tracking* system, sometimes called a "cradle to grave system," for all generators, transporters, storage sites, disposal sites, and treatment facilities for waste defined as hazardous, and specifically defines what comprises hazardous waste.
- The Law requires all *new* landfills to install a groundwater monitoring system.
- The Law requires all *existing* sites to be *retrofitted* with groundwater monitoring systems.
- Hazardous waste disposal sites are required to have two impermeable liners installed to trap and collect *leachate,* liquid that passes through the waste and may contain hazardous chemicals.
- Land disposal of liquid hazardous waste and certain categories of solid hazardous waste is prohibited.

We should note here that the legal definition of "hazardous waste" is complex and beyond the scope of this chapter. You will find a discussion elsewhere in this text.

Safe Drinking Water Act

First passed in 1974, the Safe Drinking Water Act (SDWA) was amended in 1977 and 1986. It sets minimum standards for drinking water in the United States. It also seeks to protect what it describes as "sole source" aquifers and certain other aquifers from pollution, resulting from subsurface injection of waste. The 1986 Amendment authorized the EPA to designate certain aquifers as of paramount importance due to their use as a "sole source" for drinking water in an area. The Amendment further mandated the states to protect the surface area around municipal drinking water well sites from contamination.

Concept Check 8-12. Do you see any authority in SDWA for the actions of Pennsylvania to close the Blue Bell Gulf station? Explain.

- -

Finally, SDWA required EPA to set "maximum contamination levels" (MCL) for a number of substances deemed hazardous. We have given some examples in the Blue Bell Gulf case study.

Federal Insecticide, Fungicide and Rodenticide Act

The Federal Insecticide, Fungicide and Rodenticide Act (FIFRA) was passed in 1972 and amended in 1978. One of its provisions sought to protect groundwater from improper pesticide use, storage, and disposal. Through an agonizingly long, tedious, and complicated process, EPA tests and assesses the safety of pesticides and can prohibit their use. In general, due to lack of funding and inadequate staffing, EPA cannot test new pesticide products as fast as they are introduced onto the market. The result is that many pesticides are in use that have not been adequately tested.

The Law sets two categories of pesticides: *general use* and **restricted use**. Restricted use pesticides can be banned in areas where groundwater has been contaminated, or where the potential exists for such contamination.

Federal Water Pollution Control Act

The Federal Water Pollution Control Act (FWPCA) was passed in 1972 and has been amended several times. It indirectly protects groundwater by regulating waste disposal into surface waters. We discuss the Act in more detail elsewhere in this text.

Comprehensive Environmental Response, Compensation and Liability Act

Passed in 1980 and amended several times since, the Comprehensive Environmental Response, Compensation and Liability Act (CERCLA) is perhaps the most controversial of all "clean water" legislation. This is because it can hold any owner of a property liable for the property's remediation, even if that party was responsible for only a small part of the pollution, or none of it at all! CERCLA's objective was the remediation of hazardous waste sites ("Superfund" sites) that existed before passage of RCRA (see above). CERCLA requires EPA to set up a National Priority List (NPL) for hazardous waste sites to be remediated, paid for in part by fees leveled on industries such as chemical plants and refineries. As of 2004, the fees were no longer being collected, and the "trust fund" was being rapidly depleted. After more than two decades of authority under the Law, over 1000 sites remain on the NPL, prompting critics to call for alteration of the statute to reduce lawsuits over liability for pollution cleanup.

■ POSTSCRIPT

We began this chapter with a dispute over groundwater withdrawals in Arkansas. As you have seen, humans usually assume resources are infinite, and use them as if there were no tomorrow. Groundwater is a prime example of a resource that has great value, but is easily rendered useless or too expensive to use economically by overuse or by contamination. Users then call on "government" to bail them (or us) out.

Concept Check 8-13. Think again about the Arkansas case. (A) Is it better to use resources sustainably, or to maximize their use to maximize short-term gain? (B) On what basis did you reach your conclusions? (C) What importance did "point of view" have in your decision?

Concept Check 8-14. Who is speaking for future generations in the Arkansas case? Cite evidence for your position.

Concept Check 8-15. Do you see cause for optimism in the material you read in this chapter? Explain.

Concept Check 8-16. How important is education about groundwater in ensuring availability both for now and future generations?

CHAPTER SUMMARY

1. The Alluvial Aquifer underlies 32,000 square miles of Arkansas, Tennessee, Missouri, Mississippi, and Louisiana. The aquifer is used mainly for rice and fish farming.
2. Withdrawing water faster than it can be replenished is called mining. Prolonged mining can irreversibly drain aquifers.
3. In arid regions, much of the groundwater being used today was emplaced during wetter conditions thousands of years ago and is not presently being replaced at a rate equal to consumption.
4. Israel and the Palestinian Authority are equally dependent on groundwater for irrigation and public water supplies.
5. Global groundwater depletion has been significant enough to contribute to sea level rise during the past century.
6. Aquifers underlying nuclear-weapons production sites are some of the most contaminated in the world. Such sites represent the "avoided costs" of nuclear weapons programs over the past half-century.
7. Water that seeps below the root zone of plants becomes groundwater; this process is called recharge.
8. Porosity is a measure of the amount of pore space in material.
9. Permeability expresses the ability of material to transmit a fluid.
10. Layers of rock or sediment that store and transmit water are called aquifers. Aquitards are impermeable units that may be able to hold water but not readily transmit it.
11. There are two types of aquifers: unconfined and confined. Recharge of unconfined aquifers can be diffuse or localized.
12. The unsaturated zone is the surface zone where pores are empty or where water clings to the sides of pores by capillary attraction. Pore spaces in the underlying saturated zone are filled with water. Separating these two zones is the water table.
13. The elevation of the water table fluctuates with precipitation in humid regions.
14. Aquifers that are sandwiched between impermeable layers are called confined aquifers. Confined aquifers are characterized by a potentiometric surface.
15. A well drilled into a confined aquifer allowing water to rise above the top of the aquifer is called artesian.
16. Because water will flow towards low pressure, wells drilled into aquifers can create cones of depression.
17. A plume is a volume of groundwater, with defined boundaries, containing some contaminant.
18. Recharge rates for surface aquifers depend on precipitation, bedrock geology, soil type, and surface topography.
19. Groundwater flows through aquifers from recharge areas to discharge areas.
20. Confined aquifers are recharged via precipitation in discrete recharge areas. Land-use practices in recharge areas can adversely affect water quality and can impact users far away.
21. Although surface water supplies are renewable, groundwater supplies may be renewable or not. The safe yield is the volume of water that can be withdrawn without reducing its ability to provide water in the future.
22. Groundwater and surface water may be complexly interconnected.

23. Agriculture is the nation's biggest user of groundwater. Two thirds of groundwater withdrawal occurs in Western states, and there, nearly 80% of the groundwater is used to irrigate crops.

24. Groundwater withdrawal threatens the future of agriculture in the region of the United States known as the "High Plains." One of the most extensive aquifers in North America underlies this region: the High Plains, or Ogallala Aquifer.

25. More than half of the U.S. population depends on groundwater for their drinking water. Water that contains abundant calcium (Ca) and magnesium (Mg) is said to be hard.

26. Contamination of groundwater is usually the result of human activities on the land surface. Areas with carbonate bedrock, called karst, can experience severe groundwater contamination.

27. Herbicides, pesticides, and fertilizer from industrial agriculture are among the most widespread contaminants in shallow groundwater of rural areas in the United States. Application of pesticides to cotton may be contaminating shallow groundwater over much of the southern United States.

28. Surface subsidence can sometimes result from excessive pumping of groundwater.

29. Many coastal areas of the United States are experiencing saltwater incursion.

30. Ninety percent of Florida's population relies on groundwater for drinking water supplies. The two major aquifers are the Biscayne Aquifer and the Floridan Aquifer. The Floridan aquifer is one of the largest and most productive aquifer systems in the world. The Biscayne aquifer is the principal source of water for Dade and Broward Counties and the southeastern part of Palm Beach County in southern Florida.

31. Groundwater depletion is a serious problem for the Albuquerque, New Mexico area. The city's water use ranks among the highest in the Southwest, while water rates are among the lowest.

32. Arsenic occurs naturally in rocks and sediment in many areas of the Western United States, and thus in drinking water.

33. There is no national groundwater law or doctrine. A body of federal and state law requires cleaning of polluted groundwater. The technologies for remediating polluted groundwater fall into *ex-situ* methods, in which the groundwater is pumped out (or soil is removed) and treated, and *in-situ* methods, in which pollutants are treated within the aquifer.

34. Hundreds of thousands of underground storage tanks have been used to store petroleum products and other materials in the United States for nearly 100 years. A large portion of older stainless steel tanks leak. Remediation of contaminated sites has been required since passage of the Resource Conservation and Recovery Act (RCRA) in 1976.

35. The Blue Bell Gulf station, Pennsylvania, was the site of an underground tank leak, discovered in 1998. The remediation plan is discussed in the text.

36. There are several doctrines, or traditional rules, which state courts may follow when dealing with conflicting claims on groundwater. Common Law Doctrines hold that the landowner has the right to groundwater under his land. By contrast, some states regulate groundwater use.

37. Authority to protect groundwater is divided and fragmented between federal agencies. We cite and discuss examples in the text.

KEY TERMS

aquiclude

aquifer

aquitard

base flow

cone of depression

confined aquifer

consumptive use

ex-situ remediation
vs. in-situ
remediation

groundwater maps

hard water

hydrologic cycle

karst

leachate

perched water table

permeability

plume

porosity/effective
porosity

potentiometric
surface

recharge

remediation

restricted-use
pesticides

runoff

safe withdrawal rate

safe yield

salinization/
salination

saltwater incursion/
invasion

saturated zone

springs

substrate

unconfined aquifer

unsaturated zone

water table

REVIEW QUESTIONS

1. The Alluvial Aquifer of the Mid-South is primarily used for
   a. municipal supplies.
   b. rice farming.
   c. cooling power plants.
   d. it is presently unused.

2. Global groundwater depletion over the past century.
   a. has contributed to sea-level rise.
   b. has contributed directly to global warming.
   c. has had no discernible effects.
   d. has contributed to ozone depletion in the upper atmosphere.

3. The most contaminated subsurface site in the United States is
   a. Love Canal, New York.
   b. Three Mile Island, Pennsylvania
   c. Yucca Mountain, Nevada.
   d. Hanford, Washington.

4. Layers of substrate that store and transmit water are called
   a. aquicludes.
   b. aquitards.
   c. aquifers.
   d. artesian systems.

5. A measure of the amount of pore space in material is
   a. permeability.
   b. porosity.
   c. transmissivity.
   d. range.

6. The ability of material to transmit a fluid is expressed using the term
   a. permeability.
   b. porosity.
   c. effective porosity.
   d. piezometry.

7. Separating the saturated zone from the unsaturated zone is
   a. the cone of depression.
   b. the cone of saturation.
   c. the water table.
   d. the piezometric plane.

8. Confined aquifers may also be called
   a. artesian aquifers.
   b. subtended aquifers.
   c. transmissive aquifers.
   d. occluded aquifers.

9. Cones of depression result from
   a. excessive water withdrawal in an aquifer.
   b. water contamination in an aquifer.
   c. the formation of sinkholes in an aquifer.
   d. the introduction of pollution into an aquifer.

10. A volume of groundwater with defined boundaries containing some pollutant is called a/an
    a. aquiclude.
    b. locus of contamination.
    c. plume.
    d. hot spot.

11. The contribution of diffuse recharge decreases as
    a. the humidity of a region increases.
    b. the aridity of a region increases.
    c. the aridity of a region decreases.
    d. the amount of carbonate bedrock increases.

12. Confined aquifers depend on which of the following for recharge?
    a. the presence of carbonate bedrock
    b. injection from man-made wells
    c. stream runoff
    d. precipitation in discrete recharge areas

13. Which is true about recharge rates of confined aquifers?
    a. Some confined aquifers are recharged at such slow rates that withdrawal of large volumes of water is irreversible.
    b. Recharge is not possible in confined aquifers.
    c. Recharge in confined aquifers depends on precipitation on the land overlying the aquifer.
    d. Recharge of confined aquifers is so rapid that they may be viewed as a renewable resource.

14. The two components of stream flow are
    a. confined flow and unconfined flow.
    b. base flow and recharge flow.
    c. runoff and confined flow.
    d. runoff and base flow.

15. The amount of a stream's discharge maintained by groundwater is called
    a. base flow.
    b. recharge.
    c. transmissivity.
    d. confined flow.

16. About how much of Chesapeake Bay's nitrogen pollution is from contaminated groundwater?
    a. less than 1%
    b. more than 90%
    c. about one-third
    d. more than half

17. The source of the contamination mentioned in 16 is mainly
    a. air deposition.
    b. sewage treatment plants.
    c. coal-fired power plants.
    d. agricultural runoff.

18. The biggest user of groundwater in the United States is
    a. municipalities.
    b. agriculture.
    c. power plants.
    d. private wells.

19. Groundwater is used for what percent of municipal water in Florida?
    a. more than 90%
    b. less than 10%
    c. about half
    d. less than 1%

20. Hard water contains excessive
    a. pollution.
    b. iron and sulfur.
    c. calcium and magnesium.
    d. sodium and chlorine.

21. Contamination of groundwater is usually the result of
    a. human activities.
    b. volcanic activity, either past or present.
    c. earthquakes, which allow subsurface movement of contaminated plumes.
    d. excessive forest cover in a region.

22. Which bedrock type underlies karst topography, regions with high potential for groundwater contamination?
    a. bedded salt deposits
    b. volcanic rocks with high arsenic levels
    c. carbonate
    d. silicate
    e. any of the above.

23. Septic systems are important sources of groundwater contamination in rural areas. According to the EPA, about what percent of septic systems do not work properly?
    a. more than 90%
    b. about half
    c. about a quarter
    d. about 5%.

24. Septic systems contribute mainly which type of pollution to groundwater?
    a. nutrients (N and P)
    b. HIV and SARS viruses
    c. pesticides and herbicides
    d. heavy metals (Pb, Hg)

25. Excessive withdrawal of groundwater can cause
    a. major earthquakes like the Iran earthquake of 2003.
    b. ground subsidence.
    c. direct pollution of surface water bodies.
    d. all of the above.

26. Two major aquifers in Florida are
    a. the Biscayne and High Plains Aquifers.
    b. the Ogallala and Hanford Aquifers.
    c. the Miami and Alluvial Aquifers.
    d. the Biscayne and Floridan Aquifers.

27. What form of pollution is affecting groundwater in parts of the southwestern United States, notably the Albuquerque region?
    a. mercury from coal-fired power plants
    b. pesticides from cotton
    c. arsenic from volcanic rocks
    d. radioactive waste

28. The U.S. National Research Council concluded about groundwater contamination that
    a. polluted aquifers would eventually clean themselves, given enough time.
    b. *ex-situ* methods were cheaper and more effective than *in-situ* methods.
    c. long-term contamination may be virtually irreversible.
    d. *in-situ* methods were cheaper and more effective than *ex-situ* methods.

29. Perhaps the most controversial piece of groundwater legislation established the Superfund concept. It was
    a. CERCLA
    b. SWDA
    c. RCRA
    d. FIFRA

30. The Blue Bell Gulf remediation case study was located in
    a. New England.
    b. California.
    c. Oregon.
    d. Pennsylvania.

Floods: Their Nature, Origin, and Impact

CHAPTER OUTLINE

CASE STUDY — *A Poster Child for Flooding Rivers*

Background © Remera Technologies/AbleStock.com/Thinkstock. Title © iStockphoto/Thinkstock.

The Red River of the North

The spring of 1997 will be long remembered by residents of Grand Forks, North Dakota. Fifty thousand people lived there prior to April of that year. As the crest of the Red River flood approached Grand Forks in late April, about 47,500 of them evacuated. The normal flood stage, the level at the natural stream bank is breached and damage begins to occur, for the Red River at Grand Forks was 8.5 m (28 ft). The Red River crested *at over 16.5 m (54 feet)*. The cause was a quick spring thaw of a record-breaking winter snowfall. Estimates of the clean-up cost ran to $1 billion.

Despite warnings from emergency personnel, most residents did not believe that a serious flood was immi-nent. Although 95% of homeowners knew of flood insur-ance, only 20% actually had it. Even those who thought they lived outside flood-prone areas were unpleasantly surprised. Chris Strager, a National Weather Service mete-orologist, bought a house he thought was well above record flood levels. "We thought we were being respon-sible," he said. "We thought we were safe".

The Red River of the North Basin includes parts of Minnesota, North Dakota, South Dakota, and Manitoba, Canada. Area drained by the Red River and its tributar-ies (the *watershed;* see location map) covers 48,490 mi^2 (125,600 km^2; **Figure CS 1-1A**). The Red River ends by flow-ing into Lake Winnipeg in Manitoba, Canada.

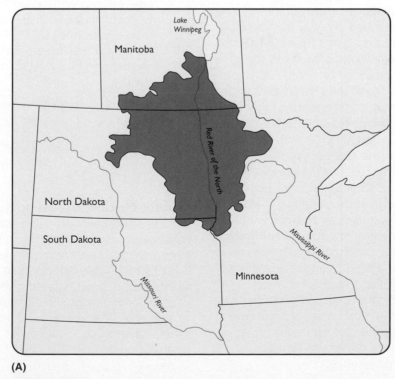

(A)

FIGURE CS 1-1 (A) Location map of the Red River of the North Basin watershed.

Continued ▶

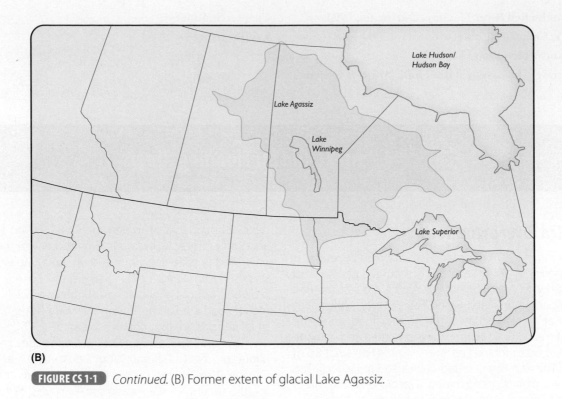

(B)

FIGURE CS 1·1 *Continued.* (B) Former extent of glacial Lake Agassiz.

Five factors contribute to "normal" Red River flooding.

1. The Red River occupies a narrow meandering channel on the flat bed of a vast prehistoric glacial lake named Lake Agassiz (**Figure CS 1-1B**). The land is so flat that flooding occurs in the same way that water spreads when poured on a table.

2. The Red River is one of the continent's youngest major streams. Hydrologists estimate its age at only 9,300 years, a long time from our perspective but insufficient for the river to have carved an efficient floodplain. When the channel floods, the water simply spills over vast areas of the former lakebed.

3. The Red River flows *north*. Spring snowmelt occurs from south to north. Thus, runoff from the southern part of the watershed combines with new melt water as the river flows north. **Infiltration capacity** of soils, that is, the maximum rate at which water can enter soils, in the region is typically low during flood season, because soils are often frozen due to the lateness of the spring thaw.

4. **Ice jams** are a common feature of north-flowing streams at this latitude. Ice from southerly thaws merges with new ice, causing ice-jam floods where the ice is blocked by bridges.

5. The **river's gradient**, a measure of how steeply it loses height, decreases downstream (northwards) from a miniscule 5 inches/mile (7 cm/km) at Fargo to an extraordinarily low 1.5 inches/mile (2 cm/km) in the vicinity of Drayton-Pembina. Melt water and precipitation thus tend to "pond" in the northern reaches.

Figure CS 1-2 illustrates the effects of the 2009 flood on the area.

Concept Check 9-1. Based on what you have read above, explain why so many floods occur on the Red River in April.

- -

Concept Check 9-2. Based on the repeated and predictable nature of flooding along the Red River, (A) what responsibility do you think the state and federal governments should assume in compensating residents for flood losses? (B) Identify means that compensation could take to reduce flood hazards in the future. (C) Discuss whether government should in general compensate citizens who suffer repeated and predictable losses from natural hazards. (D) Should public schools teach flood hazards to students?

- -

FIGURE CS 1-2 Flooding along the Red River of the North. Courtesy of Andrea Booher/FEMA.

■ INTRODUCTION

During the twentieth century, floods caused more property losses and deaths than all other natural hazards *combined*. For the twenty-first century, computer projections of global warming effects include greater severity and number of floods.

Some of the questions we address in this chapter are:

- What are the constituent parts of river systems?
- What are the causes of floods?
- How are floods measured?
- What impacts do floods have on communities?
- How does development affect flood hazard?
- What are the types of floods?
- How can flood risk be measured?
- How can flood risk be reduced?
- How can climate change increase flood hazard?

After you have read and studied this chapter, return to these questions and answer them.

Although some regions are more prone to floods than others, floods can occur almost anywhere and at any time of the day or night (**FIGURE 9-1**). Harmless-looking rivulets can turn into raging torrents within minutes during *flash floods*. Houses can be torn from their foundations and motor vehicles whirled downstream.

Most lives are lost when people are swept away by floodwaters, but most property damage occurs when structures are inundated by sediment-laden water, which often contains toxins and other pollutants such as fertilizer and pesticides, raw sewage, and animal carcasses.

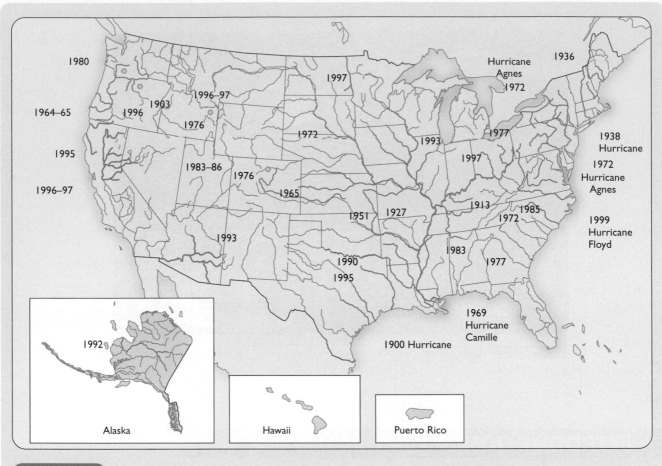

FIGURE 9-1 Floods of the twentieth century. Modified from Perry, Charles A. Significant Floods in the United States During the 20th Century—USGS Measures a Century of Floods, Fact Sheet 024-00, USGS.

Furthermore, floods transport enormous quantities of sediment and debris, and thus floods possess awesome destructive power. Erosion during floods can rapidly undermine levees and bridge foundations (**FIGURE 9-2**).

Over and over rescuers hear from flood victims, "I never expected anything like that."

■ FLOODS AND THE HYDROLOGIC CYCLE

Floods occur basically whenever water input exceeds the rate of removal. Floods are a *natural part of the hydrologic cycle*. Although floods are natural events, human interference with natural ecosystems can increase flood severity and sometimes cause floods. Even the construction of dams, to hold back water, can cause floods. The weight of water piled up behind dams has caused earthquakes, leading to dam rupture and floods. One of the major causes of floods are tropical storms (hurricanes and typhoons), which can dump as much rain over a region in 1 to 3 days as it may normally receive in a year.

The hydrologic cycle, as you have seen, relates *precipitation, infiltration, runoff, transpiration* and *evaporation*. Runoff causes floods, of course, but runoff itself can

FIGURE 9-2 A view of the 1993 Mississippi floods. Floods commonly erode stream banks and demolish "flood control" structures. Courtesy of Andrea Booher/FEMA.

be affected by the degree to which precipitation can soak into soils (called *infiltration* capacity). Clearing forests can reduce transpiration (movement of water as water vapor from leaves to the atmosphere) and increase runoff (**FIGURE 9-3A**). Planting trees can increase transpiration and decrease runoff.

Development can cover a watershed with pavement, drastically reducing infiltration and increasing runoff (**FIGURE 9-3B**).

The point is that all aspects of the hydrologic cycle are *connected* as parts of a *system*. Change one and you can change the others, often in ways not fully understood at the time. Such unappreciated changes in the hydrologic cycle by human

(A) (B)

FIGURE 9-3 (A) Clearcuts in a National Forest. List some ways removing trees en masse alters the local hydrologic cycle. © steve estvanik/ShutterStock, Inc. (B) A vast mall parking lot. How has the local hydrologic cycle been altered? © Dinga/ShutterStock, Inc.

activity can lead to, or enhance, flooding. *These connections will form a basic theme of this chapter.* Look for examples.

Consider the Mississippi River. Millennia before human interference, it had evolved into a meandering system with a floodplain of substantial width (**FIGURE 9-4A**). However, numerous changes in the drainage basin and the river itself have resulted from human interference, beginning in the nineteenth century. These changes have contributed to the severity of floods in the Basin (**FIGURE 9-4B**).

(A)

(B)

FIGURE 9-4 (A) The flood plain of the Mississippi River. It has been considerably altered during the past century and a half by human activity. Courtesy of the Gateway to Astronaut Photography of Earth. (B) A view of the devastating 1993 Mississippi River floods. Courtesy of Andrea Booher/FEMA.

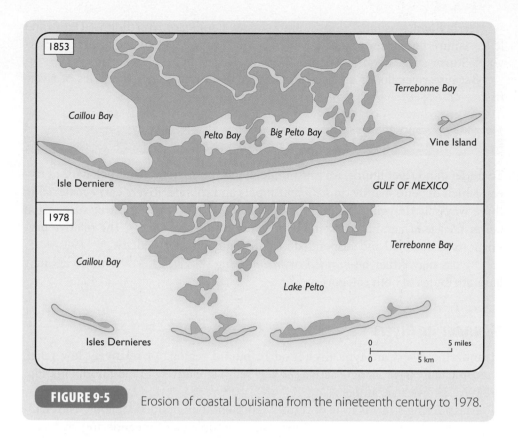

FIGURE 9-5 Erosion of coastal Louisiana from the nineteenth century to 1978.

These changes have altered not only the Mississippi but also the Gulf of Mexico and the Louisiana Coast. Damming the rivers in the watershed has reduced sediment supply to the Gulf of Mexico, causing the coast to recede along with a loss of wetlands and important fishing grounds (**FIGURE 9-5**). This imposes considerable costs on the Louisiana economy and makes the region more susceptible to damage from tropical storms and hurricanes.

Coastal land-level change is a key issue, not only for the Mississippi but also for many other areas, such as Southern California and Egypt, where sediment loss from damming rivers leads to beach erosion, among other things.

Concept Check 9-3. Explain, using examples from the previous discussion, what we mean when we assert that *changes wrought by humans, normally for well-intentioned reasons, can make floods worse and can have unforeseen consequences.*

■ CATEGORIES OF FLOODS

The U.S. Geological Survey (USGS) recognizes six main categories of floods:

- Flash floods
- Regional floods
- Dam-and-levee failure floods
- Ice-jam floods
- Floods resulting from mass movement
- Storm-surge floods

Flash floods, regional floods, and ice-jam floods are *river* floods. Dam- and levee-failure floods and mass-movement floods may affect rivers, lakes, or reservoirs. Storm-surge floods are coastal hazards, caused by tropical storms.

Because river floods are so common, we will discuss characteristics of rivers first.

■ TYPES OF RIVERS

To understand river floods, we must understand the attributes and behavior of rivers, which depend mainly on *sediment supply* and *climate*. Rivers carry material in three ways: by dragging it along the bed (*bed load*), by carrying it in suspension (*suspended load*), and by carrying it in solution (*dissolved load*). The relative proportions of bed load and suspended load depend on water velocity.

Rivers may either be *braided* or *meandering* (FIGURE 9-6). Some rivers may have attributes of both categories.

Meandering Rivers

Meandering rivers are typically found in humid regions of relatively low relief. They normally have a single active *channel* set in a relatively broad floodplain. Figure 9-6A shows an example.

Meandering streams transport finer sediment than braided streams. The channel migrates laterally across the floodplain. Floodplains of meandering rivers are normally vegetated, and the roots of the plants help constrain the river's banks at low flow, limiting sediment movement. *Flash floods* of meandering rivers are usually local events, induced by anthropogenic land-use changes.

Concept Check 9-4. In 1999, the Burroughs and Chapin Company of South Carolina purchased land along the Congaree River in Columbia, with plans to convert about 1,860 ha (4,600 ac) of farmland and forest into more than 5,000 homes, a retirement community, and a technology park, collectively called Green Diamond. According to the Federal Emergency Management Agency (FEMA), the area was classified as a floodplain.

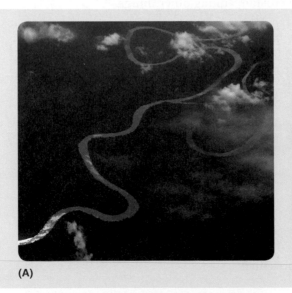

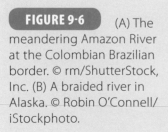
FIGURE 9-6 (A) The meandering Amazon River at the Colombian Brazilian border. © rm/ShutterStock, Inc. (B) A braided river in Alaska. © Robin O'Connell/ iStockphoto.

(A)

(B)

Opponents mounted a campaign around the slogan, "It's a Floodplain, Stupid!" Explain whether you would support or oppose the Green Diamond Development.

Concept Check 9-5. The Burroughs and Chapin Company planned to protect the development by building a levee. Explain whether this changes your support or opposition to the project.

Braided Rivers

Braided rivers, found in glacial regions and on *alluvial fans*, have many channels (**FIGURE 9-7**). The irregular nature of precipitation, runoff, and infiltration in deserts means that flow in braided rivers can be extremely variable. Many braided streams have little flow for months or years and then may carry huge amounts of water in a flash flood. They also may carry vast amounts of sediment in *debris flows* and *mudflows* (discussed elsewhere in the text).

Many of the rapidly urbanizing regions of the western United States are expanding onto alluvial fans drained by braided rivers and are increasingly susceptible to flash floods.

FIGURE 9-7   A vast alluvial fan in the desert of western China. The streams appear bright blue. Courtesy of GSFC/METI/Japan Space Systems/U.S./Japan ASTER Science Team/NASA.

Concept Check 9-6. CheckPoint: Distinguish between *braided* and *meandering* rivers.

■ RIVER FLOODS

To understand what processes cause a river to flood, you must understand (1) some of the attributes of rivers, (2) some of the natural phenomena that determine precipitation intensity and timing, and (3) some of the effects humans have on rivers.

River floods are typically caused by storms that dump more rain on a *drainage area* than can be absorbed or stored.

Remember that storms are *weather* events that are set in a regional long-term *climate* framework. Climate determines the seasonal abundance, location, and duration of storms, and the seasonal fluctuation of soil moisture, snow cover, and rate of snowmelt, all of which may contribute to flooding.

The Stream System

Individual rivers and streams are part of a *drainage system or net*: small rivulets feed larger streams, which in turn are tributary to still larger ones, and so on.

Visualize the system as composed of *upstream* and *downstream* components. Smaller streams are upstream of larger ones. Whenever intense precipitation falls over a wide region, the runoff can with time become concentrated in downstream segments of the drainage net. Regional floods therefore develop along a predictable pattern; the water level will rise gradually in the major river, and may flood well after the precipitation event is over.

Floods can develop in tributary streams without impacting larger rivers downstream, however.

Watersheds and Divides

Each individual stream has its own *watershed* or *drainage basin*. **Watersheds** *consist of a water body (a river, lake, or small stream) and the land surface that drains into that water body.* Watersheds of large rivers like the Amazon consist of many smaller tributary watersheds.

A watershed develops over long intervals of time. It is separated from other watersheds by *divides* (FIGURE 9-8).

We showed cities on Figure 9-8B to remind you that we can affect others downstream in our watershed by our land-use decisions. Virtually all of us live "downstream" of someone else, and thus we all depend on each other to maintain the quality of our watershed. (We complicate this picture further by discussing the importance of a region's *airshed* elsewhere in the text.)

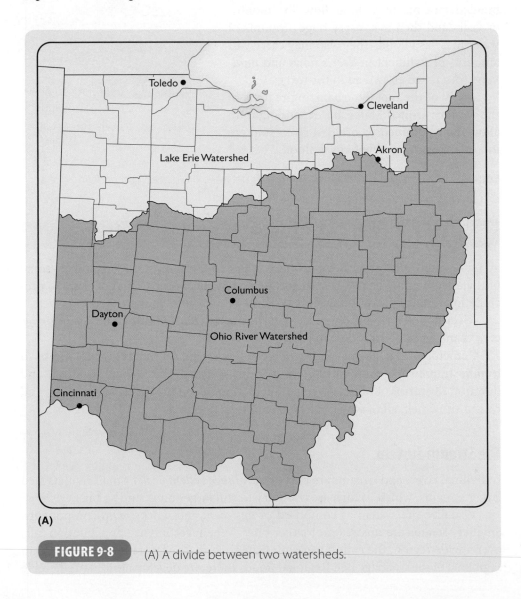

(A)

FIGURE 9-8 (A) A divide between two watersheds.

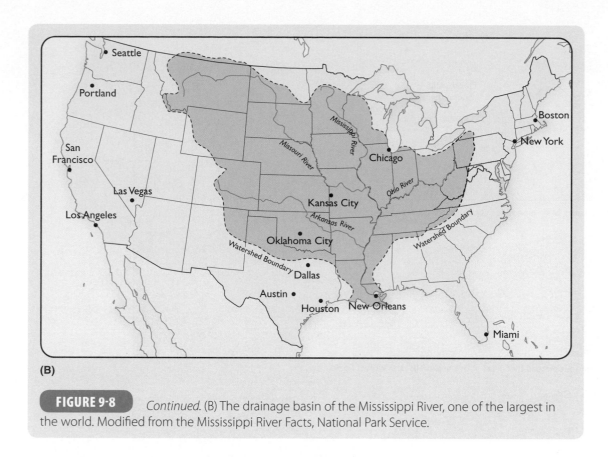

(B)

FIGURE 9-8 *Continued.* (B) The drainage basin of the Mississippi River, one of the largest in the world. Modified from the Mississippi River Facts, National Park Service.

Measuring River Flow: Discharge

Hydrologists (from *hydrology*, the study of water systems) are interested in the **discharge** of a river and the *variability* of the discharge over time. The formula for discharge (Q) is $Q = A \times v$, where A is cross-sectional area, and v is velocity (**FIGURE 9-9**).

Water velocity is an important variable, because water's ability to transport debris *increases exponentially with velocity.* According to government studies, when floodwater velocities exceed *3 ft/sec* (90 cm/sec) it becomes difficult, if not impossible, for adults to maintain their balance while walking through a flooded area (**FIGURE 9-10**). As water velocity increases, erosion of structures such as buildings, dams, and levees increases rapidly.

Discharge is volume per unit time (e.g., cubic meters per second or gallons per day) and varies seasonally.

River discharge may be measured directly using data from a *current meter* (**FIGURE 9-11A**) or it may be estimated using computer programs.

Discharge for most rivers in the United States is recorded continuously at *gauging stations* (**FIGURE 9-11B**), set up and maintained by the USGS or state agencies. Data from these stations are used to map stream flows throughout the United States. For a nationwide example, see **FIGURE 9-11C** .

Stream Flow Data: The Annual Peak Discharge, and the Array

To predict floods we need to measure the variability of stream flow over time. For each "water year" (water years begin October 1), we record the maximum discharge.

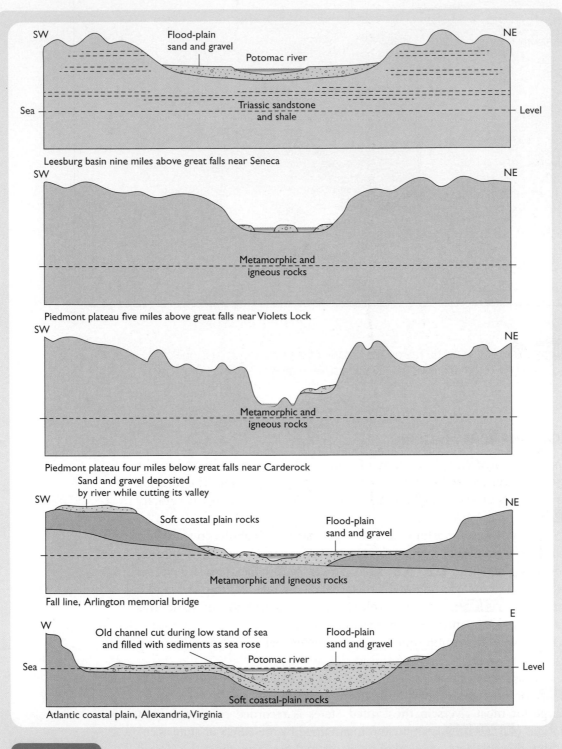

FIGURE 9-9 Cross sections through the Potomac River channel and its flood plain. Note the changes in width of the flood plain at each section. What comprises the bedrock of the Potomac River at each section? Modified from *The River and the Rocks: The Geologic Story of Great Falls and the Potomac River Gorge*, USGS/NPS.

This is the **annual peak discharge**, an example of which is shown in FIGURE 9-12. A list of the maximum floods for each water year is called an **array**.

Meandering rivers consist of a channel, in which water is confined at low flow, and a floodplain, progressively more of which is occupied as discharge increases. The height of the water above some arbitrary datum is called the *stage* (FIGURE 9-13). Stage increases as discharge increases. Above some level referred to as *flood stage* the river begins to leave its channel and occupy its floodplain, and that's where the trouble usually begins, because humans have been living and building on flood plains for millennia. This **stage–discharge relationship** is basic to evaluating flood risk.

FIGURE 9-10   People walking through a flood. If the velocity exceeds 3 ft/sec, it becomes difficult to maintain one's balance. © GOLFX/ShutterStock, Inc.

CASE STUDY   *Red River Flooding During the 1990s*

Background © Hemera Technologies/AbleStock.com/Thinkstock. Title © iStockphoto/Thinkstock.

Let's revisit the Red River. Red River floodwaters reached velocities of 10 feet per second as they passed Grand Forks. Let's see how that compares to Red River discharge during less extreme conditions (**Figure CS 2-1**). First, we need to find a discharge corresponding to the velocity of 10 ft sec^{-1}.

Value-added Question 9-1. Is there a point on the line corresponding to 10 ft sec^{-1}? If not, determine the discharge corresponding to a velocity of 10 ft sec^{-1}. What method did you use to find the discharge?

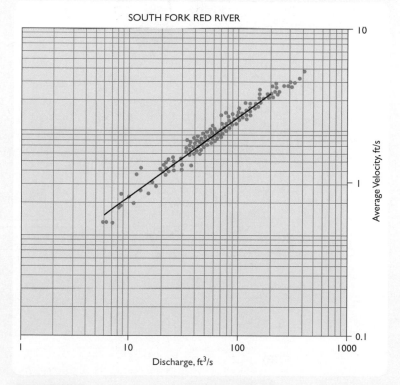

FIGURE CS 2-1   Relationship between discharge (ft^3/sec) and velocity (ft/sec) for the Red River. Data from the USDA.

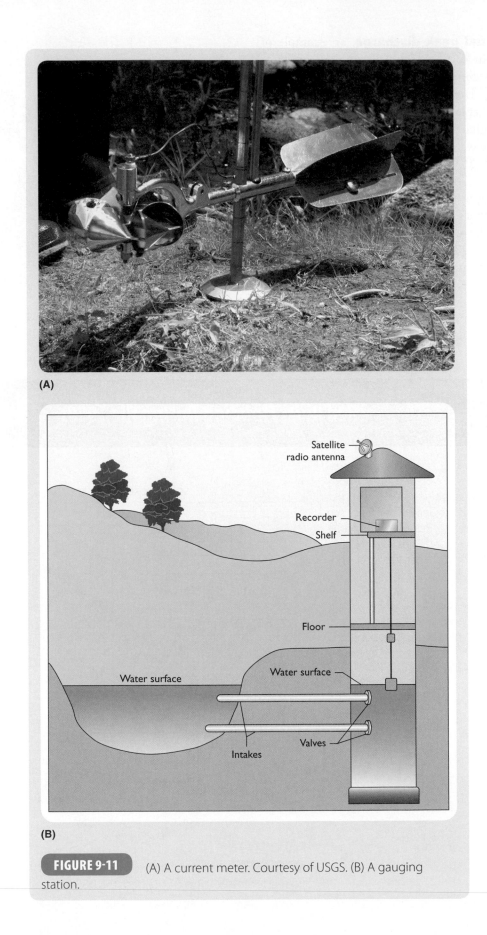

(A)

(B)

FIGURE 9-11 (A) A current meter. Courtesy of USGS. (B) A gauging station.

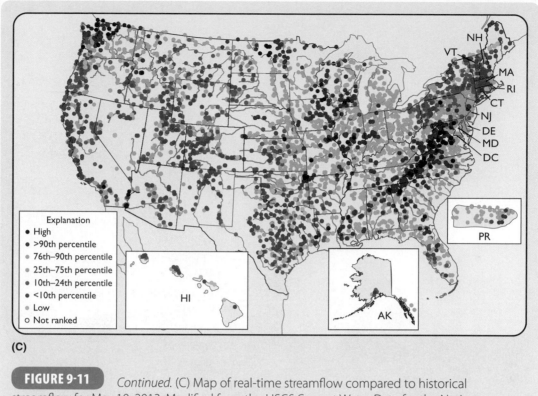

(C)

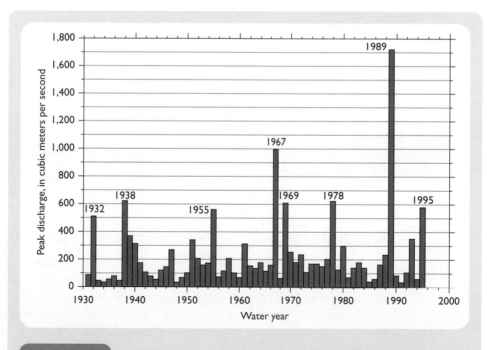

FIGURE 9-12 The annual peak discharge, Virgin River, Littlefield, Arizona. The mean annual discharge is about 5.2 cubic meters per second. During which three years were peak discharges highest? Modified from Hilmes, Marsha M. et al. Estimates of Bridge Scour at Two Sites on the Virgin River, Southeastern Nevada, Water-Resources Investigations Report 97-4073, USGS.

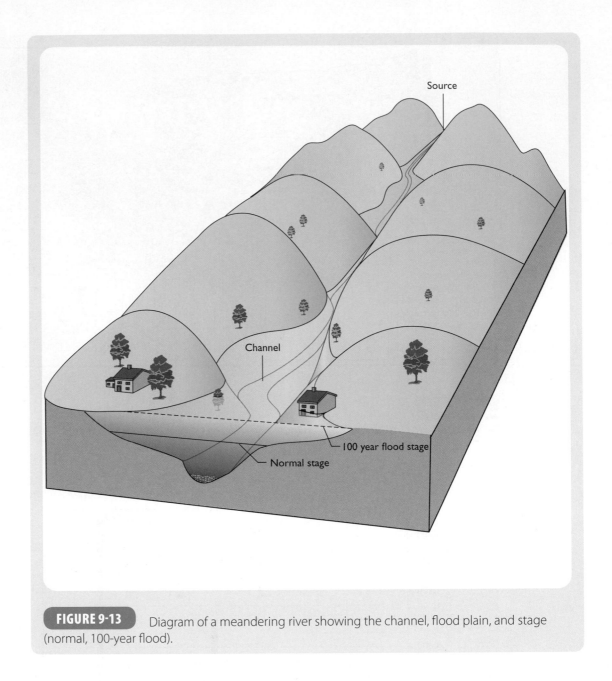

Source

Channel

100 year flood stage

Normal stage

FIGURE 9-13 Diagram of a meandering river showing the channel, flood plain, and stage (normal, 100-year flood).

Recurrence Interval, Exceedance, and Flood-Frequency Diagrams

The **recurrence interval (RI)** provides a statistical estimate of the time between recurrences of a discharge (flood) of a given magnitude. The calculation of RI for a flood of a given discharge in principle is easy. Start with the list of the maximum floods for each water year (the array). For the list of continuous years of peak discharge, the first thing to do is to *rank* the floods in order from greatest to least. If you have one hundred years of continuous record, the greatest annual flood during that interval would have a Rank of 1. The next greatest would have a rank of 2, and so forth. The RI of a flood is calculated using the formula RI = $(N + 1)/M$. where RI is recurrence interval, N is the *number of years* of continuous record (that is, in the array), and M is the rank of the flood. **BOX 9-1** is an example of how to calculate the RI for the hypothetical Ned Pepper River.

BOX 9-1   *Calculating a Recurrence Interval*

Five years of peak discharge data for the hypothetical Ned Pepper River are shown to the right. We'll determine the RI for the number 1- and number 5-ranked floods. Note that five years of data are insufficient to use for calculating a reliable RI, and thus we are using this only to illustrate the calculation. The first step, which we have done for you, is to assign a rank to each flood (**Box Table 1-1**).

The peak rate occurred in 1962. To calculate the RI for this discharge, do as follows:

$$RI = (N + 1)/M; RI = (5 + 1)/1 = 6 \text{ years}$$

For the #5 ranked discharge:

$$RI = (N + 1)/M; RI = (5 + 1)/5 = 1.2 \text{ years}$$

BOX TABLE 1-1	Ned Pepper River Data (1962-1966)	
Year	**Peak Discharge (ft³·sec⁻¹)**	**Rank**
1962	4,027	1
1963	1,019	3
1964	399	4
1965	2,778	2
1966	55	5

Concept Check 9-7. CheckPoint: Explain the concept of a *recurrence interval*. Discuss whether it is possible for two flooding events with 100-year RIs can occur in, say, a five-year period.

Concept Check 9-8. How do the two streams in Table 9-1 differ in discharge? What can you conclude about the importance of forested landscape to flood prevention?

Value-added Question 9-2.
Using the data set in **TABLE 9-1**, determine the RI of the largest flood on either stream to have occurred during the years of record.

We emphasize that the RI of a flood expresses only *statistical probability, not certainty*. Determining that a particular flood has an RI of 100 doesn't mean that such a flood will occur every 100 years, nor does it mean that only one such flood will occur every 100 years. Perhaps a better way to view the concept is as follows: given constant conditions over 100,000 years, 1,000 floods having an RI of 100 should occur. There is nothing that would prevent, say, five such floods occurring in five successive years. Many rivers have experienced more than one 100-year RI flood since 1990.

To avoid confusion, hydrologists commonly use the **annual exceedance probability** (P_e), the probability that a given maximum discharge will occur in a given year. It is the *inverse* of the recurrence interval, R, so $P_e = 1/R$.

As we pointed out in Box 9-1, *the more years of data available, the more accurate is our determination of recurrence interval*. More confidence can be placed in the results of an analysis based on 100 years of record than on an analysis based on 20 years of record.

Discharges for 100-year floods and greater can be approximated by constructing discharge-RI diagrams. These are called **flood-frequency diagrams** (**FIGURE 9-14**).

RI calculations for a flood of a given discharge, at a given location, can change if there have been significant changes in the watershed above that location, or if

Value-added Question 9-3.
If the RI for a flood is 100, what is the P_e?

TABLE 9-1				Peak Discharge From Unlogged and Logged Watersheds		

Date of Peak Discharge	Peak Discharge, Unlogged (m³/s/km²)	Peak Discharge, Logged (m³/s/km²)	Date of Peak Discharge	Peak Discharge, Unlogged (m³/s/km²)	Peak Discharge, Logged (m³/s/km²)
12/21/1955	0.89	1.04	01/21/1972	1.27	1.66
12/11/1956	1.15	1.65	12/07/1973	0.41	0.75
12/20/1957	1.29	1.62	12/20/1974	0.70	1.10
02/15/1958	0.75	0.90	12/01/1975	0.75	1.23
01/27/1959	0.65	0.92	01/08/1976	1.19	1.54
11/24/1960	0.74	0.93	12/13/1977	0.92	1.13
02/10/1961	1.03	1.23	12/04/1978	0.41	0.87
12/02/1962	0.29	0.39	02/07/1979	0.60	1.13
02/01/1963	0.48	0.70	12/25/1980	0.78	0.96
12/22/1964	1.65	1.80	12/06/1981	0.94	1.19
01/27/1965	1.35	1.50	02/14/1982	0.71	1.07
01/06/1966	0.51	1.19	12/14/1983	0.61	0.86
01/28/1967	0.60	0.80	02/13/1984	1.01	1.29
12/04/1968	0.91	1.18	12/06/1985	0.28	0.45
01/07/1969	0.43	0.54	02/22/1986	1.19	1.48
01/18/1970	0.71	0.84	12/09/1987	0.38	0.66
12/05/1971	0.79	1.35	01/10/1988	0.57	0.90

there have been changes in land use at that location since the analysis was prepared. Therefore, RI studies must be constantly revised to keep pace with development and other land-use changes. And, because many areas lack detailed discharge measurements over time, statistical analyses are not always reliable.

Hydrologists and environmental geologists are acutely concerned with the relationship between stage and discharge. Once the recurrence interval of a flood has been calculated, all that remains is to relate the stage of a particular flood to the elevation of the floodplain such a discharge will reach. Once such data are available, flood plain maps can be drawn showing the portions of the flood plain that will be occupied by floodwaters as discharge increases.

Finally, recognition and accurate measurement of high-water marks from historical floods are important to provide calibration for RI calculations, and to precisely determine water-surface profiles during floods in order to better predict future events. FIGURE 9-15 shows an example.

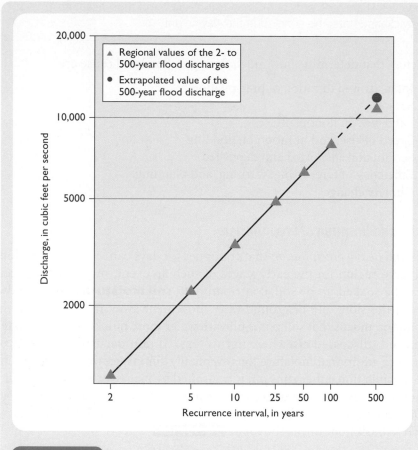

An actual flood-frequency diagram of Seneca Creek, Maryland, from the Federal Emergency Management Agency (FEMA). Question: Using the above diagram, estimate the discharge of a 100-year RI flood. Modified from FEMA.

Factors Determining Flood Damage

The great Midwest Flood of 1993 focused the attention of the Nation on the human and environmental costs associated with (1) decades of efforts to control natural flooding with structural measures, (2) unwise land-use decisions, and (3) the loss and degradation of the natural resources and functions of floodplains.

John McShane
FEMA, chair of the Federal Interagency
Floodplain Management Task Force

Concept Check 9-9. Here is how we interpret McShane's statement. "We have spent decades attempting to control floods by building dams, levees, and other structures, and by allowing or even encouraging the wrong kinds of land use in flood-prone areas by converting wetlands and forests on floodplains to farms, cities, and subdivisions. The result is that we have increased the severity and costs of floods that have occurred. Moreover, the loss of the natural filtering capacity of floodplain wetlands has led to increased water pollution from agricultural runoff."

FIGURE 9-15 High water mark sign. Courtesy of NOAA.

Explain which statement, ours or McShane's is clearest and most accurate, then speculate on why McShane stated his conclusion the way he did.

The factors that determine how much damage a flood will cause are:

- Intensity and duration of precipitation
- Season
- Land-use on floodplain
- Speed of rise, and duration of, flooding
- Sediment transported and deposited
- Efficiency of forecasting, warning, and planning
- Climate change

Intensity and Duration of Precipitation

Intense and heavy precipitation that continues for days can lead to severe flooding over a large region. In the early stages of such an event, much of the precipitation can be soaked up by soil, contributing to **soil moisture** and groundwater. However, the slow rate of groundwater movement away from a zone of heavy precipitation means that soil pores fill with water more quickly than the water can be removed (discussed elsewhere in this text). If rain persists, the ability of the soil to store additional moisture thus eventually diminishes to zero, and essentially all the precipitation either runs off, is transpired by plants, or evaporates. But transpiration and evaporation produce water vapor, which can be quickly reprecipitated. Precipitation can vary significantly over a relatively small area (**FIGURE 9-16A**), which can lead to flash flooding as shown in **FIGURE 9-16B** .

Season

In humid regions like eastern North America and the Pacific Northwest, rainfall from weather fronts is most common during autumn, winter, and early spring. Upper soil layers are more likely to be frozen or covered with snow in winter, limiting infiltration from rain. Snowmelt can augment heavy rains, most commonly in winter and early spring. This is a common cause of flooding in the Dakotas, for example, as you saw in the case of the Red River.

The hurricane season in eastern North America normally runs from early June to the end of November. Thus, late-season hurricanes can occur after frontal storms have saturated the soil, and lead to catastrophic flooding.

Land Use on Floodplain

When Europeans first settled North America, the earliest non-coastal cities generally sprang up along rivers. Where rapids or falls required boat operators to portage (carry) their cargo around the obstruction, more towns grew. Now, many of America's largest cities occupy floodplains of rivers (**TABLE 9-2**). This simple fact means that floods inevitably will damage cities.

The chance that heavy persistent precipitation will cause flooding is enhanced by **land-use changes** that increase runoff over infiltration and/or transpiration. Examples are the construction of roads, parking lots, malls and other structures that cover large areas of former natural vegetation. These impervious areas (see Case Study: Floods and Land Use) are drained by storm drains that funnel runoff directly to a stream, or to a sewage-treatment plant. Less obvious examples are

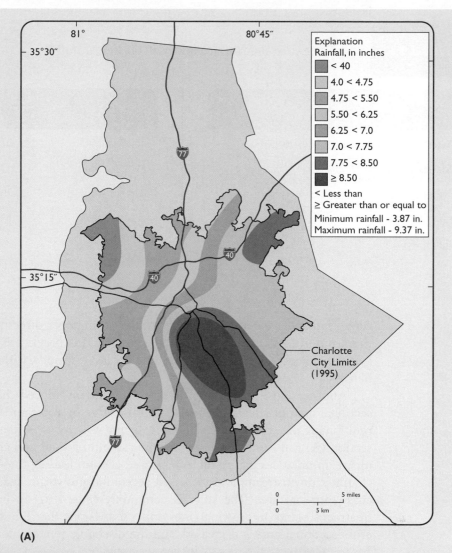

Explanation

Rainfall, in inches

■	< 40
■	4.0 < 4.75
■	4.75 < 5.50
■	5.50 < 6.25
■	6.25 < 7.0
■	7.0 < 7.75
■	7.75 < 8.50
■	≥ 8.50

< Less than
≥ Greater than or equal to
Minimum rainfall - 3.87 in.
Maximum rainfall - 9.37 in.

Charlotte
City Limits
(1995)

0 ___ 5 miles
0 ___ 5 km

(A)

(B)

FIGURE 9-16 (A) The variation in precipitation during August 26–28, 1995, in Mecklenburg County, North Carolina. Modified from Robinson, Jerald B. et al. USGS Fact Sheet FS-036-98, USGS. (B) This storm led to considerable local flash flooding. Note the amount of sediment in the water. Courtesy of Kent D. Johnson.

TABLE 9-2	Some Cities Built on Floodplains in the United States		
City	**Percent of Floodplain Urbanized**	**City**	**Percent of Floodplain Urbanized**
New Orleans, LA	100% urbanized	St. Louis, MO	67% urbanized
Great Falls, MT	97% urbanized	Denver, CO	62% urbanized
Phoenix, AZ	89% urbanized	Norfolk, VA	26% urbanized
Harrisburg, PA	83% urbanized	Richmond, VA	13% urbanized
Tallahassee, FL	83% urbanized		

roof drainage systems that transfer water from roofs directly to streets, rather than buffer areas of turf or other vegetation. This practice is often encouraged because it lessens the chance of swelling in clay-rich soils, with resulting building damage.

Another cause of flooding is related to human-induced changes in the cross-sectional area of stream channels. The discharge of a stream (Q) is calculated by multiplying the cross sectional area (A) by the velocity (v). Changes in stream cross section (A) can result from such factors as: mining sand and gravel from streams; upstream practices that result in increased erosion leading to deposition in stream channels downstream; and increased vegetation growth in channels, lowering the cross-sectional area. The latter is commonly caused by discharges from sewage treatment plants in growing cities and subdivisions that line streams. When the cross-sectional area is decreased, and the discharge remains constant, the velocity has to increase, sometimes to dangerous levels.

Development is one of the most widespread examples of contemporary land-use change in the United States. We will discuss its effect on flood hazards next.

In an undeveloped state, precipitation in a watershed follows a variety of pathways as governed by the hydrologic cycle. **Stream hydrographs** are used to show discharge over time for floods. **FIGURE 9-17** uses stream hydrographs to show the amount of storm water discharge resulting from various land uses.

In an urbanized basin, parts of this natural process can be altered. For one thing, urban areas are hotter than non-urban areas, causing a "heat island" effect. This can stimulate higher precipitation and evaporation.

Development has a greater effect on runoff for low-recurrence interval (i.e., more-frequent) floods than for high-recurrence interval floods, such as 50- or 100-year floods. During larger floods, the soil becomes saturated during long periods of intense rainfall, and loses its capacity to absorb additional rainfall. During high RI floods, runoff thus always overwhelms infiltration. Still, land conversion has a significant impact on runoff. For example, one mature tree can absorb or retard up to 200 liters per hour of precipitation. Rain can attach to leaves, stems and trunks of trees and can be transpired from roots as well. So you can see why mature forests in watersheds retard flooding.

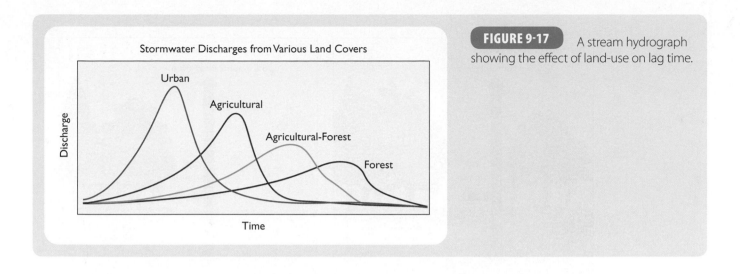

A practical example of the value of trees was exhibited when New York City residents voted to buy the land in their watershed to protect it from development.

During urbanization, impermeable surface area increases (see Case Study: Floods and Land Use). The resulting gutters, storm sewers, and drains direct water to stream channels more quickly than the infiltration mechanisms they replace (**FIGURE 9-18**).

Hydrologists use the term **lag time** to refer to the *gap between the occurrence of precipitation and the runoff from that precipitation.* As development increases, lag time *decreases;* thus, flash floods are more likely in regions that have undergone significant urbanization (see Figure 9-17). Some counties in Northern Virginia near Washington, D.C. approach 40% impervious surface as a result of paving and buildings. Even when soil is not paved over, compaction during construction can decrease the soil's infiltration capacity, in the same way that well-worn shortcuts across grassy areas on campuses compact the soil.

During urban development, stream channels are normally straightened, deepened, lined, or channeled underground within urban areas, all of which have the effect of speeding water downstream, increasing discharge. Transpiration declines as trees and shrubs are cut down and replaced by buildings and parking lots. Infiltration declines as soils are compacted and paved over. Construction itself increases sedimentation by scraping off vegetation prior to construction.

In a number of studies, urbanization was found to increase precipitation intensity and runoff, and thus flood damage (**FIGURE 9-19**).

One study in the Portland, Oregon metropolitan area found that of 1,000 homes built near floodplains between 1992 and 1995, nearly 200 were damaged in a 1996 flood. Another study found that a change from rural to urban land use in the Willamette Valley of Oregon increased storm runoff volume by 200% and peak flows by 300%. Ideally, storm water retention ponds like that shown in **FIGURE 9-20** catch and store water until it evaporates or infiltrates.

In Washington State, two studies showed that the incidence of 100-year RI floods nearly doubled when impervious surfaces covered more than 20% of the area of small basins. The number of smaller floods (up to 5 year RI) increased significantly as well.

Value-added Question 9-4.
Interpret Figure 9-17. Explain which type of land uses favor higher discharges and the possibility of floods, and which do not.

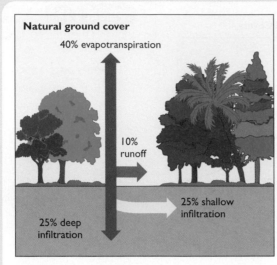

Natural ground cover

40% evapotranspiration

10% runoff

25% shallow infiltration

25% deep infiltration

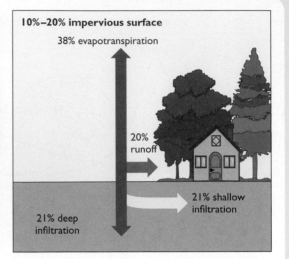

10%–20% impervious surface

38% evapotranspiration

20% runoff

21% shallow infiltration

21% deep infiltration

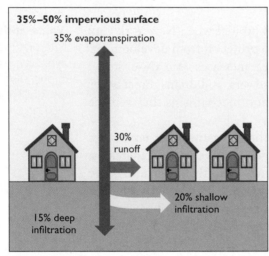

35%–50% impervious surface

35% evapotranspiration

30% runoff

20% shallow infiltration

15% deep infiltration

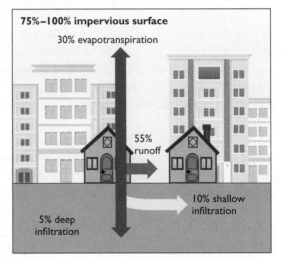

75%–100% impervious surface

30% evapotranspiration

55% runoff

10% shallow infiltration

5% deep infiltration

FIGURE 9-18   How urbanization changes the flow of water. Urbanization can change the path precipitation takes. As urbanization changes the characteristics of a river basin, water movement shifts from one in which groundwater flow is important, to one dominated by runoff.

Urbanization can contribute to flooding in downstream parts of the watershed that are not yet urbanized, because the time necessary for the runoff to flow through the basin decreases. Peak flows can increase in size and number. Increased flow velocities and sediment loads during runoff can lead to increased bank erosion and scouring of channels. *Base flows,* that part of a stream's discharge provided by groundwater, in low-flow periods commonly decrease, because of a lowered groundwater table (discussed elsewhere in this text).

For example, a 1999–2003 drought in the Mid-Atlantic States had a severe impact on groundwater levels. In some cases, groundwater fell to levels not seen since record keeping began. Many experts explain this drop in the water table by the increase in impervious surface in the watershed, which prevented groundwater recharge. Base flows in low flow periods thus decrease because of a lowered groundwater table. As a result, some rivers went nearly dry in mid-2002.

Mercer Creek near Bellevue, Wash.
1,000
800
600
400
200
0
Streamflow, in cubic feet per second
Estimated "1-in-100 chance flood" based on data from 1978–1994
"1-in-100 chance flood" based on data from 1956–1977
Highest annual flows
1960
1970
1980
1990

FIGURE 9-19 The increase in 100-year RI flood discharges as a result of urbanization in one locality in Washington State. Modified from Dinicola, Karen. The "100-Year Flood." Fact Sheet 229-96. USGS.

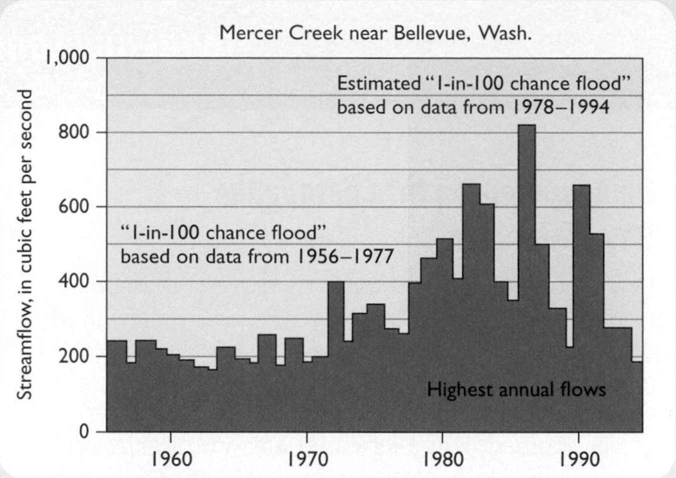

Speed of Rise, and Duration of Flooding

The faster floodwaters rise, the less chance people have to evacuate and remove or protect property. The longer floodwaters persist, the more damage that occurs to structures from sedimentation and polluted water, and the greater the loss of economic activity. A study of flooding in Bangladesh found that the duration of floods increased 25% between 1988 and 1998.

Longer intervals of flooding lead to degradation of water supplies, and sewage treatment capacity. Sewage-treatment and water-treatment plants tend to be located along water bodies for obvious reasons, so they are very prone to damage during floods.

Longer-duration floods also can undermine levees and other flood-control structures. Duration of flooding has been linked to increases in impervious surface cover. In one study in Texas, the amount of impervious cover increased from 2% to 11% between 1972 and 1995, and duration of floods doubled.

In Thailand, increased duration of floods was linked with more prolific development of waterborne parasites.

Removal of wetlands in floodplains is one of the main reasons floods have increased in severity, according to many ecologists. The Wetland Reserve Program (WRP) in Missouri has restored at least 84,000 acres of wetlands, increasing storage of floodwaters and shortened duration of floods.

Transport and Deposition of Sediment

Land-use practices, geology, and topography in the watershed determine how much sediment is available for transport during floods of a given discharge.

In general, forested land yields the least sediment, urban land more, and agriculture from row crops yields the most. Converting forest to cropland thus inevitably increases sedimentation during floods. Ways that farmers can significantly reduce sediment loss from fields is discussed elsewhere in the text.

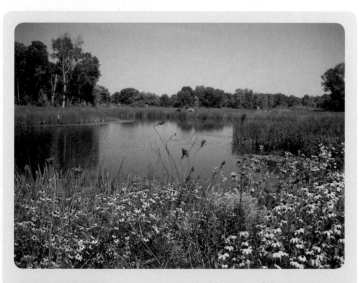

FIGURE 9-20 Stormwater retention pond in Madison, Wisconsin. Courtesy of the city of Madison, WI.

Efficiency of Forecasting, Warning, and Planning

Forecasting is generally very effective for large, regional floods if land-use changes are incorporated

Development in the Chesapeake Bay Watershed

As the population of a community grows, it is accompanied by *development*. Development typically refers to all the human-built structures in an area: subdivisions, commercial buildings (i.e., offices, shops, malls, warehouses), driveways and parking lots, and roads. One major effect of development is the increased amount of *impervious surface*, any surface material through which water cannot penetrate (e.g., roads, parking lots, and rooftops).

Concept Check 9-10. Review the diagram of the hydrologic cycle. Explain how increasing the impervious surface in an area might affect the hydrologic cycle.

Concept Check 9-11. Do you think that there is a correlation between the amount of impervious surface in an area and the likelihood of flooding? Explain your answer.

The Chesapeake Bay is the largest estuary in the United States with a watershed of more than 64,000 square miles in six states. Between 1990 and 2000, impervious surface increased from 623,628 to 780,785 ac in the watershed. Over the same period, population in the watershed increased from 14,250,226 to 15,715,448.

Concept Check 9-12. Why do you think impervious surface increased in the Chesapeake Bay watershed?

Value-added Question 9-5. Calculate the percentage increases in both impervious surface and population for the ten-year period. Report and interpret your results.

Impervious surfaces collect runoff and prevent it from infiltrating into soils and surface sediment, where rainfall can be stored and natural filtering can often remove some pollutants. These paved areas contribute significant pollution (e.g., oils, pesticides, herbicides, and nutrients) to waterways.

Approximately 100 acres of forest are lost daily in the Bay's watershed. The current forested area of 24 million acres represents 58% of the Bay's watershed. In the 1600s, about 95% of the watershed was forested.

Let's take a closer look at the relationship between impervious surface and flooding. Here, we'll examine Potomac Mills Mall, coincidentally also in the Chesapeake Bay watershed. Potomac Mills Mall is in rapidly-growing Prince William County, Virginia. According to the Prince William County Department of Public Works, a recent estimate of the impervious surface at the mall was 5,675,411 ft^2 (527, 262 m^2).

In June 1972, before Potomac Mills was built, Hurricane Agnes hit the area and dumped 14 inches of rain on the region in three days, then one of the region's most costly disasters.

Concept Check 9-13. Based on the information in this chapter, where do you think the water went during and shortly after the deluge?

What if another storm of the magnitude of Hurricane Agnes occurs (as it surely will someday) and dumps 14 inches (36 cm) of rain on the region in 72 hours. Assume further that before Potomac Mills Mall was built, the land was pasture and forest, which, given the area's subsurface geology, would have absorbed most of the storm's runoff.

Value-added Question 9-6. Calculate the maximum amount of extra runoff Potomac Mills Mall would generate when the next Hurricane Agnes hits northern Virginia. Express your answer in cubic feet, then convert it to gallons. (1 ft^3 = 7.5 gal). (*Hint*: Multiply the area of impervious surface given above by the amount of rainfall. Then multiply your answer by 7.5).

Value-added Question 9-7. Assume this runoff has to be handled over a 72-hour period. What is the resultant discharge from Potomac Mills Mall in gallons per hour? (*Hint*: Divide your answer by 72 hours; for comparison, a typical summer flow of the Potomac River is 26.4 million gal/hr.)

Of course, this number represents a maximum value, but it illustrates how development can increase a region's susceptibility to flash flooding.

To minimize the likelihood of flash flooding, as well as surface runoff, communities use *stormwater management systems*. These systems consist of a network of pipes that collect runoff from streets and large parking lots and channel it into artificial stormwater retention ponds or into creeks in the vicinity (**Figure CS 3-1**). Sometimes the runoff pipes carry the water to a sewage treatment plant, where it can be treated before discharging it into streams. Although treating runoff sounds like a good idea, it can cause serious problems during times of heavy runoff by overloading the sewage treatment plants, causing a mixture of untreated sewage and stormwater to be dumped into waterways. Many communities thus design separate systems to transmit sewage and stormwater. Although this is more expensive, it results in much less environmental degradation, especially during floods.

But stormwater management systems work only as well as the weakest link in the chain—which simply means that if anything can go wrong, it usually will. Pipes and retention ponds may get clogged with debris or fill up with vegetation and sediment, storm drains may be too small to handle runoff volumes, development since construction of the system may overload it, and so forth. This means that unless local governments take painstaking care in the design and maintenance of stormwater management systems, the more development in an area, the higher the risk of flash flooding.

One very promising development is the use of pervious paving (**Figure CS 3-2**). One type, pervious concrete, which retains most of the strength of typical concrete but has pores comprising as much as 25% of the volume, can be engineered to allow as much as 8 gal/min to pass through to the ground beneath.

FIGURE CS 3-1 Stormwater pipes, like those in Manly, Australia, release stormwater and can pollute beaches. © Thorsten Rust/ShutterStock, Inc.

FIGURE CS 3-2A Workers install Rainstone, a porous plastic product that holds rain runoff from a nearby parking lot, slowly releasing it into groundwater or a nearby stream to prevent flooding. Courtesy of Invisible Structures (www.invisiblestructures.com).

Continued ▶

FIGURE CS 3-2B A worker unrolls sod over a plastic product called Grasspave, manufactured by a Denver-based company, Invisible Structures, Inc., in a new parking lot. Grass can be grown over the plastic substrate, thus reducing surface runoff, pollution of nearby waterways, and reducing heat accumulation. Courtesy of Invisible Structures (www.invisiblestructures.com).

into forecasting models. Forecasting is much less effective for flash floods due to their irregular nature.

It is said that no chain is stronger than its weakest link. This old adage can apply to disaster planning as well. The more complex plans become, the more likelihood something will go wrong. Moreover, flood warnings are only effective if heeded by the public. Lives were lost during the 1986 Napa, California flood because residents refused to evacuate mobile homes along the rapidly rising Napa River. Warning systems that combine a network of precipitation monitors, with stream gauges that measure peak flows in real time, can provide invaluable information to authorities and can save lives and property with community support. The Passaic (New Jersey, New York) Basin flood warning system is a good example.

Climate Change

We discuss global climate change elsewhere in the text. Here, we briefly point out that climate change can have material effects on flood hazard. Computer forecasts point to changing seasonal temperature maxima and minima, which will alter precipitation and evaporation patterns. For example, if winters warm in the Pacific Northwest, snow at higher elevations will in part be replaced by rain. This will alter the hydrology of entire river systems. The region depends on stream flow from melting snow for water in the dry summer months. Warmer winters can also result in more precipitation. Tropical storms may intensify and become more numerous as sea-surface temperatures increase.

The Passaic River Basin is shown in **Figure CS 4-1**. Note that it spans the states of New York and New Jersey.

Flooding has been a problem in the Basin since the nineteenth century when industrialization of the Passaic and its tributaries began. During the 1950s and 1960s, residential and industrial development expanded significantly. By 1994, flooding cost an estimated $116 million annually according to a U.S. Army Corps of Engineers (COE) report. An April 1984 flood killed three people, caused $462 million in damage, and displaced more than 6,000 residents.

The COE considered structural methods (such as dams, levees, and floodwalls) of flood control, but concluded that because the basin was densely populated and intensely developed, a structural solution was not feasible. Following an authorization by Congress in 1976 to identify a non-structural "solution" to the Basin's chronic flooding problem, the COE recommended an expansion and modernization of the existing National Weather Service (NWS) flood warning and response system. Such an expansion would permit collection and rapid dissemination of real-time stream level and precipitation data to alert government agencies to

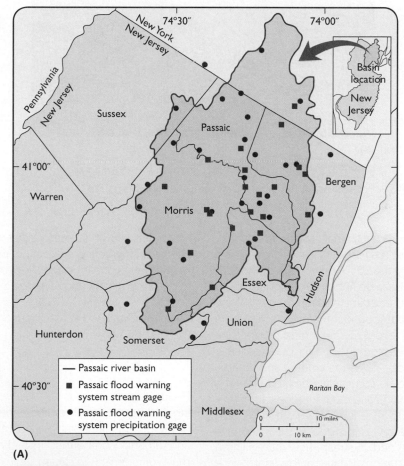

(A)

FIGURE CS 4-1   (A) The Passaic River Basin includes parts of 10 counties and 132 municipalities, covers about 2,400 square kilometers (926 square miles), and contains a population of over 2 million. It includes parts of the most developed metropolitan area in the United States. Modified from Passaic Flood Warning System, FS-092-98, USGS.

Continued ▸

(B)

FIGURE CS 4-1 *Continued.* (B) The lower reaches of the Passaic River showing the intensive development of this part of the watershed. Courtesy of Project Navigator.

impending floods, reduce or eliminate loss of life, and minimize property damage. The Passaic Flood Warning System (PFWS) was created by the USGS, the NWS, and the COE. The system collects discharge and precipitation measurements, which are used to estimate stream stages. Precipitation gauges installed throughout the drainage basin are used to measure and record the amount of precipitation that falls during a given storm. This information, together with NWS Doppler radar, allows the amount and distribution of precipitation to be accurately estimated. From this estimate, stream discharge at a given location can be predicted and converted to an estimate of stream stage.

The current system was installed in 1988, and consists of 35 rain gauges and 21 stream-flow meters linked to 12 computer base stations, located at 5 federal, 2 state, and 5 county offices. Three of the stream flow gauges and all of the rain gauges transmit real-time data to the computer base stations.

Flood watches and warnings are automatically transmitted to 15 high-risk municipalities from the base-station computers, allowing these localities to set up and carry out effective emergency evacuation procedures. Real-time knowledge of stream stage allow business and industry to reduce inventory and minimize infrastructure losses.

Concept Check 9-14. Assess strengths and weaknesses of the COE's nonstructural solution to the Passaic River's flooding.

Floods in the Passaic Basin also are responsible for spreading pollution. Since the nineteenth century, industrial development has put large quantities of toxic metals and organic compounds into the environment, and much of it remains in stream sediment. Whenever flooding occurs, some of that polluted sediment is mobilized. Polluted sediment also can be mobilized in great quantity from flooding of farmed areas, especially in regions that practice industrial agriculture, which use large quantities of fertilizers, pesticides, and herbicides.

■ STORM-SURGE FLOODS

Hurricanes, tropical depressions, and tropical storms may cause massive precipitation events that can lead to catastrophic flooding. In fact, such events are responsible for maximum-recorded flood events in regions like the eastern United States. The greatest potential for loss of life during a hurricane is from **storm surges**, which result when powerful, persistent winds "push" water onshore, as experienced by New York and New Jersey coastal residents during Hurricane Sandy in 2012. When coupled with high tides, and high sea surfaces due to extremely low atmospheric pressure, storm surges can be particularly destructive. During hurricanes, storm-surges can increase sea level by up to 15 feet (>4.5 m). Large areas of the United States' most densely populated and fast-growing Atlantic and Gulf of Mexico cities lie less than 10 feet (3 m) above mean sea level; thus, the danger there from storm surges is real and growing. In addition to the factors mentioned above, the height of a storm surge in a particular coastal area is also determined by the slope of the offshore continental shelf. A shallow slope off the coast, typical of much of the south Atlantic and Gulf coasts, will allow a greater surge to impact coastal communities.

Scientists at the U.S. National Hurricane Data Center have devised the SLOSH (Sea, Lake, and Overland Surges from Hurricanes) system to evaluate storm surges. This system uses computers fed with data on hurricane size, atmospheric pressure, forward speed, track, and wind velocity and direction, to estimate storm surge heights and winds. The forecast is reportedly accurate to ±20%. **FIGURE 9-21** shows a SLOSH simulation of storm surge magnitude along the U.S. Gulf Coast.

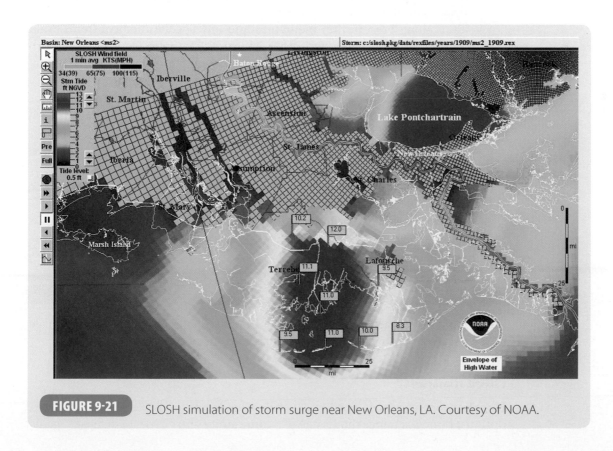

FIGURE 9-21 SLOSH simulation of storm surge near New Orleans, LA. Courtesy of NOAA.

Protecting Against Storm Surges

It is difficult to accurately forecast storm surges, because it is difficult to accurately determine how strong a tracked storm will be when it finally makes landfall. When authorities balance that uncertainty with the immense human and economic risks to their community, they often plan for a storm that is one category higher than what is forecast.

Many users of this book and/or their families will live in, or eventually move to, regions threatened by storm surges. Here is a list of rules to help citizens in storm-surge-prone areas protect themselves, their families, pets, and possessions against storm-surges.

- Pick the shortest distance you must travel to reach a safe location.
- Map out the nearest possible evacuation route. Do not get on the road without a planned route or a place to go. If possible, drive it before an evacuation order is issued.
- Select a friend or relative outside a designated evacuation zone and discuss your plan with them before hurricane season, or select a hotel/motel outside of the vulnerable area. If neither of these options is available, identify the closest possible public shelter.
- Use the evacuation routes designated by authorities.
- Notify your local emergency management office regarding anyone in your household needing special assistance to evacuate.
- Be aware that most public shelters do not accept pets.
- Secure your home prior to leaving by boarding doors and windows, securing or moving indoors all yard objects, and *turning off all utilities*.
- During hurricane season, keep your car filled with gas. Before leaving, withdraw extra money from your bank.
- Take along prescription medicines and any special medical items. Take along food and water, and other emergency supplies as well as important documents (proof of identity, home ownership, bank, family photographs, etc.)
- Recreational vehicles, boats, and trailers require extra evacuation time and help clog evacuation routes. Consider leaving them behind or start as early as possible.
- Evacuate when directed by authorities. To delay will only increase your chances of being stuck in traffic, or—even worse—not being able to get out at all.
- Expect traffic congestion and delays during evacuations.
- Stay tuned to a local radio or television station and listen carefully for any advisories or specific instructions from local officials. Monitor NOAA Weather Radio if possible.

■ FLOOD MITIGATION

A number of approaches have been tried in the struggle to reduce flood damage. Many are controversial. We list and discuss them below.

Flood Insurance

Homeowner insurance does not cover damage from floods. Federally subsidized optional flood insurance thus is one of a community's first "lines of defense" against economic losses from floods. Flood insurance rates are set by the Federal Emergency Management Agency (FEMA). The agency uses mapping of high-water marks, stream profiles, and RI determinations to prepare flood-hazard maps for a jurisdiction.

During the late 1990s, FEMA began to abandon its long-standing policy of repeatedly paying to rebuild structures situated in floodplains, so that they could be damaged again during the next flood. In one case, the same structures along the Russian River in Sonoma County, California had been destroyed and rebuilt nine times during successive floods, representing a severe waste of limited taxpayer funds.

Concept Check 9-15. What does it mean for flood insurance to be federally subsidized? Explain whether you agree or disagree with the practice.

Flood-Control Structures

Humans have been modifying river systems for millennia, but most of the early modifications were irrigation-related. In the United States, European settlers began damming rivers for power soon after their arrival.

After the 1803 Louisiana Purchase, the U.S. government set about mapping the Mississippi and its drainage system. The Lewis and Clark expedition and others were commissioned to map this drainage.

During the nineteenth century, settlements sprang up along the Mississippi, the Missouri, and other rivers of the interior, and settlers naturally began petitioning Washington for protection against the inevitable floods. The U.S. Army Corps of Engineers, through the Rivers and Harbors Act (1899) was given this responsibility, initiating a great episode of construction of *control structures* which continues today, although more and more opposition to such activity is being expressed each year.

Flood-control structures include dams, levees, revetments, weirs, and a number of smaller and more technical devices. They are built to store, confine, slow, or divert water during floods. To understand why their use is so pervasive, you should understand that many such structures have multiple purposes. For example, dams store water during floods, and that water then becomes available for recreation (houseboats, powerboats, marinas, motels, restaurants, etc). It also can be used to generate electricity for industrial and municipal use, and the water itself becomes available for agriculture, industry, and municipalities.

One usual consequence of dam construction is the alteration or destruction of native fisheries. Because dams alter stream flow, water quality, and river environments, environmentalists often oppose dam construction.

However, dams have *multiple, politically powerful constituencies* that enthusiastically support their construction, especially if those who benefit economically from the dam don't have to directly pay for it.

Cost-benefit analyses are often required prior to construction of flood-control structures. Such studies can then be used to compare effectiveness of alternatives to such structures. However, costs often include items hard or impossible to quantify, such as loss of fisheries, esthetics, or decreased water quality. Moreover, subsidies often mask the real cost of construction and operation. And benefits may be overstated, as benefits commonly depend on a complex set of economic assumptions. Much criticism of cost-benefit analyses thus has been expressed when applied to the subject of flood-control structures.

Dams

To be effective, a flood control dam doesn't actually have to have water behind it. In fact, some flood control dams in Southern California have gone three decades without storing any water (**FIGURE 9-22A**)!

Dams are typically built and maintained by Federal and State agencies, and generally benefit localities, an ideal situation from the point of view of local inhabitants. Most states are prohibited from deficit spending, a prohibition that does not apply to the federal government. The federal government therefore has ample resources to call upon to pay for flood-control structures.

Levees

Levees are natural features: low ridges formed when floodwaters overtop a river's channel and dump debris alongside the channel where velocities drop abruptly (**FIGURE 9-22B**).

Man-made levees are embankments constructed or enlarged parallel to a river to prevent flooding. Levee construction artificially deepens river channels, allowing the river to transport more water during floods. They prevent the water from occupying parts of the river's flood plain that, in the view of inhabitants, then can be used for other purposes, like agriculture and habitation.

The advantages of levee building are similar to those for dams: they protect valuable real estate and human life; they are generally paid for by those who are not the immediate beneficiaries; and they allow, at least temporarily, additional land to be used for agriculture, industry, commercial, or municipal development. They commonly increase the value of land on flood plains, because levees generate the perception that the land is now "safe" for development. However, levees do not permit floodwaters to replenish the soils on floodplains with nutrients, nor do they permit the floodplains to naturally filter floodwaters of sediment or pollutants. In addition, once levees are built, they must be maintained permanently at someone's considerable expense.

Severe floods on the Mississippi over the past decades, especially the floods of 1993, have caused many to question this reliability on flood-control structures, because the cost of floods has increased due to the increased value of structures built in harm's way.

Revetments

Revetments are structures placed on a riverbank to protect it from wave action or current. **FIGURE 9-22C** shows a few examples. Their purpose is to stabilize riverbanks, preventing lateral or downstream bank migration.

Weirs

Weirs are low dams, fences, or walls constructed across a river or stream to increase channel roughness. This increases frictional forces operating on river bottoms. In this manner, the river's velocity will decrease, reducing potential damage to structures downriver, like bridges (**FIGURE 9-22D**).

Channelization

Channelization involves the straightening and stabilization of river channels to allow rivers to transmit more runoff quicker through an area (**FIGURE 9-22E**).

(A)

(B)

(C)

(D)

(E)

FIGURE 9-22 Flood control methods.
(A) The Prado dam in Orange County, CA.
Courtesy of Space Imaging. (B) A levee built along
the 17th Street Canal, New Orleans, LA. Levees
often increase flood hazard by encouraging
development in the now "protected" area behind
the levee. Courtesy of Jacinta Quesada/FEMA.
(C) A revetment. © iStockphoto/Thinkstock.
(D) Weirs along the Almond River near Edinburgh,
Scotland. © Crepesoles/ShutterStock, Inc. (E) The
channelized Kissimmee River, South Florida.
The federal government and the State of Florida
have agreed on a multi-billion dollar plan to
begin restoration of the Everglades. Part of the
plan calls for restoring the natural channel of the
Kissimmee. Courtesy of Tony Santana/U.S. Army
Corps of Engineers.

Channel modification, sometimes including cementation, is commonly combined with flood-storage reservoirs, floodwalls, and other structures in a comprehensive flood-mitigation plan. The U.S. Army Corps of Engineers has decades of expertise in designing such comprehensive plans to address chronic flooding in built-up areas.

Channelization is an increasingly controversial practice. Channelization of streams and rivers can lead to downstream bank erosion and collapse, increased downstream sedimentation, and increasingly severe downstream flooding.

Wetlands

Wetlands have been long recognized as a powerful tool in flood mitigation. Wetlands on floodplains soak up floodwaters and filter pollutants. Unfortunately, during the past two centuries up to 90% of wetlands along rivers have been destroyed by development and converted to farms, cities, suburbs, and roads. The same is true for coastal wetlands, which have an important role in mitigating floods from tropical storms (FIGURE 9-23).

Many agencies, such as the Chesapeake Bay Foundation, are working to restore wetlands. It remains one of the nation's great environmental challenges.

Land-Use and Zoning Regulations

As more and more communities experience floods from which residents believed themselves to have been immune, communities and governments have sought to mitigate flood effects by prohibiting or regulating land use in flood-prone areas. To do this, agencies must first determine the extent to which an area is at risk of flooding. This involves the preparation of flood-hazard studies, which we briefly mentioned earlier. FEMA has responsibility for implementing the federal flood-hazard program. To that end, FEMA publishes Flood Hazard Maps, which we briefly discussed previously. Such maps divide cities and counties into zones of relatively greater or lesser flood hazard. These maps are based on flood-frequency diagrams as well as on meticulous surveys of flood plains. The intent is to relate discharge to elevation on a flood plain, and then to relate both to the likelihood of a given discharge's recurrence.

Once a jurisdiction has a flood hazard map it can regulate land use based on relative flood threat. Schools and other essential public buildings would be built far away from the most flood-prone areas, and the most threatened areas would ideally be reserved for minimal-impact use, such as parks, green space, and golf courses.

One major drawback of flood-hazard mapping is that land-use changes throughout the drainage basin that occur after the map has been prepared, can render the map's zones increasingly obsolete. To be optimally effective, therefore, such maps should be revised regularly. Unfortunately, such activities require scarce revenues to be expended, because even hydrologists and geologists like to be paid. Many localities are reluctant to pay for constant revision of flood-hazard data in the face of relentless development. And, because development's benefits are local while FEMA's costs are national, localities have little incentive to restrict development of valuable riverfront real estate.

FIGURE 9-23 (A) Wetlands, including coastal wetlands shown here, are found in many locations, along rivers or at the mouths of rivers. They are also found in low depressions where groundwater intersects with the land surface or around lakes and ponds. (B) Wetlands in the process of being destroyed for farms and industries in Brazil. © JAMIL BITTAR/Reuters/Landov.

Attempts by localities to control land use in hazardous areas has met opposition by groups using the U.S. Constitution's "Takings" Clause. The Fifth Amendment prohibits private property from being "taken for public use without just compensation." "Property rights" advocates argue that restricting a landowner's ability to sell or develop his or her property constitutes a "taking" and requires compensation, even though the property has not been seized or appropriated.

■ CONCLUSION

Floods have been impacting human settlements for millennia. For centuries, globally we have been modifying rivers to suit our own needs. We build more and more infrastructure in flood-prone areas, and profess surprise when flood damages increase.

Flood-control structures are a response to our natural desire to be protected from floods, especially if someone else can be convinced to pay for it. In recent years, the increasing costs of flood control structures and their adverse environmental impact have focused more on regulation of—and even removal of—structures or even towns from flood-prone areas.

As populations increase in flood-prone areas, flood mitigation measures will become ever more important and costly. Already, insurance companies are reducing their exposure to claims by canceling policies and refusing to write new policies in cities and entire regions prone to floods. Some projections of the impacts of climate change include more frequent and increasingly intense river floods and hurricanes. Eventually, humans may have to reevaluate whether we can afford a "come one come all" policy governing population growth in flood-prone areas.

CHAPTER SUMMARY

1. The Red River of the North Basin includes parts of Minnesota, North Dakota, South Dakota, and Manitoba, Canada. Five factors contribute to "normal" Red River flooding. It is a north-flowing, young stream, with an extremely low gradient, occupying a narrow channel on a former lake bed. Ice jams are a common feature in spring.

2. During the twentieth century, floods caused more property losses and deaths than all other natural hazards combined.

3. Most lives are lost when people are swept away by floodwaters, but most property damage occurs when structures are inundated by sediment-laden water.

4. Floods are a natural part of the hydrologic cycle. But human interference with natural ecosystems can increase flood severity and even cause floods.

5. Rivers on alluvial fans normally transport large amounts of mainly coarse sediment during rare but intense floods. Many of the rapidly urbanizing regions of the western United States are expanding onto alluvial fans drained by braided rivers and are increasingly susceptible to flash floods.

6. Flash floods of meandering rivers are usually local events, induced by anthropogenic land-use changes.

7. Climate determines the seasonal abundance, location, and duration of storms, and the seasonal fluctuation of soil moisture, snow cover, and rate of snowmelt.

8. Individual rivers and streams are parts of a drainage system or net. Each individual stream has its own watershed or drainage basin.

9. The formula for discharge (Q) is $Q = A \times v$, where A is cross-sectional area, and v is velocity. Hydrologists are interested in the discharge of a river and the variability of the discharge over time.

10. When floodwater velocities exceed three ft/sec (90 cm/sec) it becomes difficult, if not impossible, for adults to maintain their balance while walking through a flooded area.

11. The recurrence interval (RI) estimates statistically the amount of time between recurrences of a discharge (flood) of a given magnitude. The annual exceedance probability (P_e) is the probability that a given maximum discharge will occur in a year. It is the inverse of the recurrence interval. RI studies must be constantly revised to keep pace with development and other land-use changes.

12. Discharges for 100-year floods and greater can be approximated by constructing flood-frequency diagrams.

13. Factors that determine how much damage a flood will cause are: intensity and duration of precipitation, season, land-use on floodplain, speed of rise and duration of flooding, sediment transported and deposited, efficiency of forecasting, warning and planning, and climate change.

14. Urbanization has been found to increase precipitation intensity and runoff, and thus flood damage.

15. Flood warning systems that combine a network of precipitation monitors, with stream gauges that measure peak flows in real time, can provide invaluable information to authorities and can save lives and property.

16. Hurricanes, tropical depressions, and tropical storms may cause massive precipitation events, and can lead to catastrophic flooding. Storm surges result when powerful, persistent winds "push" water onshore. It is difficult to accurately forecast storm surges, because it is difficult to accurately determine how strong a storm will be when it finally makes landfall.

17. A number of approaches have been tried in the struggle to reduce flood damage. They include: federally subsidized flood insurance; installation of flood-control structures such as dams, levees, revetments and weirs; channelization of streams; and the reconstruction and protection of wetlands.

18. Flood hazard maps divide cities and counties into zones of relatively greater or lesser flood hazard.

KEY TERMS

annual exceedance
  probability

annual peak
  discharge

array

development

discharge

dissolved load

flash flood

flood-control
  structures

flood-frequency
  diagram

flood stage

hydrologic cycle

hydrology

ice-jam flood

infiltration capacity

lag time

land-use changes

natural levee

recurrence interval

river gradient

soil moisture

stage-discharge
  relationship

storm surge

stream hydrograph

suspended load

watershed

REVIEW QUESTIONS

1. The Red River of the North has an extraordinarily low gradient because
   a. it occupies an ancient failed rift.
   b. it occupies an ancient lake bed.
   c. it flows over extremely weak sediment.
   d. all of the above.

2. Braided rivers differ from meandering rivers in that braided rivers
   a. have only one channel.
   b. have no gradient.
   c. have many channels.
   d. do not develop any channels.

3. Which is true about meandering rivers?
   a. They develop in humid regions of relatively low relief.
   b. They generally have only one channel.
   c. They have a generally well-developed flood plain, compared to braided rivers.
   d. All of the above.
   e. None of the above.

4. A water body (a river, lake, or small stream) and the land surface that drains into that water body defines a(n)
   a. alluvial fan.
   b. flash flood.
   c. wetland.
   d. watershed.

5. The formula for discharge (Q) is
   a. Q = A/v.
   b. Q = v/A.
   c. Q = Av/RI.
   d. Q = n + 1/m.

6. The maximum discharge at a locality in one water year is referred to as
   a. the array.
   b. the infiltration capacity.
   c. the drainage net.
   d. the annual peak discharge.
   e. the annual peak array.

7. The height of water in a river, above some arbitrary datum is called the
   a. array.
   b. stage.
   c. discharge.
   d. flood level.

8. Recurrence interval (RI) is expressed as
   a. $RI = (N + 1)/M$.
   b. $RI = Qv$.
   c. $RI = M/(N - 1)$.
   d. $RI = M/(N - 1)$.

9. The annual exceedance probability is
   a. the inverse of the RI.
   b. the inverse of the discharge.
   c. the inverse of the stage.
   d. the reverse of the array.

10. If soil infiltration capacity is high before a rainstorm, other things being equal, flood hazard is
    a. reduced, because the soil can soak up some of the precipitation.
    b. reduced, because more precipitation will run off.
    c. increased, because more precipitation will runoff.
    d. unaffected, because the flood risk is unrelated to soil infiltration capacity.

11. Hurricanes can cause disastrous flooding. The hurricane "season" in North America is
    a. from early June through November.
    b. from March 1 to June 1.
    c. from early April through August.
    d. year-round.

12. Impervious surfaces
    a. inhibit runoff, favoring infiltration.
    b. inhibit infiltration, enhancing runoff.
    c. inhibit evaporation, favoring infiltration.
    d. enhance infiltration.

13. Which is correct concerning the effect of development on runoff?
    a. Development has a greater effect on runoff for low RI floods than for high RI floods.
    b. Development affects floods of all RIs equally.
    c. Development has a greater effect on high RI floods than on low RI floods.
    d. Evidence that development has any effect on floods is entirely anecdotal.

14. Which is the correct statement of the relative sediment yield from forest, row crops, and urban areas?
    a. Forests most, urban areas next, row crops least.
    b. Urban areas most, row crops next, forests least.
    c. Row crops most, forests next, urban areas least.
    d. Row crops most, urban areas next, forests least.

15. Floods have been responsible for spreading pollution, such as toxic metals and organic chemicals, in what watershed mentioned in this chapter?
    a. Passaic, New Jersey Basin
    b. Albuquerque, New Mexico Basin
    c. Las Vegas, Nevada metropolitan area
    d. Tucson, Arizona metropolitan area

16. The agency responsible for directly administering the federal flood insurance program is the
    a. Department of Housing and Urban Development.
    b. Environmental Protection Agency.
    c. Department of Homeland Security.
    d. Federal Emergency Management Agency.

17. Dams have been a favored means of flood control until recently because
    a. they have been universally effective at controlling floods and thus have overwhelming scientific support.
    b. they have multiple uses and multiple constituencies, so they have powerful political support.
    c. they have no effect on river biotas, so dam construction is supported by aquatic biologists.
    d. All are true.

18. Embankments constructed or enlarged parallel to a river channel to prevent flooding are called
    a. revetments.
    b. weirs.
    c. levees.
    d. fills
    e. cuts.

19. Which is true concerning wetland loss in the United States?
    a. Wetland loss along rivers and coasts has reached 90%–95% in some areas.
    b. Although wetlands were destroyed in the nineteenth century, they were largely restored during the twentieth century.
    c. More than 50% of all American wetlands have been preserved since Colonial times.
    d. Converting wetlands to other land uses is presently illegal in the United States.

20. Climate change is likely to
    a. increase the number and severity of floods in much of North America.
    b. reduce the number but not the severity of floods.
    c. reduce the severity but not the number of floods.
    d. reduce both the number and severity of floods, by reducing incidence of hurricanes and tropical storms.

Surface Failure: Mass Wasting and Subsidence

10

CHAPTER OUTLINE

■ INTRODUCTION

This chapter deals with movement of the Earth's surface material under the influence of gravity. This hazard is widely called **mass wasting**, and may also be called *slope failure*. It may be the most widespread geologic hazard on the Earth. The requirements are simple: gravity, slopes, and/or unstable surface material. Although mass wasting is often triggered by volcanic eruptions and earthquakes, the most common cause of slope failure is intense rainfall in regions with steep slopes, as we illustrate below.

Mass-Wasting from Hurricane Mitch, 1998

Hurricane Mitch (Oct. 26–Nov. 4, 1998) was probably the deadliest and most costly hurricane to hit Central America in 200 years (FIGURE 10-1). Mitch killed at least 11,000 people and cost $5 billion in Honduras and Nicaragua, about equal to one year's total Gross Domestic Product (GDP)[1] and dumped as much as 75 inches (190 cm) of rain on Nicaragua and Honduras. It set off an estimated *one million slope failures,* virtually destroying the entire infrastructure of Honduras, with 70%–80% of transportation infrastructure lost, and devastated parts of Nicaragua in the process (FIGURE 10-2).

Guatemala, Belize, and El Salvador also suffered losses of life and property. In addition, floods and slope failures in Nicaragua remobilized thousands of land mines emplaced during that country's recent civil war.

The President of Honduras, Carlos Flores, stated that Mitch eradicated "50 years of progress. We have before us a panorama of death, desolation and ruin . . . there are corpses everywhere, victims of the landslides or of the waters."

Casita Volcano, in western Nicaragua, was the site of one large slope failure. The crater area received torrential rainfall during the storm, causing a massive rock avalanche. Hundreds of thousands of cubic meters of volcanic debris washed off a fault zone down a narrow valley, followed by a mudflow (FIGURE 10-3). Nearly 2,000 people lived in the two villages. Few survived. Nicaraguan Ambassador to the United States Alfonso Ortega Urbina said, "The situation of the people is tragic. They don't have

FIGURE 10-1   Hurricane Mitch approaches Honduras. Note the size of the hurricane. Courtesy of NASA.

FIGURE 10-2 A landslide in Tegucigalpa, Honduras triggered by Hurricane Mitch: volume 6 million m³. Photo by E.L. Harp/USGS.

food. They don't have shelters. They don't have clothes. We will not have crops for the next six months so there will be hunger."

Mass Wasting in China

The following excerpt is from a report compiled by Columbia University's Earth Institute. We include it to give you an example of the awesome impact mass wasting can have when driven by rainfall.

FIGURE 10-4 shows areas affected by heavy rainfall in China, June 2002.

> *"The rainy season in China began with heavy rainfall in early and mid-June and led to severe flooding and landslides in central and northwestern China, affecting the following provinces: Shaanxi, Sichuan, Gansu, Hubei, Guizhou, Xinjiang, and Chongqing municipality. According to the Chinese government, 38 million people had been affected, 453 were killed, as many as 450,000 housing units collapsed and 1,167,000 were damaged. 1.47 million hectares of crops were damaged, 0.35 million hectares of crops were lost, and the economic cost was estimated at more than $2 billion (U.S.). Extensive damage was also done to bridges, roads, rail lines, hydropower stations, and telecommunications."*

December 15, 1998

FIGURE 10-3 Debris flow at Casita Volcano in Nicaragua after Hurricane Mitch. Mud flows covered an area 10 miles long and 5 miles wide. Courtesy of USGS.

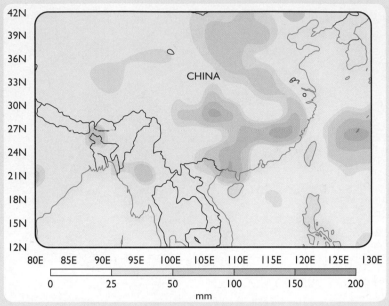

■ WHAT EXACTLY IS "MASS WASTING?"

Could these tragedies have been avoided with a greater understanding of the nature and causes of mass wasting? You decide after reading this chapter.

There is no general term for *mass wasting* of surface material under the influence of gravity, although informally many use the term "landslide." Landslide, however, refers in geologic parlance to a specific type of slope failure. We therefore will use mass wasting and slope failure interchangeably, even though the terms are somewhat cumbersome (**FIGURE 10-5**).

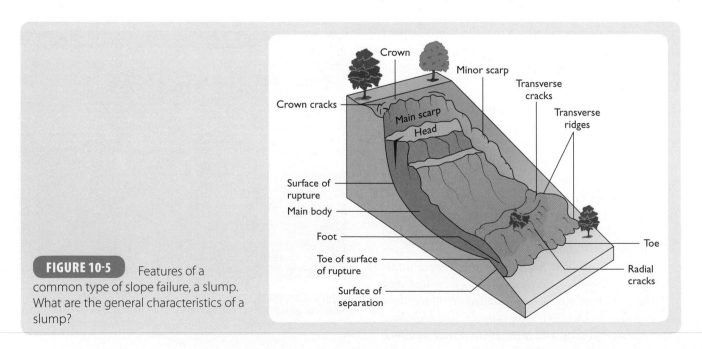

FIGURE 10-5 Features of a common type of slope failure, a slump. What are the general characteristics of a slump?

Geologists recognize several categories of slope failure based on what is moving, how it moves, whether or not it is wet, and how fast it moves. A number of classification schemes have been used.

■ CLASSIFYING MASS WASTING

Methods of classifying slope failure vary with the purpose for which the classification was designed. Some specialists even wonder whether classifications for mass wasting are useful at all, given the wide variation between individual failure events, or the lack of quantitative means to set boundaries between different types. **FIGURE 10-6** shows a simple classification.

Concept Check 10-1. Examine Figure 10-6. What are the criteria for classifying mass wasting in this scheme? What might be the advantages of this scheme?

Four widely used terms to classify slope failures are:

- Slow "flows" such as creep and solifluction
- Rapid "flows" such as mudflow, debris avalanche
- Slides including slump, rock slide, rock fall, and landslide
- Subsidence, which is the sinking of mass

Another simple classification is:

- Falls
- Flows: granular or slurry
- Slides: rotational or translational

■ WHAT CAUSES MASS WASTING?

Many geologists divide causes of slope failure into **external factors** and **internal factors**. External factors include *disturbing forces* like earthquakes and heavy traffic. Internal factors are those that change the *strength and properties* of the mass itself. Changes in both external and internal factors can lead to rapid slope failure, but failure can also be so slow as to be almost imperceptible.

The Balance of Forces Concept and Slope Failure

Almost none of the Earth's continental surface is perfectly flat. Slopes dominate surface topography. They represent a dynamic balance between *stabilizing forces,* such as *friction* and other forces that hold slope material in place, and *disturbing forces* enhanced by gravity, that pull material downslope. Geologists refer to the gravitational force that pulls material downslope as **shear stress**. Shear stress increases as weight of material and *slope angle* increases.

The resistance within the surface material to shear stress is called **shear strength**; that is, cohesional (forces that pull together) and frictional forces (those that resist movement) within the material. Cohesional forces include cements in sedimentary rock and interlocking mineral textures in many igneous and metamorphic rocks.

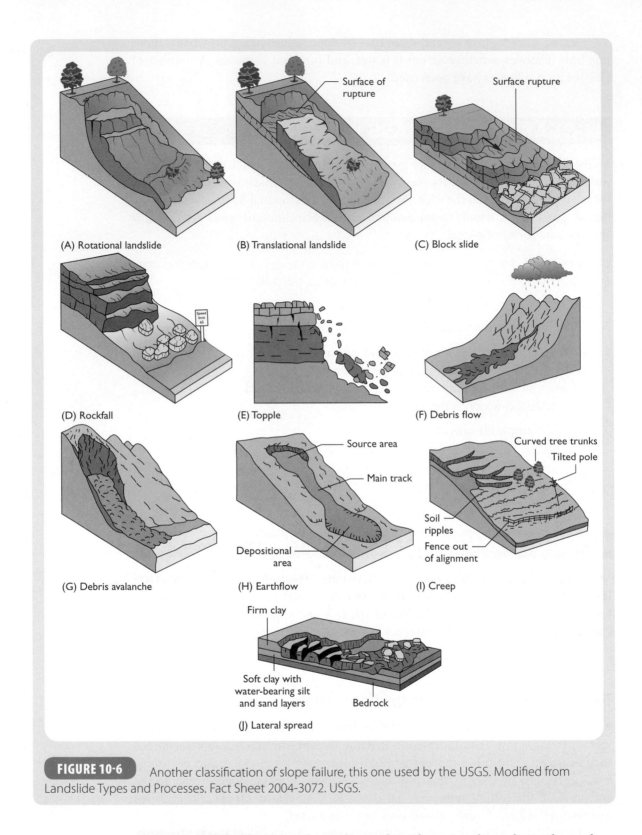

(A) Rotational landslide

(B) Translational landslide
Surface of rupture

(C) Block slide
Surface rupture

(D) Rockfall

(E) Topple

(F) Debris flow

(G) Debris avalanche

(H) Earthflow
Source area
Main track
Depositional area

(I) Creep
Curved tree trunks
Tilted pole
Soil ripples
Fence out of alignment

Firm clay
Soft clay with water-bearing silt and sand layers
Bedrock
(J) Lateral spread

FIGURE 10-6 Another classification of slope failure, this one used by the USGS. Modified from Landslide Types and Processes. Fact Sheet 2004-3072. USGS.

Shear strength holds material together and in place. Anything that reduces shear strength reduces stability.

On a stable slope, shear strength exceeds shear stress. When shear stress approaches shear strength, the slope becomes unstable. If shear strength is exceeded by shear stress, movement can occur even on gentle slopes. **FIGURE 10-7** illustrates the concept.

■ CONTROLS OF SLOPE STABILITY

The controls of slope stability are: (1) the nature of surface material; (2) abundance of water; (3) slope angle; (4) nature of disturbing forces; and (5) the subsurface geology.

Nature of Surface Material

Surface material can be *rock* (igneous, metamorphic, sedimentary), *sediment* (mixtures of coarse to fine angular to rounded debris), **regolith** (a general term for weathered rock), and *soil,* which itself may contain variable amounts of sediment of differing grain sizes with organic matter called *humus.* **Colluvium** is a general term for loose slope material. The texture (grain size, size distribution, grain orientation) of loose slope material determines its **angle of repose**, that is, the maximum angle at which unconsolidated material is stable. FIGURE 10-8 shows examples.

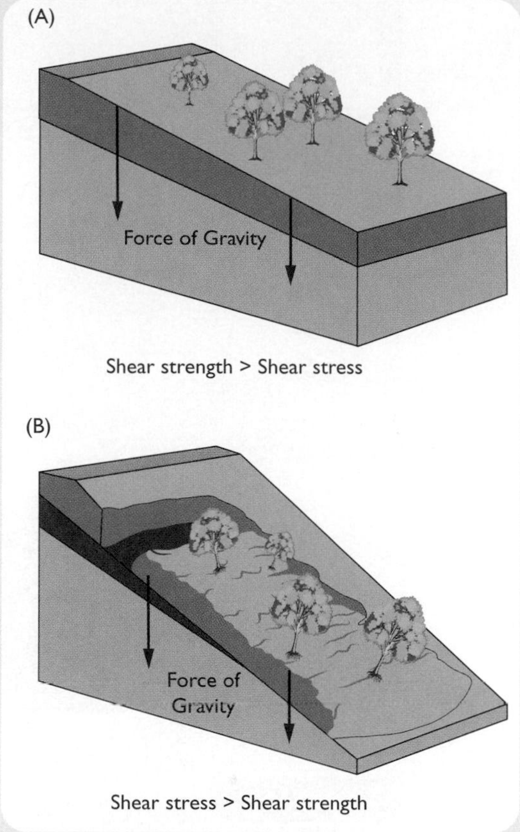

Importance of Water

Intense rainfall on steep slopes, as you saw previously, is one of the most common triggers of mass wasting. Water and/or ice also may be present in the surface material. In small amounts, water may help hold particles together (liquefaction is discussed elsewhere in this text). Here's how. Water is a **polar molecule**, that is, like a bar magnet with positive and negative poles (FIGURE 10-9). And, like magnets, the opposites—the positive of one molecule and the negative of another—attract and form a weak chemical bond called a *hydrogen bond.* Though each individual bond is weak, the sum of all the bonds together acts somewhat like cement to hold soil particles together. Thus, slight amounts of water at grain boundaries is a cohesive force.

As water content increases in bedrock or regolith, however, the slope becomes weaker for three reasons.

Concept Check 10-2. Before reading further, list reasons why you think water weakens slope material.

1. Water adds mass ("weight") to the slope material, increasing shear stress.
2. Water may move within the material, reducing friction and perhaps dissolving or chemically weathering material within the slope.
3. The **pore-water pressure** commonly builds up, especially in loose material, which can reduce material strength. The pore-water pressure is simply a force, like buoyancy, exerted by the water in pore spaces, because water cannot be compressed. Below the groundwater table, pore-water pressure is greater than atmospheric pressure and can destabilize slope material.

(A)

Force of Gravity

Shear strength > Shear stress

(B)

Force of Gravity

Shear stress > Shear strength

FIGURE 10-7 (A) A stable slope, in which the shear strength exceeds the shear stress. In other words, cohesional (forces that pull together) and frictional forces (those that resist movement) within the material are greater than gravitational forces. (B) An unstable slope, one where disturbing forces, mainly gravitational, are greater than stabilizing forces, causing slope failure.

FIGURE 10-8 (A) A pile of sediment in a sand pit that assumes a slope determined by the material's angle of repose. © iStockphoto/Thinkstock. (B) A stream flowing on a beach, eroding loose sediment. Notice that the angle of repose is virtually the same along the entire length of the stream. This is because the sediment has a nearly constant grain size. Courtesy of David Bailey.

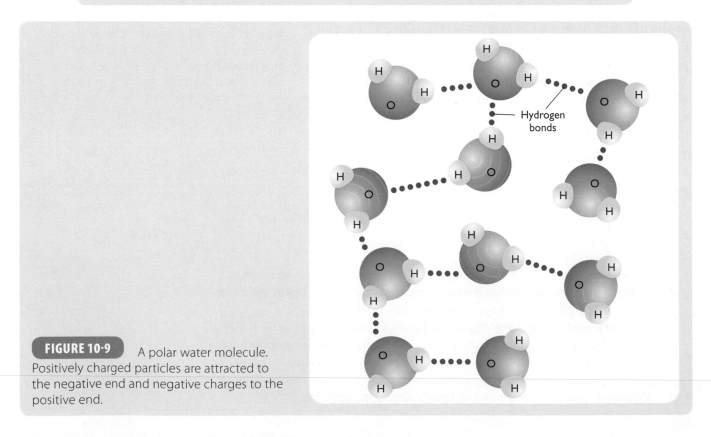

FIGURE 10-9 A polar water molecule. Positively charged particles are attracted to the negative end and negative charges to the positive end.

The content and distribution of water in slope material is related to the material's *porosity* (a measure of the openings or void spaces in a substance) and *permeability* (how easily a material transmits fluids). We discuss these concepts elsewhere in the text. Review that material now if you like. The amount and distribution of water is important if we wish to know how the slope material will behave during failure. Here is one scheme of classifying water content.

- *Dry:* Contains no visible moisture
- *Moist:* Contains some water, coating pore surfaces
- *Wet:* Contains enough water to behave in part as a fluid, has water flowing from it, or supports significant bodies of standing water
- *Very Wet:* Contains enough water to allow the mass to flow as a liquid under low gradients

Importance of Slope Angle

Other things being equal, it is intuitive that the steeper the slope, the greater the likelihood of slope failure. The reason is simple: the driving mechanism for mass wasting is *gravity*. A force acting on a body is expressed by the relationship $F = M \times A$, where F is force, M is mass, and A is acceleration. In the case of slopes, A is the acceleration of gravity. If the surface is horizontal, gravity acts at right angles to the slope, and there is no "downhill" component. If the surface is vertical, gravitational force is at a maximum, because the force is parallel to the slope. Rock and debris falls may occur, as they might on a slope if gravitational forces are strong enough (see Figure 10-7).

Causes of Slope Failure

Downslope movement can be initiated in six ways:

- *By steepening the slope, increasing shear stress.* This can happen if, for example, material at the base of a slope is excavated by mining, road building, or river erosion. This can also occur along a coast, as material at the base of cliffs is removed by wave erosion (FIGURE 10-10).
- *If the weight of surface material increases, increasing shear stress.* Placing buildings on a slope, piling material on a slope, saturating the slope during heavy rains, or even watering lawns can increase the weight of surface material. Intense rainfall commonly triggers slope failures, as you saw earlier with Hurricane Mitch. A storm in January 1982 in the San Francisco Bay Area generated nearly *18,000* slope failures.
- *When surface material is intensely shocked, reducing shear strength.* Shocks generally result from earthquakes, but even constant heavy truck traffic or blasting during mining can induce mass movement. **Sturzstrom** (German *sturz* to fall and *strom* to stream)

FIGURE 10-10 Wave erosion removes material at the base of the cliff, changing the slope. Photograph by Hilary Stockdon/USGS.

is a type of slope failure caused by a volcanic eruption, earthquake, or heavy rains, although water need not be present for the landslide. Interestingly, the horizontal movement is usually twice or more as large as its vertical drop, and its energy is much higher than a typical landslide.

- *When pore-water pressure is increased.* This happens when water saturates a slope. The pores in the slope material fill with water, displacing the air.
- *When groundwater moves through the surface material.* Groundwater can dissolve or wash away material from pore spaces, especially clays. This is important because the surfaces of clay minerals are "sticky." Mineralogists say they have *unsatisfied ionic charges.* They can attract adjacent grains and help to hold the slope material together. Over time, removing the clays can weaken the slope by reducing friction (shear strength), causing failure.
- *When surface rocks weather.* Over time, contact with the atmosphere, a process known as **weathering**, can weaken rocks structurally. For example, chemical weathering (breakdown by chemical rather than physical processes) converts feldspar, a mineral common in the Earth's crust, to clay. Mechanical weathering, in which rocks break down with no change in their chemical composition, increases rock volume and forms minute fracture surfaces, allowing fluids to enter the material more easily, thus reducing shear strength.
- *When vegetation is removed.* Clear-cutting forests can weaken slopes by increasing runoff and destroying roots that hold material in place, thus exposing the regolith to the erosive effects of rain (**FIGURE 10-11**).

FIGURE 10-11 Satellite photo of clearcuts on slopes in a one-mile square in the Olympic National Forest, Washington from 1986. Courtesy of NASA/Goddard Space Flight Center Scientific Visualization Studio.

Subsurface Geology and Slope Failure

The geology of a region strongly influences slope stability. By geology we mean the nature and *orientation* of the rock and sediment underlying an area, as well as the nature of geological contacts and planes (e.g., faults) separating, and within, rock strata. You already have seen that the slope, the nature of surface debris, and water can profoundly affect slope stability.

Here are some geological conditions that lead to unstable slopes:

- *If geological strata (layers) dip in the same direction as the slope.*

    In geological formations, separations between rock layers are known as *bedding planes.* When they are formed, these planes are horizontal (that is, parallel to the horizon). If cracks (called joints) develop at right angles to bedding, which they commonly do, water can move down the joints and flow down along bedding planes. This will decrease friction and reduce stability, causing overlying rocks to slide downslope. This is the situation that led to one of the great natural disasters of the 1960s in northern Italy, the Vaiont Dam flood. Nearly 2,000 people were killed in the flood that occurred when the landslide displaced the water out of the reservoir.

- *If horizontal beds crop out on steep slopes or cliffs, with weaker strata alternating with stronger layers.*

    The more rapid weathering of weaker layers can undercut stronger layers, leading to collapse. This situation is common in the region known as the Allegheny/Cumberland Escarpment (**FIGURE 10-12** ; an **escarpment** is a steep slope or cliff between two land areas of different elevation). It is located in the eastern United States, where alternating layers of horizontal sandstone, conglomerate, shale, and coal (see Classification of Rocks elsewhere in this text) crop out on steep slopes. We feature this region in our next Case Study.

- *Stream-dissected basalt plateaus can be prone to slope failure (**FIGURE 10-13**).*

    Examples include: the Columbia Plateau in the northwestern United States; parts of the state of Queensland in Australia; and the Isle of Skye off the west coast of Scotland. In areas such as these, especially in humid climates, basalts weather rapidly to produce abundant debris from clays (that expand when wetted), to large boulders. Groundwater movement can move down and through permeable basalts, resulting in springs issuing forth on slopes, weakening slope debris.

- *Anywhere rocks weather deeply in areas of high rainfall (**FIGURE 10-14**).*

    Examples include Jamaica and other Caribbean islands, and Yosemite National Park, USA. Here, extensive slope debris can form and creep downhill. Slope vegetation, with strong root systems, can help retard downhill creep, but if vegetation is removed by clear-cutting—for mining, agriculture, or road building, for example—rapid failure can result.

- *Slopes underlain by weak Tertiary (~65–2 million years ago) strata, at any orientation.*

    Examples include Coastal California and Australia. Here, undercutting of slopes by road building or marine erosion can quickly lead to failure.

Concept Check 10-3. CheckPoint: List and discuss three causes of slope failure.

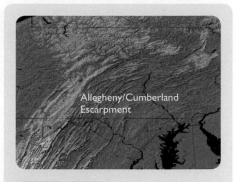

FIGURE 10-12 The Allegheny Front/Cumberland Escarpment. Ridges are underlain by strong rocks, and valleys underlain by weaker rocks. Courtesy of NOAA/NWS.

FIGURE 10-13 Basalt flows crop out on a slope in the Columbia Plateau, Northwest United States. © Kevin Schafer/Alamy.

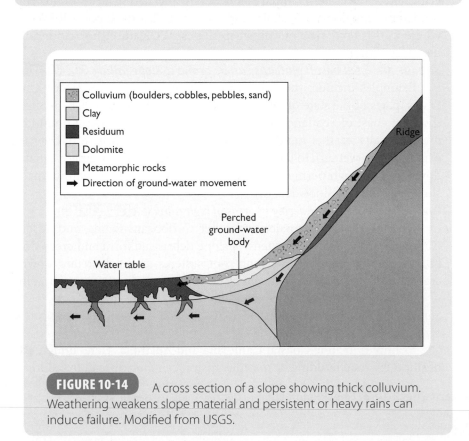

Colluvium (boulders, cobbles, pebbles, sand)
Clay
Residuum
Dolomite
Metamorphic rocks
→ Direction of ground-water movement

Ridge

Perched ground-water body

Water table

FIGURE 10-14 A cross section of a slope showing thick colluvium. Weathering weakens slope material and persistent or heavy rains can induce failure. Modified from USGS.

■ RECOGNIZING AREAS PRONE TO SLOPE FAILURE: SUMMARY

Certain characteristic landforms, geologic properties, and surface features increase the risk of slope failure. These characteristics include:

- A slope angle greater than 15°.
- Evidence of slope failure during the last 10,000 years.
- Stream or wave activity that has caused the adjacent land to be unstable.
- Buildup of snow on steep slopes.
- An *alluvial fan,* with evidence of debris flow or flood activity (**FIGURE 10-15**).
- Layers of impermeable sediment, such as silt or clay, interlayered with permeable sediment such as sand and gravel.

■ MASS WASTING HAZARDS

Impacts of mass wasting can include damage or destruction of roads and railroads, sewer and water lines, homes, and public buildings. Disruption of shipping and travel routes can cause economic losses. Slope failure may dam rivers, forming artificial

FIGURE 10-15 An alluvial fan in southern Iran. Note the channeling and the irrigated fields. Courtesy of Jesse Allen and NASA/GSFC/METI/ERSDAC/JAROS, and the U.S./Japan ASTER Science Team.

lakes, which can eventually cause flooding. Sometimes slope failure can dump huge volumes of debris into reservoirs, causing disastrous floods, like the Vaiont Dam disaster described previously. Debris flows associated with flash floods on alluvial fans can sweep away campers, vehicles, livestock, vegetation, and even small communities. Lahars generated on volcanoes can cause catastrophic damage and loss of life (discussed elsewhere in the text).

Creep

Creep refers to very slow, sometimes imperceptible downslope movement of surface material over a long period of time. Creep can affect rock, sediment, and soil, and anything on it. Even though we can't see creep, there are telltale signs,

---

CASE STUDY — Interstate Highways and Mass Wasting in Tennessee

Slope failure is commonly triggered by human activity. Here is an example from the construction of interstate highways in Tennessee.

Numerous landslides and slumps occurred during the construction of I-40, I-64, and I-75 interstate highways in Tennessee in the 1960s and 1970s, chronicled by engineer Dave Royster from the Tennessee Department of Highways.

Most slope failure occurred when roadways were cut through the Cumberland Escarpment (recall that an escarpment is a steep slope or cliff between two areas of different elevation) in east Tennessee. This escarpment separates the folded and faulted rocks of the Valley and Ridge geological province from the horizontal younger rocks of the Cumberland Plateau.

Geology of the Cumberland Escarpment

The Cumberland Escarpment is underlain by nearly horizontal to gently dipping rocks consisting of shale, sandstone, and conglomerate (see Classification of Rocks, elsewhere in this text), which in turn underlie a variable thickness of soil and colluvium, derived in part by slumping from the escarpment crest.

Sandstones and conglomerates freely transmit water, which moves downward until trapped by impermeable shale or coal layers. Springs then issue forth onto the escarpment. Springs commonly flow under colluvium, forming buried planes of weakness, as shown in **Figure CS 1-1**.

Highway Construction

Due to engineering requirements, the roadbed of the interstates could not exceed a 6% slope, which meant that the roadbeds had to gently climb the face of the escarpment (**Figure CS 1-2**). To provide a smooth roadbed over the irregular escarpment face, a series of **cuts and fills** were constructed. Cuts are excavations, and fills are piles of specially designed material, piled and compacted onto the surface of the escarpment. **Figure CS 1-3** shows cuts and fills, and the placement of cuts and fills on the highway right-of-way.

The effects of such construction practices should have been predictable. They were:

- Cuts removed support for the colluvium on the escarpment above the cuts, and increased slope angle, inducing failure.
- Fills added mass to the slope material, changing the balance between shear stress and shear strength, inducing failure.
- Fills were on occasion placed over buried springs. The extra weight of the material, coupled with *seepage erosion* by groundwater that removed supporting clays from the colluvium, induced failure. **Figure CS 1-4** summarizes these effects.

In all, more than two-dozen slope failure events occurred after construction began and before the interstate could be officially opened to traffic. Minimal cost of repairing and reconstructing failures was $12 million in 1972 dollars.

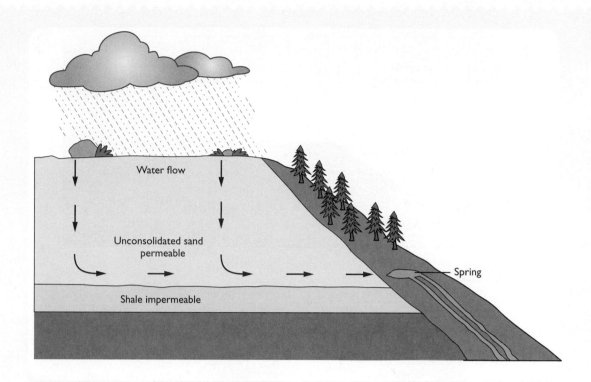

FIGURE CS 1-1 Springs form at the contact between permeable and impermeable layers.

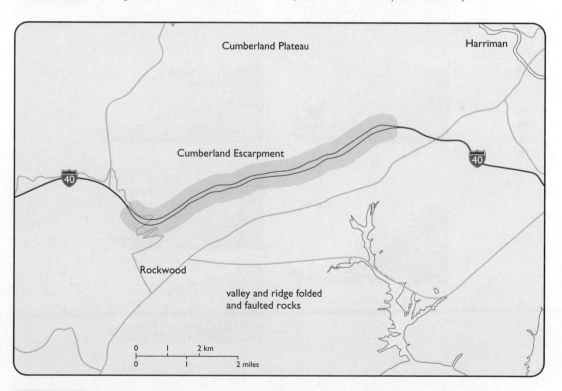

FIGURE CS 1-2 The roadbed of I-40 from Rockwood, at the base of the escarpment, to the escarpment where the roadbed turns and runs west along the top of the Cumberland Plateau.

Continued ▸

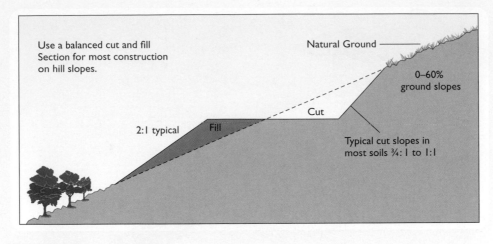

FIGURE CS 1-3 Cuts and fills.

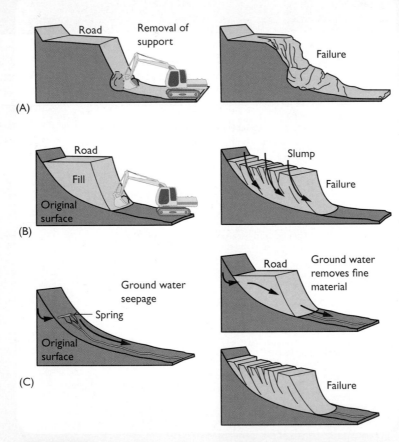

FIGURE CS 1-4 Effects of highway construction practices on slopes. (A) Removal of support induced failure. (B) Fills added weight, inducing failure. (C) Groundwater seepage removed clays and increased shear stress, inducing failure.

FIGURE CS 1-5 Revetments placed along slopes of a highway to increase shear strength. Courtesy of Scott Douglass.

Engineering a More Stable I-40

Slump and slide debris was removed and the roadbed rebuilt using common engineering practices to stabilize the roadbed. Here are examples of strategies employed.

- Groundwater seepage was controlled by wells drilled into the escarpment that pumped out groundwater, thus preventing seepage erosion that would have eventually weakened newly constructed fills below.
- Specially designed strengthened fill was used for roadbed construction.
- Cuts were reinforced with wire and rock **revetments** (structures built on slopes, bluffs, or banks to prevent erosion) to provide support for slope material above the cut (**Figure CS 1-5**).

Summary

Greater care applied to the construction of I-40, I-75 and I-24 in the first place could have avoided the delays and cost overruns experienced in constructing this leg of the highway, but it would have delayed opening the road and would have cost more up front. Highway department officials usually accepted the lowest bids in highway construction, and officials in moments of candor privately confess that it is easier to raise money to repair roads than to build new ones.

Concept Check 10-4. Best management practices (BMPs) are techniques used for a given activity that are consistent with safety and environmental regulations, cost-controls, and best-available engineering. When you see highways or other roads under construction, do you assume that BMPs are in use, or do you assume that the contract is awarded to the lowest bidder, who may "cut corners?" Would you be willing to pay higher taxes, or even a toll, to ensure that best management practices are used?

---

such as buckled and cracked walls, deformed trees, and leaning telephone poles (**FIGURE 10-16**).

Creep may cause more monetary damage annually than any other form of mass wasting. It can warp and crack walls and foundations, and damage streets and roads. One reason the cost is generally unappreciated is that homeowners' insurance usually has a high deductible. Furthermore, damage from creep to roads,

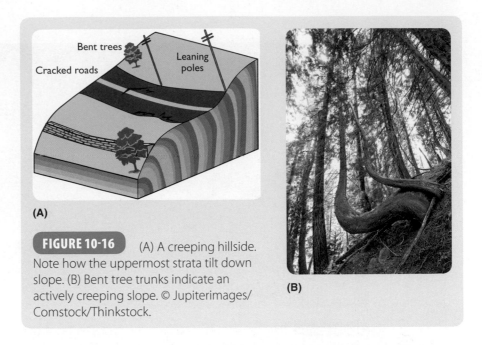

FIGURE 10-16 (A) A creeping hillside. Note how the uppermost strata tilt down slope. (B) Bent tree trunks indicate an actively creeping slope. © Jupiterimages/Comstock/Thinkstock.

streets, and public buildings may be included in routine maintenance budgets for governments.

Soils are especially susceptible to creep. Soil particles rich in clays swell when it rains or when they are wetted by ground water, causing them to expand at right angles to the slope direction. When they dry, they shrink, but gravity pulls the particle vertically down. As a result of this combination of movements, a slow creep downslope occurs; hence the term, creep. The soil can also form *terracettes:* lobes at the base of the slope, where it has begun to pile up. Impacts of creep are summarized in **FIGURE 10-17**.

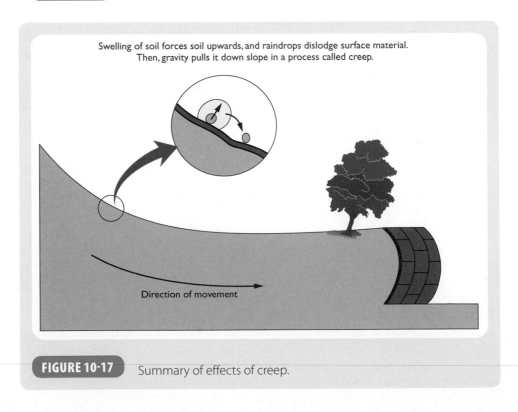

Swelling of soil forces soil upwards, and raindrops dislodge surface material. Then, gravity pulls it down slope in a process called creep.

Direction of movement

FIGURE 10-17 Summary of effects of creep.

FIGURE 10-18 Solifluction lobes in permafrost. © W K Fletcher/Photo Researchers/Getty Images.

Solifluction

Solifluction (**FIGURE 10-18**) is a type of creep in which sediment that is waterlogged moves slowly downslope over frozen ground called *permafrost*. As much as 20% of the land area of North America and Eurasia is presently underlain by permafrost, although satellite imagery has shown recent thawing as a result of anthropogenic global warming.

Solifluction lobes form on permafrost when the surface layer melts during the brief Arctic summer. The melt zone tends to be very shallow, however, and is underlain by soil in which the ice has not melted. Thus, melt water cannot penetrate deeply into the soil, meaning that, even on very gentle slopes, lobes of water-saturated soil can ooze downhill. Lobes can persist after permafrost has melted.

Solifluction can be a geohazard if structures are built on permafrost (**FIGURE 10-19**). Moreover, building structures on permafrost may induce surface melting if the structure is poorly insulated, and thus a source of heat for the underlying soil.

A proposal to construct a natural gas pipeline from Alaska's Prudhoe Bay would mean that additional areas of permafrost would be disturbed. Engineering structures and practices can ensure minimal disruption of the permafrost, however, and therefore damage from subsidence or creep can be avoided (**FIGURE 10-20**).

Warming of polar latitudes as a result of climate change is causing degradation of the permafrost in many areas, which will likely cause more solifluction.

FIGURE 10-19 A buckled road due to melting of permafrost near Yellowknife, Canada. © Ryerson Clark/iStockphoto.

Subsidence

Subsidence is the *sinking or collapse of a surface, resulting from removal of subsurface support.* **FIGURE 10-21** shows one effect of subsidence. Subsidence is a widespread type of surface failure, which we include here even though sinkholes commonly occur on horizontal surfaces, or surfaces of low relief. Subsidence has occurred in nearly every state in the United States and in nearly every country. It is an extraordinarily diverse form of ground failure, and ranges from local collapse to broad regional lowering of the Earth's surface.

The causes of subsidence are as varied as the types of failure. A few examples are: dewatering of peat or organic-rich soils, solution (that is, rock dissolving) in limestone aquifers, liquefaction, subsurface mining, and withdrawal of fluids (groundwater, petroleum, geothermal). Subsidence may accompany volcanic activity, as in the *sulfaterra* region around Naples, Italy (this and *liquefaction* are discussed in detail elsewhere in the text).

Although humans have been inducing subsidence for hundreds of years, the association between human activity and subsidence often remained obscure until modern times. For example, the relationship between groundwater withdrawal and subsidence was not recognized until 1928. We discuss the specific reasons why groundwater withdrawal causes subsidence elsewhere in the text.

Here are some brief examples of subsidence:

- Tucson, Arizona is likely to subside up to 30 cm over the next three decades due to groundwater withdrawal.
- Parts of the Las Vegas, Nevada Metropolitan area, the fastest growing area in the United States, have subsided as much as 2 m due to groundwater withdrawal.

(A)

(B)

FIGURE 10-21 (A) Collapse of a house into a sinkhole in Florida. Courtesy of NOAA. (B) A sinkhole in Winter Park, Florida. Courtesy of A.S. Navory/NOAA/NGDC.

- Parts of California's San Joaquin Valley subsided up to 10 m between 1925 and 1977 due to groundwater withdrawal.
- Much of fast-growing central Florida is subjected to ground collapse without warning due to solution (that is, dissolving) of its limestone bedrock.
- Many counties in Pennsylvania contain areas subject to subsidence from the collapse of underground coal mines.

Concept Check 10-5. CheckPoint: Define and provide two examples of subsidence.

Types, Causes, and Effects

Subsurface mining removes material from the Earth's shallow crust that formerly supported the surface. Removal of the material induces subsidence or even collapse of the surface over the mine. Widespread parts of Pennsylvania, England, and many other places are built on abandoned underground coal mines. Houses and other structures continue to collapse, sometimes hundreds of years after mining activity ceased. Timbers used during the Middle Ages in Europe to support coal workings rot, causing collapse of sometimes long-forgotten, unmapped workings. Roads built over mined areas may subside and even collapse.

In the United States, subsidence caused by mining generally is not covered under a homeowner's insurance policy. The State of Pennsylvania and other entities recommend that residents take out special insurance against subsidence. However, septic tanks, driveways, and swimming pools are not eligible for coverage.

Concept Check 10-6. What are some of the "hidden costs" of subsurface mining to society? "Hidden costs," called *externalities* by economists, are costs to society not captured in the price charged for the good extracted. For example, externalities in the automobile industry would be damage done to the environment and human health caused by tailpipe emissions.

An example of subsidence due to fluid withdrawal was illustrated above with the effects of groundwater withdrawal in California's San Joaquin Valley. But subsidence can be caused by any fluid withdrawal below the surface, if the fluid was contributing to the support of the overlying land. That fluid can be oil, gas, or water. Coastal subsidence is a special problem as it can worsen the effects of sea level rise and groundwater contamination. We discuss these subjects in other chapters. The following Case Study provides examples of the causes and impacts of coastal subsidence in two locations, California and Italy.

Subsidence can be due to solution in **karst** topography. *Sinkholes* and *disappearing streams* are common features of land in humid regions underlain by carbonate bedrock (that is, rock composed of calcium carbonate, mainly **limestones and dolostones**), resulting from the formation and collapse of shallow, commonly interconnected caverns (FIGURE 10-22). These features are characteristics of karst topography, named from the Karst region of Croatia. Widespread areas of North America, Europe, and China exhibit karst topography.

Karst areas are especially prone to groundwater contamination. The dissolving of carbonates underground may be further enhanced by acid precipitation, which affects much of Europe, eastern North America, and China. The solubility of carbonate bedrock and chemical reactions involved can also be found in the Classification of Rocks, elsewhere in this text.

Wilmington/Long Beach California

Figure CS 2-1 is a map of the Wilmington Oil Field, which began production in the early 1940s. Pumping of oil, gas and water from the field's western portion caused marked surface subsidence, resulting from **depressurization** and/or compaction of unconsolidated oil and gas reservoirs.

Concept Check 10-7. How does removal of oil cause the collapse of reservoirs? Review oil and gas reservoirs elsewhere in this text to provide a complete answer.

- -

By 1951, the rate of subsidence exceeded two feet (60 cm) per year, and total subsidence reached as much as 29 feet (9 m) in the center of the subsidence "bowl" (**Figure CS 2-2**). In the 1950s, the city of Long Beach concluded that water injection would repressure the oil reservoirs and stop the surface subsidence.

In 1964, the State of California passed enabling legislation that stated in part, ". . . the City has the power, upon any evidence of subsidence . . . to order a cessation or curtailment of production . . ." At present, the City of Long Beach measures surface elevations twice a year to prevent a recurrence of the subsidence from the 1940s and 1950s.

The Long Beach Unit began oil production in 1965 under strict subsidence control requirements. To prevent subsidence, an equal or greater amount of fluid must be injected into the reservoir compared with that removed to maintain pressure. The Long Beach Unit in 2002 produced approximately 700,000 barrels of fluid and re-injected approximately 800,000 barrels of water each day. Elevation surveys taken twice yearly since 1965 confirm that the water injection program is successfully controlling subsidence.

The Wilmington subsidence has been controlled and partly remedied by re-injection of oil-field brines (solutions of concentrated salt), in volumes of up to 1.1 million barrels (42 gal = 1 barrel) per day between 1958 and 1969. By 1969, the area of subsidence was reduced from 58 km^2 to 8 km^2. Local rebound of up to 1 foot (0.3 m) was reported. In 2012, production was 46,000 barrels a day. A March 2012 article reported a nearly one inch rise in elevation due to natural geologic processes.

Subsidence of Venice, Italy

For hundreds of years until the eighteenth century Venice was one of Europe's most powerful states. It was built in a shallow lagoon that originated over 6,000 years

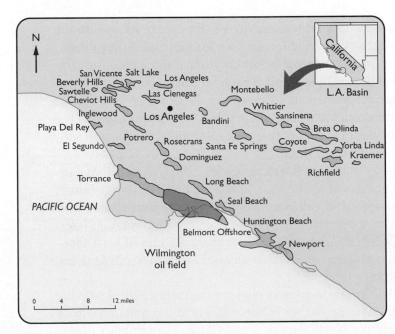

FIGURE CS 2-1 Map of the Wilmington Oil Field in California.

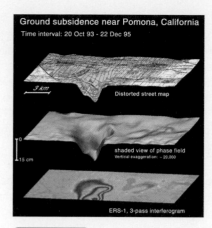

FIGURE CS 2-2 Subsidence in Long Beach, California. Courtesy of the ESA and G. Peltzer/Jet Propulsion Laboratory.

ago (average depth 1.5 m, surface area 55,000 ha; **Figure CS 2-3**). The site has been permanently inhabited since the early Middle Ages.

The city's earliest settlers built on pilings driven into the lagoonal mud on which the entire city now rests. Because of its geological setting on compactable mud, the site has been subsiding since its settlement, although at a very slow and manageable rate, which between 1550 and 1950 was about 1 mm per year.

Two factors increased Venice's subsidence rate.

1. The city government in 1550 ordered rivers to be diverted that had brought silt into the city, to avoid having to clear shipping channels reg-

ularly. Their engineering scheme worked, but *siltation* (deposition of fine sediment) in the lagoon was drastically reduced. The reduction in sediment meant that the slow natural subsidence of the site was no longer being replaced by natural sediment influx. Subsidence rate increased to about 1 mm/y as a result.

2. The nearby mainland, called the Marghera district, was industrialized during the Fascist regime of Benito Mussolini in 1930 (**Figure CS 2-4**). Industries began to extract groundwater, but withdrawal rates did not become an obvious problem until after 1950. Groundwater withdrawal caused the rate of subsidence of Venice to increase. Between 1952 and 1969 Venice subsided 9 to 10 cm, about 70% of which was due to groundwater withdrawal.

After 1970 well drilling was prohibited, and a new aqueduct provided the region with surface water. However, scientists estimated in 1977 that only 20% of the compaction due to water withdrawal of the sediments underlying Venice could be restored by eliminating groundwater withdrawal and initiating re-injection, meaning that about 8 cm of the subsidence was permanent. There is an important point here: *Sometimes the damage caused by withdrawing fluids from reservoirs cannot be repaired.*

Eight cm (3.5 in.) of subsidence may not seem like much, but you should be aware that Venice is essentially at sea level. Moreover, it is subject to tides of up to 2 m. Making the subsidence worse has been an increase in the rate of global sea level rise over the past century.

These factors in 1966 coincided with an extremely high tide to culminate in a disastrous flood, which

FIGURE CS 2-3 The site of Venice. The city is built on islands in the lagoon at the head of the Adriatic Sea. Courtesy of the ISS Crew Earth Observations experiment and the Image Science & Analysis Laboratory, Johnson Space Center.

FIGURE CS 2-4 The Porto Marghera chemical complex, across the lagoon from Venice. © Matteo Cozzi/ShutterStock, Inc.

Continued ▶

severely damaged much priceless artwork, including wall frescoes that were damaged when seawater was drawn into the walls on which the frescoes were painted. As a response, the Italian government put an end to excessive groundwater withdrawal, which was the primary cause of the increased rates of subsidence experienced since 1945. It also has proposed the construction of large, hinged dams to close the entrances to the lagoon at times of abnormally high tides. However, because the city has no sewage treatment, some waste is dumped directly into the lagoon. Environmentalists and many city officials fear a severe degradation of the lagoon's already poor water quality if circulation between the Adriatic and the lagoon is cut off. High nutrient levels result in the profuse growth of algae that must be periodically "harvested" and removed from the lagoon, and insect swarms of almost unimaginable magnitude occur throughout the year.

About 20 million tourists visit the city annually, generating revenue for the city but also huge quantities of waste and sewage. The city has only about 70,000 permanent inhabitants living among thousands of empty houses and flats. Venice continues to flood regularly (**Figure CS 2-5**).

Venice has been described as a treasure of European culture. Its fate, however, is unclear.

Concept Check 10-8. CheckPoint: Summarize the case study. Include an overview of the issue, the underlying geological causes, potential or actual solutions, and any insights you may have gained into the issue.

(A)

(B)

FIGURE CS 2-5 (A) Venice regularly floods. © Alvaro German Vilela/ShutterStock, Inc. (B) Large, hinged dams are proposed to close off the lagoon during high tides.

The formation of **sinkholes** is illustrated in Figure 10-22. With the gradual enlargement (over thousands of years) of interconnected fracture systems that readily develop in soluble carbonate strata, large underground cave systems may form in sufficiently thick carbonate bedrock. Overlying surfaces may collapse into expanding fracture systems and caverns with the slow removal of support (provided by high groundwater levels or underlying rock) by solution (**FIGURE 10-23**).

As you might imagine, sinkhole enlargement and collapse can severely damage or destroy structures built on top of or beside the structure. Collapse or subsidence also may damage or destroy infrastructure such as gas, water, sewage, and oil pipelines, and roads (**FIGURE 10-24**). Large commercial buildings are also at risk in karst regions.

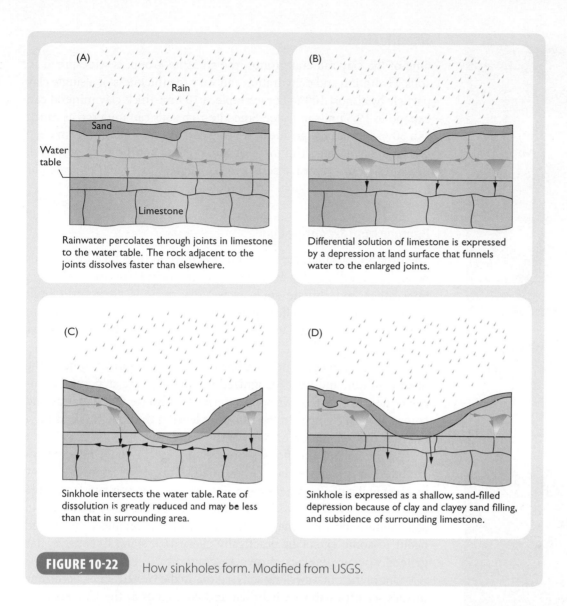

(A) Rain / Sand / Water table / Limestone

Rainwater percolates through joints in limestone to the water table. The rock adjacent to the joints dissolves faster than elsewhere.

(B)

Differential solution of limestone is expressed by a depression at land surface that funnels water to the enlarged joints.

(C)

Sinkhole intersects the water table. Rate of dissolution is greatly reduced and may be less than that in surrounding area.

(D)

Sinkhole is expressed as a shallow, sand-filled depression because of clay and clayey sand filling, and subsidence of surrounding limestone.

FIGURE 10-22 How sinkholes form. Modified from USGS.

The use of sinkholes for waste disposal is a widespread practice in some rural areas, a practice that can lead to severe groundwater contamination, because karst aquifers have relatively high rates of groundwater movement.

Finally, rapid withdrawal of near-surface groundwater can accelerate sinkhole collapse.

Expansive Clays and Swelling Soils

Clay Minerals

Swelling soils are a major geological hazard, and **expansive clays** and clay-rich shale cause extensive damage worldwide every year. In the western United States, swelling soils are called *adobe* soils or **bentonites**.

Clay minerals are among the most common minerals on earth, and are a major component of many of the planet's soils. Clays are classified into groups based on their crystal structure. Because of their layered structure and abundance of electrically charged surfaces, some clay can soak up a great deal of water.

FIGURE 10-23 A large sinkhole that formed in 1981 in Florida, during a period of abnormally low ground-water. Courtesy of USGS.

FIGURE 10-24 A street and suburb damaged by a sinkhole collapse, San Diego, California. © John Gibbins/San Diego Union-Tribune/AP Photo.

Smectite Clays

One of the clay mineral groups, the smectite group, is widespread in soils. Hydration and dehydration can vary the thickness of a single clay particle by almost 100%. For example, a 10 unit-thick clay mineral can expand to 19.5 units in water. Smectite-rich soils can undergo as much as a 30% volume change due to wetting and drying, and may form deep cracks upon drying. The force of the swelling can be very powerful: pressures from expansive soils can be as great as 15,000 pounds per square foot!

Damage from Expansive Clay

The process by which clay minerals swell when they take up water is reversible. Swelling clay expands or contracts in response to changes in weather and climate such as wet/dry periods. Structures built on soils containing swelling clays may experience major damage to foundations, driveways, etc, caused by seasonal swelling of the clays in the soil.

According to the American Society of Civil Engineers, one of the most common and least recognized causes of severe structural damage to houses is expansive soil underneath a house that expands and contracts with changes in the moisture level in the soil. This can raise or crack the foundation or push in the walls.

About half of the houses built each year in the United States are built on expansive clays. About half of these (one-fourth of new houses) will suffer some damage. The repair costs can run from 10% to 50% or more of the cost of the house. Expansive clays dominate in soils underlying some major cities such as Houston and Denver, and occur to some extent in practically every U.S. state. Expansive clay generally occurs in black, brown, or gray soils.

One of the worst swelling soils regions in the United States, according to soils scientist Sabine Chabrillat and colleagues at the University of Colorado, is the 300-km-long Front Range Urban Corridor in Colorado. This area is one of the fastest growing urban regions of North America. Expansive clays in the Cretaceous Pierre Shale and adjacent formations along the Front Range are mainly responsible for the damage.

In Colorado, by law builders must inform homeowners if expansive soils are present and provide information about them. Here are ways contractors deal with expansive soils in the Front Range and elsewhere:

- By removing, altering, or compacting expansive soils before construction. The top 1 to 3.5 meters of **regolith** is commonly treated.
- By constructing stress-resistant foundations. Foundations are commonly "pinned" to unweathered bedrock—typically sandstone—below the regolith, with special "floating" layers of material to accommodate shrinking and swelling.
- Sub-basements have sensors that operate ventilator fans to dehumidify substrate.

- By applying chemical soil treatments to reduce expansive characteristics. Due to its expense, this method is not usually applied to single-family homes.

Avoiding Problems From Expansive Soils

While expansive soils can severely damage structures, a few simple strategies, if carefully followed by homeowners, can reduce their impact. They are:

1. *Proper roof drainage:* Rain gutters with downspouts should drain to the street via pipes or durable troughs.
2. *Proper yard drainage:* All areas should drain to the street and away from the house. Even puddles can cause swelling. But note that this increases runoff at the expense of infiltration.
3. *Concrete and asphalt areas:* These also should drain to the street.
4. *Proper subsurface drainage:* Drain lines (from washers, etc.) should discharge to the street.
5. *Repair plumbing leaks:* Monitor water consumption. An unexplained increase in a homeowner's water bill could indicate a leak that should immediately be repaired.
6. *Careful attention to landscaping:* On expansive soils, the main landscaping objective should be to minimize fluctuations in soil water. Flowers and shrubs should be planted at least 1.5 m (5 ft) from building foundation. Trees can draw significant amounts of water from soils by transpiration, and should not be planted within 3 to 5 m of structures. A convenient "rule of thumb" is that a tree should be at least 1½ times its mature height away from the foundation.

   Many municipalities recommend or even mandate *xeriscaping,* which is planting drought-resistant vegetation that doesn't need watering.
7. *Watering:* This should be carefully carried out to avoid excessive water buildup in soils. Trickle irrigation is best. Minimize turf areas, because grass requires regular watering to remain green during dry seasons or periods of insufficient rainfall.

■ CONCLUSION

As you have seen, mass wasting of material at the Earth's surface is a widespread geohazard. That movement may be extraordinarily slow or catastrophically rapid. It is commonly triggered by human alteration of a natural balance between disturbing forces and resisting forces. Slope failure is enhanced by excess water buildup, and related to a wide array of disturbing forces that reduce the shear strength of slope material. As local and state revenues decline, resources to address problems of slope failure may similarly decline. This could occur at a time of unprecedented expansion of developed areas into regions more prone to slope failure. Thus, education of citizens about slopes, the criteria that govern slope stability, and the forces and activities that cause slope failure will become increasingly important in the decades ahead. Globally, unprecedented loss of property and loss of life could result from unhindered expansion of human populations into failure-prone areas, especially regions in the paths of hurricanes and typhoons. Increased impact of geohazards in developing countries could help drive up migration rates to developed countries as well.

Finally, see if you live in an area prone to mass wasting. Look at **FIGURE 10-25** for a landslide potential map of the United States.

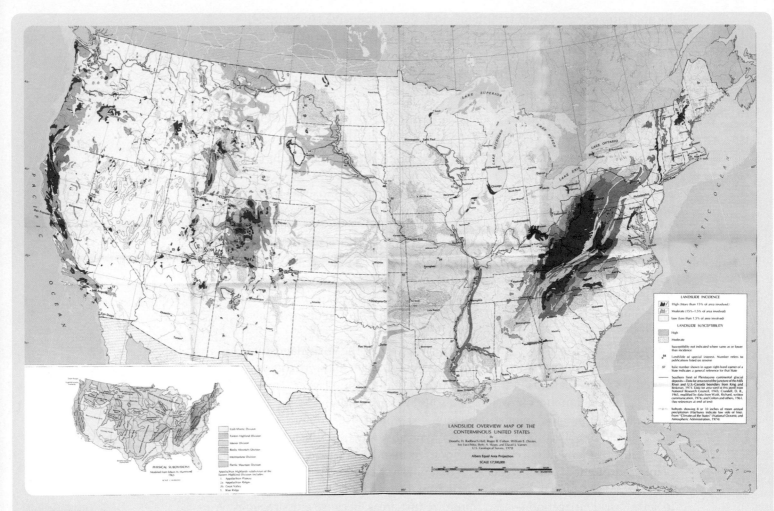

FIGURE 10-25 Map of landslide hazard potential in the United States. Red color denotes highest landslide hazard areas. Courtesy of Radbruch-Hall, Dorthy and et al. Landslide Overview Map of the Conterminous United States. Professional Paper 1183. USGS.

CHAPTER SUMMARY

1. Mass-wasting, also called slope failure, is downslope movement of material under the influence of gravity.
2. Hurricane Mitch in 1998 set off an estimated one million slope failures, virtually destroying the entire infrastructure of Honduras.
3. Heavy rainfall in early and mid-June of 2002 led to severe flooding and mass-wasting in central and northwestern China, costing an estimated $2 billion (U.S.).

4. *Surface material* consists of some combination of residual or transported unconsolidated rock material, soil, vegetation, and other weathered material (collectively called regolith), with or without water or ice.

5. Slope failures are commonly classified as falls, flows and slides.

6. Geologists consider causes of slope failure to fall into external factors and internal factors.

7. Slopes represent a dynamic balance between friction and other forces that hold slope material in place, and disturbing forces enhanced by gravity, that pull material downslope. The balance between shear stress and shear strength determines slope stability.

8. The controls of slope stability are (1) the nature of surface material, (2) the abundance of water, (3) the slope angle, (4) the nature of disturbing forces, and (5) the subsurface geology.

9. The geology of a region strongly influences slope stability. The Vaiont dam disaster, Italian Alps, is an example.

10. Certain characteristic landforms, geologic properties, and surface features increase the risk of slope failure.

11. *Sturzstroms,* or long run-out landslides, represents some of the most awesome geohazards on earth, releasing as much energy as immense volcanic eruptions.

12. Slope failure is often triggered by human activity. Numerous landslides and slumps occurred during the construction of I-40, I-64, and I-75 in Tennessee. Improper placement and composition of cuts and fills contributed to slope failure.

13. Creep refers to imperceptible downslope movement of surface material over a long period of time. Creep may cause more monetary damage annually than any other form of mass wasting.

14. Solifluction is a type of creep that develops in frozen ground called permafrost. Solifluction can be a geohazard if structures are built on or in permafrost.

15. Subsidence is the sinking or collapse of a surface, resulting from removal of subsurface support. It is an extraordinarily diverse form of ground failure, and ranges from local collapse to broad regional lowering of the Earth's surface.

16. Removal of material during underground mining can induce subsidence or even collapse of the surface over the mine.

17. Subsidence can be caused by any fluid withdrawal below the surface, if the fluid was contributing to the support of the overlying land.

18. Withdrawal of oil and water from the Wilmington, California oil field induced subsidence in the city of Long Beach, California. The subsidence was halted by fluid reinjection.

19. The city of Venice, Italy, one of Europe's great art and architectural treasures, is subsiding, due to local geological factors, global sea level rise, and fluid withdrawal below the city.

20. Karst topography develops over carbonate bedrock in humid regions. Widespread areas of North America, Europe, and China exhibit karst topography. Sinkholes and disappearing streams are common features of karst. Karst areas are especially prone to groundwater contamination.

21. One of the most common causes of severe structural damage to houses is expansive soil. One of the worst swelling soils regions in the United States is the Front Range Urban Corridor in Colorado.

22. While expansive soils can severely damage structures, a few simple strategies, if carefully followed by homeowners, can reduce their impact.

KEY TERMS

angle of repose	internal factors	shear strength
bentonite	karst	shear stress
colluvium	limestone and	sinkhole
creep	dolostone	solifluction
cut and fill	mass wasting	sturzstrom
depressurization	polar molecule	subsidence
escarpment	pore-water pressure	swelling soils
expansive clays	regolith	weathering
external factors	revetment	

REVIEW QUESTIONS

1. Three requirements for slope failure are _____, _____, and _____.

2. The most common cause of slope failure is
   a. volcanic activity.
   b. intense rainfall.
   c. earthquakes.
   d. tsunamis.
   e. global sea level rise.

3. Residual or transported unconsolidated rock material, soil, vegetation, and other weathered material is called
   a. regolith.
   b. alluvium.
   c. debris.
   d. soil.

4. An example of an external factor causing slope failure is
   a. water content.
   b. material property.
   c. earthquakes.
   d. clay mineral composition.
   e. all of the above.

5. The gravitational force that pulls material downslope is called
   a. K.
   b. shear stress.
   c. shear strength.
   d. cohesion.

6. Cohesional and frictional forces within slope material determine
   a. shear stress.
   b. water pressure.
   c. slope angle.
   d. shear strength.

7. The texture of loose slope material (colluvium) determines its
   a. shear strength.
   b. shear stress.
   c. angle of repose.
   d. weight.
   e. mass.

8. As water content increases in bedrock or regolith, the slope becomes weaker because
   a. water adds mass ("weight") to the slope material, increasing shear stress.
   b. moving water within the material can reduce friction and dissolving or chemically weather material within the substrate.
   c. the pore-water pressure commonly builds up, especially in loose material.
   d. all of the above.
   e. none of the above.

9. If a surface is vertical, gravitational force is
   a. cancelled by centrifugal force.
   b. at a minimum.
   c. cancelled by shear stress.
   d. at a maximum.

10. Which of the following increases shear stress?
    a. building on a slope
    b. increasing slope
    c. increasing cohesive forces within slope material
    d. all of the above
    e. a and b only
    f. none of the above.

11. Which are geological conditions that underlie unstable slopes?
    a. Geological strata dip in the same direction as the slope.
    b. Weak and strong horizontal interlayered strata crop out on a slope.
    c. Deeply weathered rocks in regions of high rainfall.
    d. All of the above.
    e. None of the above.

12. Slope failures that transmit large volumes of material, commonly with blocks of huge size, at high velocities, traveling kilometers within minutes are called
    a. sturzstroms.
    b. slumps.
    c. scarps.
    d. alluvial fans.

13. Creep can be recognized by observing which of the following surface features?
    a. groundwater flow
    b. deformed trees
    c. leaning telephone poles
    d. buckled and cracked walls
    e. all of the above
    f. b, c, and d only.

14. A type of creep developing in frozen ground is
    a. colluvial collapse.
    b. subsidence.
    c. solifluction.
    d. microfracturing.

15. Which of these areas have experienced subsidence due to withdrawal of groundwater?
    a. Tucson, Arizona
    b. Las Vegas, Nevada
    c. Venice, Italy
    d. all of the above
    e. none of the above.

16. Because of its geological setting, Venice, Italy had a historical (1550–1950) rate of subsidence of about
    a. 1 meter/year.
    b. 1 cm/year.
    c. 1 mm/year.
    d. 1 μm/year.

17. Karst regions are underlain by
    a. horizontal strata.
    b. carbonate strata.
    c. unconsolidated strata.
    d. strata containing abundant swelling clays.

18. Karst areas are especially prone to
    a. earthquakes
    b. groundwater contamination.
    c. overpopulation.
    d. high heat flow.

19. The process by which clay minerals swell when they take up water is
    a. reversible.
    b. irreversible.
    c. called solifluction.
    d. associated with so little pressure as to cause no damage to structures.

20. About what percent of new houses built in the United States each year are built on swelling soils?
    a. 75%
    b. 50%
    c. 25%
    d. 5%
    e. 1%

21. Many municipalities recommend or mandate xeriscaping to
    a. conserve energy.
    b. conserve water and minimize fluctuations in soil water.
    c. preserve aesthetics.
    d. reduce the spread of water-borne diseases.

FOOTNOTE

[1] The total value of goods and services in a year.

The Coast and Coastal Hazards

CHAPTER OUTLINE

POINT-COUNTERPOINT   Beach Renourishment

© Amos Struck/Shutterstock, Inc.

Point

"There is no real alternative, if you want to maintain a dry beach, than to put more sand there,"[1] says Chris Brooks, South Carolina's former top coastal regulator, about that state's beaches. "You've got intense development on the shore and you've got to keep the beach in front of them."

Counterpoint

Coastal geologist Orrin Pilkey disagrees.[2] "We're just responding to a crisis here, and a crisis there, and nobody is looking at the big picture. Sea level is rising and we expect storms to increase. Nourished beaches will get more difficult and costly to maintain. Long term, it's pointless anyway. Nature is going to win at the shoreline."

The shoreline of the United States, including Alaska, Hawaii, and the Great Lakes extends for about 95,000 mi (150,000 km). When economically important beaches

erode, they are frequently widened by the addition of sand. By 2002, there had been 333 renourishment projects involving 517 million cubic yards (MCY) of sand for the purpose of recreational enhancement and/or storm protection in the 15 Atlantic and Gulf Coast states alone, costing over $2 billion. As much as 65% of the total cost of some projects comes from the federal government.

Arguments supporting beach renourishment focus on the importance of beaches to local economies and the government's obligation to protect beachfront property, both public and private. Those opposing the activity cite overall financial costs, including the extent to which renourishment is a subsidy to beachfront businesses, and the ecological costs, for example, imperiling already endangered sea turtles by destroying nesting areas and killing adults in dredges.

Once again, two experts disagree on how to respond to a geohazard. We'll leave it to you, after reading and considering this chapter, to determine with whom you agree. *Bon chance!*

■ COASTAL GEOLOGY AND HUMANS

Chances are, you live within 100 km (62 mi) of the coast. According to the World Resources Institute, at least 60% of the planet's human population lives that close to the coastline. Coastal areas have the fastest growing populations as well (FIGURE 11-1). Not surprisingly, over half of the world's coastlines are at significant risk from development related to this population growth.

Coastal cities are reaching sizes unprecedented in human history. TABLE 11-1 shows some examples. Fourteen of the 20 most populous U.S. cities are on the

FIGURE 11-1 Examples of coastal growth. (A) Favelas in the coastal city Salvador, Brazil. (B) Fishing boats at Cape Coast, Ghana. (C) The fishing village of Tai-O, Hong Kong. (D) Shacks in the shadow of Ho Chi Minh City, Vietnam. Photos A–D © Pluff Mud Photography.

TABLE 11-1	Population of Some Large Coastal Cities	

City	2010 Estimated population
• Sao Paulo, Brazil	20.0 million
• Mumbai, India	21.8 million
• Lagos, Nigeria	11 million: growth rate 5.7%
• Dhaka, Bangladesh	14.2 million: growth rate 6%
• Shanghai, China	15.6 million

Data from: the CIA World Factbook.

coast, and all of Australia's major cities—Sydney, Melbourne, Brisbane, Adelaide, and Perth—are coastal.

Coastal counties account for only 17% of the land area of the contiguous United States; however, 16 million more people live in these counties than in the 83% of inland counties. Moreover, this population is increasing by at least 3,600 per day. U.S. coastal population is projected to increase to 166 million by 2015. Growth is projected to be most rapid along the Atlantic and Pacific seaboards, especially Florida and southern California.

Keep these statistics in mind as we discuss coasts and coastal geohazards in the following pages.

■ COASTS AND PLATE TECTONICS

One way to classify coasts is according to their plate tectonic setting. *Active* coasts occur at or near lithospheric plate boundaries; *passive* coasts do not. Coastlines along the west coast of North America are active coasts, whereas those on the east coast are *passive* coasts (FIGURE 11-2). Volcanic activity and earthquakes are common to plate boundaries. Thus, active coasts, which are characterized by mountainous terrain, narrow or no continental shelves (the seaward extension of the continent; see later in the chapter), and steep drop-offs into deep water, would be more likely to experience the geohazards associated with their setting at a plate boundary, that is earthquakes and volcanoes. Tectonic activity is minimal along passive coastlines; thus weathering and erosion are dominant, resulting in wider continental shelves, low-lying land like barrier islands, and larger sand supplies. As you will see later, the low relief of barrier islands makes them susceptible to erosion and rising sea level.

Concept Check 11-1. Figure 11-2 shows a USGS map of tectonic plates and active volcanoes. Which coastlines are active? Which are passive?

- -

Concept Check 11-2. Active coasts are also called *collision* coasts. Explain why this is so (consult the text for details about global tectonics and geohazards, if necessary).

- -

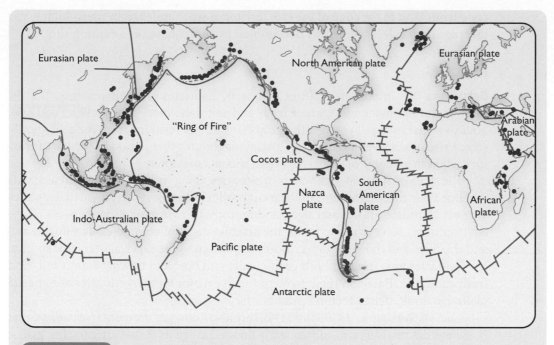

FIGURE 11-2 USGS map of tectonic plates and active volcanoes. The west coast of North America represents an active coast, since it lies at the boundary between the Pacific and North American lithospheric plates. The Atlantic coast is a passive coast. Modified from Active Volcanoes and Plate Tectonics, "Hot Spots" and the "Ring of Fire," USGS.

■ NATURE OF THE COASTLINE

Although most of us think we know what we mean when we refer to the coast, this zone has no single, precise definition. This is partially because the coast continually changes, both in space and time, and also because there are several different overlapping classification systems.

FIGURE 11-3 shows one classification of the coast. The coast includes the ocean bottom adjacent to the **shore**. The shore is essentially the zone between the tides. The coast extends inland as far as the ocean influences landforms. According to the Coastal Zone Management Act of 1972, the **coastal zone** is the transition

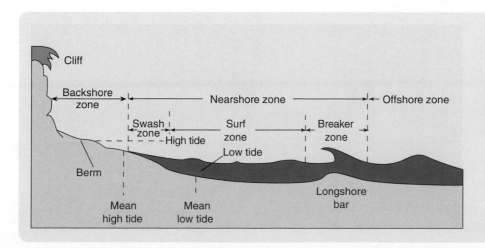

FIGURE 11-3 Subdivisions of the coastal zone. Coastal environments are often classified on the basis of the areas covered by the tide. Intertidal areas lie between high and low tide. High-energy intertidal areas are beaches dominated by sandy sediment or rocky shores with little sediment. Low energy intertidal areas are tidal flats, dominated by muddy sediment. Subtidal areas are those below low tide, and supratidal land is above normal high tide.

zone from land to the edge of the U.S. territorial sea, 3 miles from mean high water (average high tide), historically determined by the distance of a cannon shot.

Estuaries

Estuaries are important features of coasts. Estuaries are embayments, that is, semi-enclosed water bodies, where fresh-and-salt water meet and mix (**FIGURE 11-4**). Today's estuaries are all younger than 20,000 years old, resulting from the geologically recent rise of sea level since the last glacial maximum (that period in the Earth's history when the volume of global ice sheets was at a maximum).

The shallow, nutrient-rich waters of estuaries, along with the **intertidal** wetlands that line their shores, are some of the most productive ecosystems on Earth. Estuaries support abundant grasses and algae, and in tropical environments mangroves. These environments serve as nurseries for the juvenile stages of commercially valuable fish and shellfish, and thus are often critically important to the survival of coastal species.

Estuaries are classified both geologically and how and to what extent salt- and freshwater mix. Based on their geology, the five major types of estuaries are: coastal plain, bar-built, deltas, tectonic, and fjords.

Coastal plain estuaries (Figure 11-4B) are also known as *drowned river valleys,* that is, valleys cut by rivers and subsequently flooded as sea level rose after the last glacial maximum ended about 20,000 years ago. Chesapeake Bay, for example, is formed from the valley of the Susquehanna River. Other coastal plain estuaries include Delaware Bay, New York's Hudson River, Rhode Island's Narragansett Bay, and London's Thames River.

Bar-built estuaries (Figure 11-4C) form behind built-up sand bars or elongate coastal barrier islands, long, narrow sand bodies separated from the mainland by

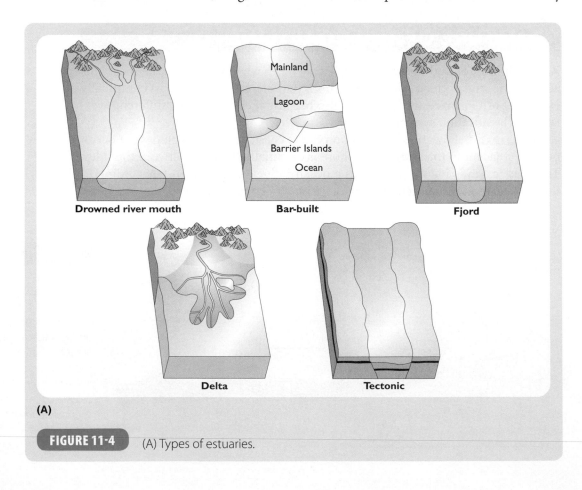

(A)

FIGURE 11-4 (A) Types of estuaries.

FIGURE 11-4 *Continued.* (B) A coastal plain estuary: Chesapeake Bay, the drowned mouths of the Potomac and James Rivers. Courtesy of NASA/GSFC/JPL, MISR Team. (C) A bar-built estuary: Pamlico Sound, behind the Outer Banks of North Carolina, an extensive barrier island complex. Courtesy of Image Science and Analysis Laboratory, NASA-Johnson Space Center. "The Gateway to Astronaut Photography of Earth." (D) A delta estuary: Nile River Delta. Courtesy of Image Science and Analysis Laboratory, NASA-Johnson Space Center. "The Gateway to Astronaut Photography of Earth." (E) Tectonic estuary: San Francisco Bay, California. Courtesy of EROS Data Center. (F) A fjord: Glacier Bay, Alaska. Courtesy of NPS.

water (which is often referred to as a *lagoon*). Bar-built estuaries commonly occur along the Gulf Coast of Texas (for example, Matagorda Bay) and Florida, in parts of North Carolina (such as, Pamlico Sound, an expansive estuary behind North Carolina's Outer Banks), and in the Netherlands.

When sediment and silt accumulate at the mouths of large rivers they may create a landform called a *delta* (Figure 11-4D). Major deltas include the Mississippi River, Santee River (South Carolina), Mékong (Vietnam), Amazon (Brazil), Nile (Egypt), and the Ganges–Brahmaputra (Bangladesh).

Tectonic estuaries (Figure 11-4E) form when folding or faulting in crustal plates causes land to subside below sea level, forming a basin that fills with seawater.

Fjords (Figure 11-4F) are typically elongated, narrow, steep-sided valleys carved by glaciers. They are also characterized by the presence of a *sill*, a ledge of glacial sediments, at their ocean terminus. Fjord estuaries commonly occur in Alaska, British Columbia (Canada), Norway, Chile, New Zealand, and Greenland.

Scientists also classify estuaries based on circulation and stratification (layering), that is, how and to what extent salt-and fresh water mix (**FIGURE 11-5**). How

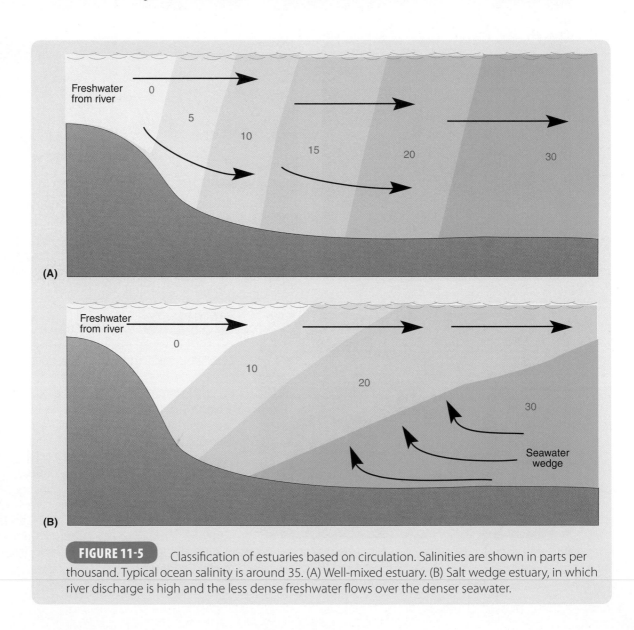

FIGURE 11-5   Classification of estuaries based on circulation. Salinities are shown in parts per thousand. Typical ocean salinity is around 35. (A) Well-mixed estuary. (B) Salt wedge estuary, in which river discharge is high and the less dense freshwater flows over the denser seawater.

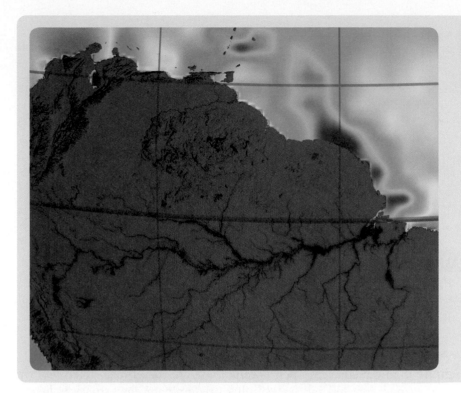

FIGURE 11-6 The freshwater plume (shown in blue) of the Amazon River. Courtesy of NASA/GSFC/JPL-Caltech.

water mixes in estuaries may sound like a purely academic distinction. However, recall that the density of fresh water is significantly less than that of seawater; thus, fresh water from rivers has a tendency to float upon a layer of heavier salt water, especially in a *salt wedge* estuary in which the volume of river discharge is high (Figure 11-5). Salt wedge estuaries, which include the Columbia, Mississippi, and Hudson Rivers, are highly stratified. If the volume of salt water is much greater than that of the freshwater, there may be partial or complete mixing (Figure 11-5). Eventually, of course, fresh water from rivers becomes completely mixed with seawater, although in the Amazon River a plume of freshwater penetrates as far as 400 km (250 mi) into the Atlantic Ocean (**FIGURE 11-6**).

How fresh and salt waters mix can determine the distribution of pollution from river sediment, and the distribution of estuarine ecosystems, among other things. *Dead zones*—areas in oceans and estuaries whose oxygen levels have been lowered to critical levels by bacterial decomposition of algal blooms caused by inputs of fertilizer and sewage—may result in fjords and other estuaries from lack of mixing (**FIGURE 11-7**) when denser salt water is "trapped" under a layer of freshwater. These *anoxic* (completely lacking oxygen) or *hypoxic* (very low oxygen levels, typically below 2 to 3 parts per million) dead zones are so-named because they cannot support oxygen-consuming marine life, such as fish, oysters, crabs, shrimp, turtles, etc.

Concept Check 11-3. CheckPoint: Explain what an *estuary* is and describe the geological types of estuaries.

One of the most basic attributes of estuaries is the balance between incoming fresh water—from rivers, groundwater, and precipitation—on the one hand, and tidal salt water influx plus evaporation on the other (Figure 11-7). If fresh water influx exceeds evaporation and salt water influx, the surface water tends to have

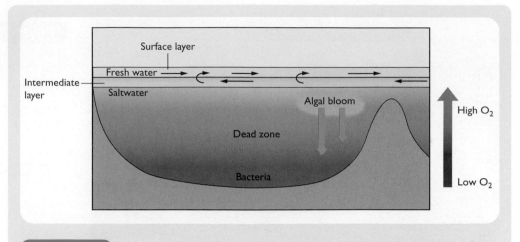

FIGURE 11-7 How a dead zone may form in an estuary or fjord. Low-density fresh water flows over higher density salt water, effectively creating a barrier between the two water types. The presence of a sill, an underwater ridge that traps the heavier salt water, further restricts circulation and facilitates development of a dead zone.

lower salinity than the adjacent open ocean. These are sometimes called *positive* estuaries. If evaporation and salt water influx predominate, the estuary is *hypersaline* (that is, possessing a salt concentration of greater than 3.5%, or 35 parts per thousand [ppt]), and these are called *negative* estuaries. **FIGURE 11-8** shows Texas' Laguna Madre, where salinities may exceed 36 ppt. Most estuaries are positive, like Chesapeake Bay (Figure 11-4).

Estuaries are some of the planet's most important real estate. Their *productivity* (the rate at which organisms convert the radiant energy of sunlight into chemical energy, primarily by photosynthesis) levels are among the highest of any ecosystem, and they are critical to the survival of many fish and invertebrate species. But estuaries are easily polluted due to their proximity to development. Pollution from hundreds of years of abuse began to severely degrade Chesapeake Bay by the 1950s. After more than 20 years of study and remediation attempts, the Chesapeake Bay Foundation's numerical estimate of Bay health as of 2010 stood at 31, on a scale of 100, which the organization characterizes as a "system dangerously out of balance." This shows how difficult estuarine remediation (that is, clean-up) is, and how important protecting estuarine health is in the first place. For details on the status of Chesapeake Bay, go to www.cbf.org/.

FIGURE 11-8 An example of a hypersaline (negative) estuary, Laguna Madre in Texas. Courtesy of Image Science and Analysis Laboratory, NASA-Johnson Space Center. "The Gateway to Astronaut Photography of Earth."

Barrier Complexes

Barrier complexes are long, narrow sand bodies separated from the mainland by lagoons and bays. Recall that barrier islands are found along passive rather than

FIGURE 11-9 Hatteras Island, a barrier island in North Carolina. Barrier islands of North Carolina are long, generally narrow, and cut by widely separated tidal inlets. Similar barrier islands are also found in parts of Louisiana, the Florida panhandle, southeast Florida, the south shore of Long Island, and the Cape Cod segment of the Massachusetts coast. The Georgia coastal barrier islands are relatively shorter, with tidal inlets spaced an average of 15 km (9 miles) apart. Similar barriers are found in northeast Florida, along most of the South Carolina coast, off the Delmarva Peninsula, in Massachusetts, and in some areas of Louisiana and east Texas. Courtesy of Bill Birkemeier/U.S. Army Corps of Engineers.

active tectonic coastlines. They contain a wealth of environments that protect and nurture countless wildlife (**FIGURE 11-9**). A system of over 400 such barriers has formed along 5,000 km (2,700 miles) of shoreline from Maritime Canada to Texas. These barrier islands are of two general types based largely on their shape: *wave-dominated* barrier islands and *mixed-energy* barrier islands (**FIGURE 11-10**). The former are long, narrow barrier islands shaped primarily by wave action. Because of their narrowness, storm waves may wash sand over the wave-dominated island, and these deposits form *washover fans* (Figure 11-10) on the back side of the island. Mixed-energy barrier islands are typically drumstick-shaped owing to influence of both waves and tides. Because barrier islands are dynamic in time and space—contracting and expanding, and migrating due to erosion and accretion—Geologist Orrin Pilkey (quoted previously in the Point–Counterpoint) considers barrier islands as almost "living organisms" that survive by adapting to their physical environment.

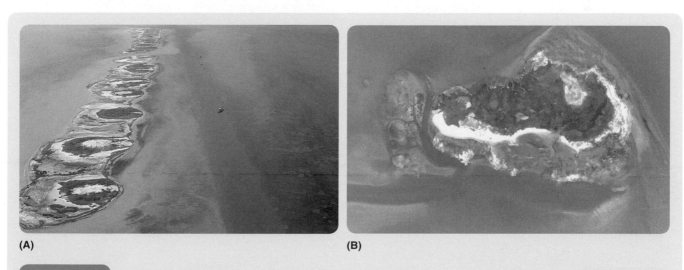

(A)   (B)

FIGURE 11-10 (A) Padre Island, Texas, a wave-dominated barrier island. Note the washover fans characteristic of these narrow barrier islands. Courtesy of Jim Wark. (B) Bull Island, South Carolina, a mixed-energy barrier island. Courtesy of the State of South Carolina and NOAA/USGS.

Barrier island complexes are the continent's main "line of defense" against hurricanes. As such, they have saved the nation billions of dollars in avoided damage.

The Coastal Barrier Resources Act was passed in 1982 to protect barrier complexes. The Act halted federal spending for infrastructure, water supplies, and disaster relief in high-risk coastal areas. One provision of the law used economic disincentives to help steer new construction away from risky and environmentally sensitive areas. The Act also eliminated federally-subsidized flood insurance on certain key parts of the barrier complexes on the Atlantic, Gulf, and Pacific Coasts, and some in the Great Lakes.

The John H. Chafee Coastal Barrier Resources System was designated under the Act to include 590,000 acres (240,000 ha) of undeveloped coastal barrier habitat along the Atlantic and Gulf coasts. In 1990, Congress included areas around the Great Lakes, in Puerto Rico and in the U.S. Virgin Islands as well. The area covered by the Act now exceeds 1.3 million acres (>500,000 ha).

Sea level rise associated with human-caused climate change may spell catastrophe for barrier islands. Few undeveloped islands remain on the U.S. east coast and many developed barrier islands may disappear in response to sea level rise.

Concept Check 11-4. Consider the options for developed barrier islands in the face of sea level rise. Select the best and worst options, in your opinion, then discuss reasons for your choices.

Reefs

Coral reefs are massive structures built of calcium carbonate by colonial corals, and a host of other organisms (**FIGURE 11-11**). They cover approximately 600,000 km² (230,000 miles²) in the tropics, an area roughly the size of Texas. The massive rock skeleton of a coral reef is the result of the actions of the coral organisms known as *polyps,* which are animals related to sea anemones that form a thin film over the reef. The hard foundation is made of calcium carbonate, which is derived from calcium ions and carbon dioxide dissolved in seawater. A reef represents the massive rock from a build-up of calcium carbonate over centuries or millennia by generations of coral polyps.

Coral reefs are important both ecologically and economically. They are among the planet's greatest storehouses of carbon, and play a critical role in the carbon cycle (**BOX 11-1**). Like barrier island complexes, coral reefs protect shorelines

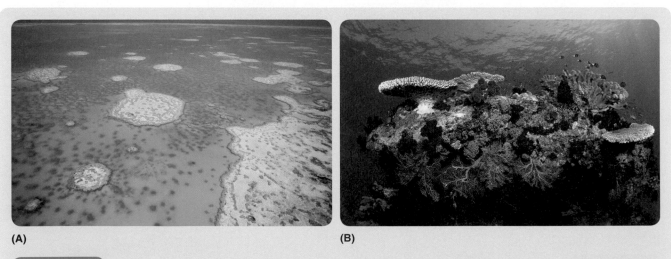

(A)　　　　　　　　　　　　　　　　　　　　(B)

FIGURE 11-11　(A) A part of the Great Barrier Reef of northeast Australia. It is one of the largest structures built by organisms on Earth, one of the planet's greatest storehouses of biological diversity, and stores (sequesters) billions of tons of carbon. © Dropu/ShutterStock, Inc. (B) A close-up showing life on the Great Barrier Reef. © Comstock/Thinkstock.

BOX 11-1 *Coral Reefs and Climate Change*

The living part of a coral reef is underlain by a foundation of calcium carbonate ($CaCO_3$) derived from calcium ions and carbon dioxide dissolved in seawater. As the coral polyps grow, they (along with calcareous algae living on or near the reef) precipitate calcium carbonate (that is, convert it from a dissolved form to a solid form), which supports the growing colony of coral and helps to cement coral rubble together. In the process, CO_2 is removed from the water. The resulting CO_2 "deficit" in the water is replenished from the CO_2 in the atmosphere.

Corals are extremely efficient at fixing carbon, that is, basically turning dissolved CO_2 into coral (remember, we are talking about the carbon cycle, and carbon is a key component of calcium carbonate). Each kilogram of $CaCO_3$ contains almost 450 grams (1 lb) of CO_2. If this CO_2 were released to the atmosphere, the Earth's surface temperature would increase significantly. If all CO_2 from coral and carbonate sedimentary rock were released, the Earth's temperature would approach that of Venus, where surface temperatures of 460°C are hot enough to melt lead!

Globally, the survival of coral reefs is threatened by increases in ocean temperatures and acidity, both direct results of increased atmospheric levels of CO_2 associated with climate change. Approximately 300 tons of CO_2 enter the ocean each second, exceeding the ocean's ability to safely absorb the gas, which is causing the ocean's pH to decrease, a phenomenon called **ocean acidification**. Acid dissolves calcium carbonate. Calcification of corals in the Great Barrier Reef has declined by over 14 per cent since 1990 as a result of acidification. According to scientists at the Carnegie Institution's Department of Global Ecology and reported in the journal *Geophysical Research Letters*[3], if atmospheric levels of CO_2 stabilize at 450 parts per million, fewer than 10% of the world's coral reefs would experience the proper ocean chemical environment necessary to sustain them. Atmospheric CO_2 levels exceeded 400 ppm in 2013.

As bad as acidification may be for coral reefs, increased ocean temperatures may be more foreboding. They can cause coral "bleaching" (**Figure B1-1**), which happens when corals expel the symbiotic photosynthetic cells (called

FIGURE B 1-1 Widespread bleaching on a coral reef in the Pacific Ocean. Courtesy of David Burdick/NOAA.

zooxanthellae) living in their tissue. Projections indicate that if current trends in greenhouse gas emissions continue, a significant proportion of the planet's remaining reefs may be lost to coral bleaching. Moreover, the polyps themselves will be directly impacted by even small changes in temperature, as a result of the physiological principle that organisms live more closely to the highest temperature they can tolerate (their *critical thermal maximum*) more so than the lowest. Animals, such as fish, capable of escaping higher temperatures frequently do so, but sedentary organisms like adult corals do not have that ability, so their mortality will doubtless increase as ocean temperatures rise.

Climate change impacts compound the stresses on reefs from local human-induced sources that we describe in the text.

In Other Words

"If current trends in CO_2 emissions continue unabated, in the next few decades, we will produce chemical conditions in the oceans that have not been seen for tens of millions of years. We are doing something very profound to our oceans. Ecosystems like coral reefs that have been around for many millions of years just won't be able to cope with the change."

Ken Caldeira
Author of the acidification study cited previously

and harbor numerous organisms. Reefs are important to the economies of many small island nations, especially in the Caribbean, by providing tourist income and fisheries for domestic consumption and export. Reefs also are major sources for the commercial tropical fish trade, in some cases with harmful consequences to the reef.

In addition to impacts of climate change described in Box 11-1, the health of coral reefs is threatened by nutrient pollution, primarily runoff of agricultural fertilizers, and undertreated sewage, which promote the growth of algae and thereby smother the reef, sedimentation (in part due to deforestation of nearby land), overfishing, and even ecotourism. Exposure of coral to increased ultraviolet radiation due to the seasonal thinning of the ozone layer may be yet another significant problem.

Approximately 75% of the world's coral reefs are classified as threatened, and if current trends continue, most could be gone by 2050.

The Continental Margin

Oceanographers recognize two overarching landforms at the continental margin: the *continental shelf* and *submarine canyons*.

Continental Shelf

The **continental shelf** is that part of the continent, underlain by continental crust, covered by the ocean. Its width ranges from a few km, to hundreds of km. Its seaward border is the **continental slope**, where slopes steepen, plunging to the ocean floor (FIGURE 11-12).

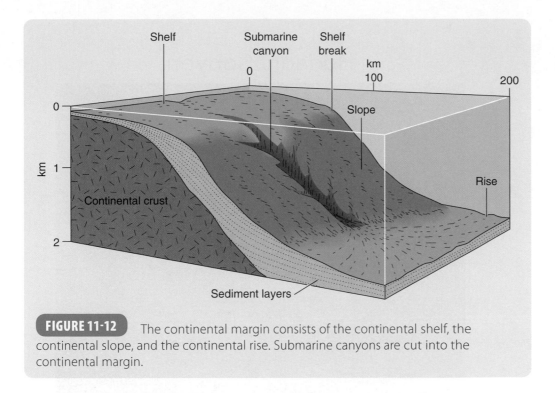

Shelf   Submarine   Shelf
          canyon   break

km
0   100   200

0
Slope

km 1
Rise

2
Continental crust

Sediment layers

FIGURE 11-12 The continental margin consists of the continental shelf, the continental slope, and the continental rise. Submarine canyons are cut into the continental margin.

The shelf is underlain by continental bedrock. At the surface is sediment and sedimentary rock eroded from the interior of the continent, which is draped like a seaward-thickening blanket over older, mainly igneous and metamorphic rocks below.

Submarine Canyons, Turbidites, and Submarine Fans

Next, we discuss features and processes of the sea floor that are not strictly coastal but result from coastal processes.

At times when the sea level was lower than today, rivers entered the sea much farther seaward than at present and were able to cut much deeper into what is now the continental shelf. This erosion formed **submarine canyons**, when rising sea level flooded the river valleys and formed the continental shelf (**FIGURE 11-13**). Canyons are important because coastal currents can transport large quantities of sand to the mouth of, and down, the submarine canyon, thus removing it from the shelf. Submarine canyons are the origin of distinctive deposits called **turbidites**, which form when unstable masses of loose sediment tumble down the canyons onto the deep ocean floor. Turbidites accumulate on **submarine fans**, or *abyssal fans,* delta-like features at the mouths of submarine canyons (**FIGURE 11-14**).

■ WAVES

Waves are travelling undulations of the sea surface. Most surface waves result from frictional effects of wind blowing across the water, creating a ripple that may grow into larger waves from the continued frictional effects.

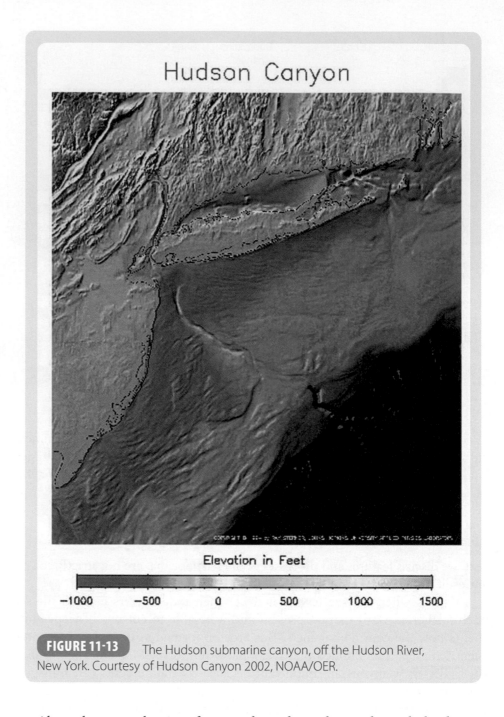

FIGURE 11-13 The Hudson submarine canyon, off the Hudson River, New York. Courtesy of Hudson Canyon 2002, NOAA/OER.

Along the coast, the size of a wave depends on the wind speed, the duration that the wind blows, and the distance over open water that the wind has blown (the wind's *fetch*). Sustained high-speed winds combined with a long fetch produces larger waves with great potential energy (**FIGURE 11-15**).

Attributes

Important attributes of waves (**FIGURE 11-16**) include:

- *Wave height:* the vertical distance from the crest to the trough
- *Wavelength:* the horizontal distances between the crests of adjacent waves
- *Wave period:* the time it takes two successive crests to pass a fixed point

(A)   (B)

FIGURE 11-14   (A) A submarine (or abyssal) fan. Courtesy of Reto Stöckli, NASA Earth Observatory. (B) Close-up of a reversely graded bed in conglomerate and sandstone of the Proterozoic LaHood Formation in Montana, U.S. © Dr. Marli Miller/Corbis.

FIGURE 11-15   Storm waves on the North Pacific Ocean. Remember that waves represent wind energy from the sun, transferred from the atmosphere to the water. Courtesy of NOAA.

- *Wave frequency:* the inverse of wave period; the number of crests passing a point per unit time
- *Wave velocity:* a wave's speed, typically in m/sec

Two other important aspects of waves are:

1. A wave transfers energy, not water. **FIGURE 11-17** shows the path traced by the individual water molecules as a wave passes. Note that this circular motion is greatest near the surface, and gradually diminishes with depth.

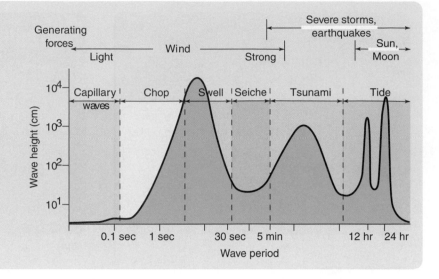

FIGURE 11-16 Important attributes of regular, symmetrical waves include height, wavelength, and period (the time it takes for successive crests to pass a fixed point).

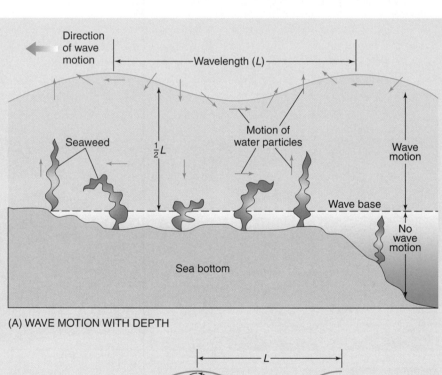

(A) WAVE MOTION WITH DEPTH

(B) ORBITS OF WATER PARTICLES

FIGURE 11-17 The motion of water particles beneath waves. (A) The arrows at the sea surface denote the motion of water particles beneath waves. With the passage of one wavelength, a water particle at the sea surface describes a circular orbit with a diameter equal to the wave's height. Water particles below the sea surface also describe orbits with the passage of each wave, as indicated by the back-and-forth motion of seaweed attached to the ocean floor. (B) The orbital diameters described by water particles beneath waves decrease rapidly with distance below the water surface. Wave-induced water motion essentially ceases at a depth (called the wave base) that is equal to one-half the wavelength.

2. The sizes and shapes of waves change as they approach the shore. Depending on the water depth, steepness of the bottom, and the wave height, the wave either begins to *break,* or reflects back out to sea.

The depth (D) to which waves affect water in coastal regions can be estimated using the formula D = L (wave length)/2 (Figure 11-17). When the water depth drops below about half of the wavelength, the friction of the wave dragging across the bottom begins to affect the entire wave (as well as sediment on the sea floor; see section on erosion). Wave height increases and wavelength decreases as the wave moves into shallower water. The wave begins to break when the top, moving faster than that part of the wave near the bottom, spills over. The energy thus released is focused on the beach, and this is a major cause of erosion.

Understanding the mechanisms of, and controls on, wave action can help us understand sediment transfer and erosion "hot spots" along shorelines.

A kind of wave known as a tsunami is discussed below (and also is covered in the discussion on earthquakes elsewhere in the text).

Longshore Transport

If a wave approaches the shore obliquely (which they usually do) or if the wave encounters an irregular bottom, the wave's direction of travel, called the wave *path,* can change. Obliquely breaking waves lead to the creation of currents that run parallel to the beach, known as **longshore currents** (FIGURE 11-18). Longshore currents can transport vast amounts of sand, some of it to submarine canyons, and are important in understanding the stability of beaches and coastal structures. The strength of the longshore current is related to wave height and angle of approach; higher waves approaching at more acute angles typically result in stronger longshore currents.

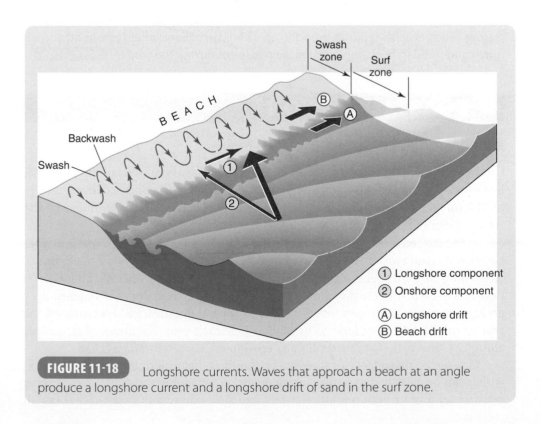

FIGURE 11-18   Longshore currents. Waves that approach a beach at an angle produce a longshore current and a longshore drift of sand in the surf zone.

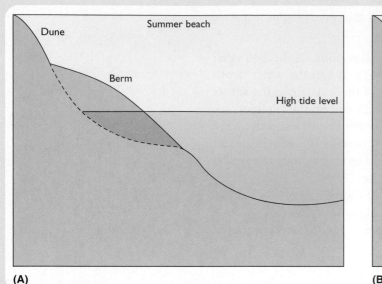

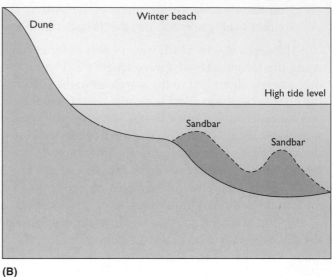

(A)

(B)

FIGURE 11-19 The cyclical differences between a summer beach (A) and a winter beach (B). More gentle summer waves move sand onward onto the beach and widen it, whereas stronger winter waves move sand of the beach and cause it to narrow. Annual fluctuation in beach size is important because scientists need to know the historic fluctuation in the beach environment in order to understand long-term impacts of sea-level change.

Seasonal Variation in Wave Energy

Wave energy, and its effect on the coast, varies seasonally. Other than erosion during extraordinary events like hurricanes and tsunami, erosion by waves is more pronounced along beaches in the Northern Hemisphere winter than during the summer, during which beaches typically are built-up. This is because weather systems other than hurricanes tend to be stronger and more persistent during the winter (**FIGURE 11-19**).

■ COASTAL EROSION

A Global Problem

Coastal erosion has become a global problem. Along the Nile Delta of Egypt, shoreline erosion averages 50 to 100 meters (~160–330 ft) per year, and occasionally rises to 200 m/year (>650 ft) due largely to the trapping of Nile River sediment by the Aswan Dam (**FIGURE 11-20**).

In the United States, all 30 states bordering an ocean or the Great Lakes are experiencing coastal erosion, and 26 are presently experiencing net loss of their shores. According to the Federal Emergency Management Agency (FEMA), about 1,500 homes will be lost to erosion annually (**FIGURE 11-21**). Losses of this magnitude or greater are projected to continue for the next several decades. FEMA estimates the cost of this loss to be at least $530 million U.S. each year. Estimates of structures along U.S. coasts that are susceptible to erosion are shown in **TABLE 11-2** .

One factor that affects local sediment supply is the underlying coastal bedrock, which can be classified as non-resistant, resistant, or highly resistant. Coastal plain beaches tend to be built on non-resistant bedrock, whereas pocket beaches (small

(A)

(B)

FIGURE 11-20 (A) Location of the Aswan Dam. (B) The dam and Lake Nasser from space. It has trapped vast quantities of sediment that previously would have been dumped into the Mediterranean and along the course of the Nile, serving to restore sediment. Courtesy of JSC PAO Web Team/NASA.

beaches between headlands, or promontories) in California are characterized by hard, resistant bedrock (**FIGURE 11-22**).

Rates of Erosion

The energy for coastal erosion comes mainly from waves. The actual *rate* of shoreline erosion depends on supply of sand, wave energy, sea-level change, and nature of coastal bedrock.

Average annual erosion rates are 2 to 3 feet/year (0.6–0.9 m/year) on the Atlantic coast and about 6 feet (1.8 m) on the Gulf coast (Figure 11-22). During intense storms, 50 to 100 feet (15–30 m) of Atlantic and Gulf coast shore can erode.

Where Does Beach Sand Come From?

A grain of *sand,* by definition, is between 0.06 and 2.0 mm (0.002–0.08 in) in diameter. By one estimate[4], there are approximately 7.5×10^{18} (7.5 *quintillion*) grains of sand on all the Earth's beaches. Where any one of those grains is going to be at any time is a question of importance to coastal dwellers, tourists, policy makers, and even insurers.

Sand can come from a variety of sources, including,

- adjacent beaches and dunes
- anywhere in a river's drainage basin
- eroded coastal landforms such as cliffs or submarine outcrops
- the continental shelf

FIGURE 11-21 Coastal home being lost to erosion. © Luc Gnago/REUTERS/Landov.

TABLE 11-2	Nationwide Estimate of Structures Susceptible to Erosion				
	Atlantic Coast	**Gulf of Mexico**	**Pacific Coast**	**Great Lakes**	**Total**
Length of coastline					
Miles	2,300	2,000	1,600	3,600	9,500
Percentage of total	24%	21%	17%	38%	
Structures within 500 feet of shoreline[a]					
Number	170,000	44,000	66,000	58,000	338,000
Percentage of total	50%	13%	20%	17%	
Structures within 60-year erosion hazard area (EHA)[b]					
Number	53,000	13,000	4,600	16,000	87,000
Structures within 60-year EHA assuming all open lots are filled					
Number	76,000	22,000	5,200	>16,000[c]	>120,000

[a]All estimates exclude structures in major urban areas. The analysis assumes these structures will be protected from the erosion hazard.
[b]The 60-year EHA is determined by multiplying local erosion rates by 60 years.
[c]Data on open lots not available in Great Lakes
Data from: *Evaluation of Erosion Hazards Summary*. The H. John Heinz III Center for Science, Economics and the Environment, April 2000. Prepared for the Federal Emergency Management Agency Contract EMW-97-CO-0375.

(A)

(B)

FIGURE 11-22 (A) Coastal plain beaches are typically found over non-resistant bedrock. © Robert Broadhead/ShutterStock, Inc. (B) Pocket beaches (small beaches between headlands, or promontories) in California overlie hard, resistant bedrock. Courtesy of Sapelo Island National Estuarine Research Reserve/NOAA.

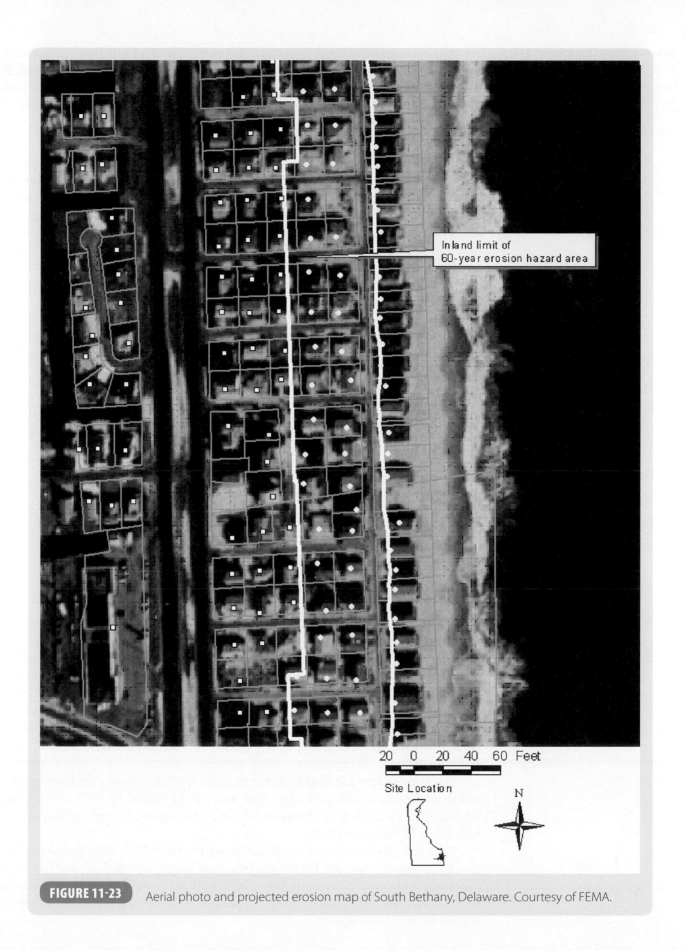

Inland limit of
60-year erosion hazard area

20 0 20 40 60 Feet

Site Location

N

FIGURE 11-23   Aerial photo and projected erosion map of South Bethany, Delaware. Courtesy of FEMA.

Value-added Question 11-1.
Study FIGURE 11-23, which shows a projected 60-year erosion hazard map for South Bethany, Delaware. Using the scale, determine how far inland the beach is projected to erode over the next 60 years. What is the average rate expected in feet per year? About how many houses will be affected?

In most areas outside the tropics, sand is of siliciclastic origin, that is, composed of grains of silicate minerals like quartz. In many tropical areas, *biogenic* sources (skeletal remains of corals, calcareous algae, etc.) also contribute significantly to sediment.

Effect of Dams, Levees, and Impoundments

Over the past century and a half, human interference with coastal processes has led to a reduction in the natural supply of sand. Thousands of dams and impoundments on rivers trap sediment. Thus, sand in vast quantities that used to naturally nourish beaches is now trapped upstream. At the same time, investments in infrastructure (sewers, roads, buildings, power lines, etc.) and growing human populations in desirable coastal settings together have built a powerful constituency supporting beach "stabilization" or "protection" practices. Coastal erosion also can be caused by natural subsidence in some areas, sea level rise globally, and loss of sediment trapped behind dams.

■ SEA LEVEL CHANGE AND COASTS

Water began to accumulate on Earth's surface before 3.8 billion years ago, a fact we know because water-laid sedimentary rocks are preserved from that time. Water now covers 71% of the Earth's surface, and the mass of water is thought to have remained nearly constant over the last several billion years. This does not, however, mean that sea level has remained constant.

What Causes Sea Level Change?

Changes in sea level, which have occurred over most of the Earth's geologic history, result from two processes:

1. *Eustatic processes* change the absolute amount of liquid water within the ocean basins. The principal mechanism driving eustatic changes in sea level is the reproportioning of water between liquid and ice, due to changes in global climate.
2. *Isostatic processes* change the underlying topography of the sea floor.

Isostatic changes can occur on *local and regional scales,* for example, the rebound of crust after glacial melting, the slow sinking of parts of coastal Virginia due to glacial isostatic adjustments, or the slow subsidence of deltas at passive continental margins; or on *global scales,* as in periods of marked changes in sea floor spreading rates. As spreading rates increase, heat flow increases, and rocks of the ocean basins swell in volume accordingly. Expanding the sea floor in turn displaces water upward and outward onto the continent.

The term *relative sea level* is also commonly used. It refers to sea level relative to the level of the continental crust and thus can be caused by both eustatic and isostatic processes, such as changes of the sea level caused by melting/freezing of ice caps or by tectonic activity.

Some scientists think that rapid falls in sea level may have contributed to mass extinction events from the Earth's geologic past. Rapid sea level decline has been hypothesized to cause changes in atmospheric oxygen levels. In this case, rapid decay (oxidation) of shallow, newly exposed organic-rich marine deposits would remove oxygen from the atmosphere.

Sea level has been rising since the end of the last glacial episode, about 15,000–18,000 years ago. The rate of global sea level rise for the last 100 years has been around 2 mm/year, and 15–20 cm in all (0.08 in/year, 6–8 in total). Over the period 1993–2003, sea level rose at an average rate of 3.1 mm/year. By 2012 the rate had increased to nearly 3.2 mm/year. USGS sea level rise animations can be found at http://cegis.usgs.gov/sea_level_rise.html.

Rate of global sea level rise for the twenty-first century is expected to rise considerably, although the magnitude of increases is subject to debate. The Intergovernmental Panel on Climate Change (IPCC) in its 2007 report concluded that sea level could rise as much as 26 to 59 cm (10–23 inches) by 2100 due to human-caused climate change (see later in the chapter). However, those projections omitted the increase in sea level caused by melting of Greenland and Antarctic ice sheets (**FIGURE 11-24**), which could at a minimum double the highest rate projected by the IPCC. A 2011 assessment projected an increase of 32 cm (13 in) by 2050.[5] It is even possible that sea level could rise by several meters by 2100 if other factors (like decreased reflectivity of the Earth's surface) come into play. The observed rate of coastal sea level rise varies from region to region because of the variability in relative sea level.

The relationship between global sea level rise and local land-level changes is a crucial one, which we revisit in our discussions of the flooding of Venice, Italy.

Impacts of Climate Change

Although ample scientific evidence now exists detailing the impacts of changing sea level upon biological communities, humans have only recently become concerned

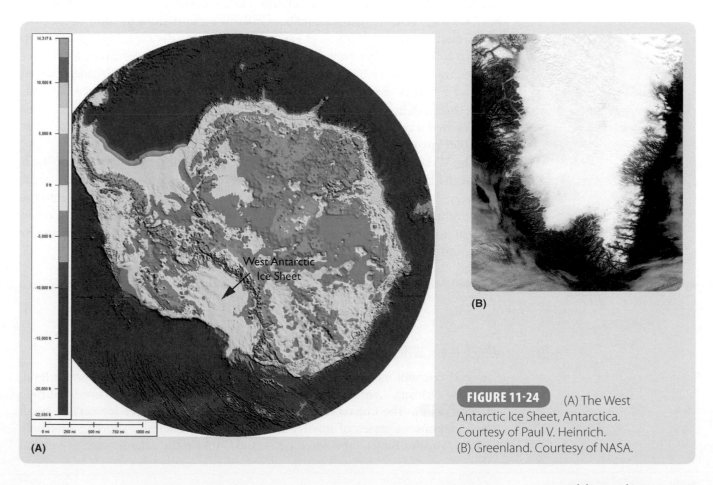

FIGURE 11-24 (A) The West Antarctic Ice Sheet, Antarctica. Courtesy of Paul V. Heinrich. (B) Greenland. Courtesy of NASA.

with the potential of such processes to affect their own lifestyles and standards of living. This newfound concern stems from the consensus among climatologists that the planet is experiencing human-induced increased global climate change.

Climate change, inducing ocean warming, will likely result in an increase in the historic rate of sea level rise due to two factors: the *thermal expansion of seawater,* and the *melting of ice on land* in glaciers and polar regions. The coefficient of thermal expansion of seawater is 0.00019 per degree Celsius, meaning that if a volume of seawater occupied 1 cubic meter of water (1000 L, 264.2 gal), after warming by 1°C, it would expand to 1.00019 m^3 (1000.19 L, 264.25 gal). Translated over the 3800-m mean depth of the ocean, an increase in temperature of 1°C will cause a sea level rise of about 0.7 m (70 cm, or 28 in) due exclusively to thermal expansion.

Thermal expansion acts upon water already in the basin. Contributions from melting ice represent water added to the current ocean volume. The complete melting of ice that is currently perched upon land, as in the ice sheets of Greenland and the Antarctic, as well as the melting of mountain glaciers, has the potential to raise sea level by as much as 80 meters (262 ft). Ice that is already floating in the ocean water, as in the Arctic ice mass, Antarctic ice shelves, and numerous smaller icebergs, may melt, but will not contribute directly to sea level rise because the mass of water contained in these features already displaces its equivalent water volume.

There is considerable evidence that the increase in melting rate has begun. Monitoring of mountain glaciers throughout the world over the last two decades indicates that 75% of them are losing mass, and that the rate of this loss has almost doubled, from 0.25 m to 0.50 m (0.82 to 1.64 ft) of water equivalent, in the last 20 years (**FIGURE 11-25**). As early as 1991 NASA reported that the extent of sea ice in the Arctic Ocean declined by 2% between 1978 and 1987. More recently, researchers have used satellite measurements of microwave emissions to show declines in permanent Arctic ice of 7% over each of the past two decades. NASA data from 2004 to 2008 show a loss in *winter* Arctic sea ice of 900 km^3/year. Some ice masses in the Antarctic have begun to melt more rapidly over the last decade, spurred by the 2.5°C increase in the average temperature on the peninsula since the mid-1940s (**FIGURE 11-26**). On September 16, 2012, NASA reported that Arctic ice mass had reached a record low for the satellite era, a reduction of 300,000 square miles (780, 000 km^2) from the previous lowest extent, in 2007.

The West Antarctic ice sheet is buttressed by an immense 500-m–thick floating ice shelf. Melting of this and other ice shelves could destabilize continental ice sheets, with the Western Antarctic Ice Sheet potentially capable of slipping into the surrounding ocean over a period of decades to centuries. If this happens, it could cause an increase in global sea level of perhaps 6 meters or more, according to scientific models, with catastrophic consequences.

Concept Check 11-5. CheckPoint: Explain ways in which global climate change will cause elevated sea level.

Which Areas Will Be Impacted?

Sea level rise will increase flooding risks in areas already near or below sea level, like New Orleans, Louisiana in the United States and coastal Holland in Europe (**FIGURE 11-27**). The impact of sea level rise will also be severe for certain small island nations. A series of low-lying islands in the Pacific, including Marshall, Kiribati, Tuvalu, Tonga, Line, Micronesia, and Cook; in the Atlantic such as Antigua

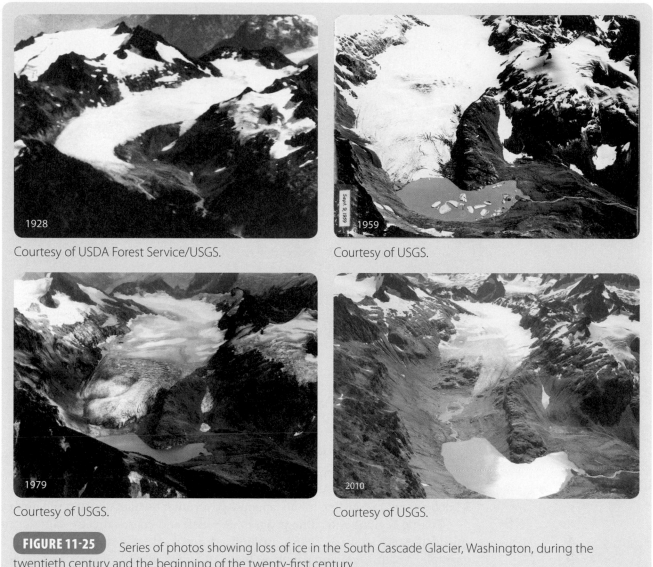

1928

Courtesy of USDA Forest Service/USGS.

Sept. 9, 1959

1959

Courtesy of USGS.

1979

Courtesy of USGS.

2010

Courtesy of USGS.

FIGURE 11-25 Series of photos showing loss of ice in the South Cascade Glacier, Washington, during the twentieth century and the beginning of the twenty-first century.

FIGURE 11-26 Collapse of a section of the Larsen Ice Shelf, Antarctica, in 2002. The area that collapsed over a 35-day period measured about 3,250 square kilometers (1,255 square miles), almost the size of Rhode Island. Courtesy of MODIS, NASA's Earth Observatory.

FIGURE 11-27 Flood control structures along Eastern Holland. Investment of the Netherlands in such infrastructure represents a considerable proportion of the country's gross national product. © iStockphoto/Thinkstock.

and Nevis; and in the Indian Ocean the Maldives, could be greatly impacted (**FIGURE 11-28**). For example, in the Maldives most of the land is less than 1 meter (3.05 ft) above sea level. A seawall built to protect Male, capital of the Maldives, cost $4,000 a meter, according to the Japanese government which donated the money.

Coastal regions with little geographic relief, and coasts already experiencing subsidence, are similarly at risk. Millions of people are at risk in Bangladesh, for example. About 17 million people in that nation live within 1 meter of sea level.

In Southeast Asia, a number of large cities, including Bangkok, Mumbai, Kolkata, Dhaka, and Manila, each with populations greater than 5 million, are located on coastal lowlands or on river deltas. Sensitive areas in the United States include the states of Florida and Louisiana, many coastal cities like New York, and inland cities bordering estuaries. **TABLE 11-3** shows a list of cities most threatened by sea level rise. By 2100, parts of an estimated 180 coastal U.S. cities could be threatened by sea level rise.

Concept Check 11-6. The year 2100 is frequently used in projections of sea level rise. Explain why this may discourage taking actions to curb or defend against sea level rise.

In addition to population centers like Miami, New York, Baltimore, Boston, smaller cities like Charleston (SC), Wilmington (NC), and Savannah (GA) also will be faced with the threat of sea level rise. Engineers claim that they can protect cities from sea level rise, but it will cost hundreds of billions of dollars.

Concept Check 11-7. Which group of cities, the major population centers or the smaller cities, if either, is likely to receive federal funding to defend against sea level rise? Discuss what you think will happen to the cities left to fend for themselves.

FIGURE 11-28 Male, capital of the Maldives. These low-lying Indian Ocean islands could be flooded by rising sea level. © Piccaya/Dreamstime.com.

If future increases in sea level (**FIGURE 11-29**) become more rapid, intertidal and some shallow **subtidal** marine communities may not be able to keep pace with sea level change and could die off, as environmental conditions change beyond their limits of tolerance. Such a change is now occurring in intertidal salt marsh environments around the Mississippi River delta in Louisiana, as coastal subsidence combined with global sea level rise exceeds the rate of vertical growth of the marsh community.

TABLE 11-3	*Major Cities Threatened by Sea Level Rise*		
U.S. Cities		**Outside of North America**	
Miami	Philadelphia	Osaka/Kobe	Cairo
New York/Newark	San Francisco	Tokyo	Guangzhou
New Orleans	Baltimore	Rotterdam	Ho Chi Minh City
Tampa-St Petersburg	Los Angeles/Long Beach	Amsterdam	Kolkata
Virginia Beach, VA	San Juan, PR	Nagoya	St. Louis, Senegal
Charleston (SC)		London	Rio de Janeiro
Boston		Shanghai	Buenos Aires
Washington		Mumbai	Lima

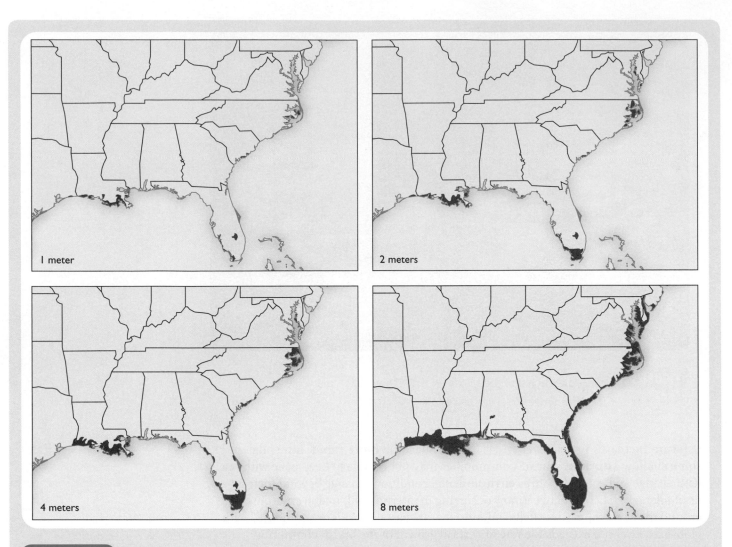

FIGURE 11-29 Four scenarios of sea level rise for the southeastern United States, based on the extent to which water temperatures increase. Modified from NOAA.

Understanding the Threat from Climate Change

Technological advances in climate research, paired with satellite-based remote sensing, have enabled scientists to more fully understand the mechanisms that cause climate change, as well sea level's response. Satellite remote sensing now permits the collection of large-scale regional and global data sets that decades ago were only a dream.

Geologists and climate scientists can use fossil, geophysical, and geochemical indicators to interpret past environmental conditions and rates of change. Examples include fossil spores and seeds from sediment that can indicate climate change episodes, and radioactive dating techniques that can precisely date the intervals containing the fossils. This information can then be used to project patterns of future climate change and sea level response in time and space. Other useful information is coming from declassified military data on sea **ice draft** (thickness under water) from the Arctic. Submarines and permanently moored devices are now continually monitoring this portion of the Earth's ice mass.

In a historical context, scientists are relatively certain that sea level has risen about 120 meters (400 ft) in the past 18,000 years, and many coastal shorelines

Bangladesh

Bangladesh (**Figure CS 1-1**), a country about the size of Illinois, Wisconsin, or Florida, lies on the northern shore of the Indian Ocean. Bangladesh was originally part of Pakistan after the Indian subcontinent gained independence from Britain in 1947. In 1970, Bangladesh seceded from Pakistan. Its population in mid 2012 was 161 million on a total area of 144,000 km^2 (55,000 mi^2). The mid 2012 population of the United States was about 314 million on a total area of 9.8 million km^2 (377,000 mi^2).

Value-added Question 11-2. What is the population density (number of people per unit of area) of Bangladesh in people per km^2?

Value-added Question 11-3. What is the population density of the United States in people per km^2?

Concept Check 11-8. Compare the population densities of the United States and Bangladesh. Do you think

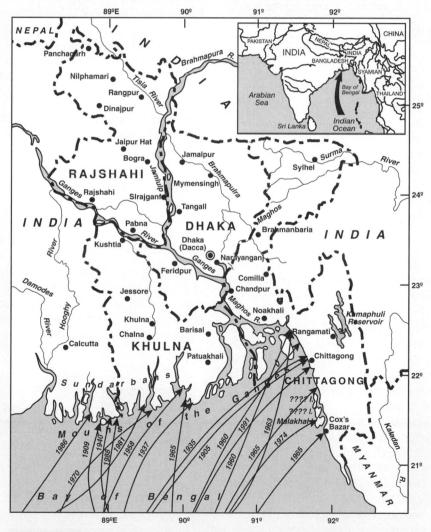

FIGURE CS 1-1 Location of Bangladesh, showing tracks of major historical typhoons.

Continued ▶

that the populations of both countries are evenly distributed throughout the countries? Explain your answer.

Floods in Bangladesh

Bangladesh is vulnerable to catastrophic flooding from rivers as well as from tropical storms. More than half the country lies at an elevation of less than 8 m (26 ft) above sea level. Moreover, more than 17 million people live on land that is less than 1 meter (3.3 feet) above sea level. About 25% to 30% of the country is flooded each year. This can increase to 60% to 80% during major floods. More than 30 million Bangladeshis live in the coastal zone and thus could become climate refugees.

River flooding in Bangladesh results from some combination of the following:

1. excess rainfall and snow melt in the catchment area, especially in the foothills of the Himalayas (an area where many of the Earth's rainfall records were set)
2. simultaneous peak flooding in all three main rivers
3. high tides in the Bay of Bengal, which can dam runoff from rivers and cause the water to "pond up" and overtop its banks

In addition, changes in land use in the catchment area can result in profound changes in the flood potential of Bangladesh's rivers. For example, deforestation of the Himalayan foothills for agriculture, fuel, or habitation makes the hills more prone to flash flooding. Then, during the torrential monsoon rains, mudslides can dump huge volumes of sediment into the rivers. This sediment can literally pile up in the river channels, reducing channel depth and thereby the channel's capacity to transport water. The result is more water sloshing over the rivers' banks during floods and more disastrous floods in Bangladesh. The silt can smother offshore reefs when it finally reaches the ocean, aided by land-use practices that clear coastal mangrove forests. Geologists estimate that the drainage system dumps at least 635 million tonnes (1.4×10^{12} pounds) of sediment a year into the Indian Ocean, which is four times the amount of the present Mississippi River.

Western-style development may be contributing to Bangladesh's increased vulnerability to natural disasters. Recently built roads blocked floodwaters, thereby increasing flooding from cyclones and the damage and loss of life.

Truly catastrophic floods can result when high tides coincide with tropical storms. In the early 1990s, such a combination took over 125,000 lives. About one tenth of all Earth's tropical cyclones occur in the Bay of Bengal (Figure CS 1-1), and approximately 40% of the global deaths from storm surges occur in Bangladesh.

Storm surges result when cyclones move onshore and the winds push a massive wall of seawater onshore with them. Surges in excess of 5 meters (16.4 feet) above high tide are not uncommon, and thus, you can imagine the impact such a storm can have on a country in which half the land is within 8 meters (26.2 ft) of sea level!

Concept Check 11-9. Bangladesh is a trivial producer of CO_2, the gas principally responsible for global climate change, compared with nations like the United States and China, and it faces potential disaster from sea level rise because of global warming and tropical storms. What responsibility, if any, do you believe the main carbon-emitting nations have to reduce the potential for environmental disaster in Bangladesh?

Miami, Florida

Miami, Florida is the playground of the super wealthy and a popular vacation spot as well as the home of large numbers of expatriates, mainly Cubans, from Caribbean countries. The fates of both places, however, are inexorably tied to global climate change. **Table CS 1-1** shows population data for Miami from 1990 to 2011. Over that period, the annual growth rate was approximately 1.2%.

Value-added Question 11-4. Determine the doubling time for Miami's population using the formula t = 70/r (where r = 1.2).

Value-added Question 11-5. Miami-Dade County had a 2012 population of about 2.56 million on a land area of 6,000 km². Calculate the city's population density in people per km². Compare it with that of Bangladesh.

Concept Check 11-10. The city of Miami is extremely vulnerable to sea level rise. Given the likelihood that a 1.5 m (5 ft) or greater sea level rise could occur as early as the turn of the century and that further increases could follow, what measures would you take now to prepare?

TABLE CS 1-1	Population for Miami, Fl from 1990-2011
2010: Miami/Dade County, 2,496,000	
2007: Miami/Dade County, 2,387,170	
2000: Miami/Dade County, 2,253,362	
1990: Miami/Dade County, 1,937,094	

have moved far inland. Off the East Coast of the United States, geologists estimate that the shoreline has migrated inland more than 80 km (50 mi) during this period, alternating between active migration and static periods. Based upon a rough average of typical beach slopes, for each unit rise in sea level, the beach will migrate 1,000 to 2,000 units inland.

■ COASTAL ENGINEERING: "MANAGING" SHORELINES

We now return to the issue raised in Point–Counterpoint at the beginning of this chapter: whether it makes sense to protect shorelines against the effects of a wave attack. As you saw previously, some take the view that we have no choice but to protect infrastructure and major urban centers along coasts, whatever the cost. Not to do so is simply "unthinkable." Others, including many marine scientists, say that humans cannot protect coasts from the ravages of nature forever, and that the more we spend on coastal engineering, the more costs we will incur in the future.

Concept Check 11-11. In light of limited government revenue and the desire of many to reduce taxes, to what extent is it an essential responsibility of government to pay, out of general revenue, to protect coastal infrastructure? If it is a responsibility of government, at what level should this responsibility lie? Explain your reasoning, or cite evidence.

We next describe the various types of coastal engineering structures and strategies used to protect coasts, and discuss the advantages and disadvantages of each. The ultimate decision to protect or not to protect will be decided through the political process.

Why Interfere with Coastal Processes?

Engineers build structures along coasts for two main reasons: (1) to trap sand to build up the shore, and (2) to protect the shore, and what we have built on it, from erosion. But it is getting more and more expensive to protect the coast. For example, the eleven most damaging hurricanes of the twentieth century all occurred after 1964 (**TABLE 11-4**). The intensity of these storms did not systematically increase compared to earlier hurricanes, so something else must have been the cause. The increased cost is most likely due to the growth in population in hurricane-prone areas, with the associated expansion in infrastructure and housing, virtually assuring that hurricanes would cause more and more damage. Thus, as more and more people move to coasts, more and more damage will likely result from storms, and more demands will be heard for protection from these natural events.

A Brief History of Coastal Protection Measures in the United States

When the Rivers and Harbors Act was passed in 1889, the U.S. Army Corps of Engineers was given the responsibility for maintaining navigable waterways, including harbor entrances. The Corps' mandate was gradually extended to protecting coasts. The building of seawalls and groins resulted. These structures temporarily protected coasts, but resulted in loss of beaches in front of the structures. Coastal dwellers then demanded beach renourishment: restoration of the lost beaches.

TABLE 11-4   Costliest U.S. Hurricanes

Rank	Name	Year	Category	Damage (U.S.)*
1.	Katrina (LA/MS/AL/SE FL)	2005	3	$105,840,000,000
2.	Andrew (SE FL/SE LA)	1992	5	$45,561,000,000
3.	Ike (TX/LA/MS)	2008	2	$27,790,000,000
4.	Wilma (FL)	2005	3	$20,587,000,000
5.	Ivan (FL/AL)	2004	3	$19,832,000,000
6.	Charley (FL)	2004	4	$15,820,000,000
7.	Irene (NC/VA/MD/DE/NJ/NY/CT/VT/NH/ME)	2011	1	$15,800,000,000
	Irene (U.S. east coast)	2011		$10,000,000,000
8.	Rita (LA/TX)	2005	3	$11,797,000,000
9.	Agnes (NE U.S.)	1972	1	$11,760,000,000
10.	Betsy (FL/LA)	1965	3	$11,227,000,000
11.	Allison (TX/LA)	2001	T.S.	$10,998,000,000
12.	Frances (FL)	2004	2	$10,018,000,000
13.	Hugo (SC)	1989	4	$9,739,820,675
14.	Camille (MS/AL)	1969	5	$9,282,000,000
15.	Floyd (NC)	1999	4	$9,225,000,000
16.	Jeanne (FL)	2004	3	$8,072,000,000
17.	Opal (NW FL/AL)	1995	3	$7,729,000,000
18.	Diane (NE U.S.)	1955	1	$7,408,000,000
19.	Frederic (AL/MS)	1979	3	$6,571,000,000
20.	New England	1938	3	$6,325,000,000
21.	Fran (NC)	1996	3	$6,140,000,000
22.	Isabel (NC/VA/MD)	2003	2	$6,112,000,000
23.	Celia (S TX)	1970	3	$5,918,000,000
24.	Great Atlantic Hurricane	1944	3	$5,706,000,000
25.	Alicia (N TX)	1983	3	$4,569,000,000
26.	Gustav (LA)	2008	2	$4,347,000,000
27.	Carol (NE U.S.)	1954	3	$4,175,000,000
28.	Georges (PR/MS)	1998	5	$3,860,000,000
29.	Juan (LA)	1985	1	$3,238,000,000
30.	Donna (FL/Eastern U.S.)	1960	4	$3,215,000,000
31.	Iniki (HI)	1992	4	$3,095,000,000
32.	Bob (NC and NE U.S.)	1991	2	$2,004,635,258

Source: National Hurricane Center Publication: *The Deadliest Atlantic Tropical Cyclones, 1492–Present.*

Coastal Protection Structures and Strategies

Engineering structures and practices can protect coasts from erosion. The following strategies or structures have been used.

Groins and Jetties

Groins and jetties are structures, usually concrete or rocks, built perpendicularly to the coast to trap sand (**FIGURE 11-30 and 31**). Jetties are specific structures built to stabilize inlets and protect harbors.

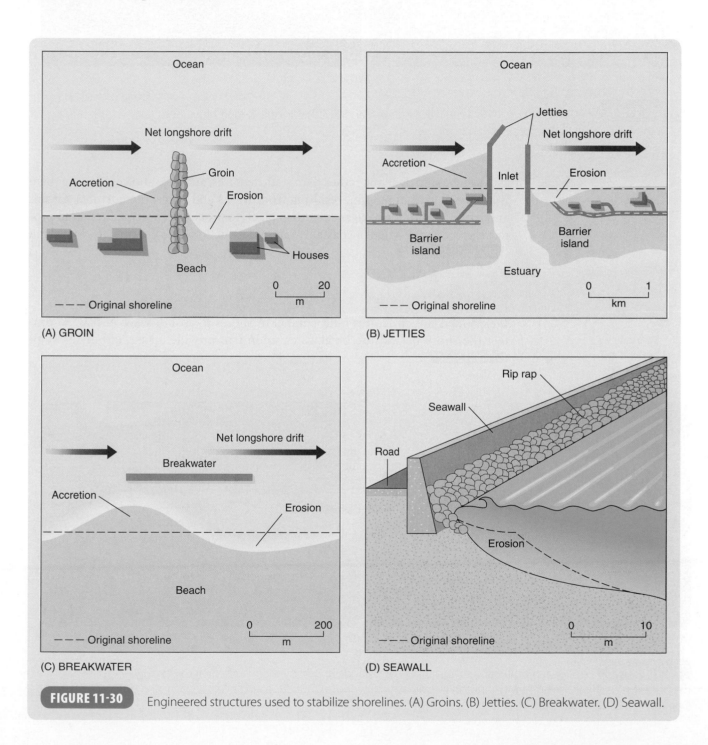

FIGURE 11-30 Engineered structures used to stabilize shorelines. (A) Groins. (B) Jetties. (C) Breakwater. (D) Seawall.

FIGURE 11-31 (A) People sitting on a groin at Pawleys Island, South Carolina. © Pluff Mud Photography. (B) Jetties. Explain how jetties work. © iofoto/ShutterStock, Inc.

Beaches are dynamic rather than static features, and the sand comprising them is almost always in motion, resulting from waves and longshore currents. Groins and jetties are built so as to keep sand in place, to protect property values, or to protect the economic value of coastal vacation destinations like Ocean City, Maryland (**FIGURE 11-32**).

Breakwaters

Breakwaters are structures built parallel to shores to absorb wave energy, to protect the shore behind the breakwater, and/or to provide a safe harbor for ships (**FIGURE 11-33**).

FIGURE 11-32 (A) Beach protection at Ocean City, Maryland. (B) Beach renourishment at Ocean City, Maryland, a major tourist destination for the Washington D.C. area and beyond. (A) and (B) courtesy of the Maryland Geological Survey, Department of Natural Resources.

FIGURE 11-33 Breakwater. Breakwaters protect the shore and provide harbor for small boats. © Jeff La Marca/Flickr.com.

Beach Renourishment

Beach renourishment, also called **soft stabilization**, refers to augmenting eroded beaches with sand either pumped from nearby sand bars or trucked in. It is practiced at numerous vacation destinations along the eastern United States (FIGURE 11-34). During winter months, waves carry sand away from beaches and move it farther offshore. Storms also may carry sand offshore or move it along the coast. In addition, submarine canyons are cut into the continental shelf when sea level was much lower than today during times of glacial maxima, as we mentioned previously. Sand in longshore transport may be intercepted by the heads of the canyon systems and carried to the deep ocean bottom by turbidity currents. Sand thus removed is no longer available to nourish beaches.

In short, beach renourishment works, but it is temporary and costly. And the sand has to come from somewhere. Removal often harms indigenous animal and plant communities.

Sea Walls and Revetments

Sea walls of various types are built usually when other means of engineering do not prevent the beach from eroding or protect hotels or valuable beach-front residences (FIGURE 11-35).

Revetments are structures built on coastal bluffs to absorb the energy of incoming waves. When properly designed, they do not significantly interfere with long-shore transport, nor do they redirect wave energy to adjacent unprotected areas (FIGURE 11-36).

FIGURE 11-34 A beach being renourished. © Todd Yates/The Facts/AP Photo.

FIGURE 11-35 A seawall stabilizing a developed shoreline along the South Carolina coast. The area behind the seawall is protected from eroding, but the beach in front disappears. © Pluff Mud Photography.

FIGURE 11-36 A revetment, left. Also note the seawall to the right. © Pluff Mud Photography.

TABLE 11-5	Advantages and Disadvantages of Coastal Engineering Structures

Advantages
They can protect coasts over short periods of time from erosion.
Structures are generally built and maintained by agencies of government, whose budgets are not dependent on local taxation.
Disadvantages
They are ineffective over time periods of several decades to centuries.
They often result in sand loss.
They encourage a sense of complacency and security, and encourage additional investment and population growth.

Advantages and Disadvantages

TABLE 11-5 summarizes the advantages and disadvantages of coastal engineering.

Other Strategies

Besides **hard stabilization** methods, other strategies include zoning, relocation of structures, and local, state, or federal legislation. Zoning ordinances may restrict building in certain areas, require certain types of building, or require structures be set back an appropriate distance from the shore. In extreme cases, structures may be relocated, as in the case of some U.S. lighthouses.

■ COASTAL GEOHAZARDS: TSUNAMI AND TROPICAL STORMS

We already have discussed some of the hazards that coasts, and the populations residing there, face from climate change. Next we discuss hazards from two geologic events: intense storms, such as hurricanes, and *tsunami,* which are seismic sea waves usually generated by earthquakes. Although we discuss tsunami in the earthquakes section of this text, the phenomena is included here as well because we understand that time may not permit coverage of all chapters in this book.

Tsunami

Tsunami, often mistakenly called tidal waves, are sea waves generated by displacement of the sea floor by volcanic or earthquake activity, or rarely by meteorite impact (**FIGURE 11-37**).

Vulnerability of the United States to Tsunami

Most of the planet's devastating earthquakes and volcanic activity occur around the Pacific Rim; thus, tsunamis are most abundant in the Pacific. But the West Coast of the

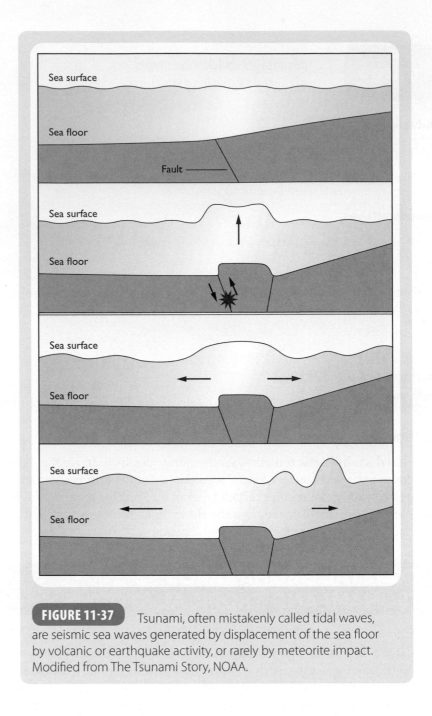

FIGURE 11-37 Tsunami, often mistakenly called tidal waves, are seismic sea waves generated by displacement of the sea floor by volcanic or earthquake activity, or rarely by meteorite impact. Modified from The Tsunami Story, NOAA.

Americas is certainly vulnerable to tsunami. In 1964, tsunami generated by the "Good Friday" Alaska Earthquake killed 11 people, caused $7.4 million of damage to Crescent City, California, and damaged coastal communities in Alaska as well (**FIGURE 11-38**).

The Pacific Coasts of Washington and British Columbia, and the inland waters of the Strait of Juan de Fuca, are vulnerable to tsunami. At greatest risk are people and property on beaches and **tidal flats**, those inhabiting low-lying areas of coastal towns and cities, and people and property near river mouths draining to the ocean.

Recently, geoscientists have warned of danger from giant tsunami that could result from volcanic eruptions on the Canary Islands (see the following Case Study) and thus could threaten the East Coast.

Damage in Kodiac, Alaska caused by the tsunami generated by the 1964 Good Friday earthquake. Figure 17, U.S. Geological Survey Professional paper 541. Courtesy of USGS.

The Great Tsunami of December 26, 2004, South Asia

On December 26, 2004 a massive earthquake struck near the island of Sumatra, Indonesia. The earthquake measured 9 on the MM scale, making it one of the largest ever measured. We discuss this earthquake in more detail elsewhere in this text. The earthquake was so powerful it changed the Earth's rotation and even shifted the position of the poles. It generated a series of tsunami that were the greatest ever recorded by humans up to that time. Here we describe briefly the extent of the damage done by the tsunami (**FIGURE 11-39**). It devastated areas of Indonesia, Thailand, India, the Maldives, Somalia, and Sri Lanka, killing up to 300,000 people. Tidal stations on the Indian coast recorded the magnitude of the waves.

The 2011 Tsunami in Japan

The March 11, 2011 earthquake that struck in the Sea of Japan devastated part of the east coast of Japan and destroyed a nuclear power plant at Fukushima (discussed elsewhere in the text). It generated a tsunami (**FIGURE 11-40**) whose wave crests reached 15 meters (46 feet), which easily overtopped the sea wall seaward of the nuclear facility that had been inexplicably designed to withstand a tsunami of only about 6 m, in one of the planet's most active earthquake zones. Waves caused by this tsunami also hit Crescent City, California and caused damage to the city's harbor.

Hurricanes and Tropical Storms

On September 15, 1999, Hurricane Floyd pounded North Carolina with 130 mph winds (**FIGURE 11-41**).

A Potential Disaster in the Atlantic Ocean: Tsunami from Cumbre Vieja Volcano, Canary Islands

Background © Hemera Technologies/AbleStock.com/Thinkstock, Title © iStockphoto/Thinkstock

The September 2001 edition of the scientific journal *Geophysical Research Letters (GRL)* contains an article by geoscientists Simon Day (University College, London) and Stephen Ward (University of California). They report on the potential for disastrous tsunami from an eruption of Cumbre Vieja volcano, located on La Palma in the Canary Islands, off Africa's west coast (**Figure CS 2-1**).

Such an enormous submarine landslide could displace colossal volumes of seawater, generating seismic sea waves in a *wave train* (a series of waves) with enough energy to force waves a maximum of 65 meters (213 ft) high as far as 16 km (10 mi) inland, causing untold destruction and loss of life in Europe, and North and South America.

Some geologists call this a "worst-case" scenario, however, as it is probably imprecise to say—as many do—that tsunami take the form of massive "walls of water." Rather, a tsunami usually appears as a rapidly rising or falling tide, a series of breaking waves, or even as a **tidal bore** when it approaches shore.

The last eruption of Cumbre Vieja was in 1971. It remains active.

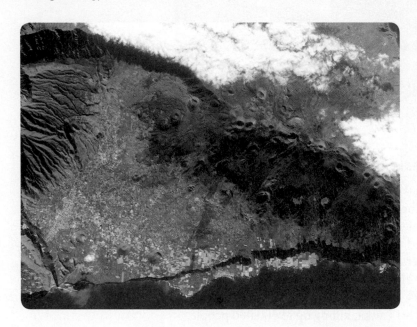

FIGURE CS 2-1 The Cumbre Vieja volcano shown at right, is located on La Palma in the Canary Islands, off Africa's west coast. Few earthquakes in the Atlantic Basin are big enough to generate tsunami with enough energy to damage coastal communities around the Atlantic. However, an eruption on Cumbre Vieja could trigger a massive landslide containing up to 700 km³ of material, which is twice the volume of the Isle of Man in the Irish Sea. Courtesy of the Expedition 17 crew and the Image Science & Analysis Laboratory, Johnson Space Center.

Two centuries earlier, much of the force of the storm could have been absorbed by the region's old-growth hardwood forest. But today, coastal North Carolina has been largely transformed by development into farms and cities, and the rain poured off the streets, plowed fields, roofs, and parking lots into its rivers, carrying untold tons of sediment and toxic waste with it. Part of Pamlico Sound was converted into a "dead zone" because the polluted runoff led to a sequence of events resulting in environmental hypoxia (extremely low dissolved oxygen levels) that caused massive die-offs of fish and other organisms. The pollution in part came from the region's thousands of CAFOs (Concentrated Animal Feedlot Operations), containing man-made "lagoons" of hog waste (Figure 11-41C). These overflowed into rivers and mixed with human sewage, hundreds of thousands of chicken, turkey, hog, and cow carcasses, agricultural chemicals, and fertilizer. Petrochemical sludge from vehicles contributed to the toxic brew.

(A)

(B)

Total precipitation reached 20 inches in places (**FIGURE 11-42**). It was one of the greatest disasters to have befallen North Carolina since European settlement.

Areas Affected

A **hurricane** is a severe tropical storm that forms in the tropical North Atlantic Ocean, most often a few degrees north of the equator, or in the Caribbean Sea, Gulf of Mexico, or eastern Pacific Ocean, where the English word for them is *typhoon*. Hurricanes need

FIGURE 11-40 Tsunami from Sea of Japan earthquake on March 11, 2011. © Mainichi Shimbun/Reuters/Landov.

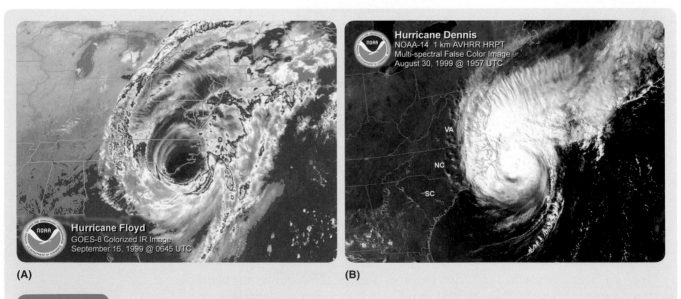

(A)   (B)

FIGURE 11-41 (A) Floyd was no ordinary hurricane. It was nearly twice as large as a "typical" Category 3 storm (wind speed 110–130 mph). (B) Just a few days earlier, Hurricane Dennis had saturated the ground, and now there was no place for Floyd's torrential rains to go. Photos (A) and (B) courtesy of NOAA.

(C)

FIGURE 11-41   *Continued.* (C) A waste "lagoon" holding excrement and urine from thousands of animals. Pigs produce more excrement than humans. Waste "lagoons" like these overflowed into surrounding bodies of water during and after Hurricane Floyd. Courtesy of Jeff Vanuga/NRCS.

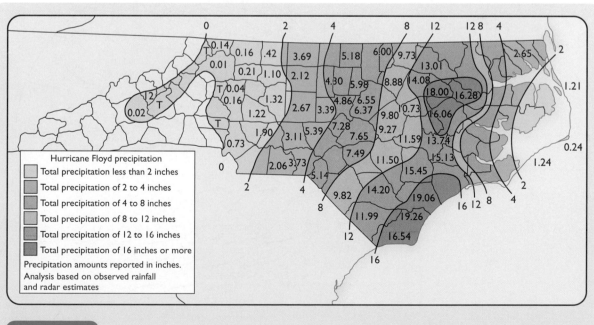

FIGURE 11-42   Precipitation from Hurricane Floyd, September 14–18, 1999. The hurricane caused dozens of deaths, killed hundreds of thousands of animals, and did $6 billion in damage. Modified from Phillip Badgett/Jonathan Blaes/NOAA.

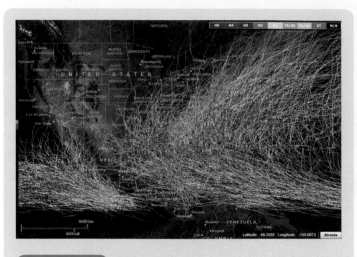

FIGURE 11-43 Tracks of tropical storms (including hurricanes) from September 16–30, 1851–2006 in the North Atlantic What part of North America is susceptible to hurricanes? When do most hurricanes occur? Courtesy of NOAA.

warm tropical oceans, high humidity, and light winds above them to originate, but it is still not clear exactly why hurricanes form in one place or time. Hurricane season in the North Atlantic Ocean extends from June through November. In fact, records show that 99% of all hurricanes occur during these months.

Hurricanes rotate in a counterclockwise direction around an "eye," a region with cloudless skies and deceptively calm winds.

Hurricane Tracks

The direction a hurricane moves is called its *track* (**FIGURE 11-43**). Hurricanes in the Atlantic generally move in a westward direction until they make land, or until they are deflected to the north/northeast.

Concept Check 11-12. **FIGURE 11-44** shows global wind belts. Notice that winds just north of the equator blow roughly to the west. At about 30 degrees north, direction changes to the north and east. Now look at the hurricane track (Figure 11-43) again, and form a hypothesis to explain hurricanes' distinctive tracks.

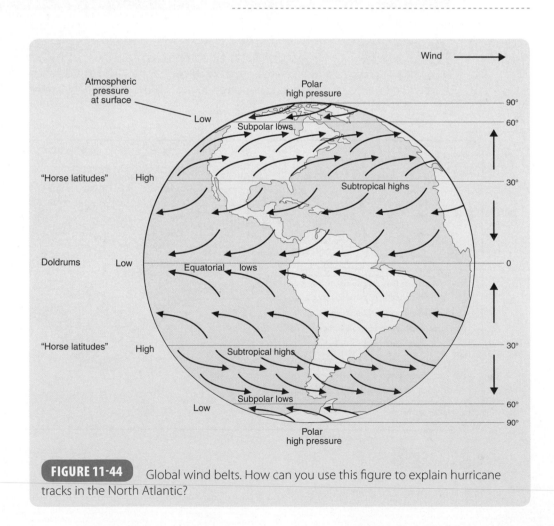

FIGURE 11-44 Global wind belts. How can you use this figure to explain hurricane tracks in the North Atlantic?

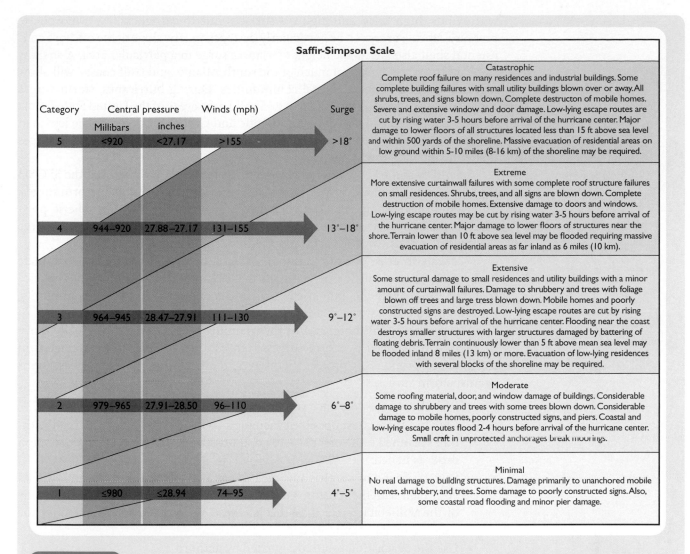

					Saffir-Simpson Scale	

Category	Central pressure		Winds (mph)	Surge	
	Millibars	inches			
5	<920	<27.17	>155	>18°	**Catastrophic** Complete roof failure on many residences and industrial buildings. Some complete building failures with small utility buildings blown over or away. All shrubs, trees, and signs blown down. Complete destructon of mobile homes. Severe and extensive window and door damage. Low-lying escape routes are cut by rising water 3-5 hours before arrival of the hurricane center. Major damage to lower floors of all structures located less than 15 ft above sea level and within 500 yards of the shoreline. Massive evacuation of residential areas on low ground within 5-10 miles (8-16 km) of the shoreline may be required.
4	944–920	27.88–27.17	131–155	13°–18°	**Extreme** More extensive curtainwall failures with some complete roof structure failures on small residences. Shrubs, trees, and all signs are blown down. Complete destruction of mobile homes. Extensive damage to doors and windows. Low-lying escape routes may be cut by rising water 3-5 hours before arrival of the hurricane center. Major damage to lower floors of structures near the shore. Terrain lower than 10 ft above sea level may be flooded requiring massive evacuation of residential areas as far inland as 6 miles (10 km).
3	964–945	28.47–27.91	111–130	9°–12°	**Extensive** Some structural damage to small residences and utility buildings with a minor amount of curtainwall failures. Damage to shrubbery and trees with foliage blown off trees and large tress blown down. Mobile homes and poorly constructed signs are destroyed. Low-lying escape routes are cut by rising water 3-5 hours before arrival of the hurricane center. Flooding near the coast destroys smaller structures with larger structures damaged by battering of floating debris. Terrain continuously lower than 5 ft above mean sea level may be flooded inland 8 miles (13 km) or more. Evacuation of low-lying residences with several blocks of the shoreline may be required.
2	979–965	27.91–28.50	96–110	6°–8°	**Moderate** Some roofing material, door, and window damage of buildings. Considerable damage to shrubbery and trees with some trees blown down. Considerable damage to mobile homes, poorly constructed signs, and piers. Coastal and low-lying escape routes flood 2-4 hours before arrival of the hurricane center. Small craft in unprotected anchorages break moorings.
1	≤980	≤28.94	74–95	4°–5°	**Minimal** No real damage to building structures. Damage primarily to unanchored mobile homes, shrubbery, and trees. Some damage to poorly constructed signs. Also, some coastal road flooding and minor pier damage.

FIGURE 11-45 The Saffir-Simpson Scale, used to classify hurricanes. Note that as pressure readings decline, the intensity of the hurricane increases. This is because as pressure in the eye drops, the gradient between the pressure in the storm's interior, and pressure outside, increases, and this is what determines potential wind speed. Note that the "surge" category denotes storm surge.

Classification

Hurricanes are classified using the Saffir-Simpson Scale (**FIGURE 11-45**). Note that as pressure readings decline, the intensity of the hurricane increases. This is because as pressure in the eye drops, the **pressure gradient** between the pressure in the storm's interior, and pressure outside, increases, and this is what determines potential wind speed. Hurricanes, tropical depressions, and tropical storms may cause massive precipitation events, and can lead to catastrophic flooding. In fact, such events are responsible for maximum-recorded flood events in regions like the eastern United States. Also note the "surge" category in Figure 11-45. This is *storm surge,* which we discuss next.

Storm Surges

The greatest potential for loss of life during a hurricane is from **storm surges**. Surges result when powerful, persistent winds "push" water onshore. When coupled with winds, high tides, and high sea surfaces due to extremely low atmospheric pressure, storm surges can be particularly destructive. The slope of the offshore continental shelf also affects the height of a storm surge in a particular area. A shallow slope off the coast, typical of much of the south Atlantic and Gulf coasts, will allow a greater surge to impact coastal communities. During hurricanes, storm-surges can increase sea level by up to 18 feet (> 5 m). Large areas of the United States' most densely populated and fast-growing Atlantic and Gulf of Mexico cities lie less than 10 feet (3 m) above mean sea level. Therefore, the danger there from storm surges is significant and growing.

Scientists at the U.S. National Hurricane Data Center have devised the SLOSH (Sea, Lake, and Overland Surges from Hurricanes) system to evaluate storm surges. This system uses computers, fed with data on hurricane size, atmospheric pressure, forward speed, track, and wind velocity and direction, to estimate storm surge heights and winds. The forecast is reportedly accurate to ±20%.

Accurate forecast of storm-surges is difficult because it is difficult to estimate how strong a storm will be when it finally makes landfall. When authorities balance that uncertainty with the immense human and economic risks to their community, they often plan for a storm that is one category higher than what is forecast.

Many users of this book and/or their families will live in, or eventually move to, regions threatened by storm surges. Here is a list of rules to help citizens in storm-surge-prone areas protect themselves, their families, pets and possessions against storm surges.

- Pick the shortest distance you must travel to reach a safe location.
- Map out the nearest possible evacuation route. Do not get on the road without a planned route or a place to go. If possible, drive it before an evacuation order is issued.
- Select a friend or relative outside a designated evacuation zone and discuss your plan with them before hurricane season, or select a hotel/motel outside of the vulnerable area. If neither of these options is available, identify the closest possible public shelter.
- Use the evacuation routes designated by authorities.
- Notify your local emergency management office regarding anyone in your household needing special assistance to evacuate.
- Prepare a separate pet plan. Most public shelters do not accept pets.
- Secure your home prior to leaving by boarding up doors and windows, securing or moving indoors all yard objects, and *turning off all utilities*.
- During hurricane season, keep your car topped up with gas. Before leaving, withdraw extra money from the bank.
- Don't forget prescription medicines and any special medical items.
- Recreational vehicles, boats, and trailers require extra evacuation time and help clog evacuation routes. Consider leaving them behind or start as early as possible.
- Do not delay your departure if requested to evacuate by authorities. To do so will only increase your chances of being stuck in traffic, or—even worse—not being able to get out at all.
- Expect traffic congestion and delays during evacuations. Take along food and water, and other emergency supplies.
- Stay tuned to a local radio or television station and listen carefully for any advisories or specific instructions from local officials. Monitor NOAA Weather Radio if possible.

CHAPTER SUMMARY

1. Active coasts occur at plate boundaries; passive coasts do not. Active coasts would be likely to experience geohazards associated with their setting.

2. The coastal zone is the transition from land, to the edge of the U.S. territorial sea.

3. Coastal environments may be classified as intertidal and subtidal.

4. Estuaries are areas where fresh-and-salt water meet and mix. They are often critically important to the survival of coastal species.

5. Barrier complexes are long, narrow sand bodies separated from the mainland by lagoons and bays. Barrier island complexes are the continent's main "line of defense" against hurricanes. As such, they have saved the nation billions of dollars in avoided damage.

6. Coral reefs are structures built by colonial corals, calcareous algae, and other organisms.

7. Reefs are very susceptible to pollution, especially excess nutrients.

8. Oceanographers recognize two overarching landforms at the continental margin: the continental shelf, and submarine canyons. The continental shelf is that part of the continent flooded by the ocean. Submarine canyons formed when rising sea level drowned coastal river valleys.

9. Coastal erosion has become a global problem. In the United States, all 30 states bordering an ocean or the Great Lakes are experiencing coastal erosion. Average annual erosion rates are 2 to 3 feet/year on the Atlantic coast and about 6 feet on the Gulf coast.

10. Sand can come from a variety of sources, including adjacent areas, a river's drainage basin, eroding coastal landforms and the continental shelf.

11. Thousands of dams and impoundments on rivers trap sediment. Sand in vast quantities that used to naturally nourish beaches thus is now trapped upstream.

12. Most surface waves result from frictional effects of wind blowing across the water.

13. Important attributes of waves include wavelength, period, height, frequency, and velocity.

14. Obliquely breaking waves lead to the creation of currents that run parallel to the beach, known as longshore currents.

15. Wave energy, and its effect on the coast, varies seasonally.

16. Changes in sea level result from eustatic and isostatic processes.

17. Sea level has been rising since the end of the last glacial episode, about 15,000 to 18,000 years ago. There is a consensus among climatologists that the planet is experiencing a human-induced increased rate of global climate change, which will increase sea level. Rate of global sea level rise for the next century could double to 4 mm/year or more.

18. Sea level rise could have a severe effect on fast-growing coastal cities, and fragile ecosystems. Based upon a rough average of typical beach slopes, for each unit rise in sea level, the beach will migrate 1,000 to 2,000 units inland.

19. Humans build structures along coasts for two reasons: to trap sand in order to protect the shore and what we have built on it from erosion.

20. U.S. Army Corps of Engineers has the responsibility for maintaining navigable waterways, including harbor entrances.

21. Groins and jetties are structures, usually concrete, built at angles to the coast to trap sand.

22. Breakwaters are structures built parallel to shores to absorb wave energy.
23. Beach renourishment works, but it is temporary and costly.
24. Revetments are structures built on coastal bluffs to absorb the energy of incoming waves.
25. Besides hard stabilization methods, other strategies to protect coasts include zoning, relocation of structures, and legislation of various types.
26. Tsunami are seismic sea waves generated by displacement of the sea floor. Tsunami are most abundant in the Pacific Ocean. Devastating tsunami occurred in 2004 and 2011 in the western Pacific.
27. The potential exists, however slight, for disastrous tsunami from an eruption of Cumbre Vieja volcano, Canary Islands, off Africa's west coast.
28. A hurricane is a severe tropical storm that forms in the tropical North Atlantic Ocean or in the Caribbean Sea, Gulf of Mexico, or eastern Pacific Ocean. Hurricanes need warm tropical oceans, high humidity, and light winds above them to originate. Hurricanes are classified using the Saffir-Simpson Scale. The greatest potential for loss of life during a hurricane is from storm surges.

KEY TERMS

barrier island complexes

coastal zone

continental shelf

continental slope

estuaries

geohazard

groins

hard stabilization

hurricane

hurricane track

ice draft

intertidal

longshore current

ocean acidification

pressure gradient

shore

soft stabilization

storm surge

submarine canyon

submarine fan

subtidal

tidal bore

tidal flat

tsunami

turbidite

REVIEW QUESTIONS

1. An estuary is
   a. a long narrow sandy beach, separated from the mainland by a bay or lagoon.
   b. a structure built by calcareous algae.
   c. an embayment where fresh-and-salt water meet and mix.
   d. a seismic sea wave.

2. Negative, or hypersaline estuaries,
   a. experience evaporation rates greater than freshwater influx.
   b. experience evaporation rates less than freshwater influx.
   c. experience no ground water influx.
   d. experience no primary productivity.

3. Barrier island complexes have saved the United States billions of dollars
   a. in avoided damage from hurricanes.
   b. in avoided damage from tsunami.
   c. in cleanup of offshore oil spills.
   d. all of the above.

4. Coral reefs are built by organisms, of
   a. calcium sulfate.
   b. calcium silicate.
   c. calcium carbonate.
   d. organic matter, like peat.

5. The reefs of the Florida Keys are being damaged by
   a. pollution from sewage treatment plants and agriculture.
   b. dust from coal plants in Asia.
   c. radioactive wastes from Mexico.
   d. all of these

6. The outer border of the continental shelf is
   a. a submarine canyon.
   b. a subduction zone.
   c. the continental slope.
   d. the subtidal zone.

7. Turbidites, distinctive deposits often originating in submarine canyons,
   a. exhibit graded beds.
   b. are always made of carbonate sediment.
   c. intercept sediment moved by longshore currents.
   d. a and b only.
   e. a and c only.
   f. a through c.

8. Average annual coastal erosion rates are
   a. a few feet/year
   b. a few hundred feet/year.
   c. a few miles/year.
   d. a few inches/year.

9. Sand is of
   a. biogenic origin.
   b. siliclastic origin.
   c. both biogenic and siliciclastic origin.
   d. volcanic origin.

10. The energy for coastal erosion comes mainly from
    a. tides.
    b. waves.
    c. tsunami and earthquakes.
    d. volcanic activity.

11. Obliquely breaking waves lead to the creation of
    a. convection currents.
    b. turbidity currents.
    c. longshore currents.
    d. all of the above.

12. Changes in sea level can result from eustatic or isostatic processes. Eustatic processes change the
    a. temperature of sea water.
    b. topography of the sea bottom.
    c. amount of water in the oceans.
    d. density of seafloor sediment.

13. Isostatic processes change the
    a. topography of the sea floor.
    b. sea water temperature.
    c. amount of water in the oceans.
    d. density of sea floor sediment.

14. An increase of average sea water temperature of 1 degree Celsius will, other things being equal, elevate sea level by
    a. 70 kilometers.
    b. 70 meters.
    c. 70 centimeters.
    d. 70 micrometers.

15. For each unit rise in sea level of the Atlantic Ocean, the shore will move inland by 1,000–2,000 units. If sea level rises by 70 centimeters, how far inland, in km, on average, will the shore migrate along the Atlantic coast (1 km = 100,000 cm)?
    a. about 1 km
    b. about 100 km
    c. about 1,000 km
    d. about 0.001 km

16. The most damaging hurricane of the twentieth century in the United States was
    a. Camille in 1969.
    b. Agnes in 1972.
    c. Floyd in 1999.
    d. Andrew in 1992.

17. The federal agency responsible for maintaining navigable waterways, including harbors is the
    a. Coast and Geodetic Survey.
    b. Coast Guard.
    c. Army Corps of Engineers.
    d. Department of Defense

18. Which is true about beach renourishment?
    a. It works but is temporary and expensive.
    b. It is impractical for Atlantic beaches.
    c. It is impractical for all beaches in the Northern Hemisphere.
    d. It is relatively cheap and permanent.

19. Which best describes the vulnerability of the United States to tsunami?
    a. The East Coast is the most vulnerable.
    b. Only Alaska is vulnerable.
    c. The Pacific Coast is the most vulnerable.
    d. Hawaii is the only vulnerable U.S. state.

20. Hurricane season in the Atlantic extends from
    a. June to September.
    b. June to November.
    c. March to December.
    d. June to November.

FOOTNOTES

[1] *Charleston Post and Courier,* May 28, 2001.

[2] "State's Options are Limited." Sammy Fretwell, *Myrtle Beach Sun News,* 5/31/01, c6.

[3] Cao, L; Caldeira, K, 2008. Atmospheric CO_2 stabilization and ocean acidification. *Geophys. Res. Lett.,* 35(19). DOI: 10.1029/2008GL035072.

[4] University of Hawaii.

[5] Rignot E.; I. Velicogna, M. R. van den Broeke, A. Monaghan, and J. Lenaerts (2011). "Acceleration of the contribution of the Greenland and Antarctic ice sheets to sea level rise". *Geophysical Research* Letters 38 (5).

12

CHAPTER OUTLINE

POINT-COUNTERPOINT Farm Subsidies

© Amos Struck/Shutterstock, Inc.

Point

In 2002 the *New York Post* reported: "If the heat surrounding former WorldCom head Bernard J. Ebbers weren't hot enough, the one-time billionaire is in the spotlight for receiving millions of dollars in taxpayer-funded subsidies through his soybean farm in Concordia Parish, Louisiana. Federal records show Angelina Plantation in Monterey, LA of which Ebbers is part-owner, received nearly $4 million in subsidies from 1996 2001."

In 2007 the European Union (EU) spent more than $62 billion in farm subsidies, making up at least a quarter of farm revenue. The comparable statistic for the United States was $16 billion. Most of those payments went to growers of rice, cotton, corn, soybeans, and sorghum, which are so-called **commodity crops**. The latter three crops are mainly fed to animals raised for human consumption. Cotton is one of the most pesticide-intensive crops grown (see later in the chapter).

While hundreds of millions of people worldwide are malnourished, the production of commodity crops exceeds demand in the United States and in the EU. Subsidies that encourage the production of crops for which there is insufficient demand generally result in greater surpluses, not fewer. There are few if any subsidies available to U.S. growers of fruits and vegetables, and no official support for organic agricultural methods.

Counterpoint

An Iowa corn (maize) and soybeans grower had this to say about farm subsidies: "If Americans want cheap food, someone's got to pay for it."

Concept Check 12-1. What do you think the farmer meant with his remark? Discuss the implications of his comment. Cite your reasons.

We begin this chapter with the substance on which agriculture depends: soil.

■ SOIL

Along with water, **soils** may be the most important natural resource on the planet. Most of our food production depends on soil, yet our knowledge of **soil biota**—their distribution, abundance, population structures, and ecological roles—are poorly understood. We need to better understand such processes as: nutrient recycling (phosphorous, nitrogen, potassium, etc.), nutrient uptake by plants, the controls on soil fertility, formation of soil organic matter, **nitrogen fixation** (conversion of atmospheric molecular nitrogen gas [N_2] to inorganic compounds like nitrates), methane (CH_4) production, CO_2 production, soil development, and soil production.

In Other Words

Soil degradation is the number one environmental problem.

Dr. Rattan Lal
Director, School of
Environment and
Natural Resources
Ohio State University

Leonardo da Vinci called soil "Earth's flesh." The functioning of this wafer-thin, dark stratum on the surface of the Earth is vital for the survival of life on land. Soils are natural resources that, like water and air, are being degraded by humans. Our activities cause soil pollution, groundwater contamination, soil erosion, and a loss of carbon and nutrients from the soil. These actions are taken with little understanding of the resilience, or stability, of soils.

Soil and the soil biota are interlocking components of the terrestrial biosphere, interacting with vegetation and global climate, and necessary for the functioning of terrestrial ecosystems.

Degradation

Rapidly growing human populations have in many areas altered or destroyed the soil's structure and countless soil species through activities such as: paving over soils, deposition of human-generated chemicals from the atmosphere, soil fumigation, dumping of waste products in soils, groundwater contamination, and soil erosion or depletion by poor farming practices.

Information on the effect of these impacts on soil biodiversity, and the role of soils in the biosphere, is largely lacking.

Some desert soils are so fragile that they can be damaged simply by walking on them (**FIGURE 12-1**).

Composition

Soils consist mainly of the decomposition and disintegration products of rock and sediment carried out by agents of chemical and mechanical weathering. Healthy soil contains countless organisms, ranging in scale from large creatures such as earthworms and moles, to microorganisms, to tinier still *microbes* (**FIGURE 12-2**). Soils also contain organic matter, called *humus,* in various stages of decomposition.

Processes that form soils are complex, but some controls on soil formation include bedrock type, slope, climate, and geological history of the site. Because there are many different kinds of rock, there are many varieties of soil. The *structure* of a soil, however, is similar worldwide (**FIGURE 12-3**). Composition of soils depends on:

- proportion of sand, silt, and clay (Figure 12-3A)
- extent to which soluble ions have been removed by groundwater
- amount of organic matter
- amount and type of *clay* minerals

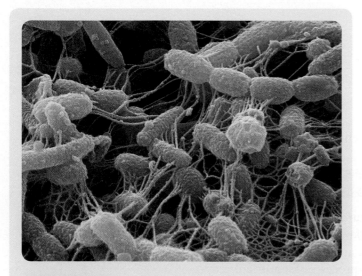

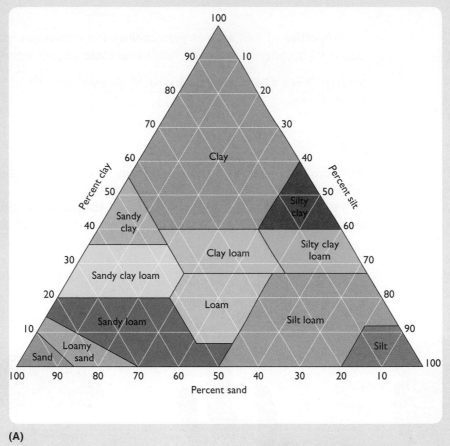

(A)

O horizon (litter) — Organic matter

A horizon (litter and topsoil) — Dark, rich in humus

B horizon (subsoil) — Light colored

C horizon (transition zone) — Varied

D horizon (parent material) — Rock or gravel

(B)

(C)

FIGURE 12-3 (A) The structure of soil depends in part on the relative proportions of sand, silt, and clay. (B) A humus layer of variable thickness (O, for organic) is underlain by a **leached zone** (A, or topsoil), with a **zone of accumulation** (B, or subsoil) below that. All of these zones eventually rest on bedrock (C), which is usually decomposing at the top. (C) Photo of soil showing humus (top) and leached zones. Courtesy of the AFCEC.

Properties of soils such as permeability, ion exchange capability, and fertility, vary as their compositions vary. We discuss these various terms and concepts later.

Concept Check 12-2. CheckPoint: What is soil composed of? What does soil composition depend upon?

Types

There are three general categories of soils, which form under differing climatic regimes. They are *pedalfers, pedocals,* and *laterites.*

Pedalfers (FIGURE 12-4A) are typical of temperate humid climates and contain few soluble ions (Ca⁺⁺, Mg⁺⁺, Na⁺, etc.), because in humid climates precipitation

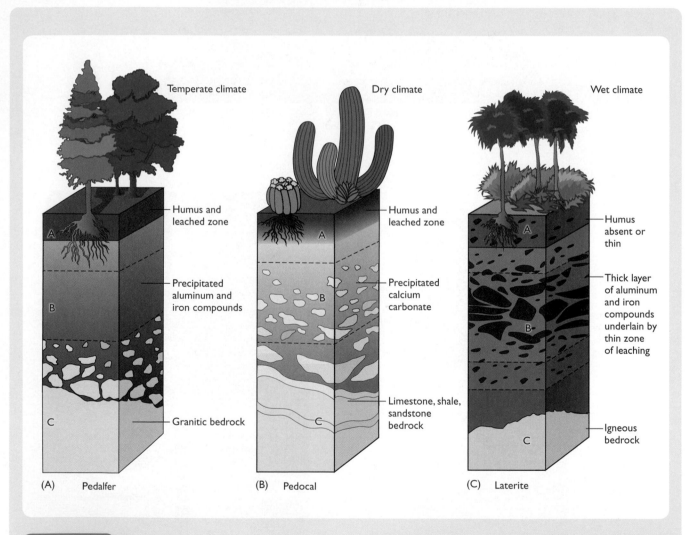

Temperate climate   Dry climate   Wet climate

- Humus and leached zone
- Precipitated aluminum and iron compounds
- Granitic bedrock

(A)   Pedalfer

- Humus and leached zone
- Precipitated calcium carbonate
- Limestone, shale, sandstone bedrock

(B)   Pedocal

- Humus absent or thin
- Thick layer of aluminum and iron compounds underlain by thin zone of leaching
- Igneous bedrock

(C)   Laterite

FIGURE 12-4 The three varieties of soil. (A) Pedalfer. These soils form mainly in moist climates where precipitation can remove many dissolved substances. (B) Pedocals. Here, lack of adequate rainfall means that some soluble materials, like calcium minerals, may remain in the soils. These deposits are called caliche (kal-ee-chee). (C) Soils produced under intense weathering conditions, favored by warm temperatures and high precipitation, have most original substances removed. Only the most insoluble materials remain, such as iron and aluminum compounds. These soils are called laterites, and are an important source of bauxite, an aluminum ore.

(e.g., rain) dissolves these ions and carries them into groundwater. Pedalfers also may contain abundant organic matter. There are many different types of pedalfer soils.

Pedocals (**FIGURE 12-4B**) usually form under dry climatic conditions. They may contain abundant soluble ions, such as Ca^{++}, B^{++}, Se^{++}, and Na^+, because precipitation is insufficient to flush them into groundwater, as the water table (discussed elsewhere in the text) is generally well below the surface.

Pedocals pose significant challenges for agriculture. The soils may be fertile, but the presence of toxic ions, like **selenium (Se)**, and arsenic (As), may render their use dangerous over the long term without expensive irrigation drainage systems, and without an "away" to flush the irrigation water containing the toxic salts (see our Kesterson Wildlife Refuge section).

Laterites are typical soils of tropical regions (**FIGURE 12-4C**). They contain few nutrients and little organic matter, and consist mainly of insoluble complexes such as iron and aluminum hydroxides. In fact, some laterites are mined for *bauxite* ore (**FIGURE 12-5**), which is refined into aluminum. Intensive agricultural practices that are successful in Western Europe and the United States (see later in the chapter) are often unsuccessful (without massive use of fertilizers) when exported to tropical countries with laterite soils, especially if rain forests are cleared and burned to make way for crops. In tropical rainforests, almost all plant nutrients are in the vegetation; thus the soils are generally of poor agricultural quality.

Concept Check 12-3. CheckPoint: Distinguish between pedalfers, pedocals, and laterites.

Soil Organisms

Although soils teem with living creatures, relatively little is known about them. Few soil organisms can even be identified to the species level. Moreover, the function of individual soil organisms is often poorly understood. Soil scientists generally speak of the functions of *communities* of soil organisms rather than individuals. Soil organisms are mainly microscopic, but some of the largest organisms on Earth—fungi—also are common in soils.

(A)　　　　　　　　　　　　　　　　　　(B)

FIGURE 12-5　(A) A bauxite mine. © Leojones/ShutterStock, Inc. (B) A close-up view of bauxite. Courtesy of NASA.

Soil microbes are essential to life on Earth.

- Microbes decompose the remains of dead plants and animals, called *residue*. Some microbes can even decompose toxic substances like the antifungal compound methyl bromide.
- Microbes create **humus**, the product of decomposition reactions.
- Microbes convert materials from one form to another (such as N_2 to nitrate and nitrous oxide).

Most organisms that need nitrogen cannot use atmospheric N_2. Thus, N must be "fixed" or converted into an inorganic compound like nitrate (NO_3). The living microbes, fungi, and invertebrates that comprise the soil food web convert carbon and nitrogen into forms available for plant growth.

- Microbes cause soils to clump, helping the surface material to resist erosion and weathering.

Our lack of precise knowledge about soil organisms and the roles they play in supporting life on Earth has troubled soil scientists (agronomists) for decades. Some worry that soil organisms could be destroyed, and thus soil functions eliminated, before their importance is even known. One thing that *is* known is that soils are one of the planet's most important storehouses of carbon, and humans alter this reservoir at our peril.

Carbon Sequestration in Soils

The storage of materials in soils is called **sequestration**. Soil organisms affect the rate of production and consumption of many substances including CO_2, methane, and nitrogen. Soil degradation has resulted in the release of greenhouse gases, contributing to climate change. The annual *flux* (movement) of CO_2 returning to the atmosphere as a result of decomposition and other soil processes amounts to approximately 68×10^9 tonnes of carbon (C) per year—at least ten times more C than that produced by deforestation and fossil fuel burning. Agronomists estimate that 1,500 billion tonnes of C is bound globally to soil. This is at least equal to, and may be greater than, the amount of C present in the atmosphere.

The environmental benefits of carbon sequestration (global climate change is discussed elsewhere in the text), and its potential to be a new income source for farmers have generated much interest in the agricultural community. As human activity releases increasing amounts of greenhouse gases like CO_2 and inorganic nitrogen, many scientists believe that soils could store more C than they presently contain. Unfortunately, C sequestration is reduced as soil temperatures rise. Soil temperatures are presently rising for two reasons: the Earth is getting hotter, and as forests are cleared, the soils below them are warmed, giving off carbon.

Concept Check 12-4. Suggest one land-use change that could result in more C sequestration, based on the above paragraphs.

- -

Soils as Geohazards: Swelling Soils

Swelling or expansive soils are soils that expand when wetted, and contract when dried. Swelling soils contain clay minerals whose crystal structures permit the large-scale introduction of water. When water enters the clay, the structure expands. This swelling exerts great outward pressure, and any structure built on swelling soils can be damaged by the expansion and contraction the soils undergo. This movement can exert enough pressure to crack sidewalks, driveways, basement floors, pipelines, and foundations. While not all expansive soils have the same swell potential, pressures can be as great as 15,000 pounds per square foot. Swelling clays can under extreme conditions double in volume when water is added, while as little as a 3% expansion can cause serious damage to structures. The clays that cause the swelling are called *bentonite,* or more specifically, *montmorillonite.* These clays commonly form as alteration products of volcanic ash deposits, which in turn are common throughout the western part of North America.

Swelling soils are one of the most expensive geological hazards in the United States, causing losses of over $2 billion each year. Regions with swelling-soil problems include:

- Santa Clara County, California (Silicon Valley)
- Las Vegas, Nevada (being developed at the rate of 1 square mile every 3 months)
- Colorado Front Range
- Los Angeles Basin, California

FIGURE 12-6 shows a house damaged by swelling soils.

Saturation of soil can occur from rainfall, lawn watering, or leaking sewer or water lines. Major damage to sidewalks, highways, utility lines, and foundations can result if the soils contain expansive clays. If construction takes place on wet materials that have high shrink-swell potential, and these materials at some later time are dried, the resulting shrinkage may cause severe cracking in structures. The City of Torrance, California describes these common effects of expansive soils on buildings. "The concentrated weight of the structure will inhibit the soil's upward expansion. Outward expansion on the other hand may continue. The footings will

FIGURE 12-6 A house damaged by swelling soils. Courtesy of the NRCS/USDA.

not be returned to their original position as the soils dry and shrink. Instead, they can 'ooze' down to a slightly lower level. This process can accumulate if the wetting and drying is allowed to continue year after year."

Ironically, droughts can provide a trigger for expansive soil damage. According to the Oklahoma Geological Survey, numerous foundation failures, which resulted from soil shrinkage, occurred in Oklahoma during the unusually hot, dry summer of 1998.

A Special Soil: Loess

Loess is a natural segue into our next two topics: soil erosion, and agriculture. Loess (pronounced "lerss" or "luhss") is German for "loose" or "crumbly." Loess is porous sediment whose composition is dominated by silt-sized grains in the 2- to 64-micron diameter range, and is composed mainly of quartz, feldspar, and mica. Loess is produced by glacial erosion. The pulverizing effect of massive kilometer-thick ice-sheets reduces solid rock to what geologists call "rock flour." This sediment is then picked up by powerful winds blowing off the ice sheets and may be deposited hundreds of kilometers away.

Loess is the source of most of the United States' best agricultural soils. Loess is distributed globally (**FIGURE 12-7**) but the thickest accumulations are in the American Midwest and Shaanxi province, China. In Shaanxi, loess erosion is a major problem. Up to half the sediment entering the Yellow River in western China is from Shaanxi. Floods in the Yellow River are one of the most serious

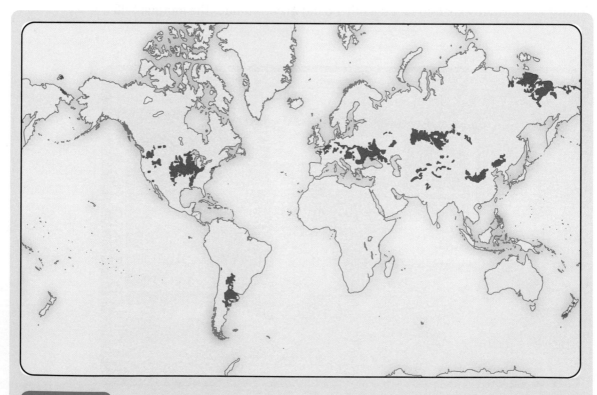

FIGURE 12-7   Global distribution of loess soils. Data from: Pidwirny, M. (2006). "Eolian Processes and Landforms". *Fundamentals of Physical Geography*, 2nd Edition.

environmental problems in China and are worsened by high sediment content. As a result, the Chinese government began a major effort to convert fragile land in Shaanxi from cropland back to forest and natural grassland, to control soil erosion. It set a target of 500,000 hectares for conversion and reportedly met that goal by 2000. The Province has approximately 3.4 million hectares (8.4 million ac) of cropland.

Although loess-based soils are extremely fertile, they are also extremely fragile, with erosion rates in the United States averaging 90 tonnes/hectare/year. Erosion from rainfall and flooding, coupled with poor agricultural practices, deposits loess into river systems, where it can harm aquatic wildlife.

Clearing of prairie land for agriculture removed long-rooted native grasses, which helped hold loess soils in place. Removal of grasses and cultivation of "row" crops accelerated gully formation, which in turn enhanced loess erosion rates (**FIGURE 12-8**). Although loess soils can be up to 60 m thick in Iowa and elsewhere, erosion has removed up to half the original soil locally.

Concept Check 12-5. CheckPoint: What is loess? Discuss its significance and problems faced by loess-based soils.

Soil Erosion

Soil erosion is widespread globally (**FIGURE 12-9**). A report by agronomist David Pimentel and colleagues at Cornell University concluded that 80% of agricultural land worldwide is moderately to severely eroded. They estimate worldwide costs of soil erosion at $400 billion annually including lost agricultural production and environmental costs. They also estimate that 75 billion tonnes of soil are eroded each year, mostly from agricultural lands.

Soil erosion is usually irreversible because, once the nutrient-rich topsoil has been stripped away, its ability to sustain plant growth is severely reduced. Run-off increases from the now-exposed and more *impermeable* subsoil, resulting in a decrease in soil water. Erosion reduces surface water quality by increasing sediment loads, and airborne dust reduces air quality.

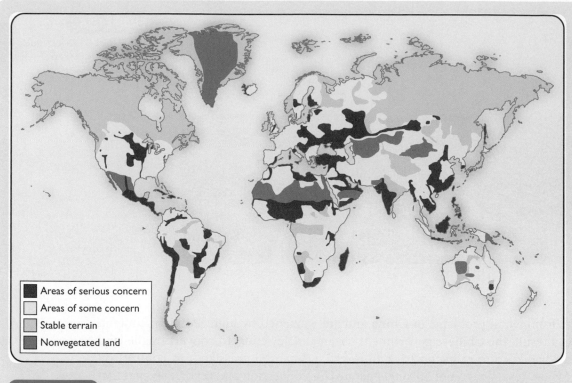

FIGURE 12-9 Global soil erosion areas of concern. What effect(s) could warming of permafrost by climate change have on polar soil stability? Modified from the International Soil Reference and Information Centre.

Legend:
- Areas of serious concern
- Areas of some concern
- Stable terrain
- Nonvegetated land

Managing Soil Erosion

Erosion from agricultural lands will generally be higher than erosion from equivalent forest or natural grasslands, but several measures can help reduce soil erosion. **FIGURE 12-10A** shows **contour plowing**, which can trap eroded soils before they can reach streams. Construction of *terraces* (**FIGURE 12-10B**) can also arrest soil movement. Planting *ground cover* can keep soil from blowing away or being rapidly removed by intense rainfall. **FIGURE 12-11** shows erosion caused by livestock. Controlling livestock damage is especially important along waterways. Changing tillage practices—for example, using **no-till plowing** involving shallower (2″) soil disturbance than traditional (8″) plowing—can reduce soil erosion, but can result in more pesticide use, because no-till is not as effective in eliminating weeds as traditional deep plowing. As well, planting lines of trees as *windbreaks* can help reduce wind erosion.

We now turn our discussion to agriculture.

■ AGRICULTURE

Humans are omnivorous. Wheat, rice, and corn—called maize outside the United States—are the three major cereal crops eaten by humans. Additional food resources include potatoes, soybeans, barley, oats, cassava, sweet potatoes, sugar cane and beets, legumes, sorghum, millet, vegetable oils, fruits, and vegetables, along with meat, milk, fish, and other seafood.

However, humans are extremely dependent on a few grains. Why do we depend so heavily on these relatively few plant species? For one thing, most wild plants that

(A)

(B)

FIGURE 12-10 (A) Contour plowing. © iStockphoto/Thinkstock. (B) Constructed terraces. © Jun Baby/ShutterStock, Inc.

we could possibly domesticate are useless to us as food. They may be indigestible, have very low food value, or have little available protein. In fact, most wild *biomass* is wood or leaves, which we can't digest.

Plant domestication has been defined by author Jared Diamond as, "growing a plant, and thereby consciously or unconsciously causing it to change genetically from its wild ancestor in ways that make it more useful to humans."

FIGURE 12-11 Soil erosion caused by livestock. © Dirk Ercken/ ShutterStock, Inc.

Grains and the Origins of Agriculture

Studies of plant remains at human habitation sites tell us when humans first domesticated them, because domesticated grains are much different from their wild ancestors. According to carbon-14 dates, grain production probably arose first in the Fertile Crescent region of the Near East, with the domestication of wheat from wild varieties occurring by 8500 BCE[1] (**FIGURE 12-12**). Grain production quickly spread along the great east–west axis of Eurasia. The domestication of rice began in China sometime around 7500 BCE. In the New World, maize was first domesticated by 3500 BCE.

Grain domestication was preceded by a major change of climate associated with the last withdrawal of glaciers about 11,000 years ago. Warming led to an expansion of the habitat for the wild grains that were ultimately to be domesticated, making it easy for humans to collect them. Technology, such as metal plows, advanced as well, resulting eventually in tools that would allow the more effective tilling of soil and the breaking up of tough sods.

Early farmers probably selected according to size: the bigger the fruit, the bigger the offspring, or so they may have thought. Cultivated peas, for example, are ten times bigger than their wild ancestors, which hunter-gatherers must have been collecting for centuries. To take another example, the oldest corncob from Mesoamerica was only 1- to 2-cm long. By 1500 farmers were growing 15 cm (6 inch) varieties, and modern cobs may exceed 30 cm.

Plant domestication developed over centuries. In fact, it is still occurring. Strawberries were not domesticated from wild forms until the Middle Ages, and pecans were not domesticated until the middle of the nineteenth century. Some of the ancestors of the plants we cultivate—eggplant, potatoes, and almonds—were poisonous! These plants have a **recessive gene** (a gene whose trait is expressed only

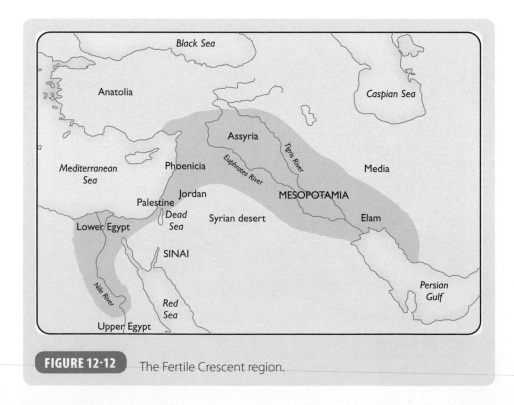

FIGURE 12-12 The Fertile Crescent region.

when both copies are identical), however, that will yield edible offspring. In the wild, such a gene is likely to stay recessive because animals will likely gobble up any edible varieties. But humans, after a bit of admittedly risky trial-and-error, would eventually select plants with the recessive gene and begin to grow only forms that displayed the nonpoisonous characteristics they needed.

Recessive genes are extremely valuable to human society. Wild wheat plants, a species of grass, have a dominant gene (genes whose trait is expressed whether or not there are two copies) that causes the stalk to disintegrate when the plant's seeds are ripe. This allows the seeds to be spilled upon the ground permitting them to germinate. But a recessive variety exists that lacks this gene; these stalks don't disintegrate, allowing for an easier human harvest, and this is the wheat plant that early farmers selected.

To summarize, here is why we became dependent on so few grains:

- Their ancestors were edible and plentiful in the wild.
- They grew quickly and were easily collected and stored.
- They contained protein and were rich in carbohydrates.

Thus, as a result of millennia of trial and error, by the days of the Roman Empire almost all modern grains were being cultivated in some form somewhere.

The Green Revolution

Over most of human history, farmers increased agricultural output mainly by plowing up forests and natural grasslands. Limits of geographic expansion were reached decades ago in densely populated parts of India, China, Java, Egypt, and Western Europe. **Intensified production**, obtaining more output from a given area of agricultural land, has thus become a "growing" necessity.

The term "*Green Revolution*" was coined to describe agriculture using enhanced irrigation, development of more productive plant species, and the increased use of fertilizers, herbicides, and pesticides. To cite one example of the effects of the "Green Revolution" it took nearly 1,000 years to increase wheat yields from 0.5 to 2 tonnes per hectare, but only 40 years to increase yields from 2 to 6 tonnes per hectare. Globally, rice and wheat yields increased nearly 100% during the second half of the twentieth century.

But our efforts to increase agricultural production are becoming more challenging and more costly. There is evidence that growth rates in grain yields have slowed in both developed and developing countries. Per capita grain production apparently began to decline in the middle 1990s, while the total area of land used for agriculture rose from 4.55 billion hectares in 1966 to 4.93 billion in 1996. According to the U.S. Department of Agriculture, in 2011 global grain harvest was up 53 million tons from 2009. Consumption, however, grew by 90 million tons over the same period.

Future increases in food production may become more difficult because a complex set of environmental and social factors must now be taken into account while developing new crop technologies. An illustration is the growing controversy over **genetically modified (GM) foods**.

Irrigated Agriculture

The earliest cities in human history, situated in what is now southern Iraq, relied on sophisticated irrigation schemes. However, **salination** (salt buildup) of soils eventually poisoned the soils and forced the sites to be abandoned.

Today, without federally subsidized irrigation projects, agriculture as it is practiced in the Western United States would not exist. Through agencies such as the U.S. Bureau of Reclamation, millions of hectares of western lands were opened to irrigated agriculture, and indeed to settlement. Beginning in 1902, and intensifying during the Great Depression of the 1930s, vast, capital-intensive federal projects dammed rivers, created lakes and reservoirs, and built and maintained wells, irrigation canals, and aqueducts. These projects provided water to farmers at rates that were typically only a small fraction of the actual cost of extraction and delivery. Federal agricultural policies created what has been called a "schizophrenic culture of entitlement and dependency" on the one hand, and a mythology of "rugged individualism" on the other. That culture, while slowly changing, persists today in many western and some southern rural communities. This culture holds that, among other things, farmers and ranchers have proprietary "rights" to water, and that the federal government is obliged to provide them water at reasonable—meaning low—cost.

Throughout the West, large corporate farms use subsidized irrigation water to produce crops such as sod, wheat, rice, potatoes, cotton, and alfalfa for which there is often insufficient demand, and in the process generate **agricultural runoff** that is often laced with toxic elements like selenium (see Kesterson NWR later in the chapter) and arsenic *leached* out of the west's arid-lands pedocal soils. Irrigation runoff usually also contains pesticides and fertilizer. Moreover, the water may be warmed by storage or flow through irrigation canals, reducing oxygen content, and may contain silt from field runoff.

A large proportion of the grains produced are used as animal feed, using most of the West's water and in the process sustaining a vast subsidized meat industry.

Conflicts among water users and concerned citizens ("stakeholders") throughout the West are growing, fueled by urbanization and a growing environmental consciousness, backed by federal endangered species and water protection legislation.

In California, agriculture's share of water has been dropping since the 1970s, but still accounts for 80% of state water use.

Concept Check 12-6. A major use of water worldwide is to irrigate feed crops for animal consumption. Identify ways in which water can be conserved.

FIGURE 12-13 shows *groundwater* use by sector. Much of the water used in agriculture is highly inefficient. Irrigation canals may be unlined, allowing water to seep into the ground, for example. And many irrigation techniques are inefficient: up to 25% of water can be lost to evaporation (**FIGURE 12-14**). Agricultural use has seriously depleted one of the most extensive aquifers in North America, the Ogallala (High Plains) Aquifer (**FIGURE 12-15**).

In China, serious depletion of groundwater has put much Chinese grain production at risk.

We discuss irrigation and agriculture elsewhere in the text. Next, we consider an example of the conflicts arising out of irrigated agriculture in the West, at the Klamath Basin.

■ THE KLAMATH BASIN, OREGON AND CALIFORNIA

FIGURE 12-16 is a location map of the Klamath Basin, an area about twice the size of Massachusetts. The **Klamath Basin Irrigation Project (KBP)** was designed, and is managed, by the U.S. Bureau of Reclamation. The KBP was established in 1905 on

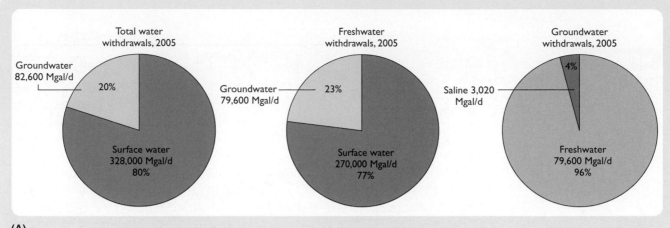

Total water
withdrawals, 2005

Groundwater
82,600 Mgal/d
20%

Surface water
328,000 Mgal/d
80%

Freshwater
withdrawals, 2005

Groundwater
79,600 Mgal/d
23%

Surface water
270,000 Mgal/d
77%

Groundwater
withdrawals, 2005

4%

Saline 3,020
Mgal/d

Freshwater
79,600 Mgal/d
96%

(A)

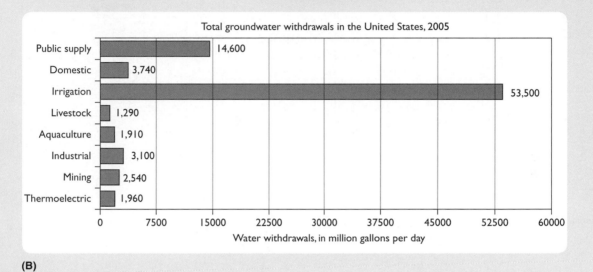

Total groundwater withdrawals in the United States, 2005

Sector	Water withdrawals (million gallons per day)
Public supply	14,600
Domestic	3,740
Irrigation	53,500
Livestock	1,290
Aquaculture	1,910
Industrial	3,100
Mining	2,540
Thermoelectric	1,960

Water withdrawals, in million gallons per day

(B)

FIGURE 12-13 (A) U.S. water use, 2005. (B) Groundwater use by sector. Which sector accounts for most water use? Parts (A) and (B) Modified from Groundwater Use in the United States, USGS.

FIGURE 12-14 Spray irrigation, an inefficient but cheap process, used mainly where water use is subsidized. © Digital Vision/Thinkstock.

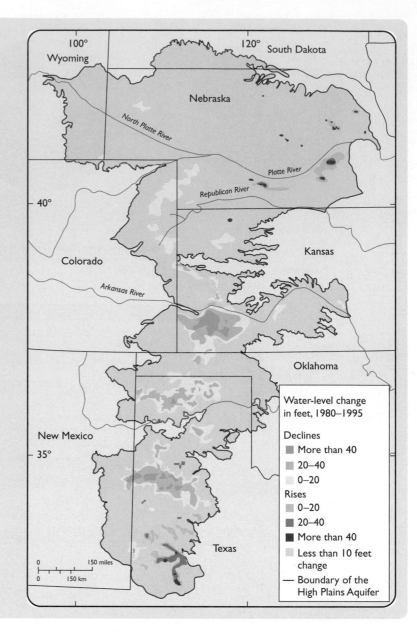

FIGURE 12-15 Depletion of the Ogallala (High Plains) Aquifer. Depletion has been so pronounced that farmers face the likelihood of abandoning irrigated farming over much of the area by 2030. Modified from U.S. Geological Survey Water-Resources Investigations Report 97-4081.

Water-level change in feet, 1980–1995

Declines
- More than 40
- 20–40
- 0–20

Rises
- 0–20
- 20–40
- More than 40
- Less than 10 feet change
— Boundary of the High Plains Aquifer

Value-added Question 12-1. Calculate the value per acre of crops in the Klamath Project as of 1999 and 2011 (assuming 210,000 acres).

land donated by California and Oregon in 1902, following the passage of the federal Reclamation Act. As of 2003 it provided approximately 400,000 acre-feet (1 AF = 1.23 million liters, or an area the size of a football field one foot deep) of irrigation water each year for about 1,600 farmers on 210,000 acres of farmland, much of which is located within wildlife refuges (**FIGURE 12-17**). As of 1999, the value of crops produced on irrigated land was $104 million. By 2011 it had risen to nearly $150 million.

Before irrigation, its few residents described the Basin as "sun-baked prairie and worthless swamps." The federal government made the land available for veterans of Word Wars I and II, and encouraged them to settle there, with promises of cheap, abundant irrigation water.

Hydrology of the Klamath Basin

The Klamath Basin can be divided into an Upper Basin, which is irrigated, and a Lower Basin, which is not. Two reservoirs, Iron Gate and Copco, mark the lower boundary of the Upper Basin. The Lower Basin begins below Copco Reservoir. The construction of Copco Dam in 1917 permanently blocked fish passage into the Upper Basin.

A key part of the watershed is Upper Klamath Lake, a vestige of once-vast glacial Lake Modoc. Today, Upper Klamath Lake has an area of about 70,000 acres (27,600 ha), and is connected to Lower Klamath Lake, which has an area of 85,000 acres (33,500 ha).

The major river in the basin is of course the Klamath, which drains out of Klamath Lake, and its largest tributary is the Trinity (see map). The two rivers stand in stark contrast for much of the year. The Trinity drains mountainous non-irrigated areas with forest cover on lower slopes, and deep cold-water reservoirs. The Trinity contains little sediment and has relatively low water temperature with high O_2 content. The bottom is visible throughout its length. The Klamath by contrast is nutrient-laden and warm, resulting in low oxygen content. It is cloudy from silt and algae, both the result of upstream irrigation.

Irrigation in the Klamath Basin Irrigation Project

Exactly how much water is used for irrigation and the number of irrigated acres is not precisely known, but a study estimated irrigation water use, after evaporation and infiltration losses, to be 400,000 acre-feet/year. Many irrigation delivery systems do not measure use, and limits are seldom strictly enforced. Within the Project, farmers pay a fee based on the number of acres they will irrigate, not on the amount of water they will use.

Center-pivot irrigation and flood irrigation methods are used within the Klamath Basin (FIGURE 12-18). Much of the water either evaporates or infiltrates without watering crops. But because users are not charged for water based on use, there is no incentive to conserve.

Water Quality in the KBP

The Oregon Department of Environmental Quality (DEQ) is required under the federal Clean Water Act to set water quality standards to protect water uses like human consumption, agricultural use, recreation, industrial water supply, and cold water fisheries. The Klamath Basin contains portions of 46 different rivers and lakes that have failed to meet these standards. All of Upper Klamath Lake is listed as "water quality–impaired" under the Clean Water Act. Its pH may reach 10 locally (that is, very basic). The lake, however, does serve as the source of a dietary supplement made from blue-green algae (cyanobacteria), and the industry provides over 500 jobs to the region.

Much of the geological substrate of the Upper Basin contains high phosphorus levels. Return flows from pastures have phosphorous (P) levels seven to ten times higher than the river's water. A cow will produce and excrete seven to ten times as much P as a human. One biologist pointed out that, with Wood River Valley cattle population somewhere between 50,000 and 100,000, it is the

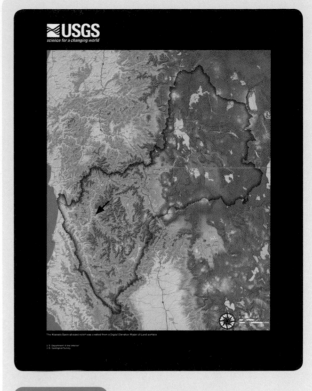

FIGURE 12-16   Location of the Klamath Basin, Oregon and California. Courtesy of USGS.

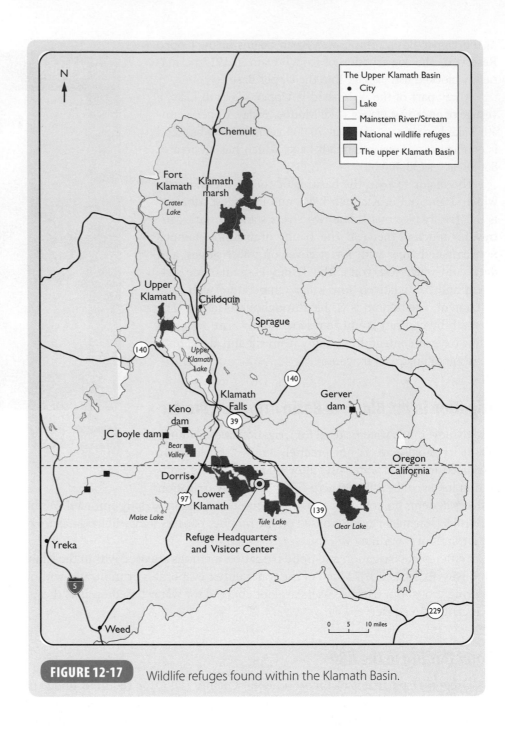

FIGURE 12-17 Wildlife refuges found within the Klamath Basin.

Value-added Question 12-2.
Review Table 12-1. What percent of the value of these crops is derived from feed crops for animals? Feed crops are barley, oats, alfalfa, pasture, grass and hay.

equivalent of having a town with 350,000 people without sewage treatment living along the river.

Agricultural Production in the KBP

Farm production in the Upper Basin is shown in **TABLE 12-1**.

The average consumptive water use in the KBP has been about 2 AF/acre.

According to the Environmental Working Group (EWG), farmers in the three Oregon counties containing Project lands received over $35 million in federal farm subsidies between 1995 and 2011. Without irrigation water provided by the U.S.

FIGURE 12-18　Flood irrigation. Fields are flooded with sheets of water. © Achim Baque/ShutterStock, Inc.

TABLE 12-1　*Agricultural Production in the Klamath Basin Irrigation Project*

	Oregon Lands in Klamath Project		Langell Valley ID and Horsefly ID	
	acreage	gross value	acreage	gross value
Barley	24,495	$7,137,843.00	692	$201,649.00
Oats	2,086	$520,666.00	854	$213,159.00
Wheat	1,670	$783,564.00	0	$0.00
Rye	1,040	$255,970.00	0	$0.00
Alfalfa	33,203	$19,722,582.00	8,670	$5,149,980.00
Grass/Hay	14,366	$2,686,442.00	5,699	$1,065,713.00
Pasture	40,927	$3,682,530.00	9,270	$834,096.00
Sugar Beets	2,744	$2,361,239 00	0	$0.00
Onions	0	$0 00	0	$0.00
Potatoes	7,349	$23,149,350.00	400	$1,250,000.00
Misc. Veg.	0	$0 00	0	$0.00
Total	127,880	$60,300,186.00	25,585	$8,724,597.00

Source: Bureau of Reclamation.

Bureau of Reclamation, farming would not be possible in the KBP, and the value of irrigators' land would drop by 75%.

Wildlife in the KBP

Eighty percent of all migrating birds in the Pacific flyway—up to two million birds—pass through the wetlands of the Upper Basin each autumn. The migrating birds are accompanied by roughly 1,000 bald eagles that spend the winter around the basin. This is the largest winter concentration of bald eagles in the contiguous (lower 48) United States.

Until around 1970, The Klamath River supported the third-largest salmon population in the United States. According to the U.S. Department of Interior, fish harvests ranged between 500,000 and 1,000,000 Chinook salmon annually off the California coast between 1947 and 1970. Suckerfish, once a staple of Native American diets, were so abundant they were used for years by irrigators as fertilizer; suckerfish have been on the federal endangered or threatened species list since 2001.

Economic Impact of Poor Water Quality

Economic impacts on the fishing-dependent communities of the coast have been significant. In 2001 testimony before Congress, the executive director of the Pacific Coast Federation of Fishermen's Associations claimed that roughly 3,780 fishing-related jobs have been lost as a result of the decline in fish. Fishers place the blame on KBP irrigation.

Before construction of the Project, the Klamath River flooded roughly 300,000 acres (118,000 ha) of wetlands during high winter and spring flows. During the last century, the Project drained most of these wetlands, turning them into farms, and water reached the remaining refuges only as runoff from cropland.

Irrigation subsidies encourage excess water use, which, according to biologists at the Oregon Natural Resources Council and others, have damaged the Lower Basin economically. Declining water quality has reduced summer tourist fishing, adversely impacting local communities depending on tourism. At one locality, the number of fishing guides fell from 25 to 6.

Several Native American tribes—the Hupa, the Yurok, the Klamath, and the Karuk—have fished the Lower Klamath and its tributaries for millennia. But the nutrient pollution, leading to algal growth, and heat combine to lower the survival rate of salmon, reducing the tribes' fish harvest (**FIGURE 12-19**). According to one hydrologist, daytime temperatures in the Klamath are so high that fish in the river's main stem can travel only at night.

Offshore, marine fishers have seen catches decline precipitously as dams, along with silt-laden, polluted, and heated water, have sharply reduced survival rates for salmon that migrate up the river system. Several species, including the once-plentiful Coho salmon, are now listed under the Endangered Species Act.

A 2002 study commissioned by the U.S. Geological Survey found that buying farmland and eliminating subsidized agriculture and restoring flows and water quality throughout the Klamath Basin would provide economic benefits from recreational activity worth 10 to 15 times the annual value of crops produced.

Adjudicating Water Issues in the KBP

Water issues here are particularly thorny because of the West's water doctrine of "**prior appropriation**," which means the first claimants have the most secure

FIGURE 12-19 An algal bloom in Elysian Lake, Minnesota. Some algae produce toxins that can harm or kill fish, and decomposing algae reduce or eliminate oxygen in the water. Courtesy of Dr. Jennifer L. Graham/USGS.

water rights. Irrigators claimed the Basin's first "water rights" in 1905. That doctrine promoted development of the arid west by encouraging settlers to divert water from streams, rivers, and lakes, but required that they put it to "beneficial use," a policy that does not permit the water to be degraded so as to render it unfit for some other user. Native Americans who had lived in the Basin for millennia were not accorded water rights under prior appropriation.

All Western states have some sort of policy for allocating water by the issuance of water rights. Additionally, all Western states use an "adjudication" to recognize historic water uses that began before water codes were enacted.

Water rights holders must use their allocation or lose it, a requirement essentially forcing farmers to waste water. Complicating the situation further is the 1909 decision by the State of Oregon to interpose itself in a water dispute between two rights holders. Consequently, Oregon adjudicates water rights allocated after 1909, but not those awarded prior to that date. Finally, ranchers and farmers who have been in the Basin for three to four generations argue their case against Native Americans who have fished the Klamath for thousands of years, but without water rights.

The severe droughts of the latter 1990s culminated in a decision by the Bureau of Reclamation to essentially shut off water to irrigators in 2001, prompting some irrigators to vandalize property and illegally, in the eyes of the Bureau of Reclamation, seize water.

The State of Oregon is presently implementing what it calls alternative dispute resolution (ADR). Specifically, The Oregon Water Resources Department is charged with administering an exhaustive negotiation process, on behalf of the state Circuit Court. The State hopes to negotiate an agreement among the numerous participants in the process (tribes, environmentalists, irrigators, mediators and other stakeholders), which will settle water rights and offer long-term guidance for watershed management. A settlement is expected to result in removal of Iron Gate, Copco 1 and 2, and J. C. Boyle Dams to enhance fish passage, and concessionary power charges to irrigators who use more efficient irrigation methods.

■ POLLUTION FROM AGRICULTURE

Modern agriculture is a major contributor to **nonpoint-source pollution**. Agricultural as well as industrial sites of pollution can be long distances from where the toxicities show up in bodies of water, including streams and rivers, lakes, groundwater, and seas. Sources include feedlots (excrement and urine from livestock), fertilizer, herbicides, and pesticides. Manure spills are usually the number one cause of fish kills (FIGURE 12-20).

Atrazine

Atrazine is the most commonly used pesticide in the United States, with around 75 million pounds (34 million kg) used annually, mainly on corn. It is extremely soluble in water, and attracts to clays under acid conditions, the usual soil pH. FIGURE 12-21 shows atrazine use in the United States.

The use of atrazine is banned in several countries in the EU, including Germany, Italy, the Netherlands, and Austria.

For years biologists have noted with growing alarm an apparent decline in global amphibian populations. In a 2002 article published in *Nature,* every water body studied that contained measurable atrazine also contained male frogs with female sex organs. Laboratory experiments showed that leopard frogs exposed to atrazine also developed female sex organs. Such chemicals are called **endocrine-disrupting toxins**. In addition, the chemical is persistent: it has been found in water and soils one year after application. Atrazine also has been implicated in high breast cancer rates in Vermont and elsewhere.

Kesterson National Wildlife Refuge

The Kesterson National Wildlife Refuge (NWR) is located in the northwest part of California's San Joaquin Valley (FIGURE 12-22). The inflow of water to Kesterson was intended to be from irrigation runoff, a situation common to almost two-dozen "wildlife refuges" throughout the American West. The refuge's ponds were completed in 1971, but until 1978 received only fresh water runoff. Beginning in 1978, irrigation runoff from farms on the San Joaquin Valley's west side flowed into the refuge. By 1981, inflow was entirely irrigation runoff. West side soils are formed on marine sedimentary rocks with high trace element content. East side soils are formed on granitic rocks containing few trace elements. By 1982, deformed wildlife were common in the refuge, and biologists had identified selenium and uranium, leached from desert soils, as the cause.

Sources associated with fish kills

- Other pesticide applications 28
- Runoff (general) 36
- Spills 69
- Industrial discharges 76
- Agriculture 139
- Sewage treatment plants 86

FIGURE 12-20 Sources of fish kills in the United States (35 states reporting). What is the largest single source of fish kills? Data from EPA, 1995.

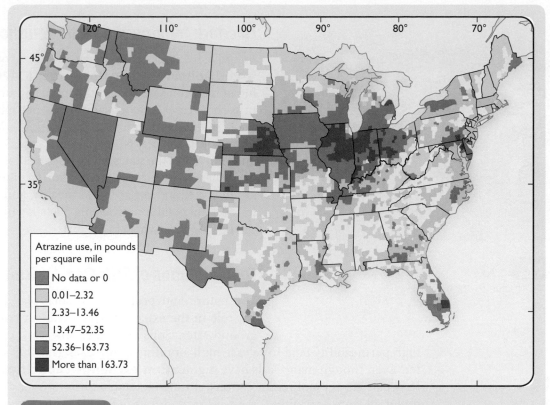

FIGURE 12-21 Atrazine use in the United States. Activity: Based on this figure, which
watershed(s) do you think would have problems with atrazine contamination?
Answer: The Mississippi is the most obvious, followed by the Ohio, the Potomac, and the
Missouri. Depending on how you interpret the data, you might have chosen others as well.
Modified from Battaglin, William A. and Goolsby, Donald A. Spatial Data in GIS format on
Agricultural Chemical Use, Land Use, and Cropping Practices in the US. USGS.

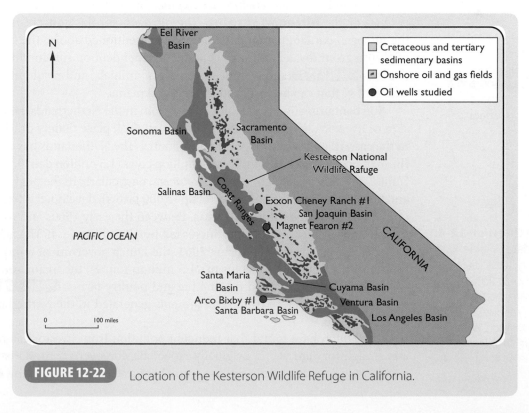

FIGURE 12-22 Location of the Kesterson Wildlife Refuge in California.

FIGURE 12-23

FIGURE 12-23 A CAFO in California. How close would you like to live to such a facility? Why? © Jim West/age fotostock.

Livestock Farming and Water Pollution

Intensive livestock raising involves the use of **Concentrated Animal Feeding Operations (CAFOs)** resulting in high concentrations of manure and urine, which in turn release ammonia, nitrogen, phosphorus, and odors into the environment (**FIGURE 12-23**). Minnesota, Iowa, South Dakota, North Carolina, and South Carolina are a few of the states where huge feedlot operations or industrial-style meat production facilities threaten groundwater and surface water quality, and contribute to local air pollution problems. Details about CAFOs can be found elsewhere in the text.

Contamination of Soils From Agriculture

The nature and properties of soils play an important role in the extent to which effluent contaminates groundwater. Soils that have low clay content and high permeability tend to release high amounts of nitrogen (N) into groundwater. Even though many soils have high **cation exchange capacities,** meaning they can attract and retain cations, nitrate (a soluble form of N and a common pollutant from agriculture) is an anion, and thus not attracted to soil clays.

Agricultural Pollution in the European Union

Value-added Question 12-3.
Review Table 12-2. On a sheet of plain or graph paper, create a graph by plotting stocking rate (on the Y-axis) against nitrogen surplus (on the X-axis), then summarize the results. Were there anomalies? If so, suggest an explanation and a hypothesis to test it, based on information in this section.

In 1995 around 43%, or 138 million hectares, of the total area of the European Union (EU) was used for agriculture. EU research has documented the extent of soluble nitrogen buildup, which can degrade groundwater. The difference between nitrogen *input* and nitrogen *retention and uptake* by crops is known as the *net-nitrogen balance* or the **nitrogen surplus.** The nitrogen-surplus concept is a measure of the nitrogen contamination *potential*. As we mentioned above, the *actual* leaching of N into groundwater and surface water is dependent on soil conditions as well as climate. **TABLE 12-2** shows concentrations of animals, and resultant N surplus, in excess of N that can be absorbed by soils and plants.

Soil contamination is a significant problem in the Netherlands. Here, the expansion of livestock, specifically pigs and poultry, took place mainly on the sandy soils in the eastern and southern part of the country. The Netherlands has 12 million pigs (human population = 16.7 million), 4 million cows (1.5 million dairy), and 90 million "units" of poultry. Dutch livestock producers, particularly in the pig (~ 10,000 farms and declining) and the poultry sectors (~ 2,000 farms), developed intensive livestock operations on farms with small areas. Between the early 1960s and the mid-1980s, the number of pigs and poultry increased by 10 million (+ 450%) and 50 million (+ 125%), respectively. But during 2001, the Dutch government announced a plan to reduce the phosphate surplus of 21.5 million tonnes, the major source of which was manure, by buying and retiring pig and poultry farms. **TABLE 12-3** shows the total amount of manure from farm animals generated in the Netherlands between 1993 and 1996.

Value-added Question 12-4.
Based on the data in Table 12-3, would reducing livestock reduce phosphorus (P) pollution? Determine the amount of P in manure of each type of livestock. Would reducing one reduce P more than others?

The Dutch have found that reducing pollution from livestock is not as simple and straightforward as they originally expected. Moreover, recent statistics on Dutch livestock production and pollution available have been distorted by epidemics of

TABLE 12-2

Stocking Rate and Nitrogen Surplus per Hectare in Parts of Western Europe

Region	Stocking Rate (animal units/ha)	N-surplus (kg N/ha)
The Netherlands	3.7	320
Belgium	2.6	170
Brittany (France)	2.4	135
Lombardy (Italy)	2.3	90
North-Rhein Westfalia (Germany)	2.0	140
Denmark	1.5	115
European Union	0.9	70

foot and mouth disease and lingering concerns over bovine spongiform encephalopathy (BSE; also known as "mad cow disease").

Organic Agriculture

By 2008, there were more than 13,000 certified *organic* producers in the United States. The U.S. Department of Agriculture (USDA) reported that U.S. organic farmland increased to 4.8 million acres (~2 million ha) in 2008 up from about 1.3 million acres in 1997. Approximately 2% of the U.S. food supply is grown using organic methods. In 2010, retail sales of organic food and beverages were approximately $39 billion. Demand for products grown "organically" is increasing at about 20% per year in the United States and the European Union.

TABLE 12-3 — **Livestock, Manure, and Phosphorus Production in the Netherlands**

	ANIMAL NUMBERS			ANIMAL WASTE PRODUCTION							
	Hogs	Cattle	Poultry	Hogs		Cattle		Poultry		Total	
		1000 head		Manure MMT	P_2O_5 Mil. kg	Manure MMT	P_2O_5 Mil. kg	Manure MMT	P_2O_5 Mil. kg	Manure MMT	P_2O_5 Mil. kg
1993	14,964	4,797	98,086	17.0	72.0	65.5	126.1	2.5	34.8	**85.0**	**232.9**
1994	14,565	4,716	93,953	16.4	65.3	63.6	119.6	2.3	33.0	**82.3**	**217.9**
1995	14,565	4,716	93,953	16.2	60.1	63.7	118.4	2.1	30.7	**82.0**	**209.2**
1996	14,419	4,551	93,552	16.9	56.9	61.7	103.6	2.4	30.5	**79.9**	**191.0**

Source: CBS/LEI-DLO, various years, and www.minlnv.nl/infomart.feiten/1997/feitlandb6.htm

In 2002, the U.S. Department of Agriculture issued rules to characterize practices that may be called "organic." Here are their criteria:

- With minor exceptions, all seeds used to grow crops must be organic.
- Animals must not be fed feed made from other animals, except fishmeal.
- The use of ionizing radiation is prohibited.
- With few exceptions, the use of genetically modified seeds or foods is prohibited.
- All fertilizer must be organic. Solid waste from sewage-treatment plants may not be used as soil conditioner or fertilizer.
- No antibiotics or growth hormones, conventional pesticides, or petroleum-based fertilizers may be used.

For details, consult the USDA website (http://www.usda.gov).

While some concern has been expressed that switching to organic methods could lower agricultural production, research indicates that long-term organic crop yields average 95% to 100% of conventional yields, after a three-year transition period from conventional to organically grown crops. This three-year interval is needed to restore soil microbes and properly condition the soil. And organic farming is carried out without the harmful environmental "side effects" from conventional agriculture, which as you have seen include air and water pollution from nitrates, phosphates, natural and artificial organic chemicals, petroleum products, growth hormones, and antibiotics; soil depletion and erosion; and polluted coastal seas and oceans.

According to the Organic Farming Research Foundation, "if USDA would increase the small proportion of its research funds currently directed toward optimizing organic farming practices, organic has the potential to produce yields fully matching or surpassing those of conventional crops."

■ CONCLUSION

There is a growing recognition that soil health is essential to sustainable agriculture. Protecting soil health begins with understanding soil processes. Agricultural subsidies in the United States, European Union, and elsewhere encourage production of crops for which there are insufficient markets. These crops are then aggressively marketed to developing countries, often undermining local agriculture in the process. Small-scale farmers are thereby driven off the land and into cities or into illegal migration to developed countries. Finally, overproduction, especially using the techniques of industrialized agriculture, can exacerbate soil degradation and water pollution.

Trends over the past 20 years show an increase in organic farming methods which, overall, have much more benign effects on soils and soil waters that industrial-style methods. A growing human population as well as global increases in meat consumption will continue to put pressure on agriculture, whether organic or conventional, and the resources on which it is based.

■ WHAT YOU CAN DO

Find out more about the impacts your food choices have on the environment. Consult the websites of the USDA (http://www.usda.gov), Environmental Protection Agency (EPA) (http://www.epa.gov), and the European Environment Agency (http://www.eea.europa.eu/) for details. Read labels! Find out what is in your food and where it comes from. Also, consider eating fewer processed foods, cutting back on your meat consumption, and eating more organically produced foods. By doing so, you are not only ingesting fewer contaminants, you are reducing the agricultural use of pesticides and other toxic compounds that adversely affect soils, streams, and groundwater.

CHAPTER SUMMARY

1. In 1999 the European Union (EU) spent $114.5 billion in farm subsidies ($72 billion in 2011): making up 49% of gross farm revenue. The United States spent $54 billion ($20 billion in 2012), equal to 24% of gross farm income. In the United States, most of those payments went to growers of rice, cotton, corn, soybeans, and sorghum: so-called "commodity" crops.

2. Along with water, soils may be the most important natural resource on the planet. But rapidly growing human populations in many areas have altered or destroyed the soil's structure and countless soil species.

3. Soils consist mainly of the decomposition and disintegration products of rock and sediment with organic matter called humus.

4. The structure of a soil consists of a top humus layer of variable thickness, underlain by a leached zone, with a zone of accumulation below that.

5. Three general categories of soils form under differing climatic regimes. They are pedalfers, pedocals, and laterites.

6. Soil microbes are essential to life on earth, but relatively little is known about them. The living microbes, fungi, and invertebrates that comprise the soil food web are responsible for changing carbon and nitrogen into forms available for plant growth.

7. Agronomists estimate that 1500 billion tonnes of C is bound globally to soil. The annual flux (movement) of CO_2 returning to the atmosphere as a result of decomposition and other soil processes, amounts to approximately 68×10^9 tonnes C/yr—at least ten times more C than that produced by deforestation and fossil fuel burning.

8. Swelling soils are one of the most expensive geological hazards in the United States. Swelling or expansive soils are soils that expand when wetted, and contract when dried. Swelling soils contain clay minerals whose crystal structures permit the large-scale introduction of water.

9. Loess, of glacial origin, is the source of most of the United States' best agricultural soils. Although loess soils can be up to 60 m thick in Iowa and elsewhere, erosion has removed up to half the original soil locally.

10. Soil erosion is usually irreversible, since once the nutrient-rich topsoil has been stripped away its ability to sustain plant growth is severely reduced. Agronomists estimate that 75 billion tonnes of soil are eroded each year, mostly from agricultural lands.

11. Humans are extremely dependent on a few grains—mainly wheat, rice, and corn, called maize outside the United States.

12. Grain domestication was preceded by a major change of climate caused by the last withdrawal of glaciers about 11,000 years ago. As a result of millennia of trial and error, by the days of the Roman Empire almost all modern grains were being cultivated in some form somewhere.

13. The term "Green Revolution" was coined to describe processes including enhanced irrigation, development of more productive plant species, and the increased use of fertilizers, herbicides, and pesticides.

14. Per capita grain production apparently began to decline in the middle 1990s.

15. Without federally subsidized irrigation projects, agriculture as it is practiced in the Western United States would not exist. Most of the grains produced are used as animal feed, using most of the West's water and in the process sustaining a vast subsidized meat industry. In California, agriculture's share of water has been dropping since the 1970s, but still accounts for 80% of state water use.

16. Many irrigation techniques are inefficient: up to 25% of water can be lost to evaporation.

17. The Klamath Basin Irrigation Project (KBP) in Oregon and California, was designed, and is managed by, the U.S. Bureau of Reclamation. As of 2003 it provided approximately 400,000 acre-feet (1 AF = 1.23 million liters) of irrigation water each year for about 1600 farmers on 210,000 acres of farmland, much of which is located within wildlife refuges. All of Upper Klamath Lake is listed as "water quality–impaired" under the Clean Water Act.

18. According to the Environmental Working Group (EWG), farmers in the three Oregon counties containing Project lands received roughly $30 million in federal farm subsidies between 1995 and 2000. Without irrigation water provided by the U.S. Bureau of Reclamation, farming would not be possible in the Klamath Basin, and the value of irrigators land would drop by 75%.

19. Agriculture as practiced in western countries is a major contributor to non-point-source pollution. Sources include feedlots (excrement and urine from livestock), fertilizer, herbicides, and pesticides.

20. Atrazine is the most commonly used pesticide in the United States, mainly on corn. A 2002 article in *Nature* found every water body studied that contained measurable atrazine also contained male frogs with female sex organs. Chemicals like atrazine are called endocrine-disrupting toxins.

21. The Kesterson Wildlife Refuge is a site of severe selenium pollution, mainly from irrigation runoff.

22. Nitrogen and phosphate contamination is a significant problem in the Netherlands and in other parts of the EU.

23. The U.S. Department of Agriculture set standards for organic foods in 2002. Sales of organic foods are growing at about 20%/year in the United States and the EU.

KEY TERMS

agricultural runoff

atrazine

CAFO (Concentrated Animal Feedlot Operation)

cation-exchange capacity

commodity crop

contour plowing

cyanobacteria

endocrine-disrupting toxins

genetically modified (GM) foods

humus

intensified production

Klamath Basin Irrigation Project

laterite

leached zone

loess

nitrogen fixation

nitrogen surplus

no-till plowing

nonpoint-source pollution

organic agriculture

pedalfer

pedocal

prior appropriation water doctrine

recessive gene

salination

selenium (Se)

sequestration

soil

soil biota

swelling soils

zone of accumulation

REVIEW QUESTIONS

1. Most U.S. agricultural subsidies are applied to
    a. organically grown crops.
    b. fruits.
    c. vegetables.
    d. commodity crops.
    e. spread evenly to all types.

2. Variably decomposed organic matter in soils is called
    a. humus.
    b. hummous.
    c. peat.
    d. carbonate.

3. Soils common to temperate humid regions are
    a. laterites.
    b. pedalfers.
    c. pedocals.
    d. karst.

4. Soils typical of arid regions are
    a. laterites.
    b. pedocals.
    c. carbonates.
    d. pedalfers.

5. Soils common to tropical regions are
    a. karsts.
    b. laterites.
    c. pedalfers.
    d. pedocals.

6. Soils with little organic matter and few nutrients, composed mainly of insoluble complexes are
    a. pedocals.
    b. pedalfers.
    c. laterites.
    d. loess.

7. As soil temperatures increase
    a. they release more carbon as CO_2.
    b. they sequester more carbon.
    c. reaction rates in soils decline.
    d. none of the above.

8. Swelling soils are hazardous because they contain certain types of
    a. clay minerals.
    b. pathogenic organisms.
    c. heavy metals.
    d. all of the above.

9. Loess originates from weathering by
   a. rivers.
   b. wind.
   c. glaciers.
   d. animals.

10. Soil erosion is usually
    a. reversible.
    b. irreversible.
    c. limited to tropical latitudes.
    d. caused by landslides generated by earthquakes.

11. A technique that can manage or reduce soil erosion is
    a. contour plowing.
    b. planting ground cover.
    c. terrace construction.
    d. all of the above.
    e. none of the above.

12. Salination of soils surrounding early cities in what is now southern Iraq, resulted from
    a. introduction of foreign seeds.
    b. irrigation.
    c. mutation of soil microbes.
    d. groundwater withdrawal.

13. Most of the corn produced in the United States is used
    a. as humanitarian aid to poor countries.
    b. to produce ethanol.
    c. as animal feed.
    d. as biomass fuel.

14. In California, agriculture accounts for what percent of the state's water use?
    a. 10%
    b. 50%
    c. 20%
    d. 80%

15. The Klamath Basin Project, a major irrigation project run by the U.S. Bureau of Reclamation, is located in
    a. Alaska.
    b. Alabama.
    c. the upper Great Plains.
    d. Oregon and California.

16. Irrigators in the Klamath Basin are charged for water on the basis of
    a. how much water they use.
    b. their ability to pay.
    c. how much land they irrigate.
    d. what type crops they grow.

17. A problem with soils in the Klamath Basin is their high
   a. phosphorus levels, which can contribute to nutrient pollution through irrigation runoff.
   b. temperature, which can liberate high amounts of CO_2.
   c. cost, which can render their products non-competitive.
   d. mercury content, which can contaminate crops.

18. Agriculture as practiced in western countries is a major contributor to
   a. point-source pollution.
   b. SOx pollution.
   c. nonpoint-source pollution
   d. lead buildup in river sediment.

19. The most commonly used pesticide in the United States as of 2000 was
   a. DDT, applied to swamps to control mosquitoes.
   b. PCBs applied to cotton to control boll weevils.
   c. atrazine, applied to corn.
   d. methyl bromide, applied to strawberries.

20. The Kesterson Wildlife Refuge has experienced severe
   a. mercury poisoning from coal-fired power plants.
   b. selenium poisoning from irrigation runoff.
   c. acid precipitation from motor vehicle emissions.
   d. all of the above.

FOOTNOTE

[1] BCE, an acronym for Before the Common Era, means the same as BC.

CHAPTER
13

Mineral Resources

Point

On February 26, 2002, the Missouri Department of Health and Senior Services (DHSS) released a report with the results of their blood-lead census, conducted in August and October 2001, on children in Herculaneum, Missouri (**Figure PC 1-1**). Tests showed that 28% of Herculaneum children six years of age and under had lead (Pb) blood levels over 10 micrograms per deciliter, a hazard level according to the U.S. Centers for Disease Control and Prevention (CDC). Within a half-mile of the Doe Run lead smelter, more than 50% of the children had unsafe levels of lead in their blood. As a result, the DHSS posted signs in Herculaneum advising citizens to avoid contact with surfaces exposed to lead dust, and the Missouri Department of Natural Resources issued a "cease and desist" order against Doe Run. The company was required to make purchase offers on about 160 homes by December 2004.

The source of the problem is the Doe Run Company's lead smelter, the largest and oldest in the United States (**Figure PC 1-2**). The smelter takes in millings from the lead mines in eastern and south-central Missouri, and refines them into lead, which is then used in shielding, batteries, and other applications where very pure lead is required.

Lead dust from the ore-hauling trucks, air emissions from the smelter, and runoff from piles of smelter waste are released to the environment. In 2000, a **Toxic Release Inventory (TRI)** published by the company listed release of more than 1 million kg (2.2 million pounds) of lead into the environment, including on-site smelter waste piles and directly to the air.

Concerned that the company may enter bankruptcy before cleaning up the health threat it created in Herculaneum, then-Missouri Governor Bob Holden, then-Senator Jean Carnahan and Representative Richard Gephardt, in whose congressional district Herculaneum was located, asked the EPA to place the site on Superfund's (the federal government's program to clean up the nation's hazardous waste sites) priority cleanup list.

A business-reporting website (http://www.Hoovers.com), describes Doe Run's parent company as, "Renco Group: a holding company for a diverse bunch of businesses including Doe Run, the world's #2 lead smelter, and coal miner Renco. Renco is owned by industrialist Ira Rennert, whose Long Island, New York, home is double the size of the White

FIGURE PC 1-1 Doe Run lead smelter near Herculaneum, Missouri.

FIGURE PC 1-2 The Doe Run smelter's smokestack, a landmark since the late 1800s, looming above Herculaneum. How close would you feel comfortable living to the smelter? © James A. Finley/AP Images.

Continued ▶

House and is said to include 29 bedrooms, 42 bathrooms, a 100-car garage, and an English pub."

The Doe Run Company operates another lead smelter in La Oroya, Peru that has been called an "ecological disaster" by local environmentalists, based on published data from the Peruvian Ministry of Environmental Health.

Counterpoint

Renco's website published the following statement: "The Doe Run Company is a privately held natural resource company focused on environmentally sound mineral production, recycling, and metals fabrication. Based in St. Louis, Doe Run is North America's largest integrated lead producer and second-largest total lead producer worldwide, employing more than 6,000 people. The company is committed to keeping its operations and communities clean and safe while producing essential raw materials—lead, zinc, copper, gold and silver—that are needed for everyday life. Doe Run has U.S. operations in Missouri, Washington, Arizona and Texas. Doe Run and its predecessors have a long, distinguished record of environmental improvement, beginning with initial processes to remove dust from the workplace as early as 1900. Since then, each successive generation of facilities has incorporated modern control technology to continue this environmental tradition."

The EPA in June 2002 approved a company plan, which includes the installation of an automatic truck wash to clean vehicles that may come into contact with lead at the smelter. In addition, Doe Run has been conducting daily street cleaning, and soil replacement along roadways and yards.

In 2002 The Doe Run Company reported that the Herculaneum, Missouri, smelter was in attainment with the Clean Air Act's National Ambient Air Quality Standard (NAAQS), that is, within levels considered harmful to public health, for the third quarter of 2002. In July 2002 the company completed a $12 million package of air pollution controls under an agreement with the State Air Conservation Commission in accordance with the federally approved State Implementation Plan (SIP) for Lead.

Results from seven air monitors in Herculaneum in 2003 ranged from 0.04 to 1.00 micrograms of lead per cubic meter of air. These values were below EPA and State Department of Natural Resources-specified air quality standards of 1.5 micrograms of lead per cubic meter of air (averaged over a three-month period).

On June 27, 2002, street signs installed by the Missouri Department of Natural Resources (DNR) were removed along lead ore haul roads in Herculaneum. The signs warned of high-lead levels on the streets and provided tips to reduce exposure. The site was set to close in 2013, but the environmental issues will likely remain.

Concept Check 13-1. Does the government have a legitimate responsibility to force private organizations to avoid polluting the environment? Do you think Doe Run would have taken such actions as it has, without government intervention?

Concept Check 13-2. Historically, is it fair to say that a part of the total cost of lead refined by Doe Run was not included in the price charged to its customers? Who or what has paid the rest? Cite your evidence.

■ MINERAL RESOURCES AND THE AMERICAN ECONOMY

The U.S. Geological Survey (USGS) keeps a database on over 50 "mineral resources" (we put these in quotes because not all the commodities listed are minerals) used in the United States. They are: aluminum, antimony, arsenic, asbestos, bauxite and alumina, beryllium, boron, cadmium, cement, cesium, chromium, clay, coal combustion products (fly ash, etc.), columbium, copper, diatomite, feldspar, fluorspar, gallium, gold, graphite, iodine, iron and steel, kyanite, lead, lime, lithium, magnesium, manganese, nitrogen, peat, perlite, platinum group minerals, pumice, rhenium, salt, sand and gravel, silicon, silver, stone, strontium, talc and pyrophyllite, tantalum, thallium, tin, vanadium, vermiculite, wollastonite, and zinc.

A detailed discussion of each of these is beyond the scope of this text. However, the United States Geological Survey offers detailed information on each of these 50 commodities on its website (http://minerals.usgs.gov/minerals/).

■ MINERAL DEPOSITS

A mineral is a naturally occurring inorganic crystalline solid with a set of physical and chemical properties (hardness, luster, etc.) and a definite chemical composition. There are more than 2,000 known minerals. The term "**mineral deposits**" refers to mineral assemblages that can be extracted and refined at a profit. Mineral deposits may be of igneous, metamorphic, or sedimentary origin. However, with rare exceptions, mineral deposits must have undergone *enrichment* by some natural process, before they are useful to humans. Iron for example makes up nearly 6% of the Earth's crust, but it must be concentrated up to about 50% to be economically viable. **TABLE 13-1** shows the abundance of the most abundant elements in the Earth's crust.

What Makes a Mineral Deposit?

The concentration at which deposits become economically valuable, and are then called **ores**, depends upon the market price of the material; the efficiency of processing to extract the valuable element(s) from the ore; any *subsidies* provided by government such as tax incentives, low-interest loans, or lax enforcement of environmental laws; shipping costs; currency fluctuations; and the costs of mitigating (that is, reducing) environmental impacts.

Mining activity has for centuries been harmful to the environment, but the scale of activity over the past century has reached levels unprecedented in human history. Even though cleaning mining-damaged landscapes, a process known as *remediation,* is now required in many countries, scientists continue to express concern about the environmental impact of natural resource extraction, including, but not limited to, gold, lead, mercury, and coal mining (see elsewhere in the text).

Concept Check 13-3. Why do you think governments provide subsidies to support mining?

TABLE 13-1	Common Elements Found in the Earth's Crust		
Oxygen (O)	46%	Sodium (Na)	2.1%
Silicon (Si)	28%	Potassium (K)	2.3%
Iron (Fe)	6%	Magnesium (Mg)	4%
Aluminum (Al)	8%	Others	<1%
Calcium (Ca)	2.4%		

Mineral Deposits and Plate Tectonics

FIGURE 13-1 shows the distribution of *porphyry copper deposits* worldwide.

Recall that subduction zones are sites of partial melting of overlying oceanic lithosphere, and/or overlying lithosphere. Dewatering of hydrous (water-containing) oceanic lithosphere within the subducting slab is a likely cause of this partial melting. Metals in the rock being melted are often preferentially dissolved in these **hydrothermal solutions**.

Geologists have known for decades that mid-ocean ridge basalts have high, but usually uneconomic, concentrations of metals such as silver, nickel, copper, and zinc. For example, metal concentrations are high along the Red Sea Rift Zone, a richly mineralized segment of the Mid-Ocean Ridge system.

When this ocean lithosphere is partially melted at subduction zones, the metals tend to become concentrated in the molten fraction (the melt), and thus may be emplaced in the shallow crust as **accessory minerals** in igneous intrusions (magma that has moved into spaces within the crust and solidified).

Types of Mineral Deposits

Mineral deposits are often divided into metals, and nonmetals. Metals have a metallic luster, are malleable and ductile (deforms under stress), and are good conductors of electricity. Metals are sometimes categorized as **base metals** (that is, metals "inferior" to the precious metals: examples are iron, copper, zinc, nickel), *precious* metals (gold, platinum group, silver), *radioactive* metals, and *rare earths,* such as niobium and iridium. Nonmetals include sand and gravel, feldspar, clays, and gemstones, to name a few.

Mineral deposits may be classified according to origin. Many valuable deposits are of igneous origin, but metamorphic ore deposits (e.g., kyanite, ruby) and sedimentary deposits (e.g., placer gold [gold fragments in streams weathered from its host rock], uranium, sand, and gravel) are also important (**FIGURE 13-2**).

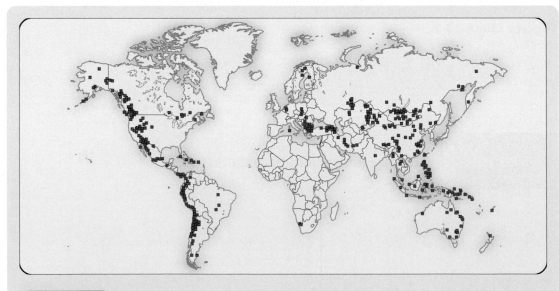

FIGURE 13-1   The distribution of younger porphyry copper deposits. Do you note any association with convergent plate boundaries? Data from: USGS.

FIGURE 13-2 Examples of mineral deposits. (A) Rubies in metamorphic rocks. © Jens Mayer/ShutterStock, Inc. (B) Sand and gravel. © Coprid/ShutterStock, Inc. and © bogdan ionescu/ShutterStock, Inc. (C) Layers of nickel-rich minerals in mafic igneous rocks. © Fletcher & Baylis/Science Source. (D) Placer gold deposits. © Accent Alaska.com/Alamy.

To give you an example of classification by origin, economic geologists commonly group *igneous* deposits as follows:

1. Segregations of valuable minerals (e.g., chromium, platinum group, nickel) in mafic intrusions (see classification of rocks, elsewhere in the text).
2. *Disseminated* deposits, in which the valuable minerals are scattered throughout an ore body (e.g., molybdenum, placer gold).
3. *Hydrothermal* deposits, formed from superheated steam solutions rising from magmas. These may be emplaced as veins in host or country rock. Examples include gold and zinc.
4. *Pegmatites,* which are small igneous intrusions consisting of very course crystals, some of which may be meters in diameter, composed of minerals containing **incompatible elements**; that is, those that don't usually form common igneous minerals. Examples are boron, lithium, uranium, and gemstones (FIGURE 13-3).

Kimberlites are intrusive bodies of enigmatic origin, some containing diamonds. The high-pressure mineral assemblage that makes up kimberlites indicates an origin within the mantle (FIGURE 13-4).

A pegmatite containing large crystals. © Jens Mayer/ShutterStock, Inc.

Rather than discuss deposits according to origin, for this general treatment we focus on the economic and environmental implications of mineral deposits. For specific deposits we will explain their geologic origin.

Human societies have made use of base metals for thousands of years. Attempts to determine when the Greek epic poems *The Iliad* and *The Odyssey* were written have been based on analyses of the poets' mention of iron and *bronze,* an alloy (a metal composed of two or more elements) of copper and tin, in weaponry.

During the past century and a half, manufacture of **steel** has resulted from *alloying* various elements with iron to provide the various characteristics exhibited by steel varieties. Stainless steel, for example, contains a thin surface layer of nickel and/or chromium. Cobalt is used to impart heat resistance to steel products, especially the engines of high-speed jet aircraft and rockets. The weapons of modern warfare have placed great demands on such relatively rare metals as cobalt, chromium and molybdenum.

Demand for metals for weaponry means that many nations, like the United States, are dependent on other nations for their entire supply of these commodities.

Concept Check 13-4. CheckPoint: Distinguish between metals and nonmetals.

Photo of a kimberlite body containing a large diamond. The mineral assemblage of kimberlites could only have formed in the Earth's mantle. © Joel Arem/Science Source.

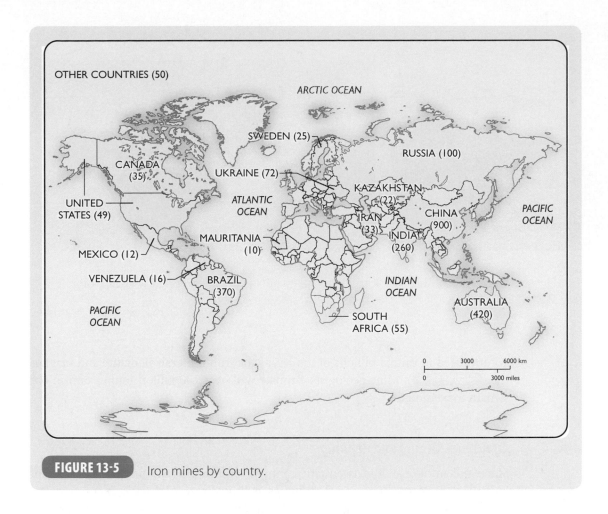

FIGURE 13-5 Iron mines by country.

Iron Deposits

Iron (Fe) is one of the planet's most abundant elements. In fact, many enormous iron mines exist, some containing ores with over 63% Fe by weight! At present more than 60 mines extract iron ore from around the world, but production is concentrated in Brazil, Australia, and South Africa (FIGURE 13-5).

Banded iron formations (BIFs) are some of Earth's most massive and important ore deposits, and they give clues to the evolution of the planet's atmosphere and life. For details about the origin of BIFs, see the website of the American Museum of Natural History (http://www.amnh.org).

BIFs are named from their physical appearance; they consist usually of alternating bands/layers of iron-rich and iron-poor minerals (FIGURE 13-6). After deposition, some of the BIFs were subjected to **secondary enrichment** processes, which helped concentrate the iron and converted it to forms easier to refine. These make up most of today's great iron deposits.

One of the great iron-ore producing regions in the world is Western Australia's Hamersley Range (FIGURE 13-7). Here, around 2.5 billion years ago, BIFs

FIGURE 13-6 A banded iron formation. © Ahmad A Atwah/ShutterStock, Inc.

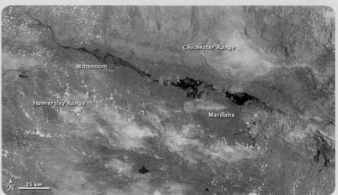

FIGURE 13-7 The Hamersley Range, West Australia. Note BIFs cropping out in right foreground. © iStockphoto/ Thinkstock and courtesy of LANCE/EOSDIS MODIS Rapid Response Team at NASA GSFC.

accumulated in untold billions of tonnes, as the iron minerals hematite and magnetite (**FIGURE 13-8**). Mine geologists estimate well over 100 billion tonnes of iron ore remain to be mined.

Aluminum Mining and Refining

Since the middle of the twentieth century, aluminum (Al) has become one of industrial civilization's essential resources, mainly due to Al's strength compared to its weight, but also for its resistance to corrosion. Refining aluminum from ore is an **energy-intensive process**. Once refined, however, the metal is extremely easy to remelt and reuse, enhancing its recyclability.

Bauxite, the main ore of aluminum, was first discovered in the French district of Les Baux in 1821, from which it took its name. Bauxite is found in extensive deposits in many countries, but mainly in the tropics and subtropics, including Australia, Surinam, Jamaica, and India. Bauxite is not really a mineral, but a leachate, typically formed where extreme weathering conditions exist, brought about by high annual temperatures and abundant precipitation. These conditions leach out all the easily soluble elements from surface rock and sediment, leaving behind only the most insoluble elements and compounds. Bauxite, along with iron oxides and hydroxides, is the most common substance produced by such extreme weathering (**FIGURE 13-9**).

Refining bauxite into ore is carried out as follows. First the bauxite must be crushed, and then processed into a white powder called *alumina* (Al_2O_3). During processing, four tonnes of bauxite ore yield two tonnes of alumina and two tonnes of waste. The

Hamersley

Turee creek Gp.

Hamersley Gp.

- [ ] Argillite
- [ ] Banded IF
- [ ] Basinal carbonates
- [ ] Quartz arenite

150 m

Fortescue Gp.

FIGURE 13-8 Stratigraphy of the Hamersley Range showing thick banded iron formation and associated sedimentary rocks.

two tonnes of alumina further reduce to one tonne of Al metal.[1]

The alumina is then dissolved in molten cryolite (Na_3AlF_6) to make the solution conduct electricity. This molten solution is placed in a large vat and a powerful electric current is passed through it. The current separates the Al from the solution, and molten aluminum metal sinks to the bottom of the vat, where it is drawn off. A byproduct of the reaction, hydrofluoric acid (HF), is extremely toxic, and may be released as a gas. A partnership with the U.S. Environmental Protection Agency (EPA) is assisting American refiners reduce HF by 50%. CO_2, a greenhouse gas, may also be emitted in the process.

Concept Check 13-5. CheckPoint: Summarize the steps involved in the transformation of bauxite into aluminum.

Concept Check 13-6. Why do you think the government doesn't simply require refiners to cease emitting HF?

FIGURE 13-9   A bauxite mine. Bauxite mining in Brazil. The red material is intensely weathered soil, with virtually all soluble material removed. Bauxite is one of the most insoluble products of such weathering, which usually takes place in the tropics. © Leojones/ShutterStock, Inc.

Even though bauxite is a rich ore, the aluminum metal is tightly locked into the structure of several minerals in the deposit and is not easily extracted. This is why refining Al is so *energy-intensive*.

A great deal of electricity is needed to refine aluminum metal, the form that is then made into foil, sheet metal, and cans (see later in the chapter). In fact, this very capital- and energy-intensive process uses 2% to 3% of the electricity produced in the United States every year! There is a significant potential for improving the efficiency of the process. The currently used, but inefficient, Hall-Héroult reduction process could be retrofitted with new technology, providing potential energy savings nationwide of 1,500 megawatts, the electricity produced by one very large power plant.

Refiners sometimes try to solve their energy supply problem by building their own hydroelectric dams. In Canada, aluminum producer Alcan owns hydroelectric dams with a total installed capacity of more than 3,500 megawatts, enough to supply a city bigger than San Francisco. Alcan smelters in Scotland and Brazil operate their own hydropower dams, but must purchase some additional power, while the company's smelters in Switzerland and Norway buy electricity from others.

Value-added Question 13-1.
How much waste is produced per tonne of Al refined?

Porphyry Copper Deposits

A good example of a mineral deposit whose origin is related to plate tectonics processes, and the geology of surrounding rocks, is the **porphyry copper deposit**. Porphyry coppers are the source of at least half the world's copper. A porphyry copper deposit is named from the porphyritic igneous intrusive (see classification of rocks elsewhere in this text) found at the deposit's center. A *porphyry* is an igneous rock which contains large crystals set in a finer, but still visible, groundmass (FIGURE 13-10). Porphyry copper deposits usually form with, and below, composite volcanoes (one made of composites of lava and ash deposits) like Mount Rainier. However, because most of the deposits were emplaced million of years ago, the surface volcanoes have been eroded away. The deposits have thus been exposed by erosion, and enriched by percolating ground water, making them easy to mine.

FIGURE 13-10 A porphyritic igneous rock. © Siim Sepp/ShutterStock, Inc.

Recall that economic mineral deposits must be *enriched* above the composition of the valuable element mined by natural processes. In the case of porphyry coppers, percolation of rainwater through the rocks dissolves copper-bearing minerals near the surface and redeposits copper belowground. In fact, it is the host, or surrounding, rock that contains the valuable mineral concentrations. The porphyritic body is only the source of the elements that are concentrated by the percolating rainwater. Each of these zones may contain a specific mineral assemblage.

Most of the valuable ore minerals are copper oxides and hydroxides, but silver, molybdenum, and even gold minerals also may be found, as *accessory* minerals. These accessory minerals can considerably enhance the economic value of a deposit. However, the economic value of a porphyry copper deposit is most strongly influenced by the country rock surrounding the intrusion. If the mineralogy of the host rock is not suitable, no economically viable resource will form.

Precious and Radioactive Metals

The precious metals include gold, silver, and platinum group elements (PGE). We will include two examples: gold, and the platinum group elements. We include the PGEs because of their strategic importance, their importance as catalysts in catalytic converters, and their growing demand as jewelry. Finally, we discuss deposits of radioactive metals.

For centuries if not millennia, humans have lusted after precious metals. Gold was the force behind the invasion of the New World by Spanish conquistadores in the sixteenth century, and gold and silver looted from Native Americans financed Spain's wars during that time and for decades after. Precious metals stolen from Spanish galleons by privateers chartered by Queen Elizabeth I made a sizable addition to the English treasury as well. So much gold and silver was brought back from the Americas that it set off an extraordinary period of inflation in Spain, and may have contributed to the collapse of Spain as a great power, a classic example of too much money chasing too few goods.

Gold flakes weathered out of gold-bearing veins and rocks are so resistant to weathering, and so much heavier than other minerals, that they typically accumulate in stream sand and gravels, where they may be collected by swishing a pan full of water and sediment around, gradually allowing the lighter material to spill over the lip. These are called **placer deposits**. Unfortunately for miners, pyrite (FeS_2), a worthless mineral that superficially resembles gold, is usually found in the same sediments, and many a tenderfoot miner has been taken in by pyrite, (hence the term "fool's gold"; FIGURE 13-11).

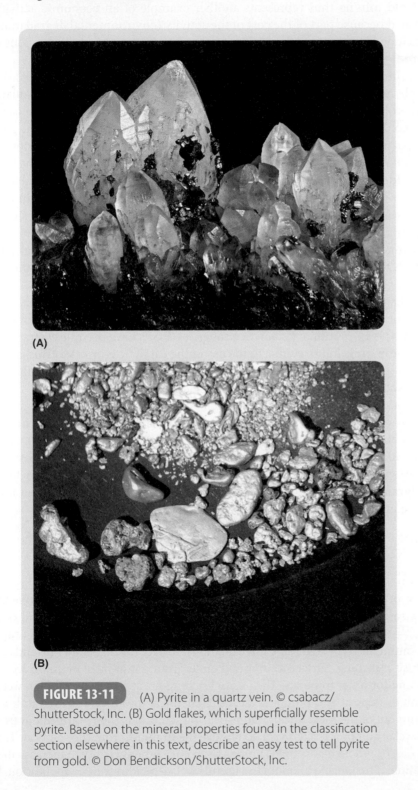

(A)

(B)

FIGURE 13-11 (A) Pyrite in a quartz vein. © csabacz/ ShutterStock, Inc. (B) Gold flakes, which superficially resemble pyrite. Based on the mineral properties found in the classification section elsewhere in this text, describe an easy test to tell pyrite from gold. © Don Bendickson/ShutterStock, Inc.

Beginning in the second half of the nineteenth century, industrial-scale gold mining used high-pressure hoses to pry gold-bearing veins free from igneous rocks in the Sierra Nevada of California. This activity unalterably changed river dynamics for rivers throughout the northern Sierra, by depositing millions of tons of sediment in rivers over the course of a few decades.

Today, the cost of remediation is left to the State of California and federal agencies. Gold mining thus represents another example of an economic activity that flourished because part of the cost of the activity (environmental degradation) was not included in the price. Gold mining has a similar impact today, in regions where the **amalgamation method** of gold recovery is practiced.

Rocks containing gold (usually also with silver) are crushed to recover the valuable metals. Because metals are present in host rocks in very small quantities, millions of tons of host rock must be processed to yield economically viable quantities of gold and silver. Amalgamation is the alloying and collection of fine gold and silver particles with droplets or coatings of mercury. The mercury-precious metal *amalgam* is then collected and heated to vaporization, to separate out the precious metals. Ideally, the mercury should be distilled for reuse, but considerable loss of mercury (and precious metals) is inevitable in the many steps involved in processing the ore. Vaporization of mercury (Hg) during amalgamation is a significant source of mercury pollution in Brazil.

Spanish colonizers named the *platinum group elements (PGEs)* metal "platina" (little silver), and thought it an impurity when they encountered it in their silver mines in Colombia.

PGEs' wear and tarnish resistance make them ideal for use in jewelry. Further useful characteristics include resistance to chemical reactions, high-temperature stability, and extraordinary catalytic properties. PGEs have become essential metals to the automobile industry, due to the use of the catalytic converter. This is a device attached to the exhaust system, which converts some pollutants to less harmful or benign substances.

Uranium is a silvery metal with toxicity comparable to lead. It is found in three naturally occurring *isotopes* (atoms of the same element with the same number of protons but a different number of neutrons): U^{238} (~95%), U^{235} (~0.3%), and U^{233}. U^{235} is capable of releasing energy by *fission,* a nuclear reaction in which an atom's nucleus splits. Uranium is widespread as a trace element; for example, there are 3 milligrams of uranium per tonne of seawater, approximately 4 grams per tonne of granite, and up to 400 grams per tonne of coal (1 g/tonne = 1 ppm).

Concept Check 13-7. What happens to the uranium when coal is burned to produce electricity?

Ores of uranium typically contain about 3 kilograms of uranium per tonne of host rock. Uranium is usually found in nature in two minerals: *pitchblende* and *carnotite* (FIGURE 13-12). Canada and Australia are the top two global producers of uranium.

Uranium is recovered using three methods: surface mining, underground mining, and *in-situ* leaching. Surface mining and underground mining of uranium are similar to mining for other minerals. The *in-situ* leaching process involves drilling wells into a uranium deposit and pumping a caustic chemical into a central well to leach the uranium out of the rock. The uranium-bearing solution is then pumped out from the other wells, and processed on site to remove the uranium from the solution. The uranium ore, or *yellowcake,* is then shipped to an enrichment facility to increase the concentration of fissionable uranium to a useful level, depending on the application. Waste

(A)

(B)

from mining and milling uranium ore is placed in *tailings impoundments* (**tailings** are the residues left over after the desired metals are removed; **FIGURE 13-13**).

Mining uranium doesn't produce any "new" radioactivity. Much of the natural radioactivity associated with uranium ore bodies is from the *intermediate decay* (breakdown) *products* of the U^{238} and U^{235} series, typically radon (**FIGURE 13-14**). Thus, uranium must be mined carefully, and mine tailings must be handled with extreme caution. After mining, tailings should be permanently covered with enough clay and soil to reduce both **gamma radiation** levels and **radon** emission rates, to levels near those naturally occurring in the region. The site may then be covered with vegetation. Erosion can remove the cover over the thousands of years the site remains abnormally radioactive, however. Essentially permanent monitoring

is essential to ensure the tailings do not pose an unacceptable risk to the public, representing another "deferred cost" to the use of uranium to generate electricity.

Concept Check 13-8. Do you agree that essentially permanent monitoring is necessary? Why or why not? Who should have responsibility for such monitoring? Who should pay? Assess the feasibility of monitoring uranium tailings for thousands of years.

The most significant threats posed by uranium mining are radon exposure to mine workers, and an increased cancer risk to the general public. According to a report by the British Columbia (Canada) Medical Association (BCMA),

- Radon exposure in underground mines was as much as 100 to 1,000 times higher than background radon.
- Radon is 7.8 times heavier than air; thus radon levels in open pit mines can be 2000 to 10,000 times higher than background radon levels for medium to high-grade ore bodies.
- Workers in uranium mills can receive up to 1000 times the gamma radiation compared to background gamma radiation levels over their working lives.

The BCMA further concluded that radiation from uranium tailings in Canada could result in 200 to 300 additional cases of lung cancer per 10,000 persons over a lifetime; in their words, "tantamount to allowing an industrially induced and publicly sanctioned epidemic of cancer."

As you can see, the issues in uranium mining, milling, and waste disposal are contentious and the subject of considerable public debate.

Finally, artillery shells in the United States are routinely made from depleted uranium; that is, uranium with almost all of the fissionable isotopes removed, because it is extremely dense (more dense than lead!). Some scientists, especially

Europeans, question the environmental impact of this practice. Depleted uranium has an environmental impact, as we mentioned previously, similar to that of lead.

Deposits of Nonmetals

Among the most important and valuable nonmetals are: sand and gravel deposits, phosphates, and clays; but perhaps the most alluring examples of nonmetals are the gemstones.

The USGS defines a *gemstone* as any organic or inorganic mineral used for personal adornment, display, or "objet d'art" because the mineral possesses beauty, rarity, and durability. In a 2004 survey, six American states accounted for more than 75% of gemstone production: Tennessee, Oregon, Kentucky, Arkansas, Arizona, and Alabama.

Due to high demand, most gemstones consumed in the United States are imported. The U.S. market for natural, uncut, colored gemstones exceeds $400 million annually. This statistic is dwarfed by the domestic market for unset gem diamonds (as opposed to industrial diamonds) at about $5 billion, the largest market in the world.

Foreign countries with major gemstone deposits other than diamond are shown in **TABLE 13-2** .

Diamonds are the public's most prized gemstone. In 2010, world diamond production was 117 million carats with an estimated value of $72 billion (1 cubic centimeter of diamond weighs 16 carats). Although diamonds are mined in many countries, production is concentrated in parts of Africa, Siberia, Australia, Venezuela, and Brazil (**FIGURE 13-15**).

The marketing of uncut diamonds is controlled by a private company, De Beers Centenary AG, with headquarters in Switzerland. De Beers carefully manages demand by extensive advertising, that on the one hand recommends diamonds be kept "forever," by linking them inextricably with romantic love, and on the other implies that diamonds gain value over time and so are an excellent investment.

The value of diamonds, other than that maintained by advertising, lies in the extreme hardness of the stone, its relative scarcity, and its optical ability to disperse light in spectra, called "fire."

Kimberlites are rocks that, by their mineral association and structure, are known to have formed deep in the Earth's mantle, and to have been explosively emplaced into the lower crust associated with volcanic activity. There, they are slowly exposed by erosion in very old regions called *shields,* over many millions of years. The geology of kimberlites is discussed elsewhere in the text.

Some South African mines can make a profit mining ores with 25 *carats* of diamond (1 cubic centimeter of diamond weighs 16 carats) per 100 cubic meters of rock, or about *2 grams of diamonds per 100 tonnes*. Diamond has a specific gravity of 3.5 grams per cubic centimeter (**FIGURE 13-16**).

The environmental and social impact of diamond mining is significant. We discuss this in detail later in the chapter.

Uranium 238 (U238) Radioactive Decay

Nuclide	Half-life
uranium-238	4.47 billion years
thorium-234	24.1 days
protactinium-234	1.17 minutes
uranium-234	245000 years
thorium-230	8000 years
radium-226	1600 years
radon-222	3.823 days
polonium-218	3.05 minutes
lead-214	26.8 minutes
bismuth-214	19.7 minutes
polonium-214	0.000164 seconds
lead-210	22.3 years
bismuth-210	5.01 days
polonium-210	138.4 days
lead-206	stable

FIGURE 13-14 The decay process of uranium is a complex process, yielding many intermediate by-products, including radon and radium.

TABLE 13-2	Gemstone Producing Countries
Country	**Gems Produced**
Afghanistan	beryl, ruby, tourmaline
Australia	beryl, opal, sapphire
Brazil	agate, amethyst, beryl, ruby, sapphire, topaz, tourmaline
Burma	beryl, jade, ruby, sapphire, topaz
Colombia	beryl, emerald, sapphire
Kenya	beryl, garnet, sapphire
Madagascar	beryl, rose quartz, sapphire, tourmaline
Mexico	agate, opal, topaz
Sri Lanka and India	beryl, ruby, sapphire, topaz
Tanzania	garnet, ruby, sapphire, tanzanite, tourmaline
Zambia	amethyst, beryl

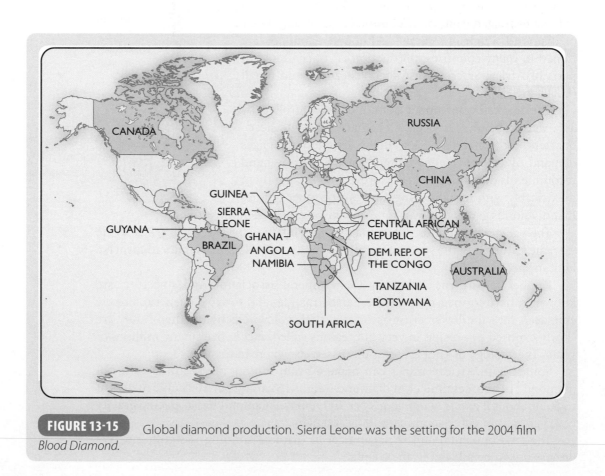

FIGURE 13-15 Global diamond production. Sierra Leone was the setting for the 2004 film *Blood Diamond*.

Aggregate is the only mineral resource produced in all 50 states. While aggregates are not minerals as strictly defined, the USGS considers these materials "mineral resources" so we will include them here. Around 90,000 Americans work in the aggregate industry. Aggregate is mainly used in construction, to make cement, road base, and *fill* of various types. Eighty percent of concrete roads, and 90% of asphalt surfaces, are aggregate. Aggregate mining in the United States totaled about 2.8 billion tonnes in 2007, about two thirds of the total nonfuel mineral resource production in the United States.

Other uses of aggregate include paint, paper, and glass manufacture. Here are some specific examples:

- Millions of tonnes of limestone are used each year in SO_2 scrubbers, at coal-fired electric power plants.
- *Industrial grade sand* is a high-value commodity. Uses include glass making, abrasives, and use as molds in foundries.
- *Construction sand and gravel* is a high-volume, low-value commodity, which means costs of transport are a significant part of the price. In fact, the cost of moving construction sand and gravel from the plant to the market often exceeds the sales price of the product at the plant. This is why there are tens of thousands of sites around the country where sand and gravel is mined. Sand is also used in beach *renourishment*.

FIGURE 13-16 Giant ore trucks loading ore. In South African mines, such a truck would contain only half of a cubic centimeter of diamonds! Only about one third of those diamonds would be of gem quality. © kaband/ ShutterStock, Inc.

Value-added Question 13-2.
Calculate the per capita U.S. consumption of aggregate in 2007. Use a population of 302,000,000.

Production of construction sand and gravel in the United States increased from 1950 (320 million tonnes) to 1990 (826 million tonnes), in 2008 was 1.04 billion tonnes, but fell to 790 million tonnes in 2011 due to the decline in economic activity.

The single greatest impact on growth in production in the United States was construction of the Interstate Highway system between the early 1960s and the late 1970s. Since then, sand and gravel production has fluctuated with construction activity.

While the price of construction sand and gravel has fluctuated little over the past three decades, analysts expect the price to rise in the future due to lower deposit quality, more stringent environmental regulations, and land use restrictions.

Trucks account for 80% to 90% of shipments. Because most trucks are diesel-powered, price of sand and gravel could be impacted by new regulations that will sharply reduce diesel emissions, which have been found to contain at least 40 carcinogenic or mutagenic compounds, as well as rising global prices for diesel fuel.

Sand and gravel deposits can result from fluvial (river), lacustrine (lake), glacial, marine, or residual (weathering) processes. In Virginia for example, sand and gravel deposits are mainly fluvial, while in North Dakota the deposits are mainly of glacial origin.

After 1980, about 70% of all landfill sites in the United States closed rather than adhere to the stricter requirements set by the Resource Conservation and Recovery Act (RCRA). They were mainly replaced by large regional landfills, often requiring long transport distances for waste. As a result, recycling of aggregate began to be supported by some localities. For example, in 1995 Los Angeles required recycled aggregate to be used in all road projects unless specifications required virgin product. King County, Washington reported that cost of recycled aggregate matched that for virgin aggregate.

Clays and clay minerals have been mined since the Stone Age. They are among the most important minerals used today by manufacturing and environmental industries. The United States produced 26 million tonnes in 2011.

The attributes common to all clay minerals relate to their chemical composition, layered structure, and size.

Water molecules are strongly attracted to clay mineral surfaces. When a little clay is added to water, a *slurry* forms, because the clay distributes itself evenly throughout the water. This property of clay is used by the paint industry to disperse pigment (color) evenly throughout paint. Without clay, it would be difficult to evenly mix the paint base with pigment. Mixing clay with a little water results in a mud that can be shaped and dried to form a relatively rigid solid. This property allows potters and the ceramics industry to produce dinnerware, and many other items.

Environmental industries use clays to produce impervious liners for containment of waste in landfills, and at industrial sites. Waste disposal is discussed elsewhere in the text.

Swelling clay expands or contracts in response to changes in environmental factors (wet and dry conditions, temperature). *Hydration and dehydration* can vary the thickness of a single clay particle by almost 100%. Structures built on soils containing swelling clays may be damaged by seasonal swelling of the clays in the soil. Swelling soils as geohazards is discussed elsewhere in the text.

Most clays have the ability to soak up ions from a solution and release the ions later under different conditions. The ability to exchange ions is due to the *electrically charged surface* of clay minerals. Ions can be attracted to the surface of a clay particle or taken up within the structure of these minerals. Many types of ions in solution can adhere to clay surfaces, or be taken internally, including organic molecules like pesticides. This fact allows clays to be important tools in remediating polluted sites. This property is examined further in discussions of groundwater elsewhere in the text.

Important clay resources include bentonite, kaolin, ball clay, fuller's earth, and common clay. Major domestic uses for these clays are shown in **TABLE 13-3**.

TABLE 13-3	Uses of Clays
Type of Clay	**Uses**
bentonite	pet waste absorbent, foundry sand bond, drilling mud, iron ore pelletizing
common clay	brick, cement, lightweight aggregate
ball clay	floor and wall tile, sanitaryware, and many other uses
fire clay	mainly as refractories
fuller's earth	as an absorbent
kaolin	mainly in high-quality paper manufacture, and in refractories.

Data from: *Minerals Yearbook: Volume 1:* Metals and Minerals: 2008: USGS.

Many metal ores are *sulfides*. Examples include sphalerite (ZnS, zinc sulfide), chalcopyrite ($CaFeS_2$), galena (PbS), and argentite (AgS). Refining of these ores is called **smelting**, and refining plants are called smelters. Smelting of these ores involves "roasting" powdered ore minerals in air, to produce metal *oxides* or elements. By-products of refining can be serious sources of toxic air pollution, but modern refining methods can reduce emissions drastically, if laws and regulations are properly enforced.

Because many ores are sulfides, sulfur dioxide (SO_2) is a major gaseous emission of smelters. Untreated emission of sulfur dioxide can produce strongly acid precipitation as well as acid aerosols, which can result in severe local and regional air pollution. In addition, various metals are present in trace amounts in the ores. When the ore is roasted, these metals can be volatilized. Examples are cadmium (often found in copper ores), arsenic (a widespread trace element in many ores), and lead (from lead smelting and copper/silver processing). A variety of other trace elements may be emitted as well, depending upon the ore being refined. Mercury is a widespread air contaminant produced by coal burning, for example.

Sulfur Dioxide Emissions

Consider the following oxidation reaction. Note that eight molecules of SO_2 are produced for each four original FeS_2 units.

$$4\ FeS_2 + 11\ O_2 \rightarrow 2\ Fe_2O_3 + 8\ SO_2\left(gas\right)$$

Sulfur dioxide (SO_2) has two kinds of adverse impact: health impacts on humans, and environmental impacts on ecosystems. Air pollution impacts are discussed elsewhere in the text.

We illustrate modern methods to reduce emissions by describing operations at the Mount Isa mine complex in Australia (**FIGURE 13-17**). Mount Isa, operated by Xstrata plc, is the largest underground mine in Australia and one of the largest in the world. It began operating in 1931, and ranks among the world's top three producers of lead. In addition, it is one of the five largest global producers of silver, it is the 10th largest source of zinc, and is one of the world's top twenty copper producers.

The lead, zinc, and silver mineralization occurred in sedimentary rocks around 1600 my ago, when hydrothermal fluids circulated through sedimentary rocks. Hundreds of millions of years later, copper mineralization occurred.

Emission control gradually has improved over the past several decades as a response to government mandate. Sulfur dioxide is approximately 80% captured by scrubbers and then converted to sulfuric acid for use in fertilizer production. The company reports that a new scrubber will remove 98% of SO_2 by 2013. The reaction is:

$$2\ SO_2 + O_2 + 2\ H_2O \rightarrow 2\ H_2SO_4\left(\text{sulfuric acid}\right)$$

Ground level SO_2 in 2001 averaged less than 11 micrograms (μg)/m^3 at Mt Isa, down from an average of 17 in 1997. The company further reduced SOx by more than 70% by 2010.

FIGURE 13-17 (A) Location of the Mt. Isa mine, Queensland, Australia. (B) View of the Mt. Isa complex. © Bill Bachman/Alamy.

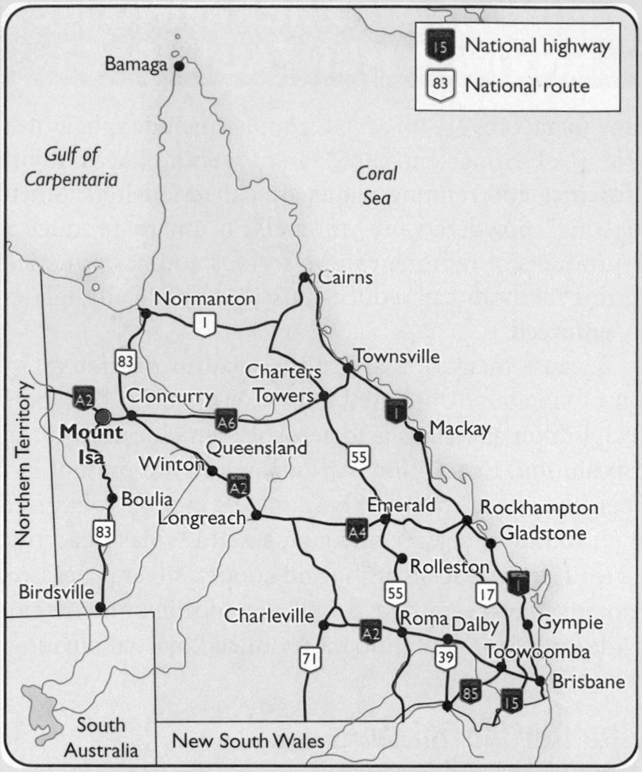

(A)

(B)

Lake Coeur d'Alene, Idaho Watershed

A legacy of metal mining is Lake Coeur d'Alene, which rests in one of the most picturesque watersheds in the Northwest United States (**FIGURE 13-18A**). Between 1880 and 1910, a combination of railroad construction, clear-cut forest practices, and metal mining severely degraded the lake and its watershed. A century later, government agencies and environmental groups continued to try to undo the effects of these activities. By the 1970s, an estimated 72 million tons of mining and processing wastes had been dumped into the 21 square mile natural lake and its watershed.

FIGURE 13-18 (A) Lake Coeur d'Alene in Idaho. © WellyWelly/ShutterStock, Inc. (B) Map showing the Bunker Hill Superfund site.

In addition to nutrients nitrogen (N) and phosphorus (P), toxic levels of metals pollute the lake. Bunker Hill, Idaho was the site of one of the nation's richest Pb-Zn deposits. It is now a Superfund site (**FIGURE 13-18B**). Rivers draining the Bunker Hill Superfund site dissolve appreciable quantities of zinc, which inhibits growth of some aquatic plants. Mining and smelting of ores have left the soils surrounding the smelters highly contaminated with lead, zinc, arsenic, and mercury. The high S content of the ores has also acidified the contaminated soils.

Between 1975 and 1995, levels of nutrients in the lake fell 50%, due in part to (1) elimination of direct discharges of mining and smelting wastes into one of the lake's tributaries, (2) diversion of untreated sewage to municipal wastewater-treatment plants, and (3) application of **best-management-practices (BMPs)** by timber and agricultural industries.

A remediation project was started in 1997 to control slope erosion, neutralize soil acidity, and stabilize metals in soils. To that end, some steep slopes were covered with municipal **biosolids** to enhance soil for vegetation and soil microbes. The slopes were also treated with wood ash and sawmill residue, to help reduce acidity and add nutrients, especially potassium and phosphorus (**FIGURE 13-19**).

Erosion control is critically important in the watershed, because runoff during episodes of intense rainfall can dump immense quantities of toxics into streams: pollution that will then be carried into the lake. For example, approximately

FIGURE 13-19   (A) Before and (B) after photos showing revegetation at Bunker Hill, Idaho. Parts A and B courtesy of Cami Grandinetti/EPA.

720 tonnes of lead and 180 tonnes of zinc entered Coeur d'Alene Lake on February 10, 1996, one day after the peak discharge of the second largest flood ever recorded. This represented nearly three times the amount of Pb transported during the entire water year 1991, and more than 12 times the amount transported during water year 1992. Flood discharges also elevate trace element concentrations by as much as several orders of magnitude. Cadmium, copper, lead, zinc, and mercury concentrations during the February 1996 floods exceeded criteria established for the protection of aquatic life. Concentrations of lead exceeded drinking water standards.

The Carson River, Nevada Watershed

Paul J. Lechler, a chief scientist with the Nevada Bureau of Mines and Geology, has researched mercury contamination in the Carson River, Nevada from *amalgamation* milling of gold-silver ores within the watershed. Mining of the Comstock Lode in Virginia City, Nevada mainly took place between 1860 and 1895 (**FIGURE 13-20**). The Carson River is a designated Superfund site by the Environmental Protection Agency because of high levels of mercury in river channel and floodplain sediments.

 Approximately 15 million pounds of mercury were dumped into the Carson River drainage system during the milling of ore. The Comstock Lode yielded 8 million

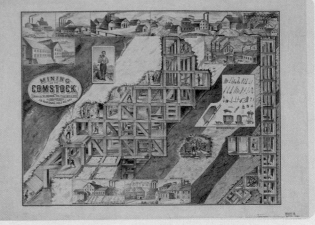

(A)   (B)

FIGURE 13-20   (A) The Carson River watershed. © LatitudeStock/Alamy. (B) The Comstock Lode deposit. Courtesy of Library of Congress, Prints & Photographs Division [reproduction number [LC-DIG-pga-01999].

ounces of gold and 192 million ounces of silver. Lechler estimates that 3 million ounces of gold and 64 million ounces of silver were lost with the mercury. These amounts of mercury, gold, and silver now reside in mill tailings, channel sediments, and flood plain deposits of the Carson River, largely along the 70-mile (120 km) stretch between Carson City and Fallon, Nevada.

Up to 200 mills may have processed Comstock ore between 1860 and 1895. According to the USGS, high mercury levels in fish, amphibians, and invertebrates downstream of the hydraulic mines are a consequence of amalgamation.

Although most milling was done in Six Mile Canyon and along the banks of the Carson River, some Comstock ore was shipped to mills in Washoe Valley. Studies detected some mercury contamination in Washoe Lake (now dry), and Little Washoe Lake sediments (**FIGURE 13-21**). These lakes drain into the Truckee River. Concentrations of 200 to 300 ppb total mercury are common in the upper 30 to 40 cm of lake sediments in the middle of Washoe Lake. Closer to the edges of the playa, 100 to 150 ppb total mercury is common. At the north end of Little Washoe Lake, muds contain 2 to 14 ppm mercury.

The use of NaCl (salt) in the mining process led to the formation of mercuric chloride ($HgCl_2$). This is a very soluble form of mercury and may have contributed to the widespread nature of mercury contamination. Today in the region, mercury vapor continues to seep out of the sediment. An evaluation of the potential toxicity of mercury can be found elsewhere in this text.

Most of the mercury used in amalgamation mining in the Sierra Nevada was derived from mines in the California Coast Ranges. Study (**FIGURE 13-22**), which shows the impact of mercury mining in this area.

Value-added Question 13-3.
At $1,400 U.S. per ounce (or determine the current price from a newspaper or website), calculate the value of the gold "pollution" in the tailings and the Carson River.

FIGURE 13-21   A view of Washoe Lake. Courtesy of State of Nevada; Department of Conservation & Natural Resources.

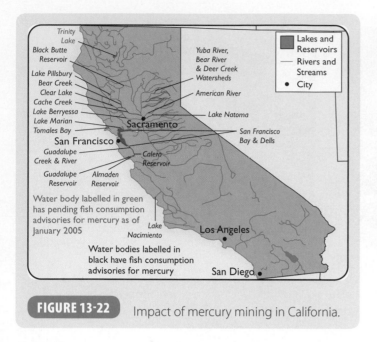

FIGURE 13-22 Impact of mercury mining in California.

Social and Environmental Impact of Diamond Mining and Trade

We illustrate an example of the environmental and social impacts of the diamond trade using the Democratic Republic of the Congo (DROC; **FIGURE 13-23**).

Fighting between DROC rebel forces and government troops has caused large-scale internal displacement of the Congolese people. The diamond mining industry was used to finance the DROC's civil war. Most of the diamond mining in the DROC takes place in the eastern part of the country, which is controlled by anti-government factions that historically were supported by Uganda and Rwanda. The United Nations Human Rights Field Operation in the DROC, which began in 1996, was forced to evacuate in 1998 due to a renewed outbreak of the conflict.

Diamonds in Angola

South of the DROC, Angola has diamond reserves estimated at 180 million carats, principally in the provinces of Lunda Norte, and Lunda Sul in the central and northeastern parts of the country (**FIGURE 13-24**). Diamonds are produced from kimberlites and stream deposits from the weathering of kimberlites. Most of the diamond-rich kimberlite ore bodies are located along a northeast–southwest trend that extends into the neighboring DROC. Diamond production in Angola generated over $1.2 billion in 2008 alone, although exact numbers are uncertain due to the amount of illegal diamond mining and smuggling. Angola produced an estimated 5.3 to 6.0 million carats in 2003 officially.

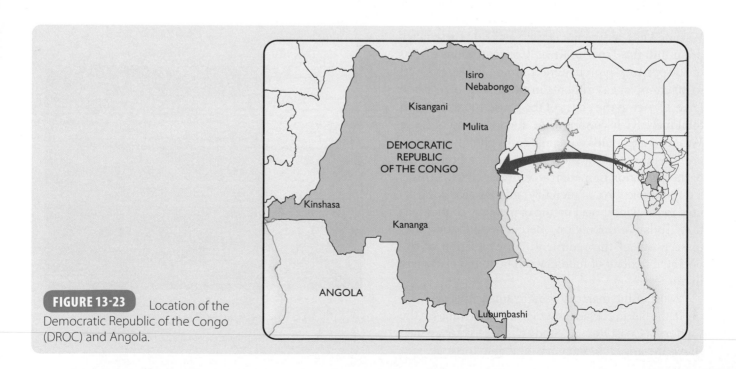

FIGURE 13-23 Location of the Democratic Republic of the Congo (DROC) and Angola.

FIGURE 13-24 (A) Location of diamond reserves in Angola. (B) Satellite photo of Catoca Diamond Mine, Angola. Courtesy of NASA/GSFC/METI/ERSDAC/JAROS, and the U.S./Japan ASTER Science Team. (C) Diamond mine in Angola. Courtesy of Japie Duminy.

Individual miners account for more than half of Angola's production. The Catoca mine, Angola's only commercial kimberlite mine, began production in early 1997 and produced more than 9.5 million ct in 2007. Geologists estimate reserves at 135 million carats.

Most of the diamond-rich region is located within territories traditionally controlled by rebel Unita forces. Central government and rebel groups frequently use mineral and nonmineral commodities to purchase military equipment.

Diamonds also have been used to finance civil war in Sierra Leone, where as many as 75,000 civilians were murdered during the 1990s, as described in the film, *Blood Diamond*.

If order were to be established in the Congo and throughout other mineral-rich countries of Africa, that unfortunate region could benefit enormously from its diamond resources as well as many other valuable commodities.

Concept Check 13-9. Would the impact of the mining of a commodity influence your choice to buy it? Why or why not?

■ RECYCLING OF MINERAL RESOURCES

Recycling of mineral resources varies greatly worldwide, as does the reliability of reporting. In the United States, statistics on recycling are regularly published by the USGS and by metal industry trade associations.

Amount of recycling generally depends on the value of the commodity, ease of extraction of the valuable commodity from the used product, national laws and regulations, access to and price of virgin material, and subsidies. In the United States, some examples of recent recycling rates range from over 60% in the case of lead (mainly from lead-acid batteries), to around 50% for iron and steel, to over 19.4 million tonnes of Al recovered and recycled in 2009.

CHAPTER SUMMARY

1. The U.S. Geological Survey (USGS) keeps a database on over 50 mineral resources used in the United States.
2. "Mineral deposits" refers to mineral assemblages that can be extracted and refined at a profit.
3. Mineral deposits may be of igneous, metamorphic, or sedimentary origin.
4. With rare exceptions, mineral deposits must have undergone enrichment by some natural process, before they are useful to humans.
5. The concentration at which deposits become economically valuable, and are then called ores, depends upon the market price of the material; the efficiency of processing to extract the valuable element(s) from the ore; any subsidies provided by government such as tax incentives, low-interest loans, or lax enforcement of environmental laws; shipping costs; currency fluctuations; and the costs of mitigating (that is, reducing) environmental impacts.
6. Mining activity has for centuries been harmful to the environment, but the scale of activity over the past century has reached levels unprecedented in human history.
7. Remediation is now required in many countries because of the environmental impact of natural resource extraction, including, but not limited to, gold, lead, mercury, and coal mining.
8. Mineral deposits are often divided into metals, and nonmetals.
9. Mineral deposits may of igneous origin, but metamorphic ore deposits and sedimentary deposits are also important.
10. The manufacture of steel has resulted from alloying various elements with iron to provide the various characteristics exhibited by steel varieties.

11. Banded Iron Formations (BIFs) are some of Earth's most massive and important ore deposits, and they give clues to the evolution of the planet's atmosphere and life.

12. Aluminum (Al) is one of industrial civilization's essential resources, mainly due to its strength compared to its weight, as well as its resistance to corrosion.

13. Refining aluminum from ore is an energy-intensive process. However, once refined, the metal is extremely easy to remelt and reuse, enhancing its recyclability.

14. A great deal of electricity is needed to refine aluminum metal, the form that is then made into foil, sheet metal, and cans.

15. A mineral deposit whose origin is related to plate tectonics processes, and the geology of surrounding rocks, is the porphyry copper deposit.

16. Most of the valuable ore minerals are copper oxides and hydroxides, but silver, molybdenum, and even gold minerals may also be found, as accessory minerals.

17. The precious metals include gold, silver, and platinum group elements (PGE).

18. Gold flakes weathered out of gold-bearing veins and rocks are so resistant to weathering, and so much heavier than other minerals, that they typically accumulate in stream sand and gravels.

19. Uranium is a silvery metal with toxicity comparable to lead. It is found in three naturally occurring isotopes.

20. Uranium is recovered using three methods: surface mining, underground mining, and in-situ leaching.

21. The most significant threats posed by uranium mining are radon exposure to mine workers, and an increased cancer risk to the general public.

22. Among the most important and valuable nonmetals are: sand and gravel deposits, phosphates, clays, and gemstones.

23. The environmental and social impact of diamond mining is significant.

24. Clays and clay minerals have been mined since the Stone Age and are among the most important minerals used by manufacturing and environmental industries.

25. Important clay resources include bentonite, kaolin, ball clay, fuller's earth, and common clay.

26. Refining of metal ores is called smelting, and refining plants are called smelters. Since many ores are sulfides, sulfur dioxide (SO_2) is a major gaseous emission of smelters.

27. In developed countries, most sulfur dioxide emissions are captured by scrubbers, and converted to sulfuric acid for use in fertilizer production.

28. In the United States, recycling rates range from over 60% in the case of lead (mainly from lead-acid batteries), to around 50% for iron and steel, to over 19.4 million tonnes of Al recovered and recycled in 2009.

KEY TERMS

accessory mineral(s)

aggregate

amalgamation method

base metals

banded iron formation (BIF)

best management practices (BMPs)

biosolids

clay and clay minerals

energy-intensive process

gamma radiation

hydrothermal solutions

incompatible elements

kimberlite

mineral deposit

ore

placer deposits

porphyry copper deposit

radon

secondary enrichment

smelting

steel

swelling clays

tailings

Toxic Release Inventory (TRI)

REVIEW QUESTIONS

1. Mineral deposits may be of what origin?
   a. igneous only
   b. metamorphic only
   c. sedimentary only
   d. all of the above
   e. none of the above

2. All of the following determine the concentration at which a mineral deposit becomes economically viable except
   a. the market price of the material.
   b. whether it is igneous, sedimentary, or metamorphic.
   c. the efficiency of processing to extract the valuable element(s) from the ore.
   d. lax enforcement of environmental laws.
   e. none of the above.

3. *Remediation* refers to
   a. subsidies provided by government such as tax incentives and low-interest loans.
   b. dewatering of hydrous (water-containing) oceanic lithosphere.
   c. refining bauxite ore into aluminum.
   d. the process of magma that moves into spaces within the crust.
   e. cleaning mining-damaged landscapes.

4. Metals
   a. differ from nonmetals in that metals do not deform under stress.
   b. include sand, gravel, and gemstones.
   c. may be classified as base metals, precious metals, radioactive metals, and rare earths.
   d. are poor conductors of electricity.
   e. all of the above apply to metals.

5. Stainless steel
   a. is an alloy of iron with nickel and/or chromium.
   b. is a form of cobalt.
   c. is an alloy of copper and tin.
   d. was used to determine when the Greek epic poems *The Iliad* and *The Odyssey* were written.
   e. is a naturally-occurring metal.

6. Iron ore production is concentrated in
   a. the United States and Mexico.
   b. Jamaica and Surinam.
   c. Japan and China.
   d. India.
   e. Brazil, Australia, and South Africa.

7. BIFS
   a. is an acronym for banded iron formations.
   b. give clues to the evolution of the Earth's atmosphere and life.
   c. are named for alternating bands/layers of iron-rich and iron poor minerals.
   d. include Hamersley Range, West Australia.
   e. all of the above.

8. Aluminum
   a. is refined from iron ore.
   b. has become an industrial staple because of its strength compared to its weight.
   c. requires a lot of energy to refine.
   d. a and c only.
   e. b and c only.

9. Refining aluminum takes what percentage of the electricity produced each year in the United States?
   a. <1%
   b. 2% to 3%.
   c. 10%
   d. 5%
   e. This is a trick question because bauxite is not mined in the United States.

10. A good example of a mineral deposit whose origin is related to plate tectonics processes is
    a. porphyry copper.
    b. bauxite.
    c. stainless steel.
    d. all of the above.
    e. a and b only.

11. The PGE mineral group
    a. includes gold and silver.
    b. are used as catalysts in catalytic converters.
    c. are used in jewelry.
    d. are radioactive.
    e. b and c only.

12. Placer deposits are
    a. assemblages of bauxite in Jamaica.
    b. refer to pyrite.
    c. gold flakes in stream sand and gravel.
    d. should be avoided due to their radioactivity.
    e. none of the above.

13. Remediation of industrial-scale gold mining is currently the responsibility of
    a. the gold miners.
    b. jewelers.
    c. a tax on wedding bands.
    d. the states and federal agencies.
    e. bauxite refiners.

14. Uranium
    a. is a nonmetalic, nontoxic element.
    b. has two naturally occurring isotopes.
    c. is not widespread in its distribution.
    d. is not capable of releasing energy by fission.
    e. none of the above are true.

15. Among the most important and valuable nonmetals are:
    a. sand and gravel deposits.
    b. phosphates.
    c. clay.
    d. gemstones.
    e. all of the above except d.

16. Aggregate refers to
    a. a mineral resource exclusively from South and North Carolina.
    b. a mineral resource whose industry employs 90,000 Americans.
    c. is mainly used as a rare gemstone in wedding rings.
    d. all of the above.
    e. b and c only.

17. The single greatest impact on growth in production of aggregate in the United States was
    a. the construction of the interstate highway system.
    b. the increase in demand for wedding rings.
    c. the increased demand for stainless steel.
    d. nothing, because aggregate production decreased.
    e. the demand for aluminum cans.

18. Sand and gravel deposits arise from
    a. fluvial (river) processes.
    b. lacustrine (lake) processes.
    c. glacial processes.
    d. marine processes.
    e. all of the above.

19. Clay
    a. is used to produce impervious liners for containment of waste in landfills, and at industrial sites.
    b. repels water molecules.
    c. minerals have been mined only in the last 250 years.
    d. minerals include pyrite.
    e. is alloyed with nickel to iron to produce stainless steel.

20. Sulfur dioxide (SO_2)
    a. is a major gaseous emission of smelters.
    b. has health impacts on humans.
    c. causes environmental impacts.
    d. cannot be captured by emission scrubbers.
    e. all of the above except d.

FOOTNOTE

[1] Source: Rio Tinto Alcan. Accessed September 11, 2011 from http://www.alcan.com/.

Energy and Emissions: Global Trends

POINT-COUNTERPOINT

Point

J. R. Whitehead, formerly a member of the **Club of Rome**,[1] had this to say about global energy: "Energy is only one of many problems which threaten the global future, although it ranks with population growth as one of the most serious. The world is due to meet the greatest energy crisis ever within the next half century—the progressive decline of fossil fuels, but not before their excessive use has irreversibly damaged the ecosystem."

Counterpoint

According to Edison Electric Institute[2] (EEI): "To ensure adequate energy supplies for the future (in the United States) we will need more power plants. One thousand three hundred ten new power plants with a total of 393,000 **megawatts** of capacity are projected to be needed by 2020. Most of the new capacity is expected to be fueled by natural gas."

Furthermore, EEI, citing Department of Energy projections, suggests that the United States will consume 23% more electricity by 2030 compared to 2005.

Value-added Question 14-1. Based on EEI's estimate, what is the average capacity in megawatts of each of the new power plants?

Value-added Question 14-2. Based on their estimate, how many power plants must be built in the United States each month between 2002 and 2020?

E nergy is the ability to do work, and has always been essential to humans, and indeed to the functioning of planet Earth. Until the eighteenth century, most energy used by humans was from muscle power, waterpower, wood, or dung.

In Western Europe, the demand for coal as an **energy** source began to accelerate with the invention of the steam engine and with the deforestation of the region. Wood as a fuel source became limited as trees were used to make charcoal, which was used to refine and smelt metals. Mature trees were cut down for ship masts. A major agent of deforestation was land clearances for expanded agricultural production.

How is the world fixed for energy? The International Energy Agency (IEA) reports,

- The world has energy reserves to meet its needs for at least the next several decades.
- Oil fields in developed countries (mainly, Australia, Canada, Norway, and Britain) have peaked or soon will peak and decline. New hydrofracturing techniques could significantly increase the United States' oil reserves, as it has already done for natural gas. However, consumers such as China and India may become more dependent on a small number of Middle East oil suppliers.
- Significant *infrastructure* additions (pipelines, fields, scrubbers, etc.) will be needed to bring natural gas to market and to reduce coal emissions. Due partly to subsidies provided to fossil energy, renewable alternative energy sources are not yet competitive with fossil energy sources. The cost of renew-

able energy sources must continue to fall so that they can compete with fossil fuels, or fossil fuel subsidies must be reduced or eliminated.

- Uranium, together with the plutonium produced by uranium use, could fuel nuclear plants for the foreseeable future. Here the issue is one of public trust. The public's concern about plant safety and waste disposal still inhibits the use of nuclear energy. To meet these challenges and exploit abundant reserves, massive investments would be required in nuclear plants and technology. But even with such investments, nuclear energy could prove unacceptable to many nations with democratic institutions due to a lack of public trust. And the 2011 Fukushima disaster threatens to destroy what little public trust remains.

■ FOSSIL FUELS

Burning fossil fuels—coal, oil, and natural gas—accounts for nearly 90% of global energy production, although renewable energy investment is growing rapidly (**FIGURE 14-1**). Advantages of fossil fuels are: massive subsidies are provided for the extraction and use of fossil fuels; these fuels are easily transported; they can be used to produce electricity if desired; liquid fuels are essential to modern transportation; and the technology is well known.

After World War II, petroleum replaced coal as the world's major source of energy. Oil provides the main source of energy for the *transportation* sector of economies. Transportation in turn accounts for the largest demand on the world's oil reserves and contributes to some of humanity's most pressing environmental problems, such as climate change and pollution of water and soils (**FIGURE 14-2**). Fossil fuels contribute significantly to greenhouse gas buildup when burned, a major factor in global climate change. Finally, methane, the major constituent of natural gas, further contributes to climate change as it leaks from pipelines and coal seams.

Here is the world's present situation relative to fossil fuels.

1. World coal reserves are sufficient to cover global demand for at least one hundred years, even at slow rates (~1% per year) of increase. But if coal consumption is to grow without irreversible climate change and unacceptable environmental degradation, new technologies must be developed to limit emissions of carbon dioxide and other toxic gases. Moreover, remaining coal reserves are less accessible than those already mined. Emissions from coal plants remain one of the planet's major sources of toxic pollutants—including radioactive particles, mercury, SOx (sulfur oxides), NOx (nitrogen oxides), lead, and arsenic. Underground coal mining is one of the world's most dangerous professions, and degradation from surface mines (e.g., mountaintop removal) is a severe regional problem.

2. The 21st century has seen a rapid increase in the discovery rate and use of natural gas in the United States (Figure 14-1). Heightened environmental concerns made gas more desirable

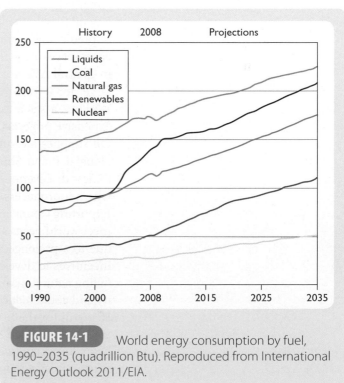

FIGURE 14-1   World energy consumption by fuel, 1990–2035 (quadrillion Btu). Reproduced from International Energy Outlook 2011/EIA.

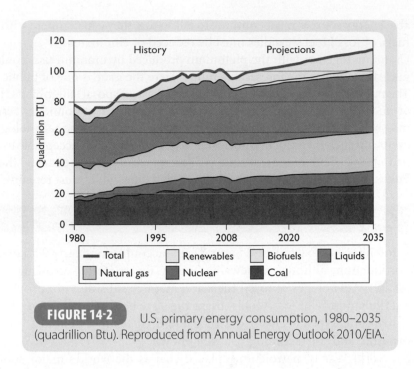

FIGURE 14-2 U.S. primary energy consumption, 1980–2035 (quadrillion Btu). Reproduced from Annual Energy Outlook 2010/EIA.

as a power-generation source. But gas reservoirs are often situated far from markets, as in Saudi Arabia, Latin America, and Alaska, requiring the construction of long and expensive pipelines or the development of other transportation methods, such as liquefied natural gas (LNG). This fact bedevils European attempts to secure more natural gas reserves at low prices because there is little indigenous gas. Gas prices may not be high enough to justify developing far-off gas fields. For example, in 2010 British Petroleum (BP) and other energy firms tentatively decided not to build a natural gas pipeline from the giant Prudhoe Bay oil field to the lower 48 states, citing the high investment required and the unpredictable world price for gas. Additionally, many regions where natural gas reserves are plentiful, like the Middle East and the former Soviet Republics in Central Asia, are politically unstable, and gas infrastructure could be an easy target for terrorists. Moreover, the world price of gas fluctuates continuously. Without a stable or guaranteed price, investors are often unwilling to fund expensive projects that can take years to bring to market.

3. Oil supplies are sufficient for several decades at present prices and levels of global demand. Should demand increase, prices could rise significantly, leading to new discoveries and additions to reserves. There is considerable debate over whether producers, mainly members of OPEC (the Organization of Petroleum Exporting Countries) and Russia, would be willing to increase oil production to meet world needs at fluctuating prices. Should prices *fall* significantly, perhaps due to a prolonged global economic downturn or major increases in supply, incentives to develop alternatives to oil could disappear. For example, rapidly falling natural gas prices in the U.S. in 2012 lowered domestic exploration activity and could reduce available supplies in the near term.

4. In 2013, for the first time in three million years, atmospheric CO_2 concentrations exceeded 400 pp. Availability of energy in the long run may not be as important a concern as the growing *environmental impact* of burning ever-increasing amounts of fossil fuels, coupled with a public unwilling to trust an expansion of the nuclear industry.

■ TRENDS IN GLOBAL ENERGY USE AND EMISSIONS

Global energy demand has risen more than 80% since 1971. It is likely to increase at low rates (1%–2% per year) over the next several decades, fueled by economic expansion and industrial development in countries like China, India and Indonesia, according to the IEA, the United States Energy Information Administration (EIA) and the World Resources Institute (WRI) (**FIGURE 14-3**).

Burning fossil fuels provides nearly 90% of global energy. Thus, along with rising energy use will come rising greenhouse gas emissions from fossil fuels, resulting in a projected acceleration of rate of global climate change. Carbon dioxide is a major greenhouse gas (GHG), and fossil-fuel combustion accounts for about 80% of global carbon dioxide (CO_2) emissions each year (see Figure 14-1), with the rest mainly from deforestation and cement manufacture. Despite widespread global acceptance of the Kyoto Protocol (with the notable exception of the United States), global energy consumption—and annual CO_2 emissions—have risen by almost 50% since 1993. And EIA forecasts that global oil production could double by 2050. China adds one large coal-fired power plant each week (**FIGURE 14-4**).

Developing nations account for more than 80% of world population but consume only about one third of the world's energy. That will likely change quickly. For example, the developing nations' share of commercial energy consumption grew to nearly 40% by 2010. By 2008 China overtook the United States as the world's greatest total emitter of GHGs, but on a *per capita* (per person) basis the individual Chinese lags behind the average American (or Western European; Figure 14-4).

Concept Check 14-1. Is total GHG emissions or *per capita* emissions more important? Why?

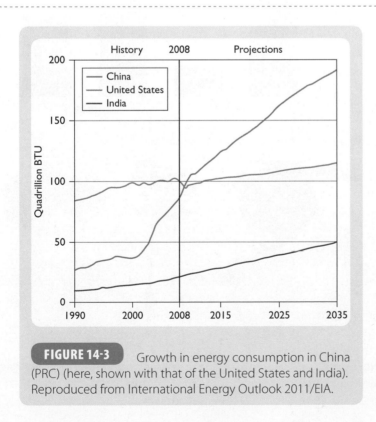

FIGURE 14-3   Growth in energy consumption in China (PRC) (here, shown with that of the United States and India). Reproduced from International Energy Outlook 2011/EIA.

(A) Total emissions of CO_2 for major emitters. Data from: EIA. (B) 2009 Per capita CO_2 emissions. (C) Per capita CO_2 emissions from a different perspective: A map of the world distorted to represent per capita carbon dioxide emissions rather than land area. © Copyright SASI Group (University of Sheffield).

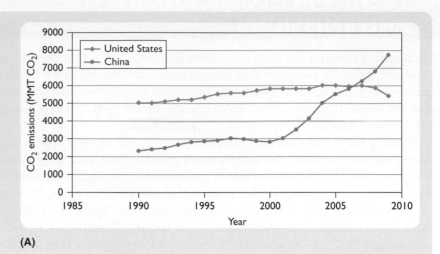

(A)

(B)

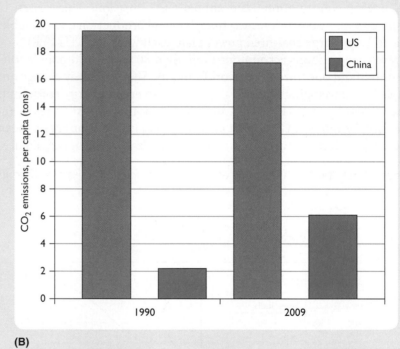

(C)

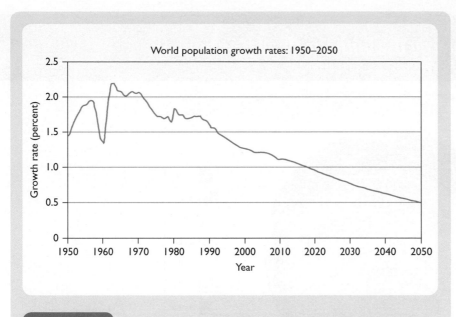

FIGURE 14-5 The U.S. Census Bureau's projection for rate of future global population growth. What was the rate as of 2012? What is the rate projected to be by 2050? When is the global growth rate projected to fall to 1%? At this rate, how long would it take global population to double? Reproduced from International Data Base World Population Growth Rates: 1950–2050/U.S. Census Bureau.

Factors driving this increased energy demand in the developing world include: rapid industrial expansion and infrastructure development; population growth (**FIGURE 14-5**) and urbanization; rising incomes that enable the purchase of energy-consuming appliances and cars that could not be afforded previously, especially in China and India; lax enforcement of emission regulations that makes energy less expensive in developing countries; and subsidies that encourage the use of fossil fuels, such as subsidized cooking oil in Indonesia.

BOX 14-1 *Measuring Energy Use with the Quad*

Energy use is typically expressed in **quads**, where 1 quad = 10^{15} **Btu** (British Thermal Units) (**Figure B 1-1**).

One quad is approximately equal to the energy obtained from each of the following sources:

- 36 million tonnes of coal (400,000 rail cars)
- 24 million tonnes (167 million barrels) of oil
- 28 billion cubic meters of natural gas
- 33 1000-megawatt electric power plants, operated for one year

- 660 square kilometers total surface area of optimally oriented, 10%-efficiency photoelectric solar collectors at sea level in clear conditions, operated for one year
- 44,000 750 kW (**kilowatt**) optimally-sited windmills, operating at peak output for one year

(continued on next page)

BOX 14-1 *Measuring Energy Use with the Quad (Continued)*

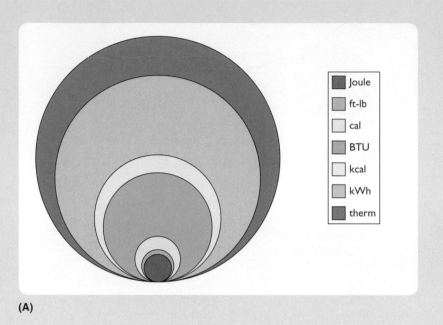

(A)

Energy Units	equals about				
	Joule	**ft-lb**	**kWh**	**BTU**	**kcal**
1 Joule (J)	1	0.737	2.78×10^{-7}	9.48×10^{-4}	2.39×10^{-4}
1 foot-pound (ft-lb)	1.36	1	3.77×10^{-7}	1.29×10^{-3}	3.24×10^{-4}
1 kilowatt hour (kWh)	3.60×10^6	2.66×10^6	1	3,415	861
1 British Thermal Unit (BTU)	1,055	778	2.93×10^{-4}	1	0.252
1 kilogram calorie (kcal)	4,185	3,087	1.16×10^{-3}	3,97	1

(B)

FIGURE B 1-1 (A) Units used to measure energy. Assuming the circles are to scale, how much larger is a therm than a joule? (B) Some conversion factors. How many BTUs are in a kilowatt-hour?

■ ENERGY USE AND POPULATION

As we have seen, both energy use and human population are growing and projected to continue to do so for some time (**FIGURE 14-6**). Population growth alone, however, is not a satisfactory criterion by which to predict future energy consumption, because of the large range in energy consumption per capita from country to country. On a worldwide basis, the *mean* annual global per capita energy consumption is about 60 gigajoules. One gigajoule = 0.28 megawatt-hours of electricity, = 947,800 Btu, = 0.165 barrel of oil. However, the *median* per capita value (that is, the middle value) is much lower than 60 gigajoules, because the developed nations consume energy at a rate that is one or two orders of magnitude greater than that of the developing countries, where most people live. For example, Mexico, Malaysia, and Portugal have per capita energy consumption near the mean value (59, 52, and

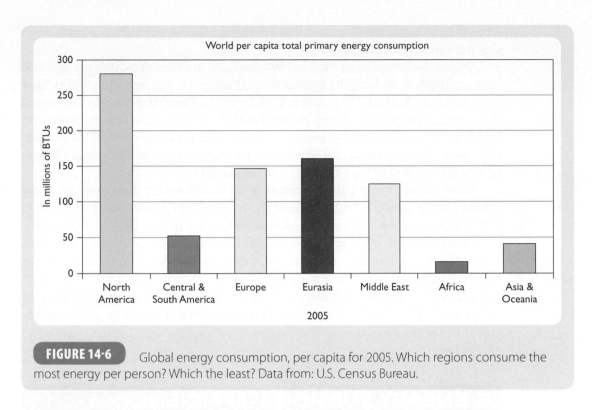

FIGURE 14-6 Global energy consumption, per capita for 2005. Which regions consume the most energy per person? Which the least? Data from: U.S. Census Bureau.

61 gigajoules, respectively) but the per capita consumption in the United States and Canada is 317 to 319 gigajoules. At the lower end of the scale are India (10 gigajoules), Nigeria (7 gigajoules), and Haiti (1 gigajoule).

■ THE GROSS DOMESTIC PRODUCT AND ENERGY USE

A country's **gross domestic product (GDP)**, though increasingly criticized by environmentalists and some economists, is a widely used economic indicator that correlates with many parameters often associated with **standard of living**. Data from many countries show GDP to be *positively* correlated with adult literacy rate, life expectancy, and percentage of the population with access to sanitation. Many thus use GDP as a proxy for evaluating *standard of living*. Economists have adopted a form of GDP that allows for comparison of a currency's *purchasing power* in that country, because purchasing power varies widely among countries. This is GDP based on purchasing power parity (PPP), and it represents *the amount of a common "market basket" of goods each currency can purchase locally*, and is thus a better economic indicator.

The use of GDP to measure standard of living is controversial outside conventional economics, because GDP includes economic activity of all kinds: money spent to clean up oil spills, to reduce air pollution, and money spent on medication for miners with black lung disease, for example. In other words, the Exxon Valdez oil spill increased U.S. economic activity, and thus contributed positively to GDP. Also, GDP assigns no value to environmental "goods," such as the value of wetlands in cleaning polluted water. In the context of this discussion, however, we will continue to use GDP as one measure of standard of living, always realizing its limitations.

Concept Check 14-2. List alternative ways to measure quality of life. Here are some examples to get you started: access to clean water, peace, low crime rates, and so forth.

According to economists, growth in GDP should exceed population growth, such that per capita GDP increases.

During the period from 1980 to 1990, the GDP in China grew 10% per year and the consumption of electricity grew at approximately 9% per year. (The World Bank estimated real Chinese GDP growth at 9.3% for 2011.) These rates considerably exceed the increase in population during the same period, which was less than 1.5% per year. Under such conditions, and *ignoring environmental impacts,* per capita GDP can rapidly rise, providing a chance for a better life for many of a country's citizens. Chinese GDP was estimated by the CIA at $11.3 trillion in 2011.

Many nations show parallel growth of GDP and energy consumption. Some exhibit energy growth rates that *exceed* growth in GDP. For example, from 1970 to 1992 the energy consumption in Mexico increased 86% and the GDP increased 57%.

Such conditions often apply to countries like Mexico that subsidize energy use. Many developing countries, for example, subsidize the price of kerosene, which is widely used for cooking. During approximately the same time period (1977–1992), a 102% increase in GDP in Malaysia was accompanied by an energy consumption increase of 147%. Many environmentalists believe that it would be advantageous globally to set as a goal decoupling the two variables, by increasing the *efficiency* of energy use.

■ ENERGY SUBSIDIES, EFFICIENCY, AND EMISSIONS

Concept Check 14-3. How would you behave if your state or local government decided to subsidize gasoline prices? Electricity prices? Perhaps they could allow you to deduct all of your gasoline expenses from your state taxes, or the state would simply agree to rebate half your costs at the end of the year. Perhaps your state, as California tried to do, would mandate a maximum price consumers could be charged for electricity. Would this affect how much you drove? What kind of car you bought? How much electricity you used? How careful you were about turning off the lights? Evaluate the advantages and disadvantages of subsidizing energy and fuels prices.

Here are three definitions of an **energy subsidy**. (1) The Organisation for Economic Cooperation and Development (OECD) defines it as, "*any measure that keeps prices for consumers below market levels, or for producers above market levels or that reduces costs for consumers and producers.*" (2) The U.S. Energy Information Administration (EIA) uses, "*any government action designed to influence energy market outcomes, whether through financial incentives, regulation, research and development, or public enterprises, energy producers or lowering the price paid by energy consumers.*" (3) According to the International Energy Agency (IEA), "*any government action that concerns primarily the energy sector that lowers the cost of energy production, raises the price received by energy producers or lowers the price paid by energy consumers,*" constitutes an energy subsidy.

The IEA in 1999 conducted an analysis of the impact of energy subsidies in eight developing countries. By keeping prices below the actual cost of producing energy, energy subsidies impose heavy burdens on economic efficiency, environmental quality, and government budgets. While some countries around the world have made progress in reducing or eliminating such subsidies (the U.K. has ended coal subsidies), much remains to be done, particularly in developing countries.

Concept Check 14-4. The Clean Air Act and subsequent legislation requires coal-burning power plants to remove some toxic pollutants from their emissions. How would *not* requiring coal-burning power plants to clean their emissions of toxic pollutants and CO_2 constitute an energy subsidy? How would such a decision affect how utility executives made decisions about new power plants' fuel type?

Energy Subsidies and Sustainable Development

Removing energy subsidies would support the *three principal aims of* **sustainable development**: social welfare, environmental protection, and economic growth. Money used to support subsidies could be redirected to social benefits, such as universal health care, primary education, clean water, and environmental remediation. Environmentalists insist that environmental benefits accrue from strict market pricing, which means including all the costs in the price of a product. This in turn could reduce both local and global pollution, including CO_2 emissions.

Concept Check 14-5. If subsidies to support fossil fuel use are "bad", are subsidies to encourage the development and use of renewable energy "good?" Why or why not?

Here are some examples of fossil fuel energy subsidies worldwide:

- Germany provided subsidies to its coal industry amounting to US$ 115 per tonne for 34 million tonnes of coal mined in 2000, a subsidy totaling US$ 3.9 billion. It stood at 2.3 billion euros ($3.3 billion) by 2006.
- Bangladeshi government energy subsidies in 2010 amounted to $5 billion, according to the World Bank.
- The U.S. Agency for International Development estimated that total energy subsidies in Indonesia cost $16 billion in 2010.

Eliminating subsidies can have a longer-term positive impact. Subsidy elimination can contribute to individual welfare by eliminating a stimulus for excess consumption, which leads to the rapid depletion of the stock of **natural resources**, such as fossil fuels, mineral deposits, and wood. Strict market pricing can stimulate recycling and renewable energy, in turn enhancing sustainable development.

However, eliminating subsidies can have profound negative impacts on local communities. In the Ukraine, hundreds of thousands of residents depend on employment in coal mining. Should subsidies be removed without comprehensive and expensive retraining, the local population could suffer an economic depression, which in many countries has resulted in political unrest. This is one reason why many countries, especially democracies and poor countries with large and growing populations, find it expedient to continue subsidies for fossil fuels.

Under-Pricing of Energy Resources

The IEA found that, in addition to energy subsidies in developed countries like Australia, Canada, and the United States, widespread under-pricing of energy resources occurs in many of the largest economies outside the developed world: China, India, Indonesia, Iran, Kazakhstan, Russia, South Africa, and Venezuela.

These price subsidies result in substantial burdens on the environment. According to IEA, the removal of energy price subsidies in these countries would:

- Reduce primary energy consumption by 14%.
- Increase GDP through higher economic efficiency by almost 1%.

- Lower CO_2 emissions by 17%.
- Generate domestic environmental benefits, including reduced local air pollution.

Global energy use is growing. Environmental consequences are significant, predictable, measurable, and possibly avoidable should energy producers use renewables and other nonfossil fuels to replace fossil fuels. Subsidizing the price of energy from fossil fuels distorts markets, eliminates incentives for nonpolluting renewable energy sources, results in more rapid depletion of resource stock, and produces avoidable pollution, which adds to global health costs. Increasing energy efficiency is cost-effective, and can have positive effects on global emissions, citizens' health, and government budgets.

Reducing subsidies that encourage energy use, however, can have severe consequences on the poor and on localities dependent on fossil fuels for income. It can lead to social instability as well.

CHAPTER SUMMARY

1. Energy is the ability to do work.
2. The world has ample energy reserves to meet its needs for at least the next several decades.
3. Many giant oil fields in developed countries will soon peak and decline, forcing consumers to become more dependent on Middle East oil suppliers.
4. Due partly to subsidies provided to fossil energy, renewable alternative energy sources are not yet competitive with fossil energy sources.
5. Uranium could fuel nuclear plants for the foreseeable future.
6. Burning fossil fuels accounts for nearly 90% of global energy production.
7. World coal reserves are sufficient to cover global demand for at least one hundred years, even at slow rates (~1% per year) of increase.
8. Heightened environmental concerns make natural gas desirable as a power-generation source. But without a stable price, investors are often unwilling to fund expensive natural gas delivery projects.
9. Oil supplies are sufficient for several decades at present prices and levels of global demand. However, global energy demand is likely to increase at rates of 1% to 2% per year over the next several decades.
10. Along with rising energy use will come an inevitable increase in greenhouse gas emissions from fossil fuels.
11. Developing nations account for more than 80 percent of world population, but consume only about one third of the world's energy. Developed nations consume energy at a rate that is one or two orders of magnitude greater than that of developing countries, where most people live. That will likely change quickly.
12. Energy use is typically expressed in quads: 1 quad = 10^{15} Btu.

13. Many nations show parallel growth of GDP and energy consumption. Some exhibit energy growth rates that *exceed* growth in GDP.

14. Energy subsidies impose heavy burdens on economic efficiency, environmental quality, and government budgets. Removing energy subsidies would support the three principal aims of sustainable development: social welfare, environmental protection, and economic growth.

15. Eliminating subsidies can have profound negative impacts on local communities.

KEY TERMS

BTU	kilowatt (kW)	standard of living
Club of Rome	megawatt	sustainable
energy	natural resources	development
energy subsidy	quad	
gross domestic product (GDP)		

REVIEW QUESTIONS

1. Fossil fuels account for about what percent of global energy consumption?
   a. 90%
   b. 50%
   c. 25%
   d. 10%

2. World coal reserves are sufficient to meet global demand for
   a. two or three decades.
   b. at least 500 years.
   c. at least 100 years.
   d. the next ten years.

3. The 1990s saw an increase in the use of
   a. tidal power.
   b. wave power.
   c. natural gas.
   d. all of these.

4. Global energy use is likely to increase at what rate over the next several decades?
   a. 10%/year
   b. 20%/year
   c. 1%–2%/year
   d. 12%–15%/ycar

5. Fossil fuel burning accounts for about what percent of human-generated CO_2 emissions?
   a. 10%
   b. 100%
   c. 80%
   d. 25%

6. 1 quad is equal to
   a. 10^{15} kilowatts.
   b. 10^{15} Btu.
   c. 10^{15} gallons.
   d. 10^{15} tonnes.

7. Oil production is expressed in
   a. liters.
   b. pints.
   c. gallons.
   d. barrels.

8. The Exxon Valdez oil spill
   a. increased economic activity, and thus U.S. GDP.
   b. reduced economic activity, increasing GDP.
   c. increased economic activity, reducing GDP.
   d. reduced economic activity, reducing GDP.

9. A major German energy subsidy supports
   a. oil production.
   b. wave power production.
   c. tidal power production.
   d. coal mining.

10. Removing energy subsidies
    a. can have major negative impacts on local communities.
    b. can lead to political instability.
    c. can increase economic efficiency.
    d. all of these.
    e. b and c only.

FOOTNOTES

[1] The Club of Rome is a group of scientists, economists, businessmen, international high civil servants, heads of State, and former heads of State. Details on its membership are on their website (http://www.clubofrome.org).

[2] Edison Electric Institute is the trade organization representing the electric power industry. More information is on their website (http://www.eei.org).

Fossil Fuels: Coal

CHAPTER OUTLINE

POINT-COUNTERPOINT　　*Coal as a Viable Fuel*

© Amos Struck/Shutterstock, Inc.

Point

As far back as 1999, the Worldwatch Institute issued a report calling for a global phase-out of coal use, citing coal burning's impact on climate change, air and water pollution, ecosystem contamination, and human health. The report argued that particulate and sulfur emissions from burning coal caused more than 500,000 premature deaths a year globally, and untold millions of new respiratory illnesses. The authors further alleged that, "piecemeal attempts to deal with coal's health and environmental effects have created new, more chronic problems." For example, high smokestacks built to transport emissions far from the source led to widespread acid rain, while declining sulfur levels, resulting from improved pollution control methods or burning lower sulfur coal, have in many places been offset by increased nitrogen emissions, polluting numerous European and North American lakes. Coal use in China is laying waste to that country's environment in ways not seen since Dickensian England.

Fossil fuel subsidies amounted to over $500 billion globally as of 2011, according to the International Energy Agency (IEA). Polluted dust clouds from Asian coal plants now reach the American west coast.

The U.S. Centers for Disease Control and Prevention (CDC) estimated that each year up to 2,000 former miners in the United States die from lung diseases caused by exposure to coal mine dust. According to the U.S. Department of Labor, more than 100,000 beneficiaries and 18,000 dependents received benefits in 2011 totaling $203 million arising out of exposure to coal dust in mines.

Counterpoint

The World Coal Institute asserts that "coal bears the [unjust] stigma of being a 'dirty' polluting fuel. Today, coal can be burned cleanly and effectively throughout the world, using constantly improving technologies. Technologies now being developed allow even greater reduction of emissions during combustion." The Edison Electric Institute reports that "(b)y 2010, sulfur dioxide (SO_2) emissions in the United States [were] at their lowest level in one hundred years (except for a few years during the Great Depression)," largely due to utility reductions and a global economic slowdown. They point out that sulfur emissions from coal plants have declined 40% from 1980 levels, and NOx (nitrogen oxide, sometimes pronounced "nox") emissions have dropped 30%. Finally, they assert that, "electric utilities are leading all U.S. industries in voluntary actions to mitigate greenhouse gases."

Concept Check 15-1. Are the positions by the two groups contradictory? Do they deal with the same subjects? If so, then how do their subject matters differ?

■ COAL'S CHALLENGES

The modern era of environmental protection probably began in 1952, when unusual weather conditions led to air stagnation over London, England (FIGURE 15-1). At least 4,000 people were killed by air pollution from millions of coal fires. This disaster led to the passage of the 1956 British Clean Air Act. The Act gave local governments the right to establish "smoke-free" zones. London as a result banned traditional coal fires, and mandated the conversion of most domestic heating to natural gas.

Mine Safety

The U.S. coal industry is the safest in the world; nevertheless, during the period 2002–2012, an average of 30 miners were killed yearly in accidents, mainly in underground mines, and mostly due to methane explosions (see later in the chapter).

The federal Mine Safety and Health Administration (MSHA; http://www.msha.gov) has the responsibility of industry oversight in the United States. As electric utilities, the major coal consumers in the United States, conform to the requirements of the Clean Air Act (discussed elsewhere in the text), they often choose to burn coal mined in open pits in Western states like Wyoming and Montana (**FIGURE 15-2**). In open pit mines, hundreds of millions of tons a year are produced with extraordinary safety. Many mines operate an entire year without an injury, much less a fatality.

FIGURE 15-1   London smog in 1952, resulting in large part from coal fires. © UK History/Alamy.

While safety is a primary concern in the United States, the same cannot be said for other countries that have large scale coal mining industries. We begin with examples of the impact of coal mining in Ukraine (a former Soviet "Republic") and in China. The reason we do this is that developing countries, where 5/6 of the global human population lives, often seek the most expedient ways to increase energy production and exports, regardless of the environmental and social impacts that developed countries are able to address more comprehensively because of greater wealth. Coal use is rising fastest in developing countries China and India.

FIGURE 15-2   An enormous open pit coal mine in Wyoming. Note the size of equipment for scale. © Andrey N Bannov/ShutterStock, Inc.

▲ In Other Words

"May this mine be cursed. It has taken my little son forever. First my husband and now my son."

Mother of dead miners after Ukraine mining disaster, 2000.

Coal has been mined in the Ukraine since at least 1721. Ukraine has about 400,000 coal workers and around 200 mines, concentrated in the Donbass coal region in the east of the country around Donetsk (**Figure CS 1-1**). Ukraine had 50 to 100 billion tonnes in proven **coal reserves** (see below) as of 2010, depending on the source of the estimate. Coal supplies about half of Ukraine's domestic energy, even though output fell by over 40% between 1992 and 2002. The country expanded its reliance on coal after the Chernobyl nuclear disaster in 1986. During 2010, Ukraine produced about 60 million tons of coal.

The Ukraine coal industry has been characterized by low productivity, insufficient investment in infrastructure, poor labor relations, and a bad safety record, all by Western standards. Low productivity means many more miners are employed than in industries where productivity is high. Here's an example: In the early twentieth century, it took 130,000 miners to produce nearly 130 million tons of coal per year in West Virginia.

Value-added Question 15-1. What was the annual production per miner in West Virginia?

In 1979, it took 58,500 miners to produce 112 million tons. In 2009, fewer than 15,000 miners produced almost 145 million tons.

Value-added Question 15-2. Calculate the annual production per miner in 1979 and 2009.

Concept Check 15-2. What do you conclude from the calculations you just made?

Many of the Ukraine mines operate at depths in excess of 300 m (1,000 ft), and in a third of the mines the coal seams are so steep that they can only be mined by hand. Many of the mines actually operate at a loss. In spite of, or perhaps because of the above inefficiencies, the Ukrainian govern-

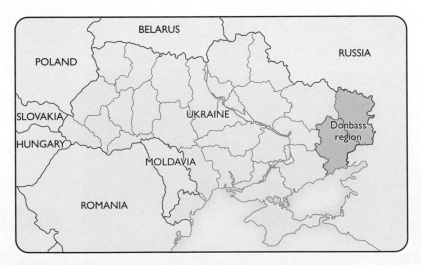

FIGURE CS 1-1   Map of the Donbass region of the Ukraine.

TABLE CS 1-1	Coal Mine Deaths in Ukraine (1996-2001)

Year	Miners Killed
2000	318
1999	280
1998	360
1997	260
1996	340

Data from: World Nuclear Association.

ment continues to pour subsidies into the industry. These subsidies averaged about $15 million (U.S.) per month in 2010, according to the UN Environment Programme.

Working in Ukrainian mines is extremely dangerous (**Table CS 1-1**). As of 2011, more than 2 miners died for every million tonnes coal mined. In late 2007, nearly 100 miners were killed in a Donbass mine explosion. In 2008 the new government proposed to inspect every operating mine.

Miners' wages are determined by tonnage of coal mined, thus discouraging strict adherence to safety practices. Health care for miners was formerly provided by the Soviet government. With the collapse of the USSR, the miners were increasingly forced to rely on private health care.

Value-added Question 15-3. Assume the life of each of the 100 miners was worth $3 million United States, which is the approximate value placed on miners killed in accidents in the United States by insurance companies. How much extra would each tonne of coal cost, assuming the price of coal had to include this cost? Note: It is difficult to compare the price per tonne of coal in the Ukraine to that in the United States.

CASE STUDY　Coal Use in China

Background © Hemera Technologies/AbleStock.com/Thinkstock. Title © iStockphoto/Thinkstock.

Guizhou Province, China: The Impact of Domestic Coal Use

Residential coal use was phased out in the United States during the mid-twentieth century. For an analysis of the impact of residential coal use, we will consider Guizhou (pronounced Gway-JOE) Province in south-central China (**Figure CS 2-1**). Coal replaced wood for fuel towards the end of the nineteenth century, after the cutting of the region's extensive forests. The province has proven reserves of over 50 billion tonnes, with over 3 trillion cubic meters (~90 trillion cubic feet [TCF]) of associated methane.

Two crops important to village diets are chili peppers and corn. To dry the peppers, villagers hang them in bunches over unvented stoves, where the peppers absorb contaminants from the coal smoke. One such contaminant is arsenic (chemical symbol As), which is endemic in the region's coals in concentrations of up to 35,000 parts per million (ppm).

Continued ▶

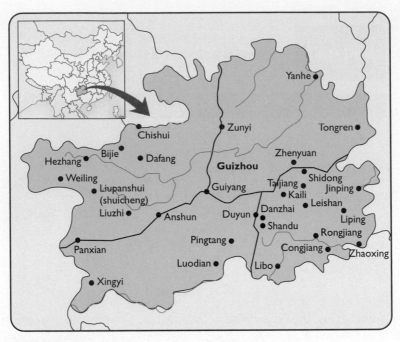

FIGURE CS 2-1 Location of Guizhou Province, China.

Value-added Question 15-4. At a concentration of 35,000 ppm by weight, how much of the coal consists of arsenic? Express this in grams per kg. 10,000 ppm = 1%; 1,000 g = 1 kg.

Peppers can absorb As up to a concentration of 500 ppm when left to dry over the smoking stoves. For comparison, in 2001, the U.S. Environmental Protection Agency (EPA) lowered its safe exposure level to arsenic in drinking water from 50 parts per **billion** (ppb) to 10 ppb. One ppb is 1000 times less concentrated than one ppm. The result: tens of thousands of Guizhou residents exhibit symptoms of arsenic poisoning, consisting mainly of pre-cancerous and cancerous skin lesions.

Far more common in the region, however, is a condition called **fluorosis**, or fluoride poisoning. Fluorosis can manifest itself in discolored teeth, but far more serious are malformed bones (**Figure CS 2-2**). In 1989, Chinese scientists demonstrated that corn dried over unvented coal stoves burning briquettes made with high-fluorine coal (with As concentrations sometimes exceeding 200 ppm as well) was the probable source of the fluorosis. The condition affects up to ten *million* citizens of Guizhou and surrounding regions.

Burning coal containing the trace metal selenium (Se) also may be responsible for human **selenosis** (selenium poisoning) in the region.

Concept Check 15-3. Assess the impact of home coal stoves on the healthcare system of the region. How is the cost of health care adversely impacted by residential coal use? Does this constitute a "subsidy" for coal? If so, then how?

Concept Check 15-4. Suggest some simple ways these problems can be addressed, assuming that immediately replacing coal as a home heating fuel is impractical. *Hint*: Do you think that residents would continue burning coal in unvented stoves if they were fully informed of the hazards? Are there any simple construction measures that could help address the problem of pollution from unvented stoves?

Coal Mining "Accidents"

Concept Check 15-5. Why do you think we place the word "accidents" in quotation marks? Identify the logical conditions under which it is accurate to call disasters that kill miners "accidents."

In addition to toxic contaminants, mining of coal in Guizhou has taken a toll on the region's miners. By 2007, one dollar in eight spent by Americans on goods was

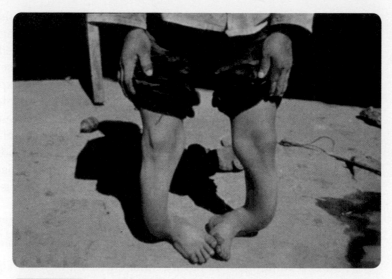

FIGURE CS 2-2 Example of fluorosis, Guizhou. Reproduced from *PNAS* March 30, 1999 vol. 96 no. 7 3427–3431.

spent on items imported from China. During 2000–2007, the Chinese government reported that an average of 5,000 miners was killed across China annually. Although official figures reported around 2,000 miners killed in 2011, labor activist Han Dongfang has claimed the actual statistic is at least twice the official death toll. Even though the death toll has fallen, production has nearly doubled, such that approximately three miners die for every 1,000,000 tonnes of coal output. This death rate is at least 5 times that of U.S. mines. While China has experienced tragic loss of life from coal mine accidents at a rate that would likely not be tolerated in the United States, little has been done to mitigate risk to miners. Such impacts are termed **externalities** by economists, as we discussed elsewhere in the text, if the cost of the disasters is not a part of the price users pay for the coal. Were Chinese authorities to impose measures to drastically reduce mine accidents, it is likely that the price of Chinese coal would rise considerably.

Concept Check 15-6. Why do you think authorities are unwilling or unable to sharply reduce mine deaths?

Value-added Question 15-5. Calculate how much, in U.S. dollars, each tonne of coal mined in 2000 in China would have to sell for, solely to compensate miners'

families for the six miners killed per million tonnes of output. Economists in the United States value a human life at $3 million.

Value-added Question 15-6. In 2011, Chinese coal averaged about 800 Chinese yuan/tonne. What is the comparable price in U.S. dollars? Use the conversion rate of 7 yuan/U.S. dollars. Calculate the final cost of per tonne of Chinese coal in U.S. dollars if miners' families were compensated.

Concept Check 15-7. Rural and small-town Chinese for decades have been migrating to large cities in search of work. There, many of them desperately try to immigrate to other countries, legally or otherwise, often the United States. Explain whether you think dangerous working conditions serve as a possible driving mechanism for illegal immigration.

Concept Check 15-8. Critics of "free trade" claim that products made in countries with lax environmental standards have an unfair advantage over those made in countries with strict environmental standards. Explain how this criticism could be applied to Chinese coal use.

Coal Mining and Pulmonary Disorders

According to the American Lung Association, *pneumoconiosis* (new-mo-con-EEÓ-sis), also known as **black lung disease**, is caused by the long-term inhalation of coal dust. An estimated 4.5 percent of coal miners in the United States are affected; with about 0.2 percent having scarring on the lungs, the most severe form of the disease.

Laws in the United States, if enforced, limit workers' exposure to coal dust, but in other countries such as China and Ukraine, such laws are nonexistent, not enforced, or ignored by miners. The result is that, in China for example, an estimated 7,500 to 10,000 new cases of pneumoconiosis are diagnosed annually. In addition, many miners smoke: China's 350 million smokers consume more than one-third of the world's cigarettes. Such activity is **synergistic** as the impact of smoking combined with black lung is far more serious than the additive effects of both.

■ COAL: A GEOLOGICAL RESOURCE

After reading about the hazards inherent in mining coal (and burning it in unvented stoves), you might be asking yourself if this scenario is overly alarmist. Why, if coal mining is so hazardous for example, is there such a demand for coal?

Despite these issues, coal is a critically important resource, due mainly to: the relative ease and low cost with which coal may be mined; the progress made in many industrialized countries to improve mine safety and control coal emissions; the lack, or non-enforcement, of such laws in many developing countries; and the energy readily extracted when coal is burned. Technologies are available to significantly reduce some emissions from coal-burning power plants, and it is theoretically possible to "sequester" (or store) the CO_2 emitted in coal combustion. In the United States new coal-fired electric power plants emit low levels of toxics compared to coal power plants built during previous decades, many of which are still in operation and still emitting high levels of pollution. And coal-mining patterns are changing as well: in the United States, open-pit (surface) coal mines, posing much less risk to workers from injury and pulmonary disorders (but ignoring for now their environmental impact), now produce more coal than underground mines as you saw above.

Concept Check 15-9. CheckPoint: Summarize reasons why coal is a critically important resource.

Consumption and Price in the United States

In 2013, the U.S. produced more than 1,000 million tons of coal. Most of that coal was burned to produce electricity (FIGURE 15-3). About two-fifths the electricity produced in the United States comes from coal-burning power plants. Coal from surface mines typically sells for about 50% of that derived from underground mines. Coal may be shipped by rail (at about $17 per ton), barge, or truck, but barge shipment, which averaged about 1.5 cent per ton-mile in 2010, is by far the cheapest method of transport.

Coal is also the preferred fuel for "heat-intensive" industries such as steel, aluminum, concrete and wallboard.

Concept Check 15-10. If coal is used to produce electricity, electricity is used to produce aluminum, aluminum is extraordinarily recyclable, and coal use generates significant adverse environmental impact, does it make sense to maximize our recycling rate for

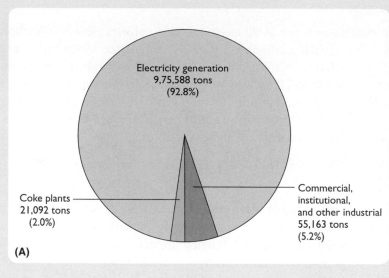

(A)

FIGURE 15-3 (A) Sectors in the U.S. economy where coal is used. Data from: EIA. (B) Sources of natural gas used to produce electricity. Reproduced from Annual Energy Outlook 2013 Early Release Overview/EIA.

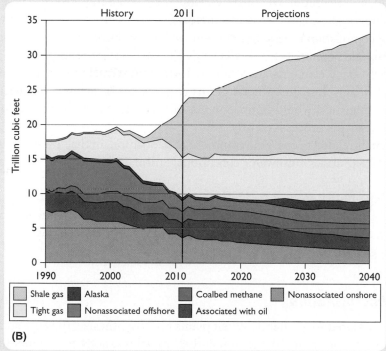

(B)

aluminum, and the other materials listed above? If it makes sense, then suggest reasons why the rate of aluminum recycling (less than 60% for cans for example) is not higher?

Global Reliance on Coal

Many countries are more reliant on coal than is the United States. Poland and South Africa rely on coal for 90% of their electricity. Ukraine, as you saw previously, became dependent on coal after the Chernobyl disaster. China burned nearly 4,000 billion tonnes of coal in 2011. More than 70% of the electricity generated in China, India, Greece, Australia, and the Czech Republic comes from coal. Recall that environmental controls on coal mining and burning may be lower than those in Western countries, or lacking altogether.

Composition

Although it can contain as many as 70 elements, coal consists mainly of carbon, hydrogen, and oxygen. Coal is a complex mix of organic matter, which burns, and inorganic material, which does not. Burning the latter produces **ash**. The inorganic materials in coal can be found (1) as mineral grains of variable size or (2) linked with organic molecules. Minerals commonly found in coal include pyrite, quartz, clays, and gypsum. A desirable **bituminous coal** may have the following composition (but ash and moisture contents are highly variable):

> C, 78%; H, 5%; O, 6%; N, 2%; H_2O (moisture), 3%; S, <1%; Other (ash when burned) <5%

Coal also contains many **trace elements** (those whose concentration is less than 100 ppm). Trace elements in coal are becoming increasingly of concern to coal-burning electric utilities due to environmental impacts. Trace elements include antimony, arsenic, beryllium, cadmium, chromium, cobalt, lead, manganese, mercury, nickel, lead, and selenium, which are variably toxic when ingested (discussed elsewhere in the text). These elements and minerals may be volatilized (converted to gas) during coal combustion unless they are removed before burning. Burning coal without removing such materials from exhaust gases leads to serious air pollution.

Many of these trace elements are found as **ionic substitutions** in minerals within the coal. Removal of the mineral grains (e.g., pyrite [FeS_2] and sphalerite [ZnS]) before combustion would remove much of the potential for air pollution from the coal, but such a process is costly.

Sulfur content is related to coal's environment of deposition, in that lower-sulfur coals were deposited under freshwater conditions. Sulfur is found as FeS_2 and as organic sulfur.

Coal-burning utilities are required by law to publish **Toxic Release Inventories (TRI)** yearly. In the TRI, they list toxic substances released by combustion, by weight. You can find TRIs on most coal-burning electric utilities' web pages or at http://www.epa.gov/tri/.

Geology of Coal Formation

Coal is essentially stored solar energy. When plants photosynthesize, they use solar energy to produce tissue (growth) and to fuel their metabolic activities. When the plants die, organisms often convert plant tissue back to simple compounds like H_2O and CO_2. Under certain circumstances, however, which have been replicated in many places on the planet over the past 350 to 400 million years, plant tissue has been buried under anoxic (oxygen-free) conditions, allowing the plant tissue to remain undecomposed. Over millennia, the buried plant tissue may be converted through a series of stages involving microbial activity, pressure, and heat, into various *grades* of coal; they are *peat*, **lignite**, *sub-bituminous*, *bituminous*, or *anthracite* coal, the form of coal with the highest amount of energy per kg (**FIGURE 15-4A**). **Anthracite** is relatively rare, however, as it usually results from metamorphism of bituminous coal (see **FIGURE 15-4B** for the attributes of each grade of coal). To form, coal requires an oxygen-free environment in which to accumulate, and therein lies much of the source of its environmental contaminants. Oxygen-free groundwater may transport significant quantities of metal cations that may be deposited along with the coal vegetation.

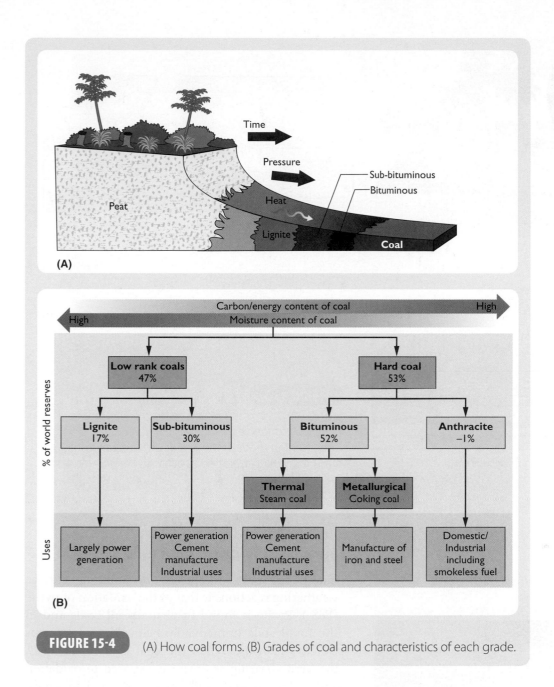

FIGURE 15-4 (A) How coal forms. (B) Grades of coal and characteristics of each grade.

Coal is formed from **peat** (**FIGURE 15-5**). Peat is simply undecayed plant matter. Although peat can accumulate in any climatic zone, for thick coal deposits to accumulate tropical and subtropical conditions provide the rate of plant growth needed. A general rule is ten to twelve meters of plant material yield one meter of coal.

Most coals are found in sedimentary rocks interpreted to have formed in deltas, or behind barrier island complexes, and that exhibit much fluctuation of sea level (Figure 15-5). For thick coal seams to accumulate, the area of coal accumulation must persist for geologically significant intervals (e.g., hundreds of thousands of years). In addition, the area must slowly subside, with the water table maintained close to the surface of the vegetation, to maintain a hypoxic or anoxic (low-oxygen) environment.

Concept Check 15-11. CheckPoint: Explain the process by which coal forms.

(A) Formation of a coal swamp behind a barrier reef.

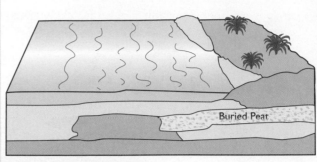

(B) Peat becomes buried as sea level rises

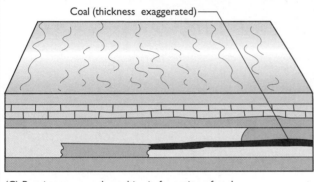

Coal (thickness exaggerated)

(C) Peat is compressed, resulting in formation of coal.

FIGURE 15-5   One scenario under which coal forms. The coal swamp could be in an environment behind a barrier island, for example. Fluctuation of sea level over hundreds of thousands of years can result in many individual layers of coal accumulating.

FIGURE 15-6   Acid-mine drainage in Idaho, stored in a "holding pond." Courtesy of EPA Region 10 (Pacific Northwest).

Oxidation of Pyrite and Water Pollution

One of the most widespread, and polluting, surface weathering reactions is that of the oxidation of pyrite (FeS_2). Pyrite is found in virtually all coals and in the fossil soils immediately below the coal seams called *underclay*. When the coal is stripped away, the pyrite-bearing sediment is exposed to oxidation and hydration, and produces sulfuric acid solutions. Strip-mining coal may result in oxidation and hydration of associated materials, leading to acidification of surface water (**FIGURE 15-6**).

The following formula illustrates pyrite weathering and the production of acidic substances that can contaminate surface waters.

$$FeS_2 + \frac{15}{2O_2} + 4H_2O \rightarrow Fe_2O_3 + 4SO_4^{(-2)} + 8H^+$$

Note the large amount of H^+ (a measure of acidity) produced as well as the SO_4. These form sulfuric acid in natural waters.

Classification

Coal may be classified using any of the following criteria:

- Carbon content
- Energy content (how much heat can be produced when burned)
- Sulfur content
- Ash (insoluble) content
- Water content

Coal is *ranked* according to its organic carbon content. Organic carbon generally increases with depth, duration of burial, and heating, which can drive off other substances, like water, thus concentrating the C. The rank of a coal therefore reflects changes in plant matter resulting from burial and increased temperature, which drives off "volatiles" such as water, CO_2, CH_4 (methane), and H (hydrogen). Most coal burned in the United States is **bituminous**.

TABLE 15-1 shows the relationship between rank and C content.

TABLE 15-1	Relation Between Rank and Content of Volatiles
Rank	**Fixed Carbon (dry wt %)**
anthracite	86–98
bituminous	69–86
lignite	<69
peat	<50

Coal Reserves

The total amount of a resource that can (legally) be extracted at a profit *using present technologies* is referred to as *reserves* (**FIGURE 15-7**).

Reserves may change with technological innovation and with higher prices as well.

In fact, China has been increasing production at about 9% annually, significantly reducing the lifespan of the reserves.

Value-added Question 15-7.
Chinese coal production is on the order of 4,000 million tonnes annually. Assuming no increase, how many years would China's coal last?

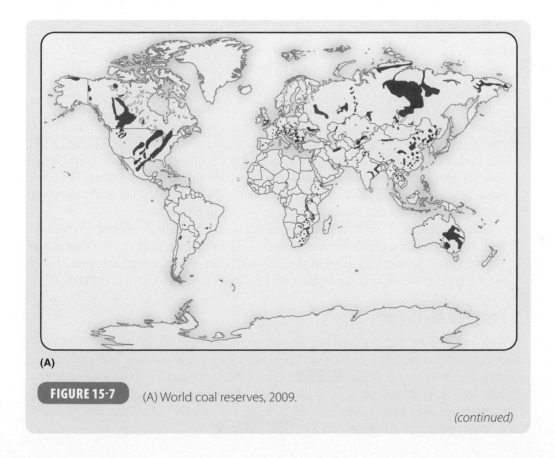

(A)

FIGURE 15-7   (A) World coal reserves, 2009.

(continued)

| Region/Country | Recoverable reserves by coal rank | | | | 2008 production | Reserves-to-production ratio (years) |
	Bituminous and anthracite	Subbituminous	Lignite	Total		
World total	**445.7**	**287.0**	**215.3**	**948.0**	**7.5**	**126.3**
United States[a]	119.2	108.2	33.2	260.6	1.2	222.3
Russia	54.1	107.4	11.5	173.1	0.3	514.9
China	68.6	37.1	20.5	126.2	3.1	40.9
Other non-OECD Europe and Eurasia	42.2	19.1	40.1	101.4	0.3	291.9
Australia and New Zealand	40.9	2.5	41.4	84.8	0.4	191.1
India	61.8	0.0	5.0	66.8	0.6	117.5
OECD Europe	6.2	0.8	54.3	61.3	0.7	94.2
Africa	34.7	0.2	0.0	34.9	0.3	123.3
Other non-OECD Asia	3.9	3.9	6.8	14.7	0.4	34.4
Other Central and South America	7.6	1.0	0.0	8.6	0.1	95.8
Canada	3.8	1.0	2.5	7.3	0.1	97.2
Brazil	0.0	5.0	0.0	5.0	0.0	689.5
Other[b]	2.6	0.6	0.1	3.4	0.0	184.5

[a]Data for the U.S. represent recoverable coal estimates as of January 1, 2010.   Sources: World Energy Council and EIA.
[b]Includes Mexico, Middle East, Japan and South Korea.

(B)

FIGURE 15-7   Continued. (B) Recoverable reserves by coal rank (billion tons). Reproduced from International Energy Outlook 2011/EIA.

U.S. Reserves

Coal resources of the United States are shown in **FIGURE 15-8**. Ten states contain most of the United States' coal reserves as you can see. Coal production is shown in **FIGURE 15-9**.

Concept Check 15-12. Compare the list of coal reserves by state and coal production by state. Propose explanations for any major differences.

Value-added Question 15-8.
What was the United States' production for 2010? Based on the reserves figures in Figure 15-9, how long would these reserves last, assuming no increase?

Sulfur content is one of the most important attributes of coal. Low-sulfur coals are highly valued because burning them produces less sulfur oxide (SOx, sometimes pronounced "sox") pollution than coals with higher sulfur content (**FIGURE 15-10**). The Powder River Basin contains the largest reserves of low-sulfur coal in the United States. Campbell Co., Wyoming alone produces one third of all U.S. coal. The Eagle Butte Mine (**FIGURE 15-11**) alone has more than 810 million tons of coal reserves in a coal seam 120 feet thick with an average S content of 0.35%.

Low-sulfur coal like that from Wyoming is increasingly important to meet the requirements of the Clean Air Act. However, as we discuss later in the chapter (see "Clean Coal" Technology later in the chapter), cost-effective technologies exist to permit higher-sulfur coals to be burned with low emissions of toxic substances.

Value-added Question 15-9.
Until 2008, U.S. coal production had been increasing at around 1% annually. What is the doubling time for this rate? Recall that doubling time = 70/% increase (see population math made easy, elsewhere in the text). How would increased production at this rate affect the estimate you got in the question above?

Coal Bed Methane

Associated with coal deposits are variable amounts of *methane* (the primary component of natural gas), which accumulates in fracture systems and on internal surfaces in the coal (**FIGURE 15-12A**). The U.S. Geological Survey estimates that 100 trillion (one trillion = 10^{12}) cubic feet of coal-bed methane is economically recoverable using present technology, which is nearly five times the United States' annual consumption of natural gas (**FIGURE 15-12B**).

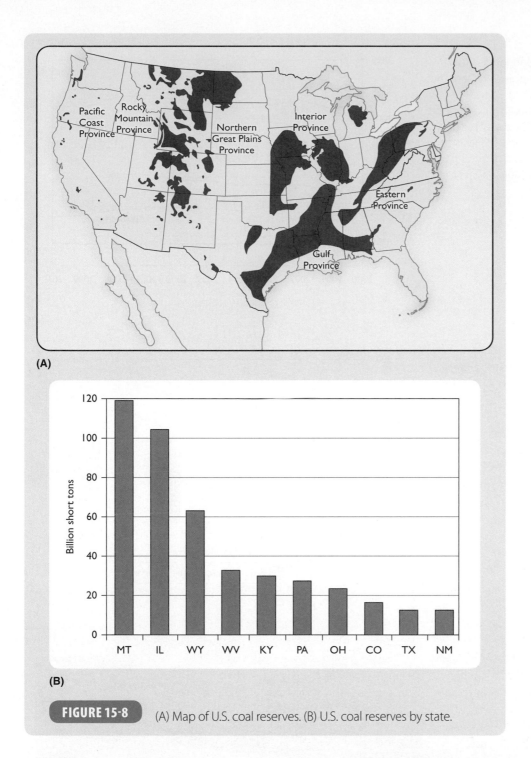

(A) Map of U.S. coal reserves. (B) U.S. coal reserves by state.

Extracting coal bed methane has two advantages besides the value of the methane. First, recovering the methane reduces the likelihood of a mine explosion should underground mines be developed. Second, the methane if allowed to vent to the atmosphere is a powerful greenhouse gas (global climate change is discussed elsewhere in the text). Burning the methane yields heat that can be used to heat buildings, for example, or to produce electricity. In either case, far fewer emissions result when compared to burning coal or oil. No SOx is emitted, CO_2 levels are about half those emitted by burning an equivalent amount of coal, no toxic metals or particulates result, and natural gas burning produces less NOx than other fossil fuels.

FIGURE 15-9 Coal production to 2010. Reproduced from U.S. Coal Supply and Demand: 2010 year in review. Data from: U.S. Energy Information Administration, Quarterly Coal Report, October–December 2010, DOE/EIA-0121(2010/Q4) (Washington, DC, April 2011); Coal Industry Annual, DOE/EIA-0584, various issues; Annual Coal Report, DOE/EIA-0584, various issues/EIA.

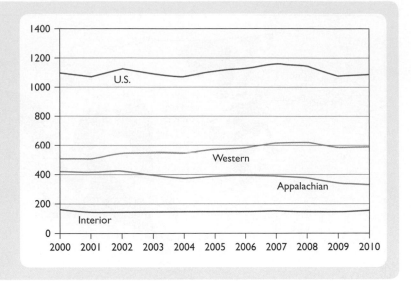

FIGURE 15-10 Low sulfur coal reserves of the United States. Data from: EIA.

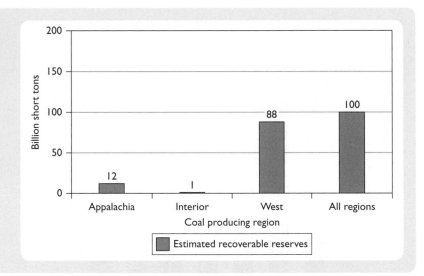

FIGURE 15-11 The Eagle Butte Mine, Wyoming. © Jim Parkin/ShutterStock, Inc.

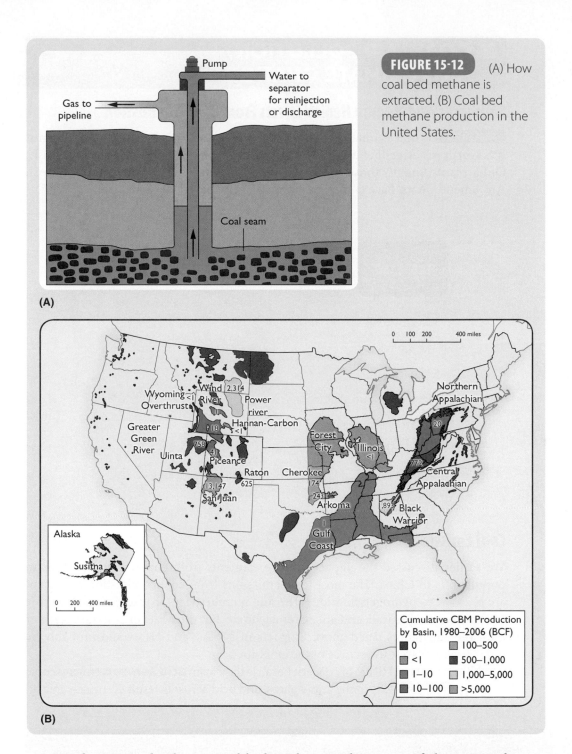

FIGURE 15-12 (A) How coal bed methane is extracted. (B) Coal bed methane production in the United States.

One barrier to developing coal-bed methane is that many of the potential deposits are in public lands prized for their recreational and habitat value. Development of the reserves potentially could compromise these relatively pristine areas. Another disadvantage is the amount of water in coal that would be produced along with the methane. For example, in Wyoming's Powder River Basin, wells produce 5 to 20 gallons of water per minute each. If the water were of lower quality than surface water—which it usually is—surface disposal of coal bed water would degrade streams. The water may be reinjected, but that increases cost. It is also feasible to evaporate the water and sell the saline residue. This may be most applicable to western regions with high evaporation potential, but the economics are uncertain.

■ COAL, AIR POLLUTION, AND CLIMATE CHANGE

Emissions and Human Health Effects From Coal Combustion

TABLE 15-2 is one estimate of average emission values for coal-fired power plants. It is worth restating that emissions from coal-fired power plants vary enormously. Older plants, mainly those excluded from immediate coverage by the 1990 Clean Air Amendments, have emissions far higher than new coal power plants.

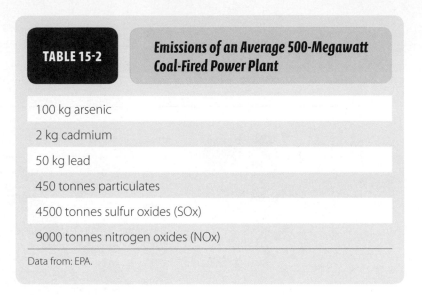

TABLE 15-2 — **Emissions of an Average 500-Megawatt Coal-Fired Power Plant**

100 kg arsenic
2 kg cadmium
50 kg lead
450 tonnes particulates
4500 tonnes sulfur oxides (SOx)
9000 tonnes nitrogen oxides (NOx)

Data from: EPA.

Coal Emissions

We earlier discussed the impact of trace elements from coal burning on human populations in China. The most toxic emissions from coal combustion at present are NOx, SOx, hydrogen fluoride (HF), and mercury (Hg). Nickel (Ni) and arsenic (As) are emitted as well, and are extremely toxic, but their rate of emission is lower than the substances listed above. Impacts of Ni, As, and HF would probably be most severe nearest the source of the emission.

NOx, SOx, and HF are of concern because they form **acid aerosols** (microscopic droplets; **FIGURE 15-13**) in the atmosphere, and acid aerosols result eventually in acid

---

BOX 15-1 *It's Your Grandpa's Fault: "Protected" Coal Power Plants*

Under 1977 Amendments to the Clean Air Act, plants operating or in construction before 1977 were not required to meet the pollution standards applied to newer plants. The reason is that plants "**grandfathered**" under the "existing plant" provision were expected to be replaced at the end of their 30-year projected lives by new plants. Yet many continue to operate, releasing four to ten times more pollution than new coal-fired power plants.

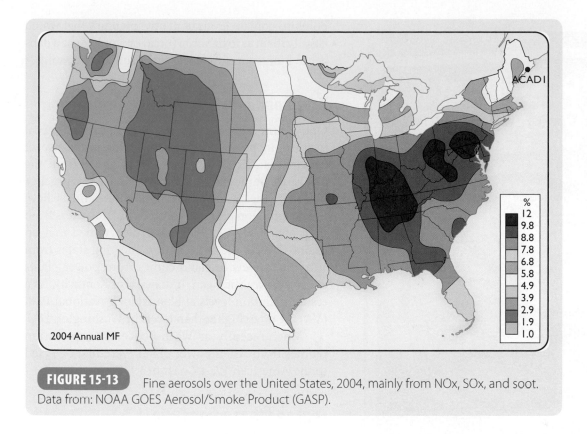

2004 Annual MF

%
12
9.8
8.8
7.8
6.8
5.8
4.9
3.9
2.9
1.9
1.0

FIGURE 15-13 Fine aerosols over the United States, 2004, mainly from NOx, SOx, and soot. Data from: NOAA GOES Aerosol/Smoke Product (GASP).

precipitation (known as acid rain, acid snow, etc.). A typical coal-fired power plant in the United States emits around 1,000 tonnes of hydrochloric acid (HCl), 450 tonnes of sulfuric acid (H_2SO_4), and 80 tonnes of hydrogen fluoride (HF) per year. Information on the impacts of specific pollutants can be found at the Centers for Disease Control website (http://www.cdc.gov) We summarize the most serious impacts here.

- *H_2SO_4 aerosols* can erode buildings and irritate lungs even in minute quantities. SOx aerosols can reduce agricultural yields and increase acidity of soils, eventually freeing toxic aluminum (Al^{3+}) from soils, where it may enter streams and lakes and poison aquatic life.
- *NOx* contributes to smog-forming reactions, and can pollute waterways by enriching them in soluble nitrogen, causing the explosive growth of surface algae, eventually leading to **hypoxia** (low dissolved oxygen levels).
- *HF* is an extremely reactive and corrosive substance; its concentration is small enough to ensure that it is of minor concern except immediately downwind of an emission plume. HF is also emitted by industries like aluminum refining.
- *Mercury (Hg)* can be converted to *methyl mercury* by native soil bacteria. In this form it is extremely toxic to living organisms even in minute quantities (ppb concentrations). Mercury in other forms (elemental mercury, ionic mercury) is also toxic, but not nearly as much as methyl mercury.

Mercury emissions from coal plants have led to global pollution of waterways worldwide, including every major river in South Carolina, such that warnings are posted cautioning residents about consumption of fish. Similar warnings have been issued for the Great Lakes and many other waterways in the eastern United States (**FIGURE 15-14**). Mercury contamination of the oceans has reached alarming levels, and most large oceanic fish and mammals, including bluefin tuna, grouper, orca, and

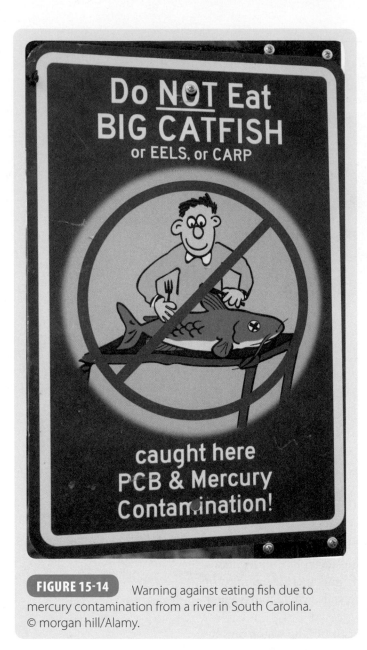

Value-added Question 15-10.
If utilities, steel mills, etc. in the U.S. burn approximately 770 million tonnes of coal per year, how much CO_2 is emitted?

dolphins contain levels of methyl mercury that would classify them as toxic waste (orca) or would pose a danger to human health if ingested in sufficient quantity.

Transboundary Effects of Coal Combustion

One of the severe problems associated with coal burning in power plants is the widespread use of very tall stacks to disperse emissions far from the local environment (**FIGURE 15-15**). Since this practice began in the 1970s, environmental scientists have traced emission plumes hundreds of kilometers from their source. The U.S. EPA has determined that up to one third of the air pollution affecting states from Virginia to New England comes from power plants as far away as Michigan, Indiana, and Kentucky. As a result, pollution levels at Shenandoah National Park (VA) are often higher than levels in Washington D.C.

More recently as we noted earlier, pollution plumes from Asian power plants were first detected in the American West Coast in 1997.

Coal and the Carbon Cycle

The **carbon cycle** describes how atoms of carbon move between the atmosphere, biosphere, hydrosphere, and lithosphere. Coal combustion is a major anthropogenic source of CO_2. The typical bituminous coal burned in power plants contains about 85% carbon by weight. Thus, each kg of "average" coal contains about 850 g of carbon by weight. When the coal is burned, the carbon combines with oxygen to form CO_2. Given the atomic weights of C and O, one can calculate the amount of CO_2 produced per kg of average coal burned. For each kg of coal burned containing 85% C by weight, approximately 3.5 kg of CO_2 is emitted.

While CO_2 is not a toxic substance in the sense of SOx or NOx, its emission in large quantities may cause irreversible climate change, and the Supreme Court has ruled that it is a "pollutant" under the terms of the Clean Air Act.

"Clean Coal" Technology

"Clean" coal technology, while an oxymoron to some, is the subject of significant research funded both by government and private industry. Clean coal technology projects presently funded by industry and the Department of Energy (DOE) fall into three categories. They are:

- *Environmental control technologies*, of which there are three types: SOx control systems, NOx control systems, and combined control systems. These include advanced "scrubbers" that remove pollutants before they are vented to the atmosphere. We illustrate these technologies below.

- *New combustion technologies*, including low-NOx burners and fluidized bed combustion systems, which remove pollutants, or prevent pollutants from forming (**FIGURE 15-16**).
- *Advanced coal-processing systems* that remove potential pollutants before combustion, including technologies that convert coal to other substances like natural gas (called **gasification**), removing pollutants in the process

Ultimately, advocates of coal use describe *gasification* combined with *carbon sequestration* as the ultimate environmentally friendly way to utilize the world's enormous coal reserves. Gasification could convert coal to a number of products, including hydrogen. One such system is Integrated Coal Gasification Combined Cycle (ICGCC) power generation. In this process, coal is powdered and reacted with oxygen and other gases at high temperatures, producing a gas composed mainly of carbon monoxide and hydrogen. Sulfur compounds and other contaminants, including ash, are removed at this stage. The energy released from gas combustion powers a gas turbine. The thermal exhaust from the gas turbine is used in turn to power a steam turbine, resulting in a "combined cycle power system." Such a system could achieve thermal efficiencies approaching 50%, compared with conventional coal-fired thermal power generation systems that can achieve 35% to 40%, and nuclear plants that may achieve 33% to 35% efficiencies.

Commercial operation would require a system to economically gasify 1,500 to 2,000 tons of coal per day. According to the Department of Energy (DOE), gasification-based power plants would cost about $1200 per kilowatt to build, compared to conventional coal plants at around $900 per kilowatt.

FIGURE 15-15 Tall stacks that disperse pollution far from a coal-fired power plant. *Question:* Should users of electricity from such coal-fired plants compensate those harmed by their pollution far from the source? Why or why not? © AbleStock.

Concept Check 15-13. Would this constitute an appropriate use for government subsidies, that is, to pay utilities the difference in construction costs, or to provide tax advantages to encourage such plants to be built? Why or why not? What evidence did you use?

Concept Check 15-14. The U.S. DOE's *Vision 21* program seeks to develop "virtually pollution-free energy plant." The program is described on the Agency's website (http://www.fossil.energy.gov/programs/powersystems/vision21/). Why do you believe such a system is not already in place? Why do you believe such a proposal depends largely on government-funded research to be viable?

Value-added Question 15-11. How much more would a 1,000-megawatt ICGCC plant cost to build than a conventional 1000-megawatt coal-fired station?

A Last Look at Coal: Mountaintop Removal

On July 7, 2001, water and debris from the Kayford Mountain **mountaintop removal (MR)** project inundated Dorothy, West Virginia during a heavy rainstorm (**FIGURE 15-17**). Two people were killed and some 3,500 homes were damaged or

FIGURE 15-16 A diagram showing how fluidized bed combustion works. Powdered coal and water are injected into a pressurized cylinder, along with crushed magnesium-rich limestone and steam. The sulfur gases combine with the limestone to form calcium sulfate. Reproduced from Coal Becomes a "Future Fuel"/NETL/DOE.

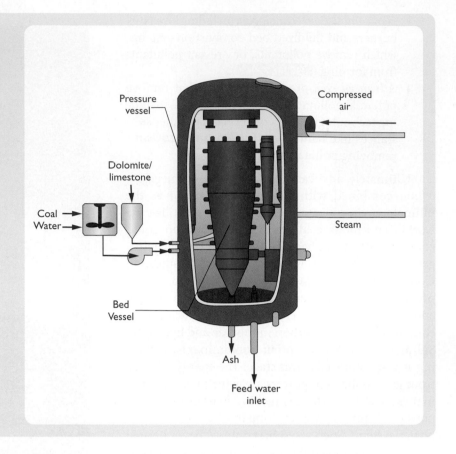

FIGURE 15-17 The Kayford Mountain mountaintop removal site. © ZUMA Press, Inc./Alamy.

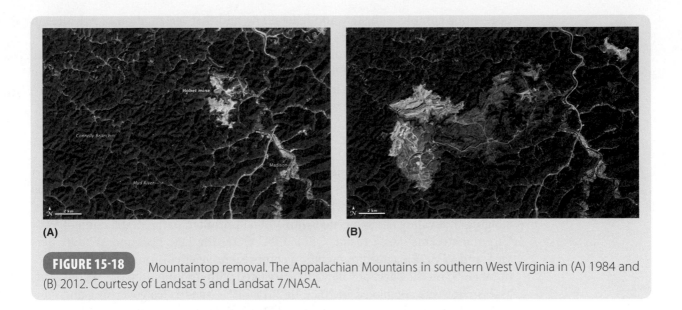

(A)　　　　　　　　　　　　　　　　　　　　　　(B)

FIGURE 15-18 Mountaintop removal. The Appalachian Mountains in southern West Virginia in (A) 1984 and (B) 2012. Courtesy of Landsat 5 and Landsat 7/NASA.

destroyed. The mine's drainage structures failed, allowing a torrent of water and detritus to overwhelm runoff ponds.

As you recall, coal mining is carried out in one of two ways: underground mining or surface mining. A relatively new type of surface mining is mountaintop removal (**FIGURE 15-18**). MR refers to the practice of removing large quantities of rock and soil above horizontal coal seams underlying mountaintops. Most MR takes place in a geological province called the Allegheny/Cumberland Plateau in the eastern United States. Excess "waste" material (called *spoil*) is dumped or bulldozed into adjacent valleys, irreversibly altering surface drainage patterns.

Surface mines require the construction of a *slurry pit*. A dammed creek is used to provide water by which coal is washed before shipment to market. The waste material (slurry) is dumped into a slurry pit. Slurry pits have failed, sometimes resulting in severe destruction and loss of life.

Coal in the West Virginia Economy

By far the leading energy product of West Virginia and one of the state's most economically important industries is coal. Each year West Virginia produces more than 170 million tons of coal of which 60% is exported. The bulk of the portion that remains in the state is used to generate around 92 million megawatt-hours of electricity of which 66 million megawatt-hours are exported. West Virginia is ranked fourteenth in the United States in electricity production.

Lack of economic opportunity often drives ecologically destructive practices. West Virginia is a focus of MR, perhaps in part due to that state's ranking as fiftieth in per capita income.

The number and size of MR operations in West Virginia, Kentucky, and to a lesser extent Virginia, have increased due to a more permissive regulatory environment, economies of scale, and the availability of larger, more efficient earth-moving equipment.

A Yale University study reported that by 2010 MR had leveled more than 25 percent of southern West Virginia's mountains, resulting in the burial of nearly 2,000 miles of streams, and clear-cutting of over 300,000 acres of hardwood forests. The National Mining Association (http://www.nma.org) argues that all MR operations comply with state and federal law, and that reclamation of MR sites in many ways increases the area's biological diversity.

MR operations comprise one fourth to one third of all coal mining in the region. According to West Virginia's Office of Surface Mining Reclamation and Enforcement (OSM), "the citizens of West Virginia and Kentucky appear to be justified in their concerns about the cumulative environmental impacts from mountaintop operations."

Regulatory and Legal Issues

In July 1998, the West Virginia Highlands Conservancy filed suit against the West Virginia Department of Environmental Protection (WVDEP) and the U.S. Army Corps of Engineers (COE), which has assumed permitting authority of MR sites, alleging that valley fills associated with MR had resulted in the destruction of West Virginia streams[1]. As a result, OSM, EPA, COE, the U.S. Fish and Wildlife Service, and WVDEP agreed to prepare a comprehensive environmental impact statement (EIS) on the effect of valley fills associated with mountaintop operations. This assessment was completed in 2005. In 2011 EPA issued rules governing and controlling MR, but the decision was stayed by a U.S. District Court.

Surface Mining Control and Reclamation Act

Under the Surface Mining Control and Reclamation Act of 1977 (SMCRA), surface mines were required to restore disturbed land to near its "original contour." A major exception was in cases where mining would leave an essentially flat or gently rolling area that could theoretically be used for "economic development." The idea was that an activity that produced flat land added value to the region, as flat land was more valuable than steeply sloping land for agriculture, or residential or commercial development. In any case, operators were prohibited from placing *spoil* (any material removed during mining that is not replaced) so as to "damage natural watercourses."

MR removes the entire top of mountains to access underlying coal deposits. These operations may remove up to a dozen or more coal seams as well as the intervening overburden (**FIGURE 15-19**). Because capital costs for the type of equipment used in MR are extraordinarily high, MR is usually not practical unless large reserves can be mined. As the use of MR in Appalachia has increased, so has the size of the permits issued. WVOSM reports that permit applications for a mine area in excess of 3,100 acres are not unusual.

Valley Fills

The material covering coal seams is called **overburden**. Mining methods frequently use heavy explosives to break up overburden into smaller pieces, causing the overall volume of the material to increase. MR in Appalachia causes the amount of spoil produced to typically exceed the amount that can safely be placed back into the mining excavation. This extra overburden volume is called *excess spoil*. In Appalachia the excess spoil is placed in adjacent valleys. These are then called **valley fills**.

According to OSM research, valley fills result in:

- altering the aesthetics of the landscape;
- adverse impacts on streams and associated fish and wildlife habitats;
- diminished quality of life for citizens living near the operations;
- lack of economic development following mining, and
- lower property values.

FIGURE 15-19 Multiple coal seams in carboniferous rocks, located in West Virginia. © Universal Images Group Limited/Alamy.

OSM concluded that, "these problems are aggravated by dust from blasting and haulage trucks, noise and vibration from blasting, truck traffic, loss of stream headwaters from spoil disposal in valley fills, and related impacts."

Concept Check 15-15. Mountaintop removal has been called "violently decapitating majestic mountains" and a threat to an entire way of life in Appalachia. Explain whether you agree with that characterization.

■ CONCLUSION: IS COAL A FAUSTIAN BARGAIN?

In Goethe's *Dr. Faustus*, the devil offered the play's protagonist his every wish fulfilled in return for his immortal soul. Environmental groups assert that coal retains a price advantage over renewable or pollution-free sources of energy like solar, wind, or hydrogen because the total impact and cost of coal mining and use is not included in the price consumers pay for electricity. An obvious example is the death and injury of thousands of coal miners each year in the Ukraine, China, and elsewhere. Another example cited is acid precipitation, which imposes measurable costs on agriculture and buildings. In addition, mercury emitted by coal plants has become a global pollutant. We cannot forget that coal burning is the largest global source of CO_2 and a significant source of atmospheric methane.

Is coal a **Faustian bargain**, or is such an analysis unduly alarmist? Industry and DOE have asserted that coal can "eventually" be burned with virtually no emissions. What will happen in the meantime?

Research the issue and decide for yourself.

CHAPTER SUMMARY

1. In 1999, the Worldwatch Institute issued a report advocating a global phase-out of coal use.

2. The World Coal Institute issued a report in 2000, which asserted that "coal bears the unjustifiable stigma of being a 'dirty' polluting fuel."

3. The modern era of environmental protection probably began in 1952, when unusual weather conditions led to air stagnation and at least 4,000 people were killed by air pollution from millions of coal fires in London, England.

4. The U.S. coal industry is the safest in the world. In open pit mines, hundreds of millions of tons a year are produced with extraordinary safety.

5. By contrast, the Ukraine coal industry is characterized by low productivity, insufficient investment in infrastructure, poor labor relations, and a bad safety record. On average about 300 miners are killed each year mining coal.

6. Tens of thousands of Guizhou, China, residents exhibit symptoms of arsenic poisoning, mainly from burning coal in houses with unvented stoves. Fluorosis and selenosis, from the same source, affect up to ten million Chinese in Guizhou and surrounding regions.

7. During 2000 to 2007, the Chinese government reported an average of 4,000 miners killed across China yearly. Labor activists assert the real figures are twice as high. By 2012, official figures put miners deaths at about 2,000.

8. In addition to methane gas explosions, underground mines are susceptible to collapse, and flooding from groundwater.

9. An estimated 4.5 percent of coal miners in the United States are affected with black lung disease.

10. In spite of safety and environmental factors, coal is a critically important resource. About two-fifths of the electricity produced in the United States comes from coal-burning power plants, due mainly to its low cost. Many countries are more reliant on coal than is the United States.

11. Coal is a complex mix of organic matter, which burns, and inorganic material, which does not.

12. Trace elements in coal are becoming increasingly of concern to coal-burning electric utilities, due to environmental impacts. Trace elements include As, Be, Cd, Cr, Co, Hg, Mn, Ni, Pb, Sb, and Se.

13. Sulfur content is related to coal's environment of deposition, in that lower-sulfur coals are deposited under freshwater conditions. Sulfur is found as FeS_2 and as organic sulfur. Sulfur content is one of the most important attributes of coal. Low-sulfur coals are at a premium because their combustion produces less SOx pollution than coals with higher sulfur content.

14. Most coals are found in sedimentary rocks interpreted to have formed in deltas or behind barrier island complexes exhibiting much fluctuation of sea level. Over millennia, buried plant tissue may be converted through a series of stages involving microbial activity, pressure, and heat, into various grades of coal; they are peat, lignite, sub-bituminous, bituminous, or anthracite coal.

15. Strip-mining coal may result in oxidation and hydration of associated materials, leading to acidification of surface water.

16. Coal is ranked according to its organic carbon content.

17. In the United States, coal production averages around 1,000 million tons a year.

18. Low-sulfur coal, like that from Wyoming, is increasingly used to meet the requirements of the Clean Air Act.

19. Associated with coal deposits are variable amounts of coal-bed methane.

20. Emissions from coal-fired power plants vary enormously. Older plants, those excluded from immediate coverage by the 1990 Clean Air Amendments, have emissions far higher than new coal power plants.

21. The most toxic emissions from coal combustion at present are NOx, SOx, and Hg.

22. Mercury emissions from coal plants have led to pollution of major waterways worldwide, including every major river in South Carolina.

23. The U.S. EPA has determined that up to one third of the air pollution affecting states from Virginia to New England comes from coal fired power plants as far away as Michigan, Indiana, and Kentucky.

24. Coal combustion is a major anthropogenic source of CO_2.

25. "Clean" coal technology is the subject of considerable research funded both by government and private industry.

26. Ultimately, supporters of coal use describe gasification as the ultimate environmentally friendly way to utilize the world's enormous coal reserves.

27. A relatively new type of surface mining is called mountaintop removal (MR). MR refers to the practice of removing large quantities of rock and soil above horizontal coal seams underlying mountaintops. Most MR takes place in a geological province called the Allegheny/Cumberland Plateau in the eastern United States. Excess waste material (spoil) is dumped or bulldozed into adjacent valleys, irreversibly altering surface drainage patterns.

28. Under the federal Surface Mining Control and Reclamation Act of 1977 (SMCRA), surface mines were required to restore disturbed land to near its "original contour." A major exception was in cases where mining would leave an essentially flat or gently rolling area that could theoretically be used for economic development.

KEY TERMS

acid aerosols

anthracite coal

ash

bituminous coal

black lung disease

carbon cycle

clean coal
   technology

coal gasification

coal reserves

grandfathered coal
   plants

externalities

Faustian bargain

fluorosis

hypoxia

ionic substitution

lignite

mountaintop
   removal (MR)

overburden

peat

selenosis

sequester

spoil

synergistic

Toxic Release
   Inventory

trace elements

valley fill

REVIEW QUESTIONS

1. The U.S. Center for Disease Control and Prevention estimated _____ former miners died each year from exposure to coal mine dust.

   a. none

   b. 500,000

   c. 100

   d. 2,000

2. The great "killer smog" of London, England, which ushered in the modern era of environmental protection, occurred in
   a. 1886.
   b. 1918.
   c. 1952.
   d. 1991.

3. The main cause of mining deaths from "accidents" in coal mines over the past three decades has been
   a. roof collapse.
   b. underground floods.
   c. machinery malfunction.
   d. methane explosions.

4. Which segment of the U.S. economy has the greatest coal consumption?
   a. the electric utility industry
   b. the iron and steel industry
   c. the aluminum refiners
   d. direct home and residence heating

5. Arsenic poisoning, selenosis, and fluorosis affect thousands to millions of residents of Guizhou, China. The source is
   a. groundwater from polluted coal mines.
   b. air pollution from coal-fired power plants.
   c. smoke from residence stoves without outdoor vents.
   d. aluminum refineries powered by local coal.

6. According to official Chinese government sources, on average how many miners are killed in coal mine "accidents" in China each year?
   a. 250–300
   b. 20–40
   c. 5,000
   d. 2,000

7. Most coal in the United States is produced
   a. in totally automated underground mines, using industrial robots.
   b. in underground mines.
   c. in surface mines.
   d. from solidification of methane.

8. The United States consumes about how much coal each year?
   a. about 1,000 million tons
   b. about 100 million tons
   c. about 1,000 billion tons
   d. about 10 million tons

9. Inorganic matter in coal, which does not burn, is called
   a. overburden.
   b. gangue.
   c. ash.
   d. spoil.

10. Coal consists mainly of which three elements?
    a. carbon, methane, and sulfur
    b. sulfur, iron, and carbon
    c. carbon, hydrogen, and oxygen
    d. nitrogen, silicon, and carbon

11. Most coals are found in
    a. metamorphic rocks.
    b. veins.
    c. sedimentary rocks.
    d. volcanic rocks.

12. $FeS_2 + 15/2O_2 + 4H_2O \rightarrow Fe_2O_3 + 4SO_4^{(-2)} + 8H^+$. This equation expresses
    a. the burning of coal.
    b. conversion of coal to methane.
    c. oxidation of pyrite, often found in coal deposits.
    d. elimination of sulfur pollution by scrubbers.

13. Which rank of coal has the most carbon, and is thus most valuable?
    a. peat
    b. lignite
    c. anthracite
    d. bituminous

14. Total U.S. coal reserves are estimated at about
    a. 250 million tons.
    b. 250 billion tons.
    c. 250 trillion tons.
    d. 2.5 billion tons.

15. If coal use did not increase in the United States, about how long would our reserves last, at our present rate of consumption?
    a. around 25 years
    b. around 100 years
    c. around 250 years
    d. around 1,000 years

16. The region containing the largest reserves of low-sulfur coal in the United States is
    a. the Yucca Mountain region, Nevada.
    b. the Appalachian Mountain region, West Virginia.
    c. the San Joaquin Valley, California.
    d. the Powder River basin, Wyoming.

17. An additional resource found along with coal in many deposits is
    a. methane.
    b. silver.
    c. uranium.
    d. porcelain clay.

18. The state with the worst acid rain of any in the United States also ranks second in SOx emissions. It is
    a. Texas.
    b. California.
    c. Pennsylvania.
    d. New Jersey.

19. Two pollutants from coal-fired power plants that contribute to acid precipitation are
    a. CO_2 and H_2O.
    b. methane and CO_2.
    c. NOx and SOx.
    d. FeO and HgO.

20. According to EPA, approximately what fraction of air pollution affecting mid-Atlantic and northeastern states from Virginia to New England comes from coal-fired power plants in the Midwest?
    a. 10%
    b. two thirds
    c. one third
    d. 75%

21. Which of these emissions from fossil fuel combustion is not a pollutant as defined by EPA?
    a. NOx
    b. SOx
    c. CO_2
    d. mercury
    e. all are pollutants.

22. Clean Coal Technology research is primarily funded by which government agency?
    a. the Department of Energy
    b. the Department of the Interior
    c. the Department of Fossil Fuels
    d. EPA

23. The mountaintop removal method of coal mining mainly affects what region of the United States?
    a. New England
    b. the Cumberland Plateau
    c. Wyoming
    d. the Pacific Northwest

24. Under the federal Surface Mining Control and Reclamation Act of 1977 (SMCRA), surface mines were generally required to restore disturbed land to
    a. near its original contour.
    b. as horizontal a surface as possible.
    c. old growth forest conditions.
    d. an economically valuable condition.

25. Valley fills occur because
    a. mountaintop removal operators are required by law to level the surrounding area.
    b. state regulators require the area to be reclaimed in as near a horizontal state as is "economically feasible."
    c. breaking up overburden increases rock volume, so it won't fit back into its original hole.
    d. biologists have determined that streams benefit by lowering stream gradient.

FOOTNOTE

[1] *Bragg v. Robertson*, Civ. Action No. 2:98–636 (S.D.W.V.).

16 Natural Gas

CHAPTER OUTLINE

Point

In early 2001, an Exxon executive testified before a Congressional committee that was considering a bill to promote a national energy policy. Here is part of his testimony.

"Over a year ago, the National Petroleum Council estimated recoverable natural gas resources in the Lower-48 states at over 1,400 trillion cubic feet. At the current rate of domestic consumption this is more than 60 years of gas supply. . . . Our country's economic growth is tightly linked to having reliable and competitive sources of energy . . .

"There are a number of areas of importance to the Natural Gas Supply Association (NGSA) producers.

- *First, access to natural resources that underlie public lands.*
- *Second, a balanced regulatory framework.*
- *Third, policies that encourage development of a diverse portfolio of energy supplies.*

"Now turning to access to public lands: An estimated 40 percent of undiscovered natural gas is located on land owned by federal and state governments. Access to government land is restricted. Outside of the central and western Gulf of Mexico, producers are prohibited access to virtually all federal lands offshore. About 9 percent of resource-bearing land in the Rockies is completely off limits, and another 32 percent is subject to significant restrictions. Worst of all, restrictions are increasing. The previous Administration's Executive Order to remove 60 million acres from potential development without any consideration given to possible future energy supplies is a flaw in the rule-making process. This ruling further contributes to the loss of access to federal lands in eight western states which, for instance, declined 60 percent between 1983 and 2000.

"Lastly, we need support for the development of our frontier resources. For example, natural gas from Alaska has the potential to play a significant role in meeting our nation's energy demand. However, Alaska presents huge technical and economic challenges. We need the support of the Federal Government, states, Canada, local communities and a host of other parties . . . to achieve success."

Concept Check 16-1. Summarize what you think the Exxon spokesperson was asking from Congress.

Concept Check 16-2. Do you agree that offshore areas like the California coast and the west coast of Florida should be opened to oil and gas exploration, often against the wishes of coastal residents? Do you think a tourist industry could be at risk? How?

Concept Check 16-3. Do you agree that wilderness areas should be opened up to gas exploration? Explain your reasons.

Concept Check 16-4. In mid-2011, a consortium of companies announced they were withdrawing plans to build a long-delayed natural gas pipeline from Prudhoe Bay. Do you agree with some energy companies and State of Alaska officials that the federal government should provide financial incentives to encourage construction of a natural gas pipeline (which has already been approved by a law passed in 1976)? Should the federal government set price supports for natural gas in the same way U.S. farm policy subsidizes corn, soybeans, and dairy products? Why or why not?

Value-added Question 16-1. The Exxon representative spoke of a "60-year supply" of natural gas, assuming then-constant rates of consumption. Based on a 3% rate of increase of natural gas consumption, use the formula $T = 70/R$ (doubling time formula; review the section on population math elsewhere in this text, if necessary) to determine how many years will be required for the consumption rate to double. How long would present reserves last under these more reasonable circumstances?

Counterpoint

The following are summaries of two reports done by scientists for the Wilderness Society, an environmental advocacy group founded in the 1930s. The first deals with estimates of recoverable reserves of oil and gas in *frontier areas*. Frontier areas include Alaska, the continental shelf, and public lands in the Rockies, and elsewhere. The second

Continued ▶

deals with fragmentation of habitat resulting from oil and gas exploration on public lands, like national forests and wilderness areas.

Report 1: *"This report presents the findings from two analyses conducted by The Wilderness Society in relation to gas and oil resources on public lands in the American West. The first analysis focused on potential economically recoverable gas and oil resources on National Forest roadless areas in six Rocky Mountain States, and 15 national monuments, managed by the Bureau of Land Management in the western states. Among our key findings:*

"Economically recoverable gas in National Forest roadless areas of our study area would meet total U.S. gas consumption for about 9 to 11 weeks. Economically recoverable oil in those roadless areas would meet total U.S. oil consumption for less than 24 days. The 15 national monuments contain less than six days of gas use and 15 days of oil use for the United States.

"Our second analysis focused on three reports that indicate substantial amounts of potential gas and oil resources are off limits to development, primarily because of environmental stipulations in government leases. We found that the reports failed to take into consideration a number of important criteria, including a full accounting of the costs of bringing the resources to market. Common weaknesses among the reports include: (1) inappropriate use of technically recoverable gas rather than economically recoverable gas in reaching conclusions, (2) a failure to consider improved access to gas from directional drilling and drill bit technology, and (3) a failure to examine access to existing gas reserves."

Report 2: *"Fragmentation of habitat is widely acknowledged as detrimental to wildlife and plant species. Landscape analysis is a proven method to identify fragmentation and other agents of change in a given area. Yet landscape analysis is seldom completed prior to initiation of oil and gas projects, despite considerable evidence that oil and gas extraction and transmittal are likely to cause wide-ranging disturbances in the landscape.*

"We conducted a pilot analysis of the landscape of the existing Big Piney-LaBarge oil and gas field in the Upper Green River Basin of Wyoming, a region where more than 3,000 oil and gas wells have been drilled. We measured the degree of habitat fragmentation of the field using three metrics: linear feature density (primarily roads and pipelines), habitat in the infrastructure effect zone, and the amount of habitat in core areas (interior habitat that is remote from infrastructure).

"Our results indicate an overall density of 8.43 miles of roads and pipelines per square mile. This is at least three times greater than road densities on national forests in Wyoming, South Dakota, and Colorado, and is "extremely high" based on ratings in the Interior Columbia Basin Ecosystem Management Project.

"The overall area of oil and gas infrastructure (roads, pipelines, pads, waste pits, etc.) at Big Piney-LaBarge covers 7 square miles of habitat, or 4% of the study area. But the effect of that infrastructure is much greater. The entire 166-square-mile landscape of the field is within one-half mile of a road, pipeline corridor, wellhead, retention pond, building, parking lot, or other component of the infrastructure. One hundred and sixty square miles—97% of the landscape—fall within one-quarter mile (~500 m) of the infrastructure. With respect to core area, only 27% of the study area is more than 500 feet (150 m) from infrastructure, and only 3% is more than one-quarter mile away.

"Our results, combined with a review of the scientific literature, suggest that there is no place in the Big Piney-LaBarge field where the greater sage-grouse—a potential candidate for the endangered and threatened species list—would not suffer from the effects of oil and gas extraction. And the vast majority of the study area has road densities greater than two miles per square mile, a level estimated to have adverse impacts on elk populations. Because our results clearly show that oil and gas drilling and extraction cause significant fragmentation of habitat, we recommend that similar spatial analyses be incorporated into the evaluation and monitoring of the ecological impacts of proposed oil and gas projects."

Concept Check 16-5. If exploration for oil and gas in frontier areas were to pose a threat to species covered under the Endangered Species Act (ESA), would you

■ OIL FIELDS OF THE PERSIAN GULF

We begin with a brief section intended to explain the basis of the intense geopolitical interest that industrialized countries exhibit in the nations bordering the Persian Gulf (FIGURE 16-1). The Persian Gulf region contains more than half of the world's proven *oil reserves*, and produces half of the world's oil. The Gulf region is an extraordinary source of oil because it has huge oil fields, and extremely low costs of exploration and production. For example, it can cost as little as one dollar per 42-gallon **barrel** in the Persian Gulf to produce oil. In new oil fields of North America and the North Sea of Europe, costs range from ten to fifteen dollars a barrel and occasionally more. Go to the U.S. Energy Information Administration website (http://www.eia.gov/dnav/pet/pet_pri_wco_k_w.htm) and check the current cost to *purchase* (rather than *produce*) a barrel of oil.

Let's focus on the oil fields of Iraq and Iran to illustrate their geology, their reserves (how much can be produced using modern technology at a profit), and their distribution (FIGURE 16-2).

The oil-containing structures illustrated in Figure 16-2 formed when Arabia was pushed north against Asia by lithospheric plate motion, crumpling the rocks in between (global tectonics is discussed elsewhere in the text).

Salt domes, the dark circles indicated by lines, are cigar-shaped columns of salt squeezed upward from deeply buried salt layers by pressure of overlying rocks (FIGURE 16-3). Salt domes are an important reservoir for oil and gas. Pressure can force oil and gas up against the sides of the dome, where it accumulates because oil cannot dissolve salt.

FIGURE 16-4 shows oil and gas fields of Iraq. Oil pipelines transport Iraqi oil for processing and export. Before the Gulf Wars of 1991 and 2003, over 80% of Iraqi oil was exported. Iraq has all the attributes necessary for oil and gas to accumulate: source beds, reservoir rocks, impermeable cap rocks, and traps to concentrate the oil. We explain these terms and processes later.

Geologists estimate Iraq's oil reserves at 112 to 142 billion 42-gallon barrels. For comparison, the United States has about 25 billion barrels of proven reserves.

Concept Check 16-9. Explain why Persian Gulf oil is potentially so important to the United States.

Value-added Question 16-2.
Iraq can produce as much as 3 million barrels of oil daily (MBD). By 2011 production had risen to 2.6 MBD. At the 2011 rate, how long would Iraq's oil reserves, using the lower estimate of 112 billion barrels, last?

Value-added Question 16-3.
The United States produces about 7 million barrels of oil daily. At that rate, how long would U.S. reserves last?

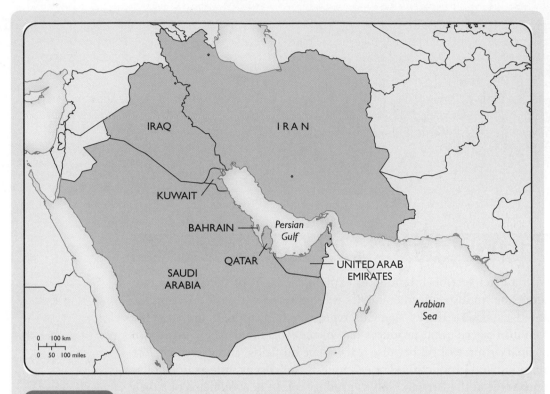

FIGURE 16-1 The Persian Gulf, home to more than half the world's proven oil reserves. Highlighted countries (Bahrain, Iran, Iraq, Kuwait, Qatar, Saudi Arabia, and the UAE), produced about 25% of the world's oil.

FIGURE 16-2 Satellite photo showing portions of southern Iran and the Persian Gulf. This photo shows the location of the major geological structures of the region, relatively young (60–50 my old) anticlines and synclines, folded structures that are known to trap oil and gas. The dark circles (arrows) are the surface expressions of vertical, cigar-shaped masses of salt deposits. The salt is impervious to oil and gas, and so traps the oil and gas. Courtesy of Lunar and Planetary Institute.

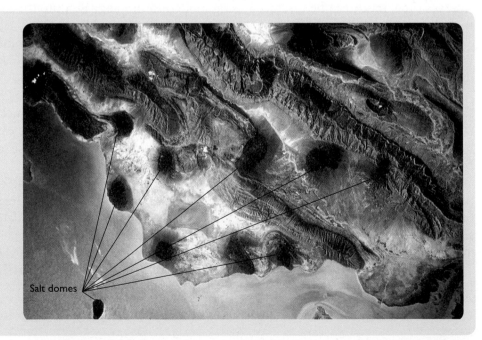

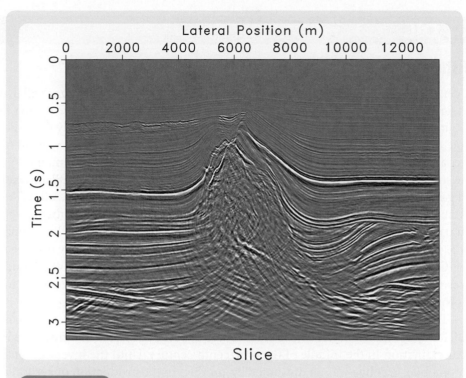

FIGURE 16-3 A salt dome outlined by seismic reflections. *Question:* Using this picture, explain how geologists can explore for potential oil and gas deposits by seismic analysis of the shallow crust. Reproduced from Fomel, Sergey, (2003)."Time-migration velocity analysis by velocity continuation" *Geophysics* 68: 1662–1672.

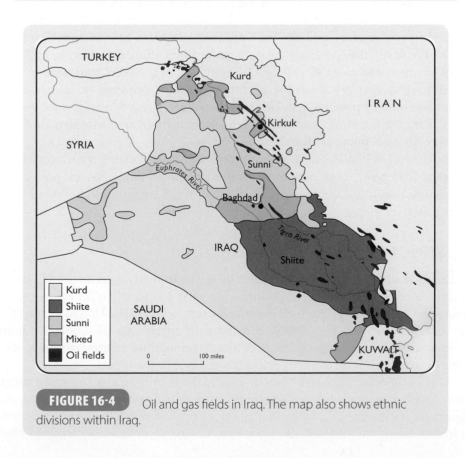

FIGURE 16-4 Oil and gas fields in Iraq. The map also shows ethnic divisions within Iraq.

■ OIL

History of Petroleum Use

A century ago, the consumption of oil was miniscule and most Americans still lived on farms. One of the great transformations in human society took place during the twentieth century: the ascendancy of oil as an energy source. Oil literally made possible the suburbanization of postwar America. As American cities have spread farther and farther from city centers, oil has become ever more essential to maintaining the suburban way of life, the life to which most Americans still aspire. Oil and natural gas are also essential to modern industrial agriculture. Not only does oil fuel our combines and tractors, but it is also the source of many pesticides and other agricultural products. Energy from oil and gas powers irrigation pumps as well. If supplies were to be shut off, industrial agriculture as we know it could eventually disappear.

Our food and most of our goods are transported using refined oil products, over highways that are maintained by government as an enormous subsidy to a motorized lifestyle.

The first product distilled from crude oil was kerosene, which was used as a fuel for lamps after whales, whose blubber was a source of whale oil, were hunted to near extinction in the middle of the nineteenth century. Until the invention of the automobile, gasoline was an unwanted by-product in the production of kerosene!

Oil Demand and Motor Vehicles

Motor vehicles with internal combustion and diesel engines began their ascendancy before World War I. An invention that helped drive America's conversion to an oil-based transportation system was the electric starter in 1911. At about that time, Henry Ford developed his mass-production system for motor vehicles, and crafted the first "hire–purchase" plan to make them affordable; that is, making a down payment and monthly payments. Before Ford, car buyers paid cash.

Officials during WWI saw the advantages gasoline-powered vehicles had over horses and battery-powered trucks, and ordered gas-powered vehicles for the war effort. The growing numbers of motor vehicles in turn drove the demand for more paved roads and "filling stations."

By the end of WWII, our transportation system was evolving from a coal-based and electric system, powering intercity trains and city trolleys, into one powered by petroleum. Now, oil alone provides nearly 40% of U.S. energy, and nearly all of the energy used by the transportation sector.

Use of Oil by Sector

FIGURE 16-5 shows the use of petroleum by sector in the United States. Oil use by electric utilities has been declining since the early 1970s, when oil prices soared, arising out of a refusal by Arab oil producers to export oil to the United States. High oil prices through the early 1980s, coupled with uncertainty of supply, led many businesses to convert to gas, or to enter into "interruptible" contracts, which allow firms to switch from oil to gas as prices favored one or the other.

One of the most important uses of oil is in the manufacture of petrochemicals. Thousands of products are manufactured from petroleum. For example, most of the plastic packaging used today is made from oil, and much of the fertilizer used in industrial agriculture is made from natural gas. Pesticides and artificial fibers

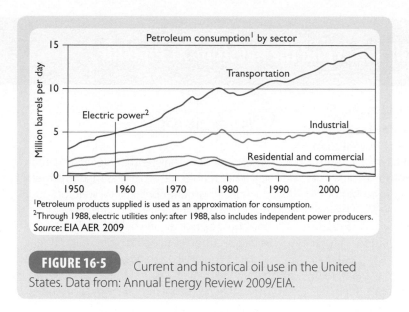

Petroleum consumption[1] by sector

[1] Petroleum products supplied is used as an approximation for consumption.
[2] Through 1988, electric utilities only; after 1988, also includes independent power producers.
Source: EIA AER 2009

FIGURE 16-5 Current and historical oil use in the United States. Data from: Annual Energy Review 2009/EIA.

like nylon are also made from oil. Medical plastics are particularly important for use as containers for intravenous/dialysis fluids, catheters, tubing, etc. In fact, some experts assert that petroleum is far too valuable to waste as a motor fuel!

Oil use by individuals for home heating is concentrated in New England and the Northeastern United States, where insufficient capacity of natural gas pipelines restricts supply. Should natural gas suppliers build new pipelines or expand existing ones, oil consumption there could drop.

The greatest source of oil use, as you can see, is in transportation, fueled by an apparently insatiable (but carefully maintained by advertising) appetite for large cars by Americans as well as by what many city planners and others call "sprawl development," that is, housing at insufficient concentration to make mass transit efficient. This subject is detailed elsewhere in the text. Many states are attempting to address inefficient development practices, which is one of the most significant energy issues of the twenty-first century. For examples of "smart growth" practices, go to http://www.smartgrowth.org.

Origin of Petroleum

Crude oil, or petroleum, is produced from organic matter in sediment by the actions of organisms in the absence of oxygen, that is, in anoxic environments. To generate oil, the sediment containing the organic matter, called a **source bed**, has to be buried and warmed by the Earth's internal heat. Such deposits, with a few interesting exceptions (see following Case Study), are now deeply buried beneath younger rocks in the North Sea and elsewhere.

When source beds are buried, complex amino acids, fats, and other organic molecules are broken into simpler, more "volatile" (that is, lower molecular weight) compounds. These lighter compounds, liquids and gases, may be displaced upwards out of their source bed by gravity, and may come to rest in layers called **reservoir rocks**. Reservoir rocks are permeable—that is, they have abundant connected pore spaces (**FIGURE 16-6**). Here is where oil, gas, and water

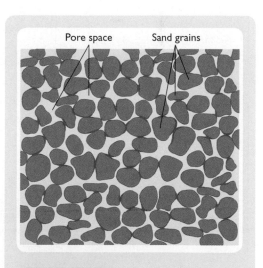

FIGURE 16-6 Close-up diagram of a permeable sandstone, a common reservoir rock in Persian Gulf oilfields, the North Sea, and elsewhere. Notice the space around and between the grains.

Most crude oil deposits are indeed deeply buried, but not all. The exceptions are vast deposits of bitumen, a viscous form of crude oil, found in sandy deposits in Venezuela and Canada. These deposits are called oil, or tar, sands. The only commercial development of these types of deposit is taking place in the Athabaska region of northern Canada, by a consortium of companies led by Shell Oil. At Athabaska, as of 2011 about 155,000 barrels of crude was being produced annually. Producing crude oil from oil sands is controversial because it requires a great deal of energy to heat the bitumen so that it may be refined, and this energy use releases greenhouse gases, such that production and use of oil from sands releases up to 15% more greenhouse gases than from conventional sources. The Keystone XL pipeline, described below, would carry about 900,000 barrels of oil daily, an amount that, when combusted, would add carbon dioxide to the atmosphere equivalent to adding more than six million new cars to U.S. roads and prevent reducing CO_2 levels to 350 parts per million, a level to which humans are thought to be able to adapt (global climate change is discussed elsewhere in the text).

In Other Words

"And make no mistake—that pipeline is a radical act. It helps unlock the planet's second-largest pool of carbon, after only the oil wells of Saudi Arabia. There's enough carbon up there that if you could burn it all off you'd raise the atmosphere's carbon concentration from its current 390 parts per million to nearly 600. Even burning a much smaller amount of these tar sands would mean that it's 'essentially game over' for the climate . . ."

Author and Environmental Activist
Bill McKibben

Another controversy is the transportation of this oil. TransCanada Corporation has proposed to extend a segment of its existing Keystone pipeline to the Gulf of Mexico near Port Arthur, Texas approximately 2,700 km (1,700 mi) through Saskatchewan, Montana, South Dakota, and Nebraska (**Figure CS 1-1**). Protestors argue that the pipeline will inevitably leak, and that this leakage could contaminate the Ogallala Aquifer (groundwater issues are discussed elsewhere in the text) as well as water for agricultural use.

FIGURE CS 1-1 (A) Map showing the existing Keystone pipeline and the Keystone XL extension. (B) Protesters opposing construction of the Keystone XL pipeline. © J. Scott Applewhite/AP Images.

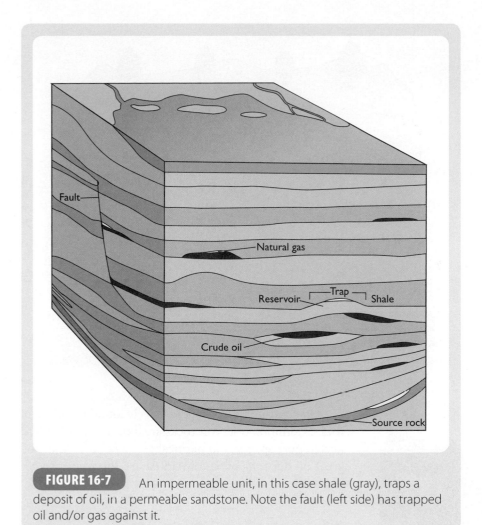

An impermeable unit, in this case shale (gray), traps a deposit of oil, in a permeable sandstone. Note the fault (left side) has trapped oil and/or gas against it.

can be found. Oil and gas deposits are formed when oil and gas is trapped in reservoir rocks by a seal or **trap**, which prevents the fluid's escape. Any impermeable barrier can potentially form a trap (**FIGURE 16-7**).

FIGURE 16-8 shows another form of trap. Here, oil may migrate upward in a permeable reservoir like the limestone in the figure, only to be trapped beneath an impermeable layer known as an unconformity. **FIGURE 16-9** illustrates a very common type of trap: a structural trap.

If no traps are encountered in its upward migration, or if the traps leak, some of the oil or gas may reach the surface, forming natural oil (or gas) seeps. The earliest discoveries of oil were at such sites. There are reportedly many natural gas seeps in the Sinai Peninsula, where the Israelites wandered after their escape from Egypt. Such seeps could be ignited by lightning strikes, or even by people. In such a way, the Biblical "burning bush" could have originated. **FIGURE 16-10** shows a natural oil and gas seep in California.

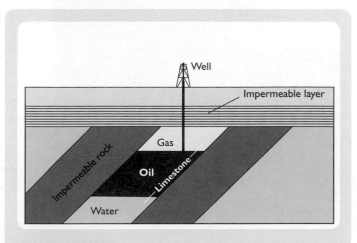

FIGURE 16-8 An unconformity, in this case where tilted rocks below intersect horizontal ones above, trapping oil and gas below.

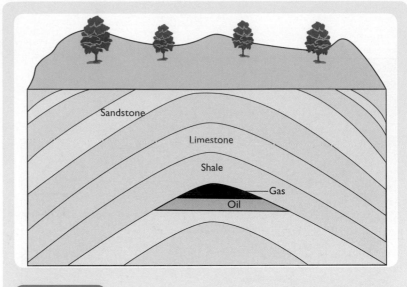

FIGURE 16-9 A structural trap. In this anticline (a fold that is convex upward, that is, in an inverted "U" shape), oil and gas migrate upward, only to be trapped in the center of the structure by impermeable shale above.

Oil deposits are almost exclusively found in sedimentary rocks. Common reservoir rocks are limestone with interconnected pore spaces, and permeable sandstones. Many of the great oil deposits of Saudi Arabia are found in permeable limestone.

What Are Petroleum Reserves?

The total amount of oil in a particular oil deposit is called a **resource**. Water and natural gas are almost always found along with the oil. Recall that gas is lighter than oil, and oil is lighter than water. Accordingly, in many reservoirs, gas is concentrated at the top, with oil below and water at the bottom.

It is not economically feasible to extract all of the oil from a reservoir rock. The amount of oil that can be extracted depends on the world price, permeability of the reservoir, size of the pores, viscosity (internal resistance to flow) of the oil, and the pressure within the rock. Oil viscosity varies considerably. The viscosity of oil is expressed by a measure known as **American Petroleum Institute (API) gravity**, and ranges from a low below 10 to a high above 50 (**FIGURE 16-11**).

Low gravity values apply to thick, waxy petroleum; higher values to more freely-flowing oil. Bitumen from oil sands are very viscous. The oil from Prudhoe Bay, Alaska (**FIGURE 16-12**) is of such low gravity that it must be heated to around 150°F (nearly 70°C) before it will flow through the 800-km-long Alaska Pipeline to the port of Valdez, where it

FIGURE 16-10 A natural oil seep at Newhall, California. Native Americans often made use of such sites to gather crude petroleum. © DigitalVues/Alamy.

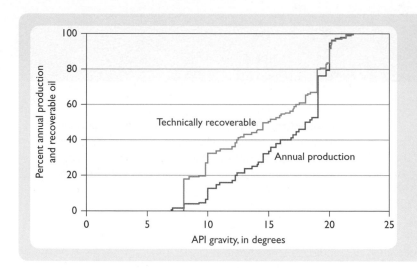

FIGURE 16-11 Annual oil production as a function of API gravity. The higher the gravity, the easier it is to extract the oil. Reproduced from U.S. Geological Survey Fact Sheet 70-03.

is loaded on tankers. The pipeline is insulated to keep in the heat, and natural gas is used as the energy source to pump the oil over two mountain ranges to Valdez.

Geologists need to know the location and distribution of oil, gas, and water in a reservoir in order to maximize the amount of oil and gas that can be extracted. Reservoir rocks with low permeability yield lower volumes of oil than those with higher permeability. This is another way of saying that low-permeability rocks retain more oil than higher-permeability reservoirs. Similar relationships pertain to many groundwater reservoirs.

The pressure in deep oil deposits is often extremely high, arising out of the weight of all that overlying rock, the extent of dissolved gas, and the incompressibility of water. Petroleum engineers seek to maintain natural pressure as long as possible and use it as effectively as possible, because using natural pressure can reduce oil extraction costs, and thus make the field more profitable.

Reserves are numerical estimates of the amount of oil that can be extracted from a reservoir, given certain assumptions. Proven reserves refer to a volume of oil that is known to exist in a reservoir that can be extracted under current prices, using current technology, without significant additional investment. Adding up all the proven reserves from all the world's known oil fields yields world proven reserves, which as of 2012 were at least 1,300 billion barrels. This statistic represents perhaps one half to one third of the oil that probably can be ultimately extracted, based on experience from individual oil fields. As much as 3,900 billion barrels of oil, or even more, thus could be ultimately extracted and used, especially if the vast oil sands of Canada and Venezuela are included.

FIGURE 16-12 A view of the Prudhoe Bay oil production facility. © Paul Andrew Lawrence/Alamy.

ANWR stands for Arctic National Wildlife Refuge.

Concept Check 16-10. What do you think is the purpose of a national wildlife refuge? Why do you believe ANWR was created?

The Arctic National Wildlife Refuge was established in 1980 by the Alaska National Interest Lands Conservation Act. According to the U.S. Fish and Wildlife (http://www .arctic.fws.gov/), ANWR *"was established to preserve unique wildlife, wilderness and recreational values; to conserve caribou herds, polar bears, grizzly bears, muskox, dall sheep, wolves, wolverines, snow geese, peregrine falcons, other migratory birds, dolly varden, and grayling; to fulfill international treaty obligations; to provide opportunities for continued subsistence uses; and to ensure necessary water quality and quantity."*

The Refuge, at 20 million acres, the approximate size of South Carolina, is home to a "full range of arctic and subarctic ecosystems . . . [and includes] 42 fish species, 37 land mammals, 8 marine mammals, and more than 200 migratory and resident bird species."

As retail gas prices in some U.S. locations exceeded $4.00/gal ($1.06/l), there were renewed calls from the American Petroleum Institute, among others, to open the ANWR on Alaska's North Slope to oil extraction. The U.S. Geological Survey (USGS) estimates between 5.7 and 16 billion barrels (mean estimate 10.4 billion barrels) ultimately might be produced over several decades.

The problem is that oil and gas leasing anywhere on ANWR is specifically prohibited by Section 1003 of the Alaska Lands Act unless the U.S. Congress decides otherwise.

First, let's discuss whether the oil in ANWR is a significant amount.

Concept Check 16-11. Does 10.4 billion barrels sound like a lot of oil to you? Explain your answer, considering the estimate of global oil reserves presented earlier.

One way to assess how meaningful 10.4 billion barrels of oil would be to the United States would be to calculate how long it would fuel our needs *assuming it was our only source of oil.*

Value-added Question 16-4. Using the U.S. annual consumption rate of oil in 2010 of 7 billion barrels per year, how many months would these new fields last? Based on your calculation, would you support drilling for oil in ANWR?

Another way of determining the importance of ANWR oil would be to compare it to U.S. proven reserves, which, according to the U.S. Energy Information Administration, were 25 billion barrels at the beginning of 2012 (excluding ANWR).

Concept Check 16-12. Explain whether this information changes your opinion about drilling in the ANWR.

As we stated earlier, oil use in the United States is about 7 billion barrels per year, of which about 4 billion barrels is imported. The U.S. Department of the Interior's Minerals Management Service and USGS have estimated that the total volume of technically recoverable oil in the United States, which includes ANWR, Gulf of Mexico, and Alaska's outer continental shelf, is about 134 billion barrels.

One argument for opening ANWR to oil drilling is that we need domestic oil immediately, both because it is cheaper than imported oil, and it frees us from dependency on imports from Middle East regimes that may be despotic.

According to the U.S. Energy Information Administration, if ANWR production begins in 2019, oil fields there would produce 780,000 barrels per day in 2027 and then decline to 710,000 barrels per day in 2030. This represents 3.1% of the projected U.S. oil consumption of 22.8 million barrels per day in 2030.

The net effect of this additional oil on gasoline prices would likely be negligible because there is an interconnected world market for gasoline. Even the most optimistic projections are a decrease of perhaps three cents a gallon in a decade or longer.

Those opposing opening ANWR for drilling also cite the area's unique biota. Environmental scientists are concerned about (1) the impact of oil exploration on the tundra environment and (2) the impact of oil exploration and production on caribou and other migrating animals. A U.S. Fish and Wildlife Service report concluded drilling in ANWR's coastal plain would cause "major environmental impacts" in an area that is the largest pristine wildlife refuge in the United States. Impacts derive from development of infrastructure (roads, buildings, etc.), pollution, and disturbance of ecosystems.

Finally, according to the U.S. EPA, every 3-mpg increase in fuel economy saves about 1 million barrels of oil per day and reduces carbon dioxide emissions by 140 million tons annually. Increasing fuel economy by 4.8 mpg eliminates the need for the 1.6 million barrels of oil a day produced in federal and state waters of the Gulf of Mexico.

Value-added Question 16-5. How much would fuel economy need to be increased to eliminate the need for the 780,000 barrels of oil a day that ANWR would provide at its peak?

Concept Check 16-13. Discuss the costs and benefits of raising fuel economy in order to avoid tapping the oil in ANWR. Discuss whether you would support raising fuel economy by this increment.

Concept Check 16-14. Summarize the issues presented in this case study.

■ NATURAL GAS

Methane (CH_4) is the main constituent of natural gas (about 75%), but natural gas usually also contains ethane (C_2H_6), propane (C_3H_8), and butane (C_4H_{10}). It is one of the planet's most important commodities. It presently meets more than 20 percent of world energy needs, and nearly a quarter of the energy needs of the United States. Because burning gas produces no sulfur oxide (SOx, or "sox") pollution, no ash or toxic emissions like heavy metals, and half the CO_2 of coal, natural gas is expected to be the fastest-growing source of fossil fuel in the twenty-first century. Natural gas may be present in reservoir rocks with oil, or may be found alone.

Nature, Origin, and Distribution

Natural gas can be produced in any geologic environment in which organic matter is decomposed by microorganisms in the absence of oxygen, with or without heat and pressure.

Natural gas deposits fall into three categories, listed below.

1. Natural gas commonly forms along with petroleum, and is usually associated with petroleum in reservoir rocks. But important reserves of gas are also found in permeable rocks with little or no oil. Until recently, the low permeability of some reservoir rocks containing gas prevented their exploitation. However, a new technique called hydrofracturing (or "fracking") uses high pressure water injected into such reservoirs through hundreds of wells to crush the reservoir rocks in place below the surface, vastly increasing the rocks permeability and allowing the gas to be extracted. While this technique can increase natural gas reserves, it also could lead to widespread pollution of surface and groundwater if not carefully monitored. In July 2011, France became the first nation to ban hydrofracturing as a method of obtaining natural gas.
2. Natural gas is produced with coal, and is typically found with coal deposits. The presence of natural gas along with coal poses the biggest hazard to miners working underground, since it is under pressure and easily ignited. We illustrate the human cost of coal mining in China elsewhere in this text.

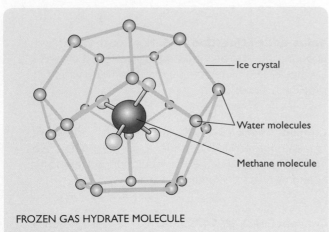

FROZEN GAS HYDRATE MOLECULE

Ice crystal

Water molecules

Methane molecule

FIGURE 16-13 A clathrate structure, with methane trapped in the center. Adapted from Suess, E., et al., *Sci. Am.* 281 (1999): 76–83. Courtesy of Deep East 2001, Naval Research Laboratory, NOAA/OER.

On the other hand, natural gas from coal mines can often be extracted prior to coal mining. Indeed, burning methane as a fuel is much preferable to "venting" the gas—that is, piping it to the atmosphere where it is dissipated—because methane in the atmosphere is a much more potent greenhouse gas than its combustion product, CO_2. Details about mining coal bed methane and global climate change can be found elsewhere in the text.

3. Methane may be formed by the action of certain bacteria in oxygen-free environments, such as water-logged soils in permafrost regions, and in deep marine sediment. In such environments, the temperature/pressure interactions may result in the formation of a structure known as a clathrate, which consists of a "cage" of ice molecules surrounding concentrated methane. Deposits of methane bound by clathrates are called **methane hydrates** (**FIGURE 16-13**). Recent research on methane hydrates in the world's oceans point to hydrates as a potential new category of natural gas reserves. Extraction of methane from clathrates is at present uneconomic.

CASE STUDY Methane Hydrates: Energy Boom or Climate Bust?

Hydrates are crystal structures made of water ice that trap natural gas (mainly methane) under conditions of high pressure and low temperatures. Hydrates could contain a vast trove of methane that potentially dwarfs other gas resources. On Alaska's North Slope alone, geologists estimate the hydrate potential at nearly 600 trillion cubic feet of methane—almost three times the known U.S. reserves.

There are two areas on Earth where methane hydrates are stored in vast quantities. One is in permafrost (permanently or seasonally frozen ground in polar regions), and the other is in marine sediment within a critical range of temperature/pressure environments.

The USGS considers the Gulf of Mexico to be the best setting on Earth to study gas hydrates. Here, they form mounds on the seafloor and can be readily sampled in sediments. **Figure CS 2-1** shows a hydrate mound on the sea floor. Hydrate mounds are often colonized by a variety of organisms. Bacteria use the hydrocarbons seeping out of the mound for energy and in turn form the base of the local food web.

Certain bacteria in marine sediments produce great volumes of methane when they feed on plant debris washed into the Gulf of Mexico from land and shore. Such biogenic methane is often trapped in layers of hydrate just below the seafloor surface.

Why Are Methane Hydrates Important?

Methane hydrates are important for two reasons. First, they may be a significant new source of a relatively "clean" fossil fuel (**Figure CS 2-2**). Natural gas is one of our most important fossil fuels, but as natural gas consumption grows in the United States and globally, many experts worry that demand may outpace supply. We have already seen price spikes in natural gas in recent years, especially in 2008, and then sharply decline in the U.S. in 2010–12 as new supplies from hydrofracturing became available.

Although estimates vary, geologists believe that the resource potential of methane in gas hydrate exceeds the worldwide reserves of conventional oil and gas deposits, plus coal and oil shale, by a considerable degree; how-

FIGURE CS 2-1 Hydrate mound on the seafloor. Courtesy of Ian R. MacDonald.

ever, there are major obstacles to producing hydrates. For example, during transport to the surface (a low-pressure environment), marine hydrate decomposes and releases its hydrocarbons.

Second, methane is an enormously powerful greenhouse gas. It is much more effective at warming the atmosphere than the most common greenhouse gas, carbon dioxide (CO_2). Any activity that releases significant new quantities of methane into the environment thus could accelerate global climate change. A rapid increase in polar and presumably global temperatures 55 million

years ago as inferred from sediment cores could have been the result of large-scale methane release.

Concept Check 16-15. Natural gas or methane is repeatedly referred to as the "cleanest fossil fuel." What do you think this means?

Concept Check 16-16. Mining methane hydrates has been called "opening a Pandora's Box." What do you think this means? Explain whether you agree or disagree.

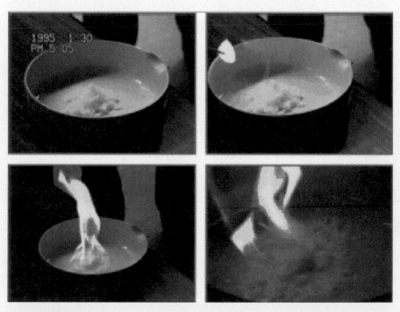

FIGURE CS 2-2 Photo showing how easily methane hydrates burn. Courtesy of Deep East 2001, Naval Research Laboratory, NOAA/OER.

Continued ▶

Formation of Natural Gas Deposits

A valuable deposit of natural gas requires a source bed, a reservoir rock, and a trap of some kind to form. Without all three, gas deposits will not accumulate. **Figure CS 2-3** shows conditions determining gas formation and accumulation in part of the North Sea.

Processing

Before it is distributed and used, natural gas usually undergoes some sort of processing. The heavier hydrocarbons propane and butane are removed and sold separately. Non-hydrocarbon gases, such as hydrogen sulfide and carbon dioxide, are also removed. Processing natural gas thus can lead to emissions of toxic hydrogen sulfide gas (H_2S) if not regulated. The "cleaned" gas may then be shipped, in the United States, the European Union, and elsewhere through tens of thousands of miles of pipeline. Natural gas is odorless, so utility companies add an "odorant" to identify leaks. Natural gas also may be transported across long distances by tanker, as we describe next.

Transportation and Storage

Natural gas may be transported in two ways: in pipelines (the Chinese built the first ones of bamboo in the sixth century bce) as a pressurized gas, or as a super-cooled liquid using specially constructed tankers. Even though natural gas commonly accumulates with oil in reservoirs, it cannot be shipped in the same pipeline used for oil. This is why the relatively large (20–35 trillion cubic feet) gas resources in Alaska's Prudhoe Bay oil field remain undeveloped. To produce and ship the gas, a new pipeline will have to be constructed. Any gas produced with Prudhoe Bay oil will continue to be either reinjected into the deep reservoir from which it came, used to heat and transport the oil in the Alaska Pipeline, or burned off in a process known as flaring (**Figure CS 2-4**).

Comparative Energy Content of Oil and Gas

To understand why wider use is not made of natural gas, consider the relative energy content of coal, oil, and gas. Each 42-gallon barrel of oil contains about 6 million BTU of energy; each kilogram of coal contains about 24,000 BTU.

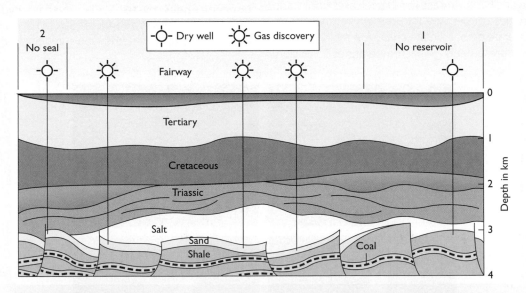

FIGURE CS 2-3 Conditions required for gas accumulation in Permian rocks of the North Sea. The sand is the reservoir rock, the coal is the source bed, and the salt layer is the necessary trap. The blue color denotes Jurassic rocks. Note that the dry wells either (1) end in shale, that is, gas does not penetrate the sandstone, or (2) end in sandstone but are not covered by impermeable salt. Thus, the necessary components of a deposit here are: (1) a coal source bed; (2) a permeable sandstone reservoir rock; and (3) an impermeable seal or cap of salt. © Geological Survey of Denmark and Greenland (GEUS).

FIGURE CS 2-4 Flaring gas from a platform in the South China Sea. Why do you think the gas is wasted in this way? Why do you think oil field operators like those in the photo above would choose to burn off such a valuable resource? © num_skyman/ShutterStock, Inc.

In volume, a kilogram of coal is about the size of a liter of bottled water. A cubic meter of natural gas contains about 30,000 BTU. Thus, it takes a relatively large volume of gas to provide the energy found in much smaller volumes of coal or oil. This helps explain why natural gas is not widely used in cars and trucks. Without pressurization, vehicles simply can't carry enough gas to match the ranges of gasoline-or-diesel-powered vehicles.

Reserves

Reserves of conventional natural gas—that is, gas in reservoir rocks—are shown in **Figure CS 2-5**. Gas reserves in coal seams are more uncertain: one estimate is shown in **Figure CS 2-6**. Extraction of methane in hydrates is not presently economically viable due to the low concentration of methane per volume of enclosing sediment, although the total amount of methane in such deposits is enormous.

Consumption

In the United States, natural gas consumption averages around 22 trillion cubic feet (TCF) per year. Global consumption of natural gas is approaching 120 trillion cubic feet and is projected to rise to nearly 170 trillion cubic feet by 2035.

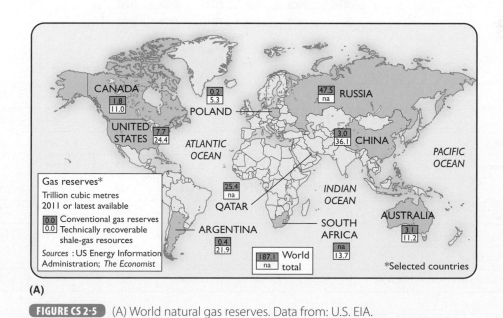

(A)

FIGURE CS 2-5 (A) World natural gas reserves. Data from: U.S. EIA.

Continued ▶

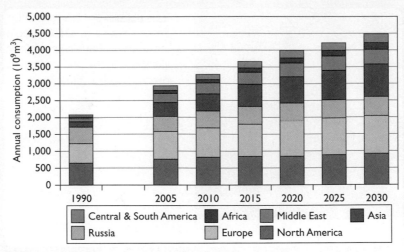

(B)

FIGURE CS 2-5 *Continued* (B) Global consumption of natural gas, actual and projected, by region. Estimate the global per capita natural gas consumption for 2010. Data from: EIA International Energy Outlook.

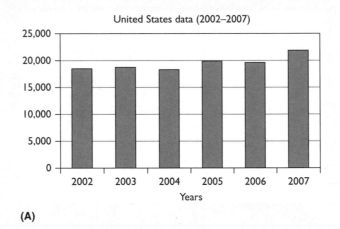

(A)

Area	2002	2003	2004	2005	2006	2007
Alabama	1,283	1,665	1,900	1,773	2,068	2,127
Colorado	6,691	6,473	5,787	6,772	6,344	7,869
New Mexico	4,380	4,396	5,166	5,249	4,894	4,169
Utah	1,725	1,224	934	902	750	922
Wyoming	2,371	2,759	2,085	2,446	2,448	2,738
Virginia					1,813	1,948
Eastern States* (IL, IN, OH, PA, WV)	1,488	1,528	1,620	1,822	273	393
Western States (AR, KS, LA, MT, OK)	553	698	898	928	1,030	1,709

(B)

FIGURE CS 2-6 (A) An estimate of U.S. coal bed methane recoverable reserves. Vertical scale is billion cubic feet. (B) U.S. coal bed methane recoverable reserves by state.

■ DISCOVERING AND EXTRACTING OIL AND GAS

Finding oil and gas deposits is based on recognizing the reservoir rocks and structures that lead to accumulation of deposits inside the Earth's crust, as you saw previously, often at depths exceeding 3 kilometers. In fact, the average oil well in the United States is around one mile (1.6 km) deep.

Exploration for oil and gas has been revolutionized over the past few decades by the use of sophisticated geophysical techniques linked to powerful computers. Since the 1970s, geologists have been able to visualize rock structures below the surface (refer to Figure 16-3, for example), and during the 1990s oil companies perfected "directional drilling," which allows exploration and production wells to literally change direction. Several oil and gas producing layers thus can be tapped by the same well, saving money, disturbing much less surface land, and allowing smaller deposits to be profitably developed (**FIGURE 16-14**). This new technology lowers costs because it requires fewer discrete wells, less casing, and less drilling rig time.

To extract oil from reservoirs, water or natural gas pressure in the reservoir itself is used, but other fluids like carbon dioxide also may be injected into the reservoir to force the oil out (**FIGURE 16-15**).

Offshore Exploration

In 1947, petroleum geologists developed the first offshore oil field using a platform, built 10 miles off the Louisiana coast. Today, a third of the world's oil is produced

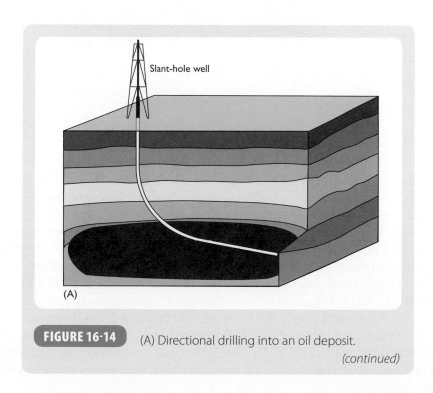

Slant-hole well

(A)

FIGURE 16-14   (A) Directional drilling into an oil deposit.

(continued)

FIGURE 16-14 *Continued.* (B) Multilateral drilling enables one master well to tap multiple reservoirs, as shown here. The technique can improve oil recovery over conventional drilling. BP estimates that one such system used in the Gulf of Mexico will remove 6 million additional barrels of oil from the reservoir compared to standard drilling techniques. (C) Horizontal well. In some reservoirs, wells can be drilled horizontally for hundreds of meters through the oil-bearing rock, making the well as much as 10 times more productive than a conventional well.

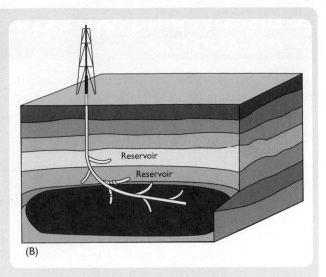

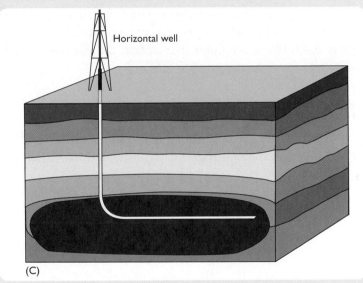

from offshore deposits, using platforms that can be 1,000 feet (330 m) high and that can cost over $1 billion to build, transport, and install. New technology allows oil to be extracted from deposits in the sea floor covered by over a mile (1.6 km) of seawater. While offshore exploration has led to the development of many additional oil fields, it also has led to considerable marine pollution. Remember that oil can be under tremendous pressure in deeply buried reservoirs. Accidents on platforms can lead to well blow-outs, like those shown in **FIGURE 16-16**. High pressure blasts oil, gas, and water out of wells, which can dump millions of gallons of crude oil into the ocean until contained. Moreover the flammable fluids can easily catch fire and explode, as happened during the early stages of the Macondo/Deepwater Horizon disaster in the Gulf of Mexico. This is one reason why many residents of the California coast and the Florida west coast have been inhospitable to offshore oil development.

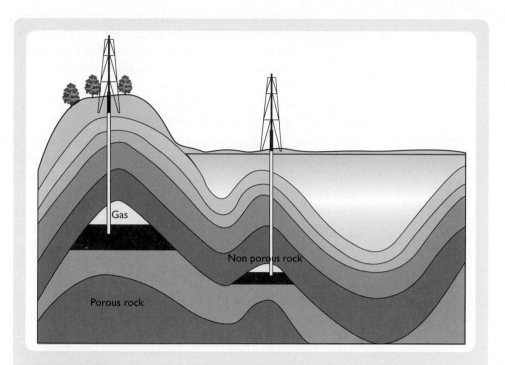

FIGURE 16-15 In these oil deposits note the natural gas, being lighter, is concentrated above the heavier oil. Drilling through the gas and into the oil can use the natural pressure exerted by the highly compressed gas to push the oil out of the reservoir. Any water in the reservoir is likely to be found below the oil as it weighs more than oil.

(A)

(B)

FIGURE 16-16 (A) The IXTOC well blowout in the Gulf of Campeche, Mexico, 1979—one of the worst oil spills in history. Courtesy of Doug Helton/NOS/ORR/NOAA. (B) The Union Oil Platform A blowout of 1969, which may have helped start the environmental movement in the United States. It fouled beaches, killed sea life for miles, and had a devastating short-term effect on the area's lucrative tourist industry. © Wally Fong/AP Images.

Improving Efficiency of Oil Recovery

The amount of oil that can be extracted from a reservoir depends on a number of factors, including the permeability of the rock, the size of the field, and the amount of natural gas and water in the reservoir. Some oil fields in the North Sea have such low permeability that as much as 95% of the oil in the field was not deemed recoverable when first developed in the 1970s.

Advanced, but expensive, techniques allow companies to significantly enhance recovery of oil from known fields. Here are a few examples:

- Bass Strait, Australia: Oil recovery increased to more than 70 percent in several Exxon/Mobil-operated fields, by applying three-dimensional (3-D) reservoir modeling, improving the performance of existing wells, drilling new wells, and installing more efficient surface processing equipment.
- Jay Field, Florida: Exxon recovered more than 60 percent of the oil from a geologically complex carbonate reservoir using advanced reservoir modeling techniques, and injection of nitrogen gas to force oil out.
- Ubit Field, Nigeria: Exxon increased the amount of recoverable oil by 500 million barrels through the application of 3-D seismic advanced reservoir modeling, by drilling of horizontal wells, and by replacing surface processing equipment.
- Prudhoe Bay, Alaska: BP increased the amount of recoverable oil by 4 billion barrels through the use of more precise reservoir models, injecting gas to improve oil recovery, installing new surface processing equipment, and by drilling horizontal wells. Operators reinject more than 6 billion cubic feet of gas each day.

Some of this **enhanced oil recovery** comes at a great energy cost, however. Critics point out that, as of 2011, 1.7 quadrillion BTUs of natural gas energy was being used to heat water to produce the steam that is injected into reservoirs containing asphalt-like petroleum. This means that the oil recovered has an inordinately large carbon cost. On the other hand, in many isolated oil fields there is no economic way to get the natural gas to a potential market.

Secondary Recovery Techniques

Oil geologists can often increase recovery of oil by injecting gases into the field, even after original gas pressure has been used up. They typically use steam, nitrogen, or carbon dioxide (which we mentioned earlier). Water is sometimes injected to push oil out of a reservoir in a process called "water flooding." Surfactants (an important ingredient in detergents) may be injected into reservoirs to 'scrub' more of the remaining oil out also. The detergent-oil mix is processed to remove the oil, and the detergent can be reinjected to recover more oil. Applying such techniques, called **secondary oil recovery**, are only considered when it is profitable to do so. Thus, higher oil prices encourage more efficient oil recovery, and lower prices discourage more efficient recovery.

Concept Check 16-17. It is clear that higher prices stimulate more oil production, but what effect do you think higher prices have on demand?

- -

Oil Refining

Oil is shipped from fields to refineries by pipeline and tanker. Nearly half the world's marine trade consists of crude oil and petroleum products. Refineries use heat, chemical reactions, and catalysts to convert crude oil into the products used by modern societies. Refineries produce gasoline, diesel fuel, kerosene, and many other hydrocarbons using a simple principle: that the hydrocarbons comprising crude oil boil at different temperatures. Impurities such as sulfur (S), nitrogen (N), and heavy metals also may be extracted from the oil during **refining**. Not all of the sulfur is refined out of the oil, because removing the S is expensive. Burning fuels thus produces sulfur dioxide (SO_2), a toxin that is also a component of acid precipitation. The United States and the countries of the European Union require refiners to remove virtually all S from refined fuels, a requirement that, while adding a few cents to the price of a gallon of fuel, will materially help clear the air in cities worldwide.

The products made from crude oil are separated in a distillation chamber. In it, the lightest molecular-weight compounds, butane and propane, are separated from the next heaviest, gasoline and kerosene, and so on through diesel fuel, home heating oil, industrial fuel oil, lubricants, thick waxes, and the heaviest, asphalt.

These products are shipped, at speeds of 3–6 miles (~5–10 km) an hour, through more than one million miles of pipelines in the United States alone.

Oil Tankers and Oil Spills

One of the most expensive forms of ecological damage results from oil spilled from tankers during tanker "accidents" (FIGURE 16-17). Large (more than a few thousand barrels) oil spills are extremely difficult, expensive, and time-consuming to remediate, and coastal ecosystems typically take years to recover from large spills. Economic losses—from lost tourist business, closed fisheries, and "cleanup" costs, for example— to some coastal areas that have suffered the effects of offshore spills, such as the Galician Coast of Spain or the Brittany Coast of France, can run into the billions of dollars. Losses to ecosystems are incalculable, using techniques of conventional economics.

FIGURE 16-17 One of the most expensive forms of ecological damage results from oil spilled from tankers during tanker "accidents." Why do you think we used quotations around the word *accidents*? Courtesy of Exxon Valdez Oil Spill Trustee Council/NOAA.

TABLE 16-1	Recent History's Most Damaging Tanker Spills		
Ship	**Year**	**Where Spilled**	**Cargo (tonnes oil)**
Prestige	2002	Galicia, Spain	77,000
Exxon Valdez	1989	Alaska	~ 30,000
Amoco Cadiz	1978	Brittany, FR	223,000
Sea Empress	1996	South Wales, UK	72,000
Torrey Canyon	1967	Cornwall, UK	117,000
Atlantic	1979	Caribbean	280,000

Data from: International Tanker Owners Pollution Federation Limited.

Worldwide, oil spills are a continuing problem, mainly because many "single-hulled" tankers are still in use. The most damaging tanker spills in recent history are listed in **TABLE 16-1**. The Exxon Valdez spill prompted the United States to prohibit single-hulled tankers from using American ports (**FIGURE 16-18**).

The greatest oil disasters, however, were not tanker spills. Iraqi damage to Kuwait's oil fields resulted in an estimated 800,000 tonne spill during the 1991 Persian Gulf War, and in 1979 an offshore platform operated by the Mexican oil company Pemex blew out, spilling an estimated 450,000 tonnes into the Gulf of Mexico. In 2010, a Transocean-owned platform leased to BP blew out when the blowout preventer manufactured by Cameron International failed to function, killing eleven workers and resulting in a minimum estimate of 200,000 tonnes of oil spilled (**FIGURE 16-19**).

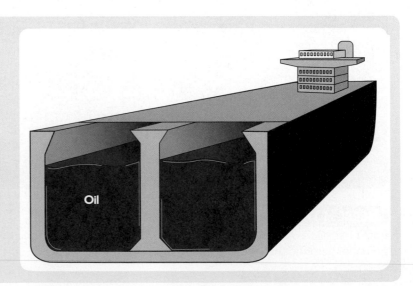

FIGURE 16-18 Structure of a double-hulled oil tanker. An outer hull is separated from an inner hull for an added margin of safely, but because there is no international law requiring decommissioning of single-hulled tankers, many are still in operation.

The 2010 *Deepwater Horizon*/BP platform explosion and oil spill (also known as the BP spill). Up to 5 million barrels (780,000 m³) of crude oil were spilled in one of the largest marine disasters of its kind. Courtesy of U.S. Coast Guard.

Focus: Deepwater Horizon 2010 Oil Spill

The Deepwater Horizon Incident in the Gulf of Mexico

On April 21, 2010, the *Deepwater Horizon* oil drilling rig, in the Gulf of Mexico, owned by Transocean and leased by BP, caught fire and sank in 5,000 feet of water, causing crude oil to gush from the well and the blowout preventer, which was built by Cameron International and failed to work. The fire continued for 36 hours until the platform sank. On September 19, 2010, a relief well was completed. The uncontrolled flow of oil, gas, and other fluids continued for 87 days, resulting in an oil spill of, in BP's words "national significance." Eleven people were killed and seventeen injured.

A BP investigative team listed eight "key findings" as causes of the incident:

1. The cement placed in the well casing by Halliburton and approved by BP was faulty and failed to "isolate the hydrocarbons."
2. The shoe track cement and the float collar, structures designed to control hydrocarbon flow at the well bottom, failed to function.

3. The Transocean rig crew and BP supervisors on the rig misread the results of a key pressure test, allowing hydrocarbons to flow uncontrolled past the blowout preventer and into the well.
4. A series of bad decisions allowed gas to rise onto the platform, which eventually caused an explosion.
5. The explosion and fire likely disabled whatever capacity the blowout preventer retained to control the well, and the blowout preventer was inoperative.

The federally-appointed Oil Spill Commission (www.oilspillcommission.gov) issued a separate report in January 2011. Here is a summary of its findings, made by the Commission's Co-Chair, William K. Reilly, formerly head of EPA: "Our investigation shows that a series of *specific and preventable human and engineering failures* were the immediate causes of the disaster. But this disaster was almost the inevitable result of years of industry and government complacency and lack of attention to safety. This was indisputably the case with BP, Transocean, and Halliburton, as well as the government agency charged with regulating offshore drilling—the former Minerals Management Service.

(continued on next page)

Only systemic reforms of both government and industry will prevent a similar, future disaster."

Here are key aspects of its findings:

1. The Macondo well blowout was the product of human error, engineering mistakes, and management failures.
2. These management failures were "not the product of a single, rogue company, but instead reveal both failures and inadequate safety procedures by three key industry players that have a large presence in offshore oil and gas drilling throughout the world."
3. Government oversight failed to reduce the risks of such a well blowout.
4. The Minerals Management Service, charged with responsibility for overseeing Gulf drilling, was critically underfunded.
5. Both industry and government were unprepared to respond to such a blowout, and unable to respond to such a vast containment operation, even though such a spill "was foreseeable."
6. The environmental damage of the spill to the Gulf will take decades to fully assess. The economic impact on the states bordering the Gulf and affected by the spill was severe. The government estimates that more than 170 million gallons of oil went into the Gulf, with "some portion" remaining in the ocean and possibly settling to the sea floor. The disaster placed further stress on coastal resources already degraded over many decades by a variety of economic and development activities, including energy production. Environmental and health effects may take years to assess.
7. "As bad as the Deepwater Horizon disaster was, it could have been much worse. At one point, *industry experts feared that a significant portion of the 4.6 billion gallon oil and gas reservoir beneath the sea floor could be released into the Gulf.*"

Further details, including the Commission's recommendations, may be found at the web address mentioned previously.

Because of the high temperature of Gulf of Mexico water, some of the oil evaporated, in contrast to the cold waters of Prince William Sound, devastated by the Exxon Valdez spill in 1988. The surfactants and dispersants BP injected into the spill, with the approval of the federal government, at the well bottom helped disperse more of the oil, making it more available to widely distributed Gulf microbes, which over time break down hydrocarbons.

BP has set aside nearly $40 billion to pay fines, compensate losses, and remediate the Gulf environment.

Around the world, in the Niger Delta of Africa, the Amazon Rain Forest, and fragile Arctic, exploration for, and production of, oil and gas pose significant risk for the marine environment. A high world price for oil ironically increases risk, as high prices result in exploration in more and more risky "frontier" areas, such as the deep ocean.

■ FUTURE OF PETROLEUM

It is clear that the world will continue to rely on oil and gas for many decades to come. But it is also clear that, while vast deposits of presently uneconomic oil are found in impermeable rocks like shales (called *oil shales*), conventional oil reserves are declining in the United States, Russia, and the European Union, where oil demand is high at present. This observation has led some to debate whether oil production has peaked, or whether it may still be possible to significantly increase supplies. Many critics assert that, because a typical oil field displays a bell-shaped curve of production and depletion, the world's oil reserves should follow the same pattern, leading them to assert that world oil production will eventually peak and begin to decline. In addition, increasing energy demand and growing world population make reducing dependency on oil more difficult. While global supplies may have been underestimated, the concentration of supplies in politically unstable regions of the world like the Middle East make dependency on oil more and more risky. Finally, continued expansion of oil and gas consumption makes addressing global climate change—which is more and more being recognized as a major national security issue for the United States—a greater challenge.

© NinaMalyna/ShutterStock, Inc.

CHAPTER SUMMARY

1. The Persian Gulf region contains more than half of the world's proven oil reserves and produces half of the world's oil.

2. Salt domes are an important reservoir for oil and gas. Pressure can force oil and gas up against the sides of the dome, where it accumulates because oil cannot dissolve salt.

3. Iraq has the attributes necessary for oil and gas to accumulate: source beds, reservoir rocks, impermeable cap rocks, and traps to concentrate the oil.

4. Geologists estimate Iraq's oil reserves at 112–142 billion 42-gallon barrels. For comparison, the United States has about 25 billion barrels of reserves.

5. One of the great transformations in human society took place during the twentieth century: the ascendancy of oil as an energy source.

6. Motor vehicles with internal combustion and diesel engines began their ascendancy before World War I. By the end of WWII, our transportation system was evolving from a coal-based and electric system, into one powered by petroleum.

7. Oil presently provides nearly 40% of U.S. energy, and nearly all of the energy used by the transportation sector.

8. One of the most important uses of oil is in the manufacture of petrochemicals.

9. Crude oil, or petroleum, is produced from organic matter in sediment by the actions of organisms in the absence of oxygen.

10. The sediment providing the organic matter for a petroleum deposit is called the source bed.

11. Organic matter is broken into simpler compounds by heat and pressure in the absence of oxygen. These lighter compounds may be displaced upward and come to rest in layers called reservoir rocks.

12. Oil and gas deposits are formed when oil and gas is trapped in reservoir rocks by a seal or trap, which prevents the fluid's escape. If no traps are encountered in its upward migration, or if the traps leak, some of the oil or gas may reach the surface, forming natural oil (or gas) seeps.

13. Proven reserves refer to a volume of oil that is known to exist in a reservoir, which can be extracted under current prices using current technology, without significant additional investment. Adding all the proven reserves from all the world's known oil fields yields world proven reserves, which as of 2012 were approximately 1,300 billion barrels.

14. Methane (CH_4) is the main constituent of natural gas (about 75%), but natural gas usually also contains ethane (C_2H_6), propane (C_3H_8) and butane (C_4H_{10}).

15. Natural gas presently meets more than 20 percent of world energy needs, and nearly a quarter of U.S. energy needs. In the United States, natural gas consumption averages around 22 trillion cubic feet (TCF) per year.

16. For a valuable deposit of natural gas to form requires a source bed, a reservoir rock, and a trap of some kind.

17. Natural gas may be transported in two ways after extraction and processing: in pipelines as a pressurized gas, or as a super-cooled liquid using specially constructed tankers.

18. Finding oil and gas deposits is based on recognizing the reservoir rocks and structures that lead to accumulation of deposits inside the Earth's crust.

19. A third of the world's oil is produced from offshore deposits, using platforms that can be 1,000 feet (330 m) high and that can cost over $1 billion to build, transport, and install.

20. Advanced, but expensive, techniques allow companies to significantly enhance recovery of oil from known fields. Ironically, higher oil prices encourage more efficient oil recovery and greater supplies, and lower prices reduce supply and discourage more efficient recovery.

21. Nearly half the world's marine trade consists of crude oil and petroleum products. One of the most expensive forms of ecological damage results from oil spilled from tankers during tanker "accidents."

22. Refineries use heat, chemical reactions, and catalysts to convert crude oil into the products used by modern societies.

23. The world will continue to rely on oil and gas for several decades to come.

24. The blowout, fire, and ultimate collapse of the Transocean Macondo drill platform in the Gulf of Mexico in April 2010 killed eleven, injured seventeen, and severely damaged the GOM environment and the economies of the states bordering the GOM. A federal commission tasked to investigate the incident reported that the disaster was foreseeable and preventable, and detailed several recommendations to minimize the likelihood of future such disasters.

KEY TERMS

American Petroleum Institute (API) gravity

barrel

enhanced oil recovery

methane

methane hydrates

natural gas

oil refining

reserves

reservoir rock

resource

salt dome

secondary oil recovery

source bed

traps

viscosity/API Gravity

REVIEW QUESTIONS

1. The Persian Gulf contains what percent of the world's reserves of oil?
   a. 90%
   b. 10%
   c. 50%
   d. 95%
   e. 75%

2. The Persian Gulf produces what percent of the world's oil?
   a. 90%
   b. 10%
   c. 95%
   d. 75%
   e. 50%

3. Motor vehicles began their ascendancy
   a. after World War II.
   b. before WWI.
   c. after 1950.
   d. before 1880.

4. Oil provides what percent of U.S. energy?
   a. 40%
   b. 90%
   c. 75%
   d. 10%

5. The sector using the most oil in the United States is
   a. electric utilities.
   b. residential and commercial.
   c. transportation.
   d. industrial.

6. The sedimentary layer containing organic matter that can be converted to oil is called a/an
   a. source bed.
   b. reservoir rock.
   c. shale.
   d. stratigraphic trap.

7. Permeable layers that can hold oil and gas are called
   a. trap rocks.
   b. reservoir rocks.
   c. source rocks
   d. aquicludes.

8. Oil and gas seeps form
   a. as a by-product of all oil deposits.
   b. whenever sandstone is the reservoir rock;
   c. when no traps are encountered.
   d. only in desert regions due to groundwater contamination.

9. Oil deposits are almost exclusively limited to
   a. the Middle East.
   b. igneous rocks.
   c. metamorphic rocks.
   d. sedimentary rocks.

10. The giant Prudhoe Bay oil field in North America is located in
    a. Alaska.
    b. Arctic Canada.
    c. the Yukon.
    d. British Columbia.

11. A volume of oil that can be extracted from a reservoir under current prices, using current technology, without significant additional investment is referred to as
    a. a demonstrated resource.
    b. a basic resource.
    c. a natural resource.
    d. a proven reserve.

12. The world's known proven reserves amount to about
    a. 1,000 million barrels.
    b. 1,000 trillion barrels.
    c. 1,000 billion barrels.
    d. 100 trillion barrels.

13. The important North Sea petroleum region has
    a. consistently lost production after 2003.
    b. doubled production after 2004.
    c. increased production by 50% after 2004.
    d. increased production by 10% per year after 2003.

14. The main component of natural gas is
    a. methane.
    b. ethane.
    c. benzene.
    d. H_2S.

15. Natural gas presently meets what % of global energy needs?
    a. about 50%
    b. about 5%
    c. about 20%
    d. about 75%

16. An advantage of natural gas over other fossil fuels is
    a. it contains the highest amount of energy per unit volume.
    b. it emits no CO_2 when burned.
    c. it emits no SOx when burned.
    d. all of the above.

17. Natural gas is transported
    a. by special tanker and by natural gas pipeline.
    b. in tankers and pipelines along with oil.
    c. mainly by tanker truck.
    d. in supercooled pipelines only.

18. Natural gas consumption in the United States in the first years of the twenty-first century averaged
    a. more than 200 trillion cubic feet per year.
    b. more than 20 million cubic feet per year.
    c. more than 20 trillion cubic feet per year.
    d. less than 5 billion cubic feet per year.

19. What event may have helped start the environmental movement in the United States?
    a. the Exxon Valdez oil spill in 1989.
    b. damage from the burned oil fields during the first Persian Gulf in 1991.
    c. the blowout of the Union Oil platform off Santa Barbara, California, 1969.
    d. the "killer smog" of London, England in 1952.

20. What percent of the world's marine trade is made up of petroleum and petroleum products?
    a. more than 40%
    b. about 10%
    c. over 90%
    d. less than 5%
    e. about 75%

21. Refineries use heat, chemical reactions, and what else to convert crude oil into useful products?
    a. CO_2
    b. oxygen
    c. mercury
    d. catalysts

22. The Exxon Valdez spill prodded the U.S. government to prohibit
    a. oil tankers from using its territorial waters.
    b. Exxon from shipping oil from Alaska.
    c. single-hulled tankers in its territorial waters.
    d. double-hulled tankers in its territorial waters.

17 Alternative Energy

CHAPTER OUTLINE

■ SUBSIDIES AND EXTERNALITIES IN FOSSIL FUEL USE

According to some energy analysts, fossil fuels represent energy's past, because so many undesirable side effects accompany the extraction and use of fossil fuels, while renewable and alternative energy represent the future.

For an example of those undesirable effects, the European Union (EU) estimates that **externalities**, or costs not included in the price of electricity production from fossil fuel and nuclear sources alone, were equal to between 1% and 2% of the EU's gross domestic product of $17,600 billion in 2010. Externalities include such things as deaths in coal mine accidents, impact of air and water pollution from fossil fuel extraction and burning, and damage from oil spills. In the United States, payments and tax breaks to fossil fuel producers exceed $5 billion per year. Two researchers in Australia estimated fossil fuel subsidies there at $AU 6.5 billion per year.

Fossil fuel subsidies to producers and consumers distort markets and reduce the competitiveness of alternative and renewable energy sources. While a few countries have cut subsidies since 1990—China eliminated $14 billion in subsidies to the coal and oil industries, and the United Kingdom eliminated subsidies to the coal industry—*total* subsidies for fossil fuels, including environmental impacts, have been estimated at $775 billion/year globally and more than $100 billion per year in the United States alone.

■ ALTERNATIVE AND RENEWABLE ENERGY

Alternative energy here refers to energy alternatives to fossil fuels. Alternatives include nuclear, hydroelectric, geothermal, hydrogen, and a variety of renewable energy sources. **Renewable energy** refers to sources that can't be "used up," so to speak. Renewable forms include energy from wind, sun (solar), biomass including wood and wood waste, tides, and waves. You have probably heard about *fuel cells* and their seemingly unlimited potential, and may wonder why they aren't included here. Fuel cells are devices that use fossil fuels or hydrogen to produce energy without combustion, thus avoiding most of the pollution problems we face when we burn fossil fuels, but they are not an energy *source*.

The main use of alternative and renewable energy today, and for the near future, is to generate electricity. In this context, you should become familiar with the term **net capacity factor**, that is, the ratio of electricity actually generated by a power plant compared to the amount that could be produced at full power with continuous operation. It depends largely on how much of the time during a year that a power plant is actually operating and producing electricity. This is one of the major shortcomings of renewable energy, at least in terms of today's power plants. Wind plants can only operate when the wind is blowing within a certain speed range. Likewise, solar plants can only produce electricity when the sun is shining. Unfortunately, users demand electricity constantly, albeit at rates that fluctuate daily and seasonally.

Concept Check 17-1. CheckPoint: Distinguish between renewable and alternative energy.

■ RENEWABLE ENERGY DEVELOPMENT

To get a sense of the state of renewable energy development at the start of the third millennium, consider the following:

- Installed wind electric-generating capacity in the United States as of December 2012 exceeded 60,000 megawatts (MW:1 megawatt is one million watts), and there are wind turbine installations in at least 26 states. American Wind Energy Association Executive Director, Randall Swisher wrote, "Wind energy installations are well ahead of the curve for contributing 20% of the U.S. electric power supply by 2030 as envisioned by the U.S. Department of Energy."
- GE Power Systems, a major supplier of wind turbines, said it expects the wind industry to grow at an annual rate of about 20%.
- FPL Energy, a subsidiary of NextEra Energy Resources, America's largest wind plant operator, had commissioned more than 10,000 MW of wind energy as of 2013.

■ ALTERNATIVE ENERGY

Total U.S. net generation of electricity in 2012 exceeded 4,050 billion kilowatt-hours (1 kW is 1,000 Watts), and was projected by the U.S. Energy Information Agency to grow to 4,908 by 2035. About 42% of the electricity was produced by coal-fired power plants, followed by 20% from nuclear, 25% from gas, 7% from hydropower, 1% from petroleum, and 5% from renewable sources. Comprehensive electricity production data can be found at http://www.eia.doe.gov/emeu/aer/txt/ptb0802a .html.

In 1949, the United States obtained nearly 10% of its total energy from alternative sources—mainly wood waste and hydropower—totaling about 3 quadrillion (**quad**) Btu out of a total production of 31 quads (1 quadrillion Btus = 10^{15}). By 2000, renewables, including hydropower, contributed over 7 quads, but energy demand had increased to 72 quads! So you can see that the proportion of U.S. energy provided by alternatives remained about the same.

Here is illustrated one of the central problems we face when we try to move to sustainable, less-polluting energy sources: *when consumption is increasing exponentially, any replacement must increase exponentially as well, but at a greater rate.*

The advantage of alternative energy sources is mainly that they are not based on burning fossil fuel. Nor are they contributors to global climate change, sometimes called *global warming,* resulting from greenhouse gas buildup, although recent trends in the use of forest biomass suggest this conclusion may need modifying. Their disadvantages, however, are varied and we will discuss them individually.

Although renewable energy is of course an alternative to fossil fuels, we will separate *alternative* energy from *renewable* energy in this chapter, purely for the purpose of packaging the discussion. We first discuss hydroelectric, nuclear, and geothermal energy.

Hydroelectric Power

Hydropower is presently the world's largest alternative source of electricity, accounting for about 15% of the world's electricity production, and 5% to 6% of total energy use. In Canada, hydroelectric power supplies 60% of domestic electricity. The United

States' largest hydropower plant is the 7,600-megawatt Grand Coulee power station on the Columbia River in Washington State (**FIGURE 17-1**). The plant's capacity is being increased to 10,080 megawatts, which will place it second in the world behind a 13,320-megawatt plant in Brazil.

Hydroelectricity is produced by damming rivers and using the kinetic energy of falling water to turn turbines (**FIGURE 17-2**). So many large dams have been constructed worldwide beginning in the middle of the twentieth century that the **global water budget** (water as a resource is discussed elsewhere in the text) has been affected. Storage of water behind dams has slightly lowered global sea level and increased evaporation from the planet's continents.

Ideal sites for hydroelectric plants are rivers with steep gradients, preferably in mountainous regions, where a river's relatively narrow channel can be easily dammed, within regions experiencing heavy precipitation. Only 20% of potential U.S. hydropower sites have been developed, but environmental concerns make most remaining sites unsuitable for **hydropower plants**.

Only around 2,400 of the United States' 80,000 dams are currently used for hydropower. Therefore, new hydropower

FIGURE 17-2 Turbine generators in a dam.
© M. Niebuhr/ShutterStock, Inc.

projects do not necessarily require building new dams—existing dams could be *retrofitted* to produce electricity.

Construction of dams has resulted in some disastrous landslides and floods. Two examples are the **Vaiont Dam** disaster of 1962 and the **Johnstown**, Pennsylvania flood of 1889. Dam construction has likewise been linked to earthquakes, for example, the 2008 magnitude 7.9 quake that killed 90,000 in southwest China. Fan Xiao, a chief engineer at China's Sichuan Geology and Mineral Bureau, suggested that the 315 million tonne weight of the reservoir's waters could have been a key factor.

Types of Hydropower Installations

There are three types of hydropower installations: conventional run-of-river, conventional storage, and pumped storage.

Most hydropower plants are *conventional* in design, meaning they use one-way water flow to generate electricity. There are two categories of conventional plants: run-of-river, and storage facilities.

Run-of-river plants use little, if any, stored water to provide water flow through the turbines. As a result, seasonal changes cause significant fluctuations in power output. There are few such plants operating in the United States.

Storage facilities use *reservoirs* behind dams to offset seasonal fluctuations in discharge (water flow), and provide a more constant supply of electricity throughout the year. Most hydroelectric plants are of this variety. Reservoirs can have the added uses of recreation, public water supply storage, and flood control. But reservoirs can also significantly increase evaporation, can trap sediment, and can alter stream ecology, resulting in local extinction of economically important species.

Pumped storage plants reuse water by constructing two reservoirs at different elevations (**FIGURE 17-3**). After water from an upper reservoir flows down a tunnel through turbines to produce electricity, it collects in the lower reservoir. During **off-peak** hours (periods of low energy demand, like evenings, as opposed to **peak** hours), the water can be pumped into the upper reservoir using spare, unwanted electricity, and reused during periods of peak demand. Pumped-storage plants are sometimes constructed with one of the reservoirs inside a mountain, and the other at the top.

Pumped-storage plants can use electricity from any other source, like a nuclear plant that can't be shut "off," in the sense that radioactive decay can't be halted. Thus, when demand for electricity from the nuclear plant is low, excess electricity can be used to pump water up to the higher reservoir. When demand is high, the pumped-storage plant can generate needed power. They could be likewise used in conjunction with solar and wind power plants.

Pumped storage systems generate about 70 units of electricity for each 100 units of electricity delivered to the plant, but without the pumped storage plant, all the surplus, unwanted electricity would be wasted.

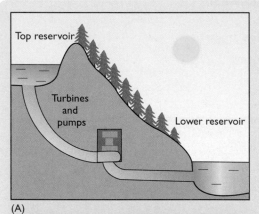

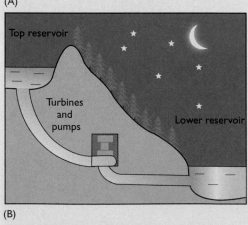

FIGURE 17-3 How a pumped-storage system works. *Questions*: How can such a facility be used to "store" electricity? What are disadvantages to pumped storage facilities?

Disadvantages

Although you may think hydroelectric energy is renewable because water in perennial streams flows all year round, conventional reservoirs behind most dams eventually fill with sediment unless the sediment is periodically removed, which is not usually cost-effective. At some point, a reservoir filled with sediment renders the dam unusable. Because most of the very large hydroelectric power dams have been built since 1930, few have as yet become unusable in this manner, but most if not all eventually will fail. In Australia, several dams built around 1890 had reservoirs filled with sediment by 1960. In the Chesapeake Bay watershed, several large power dams on the Susquehanna River may fill with silt within the next 20 years (**FIGURE 17-4**). Thus, should global energy demand continue to increase, silted-up power dams will have to be replaced, at great cost, or other forms of energy will have to be developed.

Many dams are built for flood control, and may likewise become filled over time. The Seven Oaks dam for example, built by the U.S. Army Corps of Engineers to protect areas down-gradient from Redlands, California is projected to have its storage capacity reduced from 500 feet of water to 200 feet of water, after 100 years of sedimentation.

Concept Check 17-2. Why do you think it is "uneconomic" to dredge sediment from behind dams? Where would it be placed? What impact would disposal of silt have on land? On water bodies?

A second problem with conventional hydroelectric dams is that they block rivers and thus alter entire river ecosystems. For example, construction workers on Grand Coulee Dam reported seeing thousands of 1- to 2-meter–long migratory salmon dashing themselves against the great dam in a vain attempt to get over it and reach their spawning grounds. Needless to say, the entire population was wiped out.

FIGURE 17-4 Conowingo Dam on the Susquehanna River. Sedimentation in the reservoir can reduce its capacity over time. © Philip Scalia/Alamy.

Furthermore, dams pile up vast areas of slack water, often over hundred meters deep (the world's tallest dam is 300 meters high), changing the water dynamics of the entire river **ecosystem**. Most juvenile fish that attempt to migrate downstream past the dam must pass through the dam's turbines, which kills most of them. To counter this, federal and state agencies raise fish in hatcheries. At a certain stage of development, the "fry" are then trucked around the dams in the hopes they will survive.

On the positive side, hydroelectric dams produce electricity at very low consumer rates. The capital cost of construction of the large dams of the Pacific Northwest, often carried out many decades ago, was initially borne in most cases by the federal government's taxpayers, so users don't have to directly pay this part of the electricity's cost. This subsidy means that electricity costs in regions that have access to substantial hydroelectric power, like the Pacific Northwest, historically have been some of the lowest in the United States, although this is changing as demand increases with population growth. Not surprisingly, they also have had some of the highest rates of electricity use. And energy-intensive industries—those that use a lot of electricity like aluminum refiners—usually locate in such low-cost regions.

Finally, hydroelectric power is pollution-free. No air pollution or water pollution results, and no mining waste is generated.

Many organizations are calling for dam **decommissioning** to restore watersheds, notably in California and in the Snake River in Oregon and Idaho. A number of small dams have already been removed, mainly in New England. Dam removal is also advocated for many sites in the Chesapeake Bay watershed, where more than 2000 barriers of various types to river flow have been constructed over the past 300 years.

Because dams and the reservoirs behind them have multiple constituents, and dammed rivers have many commercial and recreational uses, dam removal is very controversial.

Concept Check 17-3. Discuss whether you think more hydropower projects should be developed or that the environmental and other costs outweigh the benefits.

FIGURE 17-5 The Lardarello Geothermal Power Plant in Tuscany, the first plant of its type to be built. © Drimi/ShutterStock, Inc.

Geothermal Energy

History of Geothermal Plants

The town of Lardarello in the Tuscany region of Italy started producing electricity from geothermal resources in 1904, giving the Italians credit for developing the first *dry steam* systems (**FIGURE 17-5**). In 1958, New Zealand became the second country to produce electricity from geothermal resources, followed by the United States in 1960.

Geothermal Power in the United States

About 9,000 megawatts of geothermal electric power plants are installed around the world, with more than 3,100 megawatts of capacity in the United States. Ninety percent of that capacity is in California, with the rest in Utah, Nevada, and Hawaii. In Northern California, the Geysers geothermal field (**FIGURE 17-6**), which taps

The Geysers geothermal field in Northern California. Courtesy of NREL.

steam trapped in hot, young volcanic rocks, was built to tap the heat from recent volcanic activity north of California's Napa Valley. The Geysers produces about 750 MW, enough electricity to power San Francisco. Its capacity has been maintained by injecting sewage effluent back into the reservoir to replace lost steam.

Types of Geothermal Power Plants

Geothermal power plants (as opposed to geothermal heat pumps installed in individual residences, described later in the chapter) use steam derived from reservoirs of hot water kilometers below the Earth's surface. There are three types of geothermal power plants: dry steam, flash steam, and binary cycle.

Dry steam fields are higher-temperature fields that can produce steam under high pressure from the weight of rock overburden. Wells are drilled into the hot rocks to extract the steam. As long as steam is not withdrawn at rates that exceed groundwater **recharge**, the field will continue to produce until the excess geothermal heat is extracted, a process that could take decades or even centuries. Experts project that, barring new technologies, the Geysers geothermal plant in Northern California could lose capacity at around 10% a year due to steam depletion.

Flash steam power plants are the most common. They use geothermal reservoirs of water with temperatures greater than 360°F (182°C). As the water flows to the surface, the drop in pressure allows some of the hot water to boil, or *flash*, into steam. The steam is then separated from the water and used to power a turbine generator. Leftover water and condensed steam may be injected back into the reservoir, making this resource sustainable over long periods of time.

Binary cycle power plants operate on water at lower temperatures of about 225° to 360°F (107°–182°C). These plants use the heat from the hot water to boil a secondary fluid, usually an organic compound like isobutane with a low boiling point. This secondary fluid is boiled in a heat exchanger and used to turn a turbine. The water is then injected back into the ground to be reheated. The water and the working fluid are kept separated during the whole process, producing little or no air emissions. About 10% of geothermal capacity in the United States uses binary cycle plants.

Disadvantages

Geothermal energy is *site-specific*. Geothermal fields must be developed at the site because the steam cannot be transported far for use or it will lose its heat. Areas of high geothermal potential nearly all coincide with areas of low population in the Western United States (**FIGURE 17-7**). Moreover, toxic gases such as H_2S may be vented from the plant. These gases can usually be removed and reinjected into the reservoir at added cost. Finally, the fluids in some geothermal fields are extremely caustic, and may corrode pipes used to transmit the fluid. While not strictly a pollutant, CO_2 may also be vented from some geothermal sites, contributing to global climate change.

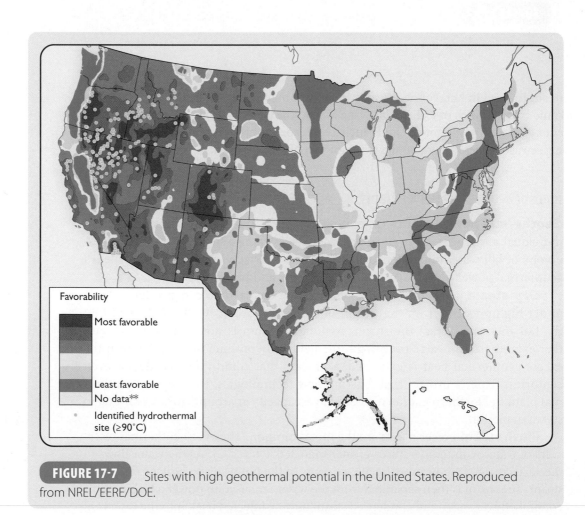

FIGURE 17-7   Sites with high geothermal potential in the United States. Reproduced from NREL/EERE/DOE.

Relatively low-temperature fields are usually uneconomic to develop unless the water in the field is of sizable quantity. Such fields may be useful for home heating and industries that need hot water, but are usually uneconomic for electricity generation.

Geothermal Heat Pumps

Geothermal heat can be tapped for residential and commercial use without generating electricity. **Geothermal heat pumps** are heat pumps that can be used for residential or commercial heating and cooling, and are 25% to 50% more efficient than conventional heat pumps. They consist of a conventional heat pump attached to 100 to 400 feet of tubing sunk into nearby ground. They take advantage of the fact that the soil a few feet below the surface is generally at the same temperature year round. During the winter the heat pump extracts heat from the soil. During the summer, the pump removes excess heat from buildings and transfers it to soils.

Although geothermal heat pumps may cost several thousand dollars more to install than regular heat pumps, they recover their cost in 2 to 10 years of operation in suitable climates, and have significantly lower maintenance costs as well. Were geothermal heat pumps to be required in new construction that use heat pumps for heating and air conditioning, the energy savings could be considerable. The disadvantage is, in addition to higher installation cost, the in-ground systems require 750 to 1,500+ square feet of ground per installed unit. **FIGURE 17-8** shows different configurations of geothermal heat pumps.

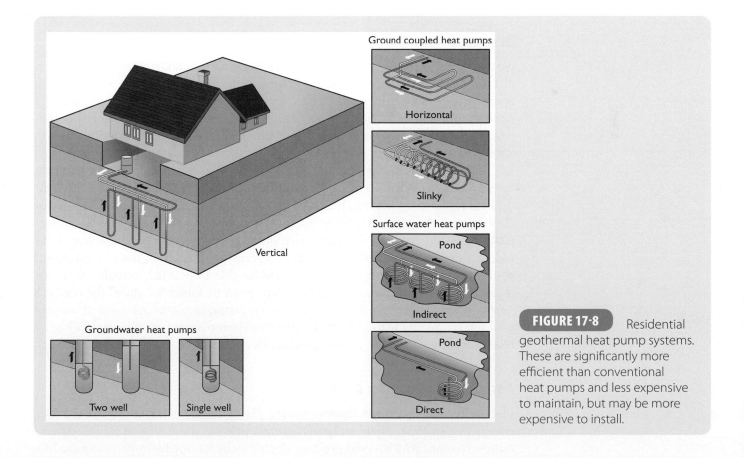

FIGURE 17-8 Residential geothermal heat pump systems. These are significantly more efficient than conventional heat pumps and less expensive to maintain, but may be more expensive to install.

Nuclear Power

As of 2011 U.S. nuclear power generation was about 780 billion kilowatt-hours (kwh), according to the Energy Information Administration (EIA). U.S. nuclear capacity of 100 Gigawatts (100,000 Megawatts) has had an annual net capacity factor of around 90% since 2000. Net capacity factor is an important criterion, since nuclear power accounted for more than 20% of the electricity generated in the United States. Because nuclear plants operate about 90% of the time, they are important **base-load plants** (see below). The top states for nuclear generation in the U.S. in 2010 were: Illinois, Pennsylvania, South Carolina, and New York.

How Nuclear Plants Work

Nuclear power plants consist of a *core* containing a *reactor*, connected to a *generator* in which a turbine spins to produce electricity. Study **FIGURE 17-9**. The core consists of **fuel rod assemblies**, which in turn contain uranium wafers. Decay of uranium atoms produces heat as well as alpha and beta particles. Heat from nuclear "decay" is transferred to pressurized water, contained in tubes surrounding the fuel rods. The water heats and produces steam. The steam, kept under high pressure, is transmitted to the generator, where the force of the steam turns a turbine, producing electricity. **Control rods**, made of neutron-absorbing material, can be lowered into the core to shut down the reactor. However, the uranium continues to decay. Therefore, unlike a gas-fired plant, nuclear power plants cannot be "turned off." This means among other things that they continue to produce heat and steam from radioactive decay whether or not there is any market for their electricity.

"Wasted" Electricity

Nuclear plants are only "disconnected" during emergencies, routine maintenance, or refueling. This illustrates one of the limitations of nuclear power. For optimum results they must be operated as base load plants. Those are plants that satisfy a region's minimal, and relatively constant, electricity needs. To illustrate this process, look at **FIGURE 17-10**, which shows the fluctuations in electricity demand in Sacramento, California.

Seasonally, electricity demand depends on climate, commercial and business activity, and the extent to which residents and commercial establishments use natural gas for heating. Regions that use natural gas have relatively lower demands for electricity during winter months because of competition from gas, which historically has been cheaper, although price fluctuations over the past few years have reduced the advantages of natural gas as a premium fuel. Thus, if a nuclear power plant is considered for a region, it should be able to provide virtually all of the region's base needs; otherwise, some means must be found to "store" the reactor's "surplus" electricity. In a few regions, pumped-storage plants are used in tandem with nuclear plants, to use surplus electricity to pump water from a lower to a higher reservoir, where it can be used to produce electricity at times of high demand.

A Brief History of Nuclear Power

Electricity from **nuclear fission** power plants began contributing to the U.S. energy mix in 1958. Nuclear plants currently provide about 19% of U.S. electricity. *Fusion* systems, which would produce electricity by forcing hydrogen atoms to fuse

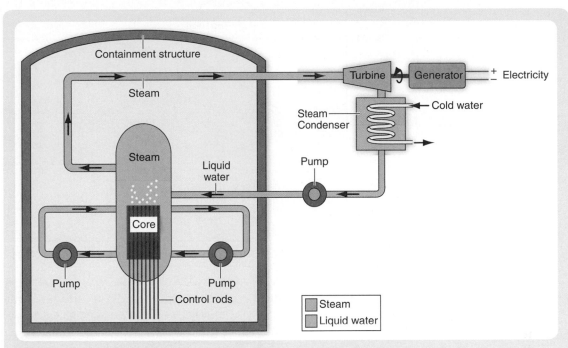

(A)

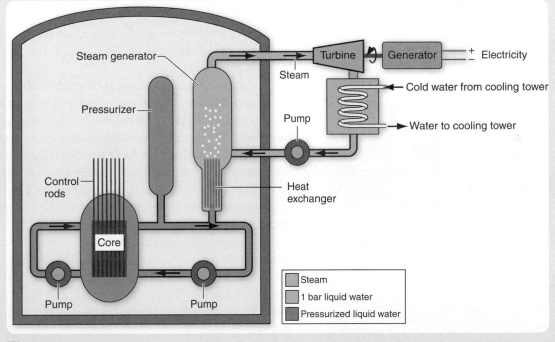

(B)

FIGURE 17-9 How the two types of nuclear power plants work. In both cases, water is boiled to produce steam. The pressurized steam turns a turbine, causing the generator to run, which produces electricity.

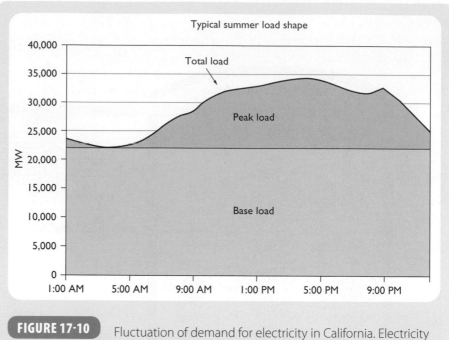

Typical summer load shape

FIGURE 17-10 Fluctuation of demand for electricity in California. Electricity use is in megawatts. Describe the daily fluctuation in demand for electricity. Data from: Transmission Agency of Northern California.

FIGURE 17-11 The meltdown of reactors at the Fukushima Daiichi Nuclear Power Plant in Japan. © Reuters/Air Photo Service/Landov.

to helium, have yet to be developed, as the technical problems associated with controlling fusion are profound, and have yet to be solved.

After the nuclear incident at Three Mile Island, Pennsylvania in 1979, demand for new nuclear plants in the United States virtually ceased. The federal Nuclear Regulatory Commission (NRC), which oversees licensing and operation of commercial nuclear plants, has extended the *operating permits* on many plants, normally awarded for 30 years, well past their projected operating lives. Therefore, nuclear plants will continue to contribute significant energy to the United States for decades to come.

The nuclear disaster at Chernobyl in the Ukraine in 1986 rendered thousands of square km around the plant uninhabitable, and generated a radioactive cloud that circled the planet. Chernobyl, Three Mile Island, and most recently the Fukushima disaster in Japan, coupled with controversy over storage of radioactive waste products, effectively drained enthusiasm for new plants in much of Europe and the United States, but plants continue to be built elsewhere, mainly in Asia, and nuclear plants continue to be important sources of electricity in the EU, Japan, South Korea, and elsewhere.

The meltdown of reactors at the Fukushima Daiichi Nuclear Power Plant in Japan (**FIGURE 17-11**), considered the second worst nuclear plant disaster after Chernobyl, following the 9.0 magnitude Tōhoku earthquake and tsunami on

March 11, 2011 has led to new efforts to curb the growth of nuclear power. In June 2011, for example, in reaction to the Fukushima meltdowns, Germany announced that it would close all 17 of its existing reactors, which provide about 23% of that country's electricity needs, by 2022.

Before the Fukushima disaster, concern about global climate change was leading policymakers to consider increasing the proportion of nuclear power. The picture has clouded significantly after Fukushima.

The Price-Anderson Act and Nuclear Liability

Operators of nuclear power plants are in large part shielded from liability in the event of a nuclear "accident" in the United States. Under the terms of the **Price-Anderson Act**, passed in 1957, the federal government assumed responsibility for monetary damages for any accident that exceeds $9 billion. Furthermore the U.S. Congress, as part of Price-Anderson, reserved the right to take "whatever actions are necessary and appropriate" to protect the public in the event of a nuclear incident where the statutory limitation on liability is exceeded. The Act was reauthorized in 2002.

Supporters of renewable energy contend that the Act provides a subsidy to the nuclear industry of about $3 billion a year, without which a private nuclear power industry would probably not exist, because scientists have calculated that the cost of an unlikely worst-case nuclear incident could exceed $300 billion. A discussion of the environmental impact of uranium mining can be found elsewhere in the text.

Nuclear Waste Issues

Concerns about waste from nuclear power center on three issues: dust dispersal at mine sites, groundwater contamination at tailings sites, and disposal of radioactive waste from reactors. Mine and tailings issues are dealt with elsewhere in the text. Here we turn to the waste from reactors. In the United States, proponents of nuclear power assert that the technology to handle reactor waste is safe, and that the small quantities of high-level radioactive waste generated since the 1950s could safely be accommodated at the proposed repository under construction at Yucca Mountain, Nevada (see following Case Study).

Concerns for the site's safety center around the length of time the waste will remain lethal—in excess of 25,000 years—which exceeds the length of times humans have had written languages by a factor of 4. Geochemists have studied fluctuations of the site's deep groundwater table and have concluded, based on analyses of groundwater flow over the past tens of thousands of years, that waste can be safely stored there. Concerns persist about the safety of spent fuel transport to the site as well as the potential for earthquake damage. We will evaluate the Yucca Mountain site later in the chapter.

Advantages of Nuclear-Generated Electricity

Nuclear decay produces no greenhouse gases (although the construction of the concrete plants produces significant greenhouse gases), and under optimum conditions emits little if any harmful radiation. Another major advantage of nuclear plants is their lack of significant conventional air emissions such as SO_x, NO_x, and mercury (Hg). And in 2007 alone, nuclear generation avoided the emission of 35 million tonnes of CO_2.

Yucca Mountain, Nevada located in the southern Nevada desert 100 miles (160 km) from Las Vegas (**Figure CS 1-1A**), was selected as the national nuclear waste repository site (**Figure CS 1-1B**), pursuant to the Nuclear Waste Policy Act of 1982. Under the terms of the National Environmental Policy Act of 1969, an environmental impact statement (EIS) was required for the proposed site. The EIS had to evaluate the site's geology, topography, plant and animal life, and all hydrologic aspects (surface streams and groundwater). Site analysis further considered the impact of time on the suitability of the site, because nuclear waste would have to be isolated from the biosphere for more than 10,000 years.

We here provide a summary of the site's geology and hydrology, and assess development of the site on the area's wildlife.

Geology

Rocks at the site are mainly Cenozoic volcanic rocks (see minerals and rocks elsewhere in the text). The repository itself is to be placed inside a welded tuff, one of the most impervious rocks on Earth (**Figure CS 1-2**).

Hydrology

Understanding the hydrology of the site involves study of surface precipitation and evaluating how much of that infiltrates to become groundwater. Moreover, scientists must determine if there are sources of groundwater outside the repository that could potentially infiltrate the repository and leach waste. (Groundwater principles are discussed elsewhere in the text.)

On average, 15 centimeters (six inches) of rain falls on Yucca Mountain each year. Due to the region's high temperature, only an extremely small fraction of that rain could soak into the ground and actually reach the water table.

Studies by scientists at Los Alamos National Laboratory and others for the Department of Energy suggest that infiltration could be diverted away from the repository level, even in the event of future climate changes.

The present water table is approximately 540 m deep, allowing placement of the repository in welded tuff 300 m underground, but still 240 m above the water table. Should precipitation increase substantially over the life of the repository—a very unlikely prospect—there would still be abundant separation between repository and the saturated zone, thus avoiding leaching of waste.

Earthquakes and Volcanic Activity at the Site

Careful monitoring of seismic activity allows underground tunnels to be designed to virtually eliminate risk from earthquakes. Moreover, although the repository site is in an ancient welded tuff, the probability of a future volcanic eruption directly impacting a repository at Yucca Mountain is estimated to be about one in 70,000,000 per year. There

(A)

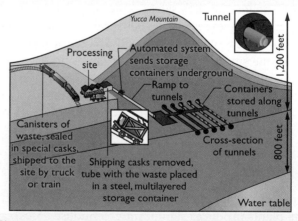

(B)

FIGURE CS 1-1 (A) A view of the Yucca Mountain Waste Repository site in Nevada. © Everett Collection Historical/Alamy. (B) A schematic diagram of the proposed facility. Reproduced from United States Nuclear Regulatory Commission.

Continued ▶

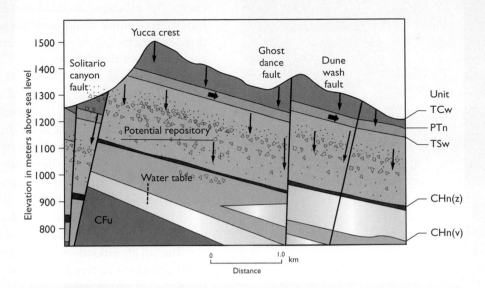

FIGURE CS 1-2 A geological cross-section through the site, showing the rocks types and the location of the water table. The repository is located within dense Tertiary age volcanic tuffs above the water table. *Question*: Based on your understanding of groundwater, is the level of the water table permanent or does it fluctuate? Based on the site's present climate, assess the likelihood of a fluctuation of the water table sufficient to engulf the repository. Because the waste can remain dangerous for more than 10,000 years, how would you determine the potential water table movement over such a time span? Reproduced from the Lawrence Berkeley Lab.

are seven inactive cinder cones near the site. Cinder cones generate among the smallest and least damaging eruptions, and thus no adverse impact is expected from volcanic activity.

In summary, scientists are confident that the site can be isolated from the biosphere for sufficient time to allow the radioactive waste to decay to harmless levels. Nothing on Earth, however, is completely risk-free. Scientists cannot therefore guarantee the site's suitability with 100% certainty (scientific "proof" is covered elsewhere in the text). The final decision to develop the site will therefore be a political one.

Every tonne of mined uranium used to produce electricity in place of coal saves the emission of 40,000 tonnes of carbon dioxide.

Nuclear fuel is "consumed" in relatively small quantities, so shipping costs are low. The fuel is solid, so tankers are not used; thus, spills are minimal in impact. Nuclear power plants theoretically can be sited anywhere power is needed as long as sufficient cooling water is available, and do not need to be close to the fuel source. Modern nuclear plants, with the notable exceptions we mentioned previously in the chapter, have good safety records.

Value-added Question 17-1.
In 2001, Australia produced 7000 tonnes of uranium for power plants. How much CO_2 emission will be avoided by using uranium instead of coal?

Disadvantages of Nuclear-Generated Electricity

Perhaps the greatest disadvantage of nuclear power is the public's perception of the risk of nuclear "accidents" based on the 1979 Three Mile Island (TMI), 1986 Chernobyl incidents, and 2011 Fukushima Daiichi Nuclear Power Plant disaster (**FIGURE 17-12**). Little radioactive material in forms harmful to life was emitted at

FIGURE 17-12 Fukushima Daiichi Nuclear Power Plant. © KYODO Kyodo/Reuters.

FIGURE 17-13 The location of the Chernobyl nuclear power plant, Ukraine. The disaster forced the government to increase dependency on coal as a power source. Ukrainian coal mines are among the most dangerous and inefficient in the developed world.

TMI; indeed Reactor 1, which was not operating at the time, continues to produce large amounts of electricity. But officials conceded that large quantities of radioactive iodine could have been emitted, putting thousands of people, mainly children, at risk from thyroid cancer within 20 miles of the plant.

FIGURE 17-13 shows Chernobyl in the Ukraine—a major planetary environmental disaster. The emissions effectively sterilized thousands of hectares of land around the plant and caused thousands of immediate or potential cancers. A plume of contaminants containing up to 40 times the radiation of the bombs the United States dropped on Hiroshima and Nagasaki at the end of World War II circled the globe.

Many experts believe the Chernobyl disaster contributed to the collapse of the USSR four years later.

The 2011 Fukushima Daiichi Nuclear Power Plant disaster occurred in one of the Earth's most tectonically active areas. The power plant is sited immediately on the coast. In anticipation of a tsunami generated by an earthquake, officials designed a seawall to defend against a 19 ft (5.7 m) wall of water. But the peak height of the tsunami was 46 ft (14 m), thus overwhelming the defenses and flooding several of the reactors, three of which experienced a full meltdown. Clean-up of the site could take a decade or longer.

Additionally, the problem of transport and storage or "disposal" of radioactive wastes has not been solved to the public's satisfaction, at least not in the United States. The controversy over the proposed long-term storage facility at Yucca Mountain (see previous Case Study) is illustrative of this concern.

Nuclear power also produces considerable amounts of "**waste heat**," that is, heat that cannot be used to produce electricity and must be dispersed into an adjacent water body or into the air. Nuclear power plants produce so much excess heat that they can transform the ecosystem of large rivers, lakes, or ocean coasts by heating the water.

Finally, for nuclear power plants to be built in the United States, the public must accept some of the risk in the form of government-subsidized insurance (see the Price-Anderson Act). This is one of the major subsidies in the entire energy industry, and supporters of renewable energy charge it puts renewables at a considerable competitive disadvantage.

Concept Check 17-4. CheckPoint: Summarize the advantages and disadvantages of nuclear energy.

BOX 17-1

Indian Point and Waste Heat

Figure B 1-1 shows the Indian Point nuclear plant on the Hudson River in Westchester County, New York. The plant's two reactors generate up to 11% of New York State's electricity. Indian Point employs 1,500 workers. In 2007, the State of New York said, "Radionuclides are leaking into the groundwater and into the Hudson River from the aging structures of Indian Point."

Scientists long have expressed concern about the effects of the plant's waste heat on the ecology of the Hudson. The plant withdraws up to 2.5 billion gallons a day from the river to cool its two reactors, and the State Department of Environmental Conservation said in a draft report that the withdrawals contributed to "significant mortality of aquatic organisms," essentially confirming earlier research which found that 1.2 billion aquatic organisms had been killed each year between 1981 and 1987 by the plant's water withdrawals and waste heat. Most of the organisms killed were eggs and larvae of various species. The plant's operators called the proposed remedy, the installation of cooling towers, prohibitively expensive, and estimated the cost at $1.6 billion. They said adding the cooling system could make the plant too expensive to operate. The operators of Indian Point filed for relicensing of the plant in 2007. Initially the state opposed the effort. The licensing process continued into 2013. The plant's present license expires in 2015.

Concept Check 17-5. Is this environmental impact an example of a subsidy for electricity production? Why or why not?

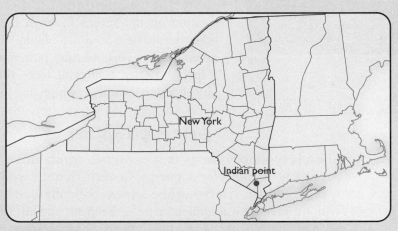

FIGURE B 1-1 The location of Indian Point nuclear power plant on the Hudson River.

■ RENEWABLE ENERGY

Tidal Power

The Earth–Moon system generates a vast amount of *tidal* energy, and where topography, tidal range, and other factors permit, it can be tapped to produce electricity. According to the U.S. Department of Energy, using present technology, for tidal differences to be harnessed into electricity, the difference between high and low tides must be at least 5 meters (16 feet). There are only about 40 sites on the Earth with tidal ranges this favorable. The greatest is in the Bay of Fundy between the Canadian Provinces of Nova Scotia and New Brunswick, with maximum tidal range of 14 meters.

Tidal power systems have been successfully producing electricity since 1966 at a favorable site along the French coast of Brittany. During a typical year, the La Rance tidal power station, a 330-m dam across an estuary fitted with 24 10-MW turbines, produces about 600 million kWh of electricity. The reversible turbines produce electricity during both incoming and outgoing tides. **FIGURE 17-14** shows the tidal power station at La Rance.

Tidal systems offer environmental advantages in that they would produce no air or water pollution, or greenhouse emissions. Drawbacks to tidal power development are similar to those for hydroelectric dams. The barriers would effectively block sites, altering the marine ecosystem. Because only a few tidal power plants have been built, very little is known about the overall impact of tidal power systems on the local environment. The effect on the environment caused by a tidal barrier is largely dependent upon the local geography and ecosystem. At the La Rance site, the environmental impact is reportedly negligible.

A major constraint to the increased use of tidal energy is the cost of building tidal generating stations. Experts estimate that the construction cost of a proposed tidal power station on the Severn River in the Western United Kingdom would be at least $15 billion. Thus, while operating and maintenance costs of tidal power plants are very low—the "fuel," seawater, is, after all, free—the overall cost of electricity generated is still very high due to high capital cost.

FIGURE 17-14 La Rance Power Station on the French coast. © PATRICK FORGET/age fotostock.

Wave Power Systems

There is enormous energy in ocean waves, which are set in motion by the wind. According to the World Energy Council (http://www.worldenergy.org), ocean waves could generate twice the electricity presently used by the entire planet! Areas with the greatest wave potential lie between 30° and 60° latitude, along the western edge of South America and the southern coast of Australia. Other favorable areas include the persistent trade winds belt north and south of the equator, and areas of high latitude (e.g., Iceland) arising from polar storms.

Total energy available depends on wave height, wave speed, wavelength, and the persistence of the waves. Using a variety of simple piston-like devices (**FIGURE 17-15**), wave systems produce energy directly from the surface motion of ocean waves or from subsurface pressure variations.

Although wave energy development is largely in the experimental stage, several nations have announced plans to encourage its development. The U.S. Department of Energy is funding research on wave power systems. Experts at the Electric Power Research Institute estimate that wave systems could produce 6% to 10% of the present U.S. demand for electricity. Wave power systems, however, could be most significant for islands like Hawaii, with no indigenous sources of fossil energy, high energy production and transport costs, and persistent waves.

But the dispersed nature of wave energy makes it difficult to harness this power in conventional large power stations. The Scottish Isle of Islay has one of the only commercial wave power stations in the world, with a capacity of 500 kilowatts

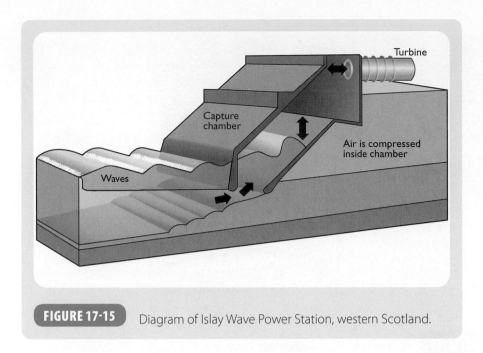

FIGURE 17-15 Diagram of Islay Wave Power Station, western Scotland.

(0.5 MW; Figure 17-15). A small wave power array is presently generating electricity off the coast of Portugal, and Pacific Gas and Electric, a California utility, has signed an agreement to purchase electricity from a wave array (Wave Connect™ Project) to be constructed off California's Humboldt County, beginning in 2012. In Oregon, a wave power device was successfully deployed in 2007 but sank when technicians tried to recover it after the test was complete (**FIGURE 17-16**).

Ocean currents also represent an enormous storehouse of energy (**FIGURE 17-17**). Development of this promising source, however, is at an early stage.

(A)

FIGURE 17-16 (A) Concept of a wave energy generating system.
Courtesy of Ocean Power Technologies, Inc. *(Continued)*

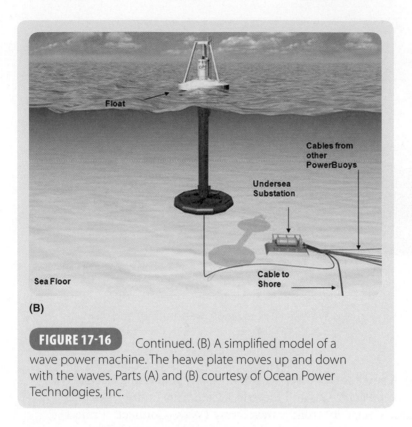

(B)

FIGURE 17-16 Continued. (B) A simplified model of a wave power machine. The heave plate moves up and down with the waves. Parts (A) and (B) courtesy of Ocean Power Technologies, Inc.

FIGURE 17-17 The energy of ocean currents may be harvested in the future with underwater turbines. Image courtesy of Marine Current Turbines, a Siemens business.

Ocean Thermal Systems

The National Renewable Energy Laboratory stated it eloquently:

> On an average day, 60 million square kilometers (23 million square miles) of tropical seas absorb an amount of solar radiation equal in heat content to about 250 billion barrels of oil (the world's proven oil reserves total about 1,100 billion barrels). If less than one-tenth of one percent of this stored solar energy could be converted into electric power, it would supply more than 20 times the total amount of electricity consumed in the United States on any given day.

Systems that convert solar energy stored in heated seawater to electricity are called Ocean Thermal Energy Conversion (OTEC). OTEC systems use the temperature differences between surface and deep water in the oceans (**FIGURE 17-18**). They operate very much like home heat pumps: using warm surface seawater to vaporize a fluid like ammonia with a very low boiling point to turn a turbine hooked to an alternator to produce electricity. At least a 20°C temperature difference in seawater is needed to efficiently produce electricity.

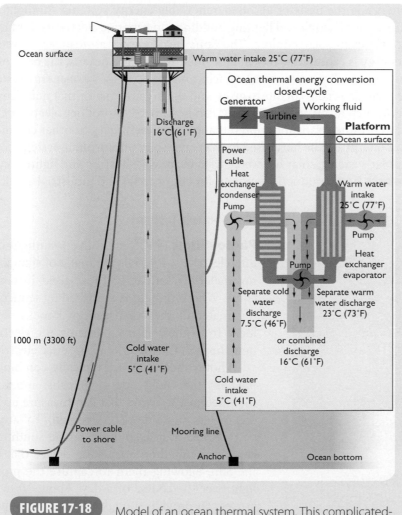

FIGURE 17-18 Model of an ocean thermal system. This complicated-looking system works much like a refrigerator in reverse. Reproduced from NOAA's Office of Ocean and Coastal Resource Management.

Cold seawater from depths of hundreds of meters would be used to condense the fluid. Warm surface seawater would vaporize it, and the process is repeated continually. The difference between power necessary to pump the seawater and the total power produced by the system is called the *net output*. A 1979 pilot project using a 50-kW OTEC system produced about 15 kW of net power. In 1981, Japanese scientists developed a 100-kW OTEC system that produced 31-kW net output.

Small-scale demonstration plants have been successfully tested. Large-scale heat exchangers must be made more efficient to produce more cost-effective electricity, and a commercial scale plant producing 40- to 100-MW net output remains to be built and operated. By the middle 2000s, researchers concluded that OTEC systems could be competitive with $60 per barrel oil and comparably priced natural gas.

According to the National Renewable Energy Laboratory, the historic low cost of energy delayed the construction of a permanent, continuously operating ocean thermal plant. Yet ocean thermal remains promising as an alternative energy source for tropical island communities, like Hawaii, that rely heavily on imported fuel, and research continues in the United States, Japan, and elsewhere.

Solar Energy

Solar energy is the basis of all other renewable energy systems. Sunlight can be used directly for heating and lighting buildings, to generate electricity, for hot water heating and even solar cooling. The Sun's heat powers atmospheric circulation, whose energy is captured with wind turbines. It heats the oceans, the basis of ocean thermal systems. It powers waves, which can be tapped by wave energy systems.

Sunlight also causes plants to grow, and the organic matter that makes up those plants is known as **biomass**. Biomass can be burned to produce electricity, or fermented to produce transportation (ethanol, biodiesel) fuels, or other chemicals.

Solar energy can be harnessed in five ways: solar photovoltaic systems, concentrating solar systems, solar hot water systems, passive heating and lighting systems, and **process heat** production systems for commercial and industrial applications.

FIGURE 17-19 A PV array at Coastal Carolina University. Installed over a bus stop shelter in 2006, the unit is capable of producing a maximum of 16 kW of electricity, an amount sufficient to power 160 100-watt light bulbs. © Pluff Mud Photography.

Photovoltaic Systems

Photovoltaic (**PV**) cells convert sunlight directly into electricity. PV cells are made of semiconducting materials similar to those used in computer chips. When sunlight is absorbed by semiconducting material, the solar energy knocks electrons loose from their atoms, allowing the electrons to flow through the material to produce electricity.

Photovoltaic cells are typically combined into *modules* that hold about 40 cells; modules are mounted in PV *arrays* that can measure up to several meters on a side (**FIGURE 17-19**). PV arrays can be mounted at a fixed angle facing south, or they can be mounted on a tracking device that follows the Sun, allowing them to capture the most sunlight over the course of a day. About 10 to 20 PV arrays can provide enough power for a household. For electric utility or industrial applications, hundreds of arrays can be interconnected to form a single, large PV system. **FIGURE 17-20** shows a house

FIGURE 17-20 A roof-mounted PV system in California. Your local utility may provide financial support for such systems. © Uwe Landgraf/ShutterStock, Inc.

in California with solar panel arrays on the roof. The performance of a PV cell is measured in terms of its efficiency at turning sunlight into electricity. The first PV cells built in the 1950s had efficiencies of less than 4%. A new commercial PV array as of 2013 had an efficiency of around 20%, but maximum cell efficiencies reportedly reached 38%, with future potential efficiencies estimated at 65%. Lower efficiencies mean that larger and more costly arrays are needed.

For comparison, a coal-fired power plant may achieve efficiencies of up to 40%, and a gas-fired plant over 50%. Nuclear plants have efficiencies of less than 35%, because of safety constraints.

Some PV cells are designed to operate with concentrated sunlight. These cells are built into concentrating collectors that use a lens to focus the sunlight onto the cells. The idea is to minimize use of expensive PV material while collecting as much sunlight as possible. But because the lenses must be pointed at the Sun, the use of concentrating collectors is limited to the sunniest parts of the country. Moreover, most require computerized tracking devices, which limit their use to electric utilities, industries, and commercial buildings.

Improving PV cell efficiencies while holding down the cost per cell is an important goal of the PV industry and government research laboratories.

Concentrating Solar Systems

Concentrating solar power systems use the Sun as a heat source. The "power tower" system uses a large field of mirrors to concentrate sunlight onto the top of a tower, heating molten salt in a heat exchanger (**FIGURE 17-21**).

The salt's heat is then used to generate electricity through a conventional steam generator. Molten salt retains heat efficiently, so it can be stored for days before being converted into electricity. The advantage of the molten salt system is that electricity could be produced on cloudy days or even several hours after sunset, addressing one of the major shortcomings of solar electricity.

A concentrating solar system in Australia. © SSSCCC/ShutterStock, Inc.

Passive Solar Heating and Daylighting

Buildings designed for passive solar heating usually have large, south-facing windows. Materials that absorb and store the Sun's heat can be built into the sunlit floors and walls, which then heat up during the day and slowly release heat at night, when the heat is needed most. This passive solar design feature is called **direct gain**. Passive solar also may provide *daylighting*, the use of natural sunlight to brighten a building's interior.

Excess solar heating and daylighting can be a problem during hot summer months. Features like overhangs can shade windows when the sun is high and help keep passive solar buildings cool. Sunspaces can be closed off from the rest of the building, and heat-blocking window films can be installed (FIGURE 17-22).

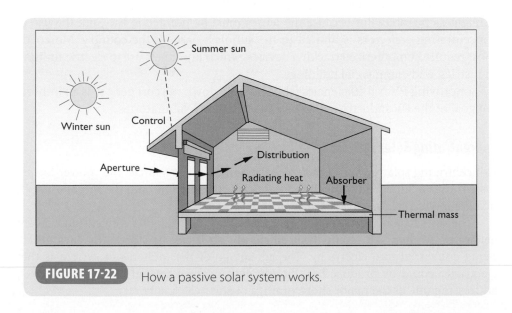

FIGURE 17-22 How a passive solar system works.

Solar Hot Water

Many parts of the world, like Israel, Greece, and China (**FIGURE 17-23A**) already use solar hot water systems extensively. Most solar water heating systems for buildings have two main parts: a solar collector and a storage tank (**FIGURE 17-23B**). The most common collector is called a flat-plate collector. Mounted on the roof, it consists

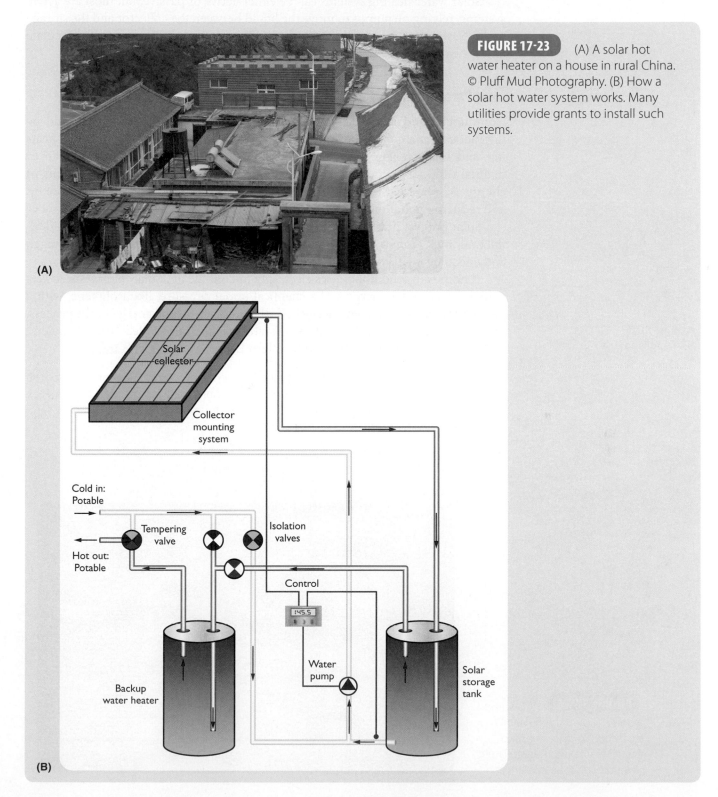

FIGURE 17-23 (A) A solar hot water heater on a house in rural China. © Pluff Mud Photography. (B) How a solar hot water system works. Many utilities provide grants to install such systems.

(A)

Solar collector

Collector mounting system

Cold in: Potable

Tempering valve

Isolation valves

Hot out: Potable

Control

145.5

Water pump

Backup water heater

Solar storage tank

(B)

of a thin, flat, rectangular box with a transparent cover facing the sun. Tubes run through the box and carry the fluid to be heated. The tubes are attached to a black absorber plate, which heats the fluid passing through the tubes.

The storage tank holds the hot liquid. It can be a modified water heater, but it is usually larger and better insulated.

Solar water heating systems can be either active or passive, but most are active systems, relying on pumps to move the liquid between the collector and the storage tank. Passive systems use gravity and the tendency for water to naturally circulate as it is heated.

Solar Process Heat

Commercial and industrial buildings can use solar technologies that would be impractical for a home. They include ventilation air preheating, solar process heating, and solar cooling. Many large buildings need large quantities of fresh air to maintain indoor air quality. Heating this air in winter can use large amounts of energy. A solar ventilation system can preheat the air, saving energy and money, and reducing pollution.

Solar process heating can provide hot water and space heating for nonresidential buildings. A typical system includes solar collectors, a pump, a heat exchanger, and adequate insulated storage tanks.

A home air conditioner uses electricity to cool the air. Solar *absorption coolers* cool air using solar energy and a chemical under pressure, like isobutene, with a low boiling point. The solar energy boils the fluid, absorbing energy in the process. Solar energy can also be used with evaporative coolers, also called "swamp coolers" in dry climates like that of the Southwestern United States (FIGURE 17-24).

Concept Check 17-6. Examine FIGURE 17-25 , which shows solar radiation maps of the United States and Germany as well as a list of the ten countries using the most solar energy. How would you respond to the statement that parts of the United States do not have sufficient solar radiation for effective solar energy?

FIGURE 17-24 A swamp cooler, a low-tech device common in dry areas like New Mexico and here in Utah. © Jaminson Moses/Alamy.

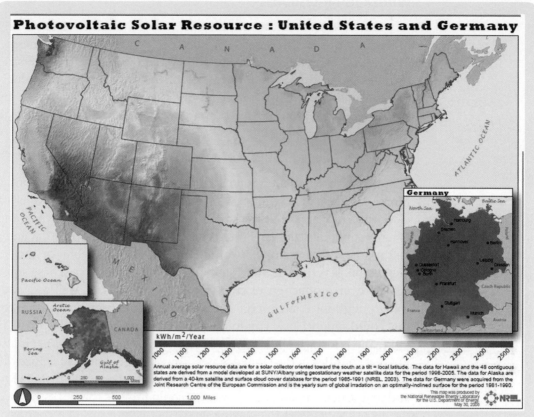

Photovoltaic Solar Resource : United States and Germany

kWh/m²/Year

1000 1100 1200 1300 1400 1500 1600 1700 1800 1900 2000 2100 2200 2300 2400 2500

Annual average solar resource data are for a solar collector oriented toward the south at a tilt = local latitude. The data for Hawaii and the 48 contiguous states are derived from a model developed at SUNY/Albany using geostationary weather satellite data for the period 1998-2005. The data for Alaska are derived from a 40-km satellite and surface cloud cover database for the period 1985-1991 (NREL, 2003). The data for Germany were acquired from the Joint Research Centre of the European Commission and is the yearly sum of global irradiation on an optimally-inclined surface for the period 1981-1990.

This map was produced by the National Renewable Energy Laboratory for the U.S. Department of Energy. May 30, 2008

(A)

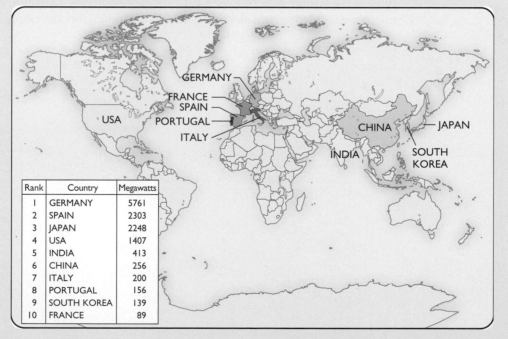

Rank	Country	Megawatts
1	GERMANY	5761
2	SPAIN	2303
3	JAPAN	2248
4	USA	1407
5	INDIA	413
6	CHINA	256
7	ITALY	200
8	PORTUGAL	156
9	SOUTH KOREA	139
10	FRANCE	89

(B)

FIGURE 17-25 (A) Solar radiation map of the United States and Germany. Courtesy of NREL/U.S. Department of Energy. (B) List of the top ten countries employing solar power. Data from: Navigation Consulting.

Wind Energy Systems

A wind turbine uses the wind's energy to generate electricity. Wind turbines, like windmills, are mounted on a tower to capture the most energy. At 100 feet (30 meters) or more above ground, they can tap faster, more persistent, and less turbulent wind. Some wind "farms" consisting of dozens of individual turbines are being installed in shallow water off the European coast in Britain and elsewhere (FIGURE 17-26). Globally, wind power systems exceeded 250,000 MW in 2012.

Subsidies for Wind Energy Development

Supporters of wind energy point to its advantages: zero pollution and free fuel. Moreover, areas of maximum wind energy potential in North America often coincide with rural areas (Nebraska, Kansas, West Texas, the Dakotas, and Wyoming) where economic stability is threatened by a declining population and a weak agricultural base. For example, two thirds of farmers in Nebraska and South Dakota depend on government price-support subsidies to survive. These subsidies in turn lead to over-production of crops already in surplus, as well as avoidable air and water pollution. Wind energy development would provide royalties to landowners that could support local communities. Loup County, NE had a population of 610 in 2011, and is losing population: in 2003, the population was 742. It is also among the poorest counties in the United States. Towns throughout the Upper Midwest could be abandoned during the next several decades without an infusion of new residents and investment. Wind energy systems could provide that investment.

In order to "level the playing field," that is, to provide a counterpoise to large subsidies for fossil-fuel production in the United States, federal and state governments are legislating considerable support for wind-energy development. The federal

FIGURE 17-26 The Danes obtain more than 10% of their electricity from wind, and expect wind to provide 20% by 2020. The Middlegrunden Wind Turbine Cooperative has an installed capacity of 40 MW. © iStockphoto/Thinkstock.

government in 2012 renewed for two additional years a 1.5 cents/kWh (kilowatt hour) tax credit for wind-generated electricity. This 1.5c/kWh tax credit helps wind compete with subsidized fossil fuels, but, because it is not a permanent feature of the federal tax code, it could be withdrawn at any time. This makes investing in wind energy risky.

Here are examples of state support for wind energy:

- *Oregon* as of 2011 provided an income-tax credit of up to $1500 a year for homeowners who produce their own wind energy.
- The *California* Energy Commission (CEC) offered a 50% rebate on the purchase price of a home- or small business wind energy system.
- In 1999, then-Gov. George W. Bush signed into law a requirement that *Texas* utilities install or contract to buy power from 2,000 megawatts (MW) of new renewable energy generating capacity by January 1, 2009. The goal was actually reached by 2002!

Potential Sites for Wind Power Development

FIGURE 17-27 shows the average annual wind energy potential for the United States. Wind speed categories of 3 or higher (WS3) are suitable for wind power development.

As you probably concluded, the potential for wind power as a source of clean, renewable electricity is enormous. Realizing that potential will depend on

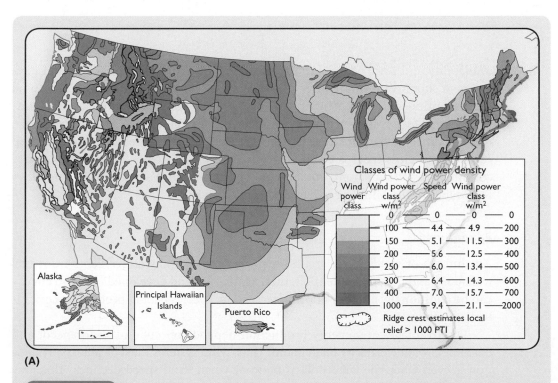

(A)

FIGURE 17-27 (A) Average annual wind energy potential for the United States. *Activity*: Identify areas in the Midwest and Northeast that are most suitable for wind power development. The greatest barrier to wind development in these favorable areas could be a lack of transmission capacity. Reproduced from Guide to Tribal Energy Development/EERE/DOE.

(Continued)

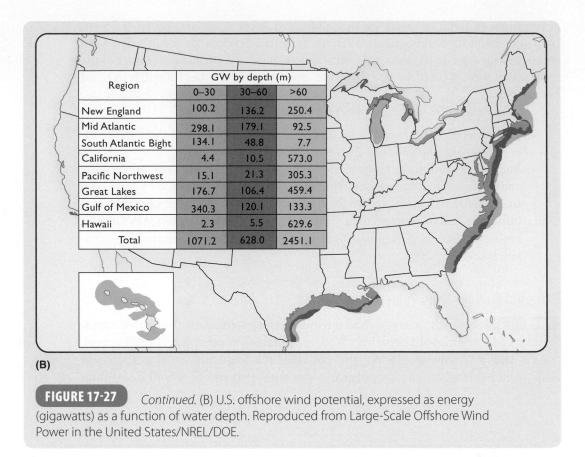

Region	GW by depth (m)		
	0–30	30–60	>60
New England	100.2	136.2	250.4
Mid Atlantic	298.1	179.1	92.5
South Atlantic Bight	134.1	48.8	7.7
California	4.4	10.5	573.0
Pacific Northwest	15.1	21.3	305.3
Great Lakes	176.7	106.4	459.4
Gulf of Mexico	340.3	120.1	133.3
Hawaii	2.3	5.5	629.6
Total	1071.2	628.0	2451.1

(B)

FIGURE 17-27 *Continued.* (B) U.S. offshore wind potential, expressed as energy (gigawatts) as a function of water depth. Reproduced from Large-Scale Offshore Wind Power in the United States/NREL/DOE.

availability and price of fossil fuels, subsidies supporting fossil fuel use, and government support for wind power at least at a level matching the subsidies for fossil fuels. Continued development of wind energy depends largely on the cost of competing fossil fuels, which in part depends on environmental standards. For example, if air quality standards for coal-fired power stations are relaxed or if mountaintop removal practices (coal is discussed elsewhere in the book) for coal mining are continued despite apparent prohibition under the Clean Water Act, wind and other renewables could be at a severe price disadvantage.

Advantages

Advantages of wind power are: no fuel charges, no spills or accidents from fuel transport, and no air or water pollution. Moreover, land developed for wind energy can be additionally used for grazing, recreation, species habitat protection, agriculture, or other uses.

Disadvantages are: energy production depends on cube of wind speed, and customers want electricity on demand. Thus, wind systems must be able to produce power even when the wind doesn't blow, or when wind speed declines. This may be accomplished at some sites by construction of a supplemental natural gas-fired generator that would be used when wind speed was insufficient, thus adding to the capital cost of the system. Research by government agencies and private wind producers is focused on means to "store" electricity produced by wind for use when required. Some methods that could prove economic are compressed air storage, pumped-storage facilities using two water reservoirs, or advanced batteries.

Future

Many experts believe that wind will be the foremost renewable source of energy by the year 2020, and an important global energy source by 2030, especially given the commitments already made to wind energy by the European Union and utilities in the United States. Furthermore, concerns about global climate change could increase demand for wind systems, as could stricter enforcement of clean air and water laws. Still, should fossil fuels prices drastically fall, lax enforcement of environmental laws become the rule, and incentives for wind power end, wind power development could be curtailed.

Biomass Energy

Humans derived most of their energy, aside from that provided by their own muscles, from *biomass* for untold thousands of years. Biomass, or energy from living matter, was still the most important planetary energy source until the eighteenth century. In the form of wood, it is still a significant energy source, especially in developing countries. Two-thirds of India's cooking energy requirement and nearly a third of its primary energy requirement come from fuelwood.

Fuelwood and other forms of traditional biomass contribute about 12.5% of Brazil's energy needs. Globally, fuelwood accounts for about 64% of estimated total world supply of combustible renewables.

But biomass is more than fuelwood. According to the U.S. Department of Energy, biomass is

> "Any organic matter that is available on a renewable or recurring basis, including agricultural crops and trees, wood and wood wastes and residues, plants (including aquatic plants), grasses, residues, fibers, animal wastes, and segregated municipal waste, but specifically excluding unsegregated wastes, painted, treated or pressurized wood; wood contaminated with plastic or metals; and tires. Processing and conversion derivatives of organic matter are also biomass."

TABLE 17-1 shows the various kinds of biomass.

Repeated energy crises beginning in the 1970s, coupled with concerns about fossil fuel depletion, environmental degradation, and global climate change have prompted a new, "higher tech" look at development of biomass energy. While biomass can be burned to produce "process" heat (heat for industrial applications), electricity, and a warm fire in the home, biomass development today focuses mainly on developing motor fuels to supplement or replace gasoline and diesel fuel.

Two of the most important technological innovations of the twentieth century were mass-production of the high-compression, internal combustion engine invented by Rudolf Diesel, and Henry Ford's mass production of the first true personal motor vehicle, the Model T, in 1908. Neither Ford nor Diesel used petroleum-based fuels for their earliest models; Ford used **ethanol** derived from corn. Diesel used peanut oil!

Both Ford and Diesel imagined that **biofuels** would power their motor vehicles in the twentieth century. But crop failures, variable prices, droughts, and competition from petroleum fuels—gasoline ("petrol") and "diesel"—led to the virtual abandonment of biofuels by 1940 in the United States.

However, the terrorist attacks in 2001 helped galvanize action on government's part to assist development of a viable biofuels industry. The U.S. Department of Energy and the U.S. Department of Agriculture are presently funding biofuels research. We will consider the state of biofuels use and the future of biofuels next.

TABLE 17-1	*Different Kinds of Biomass*

Categories of Biomass (=feedstocks)
Virgin wood
• forestry, wood processing; e.g. wood pellets from pulpwood chips (Softwoods: pine, fir, and spruce. Hardwoods: oaks, birches, beeches, willows, and poplars), sawmill residue, wood chips, split logs; Ethanol from lignocellulosic biomass that includes willows, poplars, hardwoods
Energy crops
• a form of lignocellulosic biomass, high yield crops (grasses) grown specifically for energy, incl. switchgrass, miscanthus (Chinese silvergrass), big bluestem
Food Crops
• Sugarcane, corn, soybeans, wheat, sugarbeet, vegetable oils incl. rapeseed, palm, and sunflower
Hydrocarbon-rich plants/algae
• For biodiesel, incl euphorba and jatropha
Invasive weeds/plants
• Mimosa, lantana, water hyacinth, salvinia, pistia
Agricultural residues
• residues from agriculture harvesting or processing, includes straw, vegetable/fruit peels, crop waste (e.g. corn stover), animal manure
Food waste
• from food and drink manufacture, preparation and processing, and post-consumer waste
Industrial waste and co-products from manufacturing and industrial processes

Data from: Biomass Energy Centre.

Ethanol and Biodiesel

Ethanol is an alcohol, fermented from grain, in most cases corn (called maize outside the U.S.). Since the passage of the 1990 Amendments to the Clean Air Act, ethanol's production has been subsidized by government agencies, officially to help cities meet increasingly stringent air quality requirements. Mixing ethanol with gasoline reduces carbon monoxide (CO) and hydrocarbon emissions, although it increases NOx emissions. The overall impact, however, of biofuels use is to substantially reduce the emission of ground-level-ozone-causing pollutants. However, because ethanol contains about 10% less energy per unit volume than gasoline, increasing ethanol content in motor fuels reduces fuel economy. Ethanol is also an excellent—and essentially nontoxic—octane booster, whereas alternative octane boosters such as methyl tertiary butyl ether (MTBE) are toxic and contribute to serious environmental and health problems, including groundwater pollution.

The ethanol industry in the United States produced 13.3 billion gallons of ethanol as of 2012, mainly from corn. As demand for ethanol increases, other biomass resources, such as agricultural and forestry byproducts, municipal solid wastes, industrial wastes, and crops grown solely for energy purposes, could be used to make ethanol.

Biodiesel can be made from a variety of vegetable oils. As you saw previously in the chapter, Rudolf Diesel used peanut oil to fuel his prototype engine.

Advantages

Using biodiesel instead of petroleum diesel reduces toxic emissions from diesel engines. **TABLE 17-2** shows the impact of biodiesel.

Note that NOx emissions are higher with biodiesel, but the overall effect is a significant reduction of air toxics.

Proponents of ethanol and biodiesel argue that increased production will help the agricultural community mainly by supporting prices paid to large-scale corn growers. Corn is usually produced in excess of demand largely due to government price supports. Overproduction results in low prices, and stimulates persistent calls for continued government subsidies.

Concern about the potential for rapid global climate change is adding more impetus to develop biofuels as an alternative to petroleum fuels. Because biofuels are made from plant material that took carbon dioxide out of the atmosphere in the first place, using biofuels is considered climate neutral, that is, adding no net carbon dioxide to the atmosphere. A closer examination, however, demonstrates that there is wide variability in CO_2 released from biomass, with intensively-farmed crops like corn and soybeans being far from carbon neutral.

Moreover, biofuels are nontoxic, so that groundwater contamination from leaking underground storage tanks, at present a serious national environmental problem nationwide, would be reduced. More than 46 million gallons per year leak

TABLE 17-2	*Impact of Biodiesel*	
Emission	**B100**	**B20 (20% biodiesel)**
Carbon monoxide	−43.2%	−12.6%
Hydrocarbons	−56.3%	−11.0%
Particulates	−55.4%	−18.0%
Nitrogen oxides	+5.8%	+1.2%
Air toxics	−60%–90%	−12%–20%
Mutagenicity	−80%–90%	−20%

Data from: DOE.

from thousands of small oil spills, according to the U.S. General Accounting Office. According to the U.S. Department of Energy, every gallon of nontoxic biofuels used reduces the hazard of toxic petroleum product spills from oil tankers and pipeline leaks. Such leaks average 12 million gallons per year; more than the oil dumped into Alaska's Prince William Sound by the *Exxon Valdez*, according to the U.S. Department of Transportation. Biodiesel use would also reduce toxic runoff from vehicles. Because biodiesel is biodegradable, it is an attractive alternative to petroleum diesel, especially for recreational boating and other diesel engine use in sensitive environments. For comparison, leaks and spills from the 400,000 personal watercraft operated on Chesapeake Bay are a serious source of contamination. The same is true for other public water bodies.

Should the advanced bioethanol technology being developed by DOE's Biofuels Program become commercially viable, it would be possible to produce ethanol from most plant-derived material. Many of these materials are not just underutilized and inexpensive, but create disposal problems. For example, plant residues (from rice, wheat, hay, etc.) are often burned in the field, a controversial practice in Oregon's Willamette Valley, and a practice that is becoming restricted due to air pollution concerns, notably in California. Much of the material now going into landfills is paper and paperboard that could be used for bioethanol production as well.

Drawbacks to Biofuels Production

Biofuels are produced from crops grown using fossil fuels. Examples include gasoline and diesel used as motor fuels, and natural gas used as a basis for fertilizer production. Recent research has concluded that ethanol production has an "energy efficiency ratio" of at least 1.24 (that is, 24% more energy is obtained than used from fossil fuels in production), and as much as 2.6, based on one study of biodiesel production. Another study suggested an efficiency ration of 7 would be possible if all waste material was burned to produce electricity, in addition to the material used to produce biofuels. Biofuel production could accordingly assist in reducing soil and soil water loss. The use of corn as a biofuel feedstock, however, leads to significant water pollution from pesticides and fertilizers used to grow the corn.

As discussed previously, biofuels are thought to be climate neutral. It is true that carbon emissions from plant biomass are offset by plant growth, but the regrowth may take decades, thus there is at least a short-term increase in CO_2. Burning forest biomass releases about 3,300 lb C/MW, while coal releases 2,100 pounds.

Another growing problem with biofuel production is conversion of tropical forests, most notably in Indonesia and Malaysia, to monoculture biofuel plantations (FIGURE 17-28).

Ethics

Some critics question the ethics of growing fuel instead of food, given a projected human population of up to 9 billion by the middle of the twenty-first century as well as persistent food shortages in localities worldwide.

Supporters point out that 80 to 90% of corn grown in the United States is used for livestock feed and for products such as beverage sweeteners, rather than for direct human consumption. As we mentioned in the previous section, corn is generally produced in surplus, requiring price supports, so to whatever extent ethanol supports

corn prices, taxpayer costs can be reduced. Bio-ethanol production from cellulose, which is indigestible by humans, would have even less impact on food supplies.

Wood Biomass

Increasingly, industries and power plants are burning wood and wood residue directly to produce electricity. Currently, the primary feedstock for such power plants is residue from logging and milling operations. There are thousands of small power plants burning wood waste, producing about 7,700 MW of electricity a year, and substantial "process heat," or heat that is used on site, according to the Federal Emergency Management Agency (FEMA).

Encouraged by subsidies and tax incentives deriving from new government standards for utilities to increase their share of renewable energy sources, there is enormous growth potential for woody biomass (**FIGURE 17-29**). For example, in mid-2011, Virginia Power, a subsidiary of Dominion Resources, filed a request to convert three of its smaller (50 MW each) coal-fired power stations to biomass, the fuel to come from waste forest products. The company said that, if approved, the conversion would materially reduce emissions and create about 300 permanent forest-product jobs.

However, there are major environmental drawbacks to increased use of woody biomass. First, the massive amount of wood required for new power plants cannot be provided from leftovers of forestry and milling, as they largely are currently: whole trees would be required. A 50 megawatt biomass plant in Burlington, Vermont, for example, would require 32,500 acres of forest annually, and more than 350,000 tonnes of wood. This means lost habitat (including wetlands), diminished biodiversity, threats to endangered species, water quality and quantity problems, etc. It also means new pressure to further log public forests.

In Other Words

". . . The current goal of generating 25 percent of U.S. electricity from renewable sources by 2025 would require the equivalent of clear-cutting between 18 million and 30 million acres of forest. Sufficient amounts of other biomass fuels simply don't exist in usable form. That amounts to leveling more than 46,000 square miles of forest—an area larger than the entire state of Pennsylvania."

Mary S. Booth, Ph.D.
Environmental
Working Group

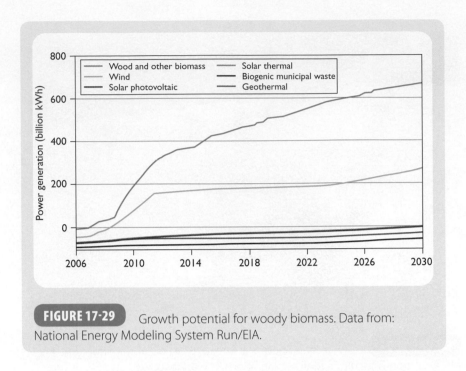

FIGURE 17-29 Growth potential for woody biomass. Data from: National Energy Modeling System Run/EIA.

"The fact is, you might get six or seven megawatts of power from residues in Massachusetts," said Chris Matera, the founder of Massachusetts Forest Watch. "They're planning on building about 200 megawatts. So it's a red herring. It's not about burning waste wood. This is about burning trees."

Moreover, natural forests do a much better job at storing the greenhouse gas carbon dioxide than tree plantations, which may use genetically modified trees engineered or bred to grow fast. Additionally, biomass is claimed to be *carbon neutral* (global climate change is covered elsewhere in the text), that is, when combusted, biomass merely returns the carbon to the atmosphere that it had previously removed, and that replanted trees then absorb more carbon dioxide. Research has shown that natural forests store more CO_2 (they are more efficient carbon *sinks*), and that more mature trees are even better carbon sinks. Forests in the southeastern United States, already the most heavily logged region in the world, are becoming even more threatened by demand for forest biomass. Ironically, much of the demand is coming from Europe, where forest biomass power plants are being constructed as part of a widespread effort to reduce use of fossil fuels! For these and other reasons, most environmental organizations are developing policies opposing forest biomass use, arguing that it should not be considered a renewable resource. For an example of such a policy, go to http://www.dogwoodalliance.org.

■ CONCLUSION

In **TABLE 17-3**, we list the alternative sources that are presently being utilized, and we list their advantages and disadvantages.

It is clear that humans eventually will have to find alternatives to fossil fuels, even if environmental concerns were to be ignored. How and over what time frame we do this poses one of the great and exciting challenges of the twenty-first century.

TABLE 17-3

The Potential for Alternative and Renewable Sources of Energy to Replace Fossil Fuels

Key:
Significance: Great, Moderate, Minor.
Future Potential: Great, Moderate, Minor.

Source	Significance Today	Future Potential	Advantages	Disadvantages
Nuclear Fission	Moderate	Moderate	No greenhouse emissions, etc	Storage and wastes
Geothermal	Minor	Moderate to great	No emissions or fuel cost	Storage
Solar	Minor	Moderate	No emissions or fuel cost	Land area Variable sunlight, storage
Wind	Moderate	Great	No emissions or fuel cost	Variable wind, storage, transmission capacity
Biomass	Minor	Moderate	Renewable No emissions No greenhouse gases	Land area needed to grow fuel crops
Hydro	Moderate	Moderate	No emissions	Effect on rivers
Tidal	Minor	Moderate	No emissions No fuel cost	Effect on estuaries and coasts
Fusion	Nil	Great	Low emissions, Vast reservoir of global hydrogen	Technical difficulties

CHAPTER SUMMARY

1. The European Union estimated that externalities, or costs not included in the price of electricity production, from fossil fuel and nuclear sources alone were equal to between 1% and 2% of the EU's Gross Domestic Product of $17,600 billion in 2010. In the United States, payments and tax breaks to fossil fuel producers exceeded $5 billion per year in the early 2000's. Two researchers in Australia estimated fossil fuel subsidies there at $AU 6.5 billion per year.

2. Alternative energy includes nuclear, hydroelectric, geothermal, hydrogen, and a variety of renewable energy sources. Renewable energy sources include wind, sun (solar), biomass, tides, and waves.

3. The main use of alternative and renewable energy today, and for the near future, is to generate electricity.

4. Forty-two percent of the electricity produced in the United States in 2011 was produced by coal-fired power plants, followed by 19 percent from nuclear, 25 percent from gas, 6 percent from hydropower, 1 percent from petroleum, and 5 percent from renewable sources.

5. The advantage of alternative energy sources is mainly that they are not based on burning fossil fuel, nor are they contributors to global climate change.

6. Hydropower is presently the world's largest alternative source of electricity, accounting for about 15% of the world's electricity production. Three types of hydropower installations are conventional run-of-river, conventional storage, and pumped storage.

7. Reservoirs behind most dams eventually fill with sediment unless the sediment is periodically removed, which is not usually cost-effective. A second problem with conventional hydroelectric dams is that they alter river ecosystems.

8. Because dams and the reservoirs behind them have multiple constituents, and dammed rivers have many commercial and recreational uses, dam removal is very controversial.

9. About 9,000 megawatts of geothermal electric power plants are installed around the world, with more than 3,100 megawatts of capacity in the United States.

10. Geothermal power plants use steam produced from reservoirs of hot water kilometers below the Earth's surface. There are three types of geothermal power plants: dry steam, flash steam, and binary cycle.

11. Geothermal heat pumps are heat pumps that can be used for residential or commercial heating and cooling, and are 25% to 50% more efficient than conventional heat pumps.

12. Nuclear power plants consist of a core containing a reactor, connected to a generator in which a turbine spins to produce electricity. Electricity from fission power plants began contributing to the U.S. energy mix in 1958.

13. The nuclear disaster at Chernobyl in the Ukraine in 1986 rendered thousands of square km around the plant uninhabitable, and generated a radioactive cloud that circled the planet.

14. Under the terms of the 1957 Price-Anderson Act, the federal government assumed responsibility for monetary damages for any nuclear plant accident that exceeds $9 billion.

15. Concerns about waste from nuclear power center on three issues: dust dispersal at mine sites, groundwater contamination at tailings sites, and disposal of radioactive waste from reactors.

16. The Earth–Moon system generates a vast amount of tidal energy, and where topography, tidal range, and other factors permit, it can be tapped to produce electricity.

17. A major constraint to the increased use of tidal energy is the cost of building tidal generating stations.

18. The energy in ocean waves could produce twice the electricity presently used by the entire planet, but the dispersed nature of wave energy makes it difficult to harness this power in conventional large power stations.

19. During the 1970s, the U.S. Government sponsored research into ocean thermal systems, but funding was cut, and commercial systems were never installed. The concept is being reinvestigated for sites like Hawai'i.

20. Sunlight can be used directly for heating and lighting buildings, to generate electricity, for hot water heating, and even solar cooling.

21. Solar energy can be harnessed in five ways: solar photovoltaics, concentrating solar systems, solar hot water systems, passive heating and lighting systems, and process heat production systems for commercial and industrial applications.

22. In the U.S., wind power capacity exceeded 60,000 MW by 2012. Supporters of wind energy point to its advantages: zero pollution and free fuel. The potential for wind power as a source of clean, renewable electricity remains enormous.

23. Biomass energy development today focuses mainly on developing motor fuels to supplement or replace gasoline and diesel fuel.

24. Since the passage of the 1990 Amendments to the Clean Air Act, ethanol's production has been subsidized by government agencies to help cities meet increasingly stringent air quality requirements. The ethanol industry in the United States produces more than 5 billion gallons of ethanol each year, mostly from corn.

25. Increasingly, industry and power plants are burning wood and wood residue directly to produce electricity, threatening forests globally.

KEY TERMS

alternative energy
base load plant
binary cycle plants
biofuels
biomass
control rods
decommissioning dams
direct gain
dry steam plants
ecosystem
ethanol
externalities

flash steam plants
fuel rod assembly
geothermal heat pump
geothermal power
global water budget
hydropower plants
Johnstown flood
net capacity factor
nuclear fission
off-peak power demand

peak power demand
photovoltaics
Price-Anderson Act
process heat
quad
recharge
renewable energy
Vaiont Dam
waste heat
Yucca Mountain Nuclear Waste Repository

REVIEW QUESTIONS

1. The main use of alternative and renewable energy today is
   a. as motor fuels.
   b. to generate electricity.
   c. to replace dams taken out of service.
   d. to replace nuclear power plants.

2. Fuel responsible for the greatest source of electricity in the U.S. is
   a. coal.
   b. nuclear.
   c. hydroelectric.
   d. renewables.

3. Nuclear power produces about what percent of total U.S. electricity?
   a. about 50%
   b. about 1%
   c. about 20%
   d. about 5%

4. Renewables (not counting dams) contribute about what percent of U.S. electricity production?
   a. about 20%
   b. about 50%
   c. about 0.1%
   d. about 4%

5. The world's largest source of alternative energy is
   a. coal.
   b. natural gas.
   c. hydroelectric.
   d. geothermal.

6. Which is a disadvantage of hydropower?
   a. hydropower installations produce electricity, which must be used on-site
   b. hydropower installations produce electricity that is much more expensive than electricity from fossil fuels
   c. hydropower operators must lease the dammed rivers, making the electricity less competitive in price
   d. reservoirs behind dams eventually become filled with sediment and must be abandoned

7. Which type of pollution is produced by hydropower?
   a. none
   b. CO_2 only
   c. ozone and CO_2
   d. ozone only

8. Most geothermal capacity in the United States is in
   a. Alaska.
   b. Texas.
   c. California.
   d. Hawaii.

9. An advantage of geothermal heat pumps is
   a. they can be installed for the same cost as a conventional heat pump.
   b. they can be installed at a lower cost than conventional heat pumps.
   c. they are substantially more efficient than conventional heat pumps.
   d. government subsidies reimburse homeowners for installation cost.

10. The U.S. state with the highest production of electricity from nuclear power is
    a. California.
    b. Washington.
    c. Maine.
    d. Illinois.

11. Demand for new nuclear plants in the United States virtually ceased after the Three Mile Island incident in 1979. Three Mile Island is located in
    a. Illinois.
    b. New York.
    c. Kansas.
    d. Pennsylvania.

12. The Chernobyl nuclear disaster in 1986 occurred in
    a. China.
    b. Russia.
    c. the Czech Republic.
    d. the Ukraine.

13. The Price-Anderson Act shields which kind of power producers against excessive liability in case of a disaster?
    a. solar
    b. nuclear
    c. coal
    d. geothermal
    e. all of the above

14. Where is nuclear waste in the United States to eventually be stored?
    a. at Yucca Mountain, Nevada
    b. at Carlsbad Caverns, New Mexico
    c. in Kansas salt mines
    d. at Sevier lake, Utah

15. Which of these forms of electricity generation produces no greenhouse gases?
    a. natural gas
    b. oil
    c. nuclear
    d. coal

16. A major drawback to the use of tidal power stations to produce electricity is
    a. there are no suitable sites around the world.
    b. the cost of construction is very high.
    c. they emit too much air pollution to compete with conventional sources.
    d. all of the above.

17. Solar photovoltaic cells are made of what type of material?
    a. thin plastic sheets, like food wrap.
    b. very pure aluminum foil.
    c. semiconducting material, like computer chips.
    d. superconductors, like liquid hydrogen.

18. Which country obtains more than 10% of their total electricity from wind?
   a. the United States
   b. Canada
   c. Denmark
   d. Russia

19. Which area has the highest wind power potential in the United States?
   a. the Upper Great Plains (Dakotas, Nebraska)
   b. Virginia and the middle Atlantic states
   c. Arizona
   d. Florida

20. Henry Ford and Rudolf Diesel expected their motor vehicles to be powered by fuels made from
   a. coal.
   b. oil.
   c. biomass.
   d. electricity.

Water Pollution: Today's Challenges

CHAPTER 18

© Rechtian Sorin/Shutterstock, Inc.

CHAPTER OUTLINE

■ AMERICAN WATERWAYS

After nearly half a century of federal legislation, at least two-fifths of the waterways of the United States are still classified as "impaired," that is, contaminated by one or more pollutants including sediment, microorganisms, nutrients (nitrogen and phosphorous), and industrial discharges. **TABLE 18-1** shows the results of the Environmental Protection Agency's (EPA) *2010 Report on U.S. Water Quality*.

In this chapter we discuss surface water pollution and remediation of contaminated groundwater. We cover methods for treating polluted water and introduce the laws on which our water treatment policies are based. You will see that the U.S. Environmental Protection Agency (EPA) is vested with enormous responsibility, meaning that the agency must be adequately funded if it is to do the job assigned to it by Congress.

Along with describing the sources and effects of water pollution, we cite three examples: chemical trace elements from industrial processes and pharmaceuticals; selenium from irrigation runoff; and water pollution associated with gold mining. Finally, one major issue, polychlorinated biphenyls (PCBs) in Hudson River, New York sediment and what to do about them, will be discussed in detail.

What Is Pollution?

FIGURE 18-1A shows satellite photographs of natural oil seeps in the Gulf of Mexico. **FIGURE 18-1B** is a satellite photo of the oil in the Gulf of Mexico arising from the 2010 BP *Deepwater Horizon* oil spill.

Concept Check 18-1. In which of these cases does the oil constitute *pollution?* Discuss the rationale for your answer.

Oil from a seep is probably chemically indistinguishable from the *Deepwater Horizon* oil spill, yet you may have called only the latter pollution, based on its

TABLE 18-1	Summary of Quality of Assessed Rivers, Lakes, and Estuaries			
Waterbody Type	**Total Size**	**Amount Assessed* (% of Total)**	**Good (% of Assessed)**	**Good but Threatened (% of Assessed)**	**Polluted (% of Assessed)**
Rivers (miles)	3,692,830	699,946 (19%)	367,129 (53%)	59,504 (8%)	269,258 (39%)
Lakes (acres)	40,603,893	17,339,080 (43%)	8,026,988 (47%)	1,348,903 (8%)	7,702,370 (45%)
Estuaries (sq. miles)	87,369	31,072 (36%)	13,850 (45%)	1,023 (<4%)	15,676 (51%)

*Includes waterbodies assessed as not attainable for one or more uses.
Reproduced from Water Quality Conditions in the United States: A Profile from the 2000 National Water Quality Inventory/EPA.

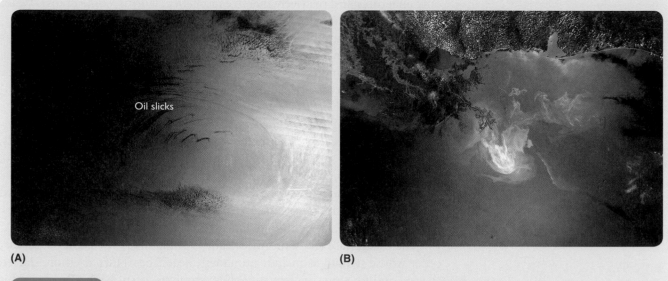

(A)   (B)

FIGURE 18-1   (A) Satellite photographs of natural oil seeps in the Gulf of Mexico. Courtesy of Goddard LAADS/NASA. (B) Satellite photo of the oil in the Gulf of Mexico arising from the BP *Deepwater Horizon* oil spill of 2010. Courtesy of GSFC/MODIS Rapid Response/NASA.

anthropogenic origin. Thus, some would argue that the same substance constitutes pollution only in cases where humans are responsible for its presence.

In general, *water pollution* refers to any hazardous or toxic material released by humans into the aquatic environment. *Marine pollution*, according to the U.N. Convention on the Law of the Sea, means the introduction by man, directly or indirectly, of substances or energy into the marine environment, including estuaries, which results or is likely to result in such deleterious effects as harm to living resources and marine life, hazards to human health, hindrance to marine activities, including fishing and other legitimate uses of the sea, impairment of quality for use of sea water and reduction of amenities.

The agent that causes the adverse effect is known as the *toxicant* or *pollutant*. The study of the toxic effects of chemicals and other substances on aquatic organisms is called *aquatic toxicology*.

Categories of Water Pollutants

There are nine general categories of water pollution:

- *Disease-causing agents*
  These include pathogenic bacteria (including the bacterium that causes cholera, discussed later), viruses, protozoans, and other parasites. These cause amoebic and bacterial dysentery, schistosomiasis, and typhoid fever, to name only a few. The main source is untreated or improperly treated human wastes, but animal wastes from feedlots or meat-packing plants are also sources.
- *Oxygen-demanding wastes*
  One of the ways of assessing the health of aquatic ecosystems is by examining the amount of dissolved oxygen (DO) in the water. Water with DO of 5 to 8 ppm is said to be **normoxic**, and can support virtually all aquatic life. Water with less than 2 to 3 ppm is said to be **hypoxic** and will support

only anaerobic and relatively inactive fauna (e.g. worms, some crabs). Water devoid of DO is **anoxic**.

Oxygen-demanding wastes are those that lead to decreases in a water body's DO level. These include sewage, paper pulp (a byproduct of manufacturing paper products), and food processing wastes. These substances can either be decomposed by oxygen-consuming microbial decomposers (bacteria or fungi) or can be chemically oxidized, in both cases reducing DO.

- *Water-soluble inorganic chemicals*

This is a very big category. It includes toxic metals, toxic metal compounds, acids, bases, and salts. Examples include heavy metals like mercury, lead, tin, and cadmium, as well as selenium, arsenic, chlorine, and nitrates. These derive from a variety of sources, including weathering of rocks accelerated by humans by mining, manufacturing processes, oil-field drilling operations, other mining, and agricultural practices.

The classic case study in this category is Minamata disease, a neural disorder whose symptoms include numbness, headaches, blurred vision, slurred speech, and for some, violent trembling, paralysis, even death. The disease is caused by mercury, and is named for Minamata Bay, in Japan, where it was first identified. In the late 1930s, Chisso Corporation dumped residues containing elemental mercury (Hg) into the bay. Metallic mercury is not water-soluble and thus sank to the bottom. However, it was converted into the soluble methyl mercury (CH_3Hg^+) by bacteria. There, through the process of **biomagnification**, in which the chemical increases in concentration in the bodies of organisms of higher trophic (feeding) levels, shellfish and fish accumulated high concentrations of mercury. These organisms constituted a major part of the diet of the local population, resulting in the observed symptoms.

- *Plant nutrients*

Plant nutrients are water-soluble inorganic nitrate and phosphate salts. Sources include sewage, runoff from livestock feedlots, fertilized fields and lawns, and industrial wastes. The result of nutrient pollution ranges from plankton blooms (massive increase in plant plankton; e.g., red tides) to dead zones (e.g., large tracts of ocean either hypoxic or anoxic). The Gulf of Mexico Dead Zone covers an area as extensive as 9,000 mi² (20,000 km²) seasonally, extending from the mouth of the Mississippi River. It is due to nutrient enrichment from agriculture in the Mississippi River watershed. The extent of the dead zone in 2012 was "only" about 3,000 mi², apparently owing to the extensive drought in the Midwest.

- *Organic chemicals*

This category includes a wide variety of both natural and synthetic chemicals such as oil, gas, solvents, and pesticides. Sources include industry, agriculture, and residential dwellings. Significant sources are improper disposal of industrial and household waste, and pesticide run-off from farms, forests, roadsides, and golf courses.

Oil receives perhaps the most attention in this category. Crude oil is a complex mixture of hydrocarbons. Some polycyclic aromatic hydrocarbons (PAH) found in crude oil are potent carcinogens.

Crude oil is sticky, insoluble, and floats on water, although some components may sink or evaporate.

Dichlorodiphenyltrichloroethane (DDT) is a chlorinated hydrocarbon (organochloride) pesticide. It is synthetic (i.e., not naturally occurring), non-biodegradable, and fat-soluble, thus it persists in the environment and bioaccumulates in organisms. As a result of DDT's interfering with reproduction and calcium deposition in eggshells, the brown pelican (*Pelecanus occidentalis*) nearly disappeared in the Southeastern United States in the early 1970s.

DDT is no longer used in the United States but is still employed in some tropical countries to control mosquitoes that carry malarial parasites.

PCBs are described later in this chapter.

- *Sediments*

  Sediment pollution is a growing problem, as fine soil particles (silt and clay) along with organic and inorganic chemicals originate as runoff from construction sites, logged forests, mining operations, and croplands. These sediments block sunlight, foul gills of fish and invertebrates, and can even smother bottom communities (e.g., coral reefs).

- *Radioactive substances*

  Radioactive substances are produced by the nuclear power industry, the Department of Defense, electrical utilities, mining industries, and as a result of medical diagnostics and treatment. In the 1960s and 1970s, radioactive waste buried in shallow trenches leached into surrounding soil and groundwater. Modern disposal techniques have been engineered to minimize the potential for leakage. Radioactive waste is discussed elsewhere in the text.

  The meltdowns and use of seawater to cool reactor cores at the Fukushima Daiichi Nuclear Power Plant after the 2011 tsunami led to renewed public interest in radiation pollution. Although the amount of radiation in the seawater was several orders of magnitude over safe levels, the half-life of the radioactivity was short and the radiation dispersed quickly, minimizing concerns.

- *Heat*

  Heat as a form of water pollution is principally produced from cooling of power plants and industrial processes, although anthropogenic climate change is beginning to heat bodies of water. Most threatened are organisms in tropics, which live closer to their upper thermal limit (e.g., corals).

- *Others*

  This final catch-all category includes plastics, packaging, disposable diapers, and discarded fishing nets.

Concept Check 18-2. CheckPoint: List and briefly discuss the nine categories of water pollution.

Endocrine Disruptors in Singapore Seawater

A category of chemicals called **endocrine disrupting chemicals (EDCs)**, or simply *endocrine disruptors*, including many pesticides, dioxin, synthetic steroids, and substances excreted in urine like birth-control medication, has been found in Singapore seawater by a team of researchers at the National University of Singapore (FIGURE 18-2). Endocrine disruptors can interfere with many basic human functions, especially development of sexual characteristics. The highest concentrations of the chemicals were found near industrial sites. Researchers found that sites with more different kinds of EDCs in solution exhibited highly *synergistic* effects, that is, the total effects were higher than the sum of the effects of each of the chemicals. Scientists concluded that the levels of hormone activity in the waters sampled off Singapore could be high enough to affect hormone "signaling" in marine life, especially after prolonged exposure or bioaccumulation in the food chain. These

results could have implications for human health as well as for marine life, as the waters off Singapore are used for seafood farming. Review this information when you read about water pollution in Puget Sound later in the chapter.

■ HISTORY OF WATER POLLUTION

Humans began to seriously degrade surface water with the growth of cities, most of which were coastal or situated on rivers. By the mid nineteenth century, water in many rivers in Western Europe was becoming too polluted to drink, and water supplies throughout Europe and North America were tainted with pathogens like cholera bacteria. Charles Dickens's novel, *Our Mutual Friend*, published in 1865, contains references to the "reeking Thames."

Widespread metal mining and smelting also contributed to water pollution. With the expansion of industrialization in the middle and late nineteenth century, water supplies rapidly became further degraded. With the expansion of industrialized agriculture after the Second World War, pollution from agricultural runoff became a serious local, regional, and even global issue.

Rivers in the United States occasionally caught fire from concentrations of flammable effluent. One such fire, on the Cuyahoga River in Cleveland Ohio in 1969 probably prompted passage of major revisions to the Clean Water Act by Congress.

Water Pollution and Chlorination

The first attempts to treat polluted water used chlorine as a disinfectant in nineteenth century England. This practice rapidly gained widespread acceptance, and today most water from municipal supplies for human consumption is chlorinated. Chlorination has probably saved as many human lives as any other health discovery, but chlorination produces toxic byproducts called **trihalomethanes (THMs)**, whose allowable concentrations in drinking water are set by government regulations.

Focus: Cholera

Vibrio cholerae is the bacterium, shown in (**Figure B1-1**), which is responsible for cholera. Cholera results from a toxin secreted by the bacterium. When a host ingests about a million bacteria, cholera can result, resulting in intense and rapid dehydration of the victim. If the patient can be quickly hydrated, cholera is rarely fatal. It does however take a heavy toll of children in rural areas of poor countries that lack adequate sanitation, medical facilities, and trained staff. India and Bangladesh are two countries where cholera is chronic. Ironically, many of their medical professionals have migrated and continue to migrate to wealthy countries like the United Kingdom, Canada, and the United States. In Bangladesh alone in the epidemic year of 1991, over 200,000 cases of cholera were reported in just three months. According to the United Nations World Health Organization (WHO) there are three to five million cholera cases and up to 120,000 deaths worldwide annually.

Cholera results when populations do not have access to safe drinking water. In Bangladesh, residents often dip drinking water out of the same bodies of water into which sewage empties. "Low-tech" approaches to reducing cholera incidence are being championed by researchers from the United States working in collaboration with local professionals.

The Director of the National Academy of Sciences until 2004, Dr. Rita Colwell reported that sari cloth, the national garb for women in Bangladesh, when folded six to eight times can filter out up to 95% of planktonic organisms in water to which *V. cholerae* attaches, cutting incidence of cholera by 50% or more (**Figure B1-2**). Though this is no substitute for access to clean water, such low-tech approaches can enormously reduce suffering as well as infant mortality, and can ease the burden on overworked health professionals.

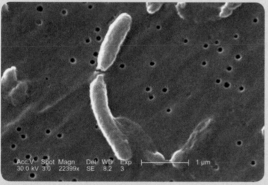

(A)

(B)

FIGURE B 1-1 (A) *Vibrio cholerae*, the bacterium that causes cholera. Courtesy of Janice Carr/CDC. (B) An open sewer in Accra Ghana where such bacteria may be found. © Pluff Mud Photography.

FIGURE B 1-2 Sari cloth. It could be used to drastically cut incidence of cholera in Bangladesh. © Aleksandar Todorovic/ShutterStock, Inc.

Point Sources

FIGURE 18-3 shows a point source, a sewer pipe in England. **Point sources** are sources of pollution that come from a discrete point. Point sources are the easiest types of water pollution to treat since they can be located, analyzed, and isolated. Furthermore, it is relatively easy to assign responsibility for the pollution. Examples of point sources include mines, industries, and sewage treatment plants. **FIGURE 18-4** shows runoff from a mine site.

Mining and Water Pollution

At mine sites, waters are often corrosive because they are often highly acidic. Acid waters can dissolve considerable amounts of metals, and many of these even at minute amounts are toxic. Examples include lead, zinc, arsenic, cadmium, chromium, and mercury.

Mines abandoned decades or even centuries ago may continue to leak toxins into surface water, much of which ends up in river sediment. Coal mines also can be sources of considerable water pollution, because surface mining often strips away coal seams to expose underclays or ancient peaty soils (**FIGURE 18-5**). Such soils, having been deposited in anaerobic environments, usually contain appreciable quantities of incompletely oxidized metal cations, and sulfur. Upon exposure to air, these materials acidify surface water, and the dissolved metals are carried into streams. This subject is treated in more detail in the mineral resources and coal sections elsewhere in the text.

Commercial and Industrial Pollution

A few examples of industries that generate potentially harmful wastes are metal plating firms, pharmaceutical firms, refineries, and chemical manufacturers. In recent years, due to the passage of laws requiring careful handling and disposal of toxic or hazardous wastes, many industries have drastically reduced their emissions, improving the environment and saving themselves millions of dollars in disposal costs and fines.

The following are some of the impacts of industrial pollution on Puget Sound, Washington.

- Copper contamination from urban areas is probably high enough to affect the hatching of floating fish eggs and the growth of phytoplankton.
- Zinc, lead, and mercury have been associated with altered bottom-dwelling communities in contaminated urban sediments. English sole collected from toxic hotspots in Puget Sound have high rates of liver tumors, which were correlated with concentrations of metals. (Catfish taken from the Anacostia River in Washington,

FIGURE 18-3 Point source of pollution. A sewer pipe. © pjhpix/ShutterStock, Inc.

FIGURE 18-4 A point source of water pollution, in this case, runoff from a mine site. Courtesy of Carol Stoker/NASA Ames Research Center.

(A)

underclay

(B)

FIGURE 18-5 (A) Coal seams. Underclays are lighter layers under the darker shiny coal seams. © Lee Prince/ShutterStock, Inc. (B) Underclays in a strip mine. Removal of coal exposes them to air and surface water, which can become highly acidic as a result. © jeff gynane/ShutterStock, Inc.

D.C. show similar effects.) The immune systems of juvenile Chinook salmon in polluted waters are more impaired than those of salmon from cleaner areas.

- Organochlorides discharged by pulp mills can bioaccumulate in organisms and cause cancer, harm to the immune and reproductive systems, and birth defects. Coho salmon near pulp mill pollution sources are more susceptible to parasites.

But industries are far from the sole source of commercial water pollution. For example, years ago New York City officials identified restaurants as sources of cooking oils, which, dumped into city sewers, cooled and formed immense clogs of fat, often impeding sewage lines and necessitated expensive cleaning operations by city employees!

Industrial Pollution in Rivers: PCBs in Hudson River Sediments

On the other side of the continent from Puget Sound lies the Hudson River (FIGURE 18-6). The Hudson River is an important source of hydroelectric power, public water supplies, transportation, and recreation. Yet the Hudson is considered by many scientists to be the most polluted waterway in New York State. The entire 200-mile stretch of the Hudson River from Hudson Falls to the Battery in New York City has been declared an EPA Superfund site. It has been the focus of an investigation by EPA since at least 1984 due to high levels of **polychlorinated biphenyls** (**PCBs**, a biphenyl with chlorine attached) in river sediment. The General Electric Co.'s Hudson Falls plant (Figure 18-6C) discharged between 209,000 and 1.3 million pounds of PCBs into the river from two capacitor-manufacturing plants located in Hudson Falls and Fort Edward. The Upper Hudson River, an approximately 40 mile reach of the river from Hudson Falls to Troy is the major focus of the investigations, and is being targeted for remediation. Previous studies had identified 40 "hot spots," defined as sediments containing greater than 50 parts per million (ppm) of PCBs (Figure 18-6D). In 1976, because of the concern over the bioaccumulation of PCBs

BOX 18-1   *Pharmaceuticals and Other Trace Substances*

In 2002, the U.S. Geological Survey published preliminary results of an extensive sampling project carried out on 139 streams in the country. The study reported that pharmaceuticals, hormones, and organic chemicals were detected. Eighty percent of the streams sampled had at least one of the contaminants. What makes the study extraordinary is the type of contaminants detected. Steroids, prescription and nonprescription drugs, insect repellents, and caffeine were widespread. Also common were disinfectants and metabolites of detergents. Metabolites are the breakdown products of an original chemical; for example, the chemical compound DDE (dichlorodiphenyldichloroethylene) is a metabolite of the organochloride pesticide DDT (dichlorodiphenyltrichloroethane), produced by the activity of organisms.

The compounds were generally present at very low concentrations; however, scientists have little experimental data by which to understand the effects of these compounds at such concentrations, nor on their persistence over time. Moreover, analyses of English rivers have shown that contaminants can react to produce new substances, some of which are unknown and thus not described by chemists. Previously, water chemists had detected the presence of estrogen from birth-control pills in the waters of Puget Sound, in Washington State.

In all, 95 chemicals were detected, but for only 14 have drinking water or safety standards been established. A 2007 review of findings revealed that 24 major metropolitan areas had water supplies with measurable quantities of such contaminants.

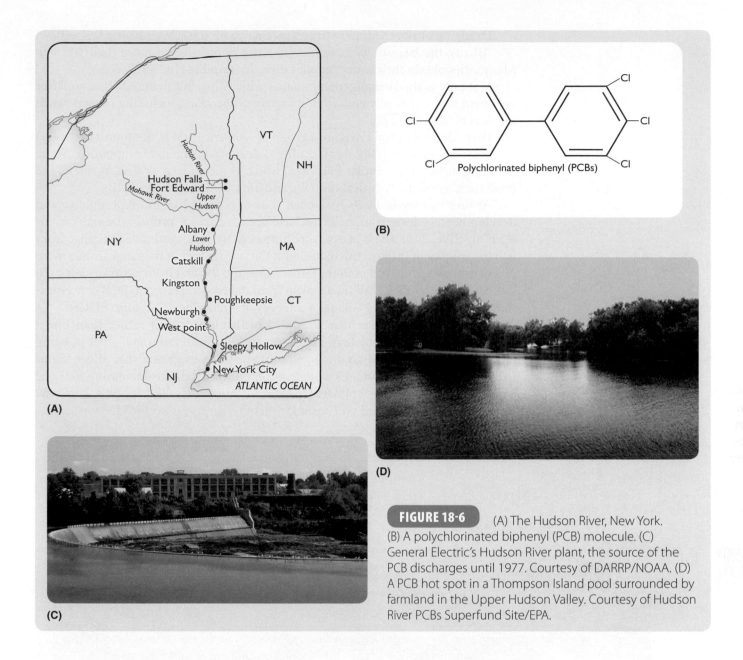

FIGURE 18-6 (A) The Hudson River, New York. (B) A polychlorinated biphenyl (PCB) molecule. (C) General Electric's Hudson River plant, the source of the PCB discharges until 1977. Courtesy of DARRP/NOAA. (D) A PCB hot spot in a Thompson Island pool surrounded by farmland in the Upper Hudson Valley. Courtesy of Hudson River PCBs Superfund Site/EPA.

in fish and other aquatic organisms and their subsequent consumption by people, the State of New York banned fishing in the Upper Hudson River and commercial fishing of striped bass and several other species in the Lower Hudson. In August 1995 the Upper Hudson was re-opened to fishing, but only on a catch and release basis. A commercial fishing ban and consumption advisories on striped bass and several other species continues in effect for the Lower Hudson River.

In February of 2002 after 18 years of study and review of stakeholder input, the EPA proposed a cleanup plan for the Hudson that involved dredging 2.65 million cubic yards (2.03 million m³) of contaminated river bottom sediment, containing about 100,000 pounds of PCBs, and would cost GE about $460 million. Although much opposition to dredging has been expressed, because dredged sediment could release buried PCBs into the river and exacerbate the problem, in 2011 Phase 1 of dredging began, with the removal of nearly 400,000 yd³ of contaminated sediment. Removal of the remaining 2.5 million yd³ was expected to take up to 7 years.

Dredging of river sediment to enhance navigation has been banned as well.

Albany, the largest city in the basin, has a population of more than 100,000. Many cities obtain their water supplies directly from the Hudson River.

Not only is the Hudson contaminated with PCBs, but polluted Hudson River sediment may eventually end up in a variety of locations, including coastal beach sediments (FIGURE 18-7).

Polychlorinated biphenyls are a group of *synthetic* (that is, not found in nature) and highly *persistent* (they don't readily decompose) organic compounds. Some PCBs are structurally similar to dioxin, which is recognized by EPA as one of the most toxic synthetic chemicals known, and are called dioxin-like PCBs.

At very low levels, PCBs have been shown to cause harmful reproductive and developmental effects in animals, and have been declared a "probable human carcinogen" by EPA. "Safe" levels of exposure have been variously estimated to range from 5 to 0.05 mg/kg of PCBs in fish tissue, with the last number in the range from a more recent estimate from studies in the Great Lakes and the first from a two-decades-old estimate made by EPA. Fish tissue taken from the Hudson range in PCB concentration from 2 to 41 mg/kg. PCBs are a category of endocrine disruptors (EDCs).

Toxicity of PCBs is primarily due to **bioaccumulation** rather than direct toxicity. PCBs bioaccumulate in organisms by (1) *bioconcentration*—that is, being absorbed from water and accumulated in tissue to levels greater than those found in surrounding water—and (2) *biomagnification*, which means increasing in tissue concentrations as smaller animals are eaten by larger ones. This is one reason why warnings have been issued on consumption of larger fish from the Hudson.

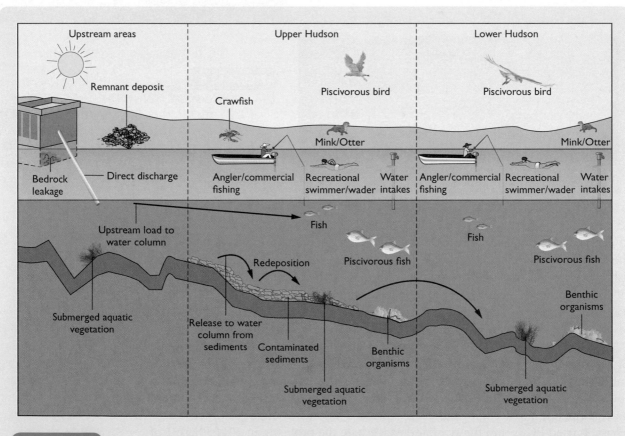

FIGURE 18-7 Exposure pathways for PCBs in the Hudson River. Reproduced from EPA.

Bioaccumulation can affect any animals that feed on fish as well. Fish in the Hudson ecosystem include yellow perch, white perch, largemouth bass, and striped bass; birds affected include tree swallows, belted kingfishers, mallards, great blue herons, and bald eagles; and mammals include bats, raccoons, mink and river otters. One study found that mink fed fish containing 5 ppm PCBs experienced a 99% decline in *fecundity* (survival of live young).

Manufacture of PCBs and other "persistent organic pollutants" (POPs) has been banned globally under terms of a 2004 United Nations treaty.

Concept Check 18-3. Stop and identify ways that industrial pollution can affect the economic health of local communities, aside from the direct cost to isolate and treat or remove the pollutant. Use the ban on sport fishing as an example, but recall the ban on navigational dredging also.

Municipal Sewage

Municipal solid waste (MSW) is discussed in detail elsewhere in this text.

FIGURE 18-8 is a map showing the location of Kolkata, India, formerly called Calcutta. The city, in the state of West Bengal, is India's largest, and its sewage system was built by the British to serve a city of 600,000. Kolkata's population now is over 15 million and growing. Figure 18-8B shows the city from space. Kolkata

(A)

(B)

(C)

FIGURE 18-8 (A) Kolkata's location. (B) Kolkata from space. Red areas are vegetation. Courtesy of JPL/NASA. (C) A view of the city after a recent flood caused by monsoon rains. *Question:* How do you think flooding affects sewage treatment in Kolkata? © Bikas Das/AP Images.

sprawls along the banks of the River Hugli at an average elevation of 9 meters, and is fewer than 100 miles from the Bay of Bengal.

Kolkata pumps untreated sewage through two main canals into thousands of hectares of ponds, where tilapia and carp, both herbivores, are grown for food. A recent study done by scientists at the University of Calgary (Alberta, Canada) found mercury and chromium levels in sewage to be "unacceptably" high, but no epidemiological studies are available that assess the risk of using untreated sewage to grow fish. Experts have concluded that even though infectious agents (diarrhea bacteria, etc.) are routinely found in fish organs, the fish is safe to eat if properly cooked and strict hygiene is maintained during preparation. However, there are no reported plans to open sashimi restaurants in the region!

U.S. sewage treatment facilities are quite different than those in most regions of India. Water pollution from sewage treatment plants depends on: (1) the degree to which sewage is treated; (2) the methods used; (3) the volumes treated and their composition; and (4) whether storm water runoff is separated from sewage.

Until the 1950s virtually no sewage in the United States was treated before disposal into the nearest body of water. After the passage of the 1956 Federal Water Pollution Control Act, sewage plants began to be built to treat sewage. With the passage of further amendments to the Act, sewage treatment gradually improved nationwide. One of the most effective means to improve treatment is to separate storm water runoff from sewage by using separate pipes. Unless separated, during storms the combined flow in sewage pipes can overwhelm the capacity of sewage plants and large volumes of untreated waste may be dumped into waterways. Examples of cities where this happens routinely are Washington, D.C., Richmond, Virginia, and Portland, Oregon. Portland is presently upgrading its storm water management system to separate sewage from storm water runoff.

FIGURE 18-9 shows a sewage treatment plant. Sewage treatment can be *primary, secondary, or tertiary*.

FIGURE 18-9   A typical sewage treatment plant. © A.B.G./ShutterStock, Inc.

Primary sewage treatment collects sewage, filters out large fragments, settles out most solids, removes surface scum (fats, etc.) and chlorinates the remaining liquids to kill potentially pathogenic organisms. The effluent is then piped to a water body: a river, lake, or the ocean. Primary treatment removes about two thirds of suspended solids, most fats and oils, up to half of any metals (which settle out with the solids), and about one third of the sewage's *biological oxygen demand* (BOD: a measure of the extent to which the effluent will degrade a water body by using oxygen). Primary treatment removes little of the waste's nutrients (N and P), and leaves the waste in a state where considerable degradation to a water body can still occur.

Secondary sewage treatment adds the following steps after primary treatment: The waste is pumped to tanks where added bacteria help reduce the volume of the *sludge* (solids), converting some organic matter to methane in the process. During secondary treatment, 80% to 90% of solids can settle out and BOD is reduced by 65% to 90%. Although P can be precipitated chemically, N content by and large is unaffected, because it is very soluble in water. The waste is then chlorinated before release.

Most municipal sewage in the United States and much in Canada receive secondary treatment. Considerable treated sludge is produced, now called "wastewater residuals" by the EPA, which is either placed in landfills or spread on fields as "soil conditioner" or fertilizer (**FIGURE 18-10**).

Tertiary sewage treatment removes most of the rest of the solids and nutrients, and bacterial action converts much organic matter to methane, which is often collected and used to power the plant, or burned on-site to minimize the risk of explosion. Ideally, all waste should undergo tertiary treatment.

FIGURE 18-10 Biosolids, or sewage sludge. If sludge is to be used on agricultural fields, it is imperative that any industrial waste be pre-treated before entering the municipal sewage system. This avoids contaminating sludge with toxic material. © Justin Kase ztwoz/Alamy.

Concept Check 18-4. CheckPoint: Differentiate between primary, secondary, and tertiary sewage treatments.

- -

Concept Check 18-5. Find the level of treatment of your own sewage by calling your local sewage treatment plant or logging on to their website. How much do they spend each year to treat sewage?

- -

Trihalomethanes (TMHs) are a family of chemicals produced when chlorine reacts with organic matter. They are one of a number of disinfectant by-products (DBPs) resulting from chlorination of sewage. **TABLE 18-2** shows the common DBPs.

Some disinfectant byproducts remain to be identified. Production of DBPs, including THMs, is usually highest in summer and early fall due to higher water temperatures. Production of THMs is directly related to concentration of organic carbon in water, from decaying plant material for example. Reducing organic carbon before chlorine application can significantly reduce THM formation. EPA regulations permit no more than 100 ppb THMs in drinking water, but there is no limit applied to chlorinated sewage.

Trihalomethanes are considered by the EPA to be probable or possible human carcinogens. Several studies pointed to a slight increase in the risk of bladder cancer

TABLE 18-2 — Common Disinfectant By-Products in Chlorination of Water

DISINFECTANT RESIDUALS
Free Chlorine
  Hypochlorous Acid
  Hypochlorite Ion
Chloramines
  Monochloramine
Chlorine Dioxide

INORGANIC BYPRODUCTS
Chlorate Ion
Chlorite Ion
Bromate Ion
Iodate Ion
Hydrogen Peroxide
Ammonia

ORGANIC OXIDATION BYPRODUCTS
Aldehydes
  Formaldehyde
  Acetaldehyde
  Glyoxal
  Hexanal
  Heptanal
Carboxylic Acids
  Hexanoic Acid
  Heptanoic Acid
  Oxalic Acid
Assimilable Organic Carbon

HALOGENATED ORGANIC BYPRODUCTS
Trihalomethanes
  Chloroform
  Bromodichloromethane
  Dibromochloromethane
  Bromoform
Haloacetic Acids
  Monochloroacetic Acid
  Dichloroacetic Acid
  Trichloroacetic Acid
  Monobromoacetic Acid
  Dibromoacetic Acid
Haloacetonitriles
  Dichloroacetonitrile
  Bromochloroacetonitrile
  Dibromoacetonitrile
  Trichloroacetonitrile
Haloketones
  1,1-Dichloropropanone
  1,1,1-Trichloropropanone
Chlorophenols
  2-Chlorophenol
  2,4-Dichlorophenol
  2,4,6-Trichlorophenol
Chloropicrin
Chloral Hydrate
Cyanogen Chloride
N-Organochloramines
MX*

*3-Chloro-4-(dichloromethyl)-5-hydroxy-2(5H)-furanone
Reproduced from Alternative Disinfectants and Oxidants Guidance Manual/EPA.

and colorectal cancer with ingestion of THMs. Other studies have found that THMs may cause organ and central nervous system damage. Still other studies link THMs to reproductive problems, including miscarriages. A California study found a miscarriage rate of 15.7% for women who drank five or more glasses of tap water containing more than 75 ppb THMs, compared to a miscarriage rate of 9.5% for women with a low THM exposure.

To reduce the THM risk, some water treatment systems are turning to *chloramines*, produced by combining ammonia and chlorine. Chloramines do not produce THMs, but because they are slightly less effective than chlorination, more chloramines must be used. Ozone and ultraviolet (UV) light also can be used to disinfect water, but they are more expensive than chlorination or chloramine use.

Trihalomethanes pose another example of the central question concerning the role of pollution in human health: is it possible to determine an exposure level *below which* there is no adverse effect? The answer for THMs is, not yet, and possibly, no. To get an idea of why this is so, consider that mutagenic and carcinogenic effects occur at the cellular level, where toxic substances interact with cells, sometimes inducing damage to DNA, sometimes other forms of cellular damage. Using Avogadro's Number ($\sim 6 \times 10^{23}$), and the molecular weight of dissolved substances, it is possible to calculate roughly how many molecules of a toxic substance are present at any given concentration. For carcinogenic substances like THMs, a 1 ppb concentration results in more than 10^{14} molecules per liter of water! As Americans are urged to drink more water, ironically for health's sake, they may be ingesting harmful amounts of carcinogens even from "pure" water.

Consumers concerned about THMs in drinking water can choose to filter their water prior to drinking. Many commercially available water filters remove THMs. But filtering water may also remove fluorine, which protects teeth from decay.

Non-Point Sources

Non-point source (NPS) pollution is the leading cause of water pollution in the United States. Non-point sources include runoff from agricultural operations (including animal feedlots), suburban lawns, leaky motor vehicles, faulty septic systems, air deposition, and urban areas. Urban runoff is illustrated in **FIGURE 18-11**.

Non-point sources from autos is believed to be the cause of tumors in fish in the Anacostia River, which runs from Maryland through Washington D.C. Up to 80% of catfish sampled had liver and/or skin tumors.

Non-point sources may also include air-deposited pollution and seepage from groundwater. (See the case study on groundwater remediation elsewhere in the text.) EPA has the authority to regulate point sources but not non-point sources, Congress having left that authority to the states under the Clean Water Act.

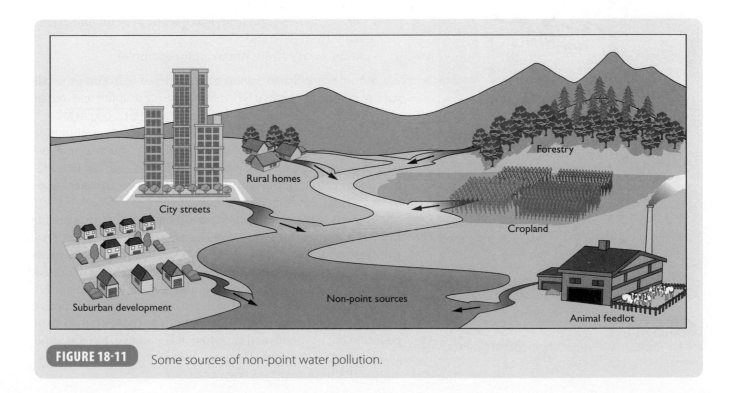

FIGURE 18-11 Some sources of non-point water pollution.

FIGURE 18-12 Santa Monica Bay, California. A 1995 research project found that four swimmers per hundred near flowing storm drains could develop fever, nausea, gastroenteritis, cold and flu-like symptoms such as nasal congestion, sore throat, fever and/or cough, compared to those who kept far away from such outflow sites. © iStockphoto/Thinkstock.

FIGURE 18-13 A sign asking citizens to be aware of storm water pollution. © Tom Uhlman/Alamy.

Urban/Stormwater Runoff

A typical city block composed mainly of rooftops, roads, and parking lots generates nine times more runoff than a wooded lot of the same size. This is called **urban runoff**. It is composed of:

- *sediment* from development and new construction;
- *oil, grease, tire dust, and toxic chemicals* from automobiles;
- *nutrients and pesticides* from turf management and gardening (over 10% of the entire state of Maryland is turf);
- *viruses and bacteria* from leaky septic systems and excrement from pets; and
- *heavy metals* (from tire wear, for example).

Storm water/urban runoff is considered a major problem nationwide; for example, it is believed to be the number one source of pollution to California's Santa Monica Bay (**FIGURE 18-12**).

To avoid unnecessary water pollution, many communities now conduct extensive education programs, and may put up warning signs on impacted waterways (**FIGURE 18-13**). In addition, many communities are building separate storm water treatment systems, as we mentioned previously.

There is a legal framework for urban runoff management. In 1987, Congress set up the *Nonpoint Source Management Program* under the Clean Water Act (CWA), authorizing EPA to help states address runoff pollution by (1) identifying waters affected by urban runoff and (2) adopting and implementing management programs to control it. These programs are meant to determine sites for **best management practices (BMPs)** to (1) keep pollution out of runoff, and (2) where it is polluted, to reduce the amount that reaches surface waters. More details about BMPs are on the EPA website (http://www.epa.gov).

Agricultural Pollution

With the advent of industrialized agriculture and factory feedlots following World War II, water pollution from agriculture became an increasing concern. Now such non-point pollution is one of the most persistent and widespread forms of water and air pollution.

Agricultural runoff includes animal waste, soil and nitrogenous dust, nutrients, soil in waterways, and pesticides, fungicides, and herbicides. It is a major source of contamination in the Gulf of Mexico. As we mentioned previously, the EPA lacks clear authority to regulate these non-point sources. Agricultural pollution could be satisfactorily addressed given sufficient public will. However, contemporary agricultural policy, through massive subsidies, encourages excess production and non-point pollution.

This subject is detailed in the flood, coastal hazards, groundwater, and soils and agriculture sections of this text.

Sources of agricultural pollution are illustrated in **FIGURE 18-14**.

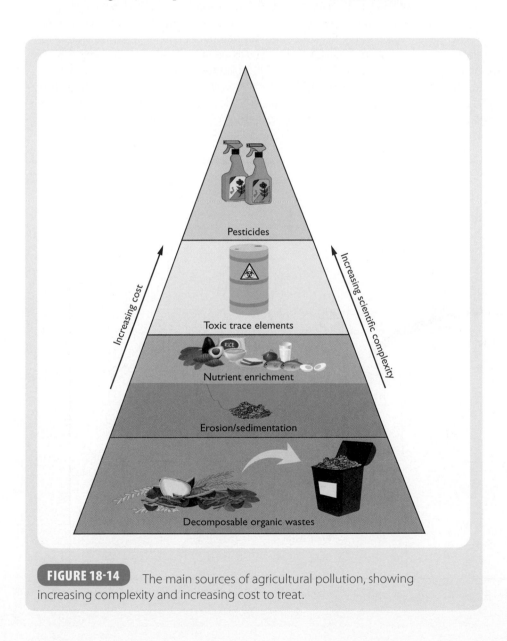

FIGURE 18-14 The main sources of agricultural pollution, showing increasing complexity and increasing cost to treat.

Focus: Selenium from Irrigation, Grand Junction, Colorado

Figure B 2-1 shows a location map of the area we will discuss. Historically, land use in the Kannah Creek, Whitewater Creek, and Callow Creek Basins has been mainly open range and irrigated agriculture. However, demand for residential housing is increasing development pressure in this area, and previously agricultural areas are becoming residential lots. Increasing residential land use has raised concerns about potentially serious water pollution, specifically concerning the potential for increased toxic selenium pollution to the nearby Gunnison River. Here's why.

Housing in the area is expected to use individual **septic systems**. Septic systems work by "flushing" wastewater into soils by way of a *leach field*. However, the area has a dry climate, meaning the soils contain soluble metals, since there is insufficient precipitation to wash the metals out of the soil (see soils and agriculture section elsewhere in the text for details). Wastewater, whether from septic systems, watering lawns or some other source, can dissolve selenium (Se) in the soil, and carry it into groundwater. In addition, areas not presently irrigated may come under residential irrigation practices. Portions of Kannah Creek and Whitewater Creek, and the entire Callow Creek basins are underlain by the Mancos Shale, which is known to contain high levels of selenium.

Increased water use in the area, septic system installation, and suburban irrigation are projected to increase the level of Se in the nearby Gunnison River. For details, go to http://www.seleniumtaskforce.org/projects.html.

Increased irrigation could also lead to the proliferation of toxic plants. Selenium (Se) is one of the few elements that plants absorb in sufficient amounts to be toxic to livestock. Because of a similar atomic structure, Se can substitute for sulfur (S) in plant tissue (in sulfur-containing amino acids, for example). This substitution of Se for S can cause chronic Se toxicity, resulting in deformity or mortality in birds, fish, and mammals. We discuss this phenomenon further in the section on Kesterson Wildlife Refuge and other impacts of Se in the groundwater section elsewhere in the text.

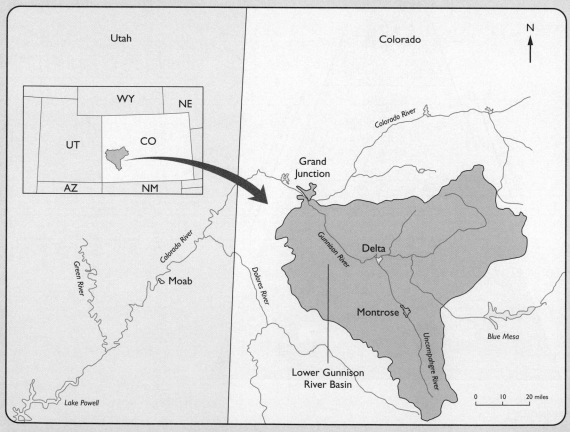

FIGURE B 2-1 Map of the basins discussed in the text.

FIGURE 18-15 A daytime photograph showing air pollution mainly from coal burning power plants in Beijing, Peoples Republic of China, in December 2004. © iStockphoto/Thinkstock.

Air Deposition

Some studies have concluded that airborne pollution can make up a significant part of the total pollution load, in a region downwind of a major source of pollution. The major sources of atmospheric pollution leading to air deposition include agriculture, coal-burning power plants, as shown in **FIGURE 18-15**, and waste incinerators. Such unlikely substances as phosphorus (a nutrient) and PCBs can be transported to waterways via air deposition. Up to one third of the pollution that degrades the air of the New York and Washington, D.C. metropolitan areas is blown in from as far away as Ontario, Canada, and the Midwest. It is therefore possible that considerable surface water contamination can result. The area that can contribute air pollution to a particular site is called the **airshed** (**FIGURE 18-16**).

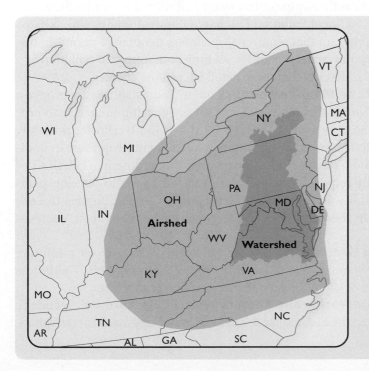

FIGURE 18-16 The airshed of the Chesapeake Bay watershed. Note that the airshed is much larger than the Bay's watershed. The airshed contains dozens of coal-fired power plants. The diagram shows that sources of NOx hundreds of km from the Bay proper can contribute to Bay pollution.

Groundwater Seepage

Recall that *infiltration* is a major component of the hydrologic cycle, and that surface water and groundwater can be intimately connected. Contamination of surface water from groundwater is therefore likely wherever contaminated groundwater enters surface water bodies like streams, lakes, and estuaries. A major source of nitrogen pollution in Chesapeake Bay, for example, is from groundwater, itself polluted by agricultural fertilizer as well as **septic tank** seepage from thousands of homes along the Bay proper.

■ ADDRESSING WATER POLLUTION

Important Legislation

The following legislation has addressed water quality in one form or another. A brief description follows each Act.

- **Solid Waste Disposal Act** (1965). This Act sought to regulate municipal waste (MSW) and to improve technologies by which MSW was handled. The federal government awarded grants to states for construction of sewage treatment plants.
- **National Environmental Policy Act** (1969, enacted 1970). This Act codified the nation's wish to restore and protect its environment, required preparation of Environmental Impact Statements (EIS) for projects with any federal funding, and set up the Council on Environmental Quality, which reports to the President.
- **Federal Water Pollution Control Act** (*Clean Water Act*: 1972 and amended). Passed after the Cuyahoga River in Cleveland caught fire from industrial pollution, the Act's purpose was to restore and protect the nations waterways. It gave EPA (Environmental Protection Agency; set up soon after passage of NEPA) authority to control water pollution. EPA under the act set up the NPDES (National Pollution Discharge Elimination System). NPDES is essentially a program that requires all polluters to obtain a permit to discharge emissions into waterways. EPA set limits on permitted pollution levels. The NPDES is described in more detail later in the chapter.
- **The Coastal Zone Management Act** of 1972. It established a program whereby states and territories could voluntarily develop programs to protect coastal waters. As a result, at least 29 states and territories presently operate federally approved coastal zone management programs. In addition, the Coastal Zone Act Reauthorization Amendments (CZARA) of 1990 specifically required coastal states and territories to develop plans to protect coastal waters from polluted runoff. CZARA applies to construction sites in these 29 states and territories where less than 5 acres is disturbed.
- **Safe Drinking Water Act** (1974 and amended). This Act set minimum drinking water standards, and sought to protect groundwater. We discuss this Act in more detail in the groundwater section of this text.
- **Resource Conservation and Recovery Act** (RCRA: 1976 and amended). RCRA authorized the EPA to take a "cradle to grave" approach to waste man-

agement, and had its main impact on the treatment of a category the act defined as "hazardous waste." The act required new landfills to have ground-water monitoring systems and required old landfills to be retrofitted. The act eventually led to the closing of 70% of all landfills in operation, and the construction of "state of the art" regional landfills, which in turn promoted the interjurisdictional transfer of wastes. It is now common for millions of tons of MSW to be transferred hundreds of km for disposal. RCRA also regulates underground storage tanks. We discuss this Act further in Chapter 11, Groundwater.

The National Pollutant Discharge Elimination System Program

The **National Pollutant Discharge Elimination System (NPDES)** permit program authorizes EPA to regulate point sources that discharge pollutants into waters of the United States.

The following activities require an NPDES permit:

- Construction sites where five or more acres are disturbed;
- Discharges from municipal separate sewer systems serving populations of 100,000 or more;
- Industrial discharges, including construction sites of five acres or more; and
- Other discharges, if identified by EPA or a state as needing an NPDES permit because they contribute to a water quality violation.

EPA is preparing regulations for other storm water discharges, such as discharges from municipal separate storm water management systems serving populations of fewer than 100,000, and discharges associated with commercial activities, light industries, and construction sites of less than five acres.

■ CONCLUSION: TRENDS IN WATER QUALITY

FIGURE 18-17 shows the status of global water quality.

The low population growth in EU countries suggests that much more progress is possible. In the United States, considerable progress has been made from the 1950s, when the first modern water quality laws were passed. Globally, however, there is less cause for optimism. As we mentioned at the top of this section, we have considerable work to do to restore our nation's waterways. The future of water pollution control will shift more toward research, to determine the impact of trace elements on environmental and human health; to improve the efficiency of existing sewage treatment plants, and to begin to deal effectively with non-point-source water pollution.

Globally, water pollution is a severe and in many areas a worsening problem. Many developing nations are reluctant to spend the considerable sums necessary to protect water supplies. Relatively little developmental aid is at present being directed toward water quality. With the population growth projected for the developing world this problem must be addressed, as it will only get worse with time.

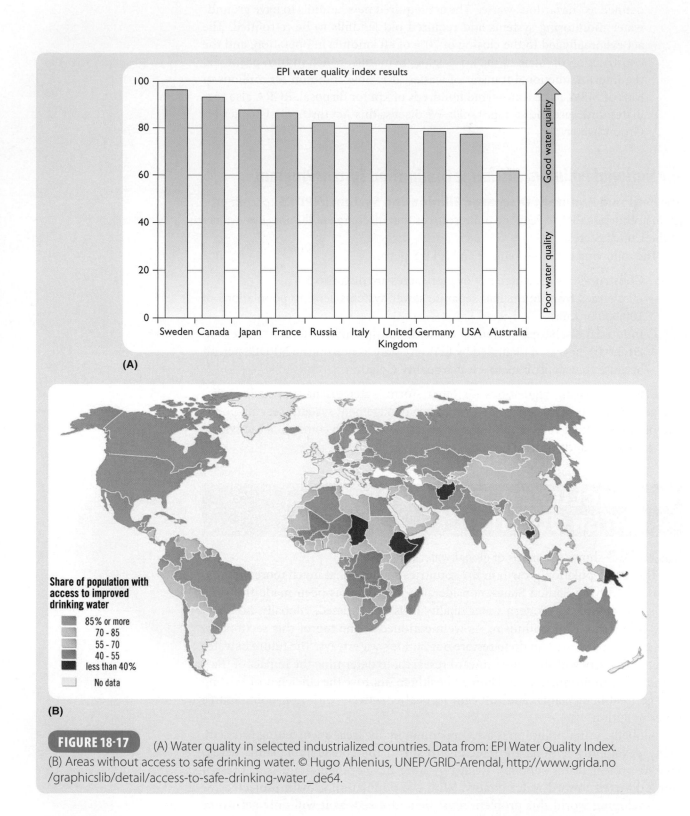

FIGURE 18-17 (A) Water quality in selected industrialized countries. Data from: EPI Water Quality Index. (B) Areas without access to safe drinking water. © Hugo Ahlenius, UNEP/GRID-Arendal, http://www.grida.no/graphicslib/detail/access-to-safe-drinking-water_de64.

CHAPTER SUMMARY

1. Nearly half of the U.S. waterways are still classified as "impaired" by EPA.
2. There are nine major categories of water pollution.
3. A class of chemicals called endocrine disruptors (EDCs) has been found in seawater off Singapore. The highest concentrations were found near industrial sites. These chemicals are found in high enough concentrations to interfere with marine life, and could have harmful effects on humans as well.
4. By the early nineteenth century, water in many rivers in industrialized countries was too polluted to drink, and water supplies throughout Europe and North America were tainted with pathogens like cholera bacteria. Metal mining and agricultural pollution have become serious sources of contamination as well.
5. The cholera bacterium is one of the most widespread forms of water pollution in developing countries.
6. Chlorination has probably improved human health as much as any practice in the last two centuries, but chlorination produces toxic by-products, called trihalomethanes (THMs).
7. Water pollution sources may be classified as point sources or non-point sources (NPS). Point sources come from a discrete point, like a sewage pipe; NPS do not.
8. Mining can produce appreciable water pollution. Water is often corrosive at mine sites, and can acidify streams and groundwater.
9. Many forms of industrial activity contribute to water pollution. Examples are metal plating firms, pharmaceutical firms, refineries, pulp mills, and chemical manufacturers. Industries can emit heavy metals like copper, zinc, lead and mercury, and organochlorines like dioxin.
10. The U.S. Geological Survey has found that 80% of 139 streams sampled contained pharmaceuticals, hormones, and organic chemicals.
11. Portions of the Hudson River, New York State, are contaminated with PCBs from industrial activity. PCBs are similar to dioxin and cause adverse reproductive and developmental effects in animals.
12. Toxic chemicals like PCBs can bioaccumulate in animals.
13. Sewage treatment plants are potential point sources of water pollution. Until the 1950s virtually no sewage in the United States was treated. Today, primary, secondary and tertiary methods of sewage treatment may be employed. Some treatment plants combine stormwater with sewage, a practice that often leads to water pollution.
14. Non-point sources of water pollution remain the leading cause of water pollution in the United States. Urban runoff is a major constituent of NPS pollution.
15. Agricultural pollution became a significant NPS after World War II. Irrigation runoff can be a significant NPS. Selenium can be mobilized from arid lands soils by irrigation runoff.
16. Air deposition of toxics into water bodies provides appreciable amounts of pollution. A noteworthy example is the Chesapeake Bay watershed.
17. Important laws to address water pollution in the United States include the Solid Waste Disposal Act (1965); the National Environmental Policy Act (1969); the Federal Water Pollution Control Act (Clean Water Act: 1972); the Coastal Zone Management Act (1972); the Safe Drinking Water Act (1974); and the Resource Conservation and Recovery Act (1976).
18. The National Pollutant Discharge Elimination System (NPDES) permit program authorizes EPA to regulate point sources that discharge pollutants into waters of the United States.

KEY TERMS

airshed

anoxic

best management practices (BMPs)

bioaccumulation

biomagnification

Coastal Zone Management Act

endocrine disrupting chemicals (EDCs)

Federal Water Pollution Control Act

hypoxic

National Environmental Policy Act

National Pollutant Discharge Elimination System

non-point source pollution

normoxic

point-source pollution

polychlorinated biphenyls (PCBs)

primary sewage treatment

Resource Conservation and Recovery Act

Safe Drinking Water Act

secondary sewage treatment

septic tank/septic system

Solid Waste Disposal Act

tertiary sewage treatment

trihalomethanes (THMs)

urban runoff

Vibrio cholerae

REVIEW QUESTIONS

1. According to the EPA's 2010 assessment, about what percent of sampled waterways were classified as impaired?
    a. Nearly 100% were impaired.
    b. Fewer than 10% were impaired.
    c. Around 40% were impaired.
    d. None was impaired.

2. Which types of pollutants were discovered by researchers in Singapore harbor water?
    a. heavy metals, like lead
    b. cholera bacteria
    c. endocrine disruptors
    d. *E. coli* bacteria
    e. all of the above

3. The disease called cholera is caused by exposure to
    a. a bacterium.
    b. radioactive waste.
    c. EDCs.
    d. fine particulates.
    e. asbestos.

4. A practice, which began in the nineteenth century, proven extremely effective in reducing death from polluted water is
    a. filtration.
    b. oxidation.
    c. ionization.
    d. chlorination.
    e. irradiation.

5. Which is an example of a water pollution point source?
   a. air deposition
   b. urban runoff
   c. a sewage treatment plant outfall
   d. agricultural runoff.

6. The pollutant in Hudson River sediment that has resulted in a 200-mile stretch of the river being declared a Superfund site is
   a. PCBs.
   b. cholera bacteria.
   c. lead from disused mines.
   d. mercury from coal mining.
   e. all of the above.

7. When smaller animals are eaten by larger ones, pollutants can become increasingly concentrated in the predator's tissue. This process is called
   a. biotoxicity.
   b. biomagnification.
   c. bioamalgamation.
   d. phytoaccumulation.

8. Sewage contains abundant nutrients. The two most abundant nutrients in sewage are
   a. silicon and carbon.
   b. nitrogen and carbon.
   c. phosphorus and carbon.
   d. nitrogen and phosphorus.

9. The most effective sewage treatment process for nutrient removal is
   a. primary treatment.
   b. secondary treatment.
   c. tertiary treatment.
   d. All are equally effective.

10. Most municipal sewage treatment in the United States and Canada is
    a. primary.
    b. secondary.
    c. tertiary.
    d. advanced.

11. Trihalomethanes are produced when chlorine reacts in treated water with
    a. organic matter.
    b. lead.
    c. halons.
    d. methane.

12. A toxic element abundant in some dryland soils (in Colorado and California, for example) that can be mobilized by irrigation water and taken up by plants is
    a. lead.
    b. selenium.
    c. nitrogen.
    d. potassium.

13. Researchers have recently identified air deposition as a major source of pollutants into the Chesapeake Bay watershed. The airshed of the Bay is _____ the Bay's watershed.
    a. larger than
    b. the same size as
    c. smaller than

14. The two major sources of air deposited pollution nationwide are believed to be
    a. coal fired power plants and motor vehicles.
    b. motor vehicles and agriculture.
    c. incinerators and coal fired power plants.
    d. mature trees and agriculture.

15. The Act with authority to regulate hazardous waste is
    a. Resource Conservation and Recovery Act.
    b. Safe Drinking Water Act.
    c. National Environmental Policy Act.
    d. Solid Waste Disposal Act.

16. The National Pollutant Discharge Elimination System (NPDES) permit program authorizes the EPA to regulate
    a. non-point sources of water pollution.
    b. emissions from strip mines.
    c. point sources of water pollution.
    d. all forms of water pollution.

Air Pollution

CHAPTER 19

CHAPTER OUTLINE

■ EFFECTS OF GLOBAL AIR POLLUTION

A ir pollution kills millions of people annually, according to the United Nations, the World Bank, and the World Resources Institute (**FIGURE 19-1**).

What are air pollutants? Where do they come from? What is their impact? Have we made progress? How do we get cleaner air? Is the air becoming cleaner or dirtier? This chapter attempts to answer these questions.

CASE STUDY — *Sulfur-Free Fuels*

Read the following statement from BP, one of the world's largest energy companies.

> *February 2002: Edinburgh became the first city in the world to offer both Sulfur Free Unleaded and Sulfur Free Diesel from 18 February 2002. BP's service stations in Scotland were the first in its worldwide network to sell both fuels.*
>
> *The new fuels are the cleanest petrol (gasoline) and diesel fuels available in the UK, containing no more than 10 ppm sulfur. This amount is so small that it is barely detectable with the most sophisticated laboratory test methods, which is why it is commonly referred to as 'sulfur free'.*
>
> *These new fuels are produced at BP's Grangemouth refinery in east Scotland, which is the only UK refinery with [the necessary] technology.*
>
> *BP Sulfur Free Petrol (Gasoline)*
>
> - *Improves the efficiency of the catalyst on car exhausts to remove pollution.*
> - *Enables engine manufacturers to further develop and release the most fuel efficient petrol engines available, which also produce less carbon dioxide.*
> - *Enables development of De-NOx technology for exhausts which reduce nitrogen oxide emissions that cause smog.*
>
> *BP Sulfur Free Diesel*
>
> - *Dramatically reduces the amount of ultra-fine particulates expelled in vehicle exhaust that can contribute to respiratory problems such as asthma. The level of particulate emissions is one of the key air quality issues that need to be addressed in cities.*
>
> *EU (European Union) legislation required sulfur-free fuels to be made available from all sites by 2008. Coinciding with this, the EU set much stricter emission limits on new vehicles manufactured from 2004.*

Concept Check 19-1. Why wasn't "sulfur-free" fuel available over the entire United Kingdom?

By 2012, the UK had made the transition to sulfur-free road fuels (defined as sulfur content not exceeding 10 parts per million or 0.001% by weight).

Concept Check 19-2. What advantage will sulfur-free diesel afford to attempts to control diesel emissions?

Value-added Question 19-1. Off-road diesels (farm equipment, etc.) burn fuel containing up to 3,400 parts per million (ppm) sulfur. How many grams of sulfur would be contained in each liter of such fuel at a concentration of 3,400 ppm? In a gallon? One ppm = 1 g/1,000 L, 1 gal = 3.785 L.

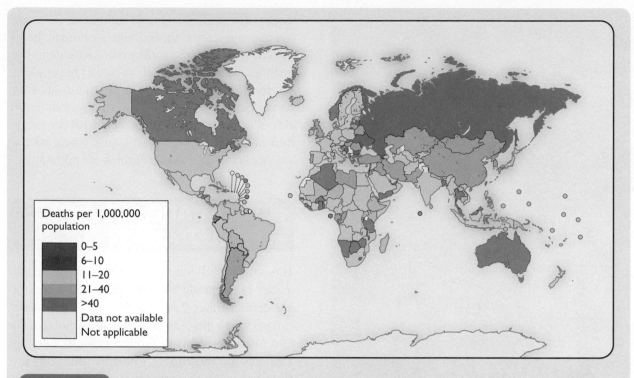

Deaths per 1,000,000
population

- 0–5
- 6–10
- 11–20
- 21–40
- >40
- Data not available
- Not applicable

FIGURE 19-1 Deaths from global air pollution, 2008, per million. Given the population of the People's Republic of China (PRC) at 1.3 billion, how many deaths are caused by air pollution annually? Reproduced from Global Health Observatory Map Gallery-Deaths Attributable to outdoor air pollution, 2008/WHO.

■ WHAT IS AIR POLLUTION?

Concept Check 19-3. How would you define **air pollution**? Take a moment and write down your definition.

- -

How did your definition compare to this one from the State of California's Air Resources Board (ARB), a division of the State's Environmental Protection Agency (EPA): "Degradation of air quality resulting from unwanted chemicals or other materials occurring in the air." **Air pollutants** are "Amounts of foreign and/or natural substances occurring in the atmosphere that may result in adverse effects to humans, animals, vegetation, and/or materials."

Let's take a moment to dissect that definition of air pollutants. Note that pollutants can be either natural or anthropogenic. In fact, volcanic eruptions can be major sources of sulfur oxide (SOx). The eruption of Mt. Pinatubo in the Philippines in 1991 spewed about 20 million tons into the atmosphere, according to the U.S. Geological Survey (USGS). And although not strictly a pollutant, carbon dioxide (CO_2) emitted by volcanoes has killed thousands of people in historic eruptions (see section on volcanic hazards elsewhere in the text).

Air pollutants can have adverse impacts on *materials*. In fact, **acid precipitation** is a major cause of building disintegration in cities around the world, especially when the building material is limestone or marble, which are common building

FIGURE 19-2 Damage from acid precipitation to Exeter Cathedral (14th century), UK. © Vanessa Miles/Alamy.

stones (FIGURE 19-2). *Ozone*, another air pollutant, can damage crops, in addition to human lungs. Finally, note that the California ARB's definition includes damage to animals as well as to people.

But the fact that some air pollutants are from natural sources does not mean that we can ignore air pollution. Quite the contrary: with the exception of episodic puffs of volcanic gases, human activity is the sole source of most air pollutants.

■ WHAT ARE AIR POLLUTANTS?

Thousands of chemicals and kinds of particulate matter can degrade the atmosphere. Air monitoring demonstrates that over 90% of Californians, and perhaps a majority of Americans—not to mention urban residents of foreign cities like Beijing, Mexico City, and Athens—breathe unhealthy levels of one or more air pollutants during some part of the year. Fortunately, relatively few pollutants are present in any quantity, even though about 200 pollutants have been labeled as *air toxics*. We describe the major air pollutants below.

Criteria for Pollutants and Air Toxics

The 1990 Clean Air Act Amendments—among the most important environmental legislation ever enacted—required the EPA to set "National *Ambient* Air Quality Standards" (NAAQS) for pollutants considered to be a threat to human health and the environment. EPA set two types of standards.

Primary standards set limits to protect public health, recognizing that there were groups of people who were more sensitive to pollution (the elderly, small children) than others.

Secondary standards set limits to protect "public welfare," treating such effects as decreased visibility, damage to animals, plants and crops, and buildings.

Pollutants

The EPA set NAAQS for what it called **criteria pollutants**. These are pollutants that occur commonly in the United States and are the EPA's indicators of air quality. They are: carbon monoxide (CO), nitrogen oxide (NOx), ozone, lead, **particulates** (**PM10** and **PM2.5**, which refer, respectively, to particles of 10 micrometers or less and particles smaller than 2.5 micrometers), and sulfur oxide (SOx). For each criterion pollutant, the EPA sets a limit for short-term exposure and (except for lead) a higher limit for long-term exposure.

Air Toxics

The EPA identified other pollutants as *air toxics*. These were pollutants for which insufficient data existed to set standards; in other words, they could identify no level below which no adverse effects could be measured.

Global Air Pollutants

TABLE 19-1 shows a list of major global air pollutants.

Heavy metals, other than lead, were not specifically covered by the 1990 Clean Air Act Amendments. Treatment of mercury under clean air legislation is controversial. Many healthcare professionals and environmentalists consider mercury in some forms to be extremely toxic. Spokespersons for the coal industry and some coal-burning utilities advocate less strict regulations. We discuss mercury in the coal, water pollution, and environmental health sections of the text.

Volatile organic compounds (VOCs) react with sunlight to produce *ozone* (O_3), a major component of smog (visible air pollution). According to the European Environment Agency, a few days' exposure to high levels of ozone can trigger a number of health problems, particularly lung and allergy problems among the very young and very old. Long-term exposure to moderate levels may also cause pulmonary distress in young children. Plants can be adversely affected as well.

A Special Case: Chlorofluorocarbons

A category of artificial chemicals called **chlorofluorocarbons (CFCs)** has been shown to destroy the atmosphere's *ozone shield*, which protects the planet's surface from high-energy ionizing radiation, for example, short-wave ultraviolet. As a result, an international treaty, the 1987 Montreal Protocol, was set up to regulate the manufacture and distribution of CFCs with a view towards their ultimate elimination. **FIGURE 19-3** shows the effect of CFCs on ozone depletion.

TABLE 19-1	**Major Global Air Pollutants**
Pollutant	**Primary Source**
Particulates <10 microns (fine)	fossil fuel combustion, fires, industry, field burning
NOx (nitrogen oxides)	(fossil) fuel combustion
SOx (sulfur oxides)	fossil fuel combustion, volcanoes
CO (carbon monoxide)	incomplete (fossil) fuel combustion
VOCs (volatile organic compounds)	gasoline, solvents, dry cleaning fluids, alcohol (many others)
Ozone (O_3)	reaction of VOC/ROG with sunlight
Heavy metals (Hg, Ni, Pb, etc)	metals refining, coal combustion

Data from: California Air Resources Board and Alaska Department of Environmental Conservation.

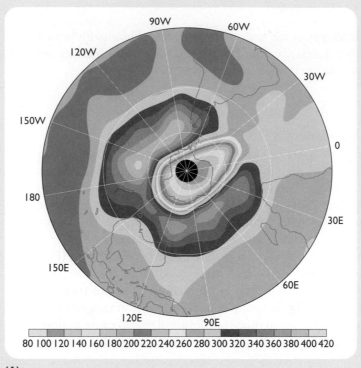

(A)

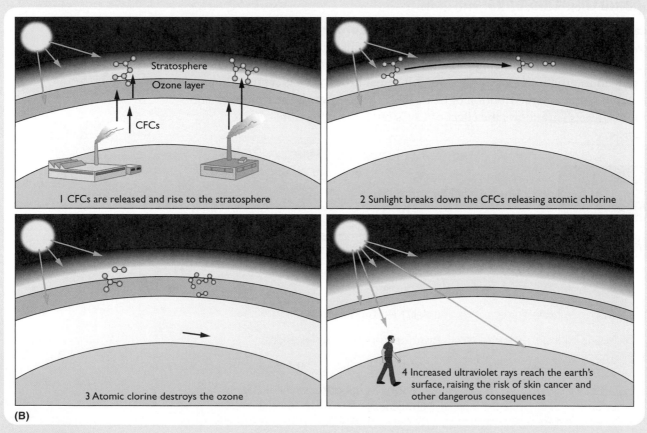

1 CFCs are released and rise to the stratosphere

2 Sunlight breaks down the CFCs releasing atomic chlorine

3 Atomic clorine destroys the ozone

4 Increased ultraviolet rays reach the earth's surface, raising the risk of skin cancer and other dangerous consequences

(B)

FIGURE 19-3 (A) The 2006 record-breaking "hole" (actually a thinning) in the ozone over Antarctica. In 2011, for the first time a thinning of the ozone layer was observed over the Arctic. Red color indicates low ozone. Reproduced from NOAA, U.S. Department of Commerce. (B) How CFCs affect ozone in the upper atmosphere.

TABLE 19-2	Atmospheric Gases and Their Concentration in the Atmosphere			
Gas	Volume (%)	Molecular Weight (Approximate)	Weight (%)	Residence Time
Nitrogen	78.000	28	75.00	—
Oxygen	21.000	32	23.00	—
Argon	00.930	40	1.30	—
Carbon dioxide	00.035	44	0.05	30–95 years
Nitrous oxide	trace	44	trace	114 years
HFC	trace	117	trace	12 years
CFC	trace	187	trace	100 years
Average atmosphere	100.000	29	100.00	—

*Percentages do not add up to 100 due to rounding, approximation, and ignoring water vapor and all minor and trace gases. Weight (%) is found by multiplying the volume (%) by the molecular weight of that gas and dividing by the average molecular weight of the atmosphere.

■ THE ATMOSPHERE

Because it was just discussed, we should give you some background on the Earth's atmosphere and its importance in understanding air pollution. The atmosphere is one of the four great, interconnected *systems* that make up the Earth's outer shell: the others are the *lithosphere*, the *hydrosphere*, and the *biosphere*.

TABLE 19-2 lists atmospheric gases with their average composition and *residence times*.

Of these, *argon* is an inert gas produced by the radioactive decay of potassium-40 (K-40), and is of no further importance in this context. **Residence time** is the average time that an atom, molecule, or complex stays in the atmosphere, before cycling into one of the other systems. Table 19-2 shows the residence times of atmospheric gases and substances implicated in global climate change.

Concept Check 19-4. Based on the data in Table 19-2, justify taking measures to reduce CO_2 emissions sooner rather than later.

Value-added Question 19-2.
What is the residence time of CO_2? This means that on average whenever a CO_2 molecule enters the atmosphere it will take _____ years for it to leave.

■ EFFECTS OF AIR POLLUTION

During the first week of December 1952, unusual atmospheric conditions prevailed across southern England. A powerful high-pressure system kept smoke and sulfur dioxide from the millions of coal fires in London and its environs from blowing "away." As the weather got colder, Londoners piled more coal on their fires. Pollution levels in the lower atmosphere quickly reached toxic levels (**FIGURE 19-4**). Visibility levels declined to one foot in many places. Authorities estimate that as many as 12,000 people may have been killed. After lengthy study, the disaster prompted

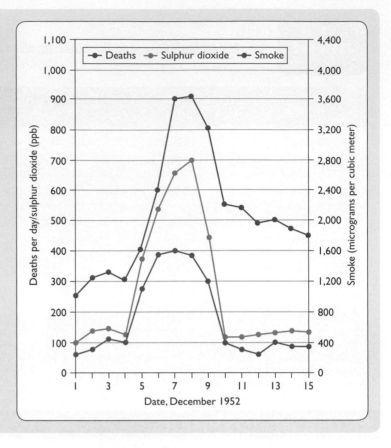

FIGURE 19-4 The buildup in air pollution over London, December 4–10, 1952, and its effects. What pollutants "spiked" in this chart? At a maximum, how many deaths per day could be attributed to the high SOx and smoke (particulate) levels? Data from: Wilkins E. T., Air pollution and the London fog of December 1952. *Journal of the Royal Sanitary Institute* 1954;74:1–21.

the Conservative governments of Winston Churchill and Anthony Eden to pass the country's first "Clean Air" law in 1956, which, among other things, banned coal fires over much of the London region.

Particulates and SOx are two forms of air pollution. **TABLE 19-3** is a chart showing some common types of air pollutants, their sources, and examples of their effects.

Toxic air contaminants (TACs) are substances from which adverse health effects may occur at extremely low levels. It is difficult or impossible to identify levels of exposure that do not produce adverse health effects. Diesel exhaust is an excellent example. In California, agricultural diesels and on-road diesel trucks each contribute about 15% of the total TAC from diesels. Diesel exhaust contains about 40 substances which are either carcinogenic, **mutagenic** (cause cellular mutations), or both. TACs differ from Criteria Air Pollutants in that according to the EPA it is impossible to identify a level of exposure to TACs that will cause undetectable effects.

■ ACID PRECIPITATION

Acid precipitation occurs when *acids* such as hydrofluoric (HF), nitric (HNO_3), hydrochloric (HCl), and sulfuric (H_2SO_4) from human activity or natural sources like volcanic eruptions enter the atmosphere, combine with water molecules, and reach the ground as **acid deposition**. This process is often called "acid rain," but any form of precipitation (snow, fog, sleet, etc.) may cause acid deposition. Natural precipitation has an acidity (pH, or hydrogen ion concentration) of around 5.6 (**FIGURE 19-5**).

In North America, major sources of acid precipitation are emissions from motor vehicles, fossil-fueled power plants, and nitrogen gases from agriculture. Both sulfur and nitrogen compounds contribute to AP.

Pollutant	Sources	Effects
Ozone. A gas that can be found in two places. Near the ground (the troposphere), it is a major part of smog. The harmful ozone in the lower atmosphere should not be confused with the protective layer of ozone in the upper atmosphere (stratosphere), which screens out harmful ultraviolet rays.	Ozone is not created directly, but is formed when nitrogen oxides and volatile organic compounds mix in sunlight. That is why ozone is mostly found in the summer. Nitrogen oxides come from burning gasoline, coal, or other fossil fuels. There are many types of volatile organic compounds, and they come from sources ranging from factories to trees.	Ozone near the ground can cause a number of health problems. Ozone can lead to more frequent asthma attacks in people who have asthma and can cause sore throats, coughs, and breathing difficulty. It may even lead to premature death. Ozone can also hurt plants and crops.
Carbon monoxide. A gas that comes from the burning of fossil fuels, mostly in cars. It cannot be seen or smelled.	Carbon monoxide is released when engines burn fossil fuels. Emissions are higher when engines are not tuned properly, and when fuel is not completely burned. Cars emit a lot of the carbon monoxide found outdoors. Furnaces and heaters in the home can emit high concentrations of carbon monoxide, too, if they are not properly maintained.	Carbon monoxide makes it hard for body parts to get the oxygen they need to run correctly. Exposure to carbon monoxide makes people feel dizzy and tired and gives them headaches. In high concentrations it is fatal. Elderly people with heart disease are hospitalized more often when they are exposed to higher amounts of carbon monoxide.
Nitrogen dioxide. A reddish-brown gas that comes from the burning of fossil fuels. It has a strong smell at high levels.	Nitrogen dioxide mostly comes from power plants and cars. Nitrogen dioxide is formed in two ways—when nitrogen in the fuel is burned, or when nitrogen in the air reacts with oxygen at very high temperatures. Nitrogen dioxide can also react in the atmosphere to form ozone, acid rain, and particles.	High levels of nitrogen dioxide exposure can give people coughs and can make them feel short of breath. People who are exposed to nitrogen dioxide for a long time have a higher chance of getting respiratory infections. Nitrogen dioxide reacts in the atmosphere to form acid rain, which can harm plants and animals.
Particulate matter. Solid or liquid matter that is suspended in the air. To remain in the air, particles usually must be less than 0.1-mm wide and can be as small as 0.00005 mm.	Particulate matter can be divided into two types—coarse particles and fine particles. Coarse particles are formed from sources like road dust, sea spray, and construction. Fine particles are formed when fuel is burned in automobiles and power plants.	Particulate matter that is small enough can enter the lungs and cause health problems. Some of these problems include more frequent asthma attacks, respiratory problems, and premature death.

Continued ▶

Pollutant	Sources	Effects
Sulfur dioxide. A corrosive gas that cannot be seen or smelled at low levels but can have a "rotten egg" smell at high levels.	Sulfur dioxide mostly comes from the burning of coal or oil in power plants. It also comes from factories that make chemicals, paper, or fuel. Like nitrogen dioxide, sulfur dioxide reacts in the atmosphere to form acid rain and particles.	Sulfur dioxide exposure can affect people who have asthma or emphysema by making it more difficult for them to breathe. It can also irritate people's eyes, noses, and throats. Sulfur dioxide can harm trees and crops, damage buildings, and make it harder for people to see long distances.
Lead. A blue-gray metal that is very toxic and is found in a number of forms and locations.	Outside, lead comes from cars in areas where unleaded gasoline is not used. Lead can also come from power plants and other industrial sources. Inside, lead paint is an important source of lead, especially in houses where paint is peeling. Lead in old pipes can also be a source of lead in drinking water.	High amounts of lead can be dangerous for small children and can lead to lower IQs and kidney problems. For adults, exposure to lead can increase the chance of having heart attacks or strokes.
Toxic air pollutants. A large number of chemicals that are known or suspected to cause cancer. Some important pollutants in this category include arsenic, asbestos, benzene, and dioxin.	Each toxic air pollutant comes from a slightly different source, but many are created in chemical plants or are emitted when fossil fuels are burned. Some toxic air pollutants, like asbestos and formaldehyde, can be found in building materials and can lead to indoor air problems. Many toxic air pollutants can also enter the food and water supplies.	Toxic air pollutants can cause cancer. Some toxic air pollutants can also cause birth defects. Other effects depend on the pollutant, but can include skin and eye irritation and breathing problems.
Stratospheric ozone depleters. Chemicals that can destroy the ozone in the stratosphere. These chemicals include chlorofluorocarbons (CFCs), halons, and other compounds that include chlorine or bromine.	CFCs are used in air conditioners and refrigerators, since they work well as coolants. They can also be found in aerosol cans and fire extinguishers. Other stratospheric ozone depleters are used as solvents in industry.	If the ozone in the stratosphere is destroyed, people are exposed to more radiation from the sun (ultraviolet radiation). This can lead to skin cancer and eye problems. Higher ultraviolet radiation can also harm plants and animals.
Greenhouse gases. Gases that stay in the air for a long time and warm up the planet by trapping sunlight. This is called the "greenhouse effect" because the gases act like the glass in a greenhouse. Some of the important greenhouse gases are carbon dioxide, methane, and nitrous oxide.	Carbon dioxide is the most important greenhouse gas. It comes from the burning of fossil fuels in cars, power plants, houses, and industry. Methane is released during the processing of fossil fuels, and also comes from natural sources like cows and rice paddies. Nitrous oxide comes from industrial sources and decaying plants.	The greenhouse effect can lead to changes in the climate of the planet. Some of these changes might include more temperature extremes, higher sea levels, changes in forest composition, and damage to land near the coast. Human health might be affected by diseases that are related to temperature or by damage to land and water.

Data from: EPA.

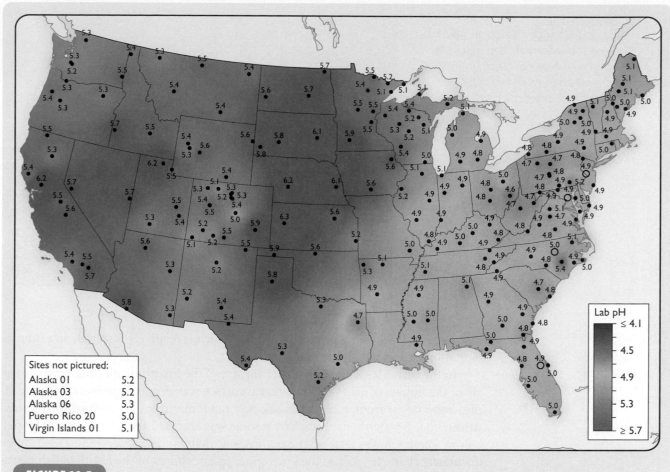

FIGURE 19-5 Average pH at sites in the United States in 2009. Acidity increases as pH decreases. Data from: EPA and National Atmospheric Deposition Program.

Impact of Acid Precipitation

Acid precipitation can corrode buildings, acidify natural waterways, and acidify soils. It can damage vegetation and at high acidities over long periods of time can harm crops, humans, and animals (**FIGURE 19-6**).

Since the 1970s, both the United States and Canada have passed legislation to control AP.

Research has demonstrated that the impact of AP depends on the *acidity of the precipitation*, the *organisms* in the affected ecosystem, the *bedrock geology*, and *soil type*. Studies in Canada, Sweden, and the United States show that regions underlain by carbonate bedrock experience less harm from AP than regions underlain by silicate bedrock. Research on biota shows that larvae of many fish species are more sensitive to acidified waters than adults.

Some soils, when acidified, can leach aluminum hydroxide (Al^{3+}) into water bodies. This hydrated alumina can selectively kill certain organisms, especially juveniles.

Acid Precipitation in Québec, Canada

Concern about acid precipitation in Québec arose by the end of the 1970s, when research demonstrated that southwestern Québec was receiving highly acidic deposition. Geologic maps indicated that the region was very sensitive to acidification

FIGURE 19-6 Dead and dying trees in Shenandoah National Park. Experts believe acid precipitation has contributed to the die-off. © Karol Kozlowski/ShutterStock, Inc.

due to its bedrock being primarily silicate, which is relatively insoluble in acid precipitation and so will not reduce its effects.

In the early 1980s, the Canadian Ministry of the Environment ordered an analysis of the impact of acid precipitation on surface waters and organisms. Water samples were taken from 1,253 lakes selected randomly from 160,000 lakes scattered throughout 5 watersheds. Each lake sample was analyzed for 19 variables (e.g., pH, color). These samples enabled the acidity, sensitivity, and acid deposition levels to be accurately characterized. At the same time, an assessment of the health of fish populations of more than 253 lakes was carried out.

FIGURE 19-7 shows the relative sensitivity of the region to acid precipitation. There are more than 450,000 (!) lakes in Québec, and many of them are acidified.

FIGURE 19-8 shows the change in species abundance as pH increases in Québec lakes. The Provincial government concluded that most of the acidification was from

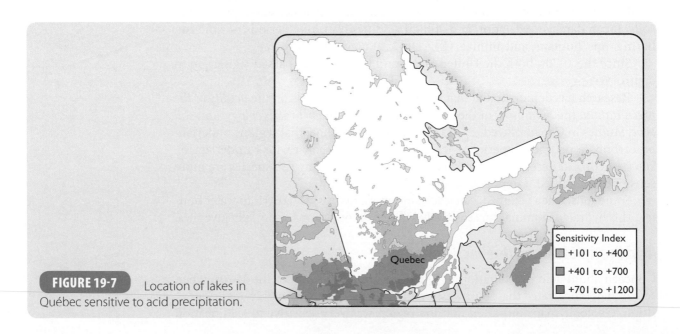

FIGURE 19-7 Location of lakes in Québec sensitive to acid precipitation.

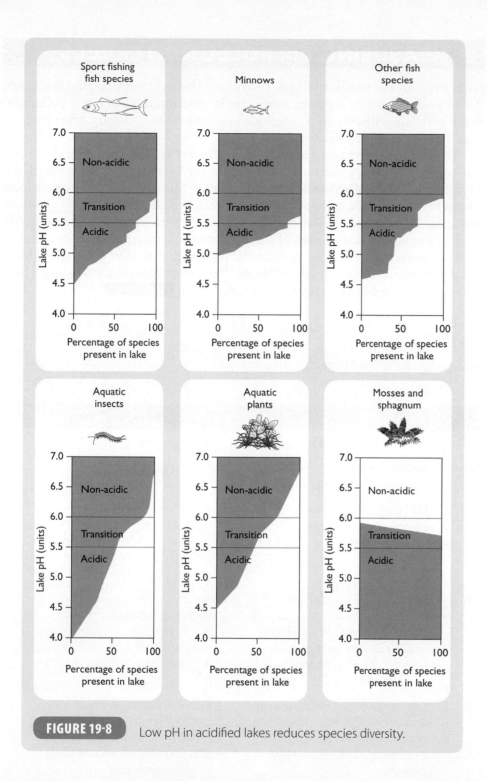

Low pH in acidified lakes reduces species diversity.

human sources, and identified power plants and smelters as major sources of acid precipitation. It further concluded, "When lake pH goes under 5, the most resistant fish species can survive, although their reproduction capacity becomes highly compromised."

Concept Check 19-5. What effects could acid precipitation have on Québec's tourism? Could this be used to actually compute a cost of pollution? If sources of the pollution could be precisely identified, should they be required to pay these costs? Should government be responsible for paying to clean up pollution, even if polluters could be identified?

■ MEASURING AIR POLLUTION LEVELS

Great progress has been made since 1970 in the United States when the first Clean Air Act was passed, in reducing atmospheric concentrations of many pollutants. Some pollutants, notably NOx, continue to increase. One category—fine particulates—has only recently been studied in detail.

The U.S. EPA has the responsibility under the 1970 Clean Air Act for determining "safe" levels of exposure to air pollutants in the United States. Accordingly, the EPA designated a **Pollution Standard Index (PSI)** with which to express local air quality. The PSI is a composite indicator calculated from atmospheric levels of ozone, nitrogen dioxide, sulfur dioxide, carbon monoxide, and particulate matter only. The concentration of each of these pollutants is measured on a continuous basis. Once each day, the *highest* concentration of each of the five pollutants is measured and assigned a value on the PSI scale. The highest of the five PSI values is then reported to the public as an Air Quality Index (AQI; **TABLE 19-4**).

TABLE 19-4	*Air Quality Index (AQI) Values and Impacts*		
Index Value	**PSI Descriptor**	**General Health Effects**	**Cautionary Statements**
Up to 50	Good	None for the general population	None required
51 to 100	Moderate	Few or none for the general population	None required
101 to 200	Unhealthful	Mild aggravation of symptoms among susceptible people, with irritation symptoms in the healthy population	Persons with existing heart or respiratory ailments should reduce physical exertion and outdoor activity. General population should reduce vigorous outdoor activity.
201 to 300	Very Unhealthful	Significant aggravation of symptoms and decreased exercise tolerance in persons with heart or lung disease; widespread symptoms in the healthy population.	Elderly and persons with existing heart or lung disease should stay indoors and reduce physical activity. General population should avoid vigorous outdoor activity.
Over 300	Hazardous	Early onset of certain diseases in addition to significant aggravation of symptoms and decreased exercise tolerance in healthy persons.	General population should avoid outdoor activity. All people should remain indoors, keeping windows and doors closed, and minimize physical exertion.
Over 400	Hazardous	Healthy people experience adverse symptoms that affect normal activity. Elderly and persons with existing diseases should stay indoors and avoid physical exertion.	General population should avoid outdoor activity. All people should remain indoors, keeping windows and doors closed, and minimize physical exertion.

Data from: EPA.

■ SUMMARY OF AIR POLLUTION SOURCES

Here we summarize information provided in the previous tables and introduce new material as well. As you saw, there are some natural air pollutants. Volcanic eruptions can be a major source of some air pollutants. SOx, HCl, HF, and hydrobromic acid (HBr) are usually present in volcanic gases as well as CO_2. During some eruptions, major quantities of some of these pollutants can be emitted. However, with some major exceptions, such as Pinatubo, the effects tend to be more local than regional and global. Forest fires can be major sources of particulates, and their effects may be regional in scope, though not usually long lasting. Human interference with forest ecosystems over the past 100 years, especially in North America, Europe, Brazil, Indonesia, and China has contributed to the severity and scale of many forest fires, supposedly "natural" sources of air pollution.

In fact, the major difference between natural and anthropogenic air pollution is that the latter tends to be *continuing and persistent*, while the former tends to be *local and episodic*. Finally, natural sources of PM2.5 particulates, CO, and heavy metals are extremely rare, and contribute virtually nothing to our air pollution problems.

The main sources of air pollutants are thus anthropogenic. They are *fossil-fueled power plants, motor vehicles, calciners* (cement manufacturers), *industries* (such as chemical manufacturers and refiners), and *industrialized agriculture* including animal feedlot operations (see the section on soils and agriculture elsewhere in the text).

■ CLEANING THE AIR

Motor Vehicles: Lead-Free Gas and Catalytic Converters

After researchers identified motor vehicles as a main contributor to Los Angeles smog in the 1950s, pressure grew to require automakers to reduce emissions and to require drivers to maintain vehicles so as to minimize emissions. The greatest single event leading to reduction in motor vehicle emissions was the phase-out of leaded gasoline in the United States in the 1970s. Other nations eventually followed suit, such that at present leaded motor fuels are available only in some developing countries.

Not only did the removal of lead help clean the air of toxic lead particulates, it also meant that **catalytic converters** could be fitted to motor vehicles for the first time (lead deactivates the catalyst). FIGURE 19-9 shows a catalytic converter, so-called because they use minute amounts of precious metals like platinum as catalysts to convert unburned hydrocarbons and CO to CO_2 and H_2O. They have drastically reduced emissions from cars, and have contributed to improvements in air quality in every city in North America since the middle 1970s. Furthermore, reductions in sulfur in gasoline and diesel fuel promise to remove tens of thousands of tonnes of SOx from the air in Western Europe and North America each year. Because sulfur eventually poisons catalytic converters, removal of S from fuels will considerably extend the life and effectiveness of those devices. While there have been some reports of deposition of catalyst metals such as rhodium and platinum by operation of a catalytic converter, so far their use has produced no known harmful side effects.

The virtual elimination of sulfur from fuel will mean that diesel engines will be candidates for converters as well, further reducing diesel emissions, especially fine particulates, an extremely toxic air contaminant.

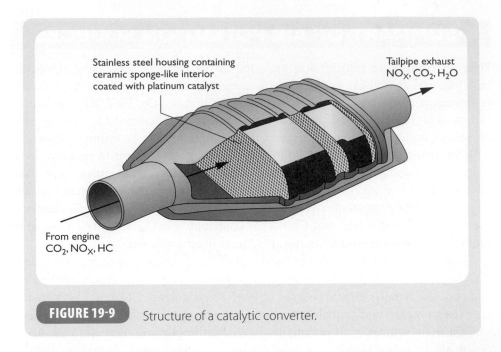

Stainless steel housing containing
ceramic sponge-like interior
coated with platinum catalyst

Tailpipe exhaust
NO_X, CO_2, H_2O

From engine
CO_2, NO_X, HC

FIGURE 19-9 Structure of a catalytic converter.

Conversion of motor vehicle fleets to hydrogen power, fuel cells, or electrics could virtually eliminate tailpipe emissions, although not necessarily overall emissions (Why?), and could help eliminate the significant air pollution caused by evaporation of petroleum-based fuels. Already a new generation of hybrid and all-electric vehicles is available with emission levels slashed as compared to low-emission gas powered cars (**FIGURE 19-10**).

FIGURE 19-10 Carol and Garvey Winans, of Georgetown, SC with their two Toyota Priuses, the first successful hybrid car. The Prius uses a small gasoline engine and an electric motor to improve fuel economy and reduce emissions. The 2009 Prius is EPA rated at 45 mpg but many owners report much higher mileage. © Pluff Mud Photography.

Pollution From Other Internal Combustion Engines

Significant amounts of air pollution result from the operation of (1) hand-held "lawn care devices," (2) off-road vehicles like personal watercraft and snowmobiles, and (3) stationary agricultural and construction diesel engines. For example, there are about 12 million marine engines operated in the United States, primarily used for recreational boating. About 400,000 ply the waters of the Chesapeake Bay. These marine engines are among the highest contributors of hydrocarbons and NOx emissions in the country, according to the EPA. The reason is that the majority of the watercraft use two-stroke (as opposed to 4-stroke) internal combustion engines. **Two-stroke engines** use a mixture of oil (for lubrication) and gasoline as a fuel. When burned, the oil emits large quantities of toxic unburned hydrocarbons.

As motor vehicles become cleaner, addressing other sources becomes increasingly important to further improve air quality. Operation of a two-stroke leaf blower for an hour can produce as much air pollution as a newer car driven for 100 miles (160 km)! Manufacturers of lawn care devices propose to cease production of two-stroke engines in favor of four-stroke engines, which produce considerably less air pollution, but are concerned that the added cost may deter purchasers.

Concept Check 19-6. Would you be willing to breathe more polluted air to allow users of these devices to save money? How do we ask these questions to children, whose systems are much more sensitive to air pollution than adults? Who speaks for plants and animals?

Tires as Sources of Air Pollution

Tire wear produces considerable fine particulate matter, consisting of (1) heavy metals from stainless steel in radials, (2) powdered polyester, and (3) latex. While latex intolerance is a significant human allergen, little research has been done on the impact of long-term exposure to tire dust. However, should it turn out to be a significant source of air toxics, tires could be designed that would wear much more slowly than those at present, helping to further reduce particulate emissions.

Cleaning Power Plant Emissions

Early (1960s) air pollution regulations required some utilities and industries to reduce emissions in the vicinity of the plant. The first antipollution devices applied to power plants were ash filters and precipitators, which trapped much of the coarse particulate matter before it could escape up the smokestack. Operators sought to further comply by building "tall stacks" up to 300-m high, which effectively dispersed stack emissions "downwind." However, residents of downwind cities were the unwilling recipients of that pollution (FIGURE 19-11). That situation in part continues today, because the 1990 Clean Air Act Amendments did not require older coal-burning power plants to upgrade or even add pollution-control equipment unless the plant itself was substantially upgraded. Many power plants, especially in the Midwestern United States where coal is the primary fuel, did not retire their old plants as the framers of the Amendments expected. Thus, old, inefficient coal-burning power plants in the Midwest (FIGURE 19-12), "**grandfathered in**" under the Clean Air Act, are the source of up to one-half of the air pollution of the Northeast and eastern Canada.

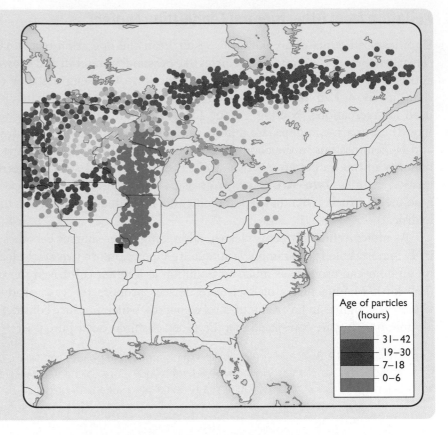

FIGURE 19-11 A computer simulation done at Washington University on July 5, 1985, showing how air pollution from St. Louis, Missouri could be dispersed around North America within 42 hours. *Questions:* What recourse do affected parties have in Canada? How could heavy metals from Missouri power plants impact organisms in Lake Superior? Modified from *Environ Health Perspect.* 2000 April; 108(4): A170–A175.

Age of particles (hours)

31–42
19–30
7–18
0–6

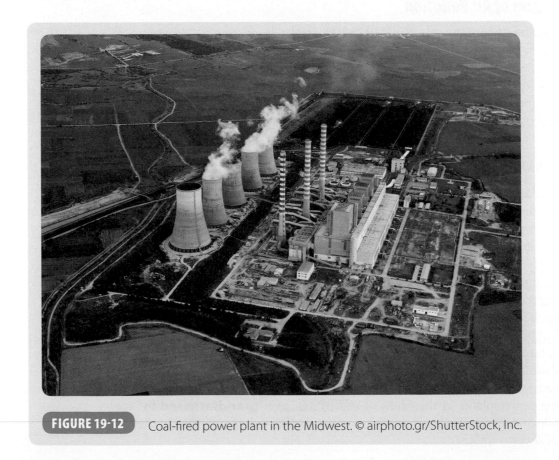

FIGURE 19-12 Coal-fired power plant in the Midwest. © airphoto.gr/ShutterStock, Inc.

Emission Control Strategies

Global trends in SOx emissions are shown in **TABLE 19-5**. Globally, operators of power plants may use a variety of means to reduce or even eliminate sulfur emissions, if required by government. They include:

1. **Fuel switching**: Many power plant operators have complied with antipollution regulations by switching fuels. Replacing coal containing high sulfur content with low-sulfur coal helped many utilities meet emissions reduction targets, but did little to reduce NOx, CO_2, particulates, or heavy metals.

     Many power producers are building natural gas-fired plants to replace or supplement coal plants. The advantage of natural gas plants is that natural gas contains no sulfur, ash, or heavy metals, and CO_2 emissions are slightly more than half those of coal plants (**FIGURE 19-13**).

     Scrubbers: Scrubbers are devices that remove sulfur gases by combining the exhaust gas with water and powdered limestone, which converts sulfur

TABLE 19-5	**Change in Global SOx Emissions Over Time**							
				Current Legislation (CLE)			**Maximum Feasible Reductions (MFR)**		
Regions of the SRES Study	**1990**	**2000**	**2010**	**2020**	**2030**	**2010**	**2020**	**2030**
OECD90:	45	29	18	14	13	5	5	5
North America	24	19	11	8	8	3	3	3
Western Europe	18	8	4	3	3	1	1	1
Pacific OECD	3	3	3	3	3	1	1	1
REF:	30	16	11	8	8	2	2	2
Central and Eastern Europe	11	6	4	2	2	1	0	0
Russia and NIS	19	10	7	6	6	2	2	2
Asia:	31	32	41	46	53	10	11	11
China	21	21	25	22	22	6	6	6
Other Pacific Asia	5	4	6	7	9	2	2	2
South Asia	5	7	11	17	22	2	3	3
ALM:	15	18	15	14	13	4	4	4
Latin America	7	8	6	5	5	2	2	2
Middle East	3	5	3	3	2	1	1	1
Africa	5	5	5	5	6	1	1	2
World Total	120	96	85	82	87	21	22	22

Describe the changes in sources of SOx after 2005. Project this trend to 2050. Is this cause for concern or optimism?
Data from: International Institute for Applied Systems Analysis.

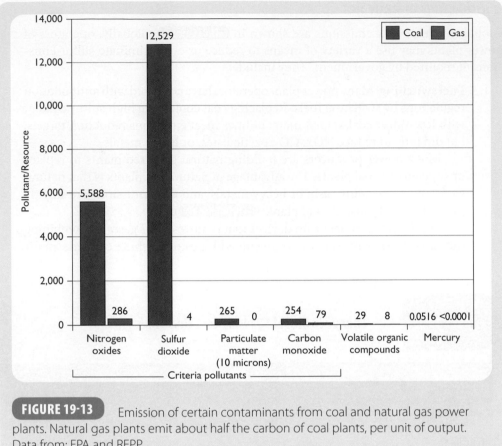

FIGURE 19-13 Emission of certain contaminants from coal and natural gas power plants. Natural gas plants emit about half the carbon of coal plants, per unit of output. Data from: EPA and REPP.

to calcium sulfite (CaS_2). This in turn can be easily converted to gypsum ($CaSO_4 \cdot nH_2O$). Gypsum is a relatively benign mineral that can be landfilled (if it has not been contaminated with toxic heavy metals), or used as drywall in building construction. Installation of scrubbers worldwide could lead to a decline in SOx. Disadvantages of scrubbers are that they cost tens of millions of dollars to install, they require power to operate, an abundant water source, and they must be carefully maintained. A coal-fired power plant with scrubbers fitted to each boiler can use up to 10% of its electricity to run the scrubbers.

2. **Cleaning coal before combustion**: Coal contains sulfur in two forms: as *pyrite* and other metals, and as *organic sulfur* in coal vegetation. Coal can be crushed and washed, which can remove much of the associated pyrite. Disadvantages are that the process is costly and requires considerable water.

Nitrogen Oxides Removal or Treatment

The 1990 Clean Air Act Amendments required states to reduce ground-level ozone. NOx is a major "precursor" to the formation of ozone. It further recognized the role played by transboundary sources in NOx and ozone pollution and required "upwind" states to reduce their NOx emissions to "downwind" states.

Nitrogen oxide gases are produced whenever fuel is burned in air. Generally, the higher the combustion temperature, the more NOx released. Strategies for reducing NOx emissions include use of low-NOx burners and switching to renewable power sources. Low-NOx burners restrict air to the hottest parts of a

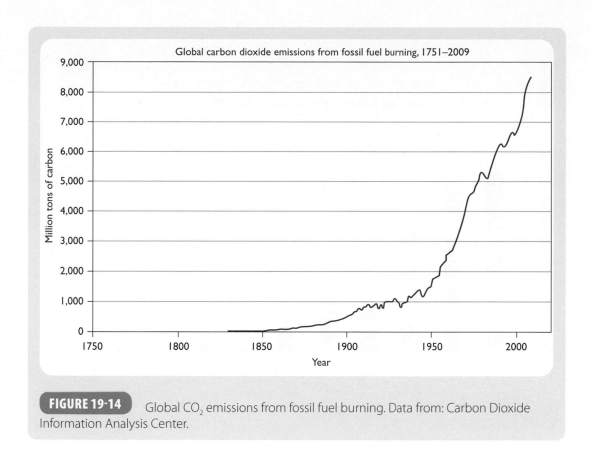

Global CO_2 emissions from fossil fuel burning. Data from: Carbon Dioxide Information Analysis Center.

combustion chamber, thus reducing formation of NOx. Reducing burn temperatures can reduce efficiency of electricity production.

Carbon Dioxide

Although the original NEPA statute may be interpreted to include CO_2 as a pollutant, higher levels of CO_2 do not threaten human health in the way that other pollutants (such as ground level ozone) clearly do. Because CO_2 was not defined by EPA regulations as a pollutant, it was not covered by the original Clean Air Act or by the comprehensive Amendments of 1990. However, increasing atmospheric levels of CO_2, along with computer forecasts of adverse impacts of global climate change, have prompted calls for measures to address CO_2 emissions (**FIGURE 19-14**), and the U.S. Supreme Court has allowed that regulating CO_2 is within the purview of the U.S. EPA. We address this issue in the global climate change section found elsewhere in the text.

■ TRENDS IN AIR QUALITY

FIGURE 19-15 shows the trends in SOx and NOx emissions from electricity generation in the United States from 1995 projected to 2030. Comprehensive data for developing countries, India, or China are either unreliable or unavailable.

Concept Check 19-7. Based on Figure 19-15, do you conclude air quality is improving or worsening? Based on your sense of the extent to which citizens in the United States and elsewhere value clean air, are you optimistic or pessimistic for the future? Cite evidence to support your conclusion.

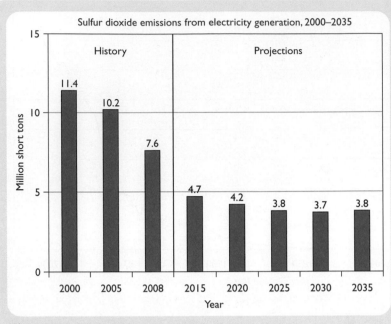

(A)

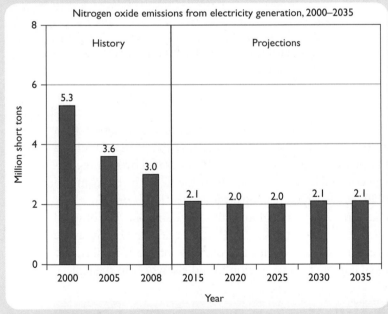

(B)

FIGURE 19-15 U.S. emissions of S and N gases from electricity generation for the years 1995–2035. (A) Sulfur oxides. (B) Nitrogen oxides. *Question*: Does this chart give you cause for optimism or pessimism? Explain. Reproduced from Annual Energy Outlook 2010/EIA.

CHAPTER SUMMARY

1. Removing sulfur from gasoline improves the efficiency of the catalyst on car exhausts to remove pollution, and enables engine manufacturers to produce the most fuel efficient engines available, which also produce less carbon dioxide, and ensures development of technology for exhausts that reduce nitrogen oxide emissions that cause smog.

2. Removing sulfur from diesel fuel reduces the amount of ultrafine particulates in vehicle exhaust that can contribute to respiratory problems such as asthma.

3. Concentration of SOx has declined significantly in the eastern United States since 1989.

4. Air pollution is poor air quality resulting from unwanted chemicals or other materials occurring in the air.

5. Air pollution can have adverse impacts on ecosystems, human health, and materials. Over 90% of Californians and millions of other Americans breathe unhealthy levels of one or more air pollutants during some part of the year. Air pollution kills millions of people worldwide every year.

6. The 1990 Clean Air Act Amendments required the EPA to set National *Ambient* Air Quality Standards (NAAQS). Criteria pollutants are CO, NOx, ozone, lead, particulates (PM10 and PM2.5), and SOx. Air toxics are pollutants for which insufficient data exist to set standards.

7. Other than lead, heavy metals were not specifically regulated by the 1990 Clean Air Act Amendments. The treatment of mercury under clean air legislation is controversial.

8. Artificial chemicals called chlorofluorocarbons (CFCs) have been shown to destroy the atmosphere's ozone shield, which protects the planet's surface from high-energy ionizing radiation.

9. Residence time is the average time that an atom, molecule, or complex stays in the atmosphere, before cycling into one of the other systems.

10. Acid precipitation occurs when acids from human activity or natural sources like volcanic eruptions, enter the atmosphere and are deposited. Major sources of AP are emissions from motor vehicles, fossil-fueled power plant emissions, and nitrogen gases from agriculture.

11. Acid precipitation can corrode buildings, acidify natural waterways, damage vegetation, and acidify soils. Some soils, when acidified, can leach Al^{3+} into water bodies. This species of Al can selectively kill certain organisms, especially juveniles.

12. Research in Québec, Canada showed that most of the acidification of lakes was from human sources, and identified power plants and smelters as major sources of AP.

13. EPA designated a Pollution Standard Index (PSI) with which to express local air quality.

14. The greatest single event leading to reduction in motor vehicle emissions was the phase-out of leaded gasoline in the United States in the 1970s.

15. Catalytic converters have sharply reduced emissions from cars, and have contributed to improvements in air quality in every city in North America since the middle 1970s.

16. Significant amounts of air pollution result from the operation of hand-held "lawn care devices"; off-road vehicles like personal watercraft and snowmobiles; and stationary agricultural and construction diesel engines.

17. Tire wear produces considerable fine particulate matter, consisting of heavy metals from stainless steel in radials, of powdered polyester, and of latex.

18. The first antipollution devices applied to power plants were ash filters and precipitators, which trapped much of the coarse particulate matter before it could escape up the smokestack.

19. Power plants have several means by which some of their sulfur emissions can be reduced, or even eliminated. They include fuel switching, installation of scrubbers, and cleaning coal before combustion.

20. NOx is a major "precursor" to the formation of ozone. NOx gases are produced whenever fuel is burned in air.

21. CO_2 is not a pollutant as defined by EPA regulations, and thus is not covered by Clean Air legislation.

KEY TERMS

acid deposition

acid precipitation

air pollutants

air pollution

catalytic converter

chlorofluorocarbons (CFCs)

criteria pollutants

Grandfather clauses in legislation

mutagenic

particulates-PM2.5, -PM10

Pollution Standard Index (PSI)

residence time

sulfur in coal

toxic air contaminants (TACs)

two-stroke engine

REVIEW QUESTIONS

1. Which is true concerning air pollution?
   a. There are no natural sources of SOx air pollution.
   b. Burning coal is the only known source of SOx.
   c. Volcanoes can be major local sources of SOx air pollution.
   d. Burning natural gas is a major source of SOx.

2. Removing sulfur from motor fuels is important because
   a. sulfur interferes with the catalyst in catalytic converters.
   b. sulfur produces fine particulates when burned, which are a pulmonary hazard.
   c. sulfur contributes to acid precipitation (AP).
   d. all of these.
   e. none of these.

3. Acid precipitation can
   a. damage crops.
   b. acidify soils.
   c. change the pH of natural waterways.
   d. all of these.
   e. none of these.

4. Which piece of legislation required EPA to set National Ambient Air Quality Standards (NAAQS) for pollutants considered to be a threat to human health and the environment?
   a. the National Environmental Policy Act
   b. the Clean Water Act
   c. the Superfund Act
   d. the 1990 Clean Air Act Amendments
   e. none of these

5. Which "heavy metal" pollutant was covered under NAAQS?
   a. mercury
   b. cadmium
   c. lead
   d. uranium

6. Volatile organic compounds (VOCs) react with sunlight to produce
   a. sulfuric acid.
   b. ozone.
   c. nitric acid.
   d. fine particulates.

7. Which chemical "family" damages the "ozone shield" in the upper atmosphere?
   a. CFCs
   b. PCBs
   c. platinum group elements
   d. all of these

8. Which contains at least 40 substances that are carcinogenic, mutagenic, or both?
   a. diesel exhaust
   b. agricultural pollution
   c. air pollution from nuclear power plants
   d. acid precipitation
   e. all of these

9. Natural precipitation has a pH of about
   a. 5.6.
   b. 7.
   c. 4.
   d. 2.5.
   e. 8.5.

10. Which bedrock type is most able to "buffer" AP and thus reduce its impact?
    a. granite and rhyolite
    b. coal and sandstone
    c. carbonate
    d. shale

11. Geologic maps of Québec, Canada indicated that the region was very sensitive to acidification due to its
    a. latitude.
    b. mean temperature.
    c. carbonate bedrock.
    d. silicate bedrock.

12. The Québec provincial government concluded that the acid precipitation the province was experiencing was mainly from
    a. methane from melting permafrost.
    b. natural sources, including birch and maple trees.
    c. power plants alone.
    d. power plants and smelters.
    e. nickel smelters alone.

13. The EPA designated a Pollution Standard Index (PSI) to express
    a. local air quality.
    b. national average air quality.
    c. the effectiveness of catalytic converters on motor vehicles.
    d. the relative pollution reduction of hybrid motor vehicles.

14. The major difference between natural and anthropogenic air pollution is
    a. natural sources tend to be continuing and persistent.
    b. anthropogenic sources tend to be continuing and persistent.
    c. both sources are continuing and persistent.
    d. both sources are local and episodic.

15. The greatest single event leading to reduction in motor vehicle emissions was
    a. the application of the catalytic converter to motor vehicles.
    b. the banning of diesel vehicles in California in 1990.
    c. the phase-out of leaded gasoline in the United States in the 1970s.
    d. the removal of mercury from motor fuels in the 1970s.

16. Hydrogen-powered vehicles would emit what type of exhaust?
    a. CO_2 and SOx.
    b. H_2O.
    c. CO_2 only.
    d. CO_2 and H_2O.

17. Operation of a typical two-cycle leaf blower for an hour can produce as much air pollution as a newer car driven for
    a. 100 miles (160 km).
    b. ten minutes at highway speeds.
    c. three minutes under urban conditions.
    d. 10,000 miles (16,000 km).

18. Tires produce what sort of emissions?
    a. none
    b. depleted uranium and lead
    c. particulates of polyester, latex, and heavy metals
    d. nutrients (N,P) and CO_2

19. Scrubbers mainly remove what pollutant from power plant exhaust gases?
    a. sulfur gases
    b. CO_2
    c. NOx
    d. ozone

20. NOx is a major "precursor" to the formation of
    a. catalytic converters.
    b. coarse particulates.
    c. heavy metals.
    d. ozone.

21. Which is true about CO_2 regulation, according to the text?
    a. CO_2 reduction is required by the Kyoto protocol, which the United States ratified.
    b. CO_2 reduction is required by the 1990 Clean Air Act Amendments.
    c. CO_2 is not a pollutant as defined by EPA.
    d. CO_2 emissions are regulated by states and local governments, as required by EPA.

Impacts of Global Climate Change

CHAPTER OUTLINE

POINT-COUNTERPOINT   *Climate Change*

© Amos Struck/Shutterstock, Inc.

Point

American Petroleum Institute, 2002:

"Amid widespread public uncertainty over the likely causes and severity of **climate change**, there is an international debate over actions the nations of the world should take to limit greenhouse emissions . . . At its core, the debate is about whether enough is known about climate change to warrant the lost jobs, higher consumer prices and a weakened U.S. economy that would come with implementing the Kyoto Protocol, an agreement which at best would make only slight progress toward solving climate change.

"There are those who believe that global warming is already causing glaciers to melt. There are also predictions for flooding of low-lying areas like Cape Cod and islands in the Pacific, drought to fertile farmlands, and increased cases of malaria and other hot weather diseases to people in some countries. . .

"Other scientists believe with equal fervor that there is at least as much evidence to suggest that warming, if it occurs, will be modest and take many decades to be felt. "Most of these climate shifts can be traced to natural phenomena, such as El Niño, sunspots or variances in the Earth's orbit . . ."

American Petroleum Institute Position, 2011:

"Oil and natural gas take us down the street and around the world. They warm and cool our homes and businesses. They provide the ingredients for medicines, fer-

tilizers, fabrics, plastics and other products that make life safer, easier and better.

"While we rely on them for most of our energy and will likely do so for years to come, emissions from their production and use may be helping to warm our planet by enhancing the natural greenhouse effect of the atmosphere. That's why oil and gas companies are also working to reduce their greenhouse emissions."

Prof. Patrick Michaels, Cato Institute, 1999

"Observed global warming remains far below the amount predicted by computer models that served as the basis for the United Nations Framework Convention on Climate Change . . .

"Moreover, carbon dioxide is increasing in the atmosphere at a rate below that of most climate-change scenarios because it is being increasingly captured by growing vegetation. The second most important human greenhouse enhancer—methane—is not likely to increase appreciably in the next 100 years. And perhaps most important, the direct warming effect of carbon dioxide was overestimated. Even global warming alarmists in the scientific establishment now say that the Kyoto Protocol will have no discernible impact on global climate."

Prof. Patrick Michaels, Cato Institute, 2010

"Most scientists agree that a substantial component of the surface warming that began in the mid-1970s is a result of changes in the earth's greenhouse effect, largely

Continued ▶

caused by increases in atmospheric carbon dioxide ... Does the most recent science and climate data argue for precipitous action? (No.) Is there a suite of technologies that can dramatically cut emissions by, say, 2050? (No.) Would such actions take away capital, in a futile attempt to stop warming, that would best be invested in the future? (Yes.) Finally, do we not have the responsibility to communicate this information to our citizens, despite disconnections between perceptions of climate change and climate reality? The answer is surely yes. If not the U.S. Congress, then whom? If not now, when? After we have committed to expensive policies that do not work in response to a misperception of global warming?"

Concept Check 20-1. Characterize the early positions of the American Petroleum Institute (API) and Prof. Michaels on anthropogenic climate change.

Concept Check 20-2. Characterize the more recent positions of the API and Prof. Michaels on anthropogenic climate change.

The Global Warming Petition Project

According to its website (http://www.petitionproject.org), over 31,000 U.S. scientists, 9000+ of whom possess Ph.D. degrees, have signed their petition, part of which reads:

> *"There is no convincing scientific evidence that human release of carbon dioxide, methane, or other greenhouse gasses is causing or will, in the foreseeable future, cause catastrophic heating of the Earth's atmosphere and disruption of the Earth's climate. Moreover, there is substantial scientific evidence that increases in atmospheric carbon dioxide produce many beneficial effects upon the natural plant and animal environments of the Earth."*

Concept Check 20-3. Before reading the remainder of this chapter, discuss your reaction to the Petition Project. Do you find the petition statement more credible in light of the number of scientists who signed it?

Counterpoint

U.S. National Academy of Sciences Position on Climate Change, 2001:

"Greenhouse gases are accumulating in Earth's atmosphere as a result of human activities, causing surface air temperatures and subsurface ocean temperatures to rise. Temperatures are, in fact, rising. The changes observed over the last several decades are likely mostly due to human activities, but we cannot rule out that some significant part of these changes are also a reflection of natural variability."

U.S. National Academy of Sciences, 2011

"Climate change is occurring, is very likely caused primarily by the emission of greenhouse gases from human activities, and poses significant risks for a range of human and natural systems. Emissions continue to increase, which will result in further change and greater risks. In the judgment of this report's authoring committee, the environmental, economic, and humanitarian risks posed by climate change indicate a pressing need for substantial action to limit the magnitude of climate change and to prepare for adapting to its impacts."

Concept Check 20-4. Summarize the 2001 and 2011 positions of the U.S. National Academy of Science. What are the differences and similarities between them?

Table PC 20-1 shows a partial list of scientific organizations with positions accepting or rejecting anthropogenic climate change as of 2011, as well as two with equivocal positions.

Concept Check 20-5. Reread and consider the comments above. If necessary, research the organizations and individual. In which do you have the most confidence? Why?

We revisit the debate on anthropogenic climate change later in this chapter.

Continued ▶

TABLE PC 20-1	Scientific Organizations with Positions Accepting or Rejecting Anthropogenic Climate Change as of 2011

A. Those Accepting Anthropogenic Climate Change

U.S. Organizations

Alliance Organization of Biological Field Stations	Association of Ecosystem Research Centers
American Association for the Advancement of Science	Botanical Society of America
American Chemical Society	Crop Science Society of America
American Geophysical Union	Ecological Society of America
American Institute of Biological Sciences	Geological Society of America
American Institute of Physics	National Academy of Sciences
American Meteorological Society	Natural Science Collections
American Physical Society	Society for Industrial and Applied Mathematics
American Society of Agronomy	Society of Systematic Biologists
American Society of Plant Biologists	Soil Science Society of America
American Statistical Association	University Corporation for Atmospheric Research

International Organizations

Accademia Nazionale dei Lincei (Italy)	International Council of Academies of Engineering and Technological Sciences
Academy Council of the Royal Society of New Zealand	Indonesian Academy of Sciences
Academy of Sciences Malaysia	Intergovernmental Panel on Climate Change
Australian Academy of Sciences	Network of African Science Academies
Brazilian Academy of Sciences	Royal Flemish Academy of Belgium for Sciences and the Arts
Caribbean Academy of Sciences	Royal Irish Academy
Chinese Academy of Sciences	Royal Society of Canada
European Academy of Sciences and Arts	Royal Society of the United Kingdom
French Academy of Sciences	Royal Swedish Academy of Sciences, and Royal Society (UK)
German Academy of Natural Scientists Leopoldina	Russian Academy of Science
Indian National Science Academy	Science Council of Japan

B. Those with an Equivocal Position

American Association of Petroleum Geologists	American Institute of Professional Geologists

C. Those Rejecting Anthropogenic Climate Change

None

There is wide agreement among government agencies, scientists, and industry that human activity is causing global climate change (TABLE 20-1). Scientific uncertainty remains on issues that include the time frame and severity of climate change impacts, where they will occur, and the cost, impact, and effectiveness of mitigation and/or adaptation strategies. You have probably heard of the Kyoto Protocol (see following section), an international treaty approved by U.S. negotiators in 1997, but not ratified by the U.S. Senate. The Protocol has been ratified by 191 nations, including the European Union, China, Canada, and by Japan, among others (FIGURE 20-1). It calls upon signatory states to set targets to reduce greenhouse gas emission, and then to meet those targets by the date(s) specified. For reference, FIGURE 20-2 shows the emissions of carbon dioxide (CO_2), the most abundant greenhouse gas, by industrialized countries from 1990, the "base year" for the Kyoto Protocol, projected to 2030. The Kyoto treaty went into effect when ratified by countries responsible for 55% of CO_2 emissions in 1990.

TABLE 20-1	The Important Greenhouse Gases and Their Primary Sources					
Gas	Atmospheric Concentration (ppm)	Annual Increase (%)	Life Span (Years)	Relative Greenhouse Efficiency ($CO_2 = 1$)	Current Greenhouse Contribution (%)	Principal Sources of Gas
Carbon dioxide (CO_2)	>400	0.4	x[1]	1	57 (44)	Burning fossil fuels in power plants and motor vehicles, mainly, and in cement production (about 6 billion tons) Deforestation, (about 1–2 billion tons)
Carbon dioxide (from biological sources)					(13)	
Chlorofluoro-carbons (CFCs)	0.000225	5	75–111	15,000	25	Foams, aerosols, refrigerants, solvents
Methane (CH_4)	1.774	1	11	25	12	Wetlands, rice, fossil fuels, livestock
Nitrous oxide (N_2O)	0.319	0.2	150	230	6	Fossil fuels, fertilizers, deforestation

Sources: Data from World Watch Institute, U.S. EPA, and *Journal of Geophysical Research*.
[1]Carbon dioxide is a stable molecule with a 2- to 4-year average residence time in the atmosphere.

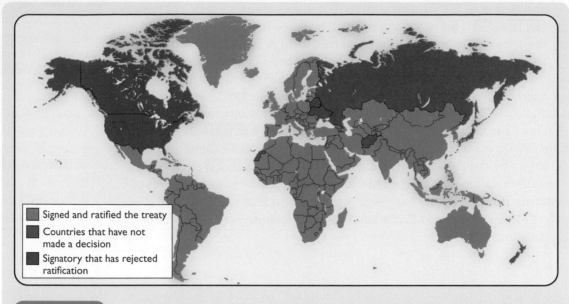

FIGURE 20-1 Nations that have ratified the Kyoto Protocol. Those in green have signed and ratified the treaty. Black depicts countries that have not made a decision. Red denotes a signatory that has rejected ratification. Data from: United Nations Framework Convention on Climate Change.

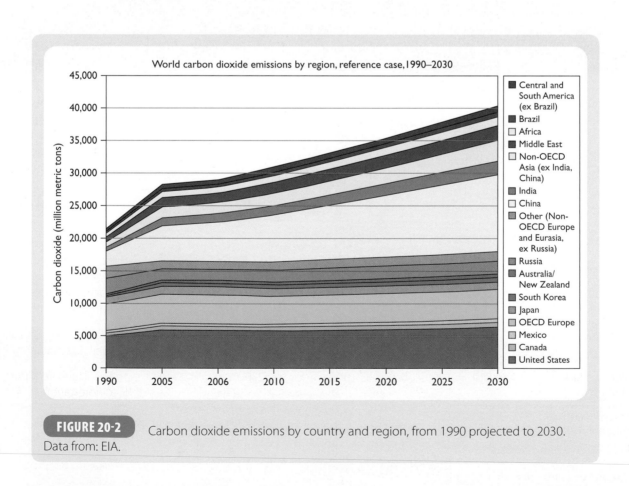

FIGURE 20-2 Carbon dioxide emissions by country and region, from 1990 projected to 2030. Data from: EIA.

■ THE KYOTO PROTOCOL

In response to a growing body of scientific evidence, and believing that to err on the side of precaution was better than to adopt a "wait-and-see" attitude, government representatives gathered in Kyoto, Japan in 1997. After much debate, most of them initialed a draft treaty that would, if approved, begin to stabilize, and eventually reduce, the amount of climate-altering greenhouse gases in the planetary atmosphere. Here are the major provisions of that treaty.

- The world's industrialized countries agreed to determine their individual CO_2 contributions during the base year, 1990.
- Signatory nations would undertake to reduce their CO_2 emissions (or their carbon equivalents) by an agreed percent by a specified date.
- Signatory nations would agree to enhance energy efficiency, promote sustainable agriculture, and protect forests, soils, and other carbon "sinks."
- They would support and promote renewable energy technologies, and research means by which CO_2 and other greenhouse gases could be "sequestered" (stored), so as not to add to climate change.
- They would seek to eliminate subsidies and market "imperfections" that encourage unnecessary greenhouse gas production.
- They would specifically promote measures to reduce greenhouse gas emission in the transport sector.
- They would agree to limit methane emissions or capture such emissions. Methane is a much more powerful greenhouse gas than CO_2.
- Signatory states would show progress towards meeting their greenhouse reductions by 2005.
- Non-industrialized countries, or those "transitioning" to a market economy, (mainly the former Warsaw Pact countries) would have other base dates than 1990.
- Countries would be allowed to "trade" emissions credits, that is, buy and sell them.
- Transfer of energy-efficiency and greenhouse-reduction technology was to be encouraged.
- Targets would be met by 2012.

Response of the United States

Although it alone emitted over one third of greenhouse gases in 1990, the United States declined to ratify the treaty, pointing out that China and India, major greenhouse emitters and countries whose emissions were growing much faster than the industrialized countries (Table 20-1), were not required to set and achieve greenhouse reduction targets.

Opponents to ratification of the treaty in the United States argued that the Kyoto Protocol's targets and timetables were arrived at arbitrarily as a result of political negotiations, and were not related to any specific scientific information or long-term objective. They protested that it could require U.S. firms to retire power plants and factories prematurely, imposing substantial and unnecessary costs on the U.S. economy. Moreover, the Kyoto Protocol did not allow countries to count legitimate mitigation activities. In fact, it restricted the use of carbon sequestration as a means of achieving its objectives.

Response of the European Union

Recognizing that climate change is one of the greatest environmental and economic threats facing the planet, the European Union (EU) has undertaken many climate-related initiatives since 1991, when it issued the first Community strategy to limit carbon dioxide (CO_2) emissions and improve energy efficiency. These include: a proposed directive to promote electricity from renewable energy, voluntary commitments by carmakers to improve fuel economy by 25% and proposals on the taxation of energy products.

A two-year study found that the cost of complying with the Kyoto targets could be limited to 3.7 billion euros (about $4 billion U.S.) per year for the 2008–2012 period, equivalent to 0.06% of the EU's expected gross domestic product in 2010.

Response of Japan

According to Japan's Minister for the Environment, "Under the Protocol, Japan has an obligation of greenhouse gas emissions reduction of 6% below 1990 levels for the first commitment period of 2008–2012. It will be devoting particular effort to the following measures.

1. The development of state-of-the-art technologies, and its introduction to the market through practical application. The Japanese government recognizes that the government itself is a major consumer, and will actively embrace the challenge and responsibilities of being a truly green consumer.
2. Economic measures which utilize market mechanisms, such as emissions trading and eco-tax systems, are known to be effective in promoting climate change mitigation measures. Japan will study and undertake such economic measures, and Japan intends to be an active participant in the formulation of the rules that would govern such measures at the global level.
3. Forests, which serve as carbon sinks, also play a vital role in the maintenance of both biodiversity and water resources. With the active participation of the citizenry, Japan will strengthen measures to preserve her forests."

Government of China's Response, 2002

"The Chinese government believes that the United Nations Framework Convention on Climate Change and its Kyoto Protocol set forth the fundamental principles and provide an effective framework and a series of rules for international cooperation in combating climate change. China hopes other developed countries will ratify or approve the protocol as soon as possible.

"China is a low-income developing country with a large population. Poverty eradication and economic development are its top priorities. Notwithstanding the foregoing, the Chinese government attaches great importance to climate change. The Chinese government at all levels has exerted tremendous efforts to address this problem. Achievements in improving energy efficiency are a supporting example."

China does not have any commitment targets within the Kyoto Protocol and is the world's largest greenhouse gas emitter (29% of global emissions in 2011). By the end of 2012, Japan was on target to meet its Kyoto Protocol target. However, Japan did not sign the Kyoto Protocol Extension (2012–2020). Whether the Kyoto Treaty will be ratified by the United States or not, it remains the only international framework for addressing climate change in any meaningful way.

■ THIS CHAPTER'S STRUCTURE

In this chapter, we first look at the history of research on global climate change. Then we discuss the phenomena called the **greenhouse effect**, and describe the latest scientific conclusions on the impacts of human-induced global climate change. Finally, we consider strategies for mitigating the adverse impacts of climate change, including carbon sequestration.

■ HISTORICAL RESEARCH ON CLIMATE CHANGE

John Tyndall

In 1857, physical scientist **John Tyndall** began research on what he called, "the radiative properties of gases" (FIGURE 20-3). He discovered that there were differences in the ability of atmospheric gases to absorb "radiant heat," which is what he called electromagnetic radiation in the *infrared* range. His research led him to conclude that oxygen and nitrogen, which together make up almost 99% of the atmosphere, had little if any ability to absorb infrared. However, two atmospheric gases, CO_2 (which he called "carbonic acid") and water vapor (H_2O_v) were powerful infrared absorbers. He concluded that without these two gases the Earth's surface would be "held fast in the iron grip of frost." And he speculated on whether changes in the concentrations of CO_2 and H_2O could affect the Earth's climate.

Svante Arrhenius

Nearly forty years later in 1895, the distinguished Swedish chemist **Svante Arrhenius** (FIGURE 20-4) presented a paper to the Stockholm Physical Society entitled, "On the Influence of Carbonic Acid in the Air upon the Temperature of the Ground." Building on the work of Tyndall and others, he argued that variations in the minor atmospheric constituents CO_2 and H_2O could profoundly affect the Earth's **heat budget**, an accounting of the balance between incoming solar radiation and re-radiation from the Earth back into space. His calculations led him to conclude that, "the temperature of the Arctic regions would rise about 8 or 9 degrees Celsius if the [CO_2 concentration] increased 2.5 to 3 times its present (1895) value." Arrhenius went on to win the 1903 Nobel Prize in chemistry.

Roger Revelle

If there is a "father of global warming" it is oceanographer **Roger Revelle** (1909–1991; FIGURE 20-5). In a pioneering

FIGURE 20-3 John Tyndall, an early researcher on climate change. © Photos.com/Thinkstock.

FIGURE 20-4 Svante Arrhenius, winner of a 1903 Nobel Prize. He argued that variations in CO_2 and H_2O could profoundly affect the Earth's heat budget. © Photos.com/Thinkstock.

article published in 1957—one hundred years after Tyndall began his work—Revelle and colleague Hans Seuss showed that CO_2 concentration in the atmosphere had increased as a direct result of the burning of fossil fuels, and that the oceans were unlikely to be able to absorb the increase. They built their work on earlier studies, including the work of British engineer Guy Stewart Callendar who, using refined analytical techniques, advanced a hypothesis that warming measured between 1900 and 1938 could be the result of rising fossil fuel burning.

Until 1957, many scientists recognized the importance of the greenhouse effect, but thought that humanity was unlikely to be able to alter global climate. Their reason: the oceans store 50 times as much CO_2 as the atmosphere. Physical laws were invoked to suggest that this ratio would remain constant, and that therefore only 2% of any additional CO_2 released by human activities would remain in the atmosphere: the rest would have been dissolved in the oceans.

In 1957 Revelle and Seuss demonstrated that the oceans could not absorb CO_2 as rapidly as humanity was releasing it. In their paper's conclusion, they used one of the most memorable phrases in the literature of science: "human beings are now carrying out a large-scale geophysical experiment."

Revelle and Seuss's hypothesis led to the establishment of monitoring stations on Mauna Loa, Hawaii, and in Antarctica, to measure worldwide trends in atmospheric concentrations. In 1965, a President's Science Advisory Committee concluded that Revelle and Seuss had been correct and officially recognized global warming from CO_2 as a possible global problem. The CO_2 concentration plot from Mauna Loa, also known as the Keeling Curve (named after Scripps Institution of Oceanography scientist Charles Keeling, an associate of Revelle), has become one of the most recognizable icons in science (**FIGURE 20-6**).

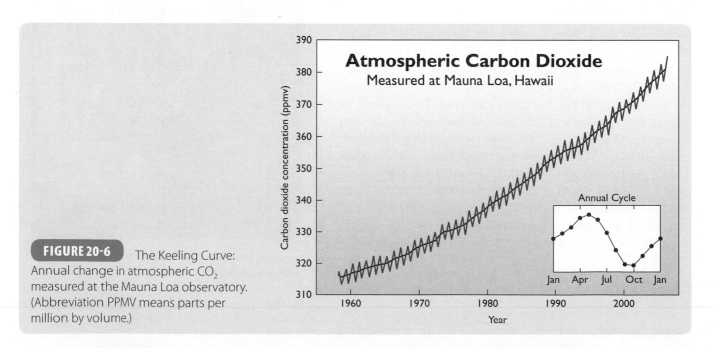

FIGURE 20-6 The Keeling Curve: Annual change in atmospheric CO_2 measured at the Mauna Loa observatory. (Abbreviation PPMV means parts per million by volume.)

By 1977 Revelle and his colleagues had concluded that 40% of the human-generated CO_2 remained in the atmosphere. Two-thirds was from fossil fuels, the rest from deforestation. They speculated that the oceans were absorbing much of the "missing" CO_2, but not nearly so much as the 50:1 ratio previously supposed.

■ CLIMATE CHANGE COMES OUT OF THE CLOSET

On June 20, 1988, during a Washington, D.C. heat wave, Dr. James Hansen, head of NASA's Goddard Institute for Space Studies, appeared before a Senate Committee. Here is part of his testimony.

> *"I would like to draw three main conclusions. Number one: the Earth is warmer in 1988 than at any time in the history of instrumental measurements. Number two: global warming is now large enough that we can ascribe with a high degree of confidence a cause-and-effect relationship to the greenhouse effect. And number three: our computer climate simulations indicate that the greenhouse effect is already large enough to begin to affect the probability of extreme events such as summer heat waves."*

Hansen's testimony created a firestorm that has only recently begun to abate with the near-universal realization that Hansen, Revelle, their colleagues, and their predecessors were right (see Table PC 20-1). *Humans are changing the climate.* The Intergovernmental Panel on Climate Change, a U.N. organization, has prepared computer projections for ranges of temperature change to 2100 based on present production of greenhouse gases (**FIGURE 20-7**).

So, what are greenhouse gases?

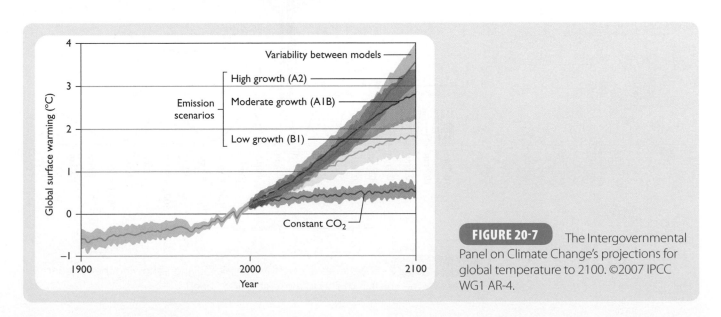

FIGURE 20-7 The Intergovernmental Panel on Climate Change's projections for global temperature to 2100. ©2007 IPCC WG1 AR-4.

■ THE GREENHOUSE EFFECT

As Tyndall inferred, were it not for the heat-retaining effects of certain atmospheric gases, causing the Earth's surface to be warmed like plants in a greenhouse walled by glass, the Earth's surface would be about 60°F colder.

Other than trivial geothermal heat, the Earth's surface is warmed by solar radiation. This solar radiation's wavelength is concentrated in the visible light and ultraviolet range (**FIGURE 20-8**). As Tyndall and Arrhenius showed, atmospheric gases are largely transparent to these wavelengths, allowing radiation in the visible range, and much ultraviolet, to pass through the atmosphere. Among other things, this is why we can "see through" the sky.

FIGURE 20-9A , the Earth's heat budget, shows the fate of this incoming radiation. The incoming solar radiation is either reflected back out to space (by clouds, snow on the surface, etc., a phenomenon known as the *albedo* effect), or is absorbed by dark surface material (rocks, the oceans, soil, vegetation, etc.). Surface material then *re-radiates* the absorbed energy; but in *a longer wavelength*, because the Earth's surface is much colder that the Sun. The radiation is emitted mainly in the **infrared**, which we experience as heat. Infrared can then be absorbed by the greenhouse gases in the atmosphere: the higher their concentration, the more effective this absorption, the warmer the lower atmosphere, and the warmer the surface. This is the greenhouse effect (**FIGURE 20-9B**).

Concentrations and Sources of Greenhouse Gases

According to the National Atmospheric and Oceanic Administration (NOAA), the concentrations of the greenhouse gases CO_2 and nitrous oxide (N_2O) are increasing at around 1% per year. The concentration of methane (CH_4), another powerful greenhouse gas, may have stabilized, at least temporarily, due to sharp declines in methane emissions from leaking gas pipelines in the former Soviet Union, among other things.

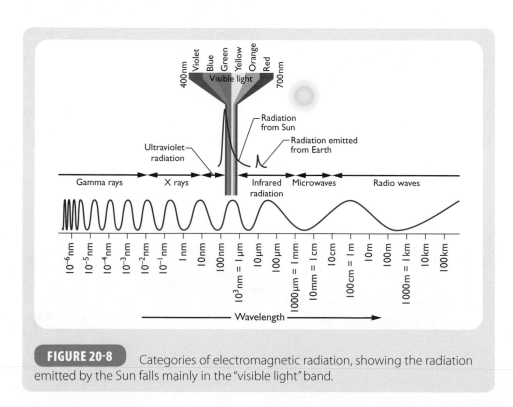

FIGURE 20-8 Categories of electromagnetic radiation, showing the radiation emitted by the Sun falls mainly in the "visible light" band.

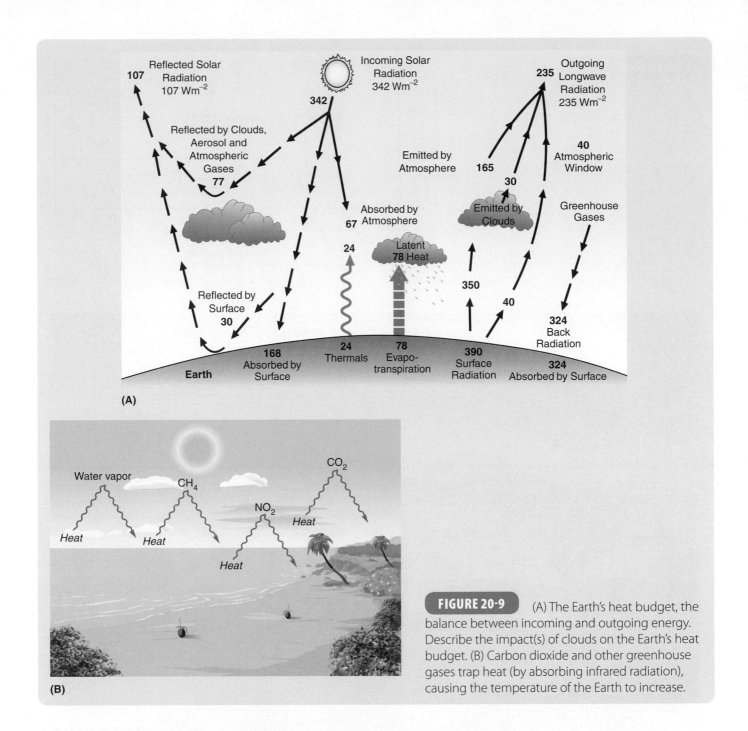

FIGURE 20-9 (A) The Earth's heat budget, the balance between incoming and outgoing energy. Describe the impact(s) of clouds on the Earth's heat budget. (B) Carbon dioxide and other greenhouse gases trap heat (by absorbing infrared radiation), causing the temperature of the Earth to increase.

Another class of greenhouse gases, the chlorofluorocarbons (CFCs), is also apparently stabilizing. CFCs are to be reduced by global agreement (The Montreal Protocol).

The current CO_2 level is higher than at any time in the past 160,000 years, and should present trends continue through the twenty-first century, the CO_2 level will reach those last seen 50 million years ago, during the Eocene Epoch of the Cenozoic (**FIGURE 20-10**).

Carbon Dioxide, Greenhouse Warming, and the Carbon Cycle

The element carbon moves from one storage reservoir to another, and this biogeochemical path is known as the **carbon cycle** (**FIGURE 20-11A**). The major reservoirs

FIGURE 20-10 Carbon dioxide levels over the past 450,000 years. These data were obtained from bubbles trapped in ice over 3,600 m deep at the Russian Vostok station in East Antarctica. Note that the trapped samples reflect four climate cycles. Note the increase in CO_2 beginning in about 1950. Data from: NASA.

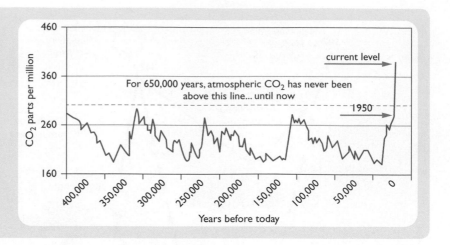

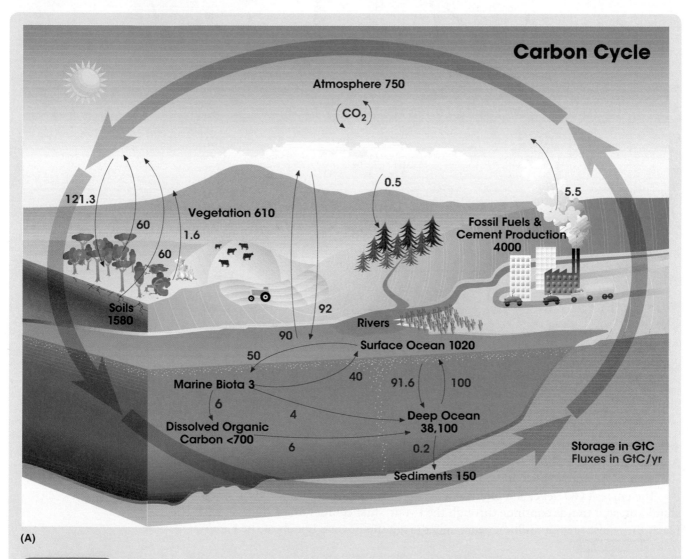

(A)

FIGURE 20-11 (A) The carbon cycle, showing inputs (sources) and sinks of CO_2. GtC is gigatonnes carbon. *Question:* Why do you think more people, who exhale carbon dioxide, are not shown? (*Hint:* Look at Table 20-1). Courtesy of NASA Earth Science Enterprise.

(Continued)

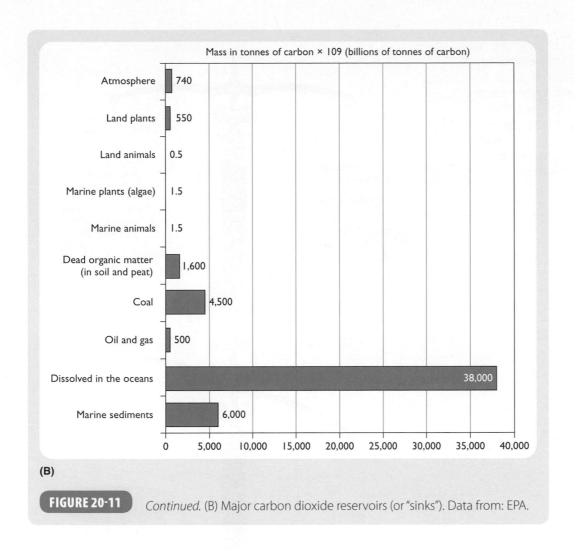

Mass in tonnes of carbon × 10⁹ (billions of tonnes of carbon)

Atmosphere	740
Land plants	550
Land animals	0.5
Marine plants (algae)	1.5
Marine animals	1.5
Dead organic matter (in soil and peat)	1,600
Coal	4,500
Oil and gas	500
Dissolved in the oceans	38,000
Marine sediments	6,000

(B)

FIGURE 20-11 *Continued.* (B) Major carbon dioxide reservoirs (or "sinks"). Data from: EPA.

of CO_2 are tabulated in **FIGURE 20-11B**. The greenhouse gas components in the carbon cycle can be thought of as an enormous balance scale, with *inputs* from natural sources on one beam, and *outputs* to natural sources on the other. For example, CO_2 from respiration, volcanic gas, and decay of vegetation would be on one scale, and CO_2 used by plants in photosynthesis, by corals to build reefs and make limestone, and carbon sequestered in sediment and soil, would fall on the other (**FIGURE 20-12A**). As long as inputs balance outputs, the CO_2 level stays roughly the same, but if any process interferes with this delicately balanced system, the amount of CO_2 in the atmosphere will change, and, other things being equal, the climate will change with it (**FIGURE 20-12B**).

Carbon dioxide is increasing as humans return the carbon in formerly stored (sequestered) fossil fuels to the atmosphere. More CO_2 is added as humans clear and burn forests, and as warming climate generates more forest fires. Soils, which previously had stored vast quantities of CO_2, now are emitting much of their CO_2 as they warm. As we mentioned, the oceans are absorbing up to half of the "new" CO_2, but how long they can continue to do this is unknown. Moreover, oceans are *acidifying*, a process that may threaten oceanic food chains, as they incorporate CO_2.

Concept Check 20-7. CheckPoint: Explain the greenhouse effect as it pertains to the temperature of the Earth.

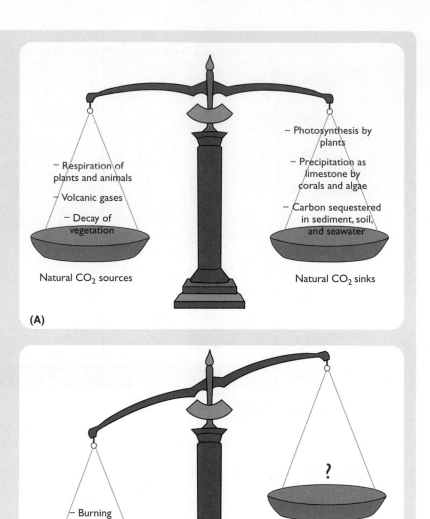

FIGURE 20-12 (A) The balance between natural sources of greenhouse gas in the carbon cycle, that is, when carbon inputs equal carbon outputs. (B) If any process (e.g., burning fossil fuels, forest fires, etc.) interferes with this delicately balanced system, the amount of CO_2 in the atmosphere will change, and, other things being equal, the climate will change with it.

Within the figure:

— Respiration of plants and animals
— Volcanic gases
— Decay of vegetation

Natural CO_2 sources

— Photosynthesis by plants
— Precipitation as limestone by corals and algae
— Carbon sequestered in sediment, soil, and seawater

Natural CO_2 sinks

(A)

— Burning fossil fuels
— Manufacturing cement
— Deforestation

Anthropogenic CO_2 sources
(B) (increases global CO_2)

?

Anthropogenic CO_2 sinks

■ SUMMARY OF EVIDENCE SUPPORTING ANTHROPOGENIC CLIMATE CHANGE

The scientific evidence supporting that the planet's temperature is increasing and humans are primarily responsible is incontrovertible. It is beyond the scope of the textbook, however, to present all of it. Here is a brief summary:

- Global temperature has risen $0.75° \pm 0.20°C$ since the beginning of the instrumental period (1880–1900). Projected increases range from 1.1° to 6.4°C (2.0°–11.5°F).
- The greenhouse effect is a well-established and understood scientific principle.
- Sources of CO_2 (burning of fossil fuels, deforestation, etc.) are undisputed.
- The rate of warming is unprecedented in recent history (and dwarfs natural variability).

- Climate models are unequivocal in demonstrating that anthropogenic CO_2 is the most significant driver of climate change.
- Volcanic contribution to warming is minor.
- Alternative explanations that withstand the test of scientific rigor are lacking.

■ IMPACTS OF RAPIDLY WARMING THE EARTH

Rising Sea Level

Some potential impacts arising out of rapid planetary warming are unmeasured, some are hypothetical, and some might even be beneficial (at least on the short term). One thing is clear: warming the Earth is causing and will continue to cause sea level to rise, and by amounts that depend (1) on the amount of ice we melt on land, and (2) on the coefficient of thermal expansion of seawater. These are discussed in the sections on coastal hazards elsewhere in the text.

A report done by scientists and consultants for the U.S. EPA attempted to quantify the costs of sea level rise to the United States. Due to funding and time constraints, the study disregarded some impacts, such as the threat to coastal water supplies from saltwater intrusion (see section on groundwater elsewhere in the text) and the severe threat to Louisiana's wetlands. Here is the abstract of that report.

> "Previous studies suggest that the expected global warming from the greenhouse effect could raise sea level 50 to 200 centimeters (2–7 feet) in the next century. This article presents the first nationwide assessment of the primary impacts of such a rise on the United States: (1) the cost of protecting ocean resort communities by pumping sand onto beaches and gradually raising barrier islands in place; (2) the cost of protecting developed areas along sheltered waters through the use of levees (dikes) and bulkheads; and (3) the loss of coastal wetlands and undeveloped lowlands.
>
> The total cost for a one-meter rise would be \$270–\$475 billion, ignoring future development. We estimate that if no measures are taken to hold back the sea, a one meter rise in sea level would inundate 14,000 square miles, with wet and dry land each accounting for about half the loss. The 1500 square kilometers (600–700 square miles) of densely developed coastal lowlands could be protected for approximately one to two thousand dollars per year for a typical coastal lot. Given high coastal property values, holding back the sea would probably be cost-effective.
>
> The environmental consequences of doing so, however, may not be acceptable. Although the most common engineering solution for protecting the ocean coast—pumping sand—would allow us to keep our beaches, [building] levees and bulkheads along sheltered waters would gradually eliminate most of the nation's wetland shorelines. To ensure the long-term survival of coastal wetlands, federal and state environmental agencies should begin to lay the groundwork for a gradual abandonment of coastal lowlands as sea level rises."

Specifically, according to the report, a rise in sea level would:

- flood wetlands and lowlands,
- accelerate coastal erosion,
- worsen coastal flooding,
- threaten coastal structures,

- elevate water tables, and
- increase the salinity of rivers, bays, and aquifers.

Other Impacts

FIGURE 20-13 shows some of the impacts scientists expect to occur as the climate warms. We consider these below as impacts on health, agriculture, forests, water supplies, coasts, and species and habitats.

Health Impacts

- *Weather-related mortalities* from heating the atmosphere, especially in summer, can increase, especially in cities in humid mid-latitudes. During heat waves in cities like Chicago, Illinois for example, hundreds of people typically die from heat-related causes, mainly the elderly poor without access to air conditioning.
- *Infectious diseases* can spread as conditions warm, if this increases the *range* of the vectors (pathogenic organisms) that spread the disease. For example, malaria can spread if conditions become more favorable for the survival of the mosquitoes that transmit the pathogen.
- *Decreasing air quality* can result from increased air temperature, which may enhance the formation and persistence of ground level *ozone*. As a result, *respi-*

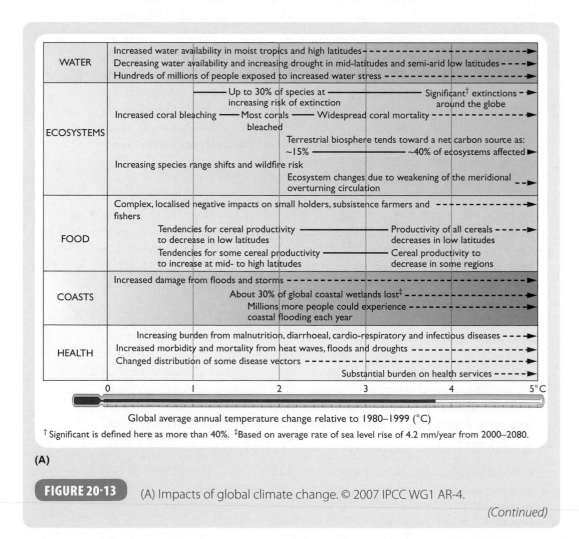

(A)

FIGURE 20-13 (A) Impacts of global climate change. © 2007 IPCC WG1 AR-4.

(Continued)

U.S. Census Region	Climate-related issues							
	Degraded summer air quality	Urban heat Islands	Snowpack/ snowmelt	Hurricanes	Heat wave	Extreme rainfall events	Drought	Wildfires
New England ME VT NH MA RI CT		✓	✓	✓				
Middle Atlantic NY PA NJ	✓	✓	✓	✓				
East North Central WI MI IL IN OH		✓			✓			
West North Central ND MN SD IA NE KS MO		✓			✓		✓	
South Atlantic WV VA MD NC SC GA FL	✓	✓		✓		✓	✓	✓
East South Central KY TN MS AL	✓	✓		✓		✓		
West South Central TX OK AR LA	✓	✓		✓	✓	✓	✓	✓
Mountain MT ID WY NV UT CO AZ NM	✓	✓	✓		✓		✓	✓
Pacific AK CA WA OR HI	✓	✓	✓		✓	✓	✓	✓

(B)

FIGURE 20-13 *Continued.* (B) Regional impacts in the United States. Data from: EPA.

ratory illnesses can increase, especially in sensitive populations like the very young and very old as well as those with chronic respiratory diseases like asthma and emphysema. Asthma is rapidly reaching epidemic proportions, especially among disadvantaged young in inner cities. This could be in part related to increases in temperature and air pollution levels during summer months.

Concept Check 20-8. How would you test the hypothesis that asthma incidence was related to increasing temperatures and pollution levels?

Agriculture Impacts

- Warming can *increase growing seasons*, and, other things being equal, increasing CO_2 can increase *plant growth rates*. However, a 2008 study headed by University of Washington atmospheric scientist David Battisti found there to be a 90% probability that, by the end of the century, the minimum temperatures in the tropics and sub-tropical regions would be higher than the present maximums, resulting in a 20 to 40% drop in key crop yields.

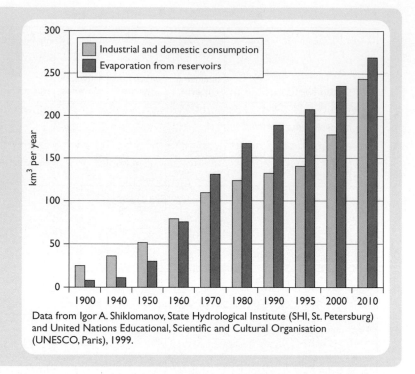

FIGURE 20-14 Loss of water by evaporation from reservoirs. Data from SHI and UNESCO, 1999.

Data from Igor A. Shiklomanov, State Hydrological Institute (SHI, St. Petersburg) and United Nations Educational, Scientific and Cultural Organisation (UNESCO, Paris), 1999.

- *Irrigation* costs could increase if atmospheric warming increases evaporation rates. About 15% of the world's crops are irrigated, but about half the crop value is from irrigated crops. Globally, over 80% of all fresh water use is for crop irrigation, and much of that water evaporates before use (**FIGURE 20-14**).
- Warming could increase evaporation rates generally, which would mean less infiltration and runoff following rainfall events. This could *reduce available soil moisture* and lower crop yields (**FIGURE 20-15**).

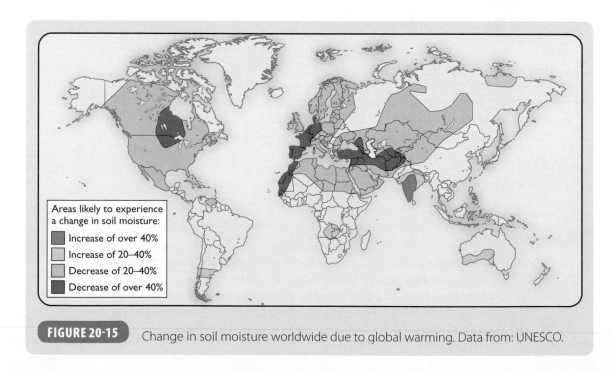

FIGURE 20-15 Change in soil moisture worldwide due to global warming. Data from: UNESCO.

Forest Impacts

- *Species composition* of forests would likely change as the climate warms (FIGURE 20-16). Some species could become extinct, succumbing to disease and insect attack for example, and some species would migrate poleward, shifting the composition of forest communities. Poleward migration of plant species would be materially affected by rate of climate change as well as land-use changes such as urbanization, which could make migration difficult or impossible locally. For example, a projected 2°C (3.6°F) warming by 2,100 could shift the ideal range for many North American forest species by about 300 km (200 mi) to the north. Many plant species lack seed dispersion mechanisms that can adjust this rapidly, and thus these species may be susceptible to regional extinction.
- *Forest productivity* could increase if rainfall were to increase along with CO_2, assuming that plant diseases did not increase in severity and extent, but

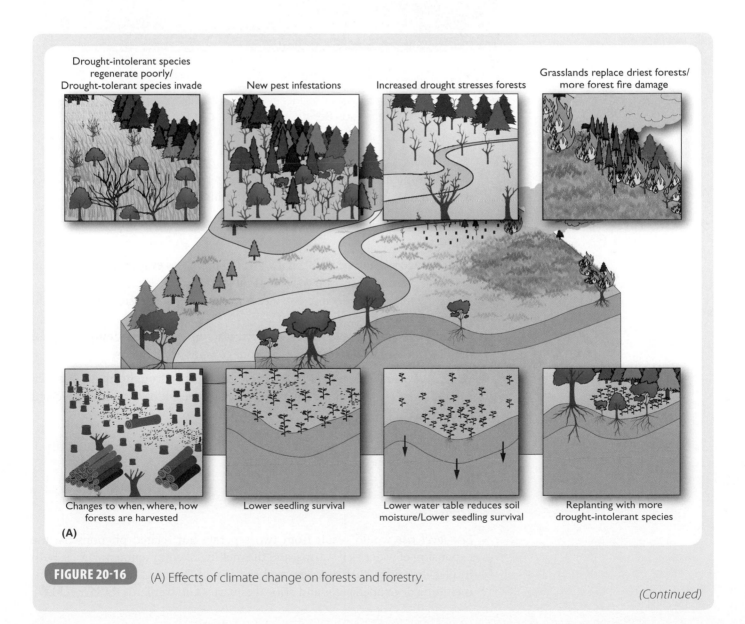

FIGURE 20-16 (A) Effects of climate change on forests and forestry.

(Continued)

(B)

many computer forecasts project a decline in soil moisture as you saw earlier in the chapter, which is linked to plant productivity.

Water Supplies

- Water *supplies* could be reduced if warming increases evaporation rates from reservoirs, or increases soil water loss rates.
- *Water quality* could decline if water temperatures increases, which would lower oxygen content, for example.
- *Water consumption* could increase as irrigators and power generators increase demand and suburban residents continue to water their lawns.

FIGURE 20-17 shows the extent of the prolonged Texas drought in 2011 as well as projected changes in water availability in Africa as a result of climate change and population growth.

Concept Check 20-9. Assess the impact of population growth on water supplies in developed as well as developing countries.

Coastal Impacts

- *Sea level rise* would result from two separate but related phenomena: the expansion of water as it warms, and the melting of continental polar ice, both in glaciers and *permafrost*. Sea level rise would flood coastal environments, including developed areas and critical estuaries and wetlands (**FIGURE 20-18**).

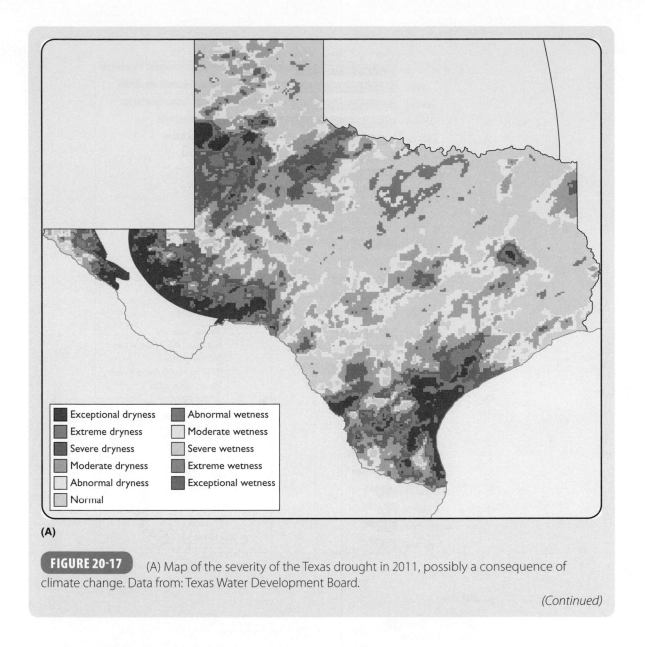

FIGURE 20-17 (A) Map of the severity of the Texas drought in 2011, possibly a consequence of climate change. Data from: Texas Water Development Board.

(Continued)

- Coastal cities' *infrastructure* would be threatened by increased severity of tropical storms in affected coastal areas.
- *Beaches* would erode as storm and wave energy increased.
- *Coastal wetland plant communities* may lose habitat because they may not be able to keep up with the predicted rate of sea level rise, or the transgression paths of these communities may be blocked by human development.

Species and Natural Areas

Climate change will cause a shift in natural community composition as plant and animal species migrate to maintain their preferred habitats.

- *Species diversity* is already declining as a result of deforestation, water pollution, and other land-use changes. This trend would accelerate.

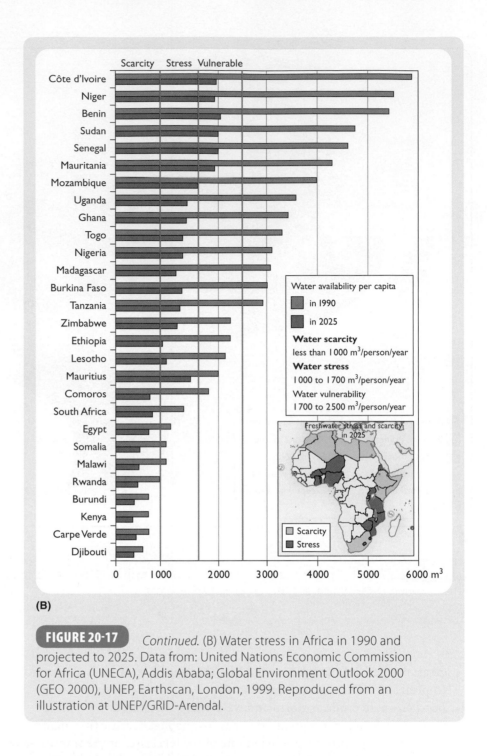

(B)

FIGURE 20-17 *Continued.* (B) Water stress in Africa in 1990 and projected to 2025. Data from: United Nations Economic Commission for Africa (UNECA), Addis Ababa; Global Environment Outlook 2000 (GEO 2000), UNEP, Earthscan, London, 1999. Reproduced from an illustration at UNEP/GRID-Arendal.

- *Habitat loss* is one of the major causes of species decline. To the extent that climate change accelerates habitat loss, species extinction rates would increase.

Concept Check 20-10. Sum up the extent of potential results of rapid climate change by listing positive changes on one side of a diagram and negative changes on the other. What is your conclusion about the overall impact of rapid climate change? What evidence did you use?

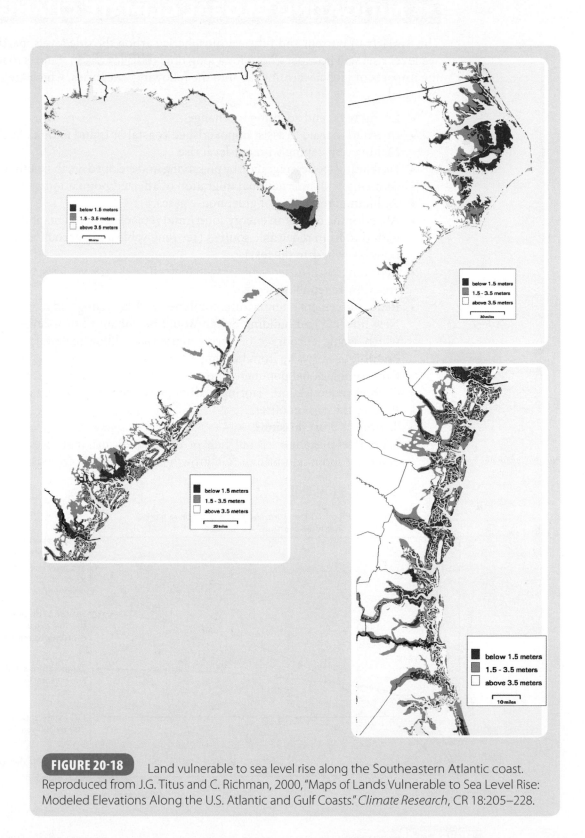

FIGURE 20-18 Land vulnerable to sea level rise along the Southeastern Atlantic coast. Reproduced from J.G. Titus and C. Richman, 2000, "Maps of Lands Vulnerable to Sea Level Rise: Modeled Elevations Along the U.S. Atlantic and Gulf Coasts." *Climate Research*, CR 18:205–228.

■ MITIGATING GLOBAL CLIMATE CHANGE

The impacts of current and future atmospheric carbon dioxide levels, particularly sea level rise, will be experienced for a long time (FIGURE 20-19). Efforts to mitigate the impacts of global warming fall into the following categories, which are variably practical.

- Living with, and adapting to, change.
- International aid to assist impoverished coastal or island nations, such as the Maldives, in coping with sea-level rise.
- Facilitating species migration by preserving undeveloped north–south corridors.
- Assisting with international migration of affected populations.
- Reducing production of greenhouse gases.
- Abandoning coal as an energy source and replacing coal-generated electricity with that from renewable sources (see renewable and alternative energy sections elsewhere in the text).
- Replacing fossil-fueled vehicles with fuel cell or electric vehicles, with fuel generated by renewable energy.
- Reducing cement manufacture, combined with recycling and reuse of cement. New housing and building design would be enhanced by changes to codes, which would encourage or permit reuse of building materials or regulate, prohibit or tax waste from building sites.
- Controlling global population growth.
- Carbon sequestration: "storing" carbon produced by burning fossil fuels.
- Purchasing carbon offsets.
- Taxing carbon emissions.
- Global reforestation. About half of the planet's original forests are gone, and only about one-fifth of "old growth" forests remain from the end of the

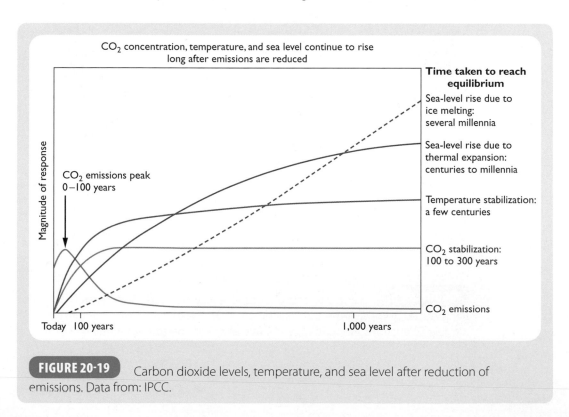

FIGURE 20-19 Carbon dioxide levels, temperature, and sea level after reduction of emissions. Data from: IPCC.

In antiquity, the Cedars of Lebanon almost certainly covered much of the mountain slopes of modern Lebanon, Israel, and Palestine (**Figure B 1-1**). The cedars (*Cedrus libani*) of the Levant (Middle East) were prized by many ancient peoples, including the Assyrians, Phoenicians, and the Egyptians, for their resins, bark, and wood. The Egyptians cut them down in large numbers for ship construction. Temples and public buildings throughout the ancient Middle East were made of cedar. By Ottoman times (1500–1700 CE) nearly all had been cut down, lastly for railroad ties by the Turks. Mature trees could reach heights exceeding 30 meters and diameters of 3 meters.

A cessation of Middle East hostilities could lead to widespread reforestation of denuded mountains.

(B)

(A)

FIGURE B 1-1 (A) Cedars on the slopes of Mount Lebanon. © Walid Nohra/ShutterStock, Inc. (B) A view of the modern range, showing effects of deforestation. © iStockphoto/Thinkstock.

Pleistocene Period, about 12,000 years ago. Forests in China, Europe, Africa, the Middle East, the United States, and elsewhere have been cut down for over thousands of years. Replanting trees, if soil conditions permit, could remove billions of tons of carbon from the surface environment. Some areas, such as parts of southern China and the Scottish Highlands, which were once extensively forested, have suffered considerable soil degradation, requiring soil enrichment before reforestation.

Carbon Sequestration Methods

World carbon emissions grew by at least 1.3% per year between 1980 and 2011. For the United States, carbon emissions grew at a 1.2% average annual rate. To address the climate change issues posed by global carbon emissions, it is theoretically possible to capture and store anthropogenic carbon gases before they reach the atmosphere.

Carbon sequestration is the capture, separation, and storage or reuse of carbon (as CO_2 or methane) from anthropogenic sources: fossil fuel burning, deforestation or industrial activity. However, the concept is usually applied to developing cost-effective measures for eliminating carbon emissions produced when coal, and to a lesser extent oil and natural gas, is burned to produce energy or electricity in power plants, because it may prove impractical to trap CO_2 from "mobile sources" like cars and trucks.

Most sequestration research is presently focused on coal, because coal is cheap, easy to mine, and contains almost twice the carbon of oil and gas. Moreover, coal reserves dwarf oil and gas reserves in total carbon content (discussed elsewhere in the text). While many environmentalists and scientists believe that "there is no such thing as clean coal," it is likely that coal will continue to be a primary fuel for electricity production for at least the next several decades, especially in countries that are presently industrializing, such as India and China. Moreover, the United States produces over 40% of its electricity from coal, as do many European countries, and analysts do not expect these economies to radically shift their electricity industries away from coal for the next few decades at least.

To be adopted, activities to sequester carbon must meet the following three requirements:

1. They must be effective and inexpensive.
2. They must provide stable, long–term storage.
3. They must be environmentally benign.

The U.S. Department of Energy (DOE) reports that, as of 2000, sequestration costs were in the range of $100 to $300/ton of avoided carbon emissions. The goal of DOE's Research and Development program for carbon sequestration was to "reduce the cost of carbon sequestration to $10 or less per net ton of carbon emissions avoided by 2015. Achieving this goal would save the U.S. trillions of dollars." By 2012 the costs were under $100 and as low as $35/ton of avoided carbon emissions.

Concept Check 20-11. Do you think it is the proper role of the United States government to assume primary responsibility for research on carbon sequestration? On what information did you form your opinion?

The following sequestration concepts may or may not be practical. For example, little research has been conducted on CO_2 capture and separation technologies, and no large-scale programs are in operation.

- Inexpensive CO_2 separation processes.
- CO_2 sequestration in geological formations, including oil and gas reservoirs, uneconomical coal seams, and in deep saline aquifers.
- Direct injection of CO_2 into the deep-ocean, and/or surface stimulation of phytoplankton growth with CO_2.
- Enhanced net carbon uptake by terrestrial ecosystems.
- Advanced chemical, biological, and decarbonization methods.

The most practical and realistic of these involve capture and injection of carbon into deep geological formations (such as coal seams that cannot be mined and deep saline aquifers) and enhancement of carbon storage in terrestrial ecosystems. We describe these next.

Oil and Gas Reservoirs

An oil or natural gas reservoir's production sometimes can be increased by pumping CO_2 gas into the reservoir to force out the remaining oil and gas. This is called *enhanced oil recovery* (EOR). U.S. oil firms already use over 30 million tons of CO_2 annually for this purpose. EOR provides an opportunity to sequester carbon at low net cost, because the corporations simultaneously obtain revenues from recovered oil/gas. Many essentially abandoned oil fields in the United States and elsewhere could theoretically store significant quantities of CO_2.

During EOR, CO_2 is stable in the reservoir, as long as the natural pressure within the reservoir is not exceeded.

This EOR application is at present economically limited to emissions of CO_2 that are near an oil or natural gas reservoir.

Coal Seams

Coal beds typically contain large amounts of methane-rich gas that is adsorbed onto the surface of the coal. Currently, coal bed methane is recovered by lowering the pressure, usually by pumping water out of the coal seam. An alternative approach is to inject carbon dioxide gas into the bed. Tests have shown that CO_2 is roughly twice as adsorbing on coal as methane, so it has the potential to efficiently displace methane and remain sequestered in the bed. CO_2 displacement of coal bed methane has been demonstrated in small field tests, but research must produce larger scale processes.

One advantage of this method is that the recovered methane can be sold, lowering the overall cost of the operation. Total U.S. coal resources are estimated at over 5 trillion tons, but 90% of it is unminable due to insufficient seam thickness, excessive depth, and danger to miners.

Another promising aspect of CO_2 sequestration in coal beds is that many large coal seams are near electricity generating stations that are themselves large emitters of CO_2 gas (**FIGURE 20-20**). The Mt. Storm, West Virginia coal-fired power plant operated by Dominion Resources is located at the mine itself. In this case, no pipeline

FIGURE 20-20 Smokestacks of the Mount Storm, West Virginia coal-fired generating plant, operated by Virginia Power, a division of Dominion Resources. © Jim West/Alamy.

transport of CO_2 gas would be required, but to be effective, CO_2 must be removed from coal before burning.

Burning coal bed methane in coal-fired power plants could provide additional power generation with low net emissions.

Geological Formations

DOE estimates that deep saline aquifers in the United States could store up to 500 billion tonnes of CO_2.

Most existing large CO_2 "point" sources, such as utilities and industries, are within easy access to a saline aquifer injection point. Sequestration in saline aquifers would thus have low transport costs.

Demonstrating that CO_2 will not escape from formations and migrate to the Earth's surface is a key objective of sequestration research. However, as part of EOR operations, the oil industry routinely injects brines from recovered oil into saline aquifers, and the U.S. Environmental Protection Agency (EPA) has allowed some liquid hazardous waste disposal injection into deep saline formations.

The Norwegian oil company, Statoil, is injecting approximately one million tonnes per year of recovered CO_2 into the Utsira Sand, a saline aquifer under the North Sea associated with a gas reservoir (**FIGURE 20-21**). The amount being sequestered is equivalent to the output of a 150-megawatt coal-fired power plant. This is the only commercial CO_2 geological sequestration facility in the world, but calculations suggest that even if only about 1% of the storage volume of the Utsira Sand were used for carbon sequestration, it could sequester the annual output of over 900 coal-fired 500 MW power stations.

Value-added Question 20-1.
Total annual CO_2 emissions from electric utilities in the United States are around 1 billion tonnes. Based on this, do you think that saline aquifers could provide a significant solution to the carbon emissions problem in the United States? If so, then for how many years?

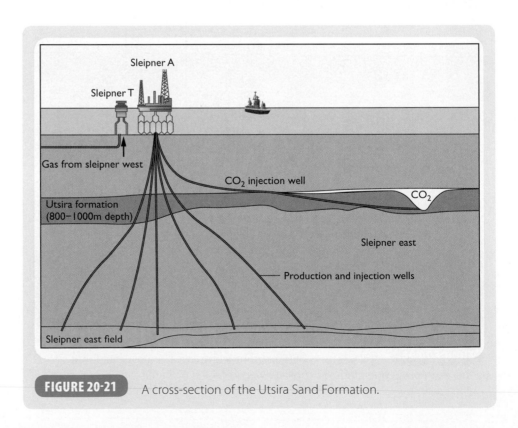

FIGURE 20-21 A cross-section of the Utsira Sand Formation.

Continental Ecosystems

There are two basic approaches to sequestering carbon in continental ecosystems: (1) *protection* of ecosystems that store carbon so that sequestration can be maintained or increased; and (2) *manipulation* of ecosystems to increase carbon storage beyond present amounts. Carbon sequestration can be enhanced either by the increased removal of CO_2 from the atmosphere by the ecosystem, or the decrease of CO_2 net emissions from the ecosystem into the atmosphere.

Enhancing the natural processes that remove CO_2 from the atmosphere could be a cost-effective way to reduce atmospheric CO_2. The terrestrial biosphere is estimated to sequester approximately 2 billion tonnes of carbon per year in vegetation and soils. Increasing this rate must consider ecological, social, and economic factors. Because soils can store enormous amounts of carbon, we will need to understand the nature and role of soil biota, which we little understand at present (see elsewhere in the text).

Encouraging reforestation in coastal regions could be adversely affected by an increased incidence of hurricanes and tropical storms predicted from warming the Earth as well as from sea-level rise. U.S. Forest Service research indicates that major hurricanes could destroy up to $1 billion worth of timber in the southeastern United States, and the rapid growth of small trees cannot compensate for loss of mature trees lost during the storm. Should numbers of hurricanes increase, carbon sequestration in southeastern forests could be compromised.

The U.S. Department of Agriculture has assessed soil carbon content of the United States. Their studies indicate that changing agricultural practices can enhance carbon storage in agricultural soils, as shown in **FIGURE 20-22**.

Over the past several hundred years, the conversion of grassland and forest to cropland and grazing lands has resulted in significant loss of soil carbon worldwide (see elsewhere in the text). A major potential therefore exists for increasing soil carbon storage through restoration of degraded soils, and aggressive application of soil conservation practices. Moreover, increasing soil carbon can significantly increase the soil's ability to trap and retain moisture, a necessity if evaporation potential increases. A Canadian study suggested planting trees across Canada could result in 2,500 million tonnes of additional carbon storage by 2050. However, the USGS has warned that the loss of soil carbon in northern high latitude terrestrial ecosystems as a result of climate warming can act as a positive feedback, enhancing additional warming. They state, "It is possible that losses of carbon storage will substantially offset carbon sequestration efforts . . . in the United States."

A 2005 British study reported that the nation lost soil carbon at the rate of 0.6%/year over the twentieth century, amounting to about 13 million tonnes annually. This soil loss offset the total carbon reduction by industry over the 1980–2002 period. Thus, soil storage of carbon may simply restore that lost by poor land use practices over the preceding several hundreds of years.

Removing marginal land from production and replacing it with an ecologically compatible land use such as wildlife habitat, can lead to increases in total biomass production as well as an increase in soil carbon content. However, a rapidly growing human population, coupled with increased meat consumption, is placing increasing demands on croplands for additional production.

In any case, researchers have concluded that terrestrial ecosystem sequestration by itself probably cannot store sufficient carbon to reverse climate change. It could, however, make a major contribution to a solution.

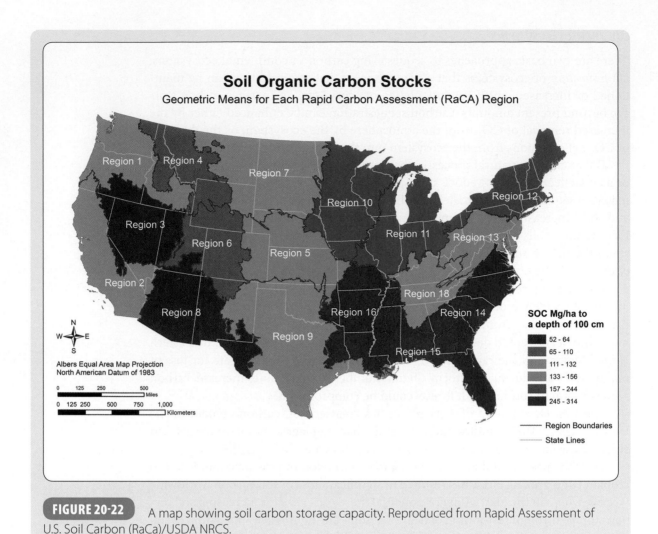

Soil Organic Carbon Stocks
Geometric Means for Each Rapid Carbon Assessment (RaCA) Region

FIGURE 20-22 A map showing soil carbon storage capacity. Reproduced from Rapid Assessment of U.S. Soil Carbon (RaCa)/USDA NRCS.

FIGURE 20-23 Outcrops of serpentinite rocks in California. Courtesy of NPS.

Mineral Carbonation

Enhanced mineral carbonation, that is, increasing the rate at which *carbonate* rocks form, is being investigated by researchers at Los Alamos National Laboratory, New Mexico. Carbonates store more carbon on Earth than any other reservoir, and organisms that make carbonate have been removing carbon from the atmosphere for the past 3.5 billion years.

Researchers react carbon dioxide with magnesium- and calcium silicate minerals to fix carbon. Enormous deposits of such silicates in the form of serpentinite rocks are found in a number of locations, notably near the high-energy-using coasts of the United States (FIGURE 20-23). Future power plants could be located near these deposits, allowing for immediate, permanent storage of their carbon dioxide emissions *if* the process can be made cost-effective.

In addition to Los Alamos, Oak Ridge National Laboratory, the Massachusetts Institute of Technology, and the Tennessee Valley Authority are among agencies and organizations actively pursuing carbon sequestration research.

■ CONCLUSION

Ignoring more than a century of scientific research, a few scientists, many politicians, and a distressingly large number of citizens say they remain unconvinced that rapid climate change, induced by human activity, poses a severe enough threat to warrant a "Manhattan Project" to mitigate its effects.

According to the Petition Project's website, signers include: "9,029 PhDs; 7,157 MS; 2,586 MD and DVM; and 12,715 BS or equivalent academic degrees." In contrast, a recently published study concluded that 97% of more than 1,372 climate researchers most actively publishing in their field supported the scientific conclusion of anthropogenic climate change.

Concept Check 20-12. Which group has the most credibility? Why?

Concept Check 20-13. Whose conclusion do you accept? Why?

It is very likely unavoidable that humanity and the planet will experience impacts, some horrific, of global climate change. Researchers at McGill University's Department of Natural Resource Science have produced a map showing that these will not be experienced equally (FIGURE 20-24).

What is clear is that many of the "solutions" to controlling greenhouse emissions would also be good for the global economy and for the biosphere. For example, the mining and burning of coal in many countries (Germany, France, China, the Ukraine) is heavily subsidized either by state aid to mining firms (France, Germany) for air and water pollution (virtually everywhere), by thousands of miners killed in mine accidents each year (Ukraine, China), or by widespread debilitating diseases like black lung among populations of miners.

Reforestation could increase income from recreation and tourism to local communities and would drastically reduce erosion rates, soil loss, and CO_2 emissions.

By 2004, sufficient nations had ratified the Kyoto Protocol such that it became the primary means by which the global community has begun to address the challenges posed by climate change. However, without action by the world's greatest agent of climate change—the United States—as well as mitigation efforts on the part of developing countries like China and India, progress will be slow and insufficient. Moreover, the 37 countries with binding targets in the Kyoto Protocol Extension (2012–2020) produce only 15% of the world's greenhouse gas emissions.

There are, however, some hopeful signs. In his 2013 State of the Union address, U.S. President Barack Obama stated, "I urge this Congress to get together, pursue a bipartisan, market-based solution to climate change . . . But if Congress won't act soon to protect future generations, . . . I will direct my cabinet to come up with executive actions we can take, now and in the future, to reduce pollution, prepare our communities for the consequences of climate change, and speed the transition to more sustainable sources of energy."

Concept Check 20-14. Did the congress act on climate change? Did President Obama issue an executive order? Research the issue and report your results.

Value-added Question 20-2.
Assess the additional cost per kWh to remove and store carbon from a 1,200 megawatt coal-fired power plant, assuming a cost of $10/tonne CO_2. Then do your analysis assuming a cost of $100/tonne. A 150-megawatt power plant emits around 1 million tonnes CO_2 annually. First you must calculate how many kWs are produced by such a plant each year, assuming an operating rate of 80% (that is, the plant is able to run and produce electricity 80% of the time). Then convert your numbers to those appropriate for a plant eight times larger.

Value-added Question 20-3.
Let's revisit the Petition Project, which, you will recall, boasts 31,000 scientists as signees. According to the U.S. Department of Education, from 1970–1971 to 2008, 10.6 million U.S. science graduates would qualify as scientists according to the Petition Project's criteria. What percent of eligible "scientists" signed the petition?

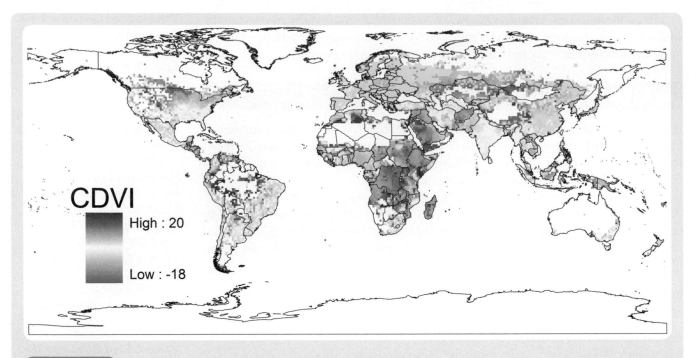

FIGURE 20-24 Global map of human vulnerability. White areas may experience climate change impacts, but human population density is very low in these areas. Reproduced from Samson, J., et al. (2011). "Geographic disparities and moral hazards in the predicted impacts of climate change on human populations." *Global Ecology And Biogeography* 20(4): 532–544.

CHAPTER SUMMARY

1. There is wide agreement among government agencies, scientists, and industry that human activity has the potential to cause, or is causing, global climate change.

2. The Kyoto Protocol, an international treaty to address global climate change, has been ratified by the European Union, China, Russia, Canada, and by Japan, among others.

3. Among other things, signatory states agreed to set targets to reduce greenhouse gas emission, and then to meet those targets by the date(s) specified. They agreed to determine their individual CO_2 contributions during the base year of 1990. They agreed to enhance energy efficiency, promote sustainable agriculture, and protect forests, soils, and other carbon sinks. They agreed to promote renewable energy technologies, and research means by which CO_2 and other greenhouse gases could be sequestered or stored. They would work to eliminate subsidies and market "imperfections" that encourage unnecessary greenhouse gas production.

4. During the nineteenth century, British scientist John Tyndall discovered that two atmospheric gases, CO_2 and H_2O, could absorb energy in the infrared range.

5. In 1895, Swedish chemist Svante Arrhenius argued that variations in CO_2 and H_2O could profoundly affect the Earth's heat budget.

6. In 1957, Roger Revelle and Hans Seuss showed that CO_2 concentration in the atmosphere had increased as a direct result of the burning of fossil fuels, and that the oceans were unlikely to be able to absorb the increase.

7. In 1965, a President's Science Advisory Committee concluded that Revelle and Seuss had been correct and officially recognized global warming from CO_2 as a possible global problem.

8. In 1988, Dr. Jim Hansen of NASA testified before Congress that the enhanced greenhouse effect was warming the climate and would likely contribute to extreme weather events.

9. The Earth's surface is warmed by solar radiation, mainly in the visible light range.

10. Surface material absorbs this energy and re-emits it, mainly in infrared, which can be absorbed by greenhouse gases.

11. The concentrations of greenhouse gases are increasing at more than 1% per year. CO_2 is increasing as humans return the carbon in formerly stored (sequestered) fossil fuels to the atmosphere in the form of CO_2.

12. Warming the Earth will cause sea levels to rise by amounts that depend (1) on the amount of ice we melt on land, and (2) on the coefficient of thermal expansion of seawater.

13. Impacts of climate change can be grouped into health impacts, impacts on agriculture, impacts on forests, impacts on water supplies, impacts on coasts, and impacts on species and habitats.

14. Efforts to mitigate the impacts of global warming include living with, and adapting to, change; international aid to assist poor coastal nations and displaced populations; facilitating species migration; reducing production of greenhouse gases; elimination of coal as an energy source; replacing fossil-fueled vehicles; global reforestation; controlling human population; carbon sequestration; reduction of cement manufacture and cement recycling.

15. Carbon sequestration is the capture, separation, and storage or reuse of carbon (as CO_2 or methane) from anthropogenic sources. Most sequestration research is presently focused on coal, because it is cheap, easy to mine, and contains almost twice the carbon of oil and gas. Sequestration in forests, soils, and in geological formations is theoretically feasible.

KEY TERMS

carbon cycle	greenhouse gases	John Tyndall
carbon sequestration	heat budget	Roger Revelle
climate change	infrared	Svante Arrhenius
greenhouse effect		

REVIEW QUESTIONS

1. Increasing global temperature will have what effect on carbon stored in soils?
   a. Carbon storage in soils will increase with increasing temperature.
   b. Soils will emit carbon as temperatures increase, increasing atmospheric CO_2.
   c. Soils have little if any stored carbon; the carbon is in vegetation.
   d. Carbon's behavior in soils is unknown and needs to be studied.

2. According to a 2001 National Academy of Sciences report,
   a. the concentration of greenhouse gases is insufficient to lead to climate change.
   b. the emission of greenhouse gases by human activity is insufficient to cause climate change at present levels.
   c. greenhouse gases are accumulating in the atmosphere as a result of human activities.
   d. greenhouse gas concentration in the lower atmosphere is actually falling, as the oceans remove carbon.

3. The nation that emitted the most CO_2 globally as of 2012 was
   a. the United States.
   b. the former Soviet Union.
   c. China.
   d. the European Union.

4. The response of the United States to the Kyoto Protocol was
   a. that the Protocol was ratified by the Senate.
   b. that the Protocol was unfair as it exempts developing nations.
   c. that the United States has agreed to reduce CO_2 levels more than the Protocol requires.
   d. that the Protocol could not be ratified until it includes agreements to reduce CO_2.

5. Until 1957, when Revelle and Seuss published their landmark paper, most scientists believed that humanity was unlikely to be able to alter global climate. Their reason was
   a. the oceans store 50 times as much CO_2 as the atmosphere.
   b. human activity did not generate greenhouse gases.
   c. the kinds of greenhouse gases humans generated were lost by the atmosphere into space.
   d. geothermal heat, and not solar radiation, was responsible for most global warming.

6. The solar radiation received by the earth is concentrated in which electromagnetic range?
   a. microwave
   b. x-radiation
   c. visible range
   d. infrared range

7. The radiation re-emitted by surface material is mainly in which range?
   a. infrared
   b. ultraviolet
   c. microwave
   d. x-radiation

8. According to U.S. government data, the concentration of greenhouse gases in the atmosphere is
   a. decreasing at about 1%/year.
   b. increasing at about 1%/year.
   c. remaining constant, because of ocean absorption.
   d. increasing at 10%/year.

9. Which of these is not a greenhouse gas?
   a. H_2O
   b. methane
   c. CFCs
   d. nitrogen

10. Which of these releases CO_2 to the atmosphere?
    a. reforestation
    b. methane release from coal mines
    c. deforestation
    d. none of these

11. According to an EPA report the total cost of a one-meter sea level rise, ignoring future development
    a. was not quantifiable.
    b. was estimated at less than $10 billion.
    c. was estimated in the hundreds of billions of dollars.
    d. was estimated in the tens of trillions of dollars.

12. Warming the air would have what effect on ozone air pollution, other things being equal?
    a. Ozone levels would increase, as higher temperatures favor ozone formation.
    b. Ozone levels would fall, as cooler temperatures favor ozone formation.
    c. Ozone formation is independent of temperature, so no effect was forecast.
    d. Warmer air would increase lung capacity, increasing resistance to ozone.

13. Warming would have what effect on agriculture?
    a. The duration of growing seasons will increase.
    b. Irrigation costs will increase.
    c. Evaporation rates will increase, thereby reducing soil moisture.
    d. All of the above are correct.
    e. None of the above is correct.

14. How would warming affect forests and forest communities?
    a. Warming could force some plant species to migrate poleward.
    b. Forest productivity could increase.
    c. Some plant species could succumb to disease and become extinct.
    d. All of the above are correct.
    e. None of the above is correct.

15. Sea level would rise with warming, from the melting of polar ice and permafrost, and
    a. the increase in volcanic activity that would accompany warming.
    b. the increase in metamorphic activity that would accompany warming.
    c. the expansion of water as it warms.
    d. None of the above is correct.

16. How would warming sea temperatures, in the equatorial Atlantic, affect hurricanes and tropical storms?
    a. Hurricanes would decrease in number.
    b. The hurricane season could become extended to more weeks/months of the year.
    c. The hurricane season would become restricted to fewer weeks/months of the year.
    d. None of the above is correct.

17. Which of these activities increases greenhouse gas concentration in the atmosphere?
    a. melting glaciers in polar regions
    b. releasing methane from coal mines
    c. evaporating water from the sea surface
    d. bacterial production of nitrogen gas

18. Most research into carbon sequestration presently focuses on
    a. cement manufacture.
    b. mobile sources like motor vehicles.
    c. coal.
    d. oil and natural gas.

19. The U.S. Department of Energy (DOE) is conducting research with the goal of reducing carbon sequestration costs to
    a. $1,000 per ton.
    b. $10 per ton.
    c. $1 per ton.
    d. 10 cents per ton.

20. The most practical and realistic plan to sequester carbon is
    a. to inject CO_2 into deep geological formations.
    b. to liquefy CO_2 and store it in space.
    c. to drastically reduce cement manufacture.
    d. to inject CO_2 into the deep ocean.

Land Use, Development, and Zoning

CHAPTER

21

CHAPTER OUTLINE

■ LAND USE

L *and use* refers to alterations of landscapes by and for human activities. Broad categories of land use are summarized on maps, such as this of the Raritan, New Jersey watershed (FIGURE 21-1).

Natural Versus Human-Altered Land Use

Natural ecosystems develop as a result of adaptations by organisms and terrain to climate and geologic change. These adaptations generally require centuries or millennia for natural ecosystems to become fully established. Human land use changes differ in that the changes occur at time spans orders of magnitude less. Thus, the "clearing" of the American West (west of the Alleghenies) began near the end of the eighteenth century and was essentially complete by the early part of the twentieth. FIGURE 21-2 shows the changes in forests for North America since European settlement.

Rapid land-use changes usually impose significant environmental costs. The clearing of Eastern old-growth forests increased sedimentation rates into rivers many-fold (Chesapeake Bay erosion rate, for example, increased from about 0.05 cm/yr to 0.60 cm/yr from before 1700 to 1750), and literally changed the

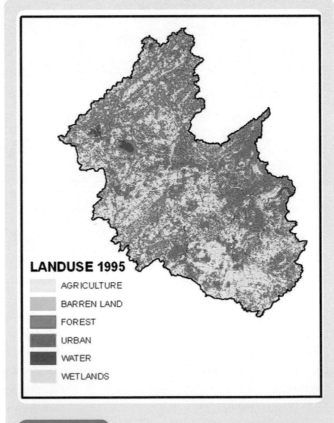

LANDUSE 1995
- AGRICULTURE
- BARREN LAND
- FOREST
- URBAN
- WATER
- WETLANDS

FIGURE 21-1 Broad categories of land use in the Raritan River Basin in New Jersey. *Question:* How would you expect these maps to change over time? Courtesy of USDA NRCS.

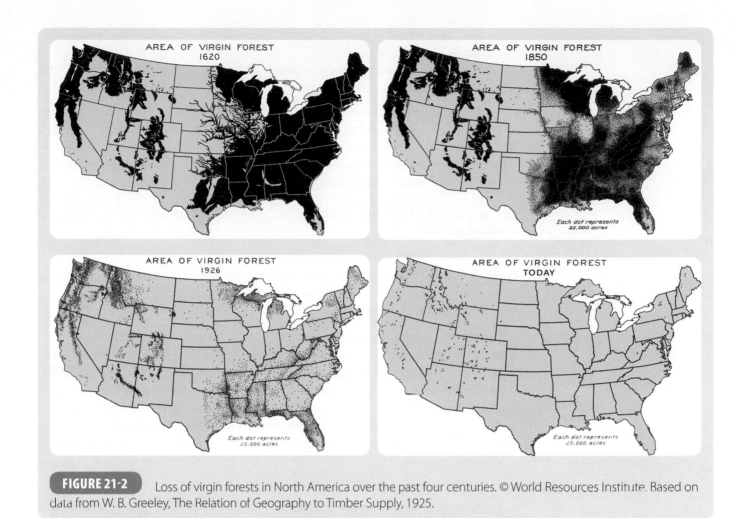

FIGURE 21-2 Loss of virgin forests in North America over the past four centuries. © World Resources Institute. Based on data from W. B. Greeley, The Relation of Geography to Timber Supply, 1925.

geography of many areas of the Atlantic Coast. It may also have contributed to global climate change by increasing carbon dioxide (CO_2) release by soils, and adding CO_2 to the atmosphere by burning vegetation.

Development

Development in a dictionary means, among other things, "a stage in growth, or advancement." But advocates of positions or activities long ago mastered the art of taking words with long-accepted meanings and adopting them for practices that can be controversial: "landfill" for trash dump is one example. In this new context, development has been used to mean *land conversion*, usually converting forest, wetland, or agricultural land to commercial, industrial, or residential use. And a *developer* is now one who facilitates land-use changes that "improve" the land; hence the term "improvements" for sewer installations, commercial buildings, subdivisions, power lines, roads, etc. Development without careful management, however, leads to a temporary or permanent increase in the amount of *pollutants* in an area. For example, sediment from construction sites can end up in streams and rivers, choking plant and animal life. According to the United States Geological Survey,

- Oil and gas from vehicles can leak onto roads and parking lots. Fertilizers and pesticides, if not applied properly, can wash off lawns. Pesticides are often found in higher concentrations in urban areas than in agricultural areas.

- Pet waste, if not properly disposed, can enter storm drains that lead to wetlands, streams, or rivers.
- Household chemicals, such as paints and cleaning products, can leak if not stored or disposed of properly.

All of these pollutants can "wash away" when it rains and end up in streams, rivers, lakes, estuaries, or ground water. Many pollutants/nutrients like phosphorus (P) also bind to the sediment, so when sediment washes away, it takes the pollutants/nutrient with it. Moreover, land-use changes can lead to increased traffic congestion, leading in turn to increased air and water pollution.

Concept Check 21-1. CheckPoint: Discuss ways in which development may lead to increased pollution.

- -

Concept Check 21-2. Why do you think we put "away" in quotes? Where is the "away" we mean when we say we threw something away?

- -

Urbanization, a land-use change, modifies the hydrologic cycle (discussed elsewhere in the text). It leads to increases in impermeable surfaces (roads, roofs, parking lots, etc.) that prevent precipitation from soaking into the ground (**FIGURE 21-3**). It usually involves compacting soil that further favors runoff over infiltration. This can increase the amount and velocity of rainwater flowing to streams and rivers, and can lower water tables. When rainwater cannot infiltrate, the result can be a loss of drinking water supplies, because many areas of the country rely on pre-

(A)

FIGURE 21-3 (A) An example of an impermeable surface. © Aaron Kohr/ ShutterStock, Inc.

(Continued)

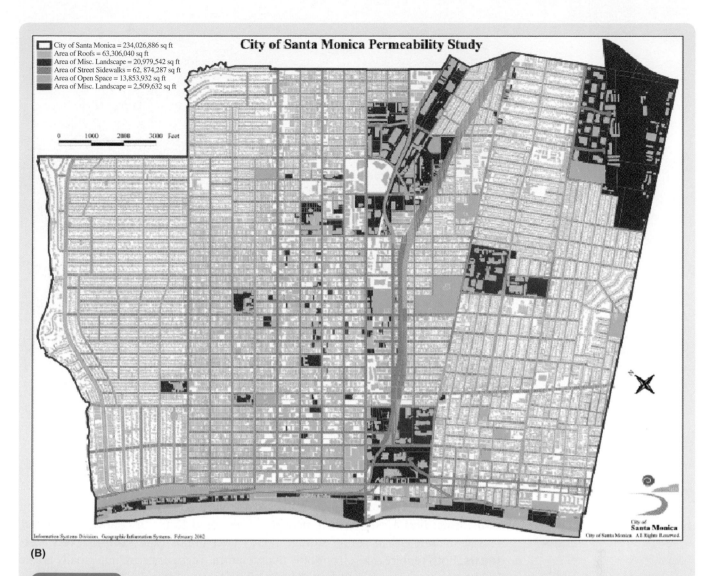

City of Santa Monica Permeability Study

City of Santa Monica = 234,026,886 sq ft
Area of Roofs = 63,306,040 sq ft
Area of Misc. Landscape = 20,979,542 sq ft
Area of Street Sidewalks = 62, 874,287 sq ft
Area of Open Space = 13,853,932 sq ft
Area of Misc. Landscape = 2,509,632 sq ft

0   1000   2000   3000 Feet

Information Systems Division, Geographic Information Systems. February 2002

City of
Santa Monica

City of Santa Monica All Rights Reserved.

(B)

FIGURE 21-3   *Continued.* (B) Permeability study of the city of Santa Monica, California. *Question:* What is your estimation of the percentage of this area that is impermeable? Compare your answer with the footnote.[1] © City of Santa Monica. All Rights Reserved.

cipitation soaking into the ground to replenish near-surface underground water supplies.

This increased speed and volume of runoff water can erode stream banks, increase turbidity and pollution, increase sedimentation, and even increase stream water temperature. All of these can have an adverse effect on the *biota* in the stream.

Urbanization also reduces tree cover, as most communities do not regulate tree loss. Loss of trees due to urbanization can have many negative impacts. Trees enhance property values. Mature trees along stream banks lower stream temperatures, helping to maintain oxygen levels. They cool their surroundings, lowering air conditioning loads, and even protect streets from deterioration! Even if "developers" replace trees destroyed during land use changes, they often replace large native species with juvenile faster-growing, shallow-rooted hybrids and non-native varieties, further negatively altering the original natural landscape (**FIGURE 21-4**).

FIGURE 21-4 The Bradford pear is a hybrid, shallow rooted tree, called "tree rats" by many horticulturalists, but much prized by developers. © Michael Hare/ShutterStock, Inc.

Stormwater Runoff

Urbanization leads to increased **stormwater runoff** and less infiltration during rainstorms. *Best Management Practices* (BMPs) used to try to control stormwater runoff include urban runoff storage ponds (wet ponds) and **constructed wetlands**. Both require careful, periodic maintenance. These costs are not usually borne by developers, but rather by local governments that are chronically short of money. A newer approach known as *Better Site Design* uses suite of strategies to *prevent* or lessen runoff.

Wet ponds, as their name implies, are runoff-holding facilities that have water in them all the time. Storm flows are trapped in the pond and released slowly to maintain healthier downstream habitats. Sediment and other pollutants settle out of the water and are not discharged to streams. Wet ponds are usually vegetated; the plants' roots hold sediment and use the nutrients (N, P) that are contained in urban runoff. The ponds are designed to be big enough to control flooding in the event of a significant storm, helping to reduce impacts on downstream habitats.

Many of the "lakes" in developments are actually wet ponds. Developers often design wet ponds to look like natural lakes. Wet ponds can be incorporated into new development site plans and can enhance the value of surrounding property.

There are several problems with this approach. One is that wet ponds are designed to transport and store a limited quantity of water, based on calculations of likely runoff from storms of a given intensity. Many are designed to handle runoff from relatively small (10-year to 25-year recurrence frequency) floods (discussed elsewhere in this text). But as development proceeds, increases in impermeable surface lead to more runoff. This extra runoff can fill retention ponds more rapidly than planned. Furthermore, ponds tend to **eutrophicate** (become devoid of oxygen) over time as nutrients from lawns and other sources fuel algae growth. This literally fills the pond with organic matter, eventually rendering it useless as a water-and sediment-storage facility.

Wetlands

Wetlands (FIGURE 21-5) serve an important function in controlling the impacts of urban runoff. In fact, many ecological economists estimate the value of services wetlands provide to humans to be in the tens of billions of dollars annually. Because wetlands are heavily vegetated, they serve as a natural filter for urban runoff. They also help to slow the flow of water to the receiving waters, and replenish groundwater. When properly designed, constructed wetlands have some advantages as an urban BMP. These advantages include: pollution removal, longevity, applicability to a range of development sites, ability to be combined with other BMPs, and wildlife habitat potential.

Contiguous Versus Isolated Wetlands

Contiguous wetlands (Figure 21-5) are wetlands immediately adjacent to a stream, river, a lake, or the ocean, and whose waters are interconnected. Contiguous wetlands serve as habitats for young fish and other fauna, help regulate water flow, protect nearby areas from floods, help recharge groundwater, and naturally filter out sediment, nutrients, and other potential pollutants. Preserving contiguous wetlands is becoming increasingly important as urbanization proceeds along water bodies, resulting in land use changes that reduce infiltration capacity, and thus filtration capacity of natural environments.

FIGURE 21-5   Examples of wetlands. © Mircea BEZERGHEANU/ShutterStock, Inc.

Isolated wetlands are wetlands completely surrounded by upland with no surface water outlet, and thus may not have obvious connections to other water bodies (FIGURE 21-6).

Isolated wetlands typically occur in depressions. Landscapes with geographic depressions may be created by glacial activity, the collapse of karst bedrock (discussed elsewhere in the text), or by wind erosion. Low-lying landscapes also include *Carolina Bays*, elliptical depressions along the Atlantic seaboard, which may represent meteorite scars.

Isolated wetlands are important habitats for wildlife, and help maintain biodiversity. They are especially important breeding habitats for amphibians, and for continental waterfowl populations. In arid and semiarid regions, isolated wetlands provide vital freshwater "oases" for wildlife and migrating birds. Their isolation has led to the evolution of unique species, specially adapted to these habitats. Endangered and threatened plants often live in wetlands, especially West Coast *vernal pools*, desert spring wetlands, and Coastal Plain ponds.

Many geographically isolated wetlands connect to other wetlands and waterways via groundwater. They thus may contribute important, but often poorly appreciated, subsurface water flows to other wetlands and surface waters. Isolated wetlands can store appreciable rainwater. Such storage is important for managing flood discharges, replenishing groundwater supplies, and providing habitat for waterfowl.

Development of isolated wetlands can have far-reaching effects on wildlife, flood control, and groundwater quality, and can even violate international agreements on protecting migratory wildfowl.

No-Net-Loss Policy

"No net loss" of wetlands—a policy goal established in 1989—has received widespread support among politicians, scientists, and environmentalists. According to the U.S. Department of Agriculture (USDA), the "no net loss" objective has been interpreted to mean that wetlands should be conserved wherever possible, and that wetlands converted to other uses must be offset through restoration and/or creation of wetlands of equal or greater area, thus maintaining or increasing the overall wetland area.

Despite the difficulty in precisely evaluating success, no-net-loss policies appear to have slowed the loss of wetlands. Federal policies that formerly encouraged the elimination of wetlands are being replaced by incentives to restore wetlands. Two agricultural wetland protection programs are the "Swampbuster Program," which denies financial subsidy for some wetland cultivation, and the Conservation Reserve Program, which pays a small subsidy to encourage farmers to stop cultivating highly erodible land, including some drained wetlands. These have produced measurable economic benefits, according to the USDA.

Mitigation Banks

Mitigation banks are a companion program to the no-net-loss policy. Using a mitigation bank, a regulatory agency will allow the destruction of a small area of isolated wetland if replaced by an equivalent or larger area of the same type, in a large tract that will be permanently maintained. A small wetland area chosen for development can be filled, or built on when an equal area with the same functions in a nearby mitigation bank property is properly built and operating. The bank can be constructed in advance by a developer or by a nonprofit organization, and credits sold or withdrawn as necessary (**FIGURE 21-7**).

While the history of mitigation banks is short, it remains unclear whether the ecological services of numerous small, isolated wetlands will ever be supplanted by a single large contiguous wetland. Also, in many areas there is little oversight of mitigation banks and virtually no enforcement. Thus, the same wetlands may be used repeatedly as mitigation, subverting the spirit of the program. But without no-net-loss policies and mitigation banks, wetlands were being rapidly depleted.

FIGURE 21-7   Photo of the Old Fort Bayou Mitigation Bank, Mississippi, maintained by the Nature Conservancy. The tract measures 1,700 acres. Courtesy of Patricia Bacak-Clements, USFWS.

Better Site Design

The approach known as *Better Site Design* (as opposed to *Conventional Design*) has been shown to be very effective in reducing runoff and protecting local watersheds from pollution associated with that runoff. Better Site Design relies more on natural approaches than structural ones. For example, when developing a site using Conventional Design, the land is typically completely cleared of trees and other vegetation, and then graded before drainage **infrastructure** (including stormwater ponds) is installed. In Better Site Design, natural areas are preserved and impervious surface is minimized. The layout and density of development vary between Conventional and Better Site Design approaches (**FIGURE 21-8**).

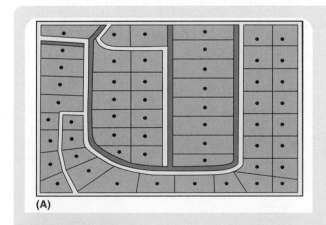

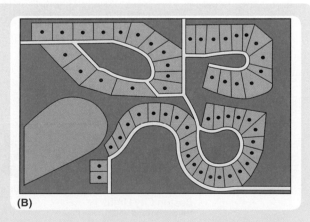

(A)

(B)

FIGURE 21-8   (A) Standard development practice using Conventional Design principles, also called sprawl. (B) Alternative approach using the Better Site Design approach.

Some of the goals of Better Site Design include:

- Prevent rather than mitigate stormwater impacts.
- Manage stormwater close to the source as possible and minimize use of wet ponds.
- Preserve natural areas and native vegetation.
- Use natural drainage pathways.

■ LAND USE AND TRANSPORTATION

The Streetcar Suburb

In the 1880s, electric streetcar lines—the descendant of the horse-pulled omnibus ("bus")—began to be extended outward from central business districts, as electricity began to be more available. This led to the growth of suburban areas ("suburbs") outside the city limits. Low fares allowed the middle classes to move to affordable housing outside the city and commute to jobs downtown. Housing could be built cheaply using inexpensive land and improved wood frame construction practices. Streetcar suburbs provided a wide range of housing, from more exclusive single-family homes to simple two- and three-family wood frame structures. These suburbs were built on small lots to guarantee an essential number of riders within walking distance of the stations and, therefore, profit for the streetcar companies. Commercial districts sprang up around the stations.

Focus: Suburban Boston, Massachusetts

Suburban development followed the construction of electric streetcar lines to previously rural areas outside Boston. By 1900, wooden detached houses filled towns such as Roxbury, West Roxbury, Dorchester, and Roslindale, all of which eventually were annexed by the city of Boston (**Figure B 1-1**).

In short, prior to the widespread adoption of the personal motor vehicle in the 1920s, growth of cities was mainly along rail and trolley lines, resulting in high-density development. After World War I, demand for motor vehicles began to accelerate, leading to demand for more paved roads. After the 1920s, passenger trips by mass-transit leveled off and then declined, while road networks and personal motor vehicles skyrocketed, leading to rapid increases in miles driven per year. For details, consult the Statistical Abstract of the United States, which is available at http://www.census.gov/compendia/statab/.

FIGURE B 1-1 Growth of Boston city limits to 1912. © Radical Cartography.

FIGURE 21-9 Abandoned buildings in Philadelphia. © Taras Kolomiyets/ShutterStock, Inc.

The Evolving Suburb

The end of World War II led to record demand for housing in the United States, because the dreaded return of the 1930s Great Depression failed to materialize. Between 1945 and 1954 birth rates soared, and nine million people moved to the suburbs in one of the great migrations of the twentieth century. Postwar suburbs were generally similar in that they were built on the outer edges of the city, had low population density per square mile, and easy access to roads. This suburban housing "boom" accelerated the decline of the central city. Abandoned central city sites are often called **brownfields**, attesting to their degraded conditions (FIGURE 21-9). Developers prefer to use "greenfields," that is, land that has not been previously used for development. The impetus for using greenfield sites accelerated with the passage of the **Superfund Act** of 1980, which held any owner, previous or present, potentially liable for cleanup costs of degraded property.

Sprawl Development

Much development in the United States today is called **sprawl**, which refers to land conversion that occurs at a rate exceeding the rate of population growth (TABLE 21-1).

Present-day American suburbs tend to form as a ring around cities that quickly transforms landscapes from rural to suburban. This ring then expands outward until the cost and time required for commuting to jobs reaches some critical point (FIGURE 21-10). Cheap land leads to cheap housing, which attracts both developers and homebuyers. Most of these communities lack access to public transportation, and low housing density does not readily support mass-transit. Residents are thus dependent upon individual vehicles for access to jobs, shopping, and recreation, all of which are usually sited outside residential areas due to zoning requirements. We discuss zoning next.

Metro Area	Population	Developed Land Area
TABLE 21-1	*Land Development and Metropolitan Area Population Growth, 1980 and 1990*	
Chicago	+ 2.5%	+ 22.5%
Cleveland	− 2.5%	+ 23%
Detroit	− 1.5%	+ 28%
New Orleans	− 1%	+ 47%
New York	0%	+ 28.5%
Phoenix	+ 40.5%	+ 58.5%
Pittsburgh	− 8%	+ 20.5%
Washington/Baltimore	+ 16%	+ 45%

Between 1980 and 1990, land development outpaced population growth in metropolitan areas. Data from: National Building Museum.

Zoning

Development is usually governed in the United States by local and state documents, such as the *comprehensive plan*, and related ordinances collectively known as **zoning**. Zoning specifies how a piece of property can be used within a city or county *zoning district*. Zoning districts are created to attract and contain certain types of development. Examples include: single-family homes, institutions such as schools and hospitals, offices, neighborhood business, and industry. Zoning designation can be changed on request.

A summary of the comprehensive plan for Boulder, Colorado is presented below. For details, go to (http://www.bouldercounty.org).

The Boulder County Comprehensive Plan (BCCP) was created to ensure that future land use decisions affecting the county should be made in a "coordinated and responsible manner."

The Plan, initially adopted in 1978, is guided by the following principles:

- Population growth is to be channeled to municipalities.
- Agricultural and ranch lands are to be protected.
- Preservation of environmental and natural resources is to be a high priority in making land use decisions.

FIGURE 21-11 shows the greenbelt surrounding the town of Boulder, a major component

FIGURE 21-10 False color infrared photo showing development around Washington DC. Courtesy of USGS/NASA.

FIGURE 21-11 The greenbelt around Boulder, Colorado. Courtesy of Boulder CVB.

of the plan to preserve green space and farms and enhance the quality of life for residents.

Why Do We Have Zoning?

Zoning originated with the passage of the Standard State Zoning Enabling Act of 1922, and the Standard City Planning Enabling Act in 1928. In a landmark 1926 Supreme Court decision, *Euclid v. Ambler Realty Co.*, the Court ruled that the exclusionary nature of zoning was appropriate, and established that the rights of the public take precedence over the rights of individual property owners. This principle continues to be the basis of zoning today. What this means is that just because you own a piece of property does not give you the right to do whatever you want with it, however unfair this might seem to some.

Early regulations dealt with minimizing public nuisances and dangers to public health. More recently, zoning regulations have attempted to focus on less easily defined public benefits and future needs.

Examples

FIGURE 21-12A is a zoning map of the town of Falmouth, Massachusetts, and **FIGURE 21-12B** is a high-altitude photo of the same locality.

Zoning regulations frequently require low-density development, and a separation of land uses: residential must be separate from retail, commercial from residential, and so forth. Furthermore, subdivisions made up of single-family houses on large lots are often separated by four-to-eight-lane highways, making walking or biking between neighborhoods difficult, if not impossible.

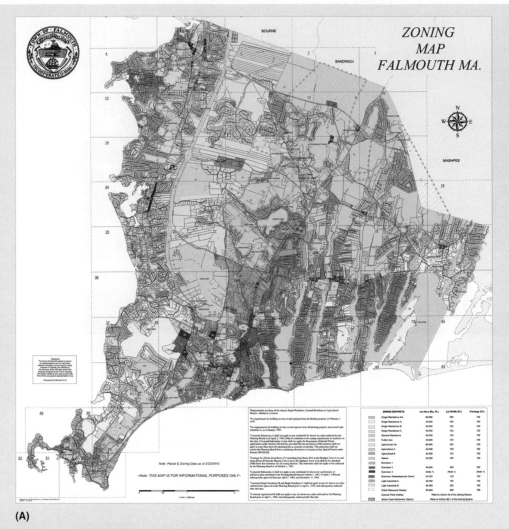

(A)

(B)

FIGURE 21-12 (A) A zoning map of Falmouth, Massachusetts. Shades of yellow mean residential zoning, purple means public use, shades of green mean agricultural, blue means marine, gray means light industrial, and shades of pink mean business zoning. *Question:* Why do you believe city officials adopted a minimum size for lots? Produced by the Falmouth GIS Department. (B) A high-altitude photo of Falmouth. Courtesy of the Town of Falmouth.

Focus: Zoning and Seismic Hazards in California

The California Department of Conservation is required by the Seismic Hazards Act of 1990 to identify and map the state's most prominent earthquake hazards in order to help avoid damage resulting from earthquakes. The Department's **Seismic Hazard Zone** Mapping Program will ultimately chart areas prone to **liquefaction** (failure of water-saturated soil) and earthquake-induced land-slides, throughout California's principal urban and major growth areas.

Look at **Figure B 2-1**, a seismic risk map of California. The unit contoured is horizontal acceleration from earthquake waves as a percent of gravitational acceleration. Study the figure and answer the questions in the caption to get an idea of how risk is associated with earthquakes.

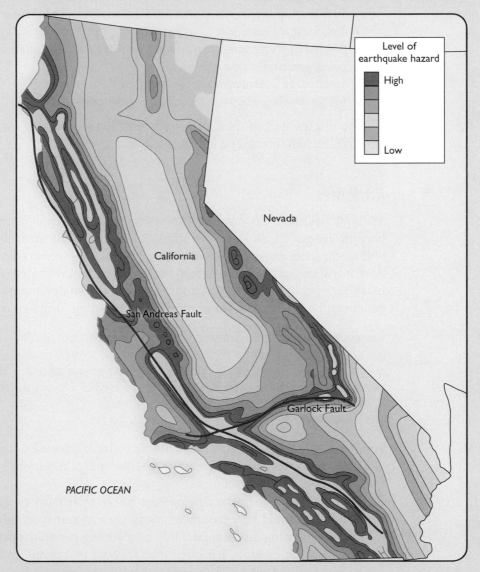

FIGURE B 2-1 A seismic risk map of California, with the San Andreas and Garlock Fault systems labeled. The contours are strength of earthquake shaking measured as percent of gravitational acceleration. *Question:* What do you think would happen at a spot where the earthquake waves had a force equal to the acceleration of gravity? How closely does the area of high acceleration follow the zones of active faults in California?

■ SUSTAINABLE DEVELOPMENT

Arising out of the growing need to support principles of sustainability, many if not most states now encourage some sort of alternative to sprawl development, sometimes called **smart growth**, which means *sustainable development*. The components of sustainable development must be applied together. Sustainable development should:

- Provide realistic transportation alternatives to the automobile.
- Plan for, rather than react to, development.
- Create pedestrian-friendly neighborhoods.
- Encourage mixed-use development within existing communities, and use **urban growth boundaries** to control sprawl.
- Preserve open space to protect critical environmental areas and services.
- Encourage a variety of housing choices.
- Support private investment using predictable, fair, quick and cost-effective government decision-making.
- Encourage "stakeholder" collaboration.
- Strike a balance between jobs and residences.

We discuss each of these principles next. Remember that for each to be most effective, the others must be included as well.

Attributes

Without *realistic transportation alternatives to the automobile*, roads will quickly become congested, air pollution will be a chronic problem, and scarce resources will be directed disproportionately to roads. Building roads will not end congestion; it will only temporarily ease it. Moreover, heavy investment in roads, the ultimate "free public good," itself encourages more sprawl development, which in turn creates still more demand for roads in a never-ending "death spiral."

Concept Check 21-3. Stop here and think about the concept of a "free public good," which is something everybody wants that is freely available to all with no cost, or for which the cost is hidden. How likely would you be to take advantage of such a "good?" Roads have been called free public goods. Why? What happens when everybody takes advantage of a free public good, like a road or a park?

- -

Some cost-effective alternatives to roads can include: sidewalks, bike paths, light rail, dedicated bus lanes, and tax incentives that discourage automobile use, or measures that encourage mass-transit.

Communities must plan development rather than react to development. New development should be encouraged along present transportation corridors and in presently existing urban areas. Only by planning can transportation choices be cost-effective, for example. It is rarely cost-effective to "retrofit" mass-transit to suburban patterns of sprawl. Urban-growth boundaries have successfully been used to contain development and protect important open space from sprawl (Figure 21-11 and **FIGURE 21-13**).

Communities must create *neighborhoods that are pedestrian-friendly*. Walking is enhanced by mixed-use development (**FIGURE 21-14**). Examples are: allowing shops to be built within walking distance of homes; providing sidewalks,

FIGURE 21-13 A false color infrared photo showing part of the urban growth boundary for Portland, Oregon as seen from space. Courtesy of USGS image from the National Land Cover Database 2001 (NLCD 2001)/NASA.

FIGURE 21-14 An urban area emphasizing pedestrian access. Stroget, a popular pedestrian area of Copenhagen, Denmark, is the longest pedestrian shopping area in Europe. *Question:* What sorts of households are most likely to find such an environment most attractive? © Sergey Goryachev/ShutterStock, Inc.

median strips, and pedestrian-friendly streets, meaning those that are easy and safe to cross; and controlling vehicle speeds. Overpasses and underpasses can be built at busy intersections, speed "humps" can be built to calm traffic, and non-dedicated bike lanes can encourage walking as well as cycling. An "urban forest" can be created by planting native trees along sidewalks and in median strips of roads, thereby reducing temperatures during summer and air pollution as well. While costly to install, these choices can actually save money over the long term, and, by enhancing property values, can broaden the tax base.

Some states, like Maryland, now channel support for development towards existing communities. If you read former-Gov. Glendening's remarks at the top of this chapter, you saw how expensive infrastructure can be. And current infrastructure costs are substantially higher. It makes little sense to abandon sewer lines in one place only to build new ones elsewhere, for example.

The federal government has begun to support brownfield development in cities. Brownfields are inner-city areas that may have been polluted by a prior use, such as a power plant, dry cleaner, or metalworking shop. By limiting liability for new owners, governments provide incentives for cost-effective development where infrastructure (sewer lines, roads, mass transit, power lines, etc.) is already in place.

It is a basic tenet of sustainable development that open-space and critical environmental areas and services must be preserved. Open space can protect air quality, recharge groundwater, and protect surface streams. Open space can also reduce flood hazard and mitigate other geohazards. It can provide esthetic value to communities, which always results in higher property values and thus a stable tax base. Water supplies can be protected through municipally-owned watersheds. This practice has been successfully implemented by New York City and Corvallis, Oregon, among other municipalities, saving money in the process.

By encouraging a *range of housing choices*, communities attract a wide range of residents. Older residents may prefer smaller more space-efficient houses. Such residents can be desirable, since they usually pay taxes and support community activities but rarely have children in public schools. Younger residents may spend more income on clothing and dining, thereby supporting local businesses.

Predictable, fair, quick and cost-effective government decision-making supports developers and reduces building cost. Litigation cost is reduced or eliminated.

Encouraging "stakeholder" collaboration ensures that all interested parties have a say in development decisions. When people feel left out of important decisions, they will be less likely to become engaged when hard choices have to be made. Local residents can best decide community values and priorities. Citizen participation can, however, be time-consuming and frustrating. On the other hand, encouraging community and stakeholder collaboration can lead to more satisfactory resolution of development issues, and greater community understanding of the importance of good planning and investment. This can vastly improve public support for smart growth.

A balance between jobs and residences is necessary if communities wish to avoid unnecessary infrastructure spending. Communities that have few jobs require residents to travel far for employment, and encourage spending away from local merchants by commuters. They also require heavy use of automobiles or high mass-transit expense. Communities that have fewer residences tend to be abandoned after the workday and on weekends, and lack support for local merchants, and a "sense of place."

CHAPTER SUMMARY

1. The phrase "land use" refers to alterations of landscapes by and for human activities.

2. Natural ecosystems develop over centuries to millennia. Human land use changes differ in that the changes occur at time spans orders of magnitude less.

3. Rapid land-use changes usually impose significant environmental costs.

4. Development is used here to mean land conversion, usually converting forest, wetland, or agricultural to commercial, industrial, or residential use. Development without careful management leads to an increase in the amount of pollutants in an area.

5. Urbanization, a land-use change, modifies the hydrologic cycle. It leads to more runoff and less infiltration. Two types of Best Management Practices (BMPs) are used to try to control stormwater runoff: urban runoff, wet ponds, and constructed wetlands.

6. Many ecological economists estimate the value of services wetlands provide to humans to be in the tens of billions of dollars annually.

7. Contiguous wetlands are those located immediately adjacent to a stream, river, a lake, or the ocean, and whose waters are interconnected.

8. Isolated wetlands are those that are completely surrounded by upland with no surface water outlet. Development of isolated wetlands can have far-reaching effects on wildlife, flood control, and groundwater quality, and can even violate international agreements on protecting migratory wildfowl.

9. According to the U.S. Department of Agriculture, the no-net-loss policy towards wetlands means that wetlands should be conserved wherever possible, and that wetlands converted to other uses must be offset by restoration and/or creation of wetlands of equal or greater area.

10. Prior to the widespread adoption of the personal motor vehicle in the 1920s, growth of cities was mainly along rail and trolley lines, resulting in high-density development.

11. Between 1945 and 1954 birth rates soared, and nine million people moved to the suburbs in one of the great migrations of the twentieth century. This suburban housing "boom" accelerated the decline of the central city.

12. Much development in the United States today is called sprawl, which refers to land conversion that occurs at a rate exceeding the rate of population growth.

13. Development is usually governed in the United States by local and state documents, such as the comprehensive plan, and related ordinances such as zoning.

14. Zoning regulations frequently require low-density development, and a separation of land uses: residential must be separate from retail, commercial from residential, etc.

15. Many if not most states now encourage some sort of alternative to sprawl development, sometimes called "smart growth." Some states (e.g., Maryland) and localities (e.g., Boulder, Colorado) concentrate support for development in existing communities.

16. To help restore inner cities states and the federal government have taken steps to encourage brownfield developments, which are, redevelopment of existing but abandoned sites.

KEY TERMS

brownfields

constructed wetlands

contiguous wetlands

development

eutrophication

infrastructure

isolated wetlands

liquefaction

mitigation banks

no-net-loss (of
    wetlands)

seismic hazard zone

"smart growth"

sprawl

stormwater runoff

Superfund Act

urban growth
    boundaries

wet ponds

zoning

REVIEW QUESTIONS

1. The cost of city services are usually unappreciated by residents. What was the cost of each mile of new sewer line, according to the former governor of Maryland?
   a. $ 2 million
   b. $200,000
   c. $2,000
   d. $200 million

2. Which is/are effects of urbanization?
   a. Urbanization can increase impermeable area, increasing runoff.
   b. It can reduce water supplies, by paving over recharge areas.
   c. It can increase erosion, by increasing runoff into streams.
   d. All of the above are correct.
   e. None of the above is correct.

3. Tree loss in urban areas can
   a. reduce surface temperature.
   b. increase property values.
   c. both of the above.
   d. neither of the above.

4. Which is/are true about wet ponds?
   a. They are built to store water in relatively small (10–25 year) flood events.
   b. They can eutrophicate over time.
   c. They can fill with sediment over time.
   d. All of the above are correct.
   e. None of the above is correct.

5. Two types of wetlands are
   a. contiguous and isolated wetlands.
   b. marine and terrestrial wetlands.
   c. isolated and terrestrial wetlands.
   d. open and closed wetlands.

6. Suburbs began to grow with the extension of electric trolley lines from cities in the
   a. 1950s.
   b. 1850s.
   c. 1940s.
   d. 1880s.
   e. 1930s.

7. Prior to the widespread adoption of the personal motor vehicle in the 1920s, growth of cities was mainly
   a. prohibited by local ordinance.
   b. along river banks.
   c. along rail and trolley lines.
   d. restricted to skyscrapers (vertical development).

8. Post World War II suburban development differed from earlier suburbs in that
   a. they had low density and depended on automobiles.
   b. they were built along rail lines and not roads.
   c. they were built on brownfields and not greenfields.
   d. they encouraged healthy growth of inner cities.

9. Sprawl generally is defined as
   a. development in which developed land area grows at a faster rate than population.
   b. development that must be accompanied by expanding mass transit by local ordinance.
   c. development in which population grows faster than developed land area.
   d. development that must maintain original tree cover under local ordinance.

10. Zoning specifies
    a. who can live on a piece of property.
    b. that schools must be within walking distance of children.
    c. how a piece of property can be used.
    d. All of the above are correct.

11. Zoning in American cities began in the
    a. 1920s.
    b. 1780s.
    c. 1950s.
    d. 1860s.

12. Urban growth boundaries, which limit development to certain areas around cities, are a feature of
    a. all modern American cities.
    b. smart growth.
    c. the National Environmental Policy Act of 1969.
    d. All of the above are correct.

13. A principle of smart growth is that communities create neighborhoods
    a. that are accessible to the automobile by controlled access roads.
    b. that have no net energy requirements.
    c. that are pedestrian-friendly.
    d. that emit no greenhouse gases.

14. "Infrastructure" refers to
    a. structures that are used to extract groundwater from under cities.
    b. features such as roads, schools, and sewer lines on which communities depend for basic services.
    c. the specific type of construction required in a zoning district.
    d. the region immediately inside an urban growth boundary.

15. A principle of smart growth related to housing is
    a. it discourages a range of housing choices, thereby reducing infrastructure cost.
    b. it requires low housing density, so children have room to play.
    c. it encourages a wide range of housing choices to attract a range of residents.
    d. it requires that parents of school-age children must pay tuition so that public schools are adequately supported.

16. Smart growth encourages
    a. residents to drive outside the area to work, to maximize pedestrian safety.
    b. communities to have as few residents as possible relative to jobs in the community, to save on school expenses.
    c. that jobs in the community be roughly in balance with residents.
    d. that buses be available to transport poor residents to work outside the community.

FOOTNOTE

[1] According to the study, permeable land comprised 43,021,138 square feet of Santa Monica's total 234,026,886 square feet, so about 18% is permeable.

Waste Production, Reduction, and Disposal

CHAPTER OUTLINE

Human society differs from natural ecosystems in that humans process resources in a *linear* fashion, like this: <u>find</u> → <u>produce</u> → <u>use</u> → <u>discard</u>. In nature, resource use is generally *cyclic*, in that one creature's waste is another's food: there is no "waste" (**FIGURE 22-1**). A good example of this is the relationship between plants and animals: plant waste (oxygen) is essential for animals, and animal wastes (nitrogen and phosphorus compounds, and CO_2) are essential for plants. Because humans do not process resources cyclically, the larger and more complex societies become, the

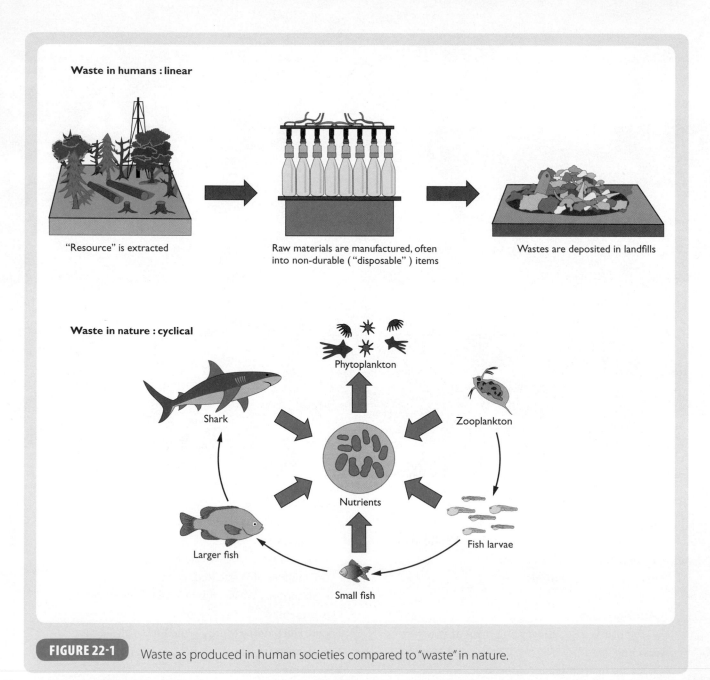

FIGURE 22-1 Waste as produced in human societies compared to "waste" in nature.

more waste is generated, and the more challenging waste handling that does not degrade the environment, becomes. Furthermore, humans produce chemicals that are not found in nature (**dioxins**, furans, PCBs, etc.) for which there is no "natural" means of disposal.

Waste may be collected and reused, recycled, buried in landfills, incinerated, exported, or dumped illegally. Laws, incentives, and higher costs have increased the amount of some waste recycling since 1980, and many states and federal government agencies have set ambitious goals (up to 50%) for further waste reduction. This does not mean the problem has "gone away," however, because even waste that is recycled must be collected, transported, and processed.

■ CATEGORIES OF WASTE

Waste generated by human activity falls into several broad, and sometimes overlapping, categories (**FIGURE 22-2**). **Municipal solid waste (MSW)** is waste generated by settlements, and today most Americans live in cities. *Industrial waste* is a broad category that includes by-products of industrial activity. Some of this is relatively benign and can be added to MSW. Other industrial waste is *hazardous* to a greater or lesser degree; examples include solvents, metal solutions, and artificial organic chemicals. This type of waste must or should be handled separately (see below). The increasing regulation and cost of hazardous waste disposal has prompted many companies to sharply reduce their waste output, an example of a rational response to an economic cost. However, the high cost of hazardous waste management can also prompt illegal dumping, or export of hazardous waste to poor countries, or to countries with corrupt governments.

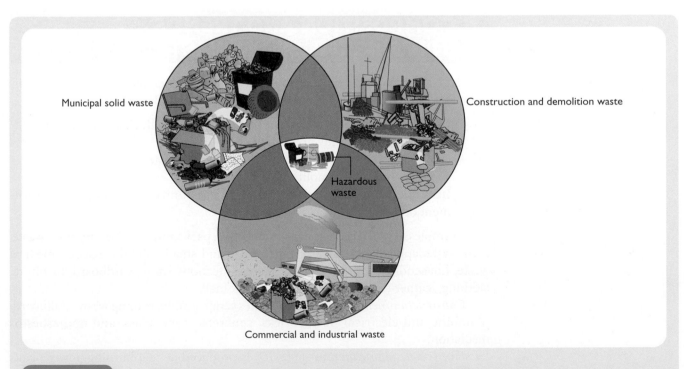

Municipal solid waste

Construction and demolition waste

Hazardous waste

Commercial and industrial waste

FIGURE 22-2 Categories of waste: municipal solid waste, construction and demolition waste, commercial and industrial waste, hazardous waste. Data from: EPA.

Human waste (sewage) is usually processed in developed countries at *sewage treatment plants*. The subject of water pollution is discussed elsewhere in the text. Some sewage plants process only residential, commercial, and industrial non-hazardous waste; others combine stormwater runoff with the above. In the latter case, heavy precipitation may overload the plant's capacity, forcing operators to shunt a mixture of untreated waste and runoff into the nearest water body. This is a serious source of water pollution. Accordingly, many communities are rebuilding their sewage systems to increase capacity, increase level of treatment, or separate storm water runoff from sewage.

The main components of human waste are nutrient- (i.e., nitrogen and phosphorous) enriched water, and solid waste, which if composted and treated, becomes **biosolids**. Biosolids, if free of hazardous contaminants, can be used as a soil conditioner. Several million tonnes per year of biosolids are in fact spread on land. Vegetative waste such as grass cuttings and tree debris can likewise be *composted* and used as mulch or fertilizer. We will consider composting later. Very concerning is that new research has shown that human waste may contain trace amounts of prescription drugs and caffeine, with unknown consequences.

Radioactive waste produced by weapons manufacture or at nuclear power plants, research institutions, and hospitals falls into several categories, from least hazardous to most hazardous. You will find radioactive waste discussed in the renewable and alternative energy sections elsewhere in the text.

Wastes from *coal mining*, a special category, are discussed in the fossil fuels section; *concentrated animal feedlot operations (CAFOs)* are covered in the soils and agriculture section; and the major category of *air and water pollution* are found elsewhere in the text.

In this chapter we focus on MSW and hazardous waste.

■ COMPOSITION AND TRENDS IN MUNICIPAL SOLID WASTE

Municipal solid waste is the term applied to refuse generated by businesses, individuals, and organizations. Major sources of MSW include:

- Non-hazardous industrial waste (e.g., flue gas desulfurization [FGD])
- Single and multifamily housing waste
- Commercial (retail and wholesale businesses) waste
- Construction and demolition waste
- Institutional waste (prisons, schools, colleges/universities, hospitals, government offices)

Examples of MSW include consumer goods, packaging, food waste, office waste, yard waste, appliances, auto parts, pet waste and small animal carcasses, medical waste, household hazardous waste, treated infectious waste, cardboard, furniture, clothing, leather (shoes, handbags), and junk mail.

Construction/demolition waste includes scrap lumber, roofing waste, wallboard, plumbing and electrical waste, bricks, concrete, stone, glass, and non-asbestos insulation.

Flue gas desulfurization waste (FGD) is produced at coal-burning power plants when limestone, water, and SOx gases combine to form calcium sulfate. FGD can be reused if made into gypsum wallboard. If disposed, FGD is usually handled in separate facilities due to its content of trace metals, but it may sometimes be handled as MSW.

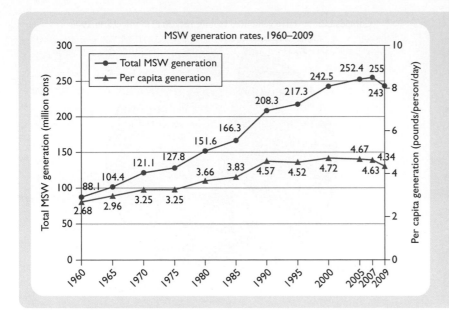

FIGURE 22-3 Municipal solid waste (MSW) in the United States. *Activity:* Evaluate how both annual and per capita values have changed over the period covered by the graph. How was the "per capita" figure determined? Do you think you "throw away" 4.5 pounds of waste a day? Keep a record of what you throw away each day in a journal and, if possible, weigh it. Report on your findings to your class. If not, how do you account for these per capita statistics? Modified from Municipal Solid Waste Generation, Recycling, and Disposal in the United States: Facts and Figures for 2009/EPA.

Municipal Solid Waste Data

Here are some facts about MSW:

- **FIGURE 22-3** shows annual U.S. MSW and per capita MSW to 2010.
- **FIGURE 22-4** shows the composition of the MSW before and after all recycled material has been removed.
- **FIGURE 22-5** shows the composition of construction and demolition waste, an important component of MSW.

One important factor governing the means of dealing with MSW is the disposal fee. Disposal fees have increased since 1980, when comprehensive new legislation went into effect governing landfill construction and operation. Waste disposal may constitute a significant part of a municipality's budget. For example, in 2005 the town of Shaftsbury, Massachusetts spent 10% of its annual budget on MSW. Non-hazardous waste disposal charges in landfills ranged from the mid-$20s to near $100/ton in 2012. Hazardous waste disposal charges averaged about four times as much.

■ WASTE TREATMENT OPTIONS

Landfills

Words convey more than just meaning. As linguists point out, they can also convey emotions. Most waste disposal sites are referred to by the neutral term **landfills** or "sanitary landfills," but they are more precisely *waste disposal sites*. Which term seems to you to be more neutral?

There are many other examples of neutral, more ambiguous names being given to practices that some find objectionable. For example, we will discuss incineration, an alternative waste disposal practice, later in the chapter. Because many found "incinerators" objectionable, industry spokespersons coined the term "waste-to-energy-facility" because incinerators could be designed to produce steam and electricity when waste was burned.

FIGURE 22-4 Composition of MSW before and after recycling. *Activity:* Critically evaluate the components. How would you reduce each of them if you were given the responsibility? Do you think price of any of the material has any impact on how much of it is "thrown away"? How would you reduce the 13.5 million tons of food "thrown away" each year? Most food is transported and delivered by diesel truck. Diesel exhaust contains at least 40 compounds that are carcinogens, mutagens, or both. Is this cause for concern? Would reducing food waste reduce air pollution? Would it reduce adverse effects on human health? Data from: Municipal Solid Waste Generation, Recycling, and Disposal in the United States: Facts and Figures for 2006/EPA.

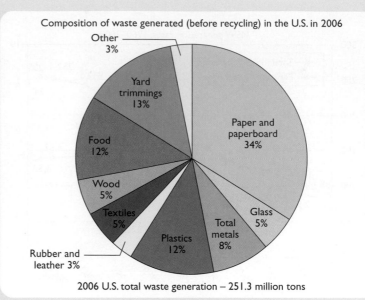

Composition of waste generated (before recycling) in the U.S. in 2006

Other 3%
Yard trimmings 13%
Food 12%
Wood 5%
Textiles 5%
Rubber and leather 3%
Plastics 12%
Total metals 8%
Glass 5%
Paper and paperboard 34%

2006 U.S. total waste generation – 251.3 million tons

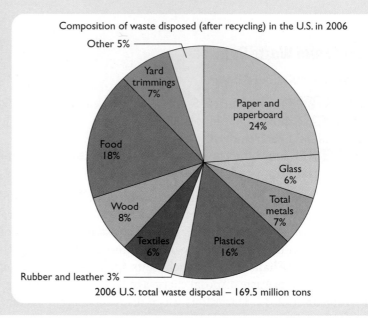

Composition of waste disposed (after recycling) in the U.S. in 2006

Other 5%
Yard trimmings 7%
Food 18%
Wood 8%
Textiles 6%
Rubber and leather 3%
Plastics 16%
Total metals 7%
Glass 6%
Paper and paperboard 24%

2006 U.S. total waste disposal – 169.5 million tons

FIGURE 22-5 Composition of construction and demolition waste. *Activity:* Do you think the price of these commodities have any effect on their discard rate? Explain your reasoning. Would taxing raw materials or providing incentives for reuse lower this amount of waste in your view? Defend your answer. Why or why not would this be a proper role of government? Data from: King County Waste Monitoring Program, February 2009.

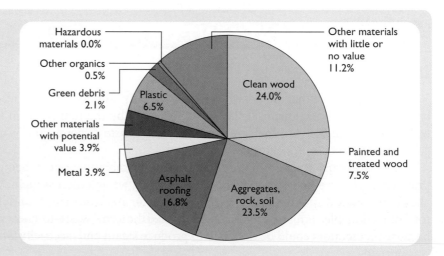

Hazardous materials 0.0%
Other organics 0.5%
Green debris 2.1%
Other materials with potential value 3.9%
Metal 3.9%
Plastic 6.5%
Asphalt roofing 16.8%
Aggregates, rock, soil 23.5%
Painted and treated wood 7.5%
Clean wood 24.0%
Other materials with little or no value 11.2%

Designing and Operating a Modern Landfill

Leachate is a fluid that contains material dissolved by the usually acid waters in the waste. It is the liquid produced when precipitation falls on a landfill and sinks into the wastes. Most landfills contain little oxygen; they are *hypoxic* or even *anoxic*. Under these conditions, potentially toxic heavy metals such as lead, nickel,

Focus: Hernando County's Facility

The Hernando County (Tampa Bay area, Florida) Waste Management Division opened a 16-acre, double-lined, disposal cell expansion at the Northwest Waste Management Facility (NWMF). Officials expected the expansion to provide eight years of solid waste disposal capacity. The NWMF sits amid 365 acres of land that is projected to provide Hernando County disposal capacity for at least 50 years.

Value-added Question 22-1. At what annual rate is land being converted to waste disposal at the 16-acre expansion?

Value-added Question 22-2. For the overall facility, what is the expected rate of land conversion?

Value-added Question 22-3. Hernando County's population grew from 165,973 to 172,778 people from 2009 to 2010. Use your answer from the question above to calculate how much land is devoted to waste disposal for each of these 6800 people.

Value-added Question 22-4. Calculate the population growth rate of Hernando County from 2009 to 2010, then use the Doubling Time formula (T = 70/R; population math made easy is found elsewhere in the text.).

Concept Check 22-1. Discuss whether the existing 365-acre landfill will be able to accommodate the garbage when the population doubles. How about when it doubles again?

Attributes of a Landfill

A "sanitary landfill" waste disposal site must be able to isolate potentially harmful liquid and solid waste for decades if not longer (**Figure B 1-1**).

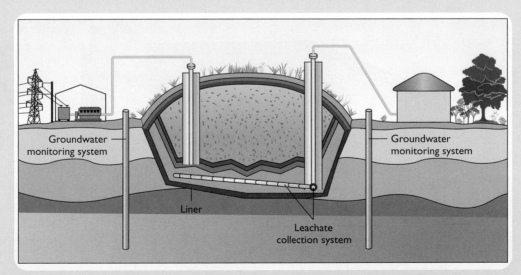

(A)

FIGURE B 1-1 (A) Attributes of a "sanitary landfill" waste disposal site. Note the liners, the leachate collection system, and the groundwater monitoring system. Many landfills vent produced methane or collect methane for use as a fuel. Given the value of natural gas (mostly methane), why don't you think this practice is universal?

(Continued)

(B)   (C)

FIGURE B 1-1 *Continued*. (B) An aerial view of a landfill. These sites can bury up to 27,000 tons of MSW each year. © Miks/ShutterStock, Inc. (C) Contrast A and B with the unlined Agbogbloshie landfill in Accra, Ghana. © Pluff Mud Photography.

To isolate the waste, **landfill cells** are underlain by one or more *liners*, made of plastic and/or clay (**Figure B 1-2**). The liner's purpose is to prevent seepage of leachate into groundwater or surface streams. *Leachate* is a name for potentially contaminated fluids in the landfill. A compacted clay layer that may be up to 3-feet (900 mm) thick is laid over the groundwater collection system. The clay is compacted to an extremely low **permeability**. The liner system is usually built with a

slight slope so that liquid will run to the lowest point in the landfill for collection and eventual treatment. A plastic liner is then laid over the clay.

To prevent damage to the clay liner, a drainage system is often placed under the liner to collect and divert groundwater flow. The direction of groundwater flow should be determined using monitoring wells constructed both upgradient and downgradient from the landfill site (Figure B 1-1A).

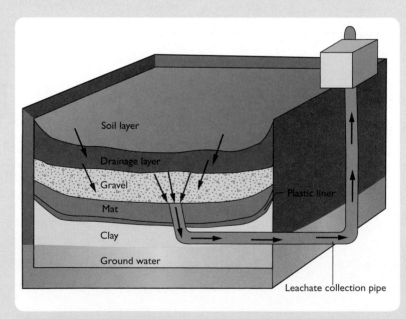

FIGURE B 1-2 A landfill cell built in 2004 for Clay County, Minnesota. Describe its attributes. The clay barrier at the base is 2 feet thick. Why do you think clay is used at the base of the cell?

chromium and mercury, to name but a few, can be dissolved (or leached, hence the name) from the waste. Some sources of such heavy metals in MSW are lead–acid batteries and flashlight batteries, which can contain mercury, and stainless steel, which can be made with nickel or chromium.

In addition to heavy metals, leachate can contain a variety of artificial chemicals, including pesticides. Although some toxic wastes may be dumped illegally in municipal landfills, the source of most toxic materials is household and commercial products like paint solvents, oils, cleaning compounds, degreasing compounds, and household pesticides. Chemical-breakdown products of plastics resulting from reactions in the waste can also be harmful.

Leachate comes mainly from precipitation seeping into the landfill. Because landfills are designed with an impermeable base, water that falls on-site can build up over time. The leachate thus must be extracted and treated or it could potentially saturate the waste and leak out.

Such a fluid if added to groundwater in any quantity could degrade it. Many landfills are situated in rural areas for obvious reasons, so groundwater contamination from leachate is a potentially serious problem for residents who depend on well water. More information on groundwater and landfills can be found elsewhere in the text.

Leachate loss from waste sites is a chronic problem (**FIGURE 22-6**). To isolate leachate from groundwater, landfills are designed to transmit surface precipitation away from the site insofar as possible. Moreover, modern landfills must have a set of monitoring wells (see above) that constantly check for groundwater contamination around the site. Finally, most landfills are equipped with leachate treatment systems that transmit leachate to collection sites, where it can be treated or shipped to a treatment facility.

One example of leachate treatment is the Conestoga landfill, located 45 miles (75 km) west of Philadelphia, Pennsylvania. Its total area is 454 acres (184 ha) and it accepts 5,000 to 10,000 tons daily. Leachate volume varies from 33,000 to 50,000 gallons per day (125,000–189,000 l/d). The facility operates an on-site leachate treatment system as well as a gas collection and treatment system (see below). Cost of leachate treatment ranged from 8 to 13 cents per gallon in 2013, according to company sources.

Concept Check 22-2. To save money and reduce taxes, should waste dump operators be free to decide whether or not to treat leachate?

Value-added Question 22-5.
For the Conestoga Landfill, using the daily waste tonnage disposed on-site, calculate the cost to treat leachate per day using the 8 cents/gal number cited above. How much, per ton of solid waste, does leachate treatment add to disposal costs?

Water quality of treated effluent was high enough to be reused at the site for irrigation.

TABLE 22-1 shows some contaminants in the leachate and their after-treatment concentrations in the effluent.

Methane

Because most landfills are hypoxic, decomposition of organic matter produces **methane**, the major component of natural gas, which is obviously flammable. Thus, an explosion could result from methane accumulation around the site. Methane could also seep into soils around the site, where it could be dissolved in groundwater. It could likewise seep into structures like basements near the site.

Hydrogen sulfide (H_2S) also can accumulate in landfills. Control and collection of landfill gas is thus an important environmental protection measure. The production of landfill gas begins soon after refuse is deposited and may

FIGURE 22-6 Leachate seeps from a landfill, Norman, Oklahoma. Courtesy of U.S. Geological Survey/Scott Christenson.

TABLE 22-1	Contaminants in Conestoga Landfill, Concentrations Before and After Treatment	
Contaminant	**Concentration (Leachate: mg/l)**	**(Effluent/max**)**
BOD#	8,700	50
Fe	183	3.5
Mn	69	0.8
Pb	0.2	0.022
Phenolics	8.5	0.05
Cyanide	0.1	0.02
Acetone	7	0.109*
1,1,1 TCE	0.887	0.1
Toluene	1.016	0.05

**Guaranteed maximum values.
*Actual measured value.
Biological Oxygen Demand, a measure of water hypoxia.
Data from: Pennsylvania Department of Environmental Protection (DEP).
Look in the glossary for an explanation of **biochemical oxygen demand (BOD)**, **phenolics**, **trichloro-ethylene (TCE)**, and **toluene**. Search elsewhere in this book for a description of the impact of lead (Pb).

continue for decades. A landfill gas collection system is usually installed to allow the recovery, control, and, if permitted, **flaring** of landfill gas from the site. To avoid methane buildup, most landfill operators drill perforated pipes into the cells and "vent" the methane to the atmosphere. Venting methane, while safer than allowing it to collect on-site or in groundwater, allows a powerful greenhouse gas to escape, where it can contribute to global climate change. Flaring methane produces CO_2, a more benign, but still climate-altering, gas. Thus, the more waste we generate, the more climate effects potentially result.

Some large landfills collect the methane, treat it to remove contaminant gases like sulfur dioxide (SO_2), hydrogen sulfide (H_2S), and CO_2, and use it to generate electricity or heat at the site.

One such is Oregon's Coffin Butte landfill, north of Corvallis. The landfill collects methane and burns it on-site to produce 5.6 MW of electricity, which is fed into the local electricity grid (FIGURE 22-7).

Some landfills even collect enough methane to sell it to natural gas suppliers.

Adverse Impacts

Landfills attract vermin, but because most are located far from habitation, this problem is minimal. But landfills also attract trucks, the major means by which waste is transported to site (FIGURE 22-8). Excess truck traffic contributes to road congestion and air pollution, since most trash haulers are diesel-powered. Some landfills are served by barge and rail, however, minimizing road impact.

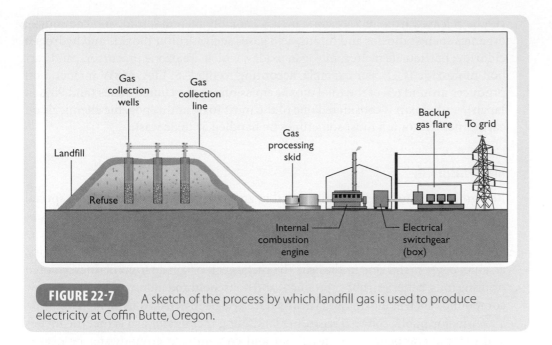

FIGURE 22-7 A sketch of the process by which landfill gas is used to produce electricity at Coffin Butte, Oregon.

Finally, landfills occupy significant space and therefore are a land use issue in many localities. Moreover, as populations grow and become wealthier, additional land for waste disposal will be essential unless per capita waste is decreased.

Advantages

Large, regional landfills are a significant source of revenue for many localities that have actively courted them as a component of economic development. According to some local officials, landfills provide jobs for residents and pay property taxes and fees, helping to keep taxes low in many rural localities.

Since 1980, about 70% of landfills have closed, meaning that the waste goes to fewer sites and thus is more easily monitored. According to some experts, it is more efficient to have waste deposited at fewer sites, thus concentrating adverse land use effects.

Incineration

Some communities in the United States, and many abroad, use **waste incineration** as a disposal strategy. Incineration reduces the ultimate volume of waste by up to 90%, and up to 35% by weight. Moreover, burning waste can produce appreciable quantities of steam and/or electricity. Concerns about incineration generally center on: the potential and real air pollution emitted; the hazardous materials in ash residues that could contaminate surface- and groundwater; impact on property values of nearby residents; and increased vehicular traffic.

According to the United Nations Environment Programme (UNEP), the following air emissions

FIGURE 22-8 A long-distance diesel-powered trash truck. The opening of regional landfills and closure of local dumpsites has increased the number of trash trucks on the nation's highways. © West Coast Surfer/age fotostock.

can be released from incinerators: metals (notably mercury, lead, and cadmium); organics such as dioxins and furans; acid gases such as sulfur dioxide and hydrogen chloride; particulate matter; nitrogen oxides, which are ozone precursors; and carbon monoxide (CO). For example, according to the U.S. EPA, MSW incineration produces around 6% of the total dioxin emissions in the United States (and 90% in Japan), and dioxin is considered one of the most toxic anthropogenic chemicals on earth. Incinerator ash must sometimes be handled as toxic waste.

Incinerator Ash

Benign incinerator ash is usually buried in an MSW landfill, preferably in a special cell, or in an ash-only landfill. Ash landfills should be carefully designed so as to prevent heavy metals from migrating from the ash into the environment.

Burning trash can concentrate heavy metals, such as lead, cadmium, mercury, arsenic, copper, and zinc in the ash. The metals can originate from plastics, colored printing inks, batteries, rubber products, and household and industrial waste. Organic compounds such as dioxins and **furans** have also been detected in incinerator ash.

A major public concern regarding landfilling incinerator ash is that the metals and organic compounds can leach out and contaminate groundwater or nearby surface water. Incinerator ash could also conceivably affect human health through direct inhalation or ingestion of airborne ash.

The risks from waste incineration have been debated by researchers, industry experts, and citizens. Critics argue that incinerator ash occasionally fails hazardous waste tests for toxicity, but industry experts counter that these tests overstress the actual environmental risks posed by the ash. In addition, field tests performed on leachate from ash landfills indicate that the metal content is below U.S. hazardous waste classification levels, and indeed usually satisfies U.S. drinking water standards.

When operated optimally, the best air pollution control equipment can remove *up to* 99+% of **dioxins**, furans, heavy metals, particulate matter, and hydrogen chloride; more than 90% of sulfur dioxide; and up to 65% of nitrogen oxides from incinerator exhaust gases. Whether these pollutants *will* be removed at these levels depends upon a number of factors, including the extent to which regulations are enforced. We discuss them briefly below.

Optimal removal of air emissions requires, not only state-of-the-art equipment, but also that MSW incinerators be properly operated and well maintained.

Optimal *combustion practices* are likewise required to control emissions by ensuring that the temperature in the combustion chamber, and the time the MSW remains in the combustion chamber, are maintained at optimal levels. Major variations in these variables can lead to a significant increase in air contaminants.

Such demanding technical considerations pose serious obstacles to incineration in most developing countries. The necessity of maintaining emissions control systems in essentially perfect order over a long period of time is challenging even in developed countries. Small mistakes in the operation of such facilities can lead to significant emissions of toxic substances, imposing health risks on local populations and ecosystems unless properly regulated.

Given the demanding conditions required to incinerate wastes with minimal air emissions, export of wastes from developed countries for incineration in developing countries could lead to a significant increase in regional air emissions, and could represent a significant avoided cost of global trade. The Basel Convention (see below) strictly regulates such shipments.

Export: Transboundary Shipment of Waste

Export of waste from one jurisdiction to another is called **transboundary shipment**. Two categories of wastes are shipped across borders: MSW and hazardous wastes. In the United States, the Supreme Court has repeatedly ruled that MSW shipments, even if valueless, constitute "commerce." The U.S. Constitution allots authority for interstate and international commerce solely to the U.S. Congress. Thus, states cannot bar or regulate the import of transboundary waste if they treat that waste any differently from identical intrastate waste. Export of waste is the preferred waste disposal method for some large cities like New York, which spends upwards of $1 billion each year for waste disposal. Export of waste, however, adds thousands of trash trucks to the nation's Interstate Highway system, increases use of imported petroleum, and adds needless pollutants to the air and water.

■ RECYCLING

According to the EPA,

- As of 2000 the United States was recycling 28% of its waste, 64 million tons, a rate that had almost doubled over the previous 15 years. By 2006, the recycling rate had increased to 32.5%, diverting over 80 million tons from waste dumps, saving the carbon emission equivalent of taking 39.4 million cars off the road, and the energy equivalent of 6.8 million households' annual energy consumption, or 222.1 million barrels of oil. Data for 2010 are shown in **TABLE 22-2**

TABLE 22-2	U.S. EPA Recycling Data for 2010

- 85 million tons composted or recycled for recovery rate of 34.1%
- 29 million tons of MSW combusted for energy recovery
- 63% of paper recovered
- 58% of yard trimmings recovered
- 35% of metals recycled
- 8 million tons of metals recycled reduced greenhouse gas emissions at equivalent of removing > 5 million cars from road for one year
- Approximately 9,000 community curbside recycling programs exist in United States, increase over reported 2002 figure of 8,875
- Approximately 3,000 community composting programs exist, decrease from reported 2002 figure of 3,227 programs
- Current amount of MSW per-person discarded in landfills lower than 1960
- Due to population growth, current total amount discarded MSW in landfills significantly higher than in 1960, yet lower than 1990

Modified from: Municipal Solid Waste Generation, Recycling, and Disposal in the United States: Facts and Figures for 2009/EPA.

- In 2000, 42% of paper, 40% of all plastic soft drink bottles, around 55% of aluminum beer and soft drink cans, 57% of steel packaging, and 52% of major appliances were recycled. By 2005, paper recycling rate reached 50%. In 2007, Americans used over 1.5 million metric tons of aluminum beverage cans but only recycled about 800,000 tons, representing a fall in recycling rate from the high of 68% reached in 1992. Aluminum manufacturers point out that it takes only 5% of the electricity to produce a new can from scrap compared to using "virgin" aluminum ore. Thus, as electricity rates increase, especially in the Pacific Northwest where many aluminum smelters are located, refiners and governments are seeking ways to increase process efficiencies and reduce use of increasingly scarce electricity from Northwest electricity producers like the Bonneville Power Authority, which gets most of its electricity from the region's carbon-free hydroelectric dams. Accordingly, the Aluminum Company of America (ALCOA) announced a goal in 2007 to increase aluminum can recycling to 75% by 2015. A company spokesperson observed that by recycling 75% of aluminum cans instead of 52% (the rate in 2007), the nation could reduce carbon dioxide emissions by nearly 12 million metric tons a year.
- "Curbside recycling" is an attempt to aid municipal recycling by picking up household trash at regular intervals. Before 1980 only one curbside recycling program existed in the United States. By 1998, 9,000 curbside programs and 12,000 recyclable drop-off centers were operating across the United States. As of 1999, 480 materials recovery facilities were processing the collected materials. These numbers fluctuate with the health of municipal budgets. To save money, cities may curtail or cease collection of recyclable material. Consider that in 2010 the number of curbside recycling programs had grown only to 9,066, serving about 71% of the residents of the U.S. By 2010, there were 633 material recovery facilities in the U.S. As of 2008, among those cities that carried out curbside recycling, municipal recycling rates varied from Houston's embarrassing 3%, to 6% in Kansas City, New York City's 17%, to 55% in Chicago, and 70% in San Francisco.

The following categories of MSW are commonly but not universally collected for recycling: aluminum cans, newsprint, corrugated cardboard, glass, steel cans, paper, and some plastics. In addition, some states have "**bottle bills**," which charge a deposit for certain beverage containers, making them valuable for recycling. As a result, in those states collecting recyclables has ironically become a source of income for many homeless and disadvantaged persons.

FIGURE 22-9A shows recycling rates for the United States. Although recycling in general has increased, recycling rates for some materials have come down over recent years, as Americans changed their dining habits (**FIGURE 22-9B**). As more time is spent in transit and at work, more meals are taken out of the home. As a result, beverage containers that used to be recycled, because they were consumed at home, are now often discarded. The recycling rate of aluminum cans declined slightly between 2000 and 2003, apparently for this reason.

Finally, recycling rates in the ten states with "bottle bills" are usually substantially higher than in states without. Bottle bills require a deposit for certain types of containers. In 2008, Oregon expanded its "bottle bill" to include plastic water bottles by initiating a 5-cents per bottle deposit.

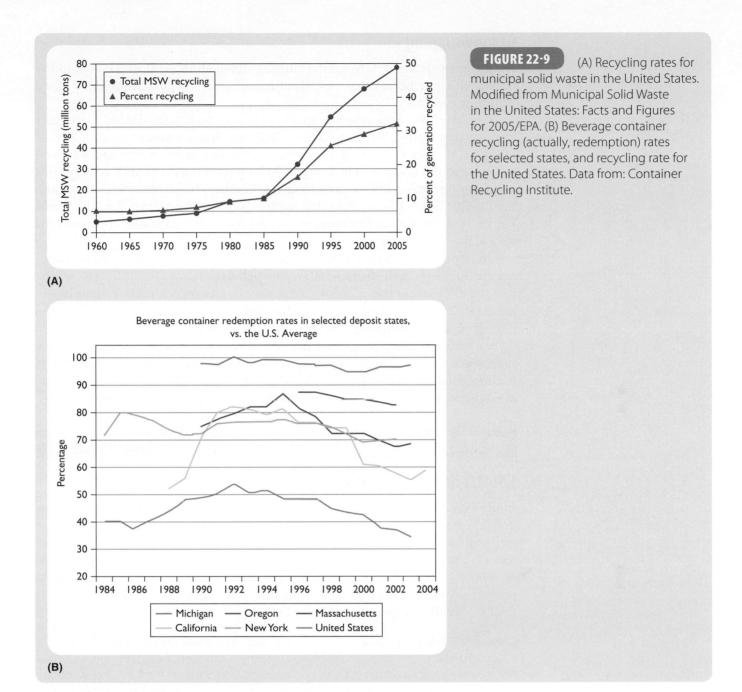

FIGURE 22-9 (A) Recycling rates for municipal solid waste in the United States. Modified from Municipal Solid Waste in the United States: Facts and Figures for 2005/EPA. (B) Beverage container recycling (actually, redemption) rates for selected states, and recycling rate for the United States. Data from: Container Recycling Institute.

Waste Injection into Aquifers

Some wastes may be injected into deep aquifers under high pressure, illustrating the saying "out of sight, out of mind" (see groundwater, elsewhere in the text). During the 1970s, Soviet authorities injected radioactive waste into deep aquifers with virtually no data on the long-term impact of such activity.

In the United States, deep injection has been implicated as the cause of earthquake swarms in the Denver, Colorado area.

Oil field *brines*, saline water produced with oil and gas, may be injected into deep saline aquifers.

Due to public concerns, injection of wastes in the United States has declined over the past few decades.

Ocean Dumping

Until the 1980s, many U.S. coastal cities routinely dumped their MSW into the ocean. This practice created vast "dead zones" where hypoxic conditions killed aerobic organisms.

Coastal cities continue to pipe treated liquid wastes into the oceans. Large cities in many developing countries, such as São Paulo, Brazil, where as much as 90% of sewage is discharged untreated, routinely dump untreated or marginally treated sewage into the ocean with little regard to long term impact of such activity. This issue may be further illustrated by Mumbai (Bombay) India. Mumbai dumps more than 2,200 million liters of untreated sewage daily into the Arabian Sea. The World Bank is underwriting an ambitious project to provide at least *primary* treatment for 80% of the city's sewage.

In Europe, three hundred million gallons (1,140 million liters) of treated sewage are dumped into British coastal waters each day.

The discharge of raw or disinfected sewage by passenger and cargo ships is permitted by international agreement on the high seas (> 12 nautical miles from the nearest land) and as close as 3 nautical miles in some cases.

As human populations grow in developing countries, sewage volumes are likely to increase, with or without massive investment in sewage treatment. Improvements in threatened ecosystems like the Chesapeake Bay are unlikely without new investment in nitrogen-removal technologies at the region's many sewage treatment plants.

■ HAZARDOUS WASTE

What Is Hazardous Waste?

In the 1976 **Resource Conservation and Recovery Act (RCRA)**, the U.S. Congress defined hazardous waste as "waste, or combination of wastes, which, because of its quantity, concentration, or physical, chemical, or infectious characteristics may (1) cause, or significantly contribute to, an increase in mortality, or an increase in serious irreversible or incapacitating reversible illness, or (2) pose a substantial present or potential hazard to human health or the environment when improperly treated, stored, transported or disposed of."

Concept Check 22-3. Pause, read again, and carefully analyze the RCRA definition, which is typical of many definitions that are found in some of our most important legislation. It at first may seem excruciatingly specific, but read deeper and answer the following questions. What does "significantly contribute to" mean in (1)? Who will define what is significant and what is not? What does "substantial" mean in (2)? Who do you think will define the word in this context?

The EPA has been given the responsibility to interpret the definition written into the law by Congress. EPA defined **hazardous waste** as (1) a waste that is *reactive, corrosive, toxic,* or *ignitable,* or (2) a waste on a list that EPA has found to be hazardous.

The agency further defined each term in (1) and set up tests by which such characteristics could be identified. For example, a waste is corrosive if it has a pH equal to or less than 2, or equal to or greater than 12.5. Look at FIGURE 22-10 . You can see that battery acid is defined as a toxic waste because of its pH. The agency went further, however, and specified that a waste was also corrosive if it corroded steel of a specific type at a rate greater than 6.35 mm per year at a temperature of 55°C (131°F).

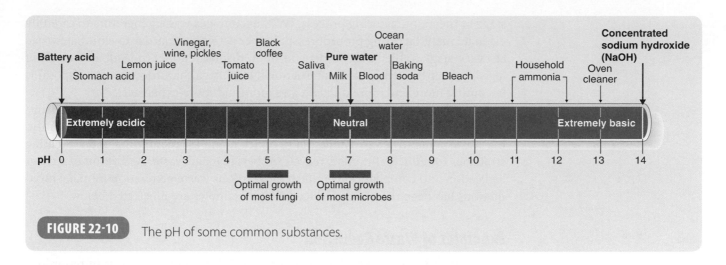

FIGURE 22-10 The pH of some common substances.

Sources

TABLE 22-3 shows examples of industries that are producers of hazardous waste.

How Is Hazardous Waste Managed?

The United States Congress declared in the text of the Hazardous and Solid Waste Amendments (1984) that it is the "national policy of the United States that, whenever feasible, the generation of hazardous waste is to be reduced or eliminated as expeditiously as possible." Accordingly, hazardous waste management strategies are now ranked, in order, from highest to lowest priority as follows:

Highest   Lowest

Waste Reduction/Pollution Prevention → Recycling → Treatment → Disposal

Application of the above strategy is called **integrated waste management (IWM)**, that is, a policy that encourages waste reduction and elimination at every

TABLE 22-3	**Some Producers of Hazardous Waste**
Industry	**Example(s) of Hazardous Waste Produced**
Petrochemicals	metals, acids, organic compounds, caustics
Metals	heavy metals, cyanide, solvents, acids, alkalis
Leather	heavy metals, sulfides
Pharmaceutical	toxic chemical residues, prescription drug byproducts, mercury thermometers
Electroplating	heavy metals
Steel	acids
Military	solvents, explosives, toxics

(Many other industry groups qualify. See Hasan for details.)

stage of the manufacturing process. IWM encourages waste generators to carefully evaluate manufacturing procedures and other activities in order to eliminate waste at every stage, rather than to depend upon disposal alone. Indeed, as a result of legislation at the federal and international level, legal hazardous waste disposal is becoming more and more difficult, and more and more expensive.

Specifically, IWM techniques rely on: (1) identification of steps in the manufacturing process; (2) identification of those steps that generate waste; (3) quantifying waste generated, and evaluation of toxicity; and (4) identifying and evaluating strategies to reduce, eliminate, recycle, or best manage waste. As a result of stricter legislation and compliance by industry, hazardous waste generation in many jurisdictions has been drastically reduced. Some examples are illustrated below.

Examples of Waste Reduction

- From 1993 to 1998, Reliant Energy, Inc. reduced the amount of hazardous waste produced at its Texas power plants by more than 80%. Waste included paint, paint thinners, gasoline, lab chemicals, and industrial cleaning wastes. In 1998, the Natural Resources Defense Council cited the company's plants as among the nation's cleanest.
- Hazardous waste initiatives in New York State reduced hazardous waste generated in 1997 by 43%, compared to 1991.
- Industries in Maine reduced hazardous waste by 41% compared to the 1987–1989 base years.

Hazardous Waste and the Basel Convention

International trade in hazardous waste other than radioactive waste is governed by the 1989 **Basel Convention**, which has been ratified by 179 countries as of 2012. Representatives of the United States, Haiti, and Afghanistan signed the Convention, but their governments have not ratified it.

In the late 1980s, stricter environmental regulations in industrialized countries led to increases in the cost of hazardous waste disposal. As a result, so-called "toxic traders" began shipping hazardous waste to developing countries and to Eastern Europe. Subsequent international outrage led to the adoption of the *Basel Convention*. Under the Convention, "transboundary movements of hazardous wastes or other wastes can take place only upon prior written notification by the State of export, to the competent authorities of the States of import and transit (if appropriate). Each shipment of hazardous waste . . . must be accompanied by a movement document from the point at which a transboundary movement begins, to the point of disposal. Hazardous waste shipments made without such documents are illegal."

During the 1989–1999 decade, the Convention set up a framework for controlling waste shipments.

The next decade's focus was on the central goal of the Basel Convention: **environmentally sound management (ESM)**. The aim of ESM was to protect human health and the environment by minimizing hazardous waste production whenever possible. ESM took an "integrated life-cycle approach," which involves tracking and regulating hazardous waste from generation through storage, transport, treatment, reuse, recycling, recovery, and final disposal.

As you saw above, many countries already had adopted "cleaner production" methods that eliminate or reduce hazardous outputs. In many cases, such methods can be both economically and environmentally efficient.

Waste handling varies tremendously across the United States and around the world. If you are attending a college or university, find out how they treat waste. If not, find out where your MSW is dumped. Do they collect and treat leachate? Do they collect and flare or reuse methane? What is the percentage of waste recycled? How much waste is deposited daily? How much leachate is treated daily? Many landfill operators take pride in their operations and will gladly arrange tours for small groups. Take a tour of your local landfill and ask to see its groundwater monitoring system, methane collection system, and leachate treatment operation. Find out what the disposal costs are. Calculate how much you personally dump each day and calculate how much your personal disposal costs are. Do you think they are excessive? Find out whether your state has a "bottle bill" and if so, how much the deposit is and what is covered. Some bottle bills are interesting as much for what they exclude as for what they include. For example, Oregon includes beer containers but not wine bottles.

Concept Check 22-4. Do you think the United States should have a national "bottle bill?" Why or why not?

CHAPTER SUMMARY

1. Human society differs from natural ecosystems in that humans process resources in a linear fashion.
2. Wastes may be collected and reused, recycled, buried in landfills, incinerated, exported, or dumped illegally.
3. Municipal solid waste (MSW) is waste generated by settlements, and today most Americans live in cities. Industrial waste is a broad category that includes by-products of industrial activity.
4. Human waste (sewage) is generally processed in developed countries at sewage treatment plants. The main components of human waste are nutrient (N, P)-enriched water, and solid waste.
5. Landfills are made up of cells underlain by one or more liners, made of plastic and/or clay. A "sanitary landfill" must be able to isolate potentially harmful liquid and solid waste for decades if not longer.
6. Leachate is a name for potentially contaminated fluids in a landfill.
7. In addition to heavy metals, leachate can contain a variety of artificial chemicals, including pesticides.
8. Because most landfills are hypoxic, decomposition of organic matter produces methane, the major component of natural gas. Some large landfills collect the methane, treat it to remove contaminant gases like SO_2, H_2S, and CO_2, and use it to generate electricity.
9. Landfills occupy significant space and therefore are a land use issue in many localities.

10. Since 1980, about 70% of landfills have closed, meaning that the waste goes to fewer sites and thus is more easily monitored. Large, regional landfills are a significant source of revenue for many localities.

11. Incineration reduces the ultimate volume of waste by up to 90%, and up to 35% by weight. Moreover, burning waste can produce appreciable quantities of steam and/or electricity.

12. Incinerating waste can emit: metals (notably mercury, lead, and cadmium); organics such as dioxins and furans; acid gases such as sulfur dioxide and hydrogen chloride; particulate matter; nitrogen oxides, which are ozone precursors; and carbon monoxide (CO).

13. A major public concern regarding landfilling incinerator ash is that the metals and organic compounds can leach out and contaminate ground water or nearby surface water.

14. Field tests performed on leachate from ash landfills indicate that the metal content is below U.S. hazardous waste classification levels.

15. When operated optimally, the best air pollution control equipment can remove up to 99+% of dioxins, furans, heavy metals, particulate matter, and hydrogen chloride, more than 90% of sulfur dioxide, and up to 65% of nitrogen oxides from incinerator exhaust gases. Whether these pollutants *will* be removed at these levels depends upon a number of factors, including the extent to which regulations are enforced.

16. In the United States, states cannot bar or regulate the import of transboundary waste, if they treat that waste any differently from identical intrastate waste.

17. As of 2009, 82 million tons composted or recycled for recovery rate of 33.8%.

18. The recycling rate of aluminum cans declined slightly since the mid-2000s.

19. Recycling rates in states with bottle bills legislation are usually substantially higher that in states without.

20. In the Resource Conservation and Recovery Act (RCRA), the U.S. Congress defined hazardous waste. EPA has been given the responsibility to interpret the definition written into the law by Congress.

21. Integrated waste management (IWM) is a policy that encourages waste reduction and elimination at every stage of the manufacturing process.

22. International trade in hazardous waste is governed by the Basel Convention, which has been ratified by 179 countries. The United States, Haiti, and Afghanistan have so far declined to ratify the Convention.

KEY TERMS

Basel Convention

biochemical oxygen demand (BOD)

biosolids

bottle bills

dioxin

environmentally sound management (ESM)

flaring

furans

hazardous waste

integrated waste management (IWM)

landfill

landfill cell

leachate

municipal solid waste (MSW)

permeability

phenolics

Resource Conservation and Recovery Act (RCRA)

toluene

transboundary shipment

trichloroethylene (TCE)

waste incineration

REVIEW QUESTIONS

1. Which nutrients comprise a significant component of human waste (sewage)?
    a. C, S
    b. C, Mg
    c. P, N
    d. Cl, Na

2. Flue gas desulfurization (FGD) waste is produced at
    a. sewage treatment plants.
    b. landfills.
    c. coal-fired power plants.
    d. nuclear power plants.

3. Landfill cells are underlain by liners, composed of
    a. plastic or clay.
    b. cement or concrete.
    c. Styrofoam.
    d. aluminum.

4. To prevent groundwater contamination, landfill liners should have very low
    a. organic content.
    b. density.
    c. clay content.
    d. permeability.

5. Liquid that can seep out of landfills, and contaminate surface and ground-water is called
    a. contaminate.
    b. liquidus.
    c. arsenate.
    d. leachate.

6. Because most landfills are hypoxic, decomposition of organic matter produces
    a. oxygen gas.
    b. chlorine gas.
    c. methane gas.
    d. CFCs.

7. Since 1980, about what percent of landfills then operating have closed?
    a. fewer than 5%
    b. more than 90%
    c. about 25%
    d. about 70%
    e. fewer than 1

8. The advantages of waste incineration are
    a. it reduces waste volume and mass, and can produce steam for power.
    b. it emits no air pollution.
    c. it eliminates waste volume and mass, and emits no air pollution.
    d. burning waste emits only oxygen and CO_2.

9. Incinerators of municipal solid waste reportedly emitted what percent of the total dioxin emission in the United Kingdom in 1996?
   a. about 10%
   b. over 90%
   c. none
   d. about 50%

10. Export of waste from one jurisdiction to another in the United States is
    a. governed by local laws and regulations.
    b. under authority of Congress, according to the Supreme Court.
    c. under authority of the states, according to the Supreme Court.
    d. under EPA authority.

11. By 2000, according to the EPA, the U.S. was recycling about what percent of its waste?
    a. over 90%
    b. less than 10%
    c. about a quarter
    d. about half

12. The rate of recycling of waste in the United States between 1985 and 2000
    a. remained constant.
    b. fell by 50%.
    c. nearly doubled.
    d. increased by about 10%.

13. The number of states with "bottle bills" in the United States is about
    a. ten.
    b. half.
    c. nearly all.
    d. all fifty.
    e. two.

14. Hazardous waste is defined in the
    a. Clean Air Act Amendments, 1990.
    b. National Environmental Policy Act, 1969.
    c. Resource Conservation and Recovery Act.
    d. Rivers and Harbors Act.

15. Under the terms of the Hazardous and Solid Waste Amendments of 1984, Congress set priorities for handling of hazardous waste in which of the following orders, from highest priority on the left, to lowest on the right?
    a. disposal → treatment → recycling → waste reduction/prevention
    b. waste reduction/prevention → recycling → disposal → treatment
    c. waste reduction/prevention → recycling → treatment → disposal
    d. recycling → waste reduction/prevention → disposal → treatment.

16. The international treaty by which hazardous waste shipments are regulated is
    a. the Kyoto Protocol.
    b. the Montreal Protocol.
    c. the Basel Convention.
    d. the United Nations Charter.

17. The international treaty mentioned in #16 above has been fully ratified by 179 countries. The three which have not are Haiti, Afghanistan and
   a. Russia.
   b. China.
   c. Iraq.
   d. the United States.

18. Many coastal cities in the United States dumped their waste untreated into the ocean until
   a. 2000.
   b. the 1960s.
   c. the 1970s.
   d. the 1980s.

19. The U.S. Congress defined hazardous waste. Which agency has the responsibility to interpret and enforce the congressional definition?
   a. EPA.
   b. Department of the Interior.
   c. the U.S. Army Corps of Engineers.
   d. the U.S. Bureau of Waste Management.

20. Most municipal solid waste is transported by
   a. air.
   b. rail.
   c. barge.
   d. truck.

CHAPTER 23

Geology and Environmental Health

CHAPTER OUTLINE

■ GEOLOGY AND HUMAN HEALTH

Several kinds of human activity produce threats to human and ecosystem health. Many of them do not strictly fit within the confines, however broad, of an environmental geology course. In this chapter, we consider but a few of the issues most relevant to our course. Other human health topics such as fossil fuels and coal, water pollution, and air pollution are examined elsewhere in the text. The subjects that we will consider in detail here are: the impacts of fine particulates, the impacts of heavy metals, and the impacts of perchlorates. We begin with perchlorates.

■ PERCHLORATES

Perchlorates ($XClO_4^-$; **FIGURE 23-1**) are highly water soluble and widespread inorganic chemicals that inhibit assimilation of iodine in the thyroid, and consequently interfere with thyroid functioning, at intake levels above 200 mg/day. Common varieties are sodium, ammonium, and potassium (the "X" in $XClO_4^-$) perchlorate. They are highly irritating to the skin and mucous membranes. Many perchlorates are naturally-occurring (see below). Other compounds, many also naturally occurring (e.g., nitrates), can affect iodine absorption.

Fetuses, infants, and children with thyroid damage can experience mental retardation, loss of hearing and speech, or deficits in motor skills. Fetuses are especially susceptible to thyroid damage. At higher levels of exposure, perchlorate is known to cause cancer.

Concept Check 23-1. How might unsafe "disposal" of perchlorates put an unfair burden on the healthcare system?

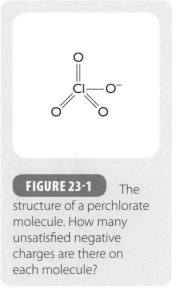

FIGURE 23-1 The structure of a perchlorate molecule. How many unsatisfied negative charges are there on each molecule?

Perchlorates can persist for decades in groundwater. Concerns regarding the presence of perchlorates in the environment have increased since 1997, when it was detected at levels above 4 parts per billion (ppb) in the drinking water of more than 15 million people in the western United States, 7 million in California alone (**FIGURE 23-2**). By 2001, perchlorates had been detected in 58 California public water supply systems and in water or soil in 17 other states, contaminating the water supply of more than 20 million people. According to the U.S. Environmental Protection Agency (EPA), perchlorate is found in just over 4% of U.S. public water systems, based on data from 3,865 public water systems between 2001 and 2005, the latest period for which such data were collected.

Two sites are worthy of special mention here. At Los Alamos National Laboratory in New Mexico, where perchlorates are used in nuclear chemistry research, perchlorates have been detected in groundwater below the lab site. And Colorado River water contains perchlorate in concentrations of up to 9 ppb, apparently derived from contaminated groundwater below the Kerr-McGee military/industrial manufacturing site at Henderson, Nevada.

Determining "Safe" Exposure Levels

The EPA tentatively set a 1 ppb limit on "safe" perchlorate levels in drinking water, subject to public comment and review. The EPA's decision could have far-reaching

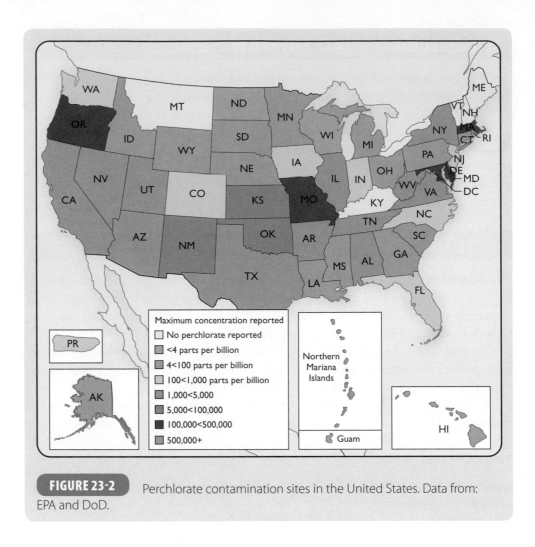

FIGURE 23-2 Perchlorate contamination sites in the United States. Data from: EPA and DoD.

impacts, because a low (~1 ppb) drinking water standard may require expensive cleanup of contaminated water by the military, other government agencies, and industry. If EPA sets the "safe" exposure level higher (~4–18 ppb) fewer cleanup and remediation costs will likely be assessed.

The California Department of Health Sciences has adopted an "action level" for perchlorate contamination at 4 ppb, pending scientific consensus on perchlorate risk. The 4 ppb level was also near the level of detection of perchlorate when the standard was adopted. Some studies indicate that the acceptable dose for children should be as low as 0.4 ppb.

Use and Sources

Perchlorates are used by the military in the manufacture of solid rocket fuel. Industrial uses include leather tanning, the manufacture of high explosives and fireworks, as a component of air-bag inflators, in electroplating and rubber manufacturing, and in batteries.

Perchlorates are found in Chilean nitrate deposits at concentrations of 1.5 to 1.8 mg/g. These deposits have been mined in the Atacama Desert, and were used worldwide as a source of nitrogen for fertilizers, explosives, and other chemicals before World War I. Nitrate production from these deposits up to 1980 was estimated at 145 million tonnes. Up to 3 billion pounds (1.36 million tonnes) of perchlorate may have been emitted into the environment from this source alone.

Concept Check 23-2. Suggest a test of the toxicity of perchlorates to humans, especially children, given that they are naturally found in nitrates in Chile's Atacama Desert. Then see the following Case Study.

Large-scale production of ammonium perchlorate for use in rocketry began in the United States in the mid-1940s. Ammonium perchlorate is unstable, and must be periodically replaced in munitions and rockets. This has resulted in the disposal of large volumes of the compound since the 1940s in Nevada, California, Utah, and elsewhere.

Studies of Perchlorate Toxicity

Studies on the toxicity of perchlorates have been funded by the aerospace and fertilizer industries. Drinking water at Taltal, Northern Chile, population ~12,000, was determined to contain 110 to 120 ppb perchlorates (**FIGURE 23-3**). The city was selected for a study to determine if thyroid abnormalities occurred in infants and school-age children. Researchers compared a population of children in Taltal with one in Antofagasta, which had no detectable perchlorates in the water supply. The study found that families in Taltal were five times more likely to have reported some type of thyroid disorder than residents of Antofagasta. There was, however, no discernable difference in the thyroid functions of children tested in the two cities.

Four other studies are of significance here. The first three were funded by industry.

- A study that fed volunteers spring water containing 3.0 mg/l perchlorate for 14 days found no change in the thyroid.
- A study of workers in an ammonium perchlorate plant found no thyroid abnormalities, even though the workers were absorbing perchlorate from the air.
- A study of volunteers' responses to variable perchlorate doses in water for 14 days found no significant response to concentrations below 180 ppb.
- The Arizona State Health Department in 2001 found a significant increase in abnormal levels of a thyroid hormone in infants whose mothers drank perchlorate-contaminated water from the Colorado River while pregnant, compared to a group in Flagstaff, Arizona whose water supply contained no detectable perchlorates.

Concept Check 23-3. Note that two of the three studies limited the test to 14 days. Comment.

Concept Check 23-4. Discuss whether industry-funding of these studies affects your confidence in their validity.

FIGURE 23-3 Location of Taltal, Chile, a site with high perchlorate levels in drinking water. Antofagasta, Chile is also shown.

Remediation of Contaminated Sites

Due to perchlorates' high solubility, they are extremely difficult to remove from groundwater. We feature one case study below: Rancho Cordoba, California.

CASE STUDY > Perchlorates at Rancho Cordova, California

Background © Hemera Technologies/AbleStock.com/Thinkstock. Title © iStockphoto/Thinkstock.

In Sacramento County, California, five drinking water wells were found to contain between 98 and 280 ppb perchlorate, much higher than the recommended maximum exposure levels (themselves based on very little data) of 4 to 18 ppb as determined by U.S. EPA. According to University of California Davis researchers, sometime before 1997 ammonium perchlorate, used in solid rocket fuel, leaked from an aerospace research site near Rancho Cordova into drinking-water wells thousands of feet away. University of California Davis researchers found that the contaminants below Rancho Cordova were lodged in ancient soil deposits called paleosols, which permitted only very slow vertical water movement (**Figure CS 1-1**). Once contaminants reach these deposits, these may be virtually impossible to remove without removing the deposit. The Environmental Working Group (EWG) estimated that perchlorate removal from contaminated groundwater below the Rancho Cordova site could cost $55 million and take more than two centuries to complete.

Summary

Studies of perchlorate toxicity must be as definitive as possible, because industry and military sources assert that perchlorates are safe at low doses. Environmental advocacy groups point to studies that show adverse impacts on infants and fetuses at extremely low doses. EPA is required by statute to decide.

Concept Check 23-5. Assess the applicability of the Precautionary Principle (found elsewhere in the text) to the use of perchlorates.

FIGURE CS 1-1 Paleosols in the Painted Hills, Oregon (paleosols are the dark reddish bands). Such deposits may have very low permeability, and thus may help trap contaminants like perchlorates. © tusharkoley/ShutterStock, Inc.

According to a *Journal of the American Medical Association (JAMA)* article, "fine-particulate and sulfur oxide–related pollution were associated with a wide range of cardiopulmonary diseases, including lung cancer. Each 10-μg/m^3 (a μg [microgram] is one millionth of a gram, or 10^{-6} g) elevation in fine particulate air pollution was associated with approximately a 4%, 6%, and 8% increased risk of all-cause, cardio-pulmonary, and lung cancer mortality, respectively. Measures of coarse particle fraction and total suspended particles were not consistently associated with mortality. Long-term exposure to combustion-related fine particulate air pollution is an important environmental risk factor for cardiopulmonary and lung cancer mortality."

The study provides a scientifically established link between fine particulates and disease. The main sources of fine particulates are (1) coal combustion in power plants, (2) diesel engines, and (3) two-cycle gas-powered lawn-care devices such as leaf blowers.

Concept Check 23-6. Describe, in your own words, the exposure-risk relationship reported by the *JAMA* article.

Concept Check 23-7. Until recently, monitoring of fine particulates was not even routinely carried out in the United States. Who should have the responsibility to monitor air for pollutants? Is this a legitimate role for government? If so, then which level (federal, state, etc.) should be responsible? Explain your answer. Recall that much air pollution is transboundary in nature. How does this affect your answer?

Concept Check 23-8. Who should pay for air monitoring? Who should pay to remove such pollutants? Explain.

Concept Check 23-9. Is it reasonable for some residents of the United States to enjoy cheap electricity from coal plants without state-of-the-art pollution removal devices, subsidized by other residents' diseases? Should a healthcare system reportedly in economic crisis have to pay for the effects of air pollution from coal plants, lawn-care devices and diesel engines? Do you think this is a "fair" question? Why or why not?

What Are Fine Particulates?

Scientists have concluded that *the finer the particulate size, the more dangerous the particle is to human health. Fine particulates* are 10 micrometers (10^{-6} meters) or less in diameter and are known as **PM10**. PM10 is roughly the same size as a bacterium or smaller. PM10 is invisible to the naked eye and small enough to be breathed into the lungs. PM10 can be composed of very small particulates of about 0.1 to 0.2 micrometers in diameter (found in automobile exhaust or fireplace emissions) and can also include particulates 10 times this size (sea salt spray or road dust). Scientists refer to a *fine and coarse fraction* of PM10, because they generally differ in chemical composition, source, and behavior in the atmosphere.

The *fine fraction* (**PM2.5**) contains particulates 2.5 micrometers or smaller. This fraction is most often produced by combustion processes and by chemical reactions, such as smog formation, occurring in the atmosphere (**FIGURE 23-4**). To an observer, an engine or a power plant emitting fine particulates may appear to

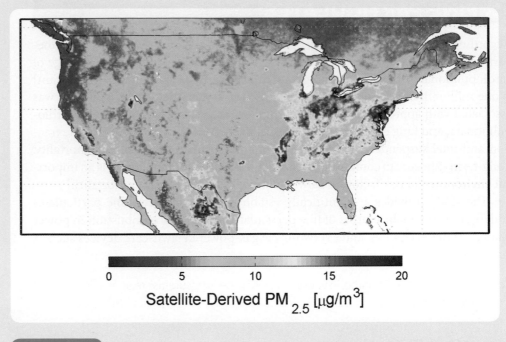

Satellite-Derived PM$_{2.5}$ [$\mu g/m^3$]

FIGURE 23-4 Distribution over the United States of atmospheric fine particulates (or soot) commonly produced by coal or petroleum combustion. Reproduced from *Environ. Health. Perspect.* 118:847–855 (2010).

be emissions-free, yet it could be emitting invisible fine particles that are readily transported through the atmosphere over distances of hundreds of kilometers, to be deposited on soil, buildings, crops, or into water bodies, or breathed into lungs. Some of the finest particles are produced by diesel engines; over 90% of particulates in diesel exhaust are less than 1 micrometer in diameter. According to the California State Air Resources Board (CALARB), these are some of the most dangerous air contaminants to which humans can be routinely exposed. In the Washington, D.C. Metropolitan Area, diesel vehicles constitute 3% of the vehicles on the road, but produce 30% of the region's smog-forming chemicals.

The *coarse fraction* consists of particles greater than 2.5 micrometers. This is the size most closely associated with natural sources. Particulates in the *coarse fraction* of PM10 are removed by the upper respiratory system, in the nose and throat: they don't penetrate far into the lung. Natural sources such as windblown dust, vegetation, and sea salt spray are major contributors. Human activities such as mining, quarrying, and cement manufacture are also important sources. These particulates settle to the ground within a matter of a few hours to a few days.

Particulates in the fine fraction (PM2.5) can remain in the air for days to weeks. They can penetrate especially deep into the lungs, collecting in the tiny air sacs called *alveoli* where the blood picks up oxygen (**FIGURE 23-5**). A growing number of potentially harmful substances have been found in PM2.5. Here are examples.

- *Sulfates*, produced from sulfur dioxide (SO_2) emissions, are acidic and may attack buildings, vegetation, and respiratory systems.
- *Elemental carbon*, as spherical aggregates and tiny solid particles produced during wood burning and engine combustion, can attach to carcinogenic combustion products like benzene and carry them into the lung (**FIGURE 23-6**).

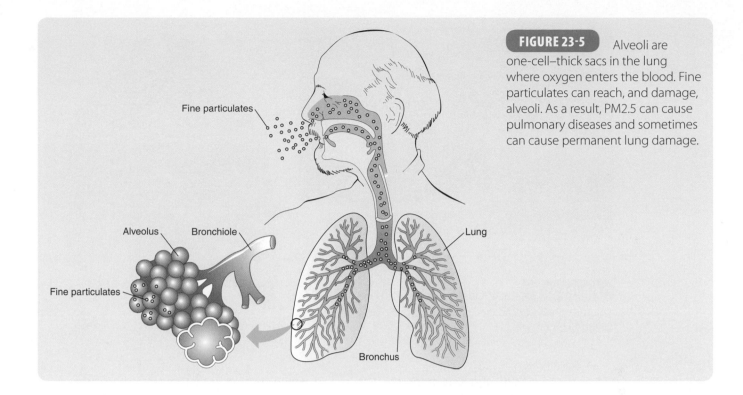

FIGURE 23-5 Alveoli are one-cell–thick sacs in the lung where oxygen enters the blood. Fine particulates can reach, and damage, alveoli. As a result, PM2.5 can cause pulmonary diseases and sometimes can cause permanent lung damage.

- A large and growing number of *soluble organic compounds* have been identified in exhaust from vehicles, combustion processes, and even barbecuing.
- *Toxic trace metals* such as lead, cadmium, and nickel (see later in the chapter) are more concentrated in PM2.5 than in larger particulates.

Assessing Hazards From Exposure to PM2.5

The State of California has tentatively proposed a Reference Exposure Level (REL) of 5 micrograms/m³ as a level below which "no *non-cancer* adverse health effects are likely" over a lifetime of exposure. Unfortunately, the estimate has a margin of error of an *order of magnitude*.

Studies have shown that exposure to PM2.5 from diesel exhaust can cause cell mutations and malfunctions in chromosomes and DNA. The California ARB considers it to be a likely carcinogen, since it contains cancer-causing compounds like benzene, 1,3 butadiene, and formaldehyde. The EPA in 1998 listed it as a "probable" human carcinogen. As you have seen, determining "safe" exposure levels are fraught with uncertainty.

Concept Check 23-10. Should humans simply accept chronic exposure to PM2.5 from diesel exhaust and other anthropogenic sources as a hazard of modern life (**FIGURE 23-7**), or should PM2.5 emitters be held accountable for their impact? Explain your answer.

Value-added Question 23-1.
Based on an order of magnitude range in the REL, what is the lowest "safe" exposure to fine particulates, like those in diesel exhaust?

FIGURE 23-6 Diesel exhaust. It contains abundant PM2.5. Most of the sulfur has been removed for on-road diesels since 2006. © Paul Marcus/ShutterStock, Inc.

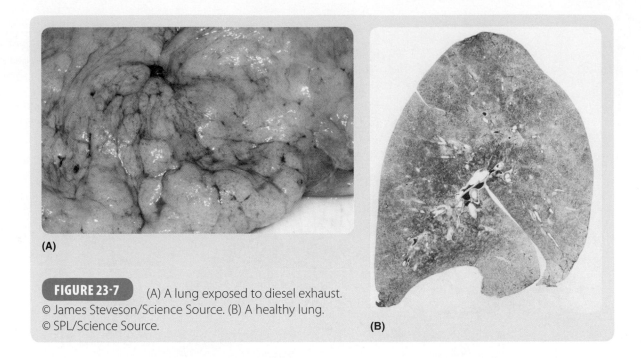

Concept Check 23-11. Who should pay to determine that impact? Should that be a responsibility of government? At what level of government should the responsibility rest?

Concept Check 23-12. Should localities have the ultimate responsibility to set exposure levels to toxics like PM2.5? Should it be determined by popular vote? Identify groups whose health may be impacted but that cannot vote.

■ HEAVY METALS AND HUMAN HEALTH

Certain **valence states** (a measure of the number of chemical bonds an atom can form based on the number of electrons in its outer shell) of *heavy* metals, as contrasted with *light* atomic-weight metals like sodium, calcium, and magnesium, have well-documented toxic impacts on animals, including humans. The metals that we consider here are arsenic (As), lead (Hg), mercury (Pb), and nickel (Ni). In the soils and agriculture section of the text, we discussed selenium (Se). Others that we do not have space to describe here, but also can have toxic effects include valence states of cadmium, chromium (the subject of the 2000 film, *Erin Brockovich*), aluminum, and the radioactive isotopes such as plutonium and uranium. We discuss the latter in the renewable and alternative energy section of the text.

Arsenic

Value-added Question 23-2.
What percentage of community water systems have a mean As concentration greater than 10 µg/L (10 ppb)?

There are about 54,000 community water systems (CWS) in the United States, and 2,302 systems are reported by an EPA study to have a mean arsenic (As) concentration greater than 10 µg/L (10 ppb).

Concept Check 23-13. As a resident, tourist, or potential resident in areas with public drinking water supplies, would you like to have access to data like this? Why or why not?

Natural Distribution

Arsenic is the twentieth most abundant element in the Earth's crust. Concentrations of As in the crust vary, but typical concentrations range from 1.5 to 5 mg/kg. Typical As concentration in coal is 4 to 13 mg/kg (4–13 ppm), but As concentrations of up to 2,000 ppm in coals are known. When the coal is burned, As can be released into the atmosphere. Volcanic activity appears to be the largest natural source of arsenic emissions to the atmosphere.

Because As occurs naturally in rock, soil, and sediment, these sources are particularly important determinants of regional levels of arsenic in groundwater and surface water. Higher arsenic concentrations, including groundwater concentrations greater than 50 µg/L, are found in sedimentary deposits derived from volcanic parent rocks. These geological conditions occur at a few locations in the West, notably near Reno, Nevada, and Eugene, Oregon.

Concept Check 23-14. Do you think residents of these cities are aware of the As concentrations in their groundwater? Would you want your government to inform you of potential high As levels in groundwater if you were a homeowner or renter using well water?

Valence States

The valence state of arsenic affects the toxicity of arsenic compounds. While (−3) is the most toxic, the (+3), and (+5) states are successively less toxic. The (+1) valence state and elemental arsenic (0) have relatively low toxicities. Inorganic arsenic, with +5 (arsenate) and +3 (arsenite) oxidation states, is more prevalent in water than organic arsenic. The distribution of these species depends largely on water oxygen levels: +5 is common in oxygenated water, and +3 is most common in hypoxic water.

Anthropogenic Sources

In addition to its release from natural sources, by weathering of rocks and minerals containing arsenic, As is released from a variety of anthropogenic sources, including:

- Refining and manufacturing of metals and alloys
- Petroleum refining
- Pharmaceutical manufacturing
- Pesticide manufacturing and application
- Chemicals manufacturing
- Burning of fossil fuels
- Waste incineration

These anthropogenic releases of arsenic can elevate environmental arsenic concentrations.

Uses

About 90% of the arsenic consumed in the United States annually is used to manufacture the wood preservative *chromated copper arsenate*. Arsenic is also a constituent of some agricultural pesticides. The most widely applied such pesticide is monosodium methane-arsonate (MSMA), which is used to kill broadleaf weeds. MSMA was the 22nd most commonly applied conventional pesticide in the United States in 1995, when between 4 to 8 million pounds of MSMA were applied. It is primarily applied to cotton but is widely applied to landscapes, especially turf. In South Florida, one study found 37% of water wells tested at municipal golf courses had elevated arsenic levels from MSMA (refer to the soils and agriculture section elsewhere in the text). Agricultural use of MSMA is being phased out in the U.S. as of 2013.

Arsenic metal is used in the production of lead-acid storage batteries. Over time, maintenance-free automotive batteries containing little or no arsenic are expected to replace lead-acid storage batteries. Thus, the demand for arsenic metal is likely to decrease.

Toxicity

Human exposure to arsenic can result in both **chronic** and **acute effects**. At low concentrations, evidence associates chronic arsenic ingestion with increased risk of skin cancer. A 1999 National Research Council (NRC) report on arsenic states that, "epidemiological studies . . . clearly show associations of arsenic with several internal cancers at exposure concentrations of several hundred micrograms per liter of drinking water." The organ systems where As-induced cancers in humans have been identified include skin, bladder, lung, kidney, nasal cavity, liver, and prostate. Considering all cancers together, the NRC states that "considering the data on bladder and lung cancer in both sexes noted in the studies . . . a similar approach for all cancers could easily result in a combined cancer risk on the order of 1 in 100" at a level of 50 µg/l.

Value-added Question 23-3.
In a city of 10,000 residents exposed to 50 µg/L arsenic, how many cancer cases could be expected?

Concept Check 23-15. Discuss the impact on the area's healthcare system. Is it better to remove the As, or treat the cancers in your view? Give your reasons. See the cost–benefit analysis below.

Other studies identified bronchitis, liver cirrhosis, and kidney problems as possibly related to chronic arsenic ingestion from drinking water.

Cost-Benefit Studies

TABLE 23-1 summarizes, insofar as possible, the costs and benefits of reducing As in public drinking water.

Concept Check 23-16. Evaluate the costs and benefits of reducing arsenic to 3 ppm based on Table 23-1.

Concept Check 23-17. Note the "potential non-quantifiable" health benefits column. Explain what this means.

Concept Check 23-18. Note that few large PWS systems have high concentrations (over 10 ppm) of arsenic; most of the problem systems are in small towns in the Western United States that draw water from groundwater sources naturally high in As. Would you be willing to pay slightly higher taxes in order for those in small communities to benefit from lowered As concentrations?

Concept Check 23-19. Is it reasonable for local residents to have a referendum to vote on whether to tolerate higher As exposure to avoid higher water fees? Explain your reasoning. *Hint*: Evaluate who would not be able to vote in such a referendum.

Arsenic Level (µg/L)	Total Annual Cost[1]	Annual Bladder Cancer Health Benefits[1]	Annual Lung Cancer Health Benefits[1,2]	Total Annual Health Benefits[1,2]	Potential Non-Quantifiable Health Benefits
3	$792.1	$58.2–$156.4	$155.6–$334.5	$213.8–$490.9	• Skin Cancer • Kidney Cancer • Cancer of the Nasal Passages
5	$471.7	$52.0–$113.3	$139.1–$242.3	$191.1–$355.6	• Liver Cancer • Prostate Cancer • Cardiovascular Effects
10	$205.6	$38.0–$63.0	$101.6–$134.7	$139.6–$197.7	• Pulmonary Effects • Immunological Effects • Neurological Effects • Endocrine Effects
20	$76.5	$20.1–$21.5	$46.1–$53.8	$66.2–$75.3[3]	• Reproductive and Developmental Effects

[1]May 1999 dollars.

[2]For 20 µg/L, the proportional reduction from the lower level risk base case is greater than the proportional reduction from the higher level risk base case. Thus, the number of estimated cases avoided and estimated benefits are higher at 20 µg/L using the risk estimates adjusted for arsenic in cooking water and food.

Costs and benefits of reducing As in public water supplies. Modified from Arsenic in Drinking Water Rule Economic Analysis/EPA.

Lead

According to the U.S. Centers for Disease Control and Prevention (CDC), "Lead poisoning affects virtually every system in the body, and often occurs with no distinctive symptoms. Lead can damage a child's central nervous system, kidneys, and reproductive system . . . Even low levels of lead are harmful and are associated with decreased intelligence, impaired neurobehavioral development, decreased stature and growth, and impaired hearing acuity."

Background

Emissions of lead in the United States have been cut by 98% since 1970, due to the national unleaded gasoline program. Does this mean that lead is no longer a health issue? In a word, no.

Housing for low-income persons remains a major source of lead contamination. In 2000, the Washington, D.C. Department of Health estimated that at least 6.5% of District children had unsafe blood-lead levels, compared to a national average of 4.4%. In 1999, the District spent over $1.5 million on lead-poisoned children. In that year at least 291 children had lead poisoning, five of them seriously enough to need hospitalization. Each case costs around $2,000 in medical costs and $4,000 in special education.

Concept Check 23-20. Is it reasonable to assume that passage of federal laws ensures that enforcement will correct this problem? Should owners of housing rented to the poor be required to ensure that their properties pose no undue risk to children from lead poisoning? Explain.

Public schools in the District of Columbia built before 1978 may themselves be sources of lead poisoning. And in recent years water dispensed from drinking fountains registered high levels of lead. The city estimated the cost to remediate these schools and to test all eligible children under the age of 6 to be $27.8 million over the first two years, if the project were to be approved.

Lead poisoning in at-risk children continues to be a problem in the District. In 2011, advocates Angela Wyan and Ralph Scott reported in the *Philadelphia Social Innovations Journal* on new initiatives and legislation aimed at identifying children with continuing high blood lead levels and stated that "the District of Columbia [has] experienced serious, documented problems with lead poisoning."

Concept Check 23-21. Make a hypothesis suggesting a possible connection between lead poisoning and a crisis in primary and secondary education. What additional information would you like to have to address the question more fully?

In 2004, high lead levels were reported in the public water supplied to thousands of residents of the District of Columbia. In 2010, according to the Centers for Disease Control and Prevention, 15,000 homes in D.C. might still have dangerous levels of lead in their water supply.

Hazards

The current minimum blood level that defines lead poisoning is 10 micrograms of lead per deciliter (1 dl = 100 ml = 0.1 liter) of blood. However, because poisoning may occur at lower levels than previously thought, various federal agencies are considering whether the standard for this level should be lowered.

Lead is one of the most toxic natural substances known, affecting virtually every system in the human body. It is similar to uranium in its effects on animals (see section on renewable and alternative energy).

Research cites lead poisoning as a major risk factor in behavioral problems and criminality. One study found that lead poisoning was the strongest predictor of disciplinary problems in school, which in turn was the strongest predictor of arrests between the ages of 7 and 22. A study of 501 boys in Edinburgh, Scotland, found that blood lead levels correlated strongly with measures of psychological deviance.

Many researchers consider high lead levels to be a risk factor for crime and delinquency, because lead poisoning causes the cognitive problems most strongly linked to criminal behavior: impulsiveness, low I.Q., hyperactivity, and low frustration tolerance.

In sum, lead can retard mental and physical development and reduce attention span. It can also retard fetal development even at extremely low concentrations. In adults, it can cause irritability, poor muscle coordination, and nerve damage to the sense organs and nerves. Lead poisoning may also interfere with reproduction, and can result in a decreased sperm count. And it may increase blood pressure.

Young children, fetuses, infants, and adults with high blood pressure are the most vulnerable to the effects of lead. It was estimated in 2000 that approximately 930,000 children between the ages of 1 and 5 had seriously elevated blood lead levels.

By 2012, the Centers for Disease Control and Prevention estimated the number at 500,000. This number has been significantly reduced from the 1980s: then, among American children 5 years old or younger, 63.3% had between 10 and 19 micrograms per deciliter; 20.5% had 20 to 29 micrograms per deciliter; 3.5% had 30 to 39 micrograms per deciliter; and 0.5% had 40 or more. Thus, the proportion of children aged 1 to 6 with lead poisoning fell to 4.4% over the period 1991 to 1994, an 80% decline from 1976 to 1980, and this number continues to decline.

Sources of Lead Emissions

TABLE 23-2 shows sources of lead emissions. TABLE 23-3 provides historical data on lead emissions in the United States.

An "average" coal-fired power plant produces 3.5 billion kilowatt-hours in a typical year of operation. One of the by-products is lead: around 114 pounds each year.

Summary

Before lead came into industrial use, human blood was estimated to have contained approximately 0.5 micrograms of lead in each tenth of a liter (deciliter) of blood, expressed in scientific shorthand as 0.5 micrograms per deciliter. (Recall that 1 microgram is a millionth of a gram and there are 28+ grams per ounce; a liter is about a quart; thus a deciliter is about half a cup.) Therefore, the "natural background level" of lead in human blood is around 0.5 micrograms per deciliter.

And finally,

- A child is estimated to lose two IQ points for each 10 mcg/dl increase in blood lead level.
- The hearing of children may be impaired by blood-lead levels below 10 μg/dl. Hearing problems can exacerbate learning disabilities.

TABLE 23-2	Sources of Lead Emissions*	
Source Category	**Emissions (short tons**)**
Electric utilities	
Coal	54
Oil	14
Industrial fuel combustion	
Coal	13
Oil	5
Miscellaneous nonresidential	
Fuel combustion	400
Chemical manufacturing	
Lead oxide and pigments	175
Metals Processing	
Nonferrous metals	
Primary lead	628
Secondary lead	505
Lead battery manufacturing	117
Ferrous metals	
Steel production	173
Waste Disposal and Recycling	
Incineration	
Municipal waste	75
Other (hospitals, etc)	546
Non-road Engines	
Aircraft	503

Sources of lead emissions. Data from: EPA.
*Statistical Abstract of the U.S.
**A *short ton* equals 2000 pounds.

Mercury

In February 2000 the Contra Costa County (California) Health Services Department issued an interim fish advisory for San Pablo Reservoir. Elevated levels of mercury (Hg) were found in largemouth bass. Pesticides, polychlorinated biphenyls, dioxins/furans, and mercury were found in channel catfish and other species of fish in the reservoir.

Mercury is widespread and persistent in the global environment, and most of its distribution is the result of human activity. Mercury is a global problem because it may be carried by atmospheric circulation to extremely remote locations, and because microbes convert a portion of mercury in aquatic ecosystems into **methyl mercury**, a powerful neurotoxin. In the United States, atmospheric deposition is highest in eastern Pennsylvania, Maryland, eastern Virginia, and coastal South Carolina, though there are "hot spots" elsewhere (FIGURE 23-8).

Value-added Question 23-4.
How much lead is emitted for each kilowatt-hour of electricity produced? Give your answer in micrograms per kilowatt-hour.

TABLE 23-3	Historical Data on Lead Emissions in the United States

Number of monitoring stations: 95 (for the entire U.S.!)
Air quality standard (micrograms/m³): 1.5

Date	1987	1990	1993	1994	1995	1996	1997
Air concentration of lead, based on 195 stations:	0.16	0.09	0.05	0.05	0.04	0.04	0.04

Date	1970	1975	1980	1985	1990	1996	1998
Total lead emitted (tons)	220,869	159,659	74,153	22,890	4,975	3,899	3,073

Data from: EPA.

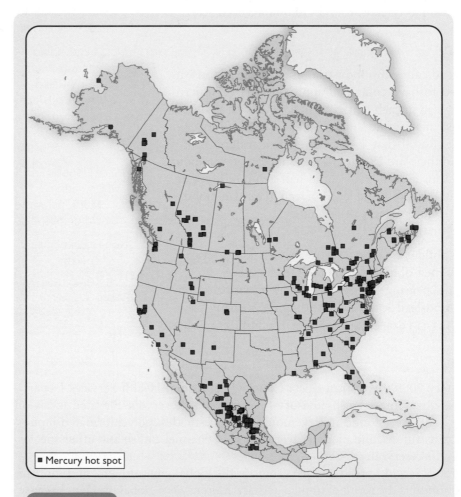

■ Mercury hot spot

FIGURE 23-8 Regions in North America with high atmospheric deposition of mercury. Data from: Commission for Environmental Cooperation of North America.

The widespread use of mercury has resulted in: (1) well-documented studies of human poisonings; (2) high-level exposures to groups of industrial workers; (3) global, chronic low-level environmental exposures; and (4) emissions as a result of agricultural production, because mercury can be contained in agricultural fungicides.

Consumption of contaminated fish, especially tuna, swordfish, shark, and whale, is believed to be the major source of human exposure at the present time. Some evidence suggests that ingestion of Hg by infants and children through vaccines containing the preservative thimerosal could be a significant source, leading to a ban of its use in most vaccines. Thimerosal has also been implicated in an upsurge of autism among children in the United States, although no scientific evidence has validated this claim.

In 1999–2000, one of six pregnant women had mercury levels in their blood of at least 3.5 parts per billion, according to EPA biologists, sufficient for levels in the fetus to reach or surpass the EPA's safety threshold of 5.8 parts per billion. In 1999–2000 this meant that 630,000 children had unsafe blood levels of mercury.

According to the EPA, between 1 and 1.6 million women in the United States of childbearing years eat enough mercury-contaminated fish to risk damaging their children's brain development. At least forty states have issued health advisories warning pregnant women or women of reproductive age to avoid or limit fish consumption. Ten states have issued advisories *for every lake and river within the state's borders*: Connecticut, Indiana, Maine, Massachusetts, Michigan, New Hampshire, New Jersey, North Carolina, Ohio, and Vermont.

Mercury in the atmosphere can be transported long distances and can contaminate isolated sites. For example, there are no natural sources of mercury in the Florida Everglades or in Minnesota's Voyageurs National Park, but mercury contamination is widespread in both places. Contamination commonly results from air deposition, but may result from groundwater contamination or surface runoff as well.

Uses

Mercury is still used in many industrial activities and is found in common items such as batteries, thermometers (although mercury in thermometers for medical use is banned in the United States and several other countries), barometers, vaccines, and even jewelry. Before 1990, paints contained mercury as an anti-mildew agent. In dentistry, mercury is still used in dental amalgams. All of these sources, and others, can release mercury into the environment. Mercury is used as a preservative in some flu vaccines.

Anthropogenic Sources

Other than natural sources from volcanoes and mineralized veins, humans generate mercury by burning coal, burning medical and other waste which contains mercury, mining gold and mercury, and from industrial activity, as noted previously. A typical 500-megawatt coal-burning power plant emits 170 pounds of mercury each year, according to the Union of Concerned Scientists. For years, however, operators of coal-fired power plants, and trash incinerators (renamed "waste-to-energy facilities"), responsible for substantial mercury emissions, have contested attempts to further regulate mercury.

Public alarm over contamination of wildlife and human health problems due to mercury toxicity has increased considerably since the mid-1960s. One issue contributing to this concern has been the number of fish consumption advisories in the majority of U.S. states, Canada, and several European countries because of high levels of mercury. Both external sources and ecosystem-specific factors can result in toxic levels of mercury in game fish. However, the Food and Drug Administration has no enforceable limit for mercury in fish—only a guideline of 1 part per million. Mercury concentrations in fish from the Everglades, for example, have often exceeded the Florida "advisory level" of 1.5 parts per million.

Mercury in Human History

Humans have been using mercury in some form for centuries. Mercury was an essential component of medicines including diuretics, antibacterial agents, antiseptics, and laxatives. By 1500 BCE the Egyptians were using mercury, as it has been found in their tombs. By the end of the eighteenth century, medicines to treat syphilis contained mercury. And during the 1800s the phrase, "mad as a hatter" was coined due to chronic mercury poisoning experienced by felters in the hat trade. In the mid-1800s hat makers used hot solutions of mercuric nitrate to shape felt hats made of wool. They typically worked in poorly ventilated rooms leading to chronic occupational exposure to mercury and neurological damage, in turn leading to sometimes psychotic behavior: the "hatter's syndrome."

Mercury in Human Medicines

Due to health concerns, most medicinal uses of mercury have been eliminated and, as a result, drug-induced sources of mercury toxicity are rare in developed countries. Among the more noteworthy incidents involving Hg are human contamination in Minamata Bay in Japan (1960; discussed elsewhere in the text), methyl mercury-treated grain in Iraq (1960 and 1970), and contamination of the Everglades. In humans, mercury poisoning is commonly misdiagnosed because of the often-delayed onset of symptoms, nonspecific signs and symptoms, and lack of knowledge within the medical profession. Indeed, without a complete history, mercury toxicity, especially in the elderly, can be misdiagnosed as Parkinson's disease, senile dementia, depression, or even Alzheimer's disease.

Mining, Energy Production, and Mercury

Mercury was used in the Sierra Nevada during the California Gold Rush to separate gold from waste. The toxic residue ended up in river sediment and stream banks. Until 2000, those who collected mercury for recycling did not have a practical way to properly dispose of it. During that year the California State Water Resources Control Board and the U.S. Department of Forestry started a pilot program to recover leftover mercury by collecting and recycling more than 200 pounds of the toxic metal. Shipping it to a recycler is costly because mercury is a hazardous substance that requires a special permit to handle, store, and transport amounts greater than 10 pounds.

Mining for mercury at some sites has gone on for centuries, and most such sites are now contaminated (FIGURE 23-9). In Slovenia, one site was mined continuously for over 500 years, only recently being abandoned. Hg wastes have contaminated a wide area in the adjacent Adriatic Sea.

FIGURE 23-9 The Almaden Mine, Spain. Mercury mining has been ongoing here for 2,000 years. © Daniel Santos Megina/Alamy.

In the United States, up to 1600 sites were mined for mercury to use in recovering gold from ore. The process, called "amalgamation," is a simple one. Crushed gold ore is passed over plates coated with mercury, to which the gold binds. The gold is then released from the amalgam by heating (or "retorting"), a process that historically released significant quantities of Hg into the atmosphere, where it ended up largely in soil and sediment. Large amounts of mercury were used, because the average concentration of gold in ores in the United States was around 6 grams per ton of rock! In the Carson River Valley of Nevada, much of the sediment is presently mercury-contaminated due to gold mining.

In the developing world, especially Brazil, and in the former Soviet Union, gold mining and processing using mercury is generating significant quantities of mercury pollution. Studies report fish in the Amazon River to be contaminated with Hg. In Brazil, labor is cheap, and the **amalgamation method** is simple: crush the ore, amalgamate it, and then burn the amalgam to vaporize the mercury and release the gold. The potential for exposure is high because mercury has a high *vapor pressure*, which means that whenever ores containing mercury are heated, the mercury is vaporized. Ironically, miners have low exposure to mercury because the Hg emitted is mainly elemental Hg, which has relatively low toxicity. However, elemental mercury may be converted to methyl mercury (MeHg: see below) in soils and waters, and may thereby enter the food chain. As a result, residents who eat fish from the Amazon are exposed to a much higher risk from mercury poisoning than miners, because MeHg is more toxic than elemental Hg.

Coal normally contains mercury. Background values in coal are typically from 50 to 70 ng/g but may be higher. Mercury content varies considerably from site to site, and even within coal seams. When coal is burned, mercury is volatilized. It enters the flue (exhaust) gases, and does not accumulate in the coal ash.

Natural Forms

Elemental mercury is liquid at surface conditions, but it is usually bound up as minerals in nature; the most common mercury-bearing mineral is HgS, cinnabar, a distinctive red mineral. Mercury may enter the environment in two forms, inorganic mercury and methyl mercury.

Inorganic mercury (metallic mercury and inorganic mercury compounds) enters the air from mining ore deposits, burning coal and waste, and from manufacturing plants. It enters the water or soil from natural deposits, disposal of wastes, and possibly from volcanic activity.

Methyl mercury (CH_3Hg^+ or MeHg) may be formed in water and soil by bacteria. The EPA has set the minimum risk level for oral exposure to methyl mercury at 0.1 μg/kg/day. Methyl mercury builds up in the tissues of fish (**bioaccumulates**). Larger and older fish (that is, those higher on the food chain) tend to have the highest levels of mercury.

How Is Mercury Toxic?

Mercury tends to bind to specific chemical structures on proteins. The result is widespread changes in the function of cells, which can damage DNA and disrupt cell division. In the nervous system, mercury interferes with development of the neuronal skeleton. Mercury also disrupts cell membranes, making them more likely to stick to one another. This may explain how cellular migration is altered during brain development. Mercury exposure also disrupts synaptic transmission, a basic brain function.

Mercury exhibits **synergistic effects** (that is, the combined effect is greater than the sum expected by adding each) with other toxics like polychlorinated

Value-added Question 23-5.
A typical 500-megawatt coal plant burns about 1,430,000 tons of coal per year. At a concentration of 60 ng/g (1 ng = 1 nanogram = 1 billionth of a gram = 10 g^{-9}), how much mercury would such a plant emit?

biphenyls (PCBs). The EPA reports that a study of children born to women who consume fish from Lake Ontario shows that prenatal PCB and mercury exposures interacted to reduce performance of 3-year-old children on certain tests. Mercury exposures in this study were quite low, yet they combined with PCB exposures to increase adverse impacts on neurodevelopment.

Concept Check 23-22. Discuss how mercury could be implicated in the "crisis" in our system of primary and secondary education. Cite evidence for your conclusions.

The nervous system of animals is very sensitive to mercury. MeHg and vapors are most harmful, because more mercury in these forms reaches the brain, where damage can be most intense. Exposure to high levels of metallic, inorganic, or organic mercury can permanently damage the brain, kidneys, and other organs of a developing fetus. Short-term exposure to high levels of metallic mercury vapors may cause effects including lung damage, nausea, vomiting, diarrhea, increases in blood pressure or heart rate, skin rashes, and eye irritation. The EPA has called mercuric chloride and methyl mercury possible human carcinogens.

Regulating Mercury Emissions

The sources of mercury emissions in Europe are similar in proportion to sources in North America.

Due to heightened public concerns, the state of mercury regulations is changing. While no statute at a federal level in the United States applies to mercury as a sole "source of concern" in the words of the EPA, a number of federal and state agencies regulate mercury in pesticides, consumer products, and mercury emissions into air and water.

The EPA regulates mercury in pesticides, and mercury releases into the environment. In 2005, the EPA issued new rules governing mercury emissions by coal-fired power plants. A "cap and trade" system was proposed that would reduce mercury emissions to 38 tons by 2010, and to 15 tons by 2018. However, a U.S. Circuit Court annulled the rules in 2008. In 2011 the EPA developed a Maximum Achievable Control Technology (MACT) standard, requiring every oil- and coal-based power plant to install mercury-specific controls rather than a cap-and trade system; however, the MACT was under review in 2012.

The Food and Drug Administration (FDA) regulates mercury in cosmetics, food, and dental products. The Occupational Safety and Health Administration (OSHA) regulates mercury exposures in workplace air.

It is reasonable to infer that reducing mercury emissions is an easier way to control mercury toxicity than attempting to remove mercury after it has lodged in the biosphere.

Mercury and mercury compounds are considered Hazardous Air Pollutants (HAPs) under the Clean Air Act. To date, the EPA has established National Emission Standards for Hazardous Air Pollutants (NESHAPs) for mercury emissions from three different sources: ore processing facilities, mercury cell chlor-alkali plants, and sewage sludge driers, but not from power plants, which are the major source of mercury emissions. Chlor-alkali plants are chemical plants that produce chlorine and caustic soda for industrial uses, using mercury in the process. Such plants are responsible for only a small percent of mercury emissions, and there are only a few of them in the United States. Thus, they are easily regulated. The EPA established a MACT for a "major source" in any listed source category. Major sources are defined as those sources that release 10 tons per year of any HAP, or 25 tons per year in total HAP emissions.

Mercury is a "toxic pollutant" under the Clean Water Act (CWA). Effluent limits for industries are based on type (metalworking, mining, etc.). EPA allows states to require effluent limits or monitoring requirements more stringent than those standards, under certain conditions.

The Clean Water Act is based on a permit system, known as the National Pollutant Discharge Elimination System (NPDES; discussed elsewhere in the text) to regulate water discharges. Facilities such as sewage plants and industries may be assigned a specific mercury discharge limit, or may only be required to monitor their discharge for mercury. Facilities report actual discharge levels in Discharge Monitoring Reports (DMRs); they are the basis for determining compliance.

Nickel

Nickel (Ni) is a silvery metal that easily forms alloys with other metals. It is widely used in steel making, electroplating, batteries, and coinage. Like the other heavy metals we have discussed previously, ingestion of Ni can have toxic effects. After a brief review of Ni toxicity, we will discuss the impacts of the Sudbury, Ontario nickel smelters.

Nickel Toxicity

About 10% of women and 2% of men are acutely sensitive to Ni, such that some of them may develop a skin rash if exposed to the metal. Otherwise, the most significant adverse Ni impact results from inhalation of Ni vapor or fine particulates of the metal, which can cause asthma, other respiratory diseases, or even cancer. Nickel can also be toxic to plants, although at low doses (<100 ppm) it is a plant nutrient.

Nickel Smelters

In 1885, a land surveyor discovered an immense deposit of copper and nickel near the town of Sudbury, Ontario, Canada, about 60 km north of Georgian Bay on the northern edge of Lake Huron (**FIGURE 23-10**). In 1886 smelting of the ores began. Initially operators used "roast yards" where fires were kept burning to drive the sulfur out of the ore minerals and concentrate the metals. The effect was to pour out vast

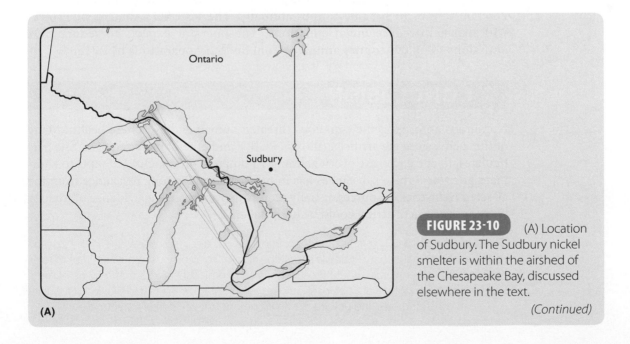

(A)

FIGURE 23-10 (A) Location of Sudbury. The Sudbury nickel smelter is within the airshed of the Chesapeake Bay, discussed elsewhere in the text.

(Continued)

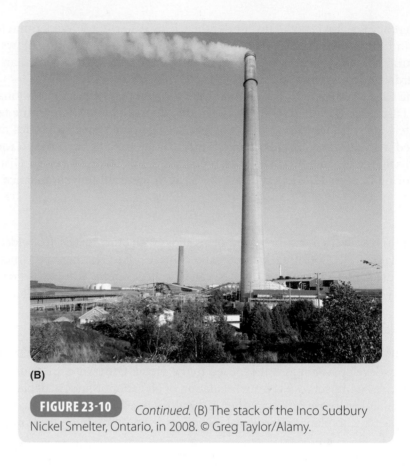

(B)

FIGURE 23-10 *Continued.* (B) The stack of the Inco Sudbury Nickel Smelter, Ontario, in 2008. © Greg Taylor/Alamy.

quantities of SOx, which denuded the vegetation for miles around the smelter, creating a veritable moonscape.

Metal smelting is one of the largest contributors to air pollution in Canada, and the smelters at Sudbury are among the most significant sources (**TABLE 23-4**). However, emissions at Sudbury are coming down (**TABLE 23-5**).

In 2006 a new fluid bed roaster (FBR) lowered SOx emissions from 265,000 tonnes to 175,000 tonnes annually. The FBR SO_2 Abatement Project will also help reduce metal emissions. The operator expects to reduce SOx emissions to 66,000 tonnes annually from Sudbury operations by 2015.

■ CONCLUSION

Numerous anthropogenic emissions threaten human and biosphere health. Many of these emissions are artificial (dioxin, PCBs), and others (mercury, SOx, fine particulates, to cite a few) are overwhelmingly human in origin. Our attempts to eliminate or control these substances has in most cases met strong resistance from the affected industries or sources, usually requiring overwhelming evidence of toxicity before government action could be undertaken.

Concept Check 23-23. The Precautionary Principle was discussed elsewhere in the text. Do you think the Precautionary Principle should be applied to all proposed new sources of toxic emissions? In other words, should those who wish to emit substances that may be harmful be required to *prove* they are not, or should government have the responsibility to regulate and clean up after those who pollute the environment? Cite evidence for your answer.

TABLE 23-4

Contributions of Metal Smelters to Selected Categories of Air Pollution in Canada

| Substance | Units | 2002 Emissions | |
| | | Base Metals Smelting | Canadian Total |
			% Emissions
Arsenic tonnes	153	201	76%
Cadmium tonnes	31	33	94%
Lead tonnes	196	223	88%
Mercury kg	1,700	2,949	58%
Nickel tonnes	ß258	475	54%
Total Particulate Matter t.	10,757	523,319	2%
Sulphur Dioxide t	669,967	1,419,520	47%

Source: Environment Canada. *Activity*: Canada has about one-tenth the population of the United States. Compare the SOx emissions from Canada to those of the United States, (U.S. 2001, 10.4 million short tons; Canada, 2000, 2.5 million metric tons). Which has higher per capita emissions? Why?

TABLE 23-5

Reduction in Sulfur Oxide Emissions at Inco and Xstrata Sudbury Nickel Smelters, 2005-2015

| Year | Emissions of SOx (metric tons) | |
	Inco	Xstrata
2005	265,000	
2007	175,000	40,250
2015 (proj.)	66,000	25,000

Activity: Calculate the total SOx emissions from the two smelters at Sudbury for 2015. What percentage reduction in SOx would this represent? For comparison, the largest stationary source of SOx emissions in Oregon is the Boardman coal-fired power plant operated by Portland GE, near Hermiston. In 2001 it emitted 18,000 short tons of SOx. Convert short tons to metric tons and compare the Boardman emissions to those from the nickel smelters at Sudbury. Which is greater? Which poses the greater hazard in your opinion? Why?

CHAPTER SUMMARY

1. Perchlorates are highly water soluble and widespread inorganic chemicals that, among other things, interfere with thyroid functioning, and may cause cancer.

2. In 1997, perchlorate was detected at levels above 4 ppb in the drinking water of more than 15 million people in the Western United States, 7 million in California alone.

3. Colorado River water contains perchlorate in concentrations of up to 9 ppb, apparently derived from military/industrial sites in Nevada. Studies of actual impact of perchlorates on human populations have been inconclusive.

4. Fine particulates have been associated with a wide range of lung diseases. Studies have shown that exposure to PM2.5 from diesel exhaust can cause cell mutations and malfunctions in chromosomes and DNA. The main sources of fine particulates are coal combustion in power plants, diesel engines, and gas-powered lawn-care devices such as leaf blowers.

5. Certain valence states of *heavy* metals have well-documented toxic impacts on animals, including humans. We consider arsenic (As), mercury (Hg), selenium (Se), and lead (Pb).

6. Anthropogenic releases of arsenic can elevate environmental arsenic concentrations. About 90% of the arsenic consumed in the United States annually is used to manufacture the wood preservative chromated copper arsenate.

7. Human exposure to arsenic can result in both chronic and acute effects, including skin and internal cancers. Cost-benefit analyses for As reduction are highly positive, according to the EPA.

8. Emissions of lead in the United States have been cut by 98% since 1970, due to the national unleaded gasoline program. Housing for low-income persons remains a major source of lead contamination.

9. Lead is one of the most toxic natural substances known, affecting virtually every system in the human body. It is similar to uranium in its effects on animals.

10. Mercury is widespread and persistent in the global environment, and most of its distribution is the result of human activity.

11. Microbes convert a portion of mercury in aquatic ecosystems into methyl mercury, a powerful neurotoxin.

12. In 1999–2000, one of six pregnant women had mercury levels in their blood of at least 3.5 parts per billion, sufficient for levels in the fetus to reach or surpass the EPA's safety threshold.

13. Humans generate mercury by burning coal, by burning medical and other waste that contains mercury, by mining gold and mercury, and through certain industrial processes.

14. In the developing world, especially Brazil and in the former Soviet Union, gold mining and processing using mercury is generating significant quantities of mercury pollution.

15. Most mercury is released by coal-fired power plants and from waste incineration.

16. Metal smelters in Canada, and notably those at Sudbury, Ontario, generate an appreciable quantity of Canada's air pollution.

KEY TERMS

acute effects

amalgamation
  method

arsenic toxicity

bioaccumulates

chronic effects

inorganic mercury

lead toxicity

methyl mercury

perchlorate

PM10

PM2.5

synergistic effects

valence states

REVIEW QUESTIONS

1. Perchlorates, which can cause thyroid problems among other things, are used mainly in
   a. the mining of gold.
   b. the preservation of wood.
   c. solid fuel for rockets and high explosives.
   d. water purification.

2. The main source of the air pollutant fine particulates is
   a. dust from construction activity.
   b. dust from cement manufacture.
   c. combustion of natural gas in power plants.
   d. coal combustion in power plants and diesel exhaust.

3. Which of these is true about the toxicity of particulates?
   a. The coarser the particulate, the more dangerous.
   b. The finer the particulate, the more dangerous.
   c. Studies have so far not established a link between size and toxicity.

4. The fine fraction of particulates is
   a. PM10.
   b. PM25.
   c. PM2.5.
   d. PM100.

5. Which is true about the toxicity of valence states of arsenic?
   a. Arsenic is not present in nature as an ion.
   b. All valence states have similar toxicities.
   c. Negative valence states are more toxic than positive valence states.
   d. The toxicity of As is insufficiently known to determine if any valence states are more toxic than others.

6. Most arsenic in the United States is used
   a. to manufacture batteries of various types.
   b. to manufacture a wood preservative.
   c. in the manufacture of certain kinds of steel.
   d. as insect and rodent poisons.

7. Which is true concerning As concentrations in public drinking water systems?
   a. Highest As concentrations are in large public supply systems that use surface water as a source.
   b. Most problem supplies are in small systems in the western United States that tap groundwater.
   c. As is a nationwide problem that has been swept under the rug for decades.
   d. All public water supply systems filter As out of water before supplying customers.

8. Which is true concerning lead poisoning?
   a. Lead affects the central nervous system only, and has distinctive symptoms.
   b. Lead affects only the central nervous system, but often has no clear symptoms.
   c. Lead affects virtually every system in the human body, often without distinctive symptoms.
   d. Lead affects every system in the body with distinctive symptoms.

9. Which is true about lead emissions in the United States?
   a. Lead emissions have been cut drastically in the past 30 years, due mainly to removing lead from motor fuel.
   b. Lead has been cut drastically, due mainly to controls on coal-burning power plants.
   c. Lead emissions have increased each year since 1970, and now constitute one of the United States' primary health problems.
   d. Lead emissions have been constant since 1970, but treatment methods have improved so that lead is no longer a health issue.

10. Lead poisoning could be an issue in public education because
    a. flaking lead paint in aging schools could poison children.
    b. lead ingestion by children in old, substandard housing could lower IQ levels.
    c. lead ingestion can lead to behavioral problems and even criminal behavior.
    d. all of the above.
    e. none of the above.

11. Which is true concerning atmospheric lead contamination in the United States?
    a. Atmospheric lead levels are on average well below the air quality standard.
    b. Lead levels are usually above the air quality standard for most metropolitan areas.
    c. Lead levels are generally above the air quality standard in rural areas.
    d. Atmospheric lead levels are no longer routinely measured by any government agencies.

12. Most of the distribution of mercury in the global environment is the result of
    a. volcanic activity.
    b. subduction zone activity.
    c. human activity.
    d. melting of permafrost in Arctic regions as a result of climate change.

13. In 1999–2000, what proportion of pregnant women tested had mercury blood levels above EPA's safety threshold?
    a. more than 90%
    b. fewer than 5%
    c. about 1 in 6.
    d. fewer than 1.

14. Mercury is used as a preservative in
    a. some vaccines.
    b. wood.
    c. some dried seafood products.
    d. most frozen seafood products.

15. How many states have issued mercury advisories for every lake and river in the state?
    a. ten
    b. none
    c. all the lower 48 states
    d. forty

16. The amalgamation process uses mercury to refine
    a. uranium.
    b. copper.
    c. gold.
    d. lead.

17. The most toxic form of mercury in nature is
    a. methyl mercury.
    b. elemental mercury.
    c. inorganic mercury.
    d. solid mercury.

18. Largest two sources of mercury emissions in the United States and Europe are
    a. autos and lawn mowers.
    b. pesticides and herbicides.
    c. active and abandoned mercury mines.
    d. power plants and incinerators.

19. Which is true concerning the regulation of mercury by government?
    a. There are no governmental regulations on mercury emissions.
    b. One federal agency, the EPA, has authority to regulate all mercury use in the United States.
    c. Regulation of mercury is left to state agencies.
    d. Several federal agencies regulate differing sources and exposures of, and to, mercury.

20. The Clean Water Act
    a. regulates mercury as a "toxic pollutant."
    b. leaves regulation of mercury to the states.
    c. has no authority over mercury because all emissions are atmospheric.
    d. has no authority over mercury because mercury levels are insufficient to cause adverse effects on human health.

CHAPTER 24

Development, Trade, and Environmental Quality

CHAPTER OUTLINE

E xperts project there will be more than one billion motor vehicles operated globally by 2025, up from 40 million in 1945. There are over 200 million motor vehicles in the United States, more than one third of the world's present total.

Concept Check 24-1. What impacts on the global environment and economy (oil demand, steel, aluminum, etc.) do you predict from a doubling of global motor vehicles by 2025?

■ WHAT IS DEVELOPMENT?

A central question in the nascent subdiscipline of *ecological economics* is to what extent **development**, leading (it is supposed) to higher per capita income; *trade*; and the state of a country's *environmental quality* are related. We consider some of these questions in this chapter.

Development can have a variety of meanings. Here we use development to refer to a complex set of changes that convert the economy of a society from one based on subsistence agriculture, to one in which most of the employed inhabitants work in manufacturing or service industries. Early stages of development depend more on the exploitation of "natural resources" and later stages on the exploitation of "human capital," that is, the creativity of the human mind.

Developed societies historically experience significant land-use changes: *deforestation*, for fuel and intensive, usually petroleum-based farming; *urbanization*, the concentration of most of a country's population into urban centers; intensive *mining* for metals and fossil fuels, and so forth. Development as a rule also has led to considerable increases in certain types of environmental pollution, such as exhaust gases from the burning of fossil fuels, and toxic factory emissions into waterways, although as some point out, pollutants such as horse urine and droppings do decline.

Deforestation usually increases *soil erosion*, which pollutes waterways. Concentrating humans in cities also concentrates waste, both human and otherwise, which if not properly treated (waste production, reduction, and disposal is discussed elsewhere in the text) can severely pollute waterways or the ocean in the case of coastal cities (human population growth is covered elsewhere in the text). Urbanization may also foster disease. The cholera and yellow fever epidemics in nineteenth century England and North America are examples.

On the other hand, urbanization also increases the concentration of waste and pollution, which can make it easier to deal with.

■ DEVELOPED VERSUS DEVELOPING COUNTRIES

The United Nations lists criteria by which countries are classified as "least developed," "developing," and "developed" among other terms. The 49 least developed countries (LDCs) are shown in **FIGURE 24-1**. The criteria for LDC were: per capita gross domestic product (GDP) under $900, weak human assets, a composite index based on health, nutrition, and education indicators; and high economic vulnerability, another composite index based on indicators of instability of agricultural production and exports, inadequate diversification, and economic "smallness." For details, consult the United Nations website (http://www.un.org).

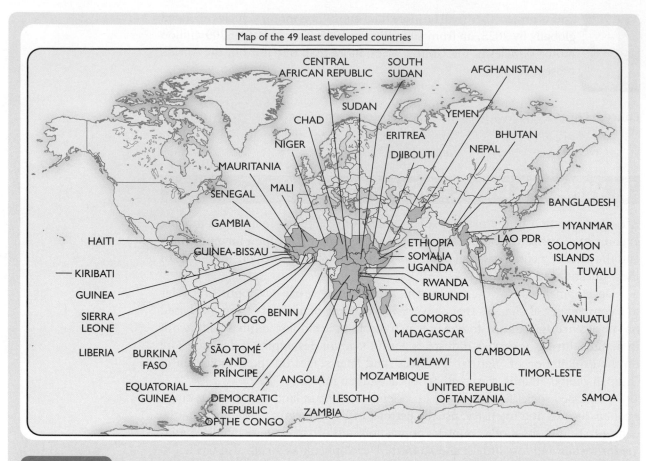

FIGURE 24-1 The 49 least developed countries. Describe the latitudinal distribution of the LDCs. Data from: United Nations Conference on Trade and Development.

■ DEVELOPMENT'S IMPACT ON THE ENVIRONMENT

Here are two hypotheses, much simplified, that purport to explain such relationships as may exist between development and the environment.

Development Harms the Environment

Many environmentalists point out that development leads to harmful land-use practices such as deforestation; harmful levels of air emissions from industry, oil-and-coal-burning power plants, and vehicles lacking emissions controls or inspections; subsidies encouraging fossil-fuel use; and water pollution resulting from deforestation, lack of sewage treatment, and wasteful agricultural practices that emphasize production above stewardship, among other things. Moreover, they cite high levels of population growth in many developing countries as a contributor to environmental decline and an encouragement of mass emigration.

Development Eventually Improves the Environment

Many economists and some environmentalists, while acknowledging harmful levels of environmental pollution in countries in early stages of development, point to the considerable empirical evidence that (1) population growth rates decline as development proceeds and (2) rates of some forms of environmental pollution decline as per capita income increases, a supposed corollary of development as we noted previously. Newer forms of technology tend to be less polluting than older forms, but also tend to be more expensive, if environmental improvements are not included in cost calculations. Moreover, countries with higher per capita incomes generally tend to have cleaner environments along with increased consumption of goods and services. Poverty and high rates of population growth are major causes of environmental degradation, such as deforestation. As nations become richer and their middle classes expand, so do demands for tougher environmental standards and regulations. Indeed, the "green" movement and "green consumerism" in the developed world evolved with the growth of the middle class after World War II.

■ ECONOMIC GROWTH AND THE ENVIRONMENT: KUZNETS CURVES

The **environmental Kuznets curve**, named for Nobel laureate economist Simon S. Kuznets, plots the *relationship between environmental quality factors and per capita income*. While controversial, Kuznets curves have been used to study emissions levels as a function of the level of economic development. The relationships that have been plotted include income against: sulfur dioxide emissions, suspended particulate matter, carbon monoxide, nitrogen oxides, and airborne lead.

In addition, researchers have developed curves for plotting the following environmental parameters against per capita income: access to safe water, presence of urban sanitation (that is, whether or not people are connected to sewage treatment plants), annual deforestation rates, total deforestation, dissolved oxygen in rivers, fecal coliform bacteria (a measure of the presence of the toxic bacterium *E. coli*) in rivers, municipal solid waste per capita, and carbon emission per capita.

Shapes of Kuznets Curves

Environmental Kuznets curves (EKC) generally exhibit one of three shapes. One shape results when an environmental benefit improves continually with increasing GDP, the total cost of a country's goods and services. **FIGURE 24-2** shows water use plotted against GDP in Sydney, Australia.

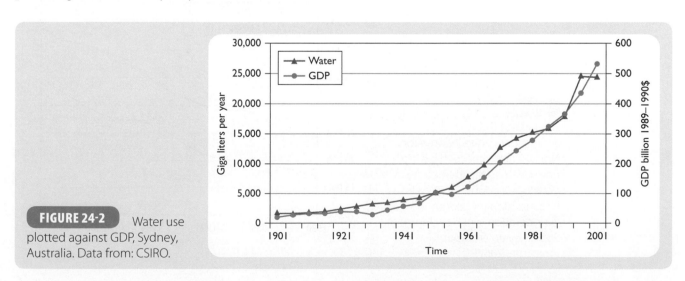

FIGURE 24-2 Water use plotted against GDP, Sydney, Australia. Data from: CSIRO.

In this case, the benefits to individuals are high (survival is at stake) and the costs of providing the service are fairly low.

FIGURE 24-3 illustrates a second shape of Kuznets curves. It shows a continuous increase in municipal solid waste (MSW) with rising incomes. Economists generally regard an increase in MSW per capita as a manageable problem for wealthy nations.

Concept Check 24-2. Do you agree that the increase in MSW is "a manageable problem" in wealthy nations? Explain your answer.

The Kuznets curve that has received the most attention, and has stimulated the most discussion, has an inverted "U" shape. The top of the curve is called the *turning*, or *tipping point*, where conditions markedly begin to change. It has been used to suggest the path air quality will follow as economic development, and presumably per capita income, increases (**FIGURE 24-4**).

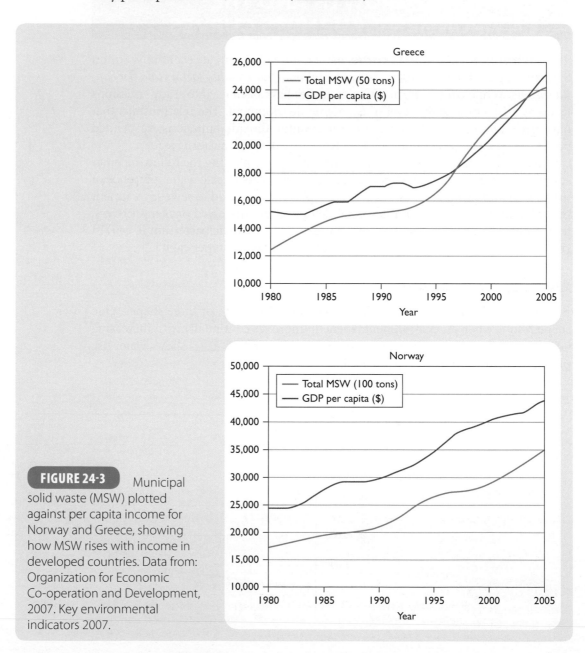

FIGURE 24-3 Municipal solid waste (MSW) plotted against per capita income for Norway and Greece, showing how MSW rises with income in developed countries. Data from: Organization for Economic Co-operation and Development, 2007. Key environmental indicators 2007.

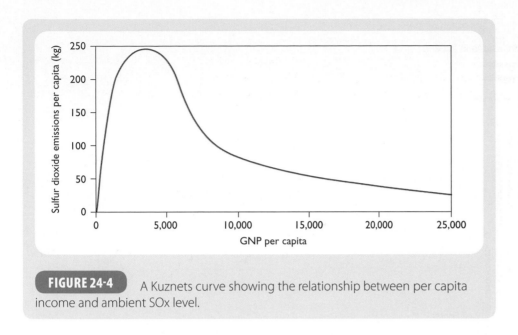

FIGURE 24-4 A Kuznets curve showing the relationship between per capita income and ambient SOx level.

The inverted "U" shape here suggests that sulfur dioxide levels increase in the early stages of a nation's economic development, eventually reach a maximum, and then decline, as per capita income increases. Similar U-shaped curves have been reported for particulates.

Researchers differ on levels of income necessary to stimulate the "turning points" in the graphs. They were found to vary between $3,000 and $8,700 for sulfur dioxide, and to range up to $10,300 for suspended particulate matter.

Concept Check 24-3. Is it reasonable to conclude that some environmental impacts of economic development are not serious because they will decline over time? Explain.

Certain types of environmental degradation can be offset at a cost. Scrubbers on power plants can remove 80 to 90% of SOx, for example, and increased per capita income gives nations the wealth with which to afford the cost. Air pollution cap and trade is discussed elsewhere.

Concept Check 24-4. How could developed countries help others to take a "short cut" around the middle of the Kuznets curve?

Carbon emissions per capita show a pattern similar to MSW (**FIGURE 24-5**). Thus, a reduction in global carbon (as CO_2) concentration could require an integrated worldwide approach.

Criticisms of Kuznets Curve Use

All economists are not as sanguine about the predictive value of Kuznets curves as the preceding section might suggest. Some economists point out that the data are usually analyzed by country, do not take into consideration international trade, nor the likelihood that wealthier countries are exporting some of their environmental problems to less developed countries. They also cannot be extrapolated to the planet as a whole from individual countries. However, even when an EKC relationship is accepted, the **turning point** on the curve (that is, where environmental degradation starts to decline with increasing per capita income) is often found to

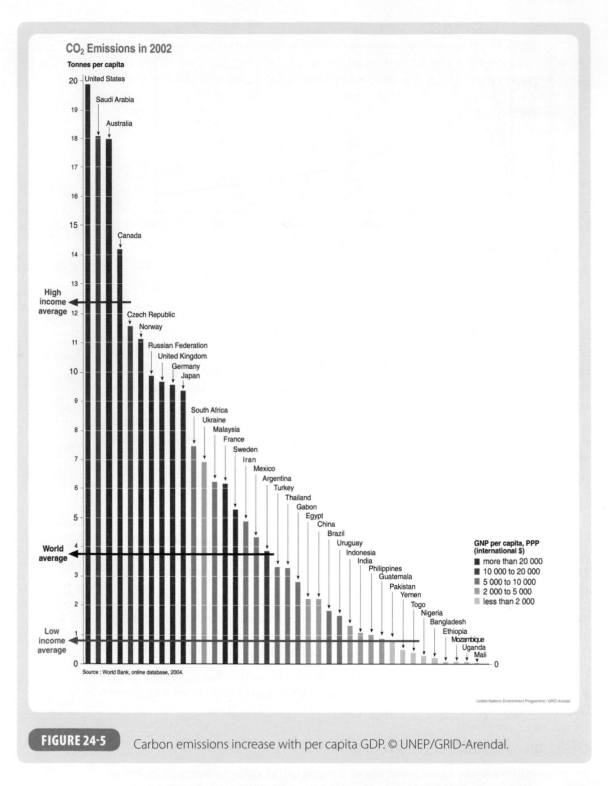

FIGURE 24-5 Carbon emissions increase with per capita GDP. © UNEP/GRID-Arendal.

be very high relative to the current per capita GDP levels of most countries of the world. For example, a 1997 study found that none of the hypothesized EKC turning points for environmental indicators is below the minimum income level of the countries studied. In the case of tropical deforestation, researchers found that per capita income levels of most countries in Latin America and Africa were well below the estimated turning point peaks, implying that deforestation could remove all forests before those societies could afford to cease the practice.

Overall, such results suggest that the majority of countries have not yet reached levels of per capita income for which environmental improvement is likely to spontaneously occur, unless this improvement is dictated by central governments. The implications are for worsening global environmental degradation even as the global economy expands and populations grow, and even as some countries make progress cleaning up specific aspects of their environment. **FIGURE 24-6** is one projection of future global SOx levels.

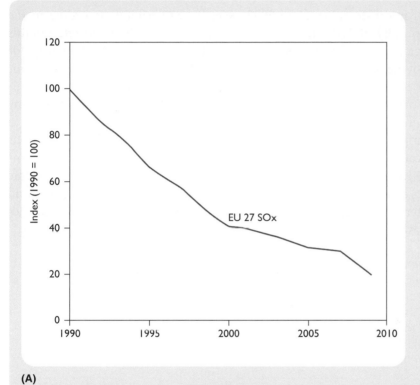

(A)

FIGURE 24-6 (A) Historical change in SOx emissions for the 27 nations of the EU (European Union). Data from: The United Nations Economic Commission for Europe (Environment and Human Settlements Division). (B) Deaths from urban air pollution, including SOx. *Question:* Could global per capita SOx decline, regardless of the absolute increase? Reproduced from Global Health Observatory Map Gallery-Deaths Attributable to outdoor air pollution, 2008/WHO.

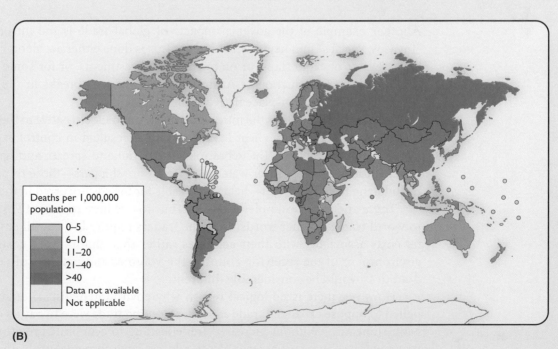

(B)

University of Nottingham (United Kingdom) economists concluded that, "meaningful EKCs exist only for local air pollutants, whilst indicators with a more global, or indirect, impact either increase . . . with income, or else have predicted turning points at high per capita income levels, with large standard errors. . . ."

GLOBAL TRADE AND ENVIRONMENTAL QUALITY

The value of international trade in 2008 exceeded $12 trillion, according to the United Nations. By 2011, this number had grown to $18.2 trillion. The effect of this amount of global trade on environmental quality is controversial. Obviously, for example, moving oil by tanker can lead to oil spills. And it is certainly possible for countries or regions to improve their own local environment by "exporting" polluting industries to other nations. For example, during the 1970s and 1980s many metal smelters closed in North America, due to the unwillingness of operators to invest in antipollution technology as required by law. Many of these operations simply relocated outside North America to developing countries, whose governments valued the jobs created by the new industry, and were willing to tolerate the resulting deterioration of their environment. Similarly, the United States and several western European countries exported toxic wastes like PCBs to countries like Nigeria, where "disposal" costs were a fraction of those in the home country.

Concept Check 24-5. Ignoring OPEC (Organization of Petroleum Exporting Countries), except for agricultural goods, all commodities have shown an increase in the proportion of exports from developing countries as compared to developed countries. Because many of these industries have historically been the source of considerable air and water pollution, is it reasonable to conclude that developed countries may be "exporting pollution" to developing countries? Why or why not? What other information would you like to have, if any, to better answer this question?

Another example of the adverse impacts of global trade is the introduction of "invasive species" into new environments. This is done either accidentally as in the ballast water of cargo ships, or on purpose as investments, or for some presumed benefit, as the introduction of Australian pines for windbreaks in the Bahamas, which have tended to displace native species (**FIGURE 24-7**).

A notorious example is the introduction of zebra mussels, native to the Black Sea, into the North American Great Lakes, which has resulted in control expenditures in the billions of dollars. The zebra mussels continue to spread, and have already infected the Mississippi River watershed. Rarely are shippers—those responsible for introducing the new species—held responsible for the cost incurred in their control.

These and other examples too numerous to cite here represent vast subsidies to world trade; in other words, were the traders required to pay all environmental costs associated with their activities rather than dump those costs onto the environment of the receiving country, the volumes and patterns of world trade doubtless would be considerably different.

However, there is no logical reason why environmentalists should feel compelled to oppose expanded trade as such, assuming trade were to pay its own "environmental way." Without increased trade leading to economic growth, countries may not have the financial resources to address environmental problems.

At the same time, unregulated "free" markets alone cannot guarantee high levels of economic growth accompanied by a clean environment. Studies in the field of environmental policy have identified many instances where unregulated markets alone do not produce an optimal allocation of resources. For example, costs associated with water pollution produced by car manufacturers may not be included in the price of cars. Or all air pollution costs resulting from electricity production may not be included in the price charged to consumers, as is the case with coal use in electricity generation. Unless some means can link the pollution abatement costs to the production process, such products will be overproduced and will "outcompete" less-polluting alternatives, like electricity from wind. Many argue that the flood of manufactured goods imported to the United States from China reflects the price advantage inherent in goods produced without regard to environmental or labor standards taken for granted in developed countries.

■ ADDRESSING ENVIRONMENTAL DEGRADATION FROM TRADE AND DEVELOPMENT

In the case of pollution that crosses national boundaries, addressing the problem usually requires international agreements, and we give two examples below. Such punitive measures as sanctions and tariffs may be counterproductive in the view of some economists, however. Tariffs on forest products, for example, have not prevented excessive logging and deforestation in Indonesia. Rather, some believe the problem to be insufficient domestic regulation. As researchers at the Progressive Policy Institute note, "Imposing trade sanctions on goods from poor countries with lax environmental standards simply lowers their economic growth, and does nothing to counter the poverty that may be contributing to the environmental problem." Instead, effective domestic regulation accompanied by "multilateral environmental agreements (MEAs)" that are both targeted and enforceable could be more effective. Technical assistance to help the poorest nations meet higher standards also can represent good investments in global environmental protection. To what extent such proposals are practical, however, continues to be debated.

Alternatively, voluntary international standard-setting organizations could develop guidelines for labeling, similar to those used to indicate sustainable forestry practices, or products of organic agriculture.

Some attempts have been made to address the most egregious problems associated with under-regulated trade. The Basel Convention of 1988 began to set controls on the export of toxic wastes from Organisation for Economic Cooperation and Development (OECD) countries to less-developed countries for disposal. In the United States, a series of environmental laws beginning with the National Environmental Policy Act of 1969 and a number of clean air and clean water legislations, culminating with the 1990 Clean Air Act Amendments, have reduced—but by no means eliminated—the amount of air and water pollution exported to Canada and Mexico.

To eliminate the adverse environmental impacts of global trade, many environmental economists, geologists, and ecologists recommend a number of new international or multilateral agreements. They include proposals to:

- Eliminate invasive species from the ballast water of vessels involved in commercial activity in international waterways.
- Eliminate international transboundary air and water pollution from industries engaged in manufacture for export.
- Require "double-hulled" tankers for the shipment of petroleum. This is already required for shipment of oil in U.S. territorial waters.
- Promote international, bilateral, or multilateral negotiations to ensure that workers engaged in export industries receive benefits and wages that are at parity with those in the countries to which the goods are shipped.
- Promote international agreements to regulate such potentially environmentally degrading activities as clear-cutting and metal smelting, so that industries cannot be subsidized by locally lax environmental standards for moving their polluting activities from one country to another.
- Promote international or bilateral agreements to eliminate or regulate the use of toxic materials or harmful practices in agriculture.

The World Trade Organization

In 1999, Seattle, Washington was rocked by sometimes-violent protests against perceived policies of the World Trade Organization (WTO; **FIGURE 24-8**). The issues raised by those protests are beyond the scope of this book. However, we will use one example to illustrate the controversy: the issue of sea turtles.

Five Asian nations challenged a U.S. law designed to protect sea turtles from certain harmful fishing practices. The U.S. law banned the importation of shrimp from countries that did not require the use of turtle-excluder devices by their fishing industry. The WTO dispute panel ruled in favor of the Asian countries, not because it disapproved of U.S. attempts to protect sea turtles, but because the panel found the United States had discriminated among members of the WTO, granting preferential treatment to Latin American and Caribbean nations, but not to Asian countries. This decision infuriated environmentalists even though it had arguably nothing to do with the desirability of saving sea turtles.

Indeed, the founding charter of the WTO formally addresses the nexus of trade and the environment. You can find it online at http://www.wto.org/. The opening paragraph states that signatories to the WTO should, "allow for the optimal use of the world's resources in accordance with the objective of sustainable development, seeking to both protect and preserve the environment." WTO rules also allow countries to impose trade regulations "necessary to protect human, animal,

FIGURE 24-8 World Trade Organization (WTO) protests in Seattle, 1999. © Beth A. Keiser/AP Photo.

or plant life, or health" or "relating to the conservation of natural resources." What's critical in the WTO rules is that measures taken to protect the environment must not be "unfair." For example, they must not discriminate. A country may not be lenient with its domestic producers and at the same time be strict with foreign producers. Nor can member nations discriminate between different trading partners.

The WTO is described as a "system of rules designed to ensure open, fair, and undistorted commerce" at the international level. It does not dictate national government policy. Sovereign nations choose to become members of the WTO and to play by its rules. So far, about 157 countries have joined, and others have applied for membership.

International Agreements to Protect the Environment

We present two examples of relatively successful international agreements to protect the global environment: the Basel Accords on Toxic Waste, and the Montreal Protocol.

Focus: The Basel Accords on Toxic Waste

This international agreement became effective in 1992. Three signatories to the Convention had as of 2013 not yet ratified it: Afghanistan, Haiti, and the United States of America.

It identified nearly 4-dozen specific categories of waste to be deemed toxic, and banned export of any of these materials from any signatory nation to any signatory nation which indicated that it did not want to receive such waste. It contained articles defining illegal activity in waste shipment, and required all signatory states to act to eliminate such activity. It stipulated that such waste should be minimized, and disposal or treatment should occur as close to the source as possible.

The Montreal Protocol on Substances that Deplete the Ozone Layer was adopted in 1987 as an international treaty whose objective was to eliminate the production and consumption of ozone-depleting chemicals. Four agencies were tasked with implementing the Protocol: The World Bank, and three United Nations (UN) agencies: the UN Environment Programme, the UN Development Programme and the UN Industrial Development Programme.

The Montreal Protocol stipulates that the production and consumption of **ozone depleting chemicals (ODCs)**, compounds that deplete ozone in the stratosphere—chlorofluorocarbons (CFCs), halons, and carbon tetrachloride, among others—were to be phased out by 2000, or 2005 in the case of methyl chloroform and methyl bromate.

Developing countries were given time to cap, and then eliminate, ODCs.

We consider three of those cases next.

China is the developing world's fastest growing consumer and producer of ozone depleting substances. Undergoing rapid industrialization and urbanization, China's ODS consumption grew around 12 to 15% annually between 1986 and 1996. China ratified the Montreal Protocol in June 1991, and was to gradually phase out ODS until total elimination was achieved. By 2012, China reports that it has phased out more than 100,000 tons of ODS since 1991

India signed the Protocol in 1992, and phased out production and consumption of ODS by the year 2010, with financial assistance from the World Bank.

In December 1991, the *Russian Federation* committed itself to continuing its membership in all international conventions, including the Montreal Protocol. Due to financial difficulties arising out of the collapse of the USSR, the Russian Federation was not able to comply with the Protocol's requirement that industrialized countries completely phase-out both production and use of ozone depleting substances by the end of 1995. With financial assistance from the World Bank, Russia set a goal to complete phase-out. Before 1995, the contraction of the former Soviet economy had reduced Russian consumption of ODS from 100,000 tons per year to 30,000 tons, only slightly larger than China's. However, Russia's inability to eliminate ODS production led to a brief "black market" in ODS in North America and Europe, where such compounds as freon were traded illegally due to the high cost of substitutes.

■ CONCLUSION

World trade has enormous potential to foster the objectives of development, but can also be the source of massive environmental degradation without multinational and international agreements to "level the playing field." The Basel Convention and Montreal Protocol are examples of the types of international agreements that could serve as templates for new agreements to address the negative impacts of underregulated international trade.

The debate about the impacts of development continues globally and locally.

Kuznets Curves are one means of analysis, howbeit controversial, by which data on the consequences of regional and national development may be evaluated.

CHAPTER SUMMARY

1. Development refers to a complex set of changes that convert the economy of a society from one based on subsistence agriculture, to one in which most of the employed inhabitants work in manufacturing or service industries.

2. Nations are divided by the United Nations into developed, developing, and least developed.

3. Many environmentalists point out that development leads to harmful land-use practices and results in increasing forms of pollution.

4. Some observers point out the considerable empirical evidence that (1) population growth rates decline as development proceeds and (2) rates of some forms of environmental pollution decline as per capita income increases.

5. The environmental Kuznets curve, named for Nobel laureate economist Simon S. Kuznets, plots the relationship between environmental quality factors and per capita income.

6. There are three types of Kuznets curves. One curve plots an environmental benefit that improves continually with increasing per capita income. A second curve, that has received the most attention, and has stimulated the most discussion, has an inverted 'U' shape. A third curve shape plots an environmental impact that worsens with increasing income.

7. Criticisms of Kuznets curves focus on the lack of reliability of some countries' data, and that the *turning point* on the curve (that is, where environmental degradation starts to decline with increasing per capita income) is often found to be very high relative to the current per capita GDP levels of most countries of the world.

8. The effect of global trade on environmental quality is controversial. Without increased trade leading to economic growth, countries may not have the financial resources to address environmental problems. At the same time, unregulated "free" markets alone cannot guarantee high levels of economic growth accompanied by a clean environment. Many countries have exported waste to others, and polluting industries have migrated from countries with strict environmental laws, to those with lax enforcement of environmental laws.

9. International agreements may provide a mechanism to address some of the environmental problems associated with global trade. Examples are the Montreal protocol and the Basel Convention on Toxic Waste.

10. The World Trade Organization (WTO) is a controversial international organization, whose founding charter formally addresses the nexus of trade and the environment.

KEY TERMS

development

environmental
 Kuznets curve
 (EKC)

ozone depleting
 chemicals (ODCs)

turning point (on
 EKCs)

REVIEW QUESTIONS

1. Kuznets curves plot
   a. lead levels against per capita income.
   b. environmental quality factors against population.
   c. environmental quality factors against per capita income.
   d. per capita income against population.

2. A Kuznets curve that plots SOx emissions against per capita income has which of the following shapes?
   a. an inverted U shape.
   b. a straight line: as income increases, SOx increases.
   c. a straight line: as income increases, SOx decreases.
   d. a curve: SOx emissions increase faster than per capita income.

3. Kuznets curves that plot carbon emissions against per capita income exhibit which of the following shapes or relationships?
   a. an inverted U shape
   b. a straight line: as income increases, carbon emissions increase
   c. a straight line: as income increases, carbon emissions decline
   d. a curve, showing carbon emissions decrease at a faster rate than incomes increase.

4. In a Kuznets curve with an inverted U shape,
   a. the "turning point" always occurs at the same per capita income.
   b. the turning point varies from country to country.
   c. the turning point occurs at a lower per capita income for developed than developing countries.
   d. there is no turning point for Kuznets curves of developed countries.

5. A widespread effect of global trade is the spread of exotic, or invasive, species. The most likely means to transmit exotic species is by
   a. private automobile.
   b. on the boots of hikers.
   c. in the ballast water of tankers.
   d. through air travel.

6. The Montreal Protocol addresses which form of global pollution?
   a. the emission of ozone-destroying chemicals.
   b. the international shipment of toxic waste.
   c. the global transport of atmospheric greenhouse gases.
   d. the global emission of mercury.

7. The Basel Accords deal with the international shipment of
   a. toxic waste.
   b. ozone-depleting chemicals.
   c. PCBs.
   d. waste from nuclear power plants.

APPENDIX I *Periodic Table of Elements*

1 IA																	18 VIIIA
lithium 3 **Li** 6.94	beryllium 4 **Be** 9.01											boron 5 **B** 10.81	carbon 6 **C** 12.01	nitrogen 7 **N** 14.01	oxygen 8 **O** 16.00	fluorine 9 **F** 19.00	neon 10 **Ne** 20.18

Element — hydrogen
Atomic Number — 1
Symbol — **H**
*Atomic Mass — 1.01

Metals
Metalloids
Nonmetals

helium 2 **He** 4.00

Lanthanide Series

| cerium 58 **Ce** 140.12 | praseodymium 59 **Pr** 140.9* | neodymium 60 **Nd** 144.24 | promethium 61 **Pm** (147) | samarium 62 **Sm** 150.36 | europium 63 **Eu** 151.97 | gadolinium 64 **Gd** 157.25 | terbium 65 **Tb** 158.93 | dysprosium 66 **Dy** 162.50 | holmium 67 **Ho** 164.93 | erbium 68 **Er** 167.26 | thulium 69 **Tm** 168.93 | ytterbium 70 **Yb** 173.04 | lutetium 71 **Lu** 174.97 |

Actinide Series

| thorium 90 **Th** 232.04 | protactinium 91 **Pa** (231) | uranium 92 **U** 238.03 | neptunium 93 **Np** (237) | plutonium 94 **Pu** (244) | americium 95 **Am** (243) | curium 96 **Cm** (247) | berkelium 97 **Bk** (247) | californium 98 **Cf** (251) | einsteinium 99 **Es** (252) | fermium 100 **Fm** (257) | mendelevium 101 **Md** (258) | nobelium 102 **No** (259) | lawrencium 103 **Lr** (260) |

*Note: For radioactive elements, the mass number of an important isotope is shown in parenthesis; for thorium and uranium, the atomic mass of the naturally occurring radioisotopes is given.

APPENDIX II *Geological Time Scale*

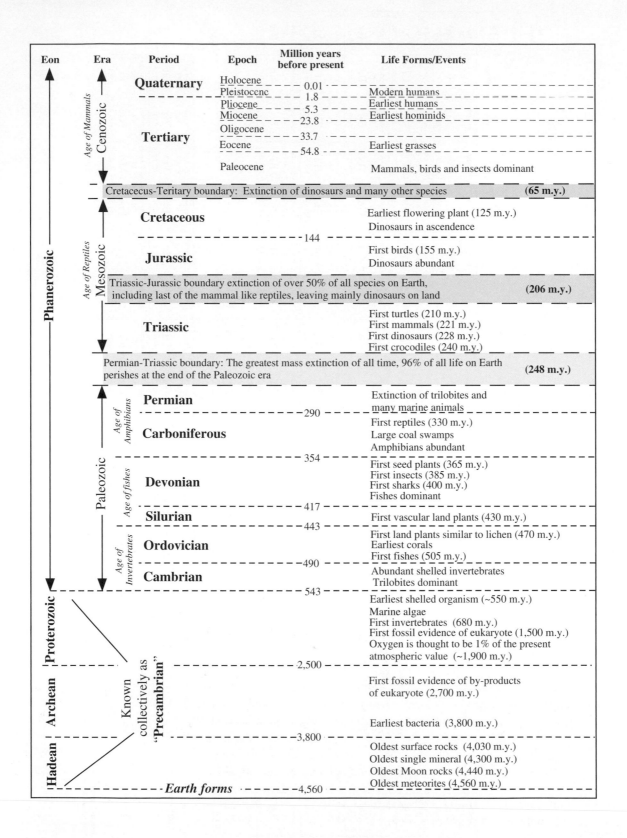

APPENDIX III
Minerals and Rocks

OUTLINE

■ INTRODUCTION: ROCK FORMATION ON HAWAI'I

The Big Island of Hawai'i contains two of Earth's most active volcanoes: Mauna Loa and Kilauea (FIGURE APIII-1). Kilauea is so active that the United States Geological Survey and other organizations have set up observatories on the island.

Eruptions on Hawai'i follow the same general pattern. Preceding the eruption by a number of weeks is an earthquake "swarm," numerous earthquakes whose focal depths (depths of origin) are as much as 60 kilometers below the volcano's summit.

Over the succeeding weeks, the earthquake swarm becomes shallower, until the volcano is observed to swell. Then, an eruption of incandescent lava occurs (FIGURE APIII-2). The lava has a low **viscosity**, meaning it flows easily and rapidly, although usually not fast enough to threaten human life. After a few minutes the red-hot surface of the new flow begins to darken and, eventually, a crust forms. Hand samples of this crust are brownish-black rock, often laced with holes called vesicles. Only rarely are any crystals visible. The holes result from escaping gas (much like the holes in bread). This gas pressure actually helps force the lava to the surface.

After several days, an entire lava flow may be solid (but still hot). Solidification takes longer with increasing flow thickness. Geologists often take samples of

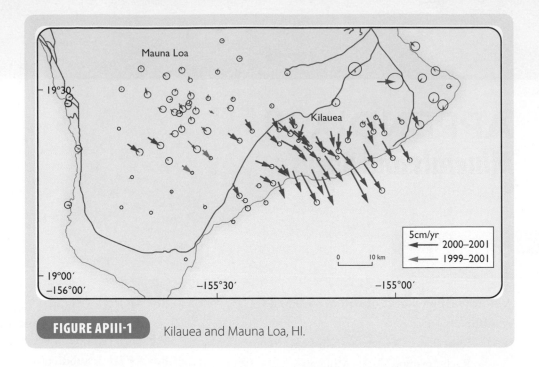

FIGURE APIII-1 Kilauea and Mauna Loa, HI.

the glowing lava before it has a chance to congeal (freeze) (**FIGURE APIII-3**). The sample is plunged into a pail of distilled water, which freezes the molten material, trapping any gases it may contain. The most common **volcanic gas** is usually, but not always, water vapor (H_2O), followed by carbon dioxide (CO_2). Volcanoes also release smaller amounts of other gases including hydrogen sulfide (H_2S), sulfur dioxide (SO_2), hydrogen chloride (HCL), hydrogen fluoride (HF), and helium (He). Then, the sample is chemically analyzed. Two analyses are shown in **TABLE APIII-1** .

FIGURE APIII-2 A lava eruption on Hawai'i. © iStockphoto/Thinkstock.

FIGURE APIII-3 Geologist collecting a hand sample of a newly-hardened lava flow, Hawai'i. Courtesy of U.S. Department of the Interior/USGS.

Lava samples from oceanic volcanoes the world over have compositions similar to, but, importantly, not exactly like, those from Kilauea. Such lavas are called **basaltic** lavas, and the rocks resulting from them, basalts, which we will define more precisely later.

Basalts are one of the most common rocks on Earth—and, indeed, one of the most common in the solar system, since samples of basalt have been brought back

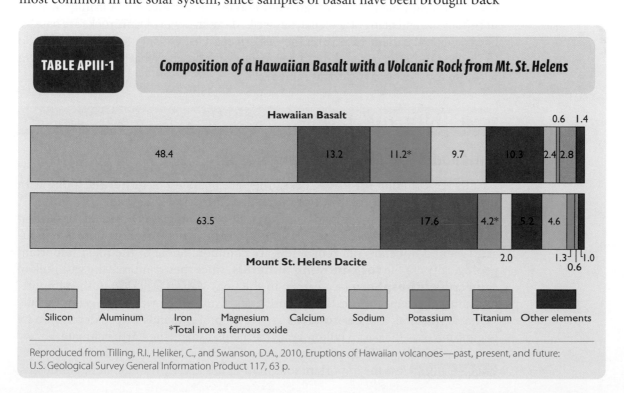

TABLE APIII-1 **Composition of a Hawaiian Basalt with a Volcanic Rock from Mt. St. Helens**

Hawaiian Basalt: Silicon 48.4, Aluminum 13.2, Iron 11.2*, Magnesium 9.7, Calcium 10.3, Sodium 2.4, Potassium 2.8, Titanium 0.6, Other elements 1.4

Mount St. Helens Dacite: Silicon 63.5, Aluminum 17.6, Iron 4.2*, Magnesium 2.0, Calcium 5.2, Sodium 4.6, Potassium 1.3, Titanium 0.6, Other elements 1.0

Legend: Silicon, Aluminum, Iron, Magnesium, Calcium, Sodium, Potassium, Titanium, Other elements
*Total iron as ferrous oxide

Reproduced from Tilling, R.I., Heliker, C., and Swanson, D.A., 2010, Eruptions of Hawaiian volcanoes—past, present, and future: U.S. Geological Survey General Information Product 117, 63 p.

Fifty million year-old basalts on the island of Skye, western Scotland. © Karola i Marek/ShutterStock, Inc.

from the Moon, and Martian Landers have analyzed rocks on the surface of Mars that look like, and have similar compositions to, basalts on Earth.

Samples of **volcanic rocks** on the extremely active island of Iceland also are overwhelmingly basalts. Basalts are common on Mount Rainier, Washington state, in Indonesia, and many other locales as well. Many ancient volcanic rocks are basalts, like these on the Island of Skye off the western coast of Scotland (**FIGURE APIII-4**). So to understand minerals and rocks, lets begin with basalt. What is basalt? To begin with, we must explain rocks and minerals.

■ MINERALS

Ionic Radii, Crystals, and Mineral Groups

Basalts are common **igneous** rocks. They are made up of **minerals**, along with masses of unordered (that is, not crystallized) material called volcanic **glass**. A mineral is *a crystalline substance with an electrically-balanced chemical formula and a defining set of chemical and physical characteristics*. There are more than two thousand named minerals, but only a few are common. Some of the most important geologically are the **rock-forming minerals**, since they comprise the vast bulk of the Earth's **lithosphere** (see earlier in the text) (**TABLE APIII-2**). Other important mineral groups include the **ore** minerals, from which we obtain economically valuable material. The **carbonates** are a group of minerals whose deposition by organisms over the past 3 billion years has helped shape the Earth's climate.

Minerals are made up of **ions**, which are electrically charged varieties of atoms. **TABLE APIII-3** lists the important ions in minerals and their ionic radii. For details,

TABLE APIII-2 — Rock Forming Minerals and Mineral Groups

Group	Mineral
Silicates	Olivine
	Pyroxene(s)
	Amphibole(s)
	Mica(s)
	Feldspar(s)
	Quartz
Carbonates	
	Calcite
	Dolomite
Sulfates	Gypsum
Halides	Halite (rock salt)

TABLE APIII-3 — Common Ions in Minerals, with Their Ionic Radii (in Units of 10^{-8} cm)

Ion (Symbol)	Charge	Radius
Oxygen (O)	−2	1.42
Silicon (Si)	+4	0.42
Aluminum/Aluminium (Al)	+3	0.51
Iron (Fe)	+2, +3	0.76, 0.64
Magnesium (Mg)	+2	0.66
Calcium (Ca)	+2	0.99
Sodium (Na)	+1	0.99
Potassium (K)	+1	1.4

(A) (B)

FIGURE APIII-5 A Si-O tetrahedron, which is the basic building block of all silicate minerals.

see the periodic table in Appendix I. Ions form when atoms gain or lose electrons, becoming charged. They are either positive (**cations**) or negative (**anions**).

How the anions and cations unite to form the regularly repeated structures we call crystals depends mainly on the size and charge of the ions. For sizes of ions, we will use here the Angstrom Unit (A), which is 10^{-8} cm. The ratio between the radius of the anion and the radius of the cation determines the crystal structure. For silicates, Si^{4+} has a radius of 0.42A, and O^{2-} has a radius of 1.4A. The most stable structure that can be built out of ions with these radii is called a **tetrahedron** (**FIGURE APIII-5**). The formula of a tetrahedron is SiO_4, which means that this structure has one small silicon cation surrounded by four large oxygen anions. Now, stop and add the positive and negative charges on the tetrahedron. Your answer should have been −4. To make a **silicate mineral**, either four positive charges must soak up these negative charges, or the tetrahedra must be linked by sharing adjacent oxygen ions, and thereby reducing the excess negative charges.

A silicate structure in which all four oxygen ions in a tetrahedron are shared with the oxygens in adjacent tetrahedra would have 4 negative charges, and 4 positive charges from the central Si, and so would be electrically neutral. This is in fact the structure of **quartz**, a common silicate mineral.

Silicate minerals are divided into groups (**TABLE APIII-4**).

Rock-Forming Silicate Minerals

We will next briefly discuss each group of rock forming silicates.

Feldspars

Table AP III-4 listed the important silicate structures and groups. The most common silicate group is the **feldspars** (**FIGURE APIII-6**). Feldspars have a framework

TABLE APIII-4	**Common Silicate Groups. A, B = Cations (Positively Charged Ions)**		
Name	**Formula**	**Examples**	**Si:O**
Isolated tetrahedra	A,B $(SiO_4)n$	olivine	1:4
Chains	A,B $(SiO_3)n$	pyroxene	1:3
Double chains	A,B (Si_4O_{11}) (OH)	amphiboles	—
Sheets	A,B $(Si_2O_5)(OH)_4$	micas, clays	—
Framework	A,B $(AlSi)_4O_8$	feldspar	1:2
	$(SiO_2)n$	quartz	1:2

FIGURE APIII-6   An example of a feldspar with formula: albite. Formula: KAl, Si3O8. May also be written 1/2K2O, 1/2Al2O3, 3SiO2. Count the numbers of the atoms and verify the formula for yourself. Courtesy of U.S. Geological Survey.

structure and exhibit another characteristic feature: **ionic substitution**. This is a phenomenon in which ions of the same charge, and similar size, can substitute for each other in a crystal structure. In lavas, the elements available are some combination of the elements you saw in Tables AP III-1 and AP III-2. So it shouldn't come as any surprise that lavas contain silicate minerals.

There are two types of feldspar: **plagioclase** and **orthoclase**, differing in composition, physical appearance, and optical properties. In the plagioclases, aluminum substitutes for silicon in the center of the tetrahedron, while sodium substitutes for calcium. In orthoclase, potassium enters the crystal structure when aluminum substitutes for silicon. This is shown in a code we use to describe a mineral, its chemical formula. The chemical formula for the type of feldspar called orthoclase is $KAlSi_3O_8$. Now remember that because feldspar is a framework silicate, its Si:O ratio has to be 1:2. To understand how to derive the 1:2 ratio from the formula, recall that both Al and Si can exist at the center of a tetrahedron. In fact, that's the only place they are found in feldspars. So in orthoclase if you add the Al in the formula to the 3Si you get 4, and if you write the ratio of Al+Si: O you get 4:8, which reduces to 1:2.

We used feldspar to illustrate ionic substitution, but ionic substitution is a feature of every common silicate mineral except quartz. In quartz, the formula is always SiO_2. **FIGURE APIII-7** shows examples of other important silicate minerals.

Olivine

If you should visit the Big Island of Hawai'i, you will see necklaces and other jewelry made from strings of green **olivine** minerals. In the olivine group of minerals,

FIGURE APIII-7 Examples of silicate minerals. (A) Olivine in peridotite rock. © Siim Sepp/ShutterStock, Inc. (B) Pyroxene crystals. © Dea A. Rizzi/age fotostock. (C) Amphibole. © Dirk Wiersma/Science Source. (D) Mica. © Tyler Boyes/ShutterStock, Inc. (E) Quartz. © Nadezhda Bolotina/ShutterStock, Inc.

the formula is $(Mg,Fe)_2SiO_4$. Virtually any proportion of Mg to Fe cations can be found in olivine, and this determines the mineral's color—more Mg, darker green; more Fe, more yellowish brown. To visualize the structure of olivine, remember the excess negative charge on the SiO_4 unit. What was it? You should remember it was −4. Now to make a mineral, the negative and positive charges have to balance, so for each −4 charge there must be +4. In olivine, the +4 comes from two iron or magnesium ions. Note here the iron has a +2 charge and not +3. If the mineral weathers, the iron may oxidize (that is, change from the +2 to the +3 ionic state) and come out of the crystal structure. This is why many olivines look rusty—the iron has started to come out of the crystal structure.

Pyroxene

Pyroxene is another very common silicate *group*, meaning that there are several common minerals that are varieties of pyroxenes. Refer back to Figure AP III-7 for details.

Amphibole

Amphiboles are hydrous silicates, meaning they have water in the crystal structure. Amphibole is not common in basalts, but is widespread in other igneous rocks, and in metamorphic rocks as well. Metamorphosed basalts can, however, contain abundant amphibole.

Mica

The **mica** group is found in a wide variety of igneous and metamorphic rocks and may also be found in sedimentary rocks as well. They have a sheet structure and contain abundant OH. Micas are valuable economic minerals as we discuss earlier in the text.

Clays

Although clay minerals do not crystallize from magmas, we introduce them here because they are so important in environmental geology. They are a **weathering product** of other silicates, especially feldspars, and volcanic glass. There are many types of clays (**FIGURE APIII-8**) and they may contain abundant water in their crystal structure. The amount of water in clays may vary depending on how much water is in the mineral's environment. Clays can absorb water when wetted, and expel it when dried. Thus, they can *change volume*. This volume change is accompanied by great pressure, which can mean that any structure built on clay-rich soils can be cracked and damaged. Such expansive clays are discussed in detail earlier in the text.

Quartz

You have already derived the structure of quartz. This mineral has widespread uses, in addition to being valued

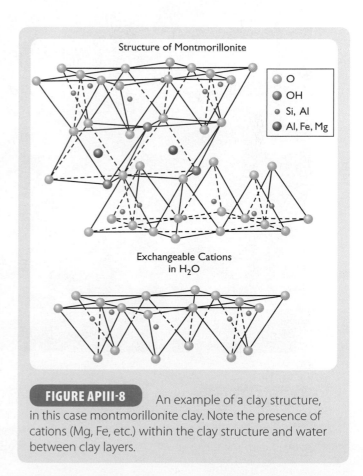

Structure of Montmorillonite

○ O
◐ OH
· Si, Al
● Al, Fe, Mg

Exchangeable Cations in H_2O

FIGURE APIII-8   An example of a clay structure, in this case montmorillonite clay. Note the presence of cations (Mg, Fe, etc.) within the clay structure and water between clay layers.

as a semiprecious stone. Very pure quartz could be the basis for a new photovoltaic (i.e., solar power cell) industry.

Other Minerals and Mineral Groups

Carbonates

Carbonates are minerals made of cations attached to the carbonate structure CO_3^{-2}. They form mainly in seawater, often precipitated by organisms like calcareous algae and corals (FIGURE APIII-9). The most common carbonates are calcite and dolomite. Calcite and the similar mineral aragonite make up limestones, and dolomite forms dolostones. They are important parts of the **carbon cycle**, since storage ("fixing") of C in rock removes it from the atmosphere (as CO_2). Carbonates are also important geologically as sources of cement, where the reaction is $CaCO_3$ + heat > CaO (lime) + CO_2. Thus, cement manufacture is an important source of the global warming gas CO_2.

(A)   (B)

FIGURE APIII-9   (A) A calcareous alga, belonging to the genus Penicillus. They are common in the Florida Keys and throughout the Caribbean. The green color is chlorophyll, and the carbonate is present as tiny needle-like crystals supporting the plant. © John M. Huisan, http://florabase.dec.wa.gov.au/. (B) A scanning electron microscope view of carbonate crystals. © Susumu Nishinaga/Science Source.

FIGURE APIII-10   Pyrite, or "fool's gold," a common sulfide mineral. © Jens Mayer/ ShutterStock, Inc.

Sulfides

Sulfides are formed when one or more cations unite with sulfur. They constitute many of the most common ore minerals, including sphalerite (ZnS), chalcopyrite (CuFeS$_2$), and galena (PbS). **FIGURE APIII-10** shows an example of these minerals. Refining of sulfide ores to recover the metals involves "roasting" the ores in air. This is an energy-intensive process and also a major source of air pollution, since the by-product is SO$_2$, a toxic air contaminant, and a major source of acid precipitation. Before passage of the first Clean Air Act in 1970, areas around metal smelters often resembled lunar landscapes, since few living things could survive the sulfide gas concentrations. Since then, smelters in the United States have either installed SO$_2$ "scrubbers" or have closed (**FIGURE APIII-11**).

Sulfates

Common **sulfate** minerals are produced by evaporating water: either seawater, or lakes in desert regions. Two examples are **gypsum** (CaSO$_4$.nH$_2$O), where the "n" refers to a variable number of water molecules, and anhydrite (CaSO$_4$). Gypsum is used extensively as wallboard in construction. Gypsum can also be produced from the by-product of SO$_2$ removal (using scrubbers) from the exhaust gases of coal-fired power plants.

Oxides and Hydroxides

These groups contain important ores as well as gems. Common minerals are shown in **TABLE APIII-5**.

(A)

(B)

FIGURE APIII-11 Views of Ducktown, Tennessee. (A) Ducktown area in the 1960s. © Ducktown Basin Museum. (B) Ducktown in the 1980s. Note the lurid green of the water from dissolved copper. © Daniel Woodrum.

■ ROCKS

Rocks are *aggregates of minerals*. There are three categories of rocks on Earth: igneous; formed from magma, a molten silicate solution; **metamorphic**, formed by heating and altering other rocks; and **sedimentary**, formed at the Earth's surface from layers of surface debris, by organisms, or by precipitates from water. The three categories of rocks are shown in FIGURE APIII-12 , a representation of the **rock cycle**.

Igneous Rocks

Volcanic (Extrusive) Rocks

We began this chapter with a look at volcanic activity on Hawai'i. For millennia volcanoes have profoundly affected humankind. They may even have to some extent controlled human migration out of Africa, by opening and closing migration routes as a result of deposition of lava in northeastern Africa, as described by geologist Steven Drury.

We discuss volcanoes and volcanic products in detail earlier in the text. We will briefly introduce the subject here.

Material that issues from volcanoes or volcanic rifts and vents is either lava or **pyroclastic** (fiery fragment) **material**. Recall that gas pressure is responsible for pushing volcanic material onto the surface. Sometimes that gas pressure is low and the material comes out as a river-like flow. This is typical of basaltic volcanoes, which have relatively low (around 50%) **silica** (Si+O) content. Other volcanoes erupt higher silica lavas such as **andesite, dacite**, and **rhyolite**. TABLE APIII-6 shows several common lava types, and their characteristics, and shows types of pyroclastic deposits. As silica content increases, the lavas become increasingly more viscous (thick and pasty). As viscosity increases, gas tends more to be trapped in the lava. Thus, gas pressure inevitably builds, often leading to great explosions, which erupt not lava streams, but incandescent particles of lava ranging in size from tiny fragments (**ash**) to large masses (bombs) (Table AP III-6).

Eruptions of silica-rich lavas and ash represent some of the most cataclysmic events in planetary history, behind only meteorite impacts in their destructive power. Fortunately, as you will see they are extremely rare.

Plutonic (Intrusive) Rocks

As we mentioned earlier, gas pressure is necessary to force lava to the surface. In addition, in the deep crust, magma is less dense than the material through which it passes, and thus magma tends to be pushed up by gravity differences. Magma cools

TABLE APIII-5	Important Mineral Groups	
Group	**Formula**	**Examples**
X = one or more cations; Y = Halide (see periodic table);		
Silicates	$x(SiO4)_n$	feldspar, quartz, etc.
Carbonates	$xCO3$	calcite, aragonite, etc.
Oxides	xO_n	corundum (ruby, sapphire)
Hydroxides	$x(OH)_n$	bauxite
Halides	xY	halite
Sulfides	$x(S)_n$	pyrite, chalcopyrite
Sulfates	$x(SO4)n$	gypsum

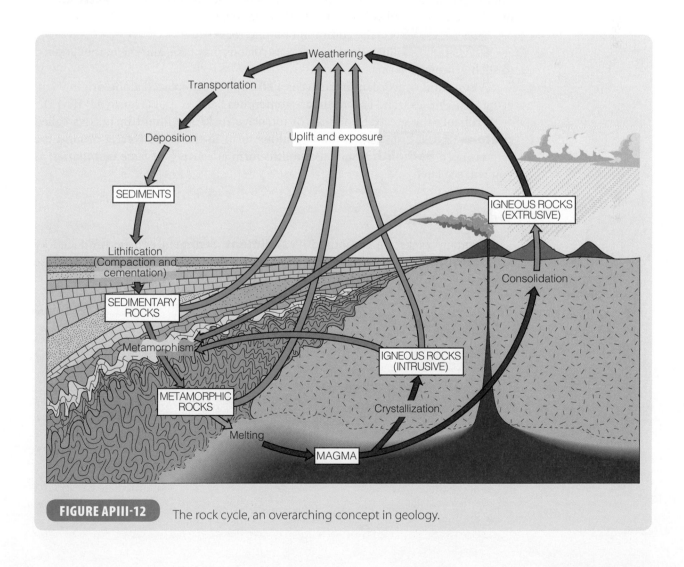

FIGURE APIII-12 The rock cycle, an overarching concept in geology.

TABLE APIII-6	Common Volcanic Rocks	
SiO₂ Content	**Rock Name**	**Comment**
Very Low nodules	Peridotite	Rare as rock, less so as in basalts
Low	Basalt	Most common volcanic rock
Medium	Andesite	Common as lava and pyroclastics
High	Rhyolite	Rare as rock, more common as fragmental and pyroclastic deposits

slowly as it rises. Often it solidifies (freezes) before reaching the surface. Magma that freezes below ground forms *intrusive or plutonic* igneous rocks. Igneous rocks that solidify at greater depths (5 to 15 km and more) tend to have well-formed crystals and no glass. Those rocks that solidify at shallower depths tend to have finer crystals. Grain size, the presence of glass, and the presence of **vesicles** can usually be used to tell volcanic from plutonic rocks.

TABLE APIII-7 lists the common plutonic rocks.

FIGURE APIII-13 illustrates the forms intrusions may take, and the names applied to such geometric forms.

A common concordant intrusive is a **sill**. Sills are emplaced along original layering in host rocks, and as a result are sometimes hard to spot (Figure AP III-13).

Probably the greatest volumes of intrusive rocks are found in masses called **plutons** (**FIGURE APIII-14**). If many plutons form in the same area at around the same time, a **batholith** results. Batholiths form at *convergent plate boundaries*, as you will see later.

Sedimentary Rocks

Sedimentary rocks are composed of **sediment**. Sediment can be produced by **weathering** of pre-existing rock, and also by organisms such as calcareous algae or corals.

TABLE APIII-7	Common Plutonic Rocks	
Common Plutonic Rock Types	**Volcanic Equivalent**	**Silica Content**
Gabbro	Basalt	Low
Diorite	Andesite	Low to Intermediate
Granodiorite	Dacite	Intermediate to High
Granite	Rhyolite	High

(A)

(B)

(C)

FIGURE APIII-13 (A) Concordant intrusive rocks are parallel to layers in surrounding rocks. These are small mafic sills. © James Steinberg/Science Source. (B) Discordant intrusive rocks cut across layers in surrounding rocks. Dikes can be quite striking in appearance. These basaltic ones are on the Island of Santorini, south of Greece. © FLPA—David Hosking/age fotostock. (C) The Great Dyke of Zimbabwe, that is several kilometers in diameter and hundreds of km long. Courtesy NASA/GSFC/MITI/ERSDAC/JAROS, and U.S./Japan ASTER Science Team.

(A)

(B)

FIGURE APIII-14 (A) A pluton in the Appalachian Mountains of western North Carolina. © J.K. York/ShutterStock, Inc. (B) The Alabama Hills, plutonic rocks in the Sierra Nevada. These are relatively young, emplaced about 100 million years ago. © Byron W. Moore/ShutterStock, Inc.

Raw materials used in the "manufacture" of sedimentary rocks are classified on the basis of origin. A common category is siliclastic material. The word siliclastic is compounded from sili- (attesting to its silicate mineral composition) and -clastic (attesting to its fragmentary nature). So the word siliclastic means fragments of silicate minerals.

Types of Siliclastic Material

Theoretically any silicate mineral, once formed into a rock, could be freed by weathering, transported by water, ice, or air, and deposited to eventually form a new sedimentary rock. But weathering at the Earth's surface is usually very efficient. This means that pre-existing silicate material is usually altered by physical processes and chemical reactions to new minerals, or is dissolved in solution. **TABLE APIII-8** shows common weathering processes acting on, and weathering products of, typical silicate minerals.

As a result, there are relatively few minerals found in **siliclastic sediment**. Common ones are quartz, orthoclase feldspars, and varieties of mica. **Clay minerals** are also common in siliclastic debris because they are a universal product of chemical weathering of feldspars and other silicates. Other weathering products are ions in solution, silica in solution, and iron oxides and hydroxides.

Under unusual circumstances (lack of water, intense cold, quick burial) unstable minerals can end up in sediment and be formed into sedimentary rocks. The very rarity of such processes and circumstances makes geologists take notice when they do see such material in sedimentary rocks.

Classification of Sedimentary Rocks

TABLE APIII-9 shows the grain size terminology used in classifying sedimentary rocks.

TABLE APIII-10 shows one classification of sedimentary rocks

For magma to form, something solid has to melt. If that melting happens in the mantle, the parent rock is **peridotite**—a magnesium-rich rock containing abundant olivine (**Figure B1-1**). Under present terrestrial conditions, there is insufficient heat from radioactive decay, the ultimate heat source, to melt all of the peridotite in the mantle in any one spot—in fact, only a small percent melts. Such a melt is called a **partial melt**. There is a rule about partial melting of silicate material that you must remember: when partial melts form, the composition of the melt (magma) depends on the original rock composition, and *what percentage of the rock is melted*. In the case of peridotite, if 100% of the rock is melted, the composition of the melt is exactly like the parent rock: peridotite. But if only 10% of the rock is melted, the magma's composition is basaltic. The reason for this is simple. Silicate melting doesn't behave like melting of the H_2O system. Stop and consider for a moment. When ice melts, no matter what percent of the ice melts, the liquid has the same composition—water. But when silicate mineral aggregates (igneous rocks) melt, the first elements to go into solution are the last ones to have entered the crystal structures when the rock originally formed.

Here's a summary of what you learned here about partial melting of silicates. The composition of the magma depends on the initial composition of the rock and the percent of the rock that melts. The early melts will be enriched in the last elements that entered crystal structures, and as more and more of the rock melts, the magma will change composition by adding those elements that entered crystal structures early. We explain why in the box below.

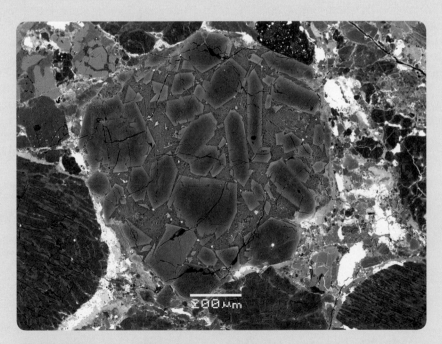

FIGURE B1-1 Zoned olivines within meteorites. Many rocks contain zoned olivine minerals, as shown here. The outer layers are iron-rich. This is a scanning electron photo. Note scale. © James Wittke.

Bowen's Research on Basaltic Rocks: Sequential Crystallization

In a lifetime of work on basaltic rocks at the Carnegie Institute in Washington, DC during the early 20th century, geochemist N. L. Bowen spent decades mixing silicate minerals, melting them, and then allowing them to cool. He would stop the cooling at different stages by plunging the mix of crystals and melt into water, thus freezing the remaining liquid. Bowen identified a **sequence of crystallization** of minerals that formed when a basaltic magma cooled. The first crystals to form were inevitably olivine and calcium-rich plagioclases. As the temperature dropped, these crystals grew, but importantly the remaining melt became enriched in whatever elements had not entered the crystals. Eventually, the difference in composition between crystals and liquid became so pronounced that reactions took place between the melt and the crystals. In the plagioclases, early plagioclase crystals responded to cooling by adding more sodium to the growing crystal. Olivine crystals *reacted* with the melt to form a new mineral, **pyroxene**. We can in fact see evidence of this in basalts. Bowen made a chart to show the sequence of **formation** of silicate minerals from basaltic melts. It is now called **Bowen's Reaction Series** (Figure B2-1). Bowen recognized two reactions going on simultaneously. One involves minerals called ferromagnesians—those with lots of iron (ferro-) and magnesium. The ferromagnesians are olivine, pyroxene, amphibole, and biotite mica. Bowen showed that these minerals form by reacting with the melt and changing crystal structures: the olivines are made of isolated tetrahedra connected by cations, the pyroxenes are made of single chains, and so on. He called this series of reactions a *discontinuous* series, because the minerals that formed changed their crystal structure.

In contrast, Bowen called the plagioclase series a *continuous* reaction series because the plagioclases did not change structure, only composition—the earliest plagioclases had mainly calcium, but as they grew they added sodium.

Now let's review the concepts of sequential crystallization and partial melting for the last time. In sequential crystallization of basaltic rocks, crystals follow a sequence when they solidify from the magma. In partial melting, the first melts consist of the last ions (elements) that solidified.

Using Bowen's rules, we can predict which rocks would form by partial melting of igneous rocks. Basaltic rocks form by partial melting of peridotites, andesitic rocks (andesites/diorites) can form by partial melting of basaltic magmas, and rhyolitic rocks (rhyolites/granite) could form by partial melting of basaltic or andesitic rocks. The processes of partial melting and sequential crystallization, accompanied by **weathering**, metamorphism, sedimentation, and the occasional meteorite impact have been responsible for the formation of the Earth's crust over the past 4 billion years.

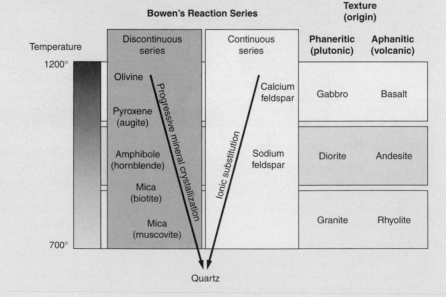

FIGURE B2-1 The two reaction series that result in the crystallization of minerals from magma, as described by Bowen. Adapted from: http://comp.uark.edu/~sboss/study2s08.htm and http://www.skidmore.edu/~jthomas/fairlysimpleexercises/brs.html.

TABLE APIII-8	Weathering Products of Common Silicates	
Mineral	**Reaction**	**Weathering Product**
Quartz	resistant	quartz grains
Feldspar	$KAlSiO + H+ = AlSiO(OH)- + K+$ $Na,CaAlSiO + H+ = AlSiO(OH) +$ $Ca_2+,Na+$	clay, K+ in solution clay, Na and Ca in solution
Olivine	$Mg_2SiO_4+H+ = SiO_4 (sol) +$ Mg^{2+}	silica, Mg ions
Ferromagnesians	$FeSiO + H+ = SiO_{sol} + Fe^{+++}$	iron oxides and hydroxides, silica

Metamorphic Rock

Metamorphic rocks form from already-existing rocks if they are subjected to increased heat and pressure. This may change original mineral size, cause new minerals to form, or both. In addition, fluids like H_2O and CO_2 may react with original minerals, especially at high temperatures, to change their composition, or even form new ones. Metamorphic rocks can form near igneous intrusions, by deep burial of rocks formed at the surface, along deep fault zones, and widely at plate boundaries. Surface rocks can be buried along faults and in regions of active sedimentation, like deltas. Common metamorphic rocks and the rocks from which they form (called protoliths) can be found in **TABLE APIII-11** .

Metamorphism and Mineral Stability

Metamorphism is a term for processes that change rocks *without melting them*. If rocks are melted by heat and/or pressure, magmas form, and the resulting rocks are

TABLE APIII-9	Grain Size Terminology Used in Classifying Sedimentary Rocks	
Category		**Size range**
Clay		1/256 mm
Silt		1/16 mm
Sand		between 1/8 mm and 2 mm
Gravel		above 2 mm

Adapted from U.S. Geological Survey Open-File Report 2011–1004.

TABLE APIII-10	Classification of Sedimentary Rocks	

Grain Size Category	Rock Name	Composition
A. Siliciclastic Rocks		
Clay	Claystone	clay minerals
Clay and Silt	Shale	clays, quartz, other
Silt	Siltstone	quartz, feldspar, etc
Sand	quartz sandstone	quartz, other
	Arkose	qtz + feldspar
	Graywacke	variable (unstable minerals)
Gravel and coarser	Conglomerate	quartz, other
	Breccia	any
B. Carbonate Rocks		
Variable	limestone	$CaCO_3$
Variable (usually fine)	dolomite/dolostone	$Ca,Mg\,CO_3$
C. Other Sedimentary Rocks		
	Coal	organic matter
	Rock salt	halite

igneous. To understand how rocks can change their structure, composition, or both without melting, recall that all minerals have fields of stability defined by temperature and pressure, outside of which they are unstable. For example, silicate minerals are stable at the pressures found in the crust and upper mantle. In the lower mantle, pressures are so high that silicate structures are not stable, and oxides, a denser arrangement of elements, are the dominant constituents.

A common silicate mineral group is the group of clays. Clays are abundant in sedimentary rocks, but are only stable at low temperatures and pressures. As T and P increase, the clays tend to recrystallize into new minerals—sometimes feldspars, sometimes micas—depending on conditions.

Nonfoliated Metamorphic Rocks

Many rocks are composed of minerals that don't form platy shapes, like sandstone (consisting of quartz and perhaps feldspar) and limestone (consisting of calcite, with or without other carbonates). When these rocks are metamorphosed, the crystals often increase in size and develop a characteristic texture. Texture is a term that refers to the size, shape and orientation of the material in a rock. Such resulting

For field study and mapping purposes, sedimentary rocks must be divided into "packages" of convenient thickness, usually on the basis of rock type. The basic unit into which sedimentary rocks are subdivided in the field is the formation. Formations are vertical successions of rocks that share some similarity, usually lithology, and which are somehow different from rocks above and below. Formations must also be thick enough to appear on a topographic map. This ordinarily means they must be at least ~20 m (~ 50 ft) thick (**Figure B3-1**).

FIGURE B3-1   Sandstone and shale from near Escalante, Utah. The sandstone layers are thin and light colored. The shales are thicker. These rocks were deposited in a coastal environment during the Cretaceous Period, about 100 million years ago. © James Steinberg/Science Source.

TABLE APIII-11   **Common Metamorphic Rocks and the Rocks from which They Form**

Protolith (or Parent Rock)		Metamorphic Rock
Shale (sedimentary)	→	Slate (metamorphic)
Slate (metamorphic)	→	Phyllite (metamorphic)
Phyllite (metamorphic)	→	Schist (metamorphic)
Sandstone (sedimentary)	→	Quartzite (metamorphic)
Limestone (sedimentary)	→	Marble (metamorphic)
Basalt (sedimentary)	→	Greenshist (metamorphic)

Progressive Metamorphism and Foliated Metamorphic Rocks

Let's follow a common sedimentary rock—shale—as it is metamorphosed by burial and increasing temperature (**Figure B4-1**). The original rock consists of clay minerals with perhaps some quartz silt, iron oxides, and even feldspar. As the temperature and pressure increase, changes begin first in the clays, which start to expel water. Pressure forces the more open clay structure into tighter arrangements, which often results in recrystallization of clays to micas. Remember that micas are sheet-structured silicates and exhibit a strong preference for a leaf-like shape. Therefore, clay-rich sedimentary rocks usually are converted to **foliated** (layered) **metamorphic rocks**. At low T/P conditions rocks called **slates** result. As T/P conditions rise, **phyllites** form. As conditions of metamorphism become increasingly extreme—as temperatures and pressures increase, the micas grow and may dominate the rock, forming a metamorphic rock called **schist**. And at very high temperatures, micas may be converted to amphiboles or even pyroxenes (if all the water is driven off—recall that pyroxenes are anhydrous). The platy or elongated minerals (micas, amphiboles) may segregate into bands separated by bands of non-platy minerals (quartz and feldspar) to form **gneiss**. This progression resulting from ever-increasing metamorphic conditions is often called increasing metamorphic grade.

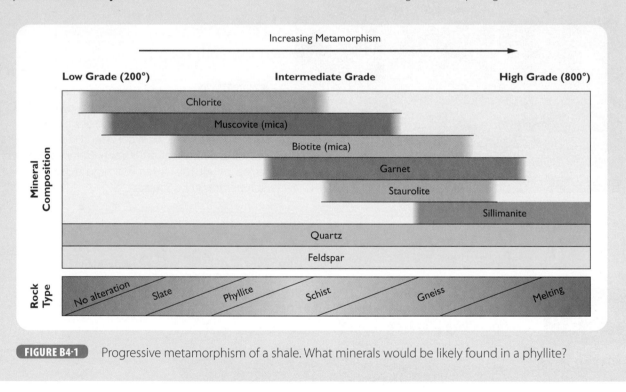

FIGURE B4-1 Progressive metamorphism of a shale. What minerals would be likely found in a phyllite?

metamorphic rocks are **nonfoliated**. **Marbles** and **quartzites** are common nonfoliated metamorphic rocks (**FIGURE APIII-15**).

Metamorphic Zones, Isograds, and Index Minerals

Geologists began to study metamorphic rocks in detail more than 100 years ago in beautifully exposed outcrops in Scotland and Scandinavia. They were able to map what they called **metamorphic zones**, based on the appearance and disappearance of minerals they called **index minerals**; that is, a metamorphic mineral characteristic of a particular zone.

(A)

(B)

(C)

FIGURE APIII-15 Examples of Non-foliated metamorphic rocks. (A) Quartzite. © Oreena/ShutterStock, Inc. (B) Marble. © Siim Sepp/ShutterStock, Inc. (C) Amphiblite. © Only Fabrizio/ShutterStock, Inc.

The boundary of a zone was drawn on a map and named an **isograd**, which represented the first appearance of an index mineral. To understand index minerals, look back at Figure B 4-1, which shows the general stability fields of common metamorphic minerals. You can see that chlorite is common in metamorphic rocks formed at low temperature, but biotite is not. At some triggering set of increasing P/T conditions, however, chlorite disappears and biotite starts to form in metamorphic rocks of appropriate compositions. Compositions are important here. Obviously if the elements needed to form biotite (Fe, Al, Si, O, H_2O, etc.) are not present in the rock being metamorphosed, biotite won't form.

Geologists recognize several metamorphic zones, thought to indicate ranges in T/P conditions under which the metamorphic rocks were formed. Recognition of T/P conditions can help reconstruct the crustal depth (low, medium, high) at which metamorphism took place. Figure B 4-1 shows the commonly recognized metamorphic zones.

Metamorphic Facies

The facies concept was proposed by the Finnish geologist Pentti Eskola around 1915. In his hypothesis, if one knew the initial composition of a rock undergoing metamorphism, any resultant mineral assemblage of the metamorphic rock

Metamorphism of Basalts

Since basalts are one of the Earth's most common rocks, you might expect that metamorphosed basalts are common, and you would be correct. In fact, even though basalts are igneous rocks and crystallize at high temperature, they are very susceptible to metamorphism, because of their glass content. Remember that glass is a geological term meaning *unordered atoms*, and does not refer to window glass, which is pure SiO_2. Glass is very unstable in nature: given a chance, glass will crystallize into minerals. The only reason glass persists is because the basalt cooled too quickly for all the ions to form crystal structures. Basalts are most easily metamorphosed where they form in the oceans—at mid-ocean ridges (MOR).

Hydrothermal Activity and Metamorphism of Basalts

Recall the concept of a model as used in geology. Beginning in the mid-1970s scientists had to change their models for the oceanic heat budget, as well as models of geochemical cycling of dissolved material (salts) in the oceans. In that year, scientists discovered **hydrothermal vents** along a segment of the mid-ocean ridge (MOR) off South America (**Figure B5-1**). Later, more and more such vents were discovered, and many of these vents were populated with startling and previously unknown communities of organisms.

The hydrothermal (hot water) vents were actually the sites of massive cation exchanges that were altering the composition of the basalts formed at the MORs. What was happening was very simple—ocean water enriched in Na (the second most abundant dissolved ion in seawater, after chlorine), was trickling down fractures in the sea floor to depths of several km. There it was heated under pressure to hundreds of degrees C. The now hot, buoyant water ascended though other fracture systems, reacting with basalt along the way. The water hydrated the rock, especially the unstable glass, and added Na to the basalts and extracted Mg, converting the basalts into metamorphic rocks called **greenstones** (which form at lower temperatures and shallower depths), or to **amphibolites** (which form at higher temperatures and greater depths. The plagioclases absorbed most of the Na, becoming more sodic in the process.

FIGURE B5-1 Black smokers, vents in the sea floor where gases and solids, consisting mainly of finely crystallized metal sulfides, pour out under high pressure and high temperature. Courtesy of OAR/National Undersea Research Program (NURP)/NOAA.

gioclases absorbed most of the Na, becoming more sodic in the process.

This metamorphism has a profound effect on oceanic crust. By adding water to the basalt, the rock becomes lighter in density, and the heat causes the rock to expand further. These effects cause rocks at MORs to stand higher than rocks away from the MORs that are colder. Colder rocks are less dense than hotter rocks. We explain the significance of this earlier in the text.

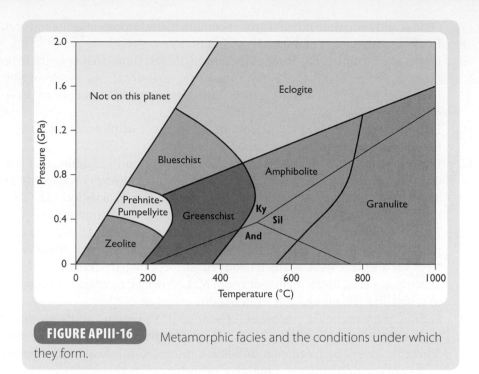

© Hemera Technologies/AbleStock.com/Thinkstock

FIGURE APIII-16 Metamorphic facies and the conditions under which they form.

would depend solely upon the T/P conditions to which the rock was raised. Thus, reasoning backward, careful determination of the complete mineral assemblage of a metamorphic rock would allow conditions of metamorphism to be worked out as well as indicate the initial chemical composition (but not necessarily mineral composition) of the "parent" rock. Today, the facies concept is widely applied to the study of metamorphic rocks. Common **metamorphic facies** and the conditions under which they form are shown in **FIGURE APIII-16**.

We end with a summary of important concepts from this important chapter.

SUMMARY OF MINERALS AND ROCKS

1. Minerals are solids with a crystalline structure, a stability range, a set of physical attributes, and a chemical formula. There are around 2,000 known minerals but only a few dozen are common, and only a few are rock-formers. Crystal structures form based on the ratio between ions and cations in the crystal.
2. Most minerals are composed of a very few elements: they are O, Si, Al, Mg, Ca, Na, K, Fe, S, and C. Geologists classify minerals into groups based on the anion or anionic complex to which cations are bound.
3. Most common minerals are silicates. Silicates are minerals consisting of SiO_4 tetrahedra, with or without cations.
4. Olivines, pyroxenes, amphiboles, micas, clays, and feldspars are important silicate groups. Quartz is a widespread and common silicate mineral.
5. Carbonates are minerals based on the CO_3 structure. Calcite and dolomite are common varieties. In the oceans, organisms manufacture much carbonate.
6. Sulfides and sulfates are minerals with widespread uses. Sulfides are important ores of metals. Gypsum is an important sulfate mineral, used to manufacture wallboard for buildings. Burning and weathering sulfides can produce acids and air pollutants.

7. Oxides and hydroxide mineral groups contain many important gems and ores.

8. Rocks are aggregates of minerals.

9. Rocks are grouped into three categories: *igneous*, those that solidify from magma; *sedimentary*, those that form from sediment; and *metamorphic*, those that form by solid-state reorganization or recrystallization from previously existing rocks.

10. Igneous rocks are classified as volcanic (extrusive) or plutonic (intrusive), based on their origin and texture.

11. Important volcanic rocks are basalt (the most common), andesite, and rhyolite. Volcanic rocks made of fragments of lava are called pyroclastic rocks. The Hawaiian Islands are made predominantly of basalts, as is Iceland. Cooling lavas emit gases, of which H_2O and CO_2 are the most common.

12. Important plutonic rocks are granite, diorite, and gabbro. An intermediate plutonic rock, granodiorite, is the most common plutonic rock. Magma is a term for a silicate melt. Many igneous rocks form by partial melting of pre-existing material in the mantle or crust. N. L. Bowen described the sequence of crystallization of minerals from basaltic magmas.

13. Sedimentary rocks are composed of sediment. Sediment can be produced by weathering of pre-existing rock, and can also be produced by organisms.

14. Siliclastic refers to fragments of silicate minerals. Siliclastic sedimentary rocks are conglomerates and breccias, sandstones, siltstones, and shales.

15. Coal and rock salt are other types of sedimentary rocks.

16. Sedimentary rocks are layered. They are grouped into formations for purposes of field study and mapping.

17. Metamorphic rocks can be classified as foliated and nonfoliated.

18. Important foliated metamorphic rocks are slate, phyllite, schist, and gneiss.

19. Important nonfoliated metamorphic rocks are marble and quartzite.

20. Seafloor basalts are easily metamorphosed to greenstones (at low T and P) and amphibolites (at higher T and P).

21. Metamorphic rocks are mapped in the field using key minerals, which define metamorphic zones. Boundaries of zones are isograds. Metamorphic facies define broad temperature, pressure regions under which metamorphic minerals come to equilibrium.

22. The origins of igneous and metamorphic rocks, and some sedimentary rocks, are linked to plate tectonics processes.

KEY TERMS

amphibole	dacite	ionic substitution
amphibolites	feldspar	ions
andesite	foliated metamorphic	isograds
anions	rocks	lithosphere
ash	formation	marble
basaltic	glass	metamorphic
batholith	gneiss	metamorphic facies
Bowen's reaction	greenstones	metamorphic zones
series	gypsum	mica
carbon cycle	hydrothermal vents	minerals
carbonate minerals	hydroxides	nonfoliated
cations	igneous	metamorphic rocks
clay minerals	index minerals	olivine

ore
orthoclase
oxides
partial melt
peridotite
phyllite
plagioclase
pluton
progressive
  metamorphism
pyroclastic material
pyroxene
quartz

quartzite
rhyolite
rock cycle
rock-forming
  minerals
schist
sediment
sedimentary
sequence of
  crystallization
silica
silicate minerals
siliciclastic sediment

sill
slates
sulfates
sulfides
tetrahedron
vesicles
viscosity
volcanic gas
volcanic rocks
weathering
weathering products

REVIEW QUESTIONS

1. The Hawaiian Islands are made up mainly of what rock type?
   a. granite
   b. basalt
   c. limestone
   d. schist

2. Another name for a silicate melt is
   a. rhyolite.
   b. granite.
   c. magma.
   d. siliciclastic.

3. Most rock-forming minerals are
   a. silicates.
   b. oxides.
   c. carbonates.
   d. sulfides.

4. The basic "building block" of silicate minerals is the
   a. octagon.
   b. octahedron.
   c. dodecahedron.
   d. tetrahedron.

5. The structure in 4 above has the ratio Si:O of
   a. 4:1.
   b. 1:4.
   c. 1:1.
   d. 1:2.

6. The most common silicate group is
   a. quartz.
   b. olivine.
   c. pyroxene.
   d. feldspar.

7. An anhydrous silicate group of minerals is
   a. feldspar.
   b. pyroxene.
   c. olivine.
   d. all are anhydrous

8. An important part of the carbon cycle is/are
   a. granite.
   b. carbonates.
   c. sulfides.
   d. silicates.

9. Volcanic rocks composed of "fiery fragments" are called
   a. granite.
   b. pyroclastic.
   c. basalt.
   d. granodiorite.

10. Magma that freezes below ground forms
    a. plutonic (intrusive) igneous rocks.
    b. extrusive igneous rocks.
    c. zoned metamorphic rocks.
    d. discontinuous igneous rocks.

11. Partial melting of peridotite produces
    a. metamorphic magmas.
    b. rhyolite.
    c. granodiorite.
    d. basalt.

12. The continuous reaction series of Bowen describes the behavior of
    a. feldspars.
    b. olivine.
    c. pyroxene.
    d. pyroclastic rocks.

13. A common mineral produced by the chemical weathering of feldspar is
    a. silicon.
    b. clay.
    c. quartz.
    d. K+.

14. A nonfoliated metamorphic rock is
    a. quartzite.
    b. phyllite.
    c. gneiss.
    d. hydrothermal.

15. A fine-grained silicilastic sedimentary rock with much clay is
    a. slate.
    b. siltstone.
    c. shale.
    d. phyllite.

16. A foliated metamorphic rock is
    a. marble.
    b. schist.
    c. quartzite.
    d. rhyolite.

17. Metamorphism of oceanic basalts makes them
    a. denser.
    b. anhydrous.
    c. less dense.
    d. magmas.

18. Index minerals in metamorphic rocks are used to define
    a. the composition of the gases in the original rock.
    b. the economic value of a metamorphic rock.
    c. zones.
    d. Formations.

19. The facies concept seeks to determine
    a. the T/P conditions under which metamorphic rocks form.
    b. the economic content of metamorphic rocks.
    c. the age of metamorphic rocks, and the time of metamorphism.
    d. all of these

20. The basic unit in the field into which sedimentary rocks are divided for purposes of mapping is the
    a. zone.
    b. facies.
    c. grade.
    d. formation.

APPENDIX IV
Basic Math and Scientific Notation

Throughout the text we occasionally ask you to make a calculation or interpret the significance of some numerical value. We call these *value-added questions*. For those of you rusty in making basic calculations, or intimidated by math, we present a brief review of two important elements: the metric system and scientific notation.

■ THE METRIC SYSTEM

The metric system is both simple and elegant. It is based on powers of ten: 10 millimeters in 1 centimeter, 1,000 meters in a kilometer, etc. This characteristic of the metric system is important to remember.

In some *Value-Added* Questions, we ask you to make conversions from the English system to the metric system and back. Some common conversions as well as metric prefixes are given in **TABLE APIV-1–APIV-4**.

Many students are able to correctly make simple conversions, e.g., lengths. However, in our experience students frequently have problems converting areas and volumes. For you, here is a useful shortcut. To convert areas, you can use the conversion factors for the units of length and square them. For example, to convert 12 square feet to square yards, do the following:

$$12\,\text{ft}^2 \times (1\,\text{yd}/3\,\text{ft})^2 = 12\,\text{ft}^2 \times 1\,\text{yd}/9\,\text{ft}^2$$

$$= 1.33\,\text{yd}^2$$

Similarly, to convert volumes, use the factors of the units of length, and cube them.

Now answer the following questions as a test your ability to manipulate and convert these units. These skills are essential to answer the *value-added questions* and understand the concepts in the issues in this book. Remember to use your conversion factors (so that units cancel each other out). For example, to determine how many liters are in 8.4 cubic meters:

$$8.4\,\text{m}^3 \times 1,000\,\text{L}/\text{m}^3 = 8,400\,\text{L}$$

Now, try the following calculations. Use appropriate abbreviations, and show all work. Answers are at the end of this appendix.

AQ-1. How many millimeters are in a 5 meters?
AQ-2. How many centimeters are in 10 kilometers?
AQ-3. How many milligrams are in a 3 tonnes? (recall that a tonne equals 1,000 kg.)
AQ-4. Express your height in feet, meters, centimeters, and millimeters.
AQ-5. Express your weight in grams, kilograms, and pounds.

TABLE APIV-1	*Metric Equivalents for Large Numbers*

One thousand = 1,000 = 10^3 (kilo or k)

One million = 1,000,000 = 10^6 (mega or M)

One billion = 1,000,000,000 = 10^9

One trillion = 1,000,000,000,000 = 10^{12}

One quadrillion (*quad*) = 1,000,000,000,000,000 = 10^{15} (commonly used in expressions of energy use)

■ SCIENTIFIC NOTATION

Numbers that are very large or very small are most conveniently added, subtracted, multiplied, and divided by expressing the numbers as powers of ten. First, you need to know is that 100 (pronounced "ten to the zero, or ten to the zero power") is defined as 1.

Fortunately, converting large numbers to scientific notation and then manipulating these numbers are both easy to learn. First, convert large or small numbers to scientific notation, a skill with which you may already be familiar. Here is an example. In scientific notation, 46,000,000 is 4.6×10^7.

AQ-6. Express 5 trillion (5,000,000,000,000) in scientific notation.
AQ-7. Express 2,470,000 in scientific notation.

You can express the same number in a variety of ways using exponents, such as 24.7×10^5, 247×10^4, and so forth, and they all will mean the same thing. It is customary, however, to express all values in the same format, by placing only one digit to the left of the decimal place (e.g., 2.47×10^6).

AQ-8. Express 42,000,000,000,000 (42 trillion) the customary way using exponents.

TABLE APIV-2	*Metric Equivalents for Small Numbers*

One hundredth = 1/100 = 0.01 = 10^{-2} (centi or c)

One thousandth = 1/1,000 = 0.001 = 10^{-3} (milli or m)

One millionth = 1/1,000,000 = 0.000001 = 10^{-6} (micro or mc or μ)

One billionth = 1/1,000,000,000 = 0.000000001 = 10^{-9} (nano or n)

TABLE APIV-3	English to Metric	
	English	**Metric**
Length:	1 inch (in)	2.54 centimeters (cm) (exactly)
	1 foot (ft)	30.48 cm (exactly) 0.3048 meters (m) (exactly)
	1 yard (yd)	91.44 cm (exactly) 0.9144 m (exactly)
	1 mile (5280 ft)	1609.34 m (exactly) 1.60934 kilometers (km) (exactly)
Mass (weight):	1 ounce, avoirdupois	28.3495231 grams (gr)
	1 pound (lb), avoirdupois	453.59237 gr (exactly) 0.45359237 kilograms (kg) (exactly)
	1 ton, net or short (2000 lb)	907.18474 kg 0.90718474 metric ton
Area:	1 square inch (in^2)	6.4516 cm^2 (exactly)
	1 square foot (ft^2)	0.092903 m^2 (exactly)
	1 square yard (yd^2)	0.836127 m^2
	1 acre (43,560 ft^2)	4046.856 m^2 0.4046856 hectares (ha)
	1 square mile	258.9975 ha 2.589975 km^2
Volume:	1 cubic inch (in^3)	16.38706 cm^3
	1 cubic foot (ft^3)	0.028317 m^3
	1 cubic yard (yd^3)	0.764555 m^3
	1 gallon (U.S.)	3.785 liters (L)
	1 liquid quart (U.S.)	0.946 L
	1 liquid ounce (U.S.)	29.573 milliliters (ml)

Temperature: 1 degree Fahrenheit equals 5/9 (.555555) degree Celsius (centigrade). To convert a temperature in degrees Fahrenheit to the equivalent temperature in Celsius, subtract 32 and divide by 1.8: degrees C = (degrees F − 32)/1.8.

TABLE APIV-4	Metric to English	
	Metric	**English**
Length:	1 centimeter (cm)	0.393701 in
	1 meter (m)	39.3701 in 3.28084 ft 1.09361 yd
	1 kilometer (km)	3280.84 ft 0.62137 mile
Mass (weight):	1 gram (gr)	0.0352739619 ounce, avoirdupois
	1 kilogram (kg)	2.204623 lb
	1 metric ton	1.10231 short tons
Area:	1 cm^2	0.155 in^2
	1 m^2	10.7639 ft^2 1.19599 yd^2
	1 hectare	2.47105 acres
	1 km^2	247.105 acres 0.3861 $mile^2$
Volume:	1 cm^3	0.06102 in^3
	1 m^3	35.3145 ft^3 1.3079 yd^3
	1 liter (L)	1.057 quarts 0.2642 gallon (U.S.)
	1 milliliter (ml)	0.0338 liquid ounce (U.S.)

Temperature: 1 degree Celsius (centigrade) equals 9/5 (1.8 exactly) degrees Fahrenheit. To convert a temperature in degrees Celsius to the equivalent temperature in Fahrenheit, multiply by 1.8 and then add 32: degrees F = (degrees C × 1.8) + 32.

■ METRIC PREFIXES AND EQUIVALENTS

For numbers summarized in scientific notation, the prefix in front of the unit (e.g., *kilo*) denotes the magnitude of the unit (e.g., kilograms = units of 1,000 g). You should memorize these equivalents (Table AP IV-1).

AQ-9. Convert 4.43 mm to (A) nm, (B) μm (micrometers), (C) cm, (D) m, and (E) km. Express your answers as decimals and in scientific notation.

■ MANIPULATING NUMBERS EXPRESSED IN SCIENTIFIC NOTATION

To add, subtract, multiply, and divide large numbers using scientific notation, you need to remember a few basic rules and check your work carefully.

Multiplication Using Scientific Notation

To multiply numbers expressed in scientific notation, multiply the bases and add the exponents.

For example, to multiply:

$$\left(7\times10^4\right)\times\left(3\times10^3\right)$$

multiply the bases (noting how we used parentheses to separate the numbers above):

$$7\times3=21$$

and add the exponents:

$$4+3=7$$

The result is 21×10^7 or, using the appropriate convention, 2.1×10^8. (This is the same as 210×10^6, which is 210 million, a convenient way to present the number in conversation. There are any number of other variations as well.)

Division Using Scientific Notation

To divide numbers expressed in scientific notation, divide the number in the numerator by the number in the denominator and subtract the exponent of the denominator from the exponent of the numerator.

For example, to divide

$$\left(12.6\times10^6\right)\div\left(4.2\times10^3\right)$$

divide the numerator by the denominator

$$12.6\div4.2=3$$

and subtract the exponent of the denominator from the exponent of the numerator

$$6-3=3$$

Thus, the answer is 3×10^3.

AQ-10. Perform the following calculations:

$(9.2 \times 10^5) \div (4.6 \times 10^{10}) =$
$(7.2 \times 10^{24}) \div (2.9 \times 10^{20}) =$
$(8.5 \times 10^{-5}) \times (3.6 \times 10^{10}) =$
$(6.2 \times 10^{24}) \times (5.9 \times 10^{20}) =$

Addition/Subtraction Using Scientific Notation

To add numbers expressed in scientific notation, you simply add the numbers (i.e., the bases) after converting both to the same exponent. To subtract numbers expressed in scientific notation, you simply subtract the numbers (the bases) after converting both to the same exponent.

For example, to add 5 billion to 9 million:

$$5,000,000,000 + 9,000,000$$

or

$$(5 \times 10^9) + (9 \times 10^6)$$

Convert both numbers to the same exponent (it does not matter which, but it is usually easier to use the smaller).

$$5 \text{ billion} = 5000 \text{ million or } 5000 \times 10^6$$

$$9 \text{ million} = 9 \times 10^6$$

Therefore,

$$(5,000 \times 10^6) + (9 \times 10^6) = 5,009 \times 10^6$$

An equally correct answer would be 5.009×10^9, which is the customary way to express the answer.

AQ-11. Perform the following calculations:

$(3.2 \times 10^5) + (4.6 \times 10^4) =$
$(7.2 \times 10^{21}) + (2.9 \times 10^{20}) =$
$(8.5 \times 10^{-5}) - (3.6 \times 10^{-6}) =$
$(6.2 \times 10^{24}) - (5.9 \times 10^{24}) =$

■ ANSWERS

AQ-1. How many millimeters are in a 5 meters?

$$5 \; \cancel{m} \times 1,000 \text{ mm}/\cancel{m} = \mathbf{5,000 \text{ mm}}$$

AQ-2. How many centimeters are in 10 kilometers?

$$10 \; \cancel{km} \times 1,000 \; \cancel{m}/\cancel{km} \times 100 \text{ cm}/\cancel{m} = \mathbf{1,000,000 \text{ cm}}$$

AQ-3. How many milligrams are in a 3 tonnes? (Recall that a tonne–also called a metric ton or *mt*—equals 1,000 kg.)

$$3 \; \cancel{mt} \times 1,000 \; \cancel{kg}/\cancel{mt} \times 1,000 \; \cancel{g}/\cancel{kg} \times 1,000 \text{ mg}/\cancel{g} = \mathbf{3,000,000,000 \text{ mg}}$$

AQ-4. Express your height in feet, meters, centimeters, and millimeters.

According to the Centers for Disease Control, in 2002 the average nineteen-year-old American male and female were, respectively, 5 ft 10 in and 5 ft 4 in.

$$\underline{\text{Female:}} \; 5 \text{ ft } 4 \text{ in} = 64 \text{ in; } 64 \; \cancel{\text{in}} \times 0.0254 \text{ m}/\cancel{\text{in}} = \mathbf{1.63 \, m}$$

$$1.63 \; \cancel{\text{m}} \times 100 \text{ cm}/\cancel{\text{m}} = \mathbf{163 \, cm}$$

$$163 \; \cancel{\text{cm}} \times 10 \text{ mm}/\cancel{\text{cm}} = \mathbf{1,630 \, mm}$$

$$\underline{\text{Male:}} \; 5 \text{ ft } 10 \text{ in} = 70 \text{ in; } 70 \; \cancel{\text{in}} \times 0.0254 \text{ m}/\cancel{\text{in}} = \mathbf{1.78 \, m}$$

$$1.78 \; \cancel{\text{m}} \times 100 \text{ cm}/\cancel{\text{m}} = \mathbf{178 \, cm}$$

$$178 \; \cancel{\text{cm}} \times 10 \text{ mm}/\cancel{\text{cm}} = \mathbf{1,780 \, mm}$$

AQ-5. Express your weight in grams, kilograms, and pounds.

According to the Centers for Disease Control, in 2002 the average nineteen-year-old American male and female weighed, respectively, 172 lb and 149 lb. Use the following example to calculate values for your weight.

$$\underline{\text{Female:}} \; 149 \; \cancel{\text{lb}} \times 1 \text{ kg}/2.2 \; \cancel{\text{lb}} = \mathbf{67.7 \, kg}$$

$$67.7 \; \cancel{\text{kg}} \times 1,000 \text{ g}/\cancel{\text{kg}} = \mathbf{67,700 \, g}$$

$$\underline{\text{Males:}} \; 172 \; \cancel{\text{lb}} \times 1 \text{ kg}/2.2 \; \cancel{\text{lb}} = \mathbf{78.2 \, kg}$$

$$78.2 \; \cancel{\text{kg}} \times 1,000 \text{ g}/\cancel{\text{kg}} = \mathbf{78,200 \, g}$$

AQ-6. Express 5 trillion (5,000,000,000,000) in scientific notation.

$$\mathbf{5 \times 10^{12}}$$

AQ-7. Express 2,470,000 in scientific notation.

$$\mathbf{2.47 \times 10^{6}}$$

AQ-8. Express 42,000,000,000,000 (42 trillion) the customary way using exponents.

$$\mathbf{4.2 \times 10^{13}}$$

(Note: In practice, exponents that are multiples of 3, e.g., 10^3, 10^6, 10^9 are often used. Thus, the above answer becomes 42×10^{12}, or 42 trillion.)

AQ-9. Convert 4.43 mm to (A) nm, (B) μm (micrometers), (C) cm, (D) m, and (E) km. Express your answers as decimals and in scientific notation.

(A) **4,430,000, or 4.43×10^6 nm**
(B) **44,300 or 4.43×10^3 μm**
(C) **0.443 cm or 4.43×10^{-1} cm**
(D) **0.00443 m or 4.43×10^{-3} m**
(E) **0.00000443 km or 4.43×10^{-6} km**

AQ-10. Perform the following calculations:

$(9.2 \times 10^5) \div (4.6 \times 10^{10}) = \mathbf{2.0 \times 10^{-5}}$
$(7.2 \times 10^{24}) \div (2.9 \times 10^{20}) = \mathbf{2.5 \times 10^4}$
$(8.5 \times 10^{-5}) \times (3.6 \times 10^{10}) = \mathbf{3.1 \times 10^6}$
$(6.2 \times 10^{24}) \times (5.9 \times 10^{20}) = \mathbf{3.6 \times 10^{45}}$

AQ-11. Perform the following calculations:

$(3.2 \times 10^5) + (4.6 \times 10^4) = \mathbf{3.7 \times 10^5}$
$(7.2 \times 10^{21}) + (2.9 \times 10^{20}) = \mathbf{7.5 \times 10^{21}}$
$(8.5 \times 10^{-5}) - (3.6 \times 10^{-6}) = \mathbf{8.1 \times 10^{-5}}$
$(6.2 \times 10^{24}) - (5.9 \times 10^{24}) = \mathbf{3.0 \times 10^{23}}$

APPENDIX V
Population Math Made Easy

Before you read this appendix, you may want to review the information in Appendix IV: Basic Math and Scientific Notation.

■ A FEW WORDS OF ENCOURAGEMENT

Two words that figure prominently in Douglas Adams' *Hitchhikers Guide to the Universe* should be committed to memory by those of us with math anxiety or who are very rusty in performing calculations: **DON'T PANIC!**

Calculating population growth can be intimidating, but it doesn't need to be. In this appendix we give you step-by-step instructions on how to make population calculations. In some cases, we even provide you with *alternative ways* to make growth calculations that involve only addition, subtraction, multiplication, and/or division. These can provide quick estimates but have some limitations, for example, they are less accurate (and in some cases, *really less accurate)*. You should use whatever method your instructor specifies.

So, get out your calculator (one that does scientific notation, commonly called a *scientific calculator*), take a couple of deep breaths, and work through this Appendix. And remember: **Don't Panic!**

■ HOW TO CALCULATE PERCENT INCREASES OR DECREASES

Value-added question 3-5 asks you to calculate the population growth rate per thousand population for each of three countries, given the birth rates, death rates, and net migration rates. Then you are asked to express this as percent. Let's make these calculations for Mexico.

Here are the data[1] for Mexico.

	Birth Rate (per thousand population)	Death Rate (per thousand)	Net Immigration (Immigration—Emigration, per thousand)
Mexico	19.39	4.83	−3.38

To make this calculation, remember that all you need to do is take the birth rate, subtract the death rate, and add net immigration (if it is negative, which means more residents leave than immigrants enter, then you subtract it). Here is the calculation, which is easy because all the data are in the same units, that is, per thousand population:

$$19.39 - 4.83 - 3.38 = 11.18 \text{ per thousand}$$

So, we have the answer expressed per thousand population, but we also want it expressed as percent. *Percent* (or *per cent*) is an abbreviation of *per centum*, which means *by the hundred*. We thus divide the answer by 10 (or multiply by 0.1, since that is the same thing) and the answer expressed as % is:

$$11.18 \times 0.1 = 1.118 \text{ per cent}$$

Recall that the decimal equivalent of 1.118% is 0.01118. We simply moved the decimal point two places to the left. For practical purposes, we can shorten this to 1.12%. We'll show you other ways of calculating the rate of population increase or decrease below.

■ HOW TO PROJECT POPULATION GROWTH USING THE COMPOUND GROWTH EQUATION

The Accurate Way

Projecting population growth means using the current population size and the rate of population growth to calculate the population at some time in the future. You'll need your scientific calculator, that is, one with an exponent key. Here is the compound growth equation:

$$\text{future value} = \text{present value} \times (e)^{rt}$$

where e equals the constant 2.71828 . . . , *r* equals the rate of increase (expressed as a decimal, e.g., 5% would be 0.05), and t is the number of years (or other unit, as long as it is the same as r) over which the growth is to be measured.

Replacing words with symbols, this equation becomes this:

$$N = N_0 \times (e)^{rt}$$

The variable N_0 represents the value of the quantity at time zero, that is, at the starting point.

This equation is central to understanding exponential growth and is one of the few worth remembering.

Sample Growth Calculation

In value-added question 3-6, you are asked to project the world population in 2025, 2050, and 2100, assuming a constant growth rate of 1.13. Let's try projecting Mexico's population in 2025, since we already calculated its population growth rate (1.118%, or 0.01118, we can round this to 1.12%). All we need to make this calculation is Mexico's population in 2010, which was 112.5 million.

Here's how to do this calculation for the year 2025, that is, for 15 years from 2010:

$$N = N_0 \times (e)^{rt}$$

$$N = (112.5 \times 10^6) \times e^{(0.0112 \times 15)}$$

$$N = 1.33 \times 10^8, \text{ which is the same as}$$

133 × 10⁶, or **133 million** (see Appendix IV for a review of manipulating numbers expressed in scientific notation)

How to Do This Calculation on a Typical Calculator

On a typical, nongraphics calculator, keystrokes are (commas are for punctuation only) as follows:

Key in *0.0112* (the decimal equivalent of 1.12%), then *x* (multiply sign), then *15*, then *=*. This gives you the exponent (the number to which *e* must be raised).

Next hit the button labeled e^x. On most calculators, e^x is above a button having another label (frequently *ln*). If this is the case on your calculator, you must hit the key labeled *2nd* or *2nd F* first, followed by the key with e^x above it. Finally, some calculators require you to hit the *=* key after the e^x key.

Next, key in *x* (multiply sign), followed by the present value (112.5×10^6; on most calculators, this is done by keying in *6.52*, then hitting the button labeled *EE* or *EXP*, and then keying in *6*. If the *EE* or *EXP* label is above the key, you must first hit the key labeled *2nd* or *2nd F* before punching *EE* or *EXP*.) Finally, hit the *=* sign.

Try projecting Mexico's population for 2050, then check the answer in the footnote at the bottom of the page.[2]

The Shorthand (Easy), but Less Accurate, Way

There is also a less accurate shortcut to make this calculation. *You should use this only if your instructor indicates that it is okay to do so.* Instead of the compound growth equation, $N = N_0 \times (e)^{rt}$, we use this equation:

$$N = (r \times N_0 \times \text{the number of years}) + N$$

Substituting numbers for symbols for the Mexico example:

$$N = [0.0112 \times (112.5 \times 10^6) \times 15] + (112.5 \times 10^6)$$

$$= 1.31 \times 10^8$$

How does this compare to the number we calculated using the compound growth equation? Whether this is good enough depends on how accurate your calculation needs to be (or which one your instructor thinks you should use).

■ USING THE COMPOUND GROWTH EQUATION TO CALCULATE THE RATE OF GROWTH OR DECREASE

Another use of the compound growth equation which, by this time you either consider magical or you never want to see again, or both, is to calculate the rate of population growth or decrease. Both the simpler growth equation and the compound growth equation can be rearranged if you want to know the average growth rate or how long it would take a population of a given size to grow (or decrease) to a different size at a specified growth rate.

The Accurate Ways

If you have the beginning and ending population values over a time period, say 10 years, and you simply want to know the growth rate over the *entire period*, you can calculate it as follows:

Growth rate over a period = (population at end of period − population at

beginning of period)/population at beginning of period

Replacing words with symbols, we get this:

$$r = (N - N_0)/N_0$$

The variable r = growth rate is expressed as a decimal (e.g., 5.2% would be entered as 0.052); N = the population at the end of the period; and N_0 (pronounced "N subzero") = the population at the beginning of the period.

This works if you want the growth rate for an entire period, say 1 year or 50 years; however, as we show you below, it is less accurate than the compound growth equation in some cases (e.g., if you want to know the average *annual* growth rate over a multiyear period, for example 50 years).

If you need an accurate value for the average annual rate of growth or decrease over time periods longer than one year, you must use this formula:

$$r = (1/t)\ln(N/N_0)$$

Let's try an example. Given a 1987 world population of 5 billion and a 1999 world population of 6.0 billion, calculate the average annual growth rate over the 12 year period.

$$r = (1/t)\ln(N/N_0)$$
$$= (1/12)\ln(6 \times 10^9)/(5 \times 10^9)$$
$$= 0.0152, \text{ or } 1.52\%$$

How to Do This Calculation on a Typical Calculator

On a typical, nongraphics calculator, keystrokes are (commas are for punctuation only) as follows:

> Key in *6* then hit the button labeled *EE* or *EXP*, and then key in *9*. Next, hit the ÷ key (division sign). Then key in *5* and hit the button labeled *EE* or *EXP*, and then key in *9*, followed by =.
> Next, hit the key labeled *ln*. Then, hit the ÷ key, enter *12*, followed by =.

The Shorthand (Easy), but Less Accurate, Way

This easy way can be far less accurate, so it should be used only if your instructor approves. For this *estimate*, we use:

$$r = [(N - N_0)/N_0]/\# \text{ years}$$

Substituting numbers for symbols:

$$r = [(6 \times 10^9) - (5 \times 10^9)/(5 \times 10^9)]/12$$

$$= 0.0167, \text{ or } 1.67\%$$

Using the compound growth equation, as you can see, is more accurate. However, this method gives us an approximation that may be appropriate for estimation.

■ USING THE COMPOUND GROWTH EQUATION TO CALCULATE HOW LONG IT WOULD TAKE FOR GROWTH OR DECREASE TO OCCUR

The Accurate Way

The compound growth equation can also be rearranged so that you can calculate how long it would take a population of a given size to grow (or decrease) to a different size at a specified growth rate. To do so, you use this formula:

$$t = (1/r)\ln(N/N_0)$$

None of this should be new to you except *ln*, which is the shorthand for an entity called the *natural logarithm*.

Value-added question 3-7 asks you how long it would take a population growing at 1.5% to increase from 5 billion to 10 billion. Let's do a similar calculation. How long would it take Mexico's population to increase from 112.5×10^6 to 225×10^6 at a growth rate of 1.12%?

$$t = (1/r)\ln(N/N_0)$$

$$= (1/0.012)\ln[(225 \times 10^6)/(112.5 \times 10^6)]$$

$$= 5.78 \times 10^1$$

$$= \textbf{57.8 years}$$

How to Do This Calculation on a Typical Calculator

On a typical, nongraphics calculator, keystrokes are (commas are for punctuation only) as follows:

Key in *225* then hit the button labeled *EE* or *EXP*, and then key in *6*. Next, hit the ÷key (division sign). Then key in *112.5* and hit the button labeled *EE* or *EXP*, and then key in *6*, followed by *=*. Your display should show the number *2*.

Next, hit the key labeled *ln*. Then, hit the ÷key, enter *0.012*, followed by *=*. Note that you do not need to first multiply by 1, since that would not change the answer.

The Shorthand (Easy), but Less Accurate, Way

Again, use this method only with approval from your instructor because it can be much less accurate. The formula is:

$$t = [(N - N_0)/N_0]/r$$

Here is an example of how long it would take a population of a given size to grow (or decrease) to a different size at a specified growth rate:

At an annual growth rate of 1.14%, how long would it take a population of 5 billion to grow to 6 billion?

This time we use this:

$$t = [(N - N_0)/N_0]/r$$

Substituting numbers for symbols:

$$\text{\# years} = [(6 \times 10^9) - (5 \times 10^9)/(5 \times 10^9)]/0.0114$$

$$= \textbf{17.5 years}$$

The answer obtained from using the compound growth equation is **15.99 years**. Demonstrate this for yourself.

We also have a useful, quick, easy, and relatively accurate rate to do similar calculations, known as doubling time.

■ DOUBLING TIME

For any population that is growing exponentially, the time it takes for the population to double can be estimated as follows:

$$t = 70/r$$

Where t = the doubling time and r = the growth rate, expressed as the decimal increase or decrease × 100 (you would enter 5 for 5%, or 0.05, increase).

For example, calculate the doubling time for a population growing at a rate of 7% per year.

$$t = 70/7 = \textbf{10 years}$$

This means that a population growing at a 7% annual rate will double in 10 years and will double again in another 10 years. Thus, at the end of 14 years, the population will be four times as large as originally.

Another useful piece of information is that it takes 10 doubling for a population growing exponentially to grow by a factor of approximately 1,000 (actually, 1,024). Doubling time is a very useful and practical concept for projecting and analyzing the implications of growth. When a population is decreasing, you can use the same equation to calculate the halving time.

■ HOW LONG WOULD IT TAKE TO FILL THE PLANET AT A DENSITY OF 1 PERSON/M² OF DRY LAND?

Let's see if we can do this calculation: Starting with the 2010 world population of 6.8 billion, and using a population growth rate of 1.13%, calculate in what year the Earth would reach this impossible density of 1 person/m^2 of dry land (in other words, when would the Earth's population grow to 1.31×10^{14}?).

$$t = (1/r)\ln(N/N_0)$$
$$= (1/0.013)\ln[(1.31 \times 10^{14})/(6.8 \times 10^9)]$$
$$= 7.59 \times 10^2$$
$$= \textbf{759 years}$$

Whew! You did it. This is about as complex as the math gets for our purposes. Pat yourself on the back, then get back to work!

FOOTNOTES

[1] Note that we wrote *Here **are** the data* instead of *Here **is** the data*. That is because the word *data* is plural and thus requires a plural verb. So, we write *data are* and *datum is*. You should as well.

[2] The answer is 1.76×10^8, or 176×10^6 (176 million).

GLOSSARY

A

Abyssal plain A flat or gently sloping seafloor area at an extreme depth (3,000–6,000 m; 10,000–20,0000 ft).

Accuracy In science, how close or far an experimental value may be from its actual value.

Acid aerosols Microscopic droplets suspended in the atmosphere until they combine with humidity and fall to Earth in the form of acid rain, polluting the soil and eventually surface waters and groundwater.

Acid precipitation (*also* acid rain) Occurs when acids such as hydrofluoric (HF), nitric (HNO_3), hydrochloric (HCl), and sulfuric (H_2SO_4) from human activity or natural sources (e.g., volcanic eruptions) enter the atmosphere, combine with water molecules, and form any type of precipitation (e.g., rain, fog, sleet, snow, etc.).

Acre-foot A unit of volume measurement for water covering 1 acre at a depth of 1 foot. It is equal to about 326,000 cubic feet.

Aftershock (1) A ground tremor caused by the repositioning of rock after an earthquake. (2) Any of several lower level ground tremors or quakes in a rupture zone after an earthquake.

Aggregate The only mineral resource produced in all 50 states, it is a mixture of rock types (sand, gravel, pebbles) that can be separable by mechanical means. Used as fill material, cement, roadbeds, and other applications.

Agricultural runoff Occurs when liquids from agricultural practices (chemical inorganic fertilizers, pesticides, etc.) leach into soils and water systems, eventually polluting them.

Air pollutants Amounts of foreign and/or natural substances in soil and water that contaminate soil, streams and other water bodies, and groundwater; substances occurring in the atmosphere that may result in adverse effects to humans, animals, vegetation, and/or materials (EPA definition).

Air pollution Degradation of air quality resulting from unwanted chemicals or other materials occurring in the air (EPA definition).

Airshed A part of the atmosphere over a specific region that contains its airflow and air pollutants. For example, the airshed of the Chesapeake Bay Watershed is that part of the surface from which pollution (from such sources as power plants, or mineral smelters) can enter the watershed.

Alternative energy Energy alternatives to fossil fuels, such as nuclear, hydroelectric, geothermal, hydrogen, and a variety of renewable energy sources (*see* renewable energy).

Amalgamation method The alloying and collection of fine gold and silver particles from ore with droplets or coatings of mercury. The mercury-precious metal amalgam is then collected and heated to vaporization, to separate out the precious metals.

Angle of repose The maximum angle at which unconsolidated material is stable (*see* shear strength, shear stress).

Annual exceedance probability A mathematical calculation of the probability that a given maximum discharge of a stream or river will occur in a given year: the inverse of the recurrence interval (R), where $P_e = 1/R$ (*see* discharge, recurrence interval).

Annual peak discharge The highest river discharge data recorded during a water-year (beginning October 1); used to predict floods by measuring the variability of stream flow over time (*see* discharge).

Anoxic Describes water that has no dissolved oxygen (DO) and cannot support life (*see* hypoxic, normoxic).

Anthracite coal (*also* hard coal) A dense, shiny coal that has a high carbon content and little volatile matter and burns with a clean flame.

API gravity The American Petroleum Institute's calculation of liquid petroleum's density compared to water to determine its value (*see* viscosity).

Aquiclude A substrate that can neither hold nor transmit water (e.g., unfractured granite).

Aquifer Underground zones of water beneath the water table; layers of rock or sediment that store and transmit water.

Aquitard A water-saturated sediment or rock (gravel, sand, silt) whose permeability is so low that it cannot transmit any useful amount of water (only slowly permeable).

Array A list of the maximum floods for each water year (annually beginning in October).

Aseismic slip (*also* creep) Occurs when rocks on either side of a fault slide more or less smoothly by each other, and no stress or minimal stress is built up that could cause an earthquake.

Arsenic toxicity Symptoms of arsenic poisoning vary with the type and concentration of the poison; inorganic arsenic may cause abdominal pains, destruction of red blood cells (hemolysis), shock, and death quickly, while lower concentrations and occasionally organic arsenic cause far less severe symptoms. Long exposure has been linked to cancers of the skin, bladder, lung, kidney, nasal cavity, liver, and/or prostate.

Ash (1) Fine pyroclastic material (pulverized rock and glass, less than 4 mm diameter) created by volcanic eruptions. A coating on vegetation makes it inedible, and when an aerosol it can take years to settle out of the atmosphere and/or may cause acid rain; great amounts can block the sun, thereby affecting global climate. (2) Noncombustible chemical compound and particulate matter released during the processing of coal.

Assumptions Statements accepted as true without proof.

Asthenosphere A zone of the Earth's mantle that lies beneath the lithosphere and consists of several hundred kilometers of deformable rock (*see* low velocity zone).

Atom The smallest unit of an element, consisting of at least one proton and (for all elements except hydrogen) one or more neutrons in a dense central nucleus, surrounded by one or more shells of electrons.

Atrazine The most commonly used pesticide in the United States; linked to the deterioration of the amphibian population worldwide, and some cancers in humans.

B

Banded iron formation (BIF) Highly metamorphosed, biochemical sedimentary rock older than 1.7 by; comprised of thin layers of iron (Fe) or iron oxides (e.g., hematite, goethite, magnetite, and limonite) as well as chert, chalcedony, jasper, and/or quartz, which form the largest deposits of iron ore.

Barrel A unit of volume equal to 42 U.S. gallons; used to measure the amount of oil to determine flow, quantity, and price.

Barrier island complexes Long, narrow sand bodies separated from the mainland by lagoons and bays with sea on the other side of the island complex.

Basaltic Describes a layer of basalt or its chemical equivalent(s). Basalt is a dark-colored, fine-grained, igneous rock found underneath sediment (*see* oceanic).

Base flow The amount of a stream's discharge provided by groundwater.

Base load plant A type of power plant that provides a more or less constant part of the total load on a local or regional electrical power-supply system.

Basel Convention An international treaty signed in 1989 to protect human exposure from hazardous waste worldwide; regulates transboundary transport between countries to protect less developed nations from becoming dumping grounds.

Base metals A classification of metals "inferior" to the precious metals (e.g., iron, copper, zinc, nickel). These metals react and oxidize more easily and are cheaper in comparison to the precious metals.

Bauxite $Al_2O_3 \cdot nH_2O$. A mineral found in Australia, Surinam, Jamaica, and India; used to produce aluminum.

Beach Area of sand, gravel, or rocks located at the edge of a shoreline between the water and a higher elevation (cliff, berm, or vegetation) with dynamic rather than static features as the sand and/or rock comprising them is almost always in motion, resulting from waves and longshore currents.

Bentonite An absorbent aluminum silicate clay formed from volcanic ash and used in various adhesives, cements, and ceramic fillers (*see* swelling soils).

Best management practices (BMPs) (Ecology) Common sense actions required by law to keep pollutants out of streams, rivers, and lakes designed to protect water quality and prevent new pollution. (Legal) Federal wastewater permit regulations initially used to refer to auxiliary procedures for industrial wastewater controls but now extended to agricultural and urban regulations.

Binary plants A type of power plant that uses water at 225°–360° F (107°–182° C) to boil a secondary fluid (usually isobutene) with a low boiling point; this secondary fluid is boiled in a heat exchanger and used to turn a turbine to generate electricity with little or no air emissions (*see* alternative energy, dry steam plants, flash steam plants).

Bioaccumulation A process that takes place within an organism when the rate of intake of a substance is greater than the rate of excretion or metabolic transformation of that substance. As these toxic chemical compounds (e.g., methyl mercury) become part of the bodies of organisms, they increase in concentration in the bodies of organisms of higher trophic (feeding) levels (e.g. marine life; *see* biomagnification).

Biochemical oxygen demand (BOD) The amount of oxygen required by aerobic microorganisms to decompose the organic matter in a sample of water (e.g., polluted by sewage), and used as a measure of the degree of water pollution.

Biofuels Fuel produced from renewable resources, especially plant biomass, vegetable oils, and treated municipal and industrial wastes. Biofuels are considered neutral with respect to the emission of carbon dioxide because the carbon dioxide given off by burning them is balanced by the carbon dioxide absorbed by the plants that are grown to produce them.

Biomagnification The increasing concentration of a substance, such as a toxic chemical, in the tissues of organisms at successively higher levels in a food chain. As a result of biomagnification, organisms at the top of the food chain generally suffer greater harm from a persistent toxin or pollutant than those at lower levels (*see* bioaccumulation).

Biomass The organic matter that makes up plants, which can be burned to produce electricity or fermented for use as fuel.

Biosolids Treated sewage sludge that meets EPA pollutant and pathogen standards for land application and surface disposal.

Birth rate Ratio of the number of live babies born to the total population in a given location over a specified period, usually expressed in 1,000s per year.

Bituminous coal (*also* soft coal) A type of coal with a high percentage of volatile matter that burns with a smoky yellow flame.

Black lung disease (*also* pneumoconiosis) A lung disease caused by the long-term inhalation of coal dust.

Blind thrusts A type of thrust fault in the Earth's crust where the plane terminates before reaching the Earth's surface (*see* fault).

Bottle bills Legislation in some U.S. states that promote recycling of glass, aluminum and plastics by giving individuals a few cents for every bottle or item recycled.

Breadth All lines of evidence that could provide some insight in addressing an issue.

Brownfields Abandoned central city sites after World War II that deteriorated and sometimes became dangerous over time.

Btu A measure of units of energy in British Thermal Units. 1 Btu = 1,055 joules or 2.93×10^{-4} kWh.

C

CAFOs (*see* Concentrated Animal Feedlot Operations).

Caldera A large, roughly circular crater left after a volcanic explosion or the collapse of a volcanic cone; they typically are much wider in diameter than the openings of the vents from which they were formed.

Carbon cycle Biogeochemical process by which the element carbon is exchanged among the biosphere, pedosphere, geosphere, hydrosphere, and atmosphere, which with the water and nitrogen cycles create the ability of Earth to sustain life.

Carbon sequestration The capture, separation, and storage or reuse of carbon (as CO_2 or methane) from anthropogenic sources (e.g., fossil fuel burning, deforestation, or industrial activity).

Carrying capacity Maximum equilibrium number of organisms of a particular species that can be supported in an environment indefinitely.

Catalytic converter A device attached to the exhaust system of an automobile, truck, or other engine to reduce or eliminate the emission of pollutants.

Cation-exchange capacity (Soils) Permeable, low clay soil particles that attract and retain cations, thereby releasing anions (e.g., agricultural pollutants such as nitrates) into the environment, which become toxic.

Cause–effect An action–reaction event. The relationship between the reason why something happens and the factor(s) or agent(s) that permits the event or leads to the result.

Chlorofluorocarbons (CFCs) An EPA class I ozone depleting substance, the manufacture of which has been regulated and in the process of being phased out since 1982.

Cinder cone Relatively small but steep conical volcanoes, emitting much pyroclastic matter of coarse size (cinder). They are predominantly of intermediate composition and benign in destruction, but can cause damage close to the crater; commonly occur on the flanks of shield volcanoes, stratovolcanoes, and calderas.

Clarity Lucidity, the clearness of thought and expression; the property of being transparent.

Clay and clay minerals Clays are hydrous aluminosilicates consisting of layers of SiO_4 tetrahedra and AlO octahedra, with cations such as Mg^{++}, K^+, and Fe^{+++}. Attributes common to all clay minerals relate to their chemical composition, layered structure, and size. Water molecules are strongly attracted to clay mineral surfaces, ranging from very little water to create a slurry (e.g., mud) to more water to shape a form that becomes solid when it dries and/or is fired (e.g., ceramics).

Clean coal technology A group of processes now under research to improve mining practices to reduce emissions and pollution; include: environmental control of NOx and SOx; new combustion technologies; advanced coal processing systems; gasification and carbon sequestering technology. Current mining and refining practices are not "clean."

Club of Rome A group of about 100 futurists from a variety of backgrounds who identified and discussed global problems in an effort towards holistic solutions to better humankind and our planet.

Coal gasification A refining process that sequesters pollution and converts coal to natural gas

Coal reserves An estimation of the amount of coal that has economic potential to be mined in the future.

Coastal zone The transition zone from land to the edge of the U.S. territorial sea, three miles from mean high water (average high tide), historically determined by the distance of a cannon shot. An extensive, low lying area bordered on one side by the sea and on the other side by a high relief land; may incorporate estuaries, barrier island complexes, coral reefs, and beaches.

Coastal Zone Management Act Enacted in 1972 to establish a program whereby states and territories could voluntarily develop programs to protect coastal waters. The Coastal Zone Act Reauthorization Amendments (CZARA) of 1990 specifically required coastal states and territories to develop plans to protect coastal waters from polluted runoff.

Colluvium A loose deposit of rock, gravel, and soil debris on a slope, or accumulated through the action of gravity at the bottom of a slope.

Commodity crop A cash crop grown for profit; often subsidized by the U.S. government (e.g., rice, cotton, corn, soybeans, and sorghum).

Compound growth equation A formula used to calculate the exponential growth of a population (or money).

Concentrated Animal Feedlot Operations (CAFOs) Animal factories or factory farms; facilities attempting to mimic high-volume manufacturing processes using livestock. Animals, feed, manure and urine, dead animals, and production operations are concentrated on a small land area, where feed is brought to the animals rather than the animals grazing or otherwise seeking feed in pastures, fields, or on rangeland.

Conduction A relatively slow process that occurs when heat is transferred by molecular motion (*see* heat transfer, convection, radiation).

Cone of depression A three dimensional valley in the water table of an aquifer surrounding a well.

Confined aquifer Aquifers that are sandwiched between impermeable layers of rock (*see* aquifer, unconfined aquifer).

Confining pressure An equal, all-sided pressure, such as lithostatic pressure produced by overlying rocks in the crust of the Earth.

Constructed wetlands Any of a variety of designed systems that approximate natural wetlands, use aquatic plants, and can be used to treat wastewater or runoff (*see* wetlands, contiguous wetlands, isolated wetlands).

Consumptive water use Water removed from a source (stream, river, lake, aquifer) and not returning an equal amount of clean water.

Contiguous wetlands Lands characterized by water saturation, located immediately adjacent to a stream, river, lake, or ocean, and whose waters are interconnected.

Continental crust (also shield or platform rock) A layer of rock in outermost zone of Earth, composed of all three rock types, but on average having a composition near that of granite.

Continental drift The movement, formation, and reformation of continents based upon the theory of plate tectonics.

Continental shelf The part of the continent that is underlain by continental crust and covered by the ocean.

Continental slope The seaward border of the continental shelf, where edges steepen, plunging to the ocean floor.

Contour plowing A method of breaking ground for crop planting by following the topography of the land, thereby helping to prevent erosion.

Control rods A neutron-absorbing material lowered and raised next to fuel rods in a nuclear reactor to control the rate of fission throughout the life cycle of the fuel rods as power is generated (*see* fuel rods, nuclear fission).

Convection Cycle of movement in the asthenosphere that causes the plates of the lithosphere to move. Heated material in the asthenosphere becomes less dense and rises toward the solid lithosphere, through which it cannot rise further. It then begins to move horizontally, dragging the lithosphere along with it and pushing forward the cooler, denser material in its path. The cooler material eventually sinks down lower into the mantle, where it becomes heated and rises up again, thus continuing the cycle.

Convergent boundary Where plates move together (*see* plate tectonics).

Core Innermost zone of a planetary body (*see* crust, mantle).

Correlation A comparison that demonstrates a relationship between two variables.

Creep (1) Occurs when rocks on either side of a fault slide more or less smoothly by each other, and no stress or minimal stress is built up that could cause an earthquake (*see* earthquake). (2) Refers to very slow, sometimes imperceptible downslope movement of surface material over a long period of time; it can affect rock, sediment, and soil, and anything on it; telltale signs include buckled and cracked walls, deformed trees, and leaning telephone poles.

Criteria pollutants Used by the EPA to determine air quality under short- and long-term exposures: carbon monoxide (CO), nitrogen oxide (NOx), ozone, lead, particulates (PM10 and PM2.5, which are particles of 10 micrometers or less and particles smaller than 2.5 micrometers, respectively), and sulfur oxide (SOx).

Critical thinking The application of analysis, assessment, and the rules of logic to the thought process (using clarity, precision, accuracy, relevancy, depth and breadth, and logic) and includes point of view, evidence, purpose, and assumptions.

Chronic effects Adverse effect on animal or human body with symptoms that develop slowly, due to long and continuous exposure to low concentrations of a hazardous substance. Such symptoms do not usually subside when the exposure stops (*see* acute effects).

Crust Outermost zone of a planetary body (e.g., Earth; *see* mantle, core).

Cumulative impact Changes in the environment from direct and indirect causes in the past, present, and foreseeable future.

Cut and fill Cuts are excavations, and fills are piles of specially designed material, piled and compacted onto the surface of the escarpment.

Cyanobacteria (*also* blue-green algae) Microorganisms related to bacteria but capable of photosynthesis.

D

Day of 6 Billion October 12, 1999, when the population of Earth reached 6 billion, according to the United Nations.

Dead zones Hypoxic regions in oceans and lakes; caused by pollution from chemical runoff into streams and rivers, mostly fertilizers that deplete oxygen so marine life cannot live (*see* nutrient pollution).

Death rate A calculation of the number of deaths in a population; can be refined by age or cause; a ratio of the deaths in an area to the population of that area, usually expressed in 1,000s per year.

Decommissioning dams Dismantling dams in an effort to restore natural river ecology and watersheds.

Density An indicator of the mass of a substance relative to its volume; weight increases with an increase in density.

Depressurization Pumping oil, gas, and/or water out of the ground destabilizes the soil above the pocket, often causing subsidence.

Depth The range of complexity; thoroughness of detail. A measure of the distance between the top (or surface) to its bottom (or end) or from the nearest to the farthest points, usually in a downward direction.

Desertification Process by which fertile land becomes desert, typically from drought, inappropriate agriculture, or deforestation.

Design earthquake The baseline earthquake whose measurements determine minimum building standards and codes in a particular region.

Development (1) A complex set of changes that convert the economy of a society from one based on subsistence agriculture, to one in which most of the employed inhabitants work in manufacturing or service industries. (2) Exploitation of land for agricultural, industrial, residential, recreational, or other purposes; a progression from a simpler or lower to a more advanced, mature, or complex form or stage.

Dilatancy-diffusion theory of earthquakes A model developed by U.S. seismologists in the 1970s to determine the probability of earthquakes based on precursor seismic waves. This theory says that just before an earthquake, stress causes the rocks about to fail to expand (to *dilate*), so that their volume increases and their density decreases, an example of strain. Numerous tiny cracks appear along the fault plane along which rupture eventually is concentrated.

Dioxin Any of several carcinogenic or teratogenic heterocyclic hydrocarbons that occur as impurities in petroleum-derived herbicides.

Direct gain A passive solar design feature using materials that soak up heat from the sun during the day and release it slowly at night.

Discharge A measure of the outflow of water volume at a given point, where discharge (Q) = (A × v), where A is cross-sectional area and v is velocity.

Discontinuity Internal zones where waves of energy abruptly change velocity; energy waves that are bent (refracted) or reflected through different densities of rock (*see* seismic waves).

Dissolved load Sediments carried in solution by moving water (streams, rivers).

Divergent boundary Where plates moved away from each other (*see* plate tectonics).

Dry steam plants A type of power plant that generates electricity from geothermal resources by using steam derived from reservoirs of hot water kilometers below the Earth's surface (*see* alternative energy, geothermal power, flash steam plants, binary plants).

Dust Veil Index A numerical index that seeks to measure the impact of a particular volcanic eruption's release of dust and aerosols over the years following the event, especially the impact on the Earth's energy balance.

E

Earthquake (*also* seismic slip) A shaking of the ground caused by movement of rocks along faults within the Earth's crust (*see* fault, creep, aseismic slip).

Earthquake forecasting An ability to predict the likelihood of an earthquake of a given magnitude occurring within a specified time interval.

Earthquake precursors Geological events that immediately precede earthquakes.

Earthquake prediction An ability to state with a degree of confidence where and when an earthquake is likely to occur.

Earthquake recurrence interval Any periodicity of earthquake activity in an area.

Ecological footprint The amount of productive land appropriated on average by each person (region, country, world, etc.) for food, water, housing, transportation, waste management, and other purposes.

Ecosystem A community of organisms together with their physical environment, viewed as a system of interacting and interdependent relationships and including such processes as the flow of energy through trophic levels and the cycling of chemical elements and compounds through living and nonliving components of the system.

Elastic rebound theory of earthquakes Forces cause plates on opposite sides of a fault to shift. These plates may move smoothly or may store energy and slowly deform. Once their rupture strength is exceeded, the energy is released as a sudden movement, and an earthquake occurs (*see* aseismic slip, earthquake, fault, seismic slip).

Endocrine-disrupting toxins Chemical compounds that mimic or disrupt the endocrine (hormone) system; many are components of pesticides and have been linked to the disappearance of amphibians and human cancers (*see* atrazine).

Energy Power that may be translated into motion, overcoming resistance, or effecting change; the ability to do work.

Energy-intensive process Where the extraction and/or refinement processes are arduous, costly, and may incur harmful emissions (e.g., greenhouse gases) or pollution.

Energy subsidies A payment from the government that keeps the price of power (electricity, coal, oil, etc.) artificially lower than the market rate for consumers or higher than the market rate for producers.

Enhanced oil recovery Any of several techniques that make it possible to recover more oil than can be obtained by natural pressure, such as the injection of fluid or gases into an oilfield to force more oil to the surface.

Environmentally sound management (ESM) An effort of the Basel Convention to protect human health and the environment by minimizing hazardous waste production whenever possible (*see* Basel Convention).

Environmental justice Exists when risks and hazards as well as investments and benefits are equally distributed without discrimination, whether direct or indirect, at any jurisdictional level; when access to natural resources and environmental investments and benefits are equally distributed; and when access to information, participation in decision making, and access to justice in environment-related matters are enjoyed by all.

Environmental Kuznets curve (EKC) Mathematic plots of the relationship between environmental quality factors and per capita income, named for Nobel laureate economist Simon S. Kuznets.

Epicenter The point on the Earth's surface directly above the focus (*also*, hypocenter), which is the point inside the Earth where an earthquake originates (*see* focus).

Escarpment A steep slope or cliff between two land areas of different elevation.

Ethanol The intoxicating agent in fermented and distilled liquors; used pure or denatured as a solvent or in medicines, colognes, cleaning solutions, and rocket fuel; proposed as a renewable clean-burning additive to gasoline.

Estuary The wide lower course of a body of freshwater that mixes with the saltwater from the oceans, and is protected from the ocean's force by barrier islands, sand spits, and mudflats (*also* known as bays, lagoons, harbors, or sounds).

Evidence Relevant factual information that can be challenged for accuracy.

Eutrophication A process by which excessive nutrients in a lake or pond cause plant life to proliferate, which depletes oxygen in the waters and causes the extinction of other organisms.

Expansive clays Types of soils containing water-absorbing minerals, thereby expanding with water and contracting when drying out.

Exponential growth A growth in which the rate is proportional to the increasing size or number in an exponential (not arithmetic) or logarithmic progression; a growth by a percentage of the starting number rather than by the addition of a fixed quantity.

ex-situ remediation (Hydrology) Refers to the method of treatment of contaminated or polluted aquifer in which the groundwater is pumped out (or soil is removed) and treated (see *in-situ* remediation, aquifer, groundwater).

External factors (Geology) Disturbing forces like earthquakes and heavy traffic that lead to slope failure (*see* internal factors, mass wasting).

Externalities (Economics) Factors not included in the costs of producing, transporting, and marketing goods (e.g., costs not included in the price of electricity production from fossil fuel and nuclear sources alone).

F

Federal Water Pollution Control Act Enacted in 1972, this legislation gave the Environmental Protection Agency (EPA) authority to control water pollution; EPA established the National Pollution Discharge Elimination System (NPDES), a program that requires all polluters to obtain a permit to discharge emissions into waterways. EPA sets limits on permitted pollution levels.

Fill (*see* cuts and fills)

Flaring of methane A safety measure to handle methane built up over time in a landfill or oil field, where the gas is captured, vented, and enflamed to burn it off, which also releases the greenhouse gas, CO_2.

Flash steam plants Uses geothermal reservoirs of water with temperatures greater than 360° F (182° C). As the water flows to the surface, the drop in pressure allows some of the hot water to boil (flash) into steam (*see* alternative energy, geothermal power, dry steam plants, binary cycle plants).

Fuel rod assembly Cylindrical or flat plates in a nuclear power plant reactor that (1) contain radioactive fuel material enriched in U-235, welded and sealed to the rod for safety; and (2) generate a nuclear chain reaction in a controlled fashion to produce heat, creating steam from the surrounding waters, which rotates turbines to generate electricity (*see* control rods, nuclear fission).

Fault (Geology) A crack in the Earth's crust from a displacement of one side in relation to the other (*see* aseismic slip, blind thrusts, creep, earthquake, normal faults, reverse fault, strike-slip faults, thrust fault).

Faustian bargain In legend, Faust trades his soul to the devil in exchange for knowledge or other earthly desire, with resulting unexpected consequences.

Fertility rate The speed or frequency of live births in an area or population in comparison with the total population.

Flood stage The level at the natural stream bank is breached by rising water and damage begins to occur.

Flood-control structures Infrastructure in flood-prone areas in an effort to prevent or mitigate flooding (e.g., revetments, dams, cuts, levees, weirs, fills, etc.).

Flood-frequency diagram A graph of the discharge versus the flood frequency of a stream or river, used to determine how often a flood of a given discharge will occur.

Fluorosis A medical condition stemming from fluoride poisoning; symptoms are discolored teeth and calcification of the bones.

Fly ash Airborne, often toxic, noncombustible chemical compound and particulate matter released most often during the processing of coal.

Focus (*also* hypocenter) The point inside the Earth where an earthquake originates (*see* epicenter, earthquake).

Furans One of a group of colorless, volatile, heterocyclic organic compounds containing a ring of four carbon atoms and one oxygen atom, obtained from wood oils and used in the synthesis of furfural and other organic compounds.

G

Gamma radiation (*also* gamma rays) One of three types of natural radiation; a very-high-frequency form of electromagnetic radiation consisting of photons emitted by radioactive elements on the course of nuclear transition.

Genetically modified (GM) foods Refers to plants modified by gene transfer to promote a particular trait.

Geohazard Any geologic event or process that presents a severe threat to humans, property, and the natural and built environment.

Geomorphology The study of landforms (origin, form, and distribution) across the landscape, and the processes (air, wind, water, earthquake, weather, etc.) that change them.

Geothermal gradient Ratio of rising temperature to increasing depth in Earth's interior.

Geothermal heat pump Used for residential or commercial heating and cooling, it is 25% to 50% more efficient than conventional heat pumps. The system utilizes the relatively stable temperatures of soils below the building to keep the structure's temperatures comfortable.

Geothermal power Thermal or electrical power produced from the thermal energy contained in the Earth (geothermal energy). Use is based on the temperature difference between a mass of subsurface rock and water and a mass of water or air at the Earth's surface. This temperature difference allows production of thermal energy that can be either used directly or converted to mechanical or electrical energy.

Global hectare Denotes one hectare of biologically productive land or water of average global productivity; underestimates productive regions because marginally and unproductive areas are not factored into the calculation.

Global water budget The amount of water transported in the water cycle between the continental mass, oceans, and atmosphere on planet Earth.

"Grandfather" clauses in legislation A provision in a statute that exempts those already involved in a regulated activity or business from the new regulations established by the statute.

"Grandfathered" coal plants Under 1977 Amendments to the Clean Air Act, utility plants operating or in construction before 1977 are not required to meet the pollution standards applied to newer plants.

Green Revolution During the 1940s–1970s, research and development of new technologies that, in agriculture, translates to using intensive fertilizer and pesticide application together with hybrid seeds.

Greenhouse effect The phenomenon whereby the Earth's atmosphere traps solar radiation, caused by the presence in the atmosphere of gases (i.e., carbon dioxide, CFCs, water vapor, and methane), which permit incoming sunlight to pass through but absorb the heat radiated back from the Earth's surface.

Greenhouse gases Any of the atmospheric gases (carbon dioxide, nitrous oxide, methane, and chlorofluorocarbons) that contribute to the greenhouse effect by absorbing infrared radiation produced by solar warming of the Earth's surface (*see* greenhouse effect).

Groins (*also* jetties) Structures, usually concrete, built perpendicularly to the coast to trap sand.

Gross domestic product (GDP) The monetary value of all the finished goods and services produced within a country's borders in a specific time period, though GDP is usually calculated on an annual basis. It includes all of private and public consumption, government outlays, investments, and exports less imports that occur within a defined territory.

Groundwater maps A detailed illustration of the elevation of a water table or the potentiometric surface, and the direction of groundwater flow. Maps are used by individuals, farmers, government agencies, and organizations to drill wells and track pollution.

H

Halomethanes Toxic or lethal compounds that are derivatives of methane (CH_4); soluble in carbons and slightly soluble in water; common as aerosol propellants, in freon, and other products.

Hardin's Third Law of Human Ecology Written by Garrett Hardin (1973), I = PAT, where I is the impact of any individual, group, region, or nation; P is population; A is affluence per capita (i.e., consumption); and T is technological damage incurred by supplying goods and materials for consumption.

Hard stabilization Structures (e.g., sea walls, revetments) added to beaches in an effort to prevent sand from eroding naturally.

Hard water Water that contains abundant calcium (Ca) and magnesium (Mg).

Hazard assessment Identifying, understanding, and quantifying the geological processes that can cause disasters in a region.

Heat budget An accounting of the balance between incoming solar radiation and re-radiation from the Earth back into space.

Heat transfer The thermodynamic movement of heat by means of radiation, convection, or conduction.

Human carrying capacity The maximum number of people that the planet can support in the long term without (1) degrading our cultural and social environments and (2) harming our physical environment in ways that would adversely affect future generations.

Humus A dark brown or black colloidal mass of partially decomposed organic matter in soil that helps retain water and improves fertility of the soil, thereby making it important for plant growth.

Hurricane (*also* typhoon) A severe tropical storm that forms in the tropical North Atlantic Ocean, most often a few degrees north of the equator, or in the Caribbean Sea, Gulf of Mexico, or eastern Pacific Ocean. Warm tropical oceans, high humidity, and light winds above them are necessary for them to originate.

Hurricane track Path along which a hurricane moves.

Hydraulic gradient Using several measurements of the groundwater's surface to determine the rate of flow from one point to another; the slope of the water table (*see* aquifer, surface map, cone of depression).

Hydrologic cycle The continuous circulation of water (H_2O) through the atmosphere and Earth.

Hydrology The study of water systems.

Hydropower plants Facilities with the technology to generate electricity from running water.

Hydrothermal solutions Mineral deposits in which hot water serves as the concentrating, transporting, and depositing agent; the most common of all classes of deposit.

Hypothesis A reasonable or proposed explanation for a phenomenon, not yet proven.

Hypoxia Low dissolved oxygen in surface water, usually caused by excess algae blooms from excess nitrogen levels; a deficiency of oxygen reaching tissues.

Hypoxic Describes water with less than 2 to 3 ppm dissolved oxygen (DO), which only supports anaerobic and relatively inactive fauna (e.g. worms, some crabs; *see* anoxic, normoxic).

I

Ice draft Upward profile of the thickness of arctic ice taken by submarines.

Ice-jam flood Occurs when ice from southerly thaws merges with new ice and piles up, causing flooding where the ice is blocked by bridges, a narrowing channel, or other cause.

Igneous A type of rock formed from magma (*see* magma).

Impacts of climate change Rising temperatures have caused many changes on Earth. In the last century there has been glacier retreat, a decrease in snow cover-extent and earlier snowmelt, and reduced duration of annual lake and river ice. Sea ice has decreased in thickness. Migrating animals have shifted polewards and in elevation in their ranges; species reproduction has altered, as have morphological and physiological adaptations. Plants have changed in abundance and diversity, phrenology, and growth. Entire ecosystems and microsystems are changing or going extinct when the changes are too rapid for their evolution. Weather patterns are in flux, causing wider areas of natural disasters such as flooding, desertification, severe storms, and/or fire.

Incompatible elements In pegmatites, elements that don't usually form common igneous minerals (e.g., boron, lithium, uranium, and gemstones).

Infiltration capacity The maximum rate at which water can enter soils; if the region is typically low during flood season, it's because soils are often frozen due to the lateness of the spring thaw.

Infrared The range of invisible radiation wavelengths from about 750 nanometers (slightly longer than red in the visible spectrum) to 1 millimeter (on the border of the microwave region).

Infrastructure The basic facilities, services, and installations needed for the functioning of a community or society (i.e., transportation and communications systems, water and power lines, and public institutions including schools, post offices, and prisons), or preservation policies, educated public, and structures and/or services necessary to preserve a natural formation.

in-situ remediation (Hydrology) Refers to the method of treatment of a contaminated aquifer in which pollutants are treated within the aquifer (see *ex-situ* remediation, aquifer, groundwater).

Instream water use Most applications of water resources (e.g., recreational uses, protection of aquatic habitats, navigation, and generation of hydropower, although the latter may have profoundly negative impacts on the dammed water body) that do not impair water's availability for future or downstream use.

Intensified production The act of obtaining more output from a given area of agricultural land due to population pressures.

Integrated Waste Management (IWM) Employs a variety of methods for waste disposal and control to minimize the environmental impact of industrial, commercial, and home/individual waste streams, including: source reduction, recycling and re-use, incineration, and landfills.

Intensity A measure of the strength of an earthquake that expresses the relative damage caused to people and structures, using a Modified Mercalli (MM) scale.

Internal factors Factors that change the strength and properties of the mass itself that leads to slope failure (*see* external factors, mass wasting).

Intertidal (*also* littoral zone) A zone between high and low tide that is exposed to both air and sea, depending on the cycle of waves and tides, seasons, and storms (*see* shore).

Intraplate earthquake A type of earthquake that occurs in the interior of a tectonic plate, far from plate boundaries (*see* earthquake, plates, plate tectonics).

Ionic substitution The replacement of one type of ion in a mineral by another that is similar to the first in size and charge.

Island arc A series of volcanic islands, distributed in a crude arc, formed by subduction at ocean-ocean boundaries that appear in an arc shape (*see* subduction, seafloor spreading).

Island chain An elongate series of upthrusts of land, usually volcanic, each of which are surrounded by sea, that are formed as a result of movement of oceanic plates over mantle plumes (*see* mantle, plates, plate tectonics).

Isolated wetlands Lands characterized by water saturation, completely surrounded by upland with no surface water outlet, and thus may not have obvious connections to other water bodies (*see* wetlands, contiguous wetlands, constructed wetlands).

Isoseismal map An illustration of the intensity of a particular earthquake using surveys of local inhabitants. Such a map uses isoseismal contours (lines of equal intensity) to represent the damage done by an earthquake.

Isostasy Equilibrium of the Earth's crust, where the forces tending to elevate balance the forces tending to depress; the state in which all pressures from every side are equal.

J

Johnstown flood (*also* Great Flood of 1889) A break in the South Fork Dam caused a flood near Johnstown, Pennsylvania, killing 2,209 people.

John Tyndall (1820–1893) A British physicist and professor who studied magnetism and thermal radiation, and scientifically identified many processes of Earth's atmosphere; discovered that CO_2 and H_2O atmospheric gases could absorb energy in the infrared range.

K

Karst Areas with humid climates underlain by carbonate (e.g., limestone, dolostone, salt, gypsum) bedrock; the landscape is characterized by sinkholes, sinking or disappearing streams, caves, springs, and underground drainage. This bedrock is soluble and with precipitation forms a weak acid rain, which widens the natural joints between rocks over time, forming caverns or sinkholes (*see* limestone and dolostone, sinkhole).

Kesterson Wildlife Refuge A federally protected area now a part of the San Luis Wildlife Refuge, located in the San Joaquin Valley, California.

Kilowatt (kW) A unit of power equal to 1,000 watts.

Kimberlite Intrusive bodies of enigmatic origin (some containing diamonds), comprised of a high-pressure mineral assemblage that indicates their origin within the mantle.

Klamath Basin Irrigation Project Dams and canals created to supply irrigation for surrounding farmers in areas of the Upper Klamath Lake and Klamath River, spanning counties in Oregon and California. Increasing population, bad agricultural processes, and pollutants subsequently caused toxic water and loss of habitat for aquatic, bird, and mammal wildlife.

L

Lag time (Hydrology) Refers to the gap between the occurrence of precipitation and the runoff from that precipitation.

Lahars (*also*, volcanic mudflows) A hot to cold mixture of water and rock fragments (clay to bolder in size) flowing fast down the slopes of a volcano, often following river valleys.

Landfill A method of solid waste disposal in which refuse is buried between layers of dirt so as to fill in or reclaim low-lying ground (*see* environmentally sound management, integrated waste management, landfill cell, municipal solid waste).

Landfill cell One of many depressions in a landfill with a plastic or clay liner to prevent seepage of leachate into groundwater or surface streams.

Land-use changes Human interactions with nature that cause increased runoff over infiltration and/or transpiration. Examples are: construction of roads, parking lots, malls, and other structures that cover large areas of former natural vegetation, agricultural practices, plus direct drainage into streams and sewer systems.

Lateral blasts Explosion that occurs when a side of a volcano collapses, releasing internal pressure on the superheated steam-and magma system within it.

Laterite A type of soil typical of tropical regions that contains few nutrients and little organic matter, and consists mainly of insoluble complexes such as iron and aluminum hydroxides (*see* bauxite).

Lava Coherent masses of hot, partially or completely molten rock that flow down slopes, generally following valleys (*see* pyroclastic material).

Lava domes Usually small and dome-shaped, these volcanic extrusions are highly viscous, silicic ($\pm$ 70% SiO_2) magma. Because of its high viscosity, this type of lava does not move far from the vent before solidifying. Domes can be deadly because of trapped gas explosions as well as pyroclastic flows.

Leachate Liquid derived from waste that may contain hazardous chemicals.

Leached zone (*also* zone of leaching) The top layer of soil characterized by the downward movement of water. Soluble minerals in this layer of regolith are dissolved by water, and the very fine clay is picked up and carried downward with infiltrating water.

Lead toxicity Even low levels of lead are associated with decreased intelligence, impaired neurobehavioral development, decreased stature and growth, and impaired hearing acuity in children; studies show it to be a major risk factor in behavioral problems, criminality, and psychological deviancy (impulsiveness, low I.Q., hyperactivity, and low frustration tolerance) in those aged 7 to 22. In adults, it can cause irritability, poor muscle coordination, damage to the sense organs and nerves, reproduction problems, and high blood pressure.

Lignite A soft, brownish-black coal in which the alteration of vegetable matter has proceeded further than in peat but not as far as in bituminous (*see* bituminous coal).

Limestone and dolostone Chemical sedimentary rocks formed usually under marine conditions containing calcium and magnesium; often comprise karst (*see* karst).

Liquefaction During an earthquake, occurs when violent shaking reduces the internal cohesiveness of sediment and causes it to collapse and flow like a fluid (*see* geohazard, mudslide, earthquake).

Lithosphere A strong, rigid zone of rock situated above the asthenosphere and includes the crust (both continental and oceanic) and the rigid uppermost portion of the mantle (*see* low velocity zone).

Loess (*also* "rock flour") A buff to grey colored, windblown loamy deposit of fine-grained, calcareous silt or clay soil, produced from glacial erosion.

Logic A series of statements or thoughts that are mutually reinforcing and exhibit intellectual standards of critical thinking.

Logical fallacy Occurs when a conclusion does not make logical sense, is internally contradictory, or not mutually reinforcing (*see* logic, critical thinking).

Long period events Seismic activity that builds up slowly and tapers off, signs of pulses of magma and gas moving into the volcano just before an eruption.

Longshore current Obliquely-breaking waves that cause currents running parallel to the beach. These currents can transport vast amounts of sand, some of it to submarine canyons, and are important in understanding the stability of beaches and coastal structures.

Low velocity zone (asthenosphere) A zone near the top of the upper mantle where seismic wave velocities are lower than the rock above and below.

M

Magma Molten material beneath or within the Earth's crust from which igneous rock is formed.

Magnetometer Instrument to measure magnetic properties of the sea floor, developed at the Scripps Institution of Oceanography.

Mantle The intermediate zone of a planetary body between the crust and the core.

Mantle plume Molten magma that rises in a narrow column towards Earth's surface.

Mass wasting (*also* landslide, slope failure) Downslope movement of loose rock and soil under direct pressures from gravitational forces. Distinct from erosion forces (wind, water, or ice and snow movement). (*See* external factors, internal factors.)

Megawatt (MW) A unit of power equal to a million watts.

Metamorphic A rock formed when a pre-existing rock is subjected to heat and pressure, causing the original minerals to align themselves, recrystallize, or both.

Methane (CH_4) A colorless, odorless gas commonly found in nature. A potent greenhouse gas.

Methane hydrates Deposits of methane gas bound by clathrates, which are "cages" of ice molecules surrounding concentrated methane.

Methyl mercury Any of various toxic compounds of mercury containing the complex CH_3Hg^- that often occur as pollutants which accumulate in living organisms (such as fish) especially in higher levels of a food chain.

Microzoning A branch of engineering seismology whose objective is to refine seismic zoning data and to determine precisely the degree of earthquake danger in regions being built up.

Mid-Atlantic Ridge A 16,000 km (10,000 mi) submarine mountain range located in a curve from the Arctic Ocean to the southern tip of Africa. Some of these mountains are tall enough to have formed the island groups of Ascension, Azores, St. Helena, and Tristan da Cunha.

Mineral deposit Any naturally-occurring inorganic crystalline solid with a set of physical and chemical properties (hardness, luster, etc.) and a definite chemical composition found on Earth.

Mine spoil Tailings from a coal or other mineral mine; these are considered waste material.

Mining groundwater Withdrawal (overdraft) of water from an aquifer faster than it can be replenished over time, thus depleting the overall supply.

Mitigation banks Where a regulatory agency will allow the destruction of a small area of isolated wetland if it is replaced by an equivalent or larger area of the same type, in a large tract that will be permanently maintained.

Model (Geology) An abstract representation of a general concept. Using data and technology, developing a topography and stratigraphy of a particular region to help predict rock changes under specific conditions (e.g., volcanic eruption or earthquakes).

Moment magnitude (MM) A more accurate measure of the absolute amount of energy released by a large earthquake than the Richter scale. MM considers the distance that the fault moved multiplied over the fault plane area (the seismic moment) that is then converted into a number like other earthquake magnitudes by a standard formula (see Richter magnitude).

Mountaintop removal The practice of removing large quantities of rock and soil above horizontal coal seams underlying mountaintops.

Municipal solid waste (MSW) All types of solid waste generated by households and commercial establishments, and collected usually by local government bodies.

Municipal waste Any garbage, refuse, sludge, sewage, or discarded material (solid, liquid, or airborne) from household, industrial, agricultural, commercial, or recreational activities within a municipality.

Mutagenic compound Any agent capable of altering a cell's genetic makeup by changing the structure of the hereditary material, DNA. Many forms of electromagnetic radiation (e.g., cosmic rays, X-rays, ultraviolet light) are mutagenic, as are various chemical compounds.

N

National Environmental Policy Act Enacted in 1970 to restore and protect our nation's environment, this legislation required the preparation of Environmental Impact Statements (EIS) for projects with any federal funding, and set up the Council on Environmental Quality, which reports to the President.

National Pollutant Discharge Elimination System A permit program that authorizes the Environmental Protection Agency (EPA) to regulate point sources that discharge pollutants into waters of the United States.

Natural gas A fuel source consisting mainly of methane produced in any geologic environment in which organic matter is decomposed by microorganisms in the absence of oxygen, with or without heat and pressure.

Natural levee Low ridges formed when floodwaters overtop a river's channel and dump debris alongside the channel where velocities drop abruptly.

Natural resources A material source of wealth (e.g., timber, fresh water) or a mineral deposit or commodity (e.g., aggregate, gemstones, coal, petroleum) that occurs in a natural state and has economic value.

Net capacity factor The ratio of electricity actually generated by a power plant compared to the amount that could be produced at full power with continuous operation.

New Madrid Fault Zone (NMFZ) An active seismic area in the United States east of the Rocky Mountains, the NMFZ is located in southern Illinois, western areas of Kentucky and Tennessee, southeastern Missouri, and northeastern Arkansas; northwestern Mississippi also has significant shaking from large earthquakes.

Nitrogen fixation The process by which free nitrogen in the atmosphere is converted by biologic or chemical methods to ammonia as well as other forms used by plants and animals.

Nitrogen surplus A measure of the nitrogen contamination potential.

No net loss A U.S. policy stating that wetlands should be conserved wherever possible; if wetlands are converted to other uses, this must be offset through restoration and/

or creation of wetlands of equal or greater area, thus maintaining or increasing the overall wetland area.

Nonpoint source pollution Contamination discharged over a wide land area, not from one specific location, or from a single source leaching fluids or projecting airborne particulates to long distances away. This diffuse pollution is caused by sediment, nutrients, and/or organic and toxic substances originating from agricultural and industrial land-use activities, which are carried to soil, lakes, streams and other water bodies by surface runoff (*see* source point pollution).

Normal fault A large crack in the Earth's crust where the side opposite the acute fault plane angle has moved down during an earthquake.

Normoxic Describes water that can support virtually all aquatic life, with dissolved oxygen (DO) of 5 to 8 ppm (*see* anoxic, hypoxic).

No-till plowing (*also* pasture cropping, direct planting) A type of farming that allows a crop to be planted without turning the soil.

Nuclear fission A nuclear reaction in which a heavy atomic nucleus, such as uranium, breaks into two lighter nuclei. Other particles, such as free neutrons and protons, as well as large amounts of energy and electromagnetic radiation, are produced during fission reactions.

Nutrient pollution Eutrophication caused by excessive nutrients that causes a cascade of events, promotes algae blooms that block light beneath water surfaces and prevents oxygenation of oceans, lakes, and other bodies of water, such that marine organisms cannot live.

O

Ocean acidification Describes the inability of the ocean to safely absorb carbon dioxide (CO_2) in the carbon cycle, thereby causing the ocean's pH to decrease as global warming progresses.

Oceanic crust A layer of rock broadly similar in composition to that of continental crust but more dense and containing higher concentrations of iron, magnesium, and calcium. Basalt is the predominant type of rock in this crust.

Ocean trench A long, narrow steep-sided depression in the Earth's oceanic crust, usually lying above a subduction zone.

Off-peak power demand Periods of low energy demand (*see* peak power demand).

Oil refining Processes using heat, chemical reactions, and catalysts to convert crude oil into the products used by modern societies.

Ophiolite sequence A layering of rocks consisting of deep-sea marine sediments overlying (from top to bottom) pillow basalts, sheeted dikes, gabbro, dunite, and peridotite; these are indicators of seafloor spreading where the oceanic crust is thinned by the forces of plate tectonics, and magma rises to the sea's floor and generates new crust.

Opinion A belief or conclusion based upon confidence or general opinion yet not substantiated by sufficient knowledge or proof.

Ore The concentration at which mineral deposits become economically valuable (*see* mineral deposit).

Organic agriculture The practice of raising crops without the use of inorganic chemical fertilizers and/or pesticides. A production system that sustains the health of soils, water, ecosystems, and people.

Overburden The material covering coal seams.

Ozone destroying chemicals (ODCs) Synthetic compounds that when aerosolized, deteriorate in the stratosphere and catalyze the ozone gas, causing holes in the ozone layer, which allows ultraviolet rays to reach the Earth's surface and leads to climate change.

P

Paleomagnetism The study of the Earth's magnetism over time. Orientation of magnetic minerals in a rock is fixed permanently unless reheated at the time of the rock's formation. Paleomagnetic study of ocean floor rocks provided early evidence for plate tectonics.

Paleoseismicity The study of prehistoric seismic movements in the Earth's crust.

Pangaea A supercontinent (from the Greek word meaning "all land") that existed before the Triassic Period, comprising all of the land masses on Earth in the Late Paleozoic Era, including Gondwanaland and Laurasia; theory is based on plant and geologic evidence.

Panthalassa A vast ocean (from the Greek word meaning "all sea") that surrounded Pangaea until the supercontinent broke apart via plate tectonics.

Particulates, PM2.5, PM10 Particles with a diameter of 2.5 or 10 micrometers or less in the atmosphere, used as a measure of air quality in specific locations or regions.

Peak power demand Periods of high energy demand (*see* off-peak power demand).

Peat Undecayed plant matter.

Pedalfer Soil typically found in temperate humid climates that contains organic material and few soluble ions (Ca^{++}, Mg^{++}, Na^+, etc.), because in humid climates, precipitation (e.g., rain) dissolves these ions and carries them into groundwater.

Pedocal A soil usually formed under dry climatic conditions that may contain abundant soluble ions, such as Ca^{++}, B^{++}, Se^{++}, and Na^{++}.

Perched water table A water table positioned over the normal water table. An accumulation of groundwater located above a water table in an unsaturated zone.

Perchlorate Highly water soluble and widespread inorganic chemical that inhibit assimilation of iodine in the thyroid gland, and consequently interfere with thyroid functioning at intake levels above 200 mg/day; at high levels may lead to cancer. Used as rocket and munitions fuel in the 1950s, this can be very persistent in groundwater.

Peridotite A magnesium-rich rock containing abundant olivine formed in the upper mantle.

Permeability The ability of material to transmit a fluid.

Persistent organic pollutant (POP) Toxic organic compounds that are not biodegradable and persist in the environment, bioaccumulate in the food web, and are a health risk to organisms (*see* bioaccumulation).

Phenolics Any of various synthetic thermosetting resins, obtained by the reaction of phenols with simple aldehydes and used to make molded products and as coatings and adhesives.

Photovoltaics Technology that converts sunlight directly into electricity.

Placer deposits Minerals and metal weathered away from their original deposits and easily collected (e.g., gold flakes).

Plate A segment of the lithosphere.

Plate boundaries In plate tectonics theory, the boundaries between lithospheric plates.

Plate tectonics A unifying principle of geology that the surface of the Earth is covered by seven major rigid lithospheric plates and a number of smaller plates, all of which are in motion, but not all in the same direction or at the same speed. This theory explains earthquakes and large scale geologic formations and processes (*see* paleomagnetism).

Plate-tectonic setting of volcanoes The type and location of a volcano strongly determines magma type produced, which in turn determines viscosity, resistance to flow (molten magma thickness), and risk to humans and property.

Point of view A particular perspective and set of bases and biases used to form an opinion.

Point-source pollution Sources of pollution that have a specific location of origin, which makes it easier to analyze, isolate, and develop a plan of remediation (*see* nonpoint-source pollution).

Polar molecule A molecule (e.g., water) with positive and negative poles.

Pollution Standard Index (PSI) A composite indicator developed by the EPA, calculated from atmospheric levels of ozone, nitrogen dioxide, sulfur dioxide, carbon monoxide, and particulate matter only; pollutants measured in local areas constantly.

Polychlorinated biphenyls (PCBs) A group of highly toxic, persistent, organic chemicals that can be odorless or mildly aromatic solids or oily liquids. Before the Clean Air and Water Act banned this class of compounds in 1984, they were used in the U.S. as hydraulic fluids, plasticizers, adhesives, fire retardants, way extenders, de-dusting agents, pesticide extenders, inks, lubricants, cutting oils, in heat transfer systems, and carbonless reproducing paper.

Population density The size of a population within a defined space.

Population pyramid The distribution of a population within a defined area by factor(s) (e.g., age, gender, economic status, etc.)

Pore water pressure A force (e.g., buoyancy) exerted by the water in pore spaces, because water cannot be compressed.

Porosity/effective porosity A measure of the amount of pore space in material, the ratio of void space to the total volume of material. (Hydrology) Pore space that can actually accommodate flow.

Porphyry copper deposit An igneous rock which contains large crystals set in a finer, but still visible, groundmass (porphyry) with copper mineralization; usually formed with, and below, composite volcanoes (one made of composites of lava and ash deposits) that have been eroded away.

Potentiometric surface Elevation to which water in a confined aquifer would rise if a well were sunk into it. The surface exists because pressures on confined aquifers are much higher than atmospheric pressure (*see* aquifer, confined aquifer).

Precautionary Principle A hypothesis whereby the scientific burden of proof is on a product or process to demonstrate that it will not cause harm.

Precision The fact, quality, or action of being exact and accurate.

Precursors Volcanism measurements suggesting signs of volcanic activity. These include: small earthquakes, changes in the position of the volcano's surface, and changes in the composition of gases emanating from the volcano.

Pressure gradient (*also* barometric gradient) The force of air (change of atmospheric pressure) per unit horizontal distance on the Earth's surface, which determines wind and weather patterns.

Price Anderson Act Legislation in the United States seeking to incentivize nuclear power by limiting the liability of nuclear plants in the event of an accident. Under the Act, the nuclear power industry is responsible for the first $10 billion of damages resulting from a spill or other accident, while the federal government is responsible for any losses exceeding that amount. It was passed in 1957 and has been renewed several times since.

Primary sewage treatment Collects sewage, filters out large fragments, settles out most solids, removes surface scum (fats, etc.) and chlorinates the remaining liquids to kill potentially pathogenic organisms. The effluent is then piped to a water body.

Prior appropriation water doctrine Legally, the first claimants have the most secure water rights.

Process heat Byproduct heat that would otherwise be ejected into the environment. In nonresidential buildings, sources of waste heat include refrigeration/air-conditioner compressors, manufacturing or other processes, data processing centers, lighting fixtures, ventilation exhaust air, and the occupants themselves.

Product life-cycle assessment (LCA) A health and environmental analysis of the products, processes, and activities involved in producing that product from extraction of raw materials to handling of wastes.

Pyroclastic flows (*also* a *nuée ardente*) A devastating, fast-moving, avalanche of molten particulates, from ash to boulders, as a result of a volcanic eruption (*see* ash, basalt, lava, pyroclastic material, tephra).

Pyroclastic material Ash, pumice, basalt, and other rock fragments found in lava from a volcano.

Q

Quad A unit of energy equal to 10^{15} Btu.

R

Radiation Occurs when electromagnetic waves travel through space; atmospheric convection. When these waves hit an object (e.g., the Earth's crust), there is a transfer of heat energy (*see* heat transfer).

Radioactivity The decay of unstable atomic nuclei, where the radioactive atom changes to a non-radioactive element by emitting particles and energy.

Radon A colorless, radioactive, inert gaseous element formed by the radioactive decay of radium.

Rate of natural increase/decrease A calculation of the birth rate minus the death rate in a population, without factoring in migration.

Recessive gene (Genetics) A gene or allele, the trait of which is expressed only when both copies are identical; opposite of dominant gene.

Recharge A process whereby surface water and groundwater is replenished by precipitation through an unsaturated zone or by surface water bodies (streams, rivers, lakes, leaky canals, etc.; *see* unsaturated zone)

Recurrence interval (1) An estimate of how regularly volcanoes erupt. (2) A statistical estimate of the time between recurrences of a river's discharge (flood) of a given magnitude.

Refining (*see* oil refining)

Regolith A general term for weathered rock; a region of unconsolidated rock and dust (including soil) that sits on top of a layer of bedrock.

Relevance Pertinence or connection to the matter at hand.

Remediation (*see* ex-situ remediation, in-situ remediation)

Renewable energy Energy that cannot be used up, such as that from wind, sun (solar), biomass including wood and wood waste, tides, and waves (*see* alternative energy).

Replacement fertility rate The ratio of live births a woman would bear to replace herself and her partner in a population. Used to determine the degree of stability in a population (*see* total fertility rate).

Reserves Amount of oil that can be produced using modern technology at a profit.

Reservoir rock Porous layers of rock filled as volatile compounds from source sediments migrate upwards by gravity to form oil, gas, and water deposits (*see* source sediments).

Residence time The average duration that an atom, molecule, or complex stays in the atmosphere, lithosphere, or hydrosphere, before cycling into one of the other systems.

Resonant frequency of buildings The frequency at which a building shakes while being disturbed by seismic violence, used to develop technology to preserve the structure despite earthquakes.

Resource Conservation and Recovery Act (RCRA) Enacted in 1976, RCRA authorized the EPA to take a cradle-to-grave approach to waste management, and had its main impact on the treatment of a category the act defined as hazardous waste.

Restricted-use pesticides Chemicals and compounds that are banned in areas where groundwater has been contaminated, or where the potential exists for such contamination.

Reverse faults A large crack in the Earth's crust where one part has moved against another part during an earthquake, in which the side opposite the obtuse angle of the fault plane moves upward relative to the other side. Reverse faults occur where two blocks of rock are forced together by compression (*see* normal faults).

Reverse osmosis A process to pump salty water through a series of filters, removing dissolved matter until it becomes potable; its cost efficiency relies on a cheap power source and a cheap, safe place to dump salty and often toxic waste products.

Revetment Structures built on slopes, bluffs, or banks to prevent erosion.

Richter magnitude A mathematical relationship used to measure the energy release of earthquakes, developed in 1935 by Charles F. Richter (*see* moment magnitude).

Risk The threat of humans suffering loss, harm (injury, disease), or death; danger.

Risk assessment The process of judgment (identification, analysis, estimation of levels) to determine the likelihood that a negative event will occur.

Risk paradigm The basis of Western torte system of civil law, where burden of proof falls not on the producers of goods, but rather on those who allege that they have suffered harm. poor water, air, or food quality or remove/consume natural resources entirely with no reuse.

River gradient A measure of how steeply the water loses height over a distance.

Rock strength Inherent structural characteristics that define the durability of a particular rock (e.g., mass, water content, brittleness, temperature, etc.).

Roger Revelle (1909–1991) Oceanographer at Scripps Institute, scientist, and administrator; in 1957 with colleague Hans Suess discovered that concentration of atmospheric carbon dioxide as a direct result of the burning of fossil fuels was unable to cycle through the oceans at a rate that would prevent global warming.

S

Safe Drinking Water Act Enacted in 1974 to set minimum drinking water standards and protect groundwater.

Safe withdrawal rate An amount of water that will not deplete the aquifer or lead to undesirable results such as formation of a cone of depression, subsidence of the surface, or saltwater incursion (*see* aquifer, cone of depression, saltwater incursion).

Safe yield (of aquifer) The amount of water that can be taken over a period of time without depleting the aquifer.

Salination The buildup of salts in water soil, thereby making plants inedible or poisonous to organisms, or causing the plants to die and water to be polluted.

Salinization/salination In an arid, poorly drained region, describes the accumulation of soluble salts in soil from the evaporation of water that bore them to the surface.

Salt dome (Geology) An anticlinal fold with a columnar salt plug at its core.

Saturated zone Zones under an unsaturated zone and water table where the pores are filled with water (see aquifer, unsaturated zone, water table).

Saltwater intrusion Occurs when a freshwater aquifer, situated next to a body of saltwater, becomes depleted, and the saltwater leaches into the freshwater aquifer, poisoning and eventually making it unusable (*see* salination, saltwater incursion/invasion).

Saltwater incursion/invasion (SWI) Where fresh water and salt water are both present in an aquifer, the fresh water, being lighter, floats on a wedge or lens of salt water. SWI occurs when wells withdraw too much fresh water, so the lens of denser salt water can infiltrate farther into the aquifer and thereby poison it.

Science An established knowledge base and the ongoing pursuit of knowledge and understanding of the natural and social world following a systematic methodology based on empirical evidence (*see* scientific method). The formation and rigorous, objective testing of hypotheses.

Scientific method A method of discovering truth by utilizing precise observations, collecting empirical and relevant evidence, developing hypotheses and testing them rigorously, and performing accurate, controlled confirmation.

Scientific theory A broadly accepted explanation for an important phenomenon.

Seafloor spreading A process by which new ocean floor is created as molten magma from the Earth's mantle through convection extrudes onto the sea floor's surface, causing ridges to form and eventually spread bilaterally, while preserving information on the current direction and intensity of the Earth's magnetic field (*see* plate tectonics, crust, convection, paleomagnetism).

Secondary enrichment Processes that help concentrate the minerals in mineral deposits and convert it to forms easier to refine.

Secondary oil recovery Technologies for retrieving as much oil as possible from a source bed.

Secondary sewage treatment After primary treatment, the waste is pumped to tanks where added bacteria help reduce the volume of the sludge (solids), converting some organic matter to methane in the process. The waste is then chlorinated before its release into the environment.

Sedimentary A type of rock formed by erosion of pre-existing rocks or by organisms, like corals, forming layers over time, and pressure.

Seiche A type of wave that occurs in a standing body (e.g., gulf, lagoon, lake).

Seismic hazard Any dangerous physical phenomenon (ground shaking, ground failure) associated with an earthquake.

Seismic hazard zoning Areas zoned as a hazard because of unstable soils or rocks during precipitation or earthquakes.

Seismic risk map A collection of historical data illustrating the probability of ground acceleration (the major cause of earthquake damage) for a given locality. Data on geologic hazards, faults, local building codes, and other tools for engineers are assembled.

Seismic wave An Earth vibration from an earthquake, volcanic eruption, or explosion, measured on a seismograph. Common types are: Primary (P) waves, Secondary (S) or Shear waves, and Surface waves.

Seismogram A record of an earthquake recorded by a seismograph used by scientists to calculate the location in intensity of an earthquake.

Seismograph Instrument that records the passing of seismic waves (*see* earthquake, seismic wave, seismogram).

Selenium An essential mineral found in foods that is a component of an antioxidant enzyme, glutathione reductase, that is key in tissue respiration. In large amounts, selenium salts are toxic.

Selenosis Selenium poisoning from eating certain plants, and/or exposure to selenium in soil and/or aerosol particulates.

Septic tank/septic system A sewage-disposal metal container in which a continuous flow of waste material is decomposed by anaerobic bacteria or in a leach field, a plot of land with series of buried pipes that drain off the filtered liquid sewage.

Sequestering (Geology) The capture of carbon dioxide and/or other pollutants during mining, refining, and/or during its use; refers to fossil fuels (oil, coal, natural gas).

Shear strength The resistance within the surface material to shear stress; cohesional (forces that pull together) and frictional forces (those that resist movement) within the material that prevent movement. Cohesional forces include cements in sedimentary rock and interlocking mineral textures in many igneous and metamorphic rocks (*see* mass wasting, sheer stress, strain).

Shear stress The gravitational force that pulls material downslope; it increases as weight of material and slope angle increases (*see* mass wasting, sheer strength, strain).

Shield volcanoes The largest type of volcano; composed of basalt, has broad, nearly flat-to-gently sloping flanks, and are shaped like an ancient warrior's shield (*see* basalt).

Shore Zone of land bordering a body of water.

Silica content of lava Concentration of SiO_2 helps to determine the classification of a volcano (*see* lava, pyroclastic material).

Sinkhole (*also* sink, doline) A topographic depression formed as the underlying limestone is dissolved by groundwater. Some are caused by the collapse of a series of caverns and cavities, and have steep walls and underground water flowing into it. Shallower ones are due to the gradual dissolving of the bedrock under the soil mantle and receive local drainage (*see* karst).

Slope failure The downward and outward movement of a mass of soil beneath a natural slope or other inclined surface; four types are rockfall, rock flow, plane shear, and rotational shear (*see* mass wasting)

"smart growth" Sustainable development promoted by many U.S. regions.

Smelting The process of heating and melting ores to extract metals.

Soft stabilization Practices such as beach renourishment in an effort to deter sandy beach erosion.

Soil Comprised of the decomposition and disintegration products of rock and sediment carried out by agents of chemical and mechanical weathering, a wafer-thin, dark stratum on the surface of the Earth, vital for the survival of life on land. Soil and soil biota are interlocking components of the terrestrial biosphere, interacting with vegetation and global climate, and vital for the functioning of terrestrial ecosystems.

Soil biota The biological life in the top layer of the Earth's surface. The mixture of soil particles and the total living organisms (animals and plants) in a defined area.

Soil moisture The amount of water absorbed by soil (the top few meters of regolith).

Solid Waste Disposal Act Enacted in 1965 to regulate municipal waste (MSW) and to improve technologies by which MSW was handled. The federal government awarded grants to states for construction of sewage treatment plants (*see* municipal solid waste).

Solifluction (*also* permafrost) A type of creep in sediment that is waterlogged moves slowly downslope over frozen ground (*see* creep, mass wasting).

Source bed Sedimentary rocks containing abundant organic matter that may provide source(s) of oil and gas deposits from which hydrocarbons have been extracted or are capable of being extracted.

Sprawl development Land conversion that occurs at a rate exceeding the rate of population growth.

Springs Form where aquifers break the ground surface; this water may then run into streams, becoming runoff.

Stage–discharge relationship The height of water in a river as discharge varies.

Standard of living A level of material comfort in terms of goods and services available to someone or some group.

Steel A generally hard, strong, durable, malleable alloy of iron and carbon, usually containing between 0.2 and 1.5 percent carbon, often with other constituents such as manganese, chromium, nickel, molybdenum, copper, tungsten, cobalt, or silicon, depending on the desired alloy properties, and widely used as a structural material.

Storm surge Occurs when powerful, persistent winds push water onshore during storms and, when coupled with winds, high tides, and high sea surfaces due to extremely low atmospheric pressure, storm surges can be particularly destructive.

Stormwater runoff Precipitation that percolates or flows, and eventually ends up in water sources (streams, rivers, groundwater, ponds, lakes, oceans).

Strain (Geology) The amount of deformation an object experiences compared to its original size and shape after being subjected to geologic or explosive forces (*see* stress, shear stress).

Stratovolcanoes (*also*, composite volcanoes) Eruption of both lava flows and ash of intermediate-to-felsic chemical composition; conical and steep in shape and are built over periods of hundreds of thousands of years; the signature type of volcanoes of ocean–continent convergent boundaries and island arcs.

Stream hydrograph A type of graph used to show discharge over time for floods (*see* discharge).

Stress (Geology) The force per unit area (e.g., tons per square inch, or dynes per square cm) that zones of rocks exert against each other. It has the same units as pressure, and is one special variety of stress, but it is much more complex because it varies both with direction and with the surface it acts on (*see* strain).

Strike-slip faults A tectonic event that occurs when two plates of the Earth's crust slide past one another during an earthquake (*see* plate techtonics).

Sturzstrom Long run-out landslides that release as much energy as immense volcanic eruptions.

Subduction zone Located at convergent boundaries where heavy, old oceanic lithosphere sinks back into the mantle (*see* plate tectonics).

Submarine canyon Narrow, steep sided valley cut into the continental shelf; formed from ancient rivers during the Ice Ages (*see* turbidite, abyssal plain).

Submarine fan (*also* abyssal fan) Delta-like features at the mouths of submarine canyons.

Subsidence The sinking or collapse of a surface, resulting from removal of subsurface support.

Subsidy Financial assistance given to industries that typically reduces the price of their product.

Subtidal Zone that is completely flooded by tidal water at low tide.

Sumatra-Andaman earthquake An undersea megathrust earthquake, magnitude 9.1, that took place on December 26, 2004 off the coast of the island of Sumatra, generating deadly tsunamis and the longest recorded faulting to date; 250,000 people in over four nations were killed.

Sunda Trench A deep sea trench located in the eastern part of the Indian Ocean.

Superfund Act The Comprehensive Environmental Response, Compensation, and Liability Act (CERCLA), commonly known as Superfund, enacted by Congress in 1980. This law created a tax on the chemical and petroleum industries and provided broad Federal authority to respond directly to releases or threatened releases of hazardous substances that may endanger public health or the environment.

Supratidal The shore area adjacent to and above high tide that can be covered with sea during storms.

Suspended load When material is carried in suspension by a river.

Sustainable development "Development that meets the needs of the present without compromising the ability of future generations to meet their own needs" (World Bank). The act of maintaining a delicate balance between the human need to improve lifestyles and feeling of well-being on one hand, and preserving natural resources and ecosystems on which we and future generations depend.

Sustainability Development that meets the needs of the present without compromising the ability of future generations to meet their own needs. Transformation from a wasteful linear model of resource use, in which natural resources (be they living or nonliving) are extracted, used, and then thrown "away," to a cyclical model built around reduction, reuse, and recycling.

Svante Arrhenius (1859–1927) Swedish physicist and physical chemist known for his theory of electrolytic dissociation and his model of the greenhouse effect; awarded the Nobel Prize for Chemistry in 1903 (*see* greenhouse gases).

Swelling clays A type of soil that expands or contracts in response to changes in environmental factors (wet and dry conditions, temperature). Hydration and dehydration can vary the thickness of a single clay particle by almost 100 percent, thus making it a hazardous foundation for structures built on top of a clay base.

Swelling soils (*also* expansive soils) Contain clay minerals (bentonites, montmorillonite) whose crystal structures permit the large-scale introduction of water, so when wetted it expands, and as it dries out, it contracts. A geohazard because it moves and thus is a danger for any structures built on top of it.

Synergistic effects (Biology) The effect when chemical substances or biological structures interact, resulting in an overall effect that is greater than the sum of individual effects of any of them.

T

Tailings piles Waste rock from a mine; may become a geohazard with weathering.

Tephra (Greek word for ash) Describes volcanic rock fragments and lava of varying sizes that are explosively blasted into the air or lifted by hot gases. Pyroclastic material of increasing size is called ash, lapilli, blocks, and bombs, respectively.

Tertiary sewage treatment After primary and secondary treatments, most of the rest of the solids and nutrients are removed, and bacterial action converts much organic matter to methane, which is often collected and used to power the plant, or burned on-site to minimize the risk of explosion.

Theory A widely accepted explanation for a phenomenon.

Thrust faults A tectonic event where one plate of the Earth's crust is compressed against another plate and the lower stratigraphic layer (older rock) is forced on top of the younger surface rock.

Tidal bore A body of water that, during exceptionally high seas, rushes up rivers faster than the tide; a type of wave that is the leading edge of a tide flowing up river.

Tidal flat (*also* mudflat) An often muddy or marshy shallow flat area covered with water during high tides.

Tonne (also, metric tonne) Equals 1,000 kg (2,200 pounds). A ton (also short ton) equals 2,000 pounds.

Topsoil loss Soils degraded by nutrient loss, erosion, and salinization (salt buildup) at rates that far exceed their regeneration or purification rates (*see* soil).

Total fertility rate Assuming no migration and static mortality rates, the average number of children born to a woman during her childbearing years in a population, calculated as the sum of age-specific fertility over a span of time.

Toluene A colorless flammable liquid, $CH_3C_6H_5$, obtained from coal tar or petroleum and used in aviation fuel and other high-octane fuels, in dyestuffs, explosives, and as a solvent for gums and lacquers; a hazardous material according to the EPA.

Toxic air contaminants (TACs) Substances from which adverse health effects may occur at extremely low levels.

Toxic Release Inventory (TRI) According to the Environmental Protection Agency (EPA), an annual report by the facility and/or manufacturer of "those chemicals or mixtures that may present an unreasonable risk of injury to human health or the environment."

Trace elements Elements found in quantities of less than 100 parts per million (ppm).

Transboundary Something that crosses a border; in pollution, a toxic substance originating in one place that migrates through air, soil, or water to another site.

Transboundary effects Occurs when a natural phenomenon, accident, or other crosses a border, province, territory, or other defined land or water mass.

Transboundary shipment The export of one area's municipal solid waste or hazardous waste to another location or nation (*see* municipal solid waste).

Transform boundary Results when plates slide laterally past one another. Also called shear boundaries.

Traps Occurs when oil or natural gas is trapped in reservoir rocks or other barrier by a seal, thereby containing the fluid (*see* reservoir rocks).

Trichloroethylene (TCE) A flammable, heavy, colorless, toxic liquid, C_2HCl_3, used to degrease metals, as an extraction solvent for oils and waxes, as a refrigerant, in dry cleaning, and as a fumigant; as a fumigant; causes cancer and liver and lung damage; classified by EPA as a hazardous material.

Trihalomethanes (THMs) A class of organic compounds, based on the methane molecule (CH_4) where the hydrogen atoms normally present are replaced by three halogen atoms (chlorine, bromine, fluorine and/or iodine). Found in many disinfectant products as well as chloroform and dibromochloromethane. THMs may form naturally in drinking water to which disinfectants have been added, which according to the EPA may, over time, lead to the development of certain cancers.

Tsunami (mistakenly termed tidal waves) A set of long, seismic sea waves created by displacement of the sea floor usually by earthquake or volcanic activity, or rarely by meteorite impacts; these powerful waves travel very fast across large distances of a lake or ocean basin, and can devastate land masses when they hit.

Turbidite Formed when unstable masses of loose sediment tumble down the submarine canyons onto the deep ocean floor (*see* submarine canyon, abyssal plain). Characterized by distinctively graded layers or beds.

Turning point on EKCs (*also* tipping point) The top of the inverted "U" shape of a Kuznets curve, where air quality markedly improves in relationship with the increase in per capita income (*see* environmental Kuznets curve).

Two-cycle engine A reciprocating internal combustion engine that requires two piston strokes or one revolution to complete a cycle.

Typhoon (*see* hurricane)

U

Unconfined aquifer A body of water in which the water is not confined by a low-permeability overlying rock bed and is recharged from the surface (*see* aquifer, confined aquifer).

Unifying principle An overarching, consolidating explanation for seemingly diverse phenomena that assembles them into a coherent whole. Plate tectonics and biological evolution are examples.

Unreinforced buildings Structures that have no metal rods or other supportive measures in their frames to help prevent building failure in an earthquake, explosion, high wind, or other event.

Unsaturated zone A surface zone where pores are empty or where water clings to the sides of pores by capillary attraction (*see* aquifer, saturated zone).

Urban growth boundaries (UGB) To ensure compact towns and cities, and decrease transportation burdens and deter pollution, a zoning tool that maintains a relatively high density of housing and commercial development inside the boundary and a rural density outside the boundary.

Urban runoff Excess rainwater that migrates into groundwater and surface waters from individual, agricultural, and industrial sources, for example: agricultural

operations (including animal feedlots), suburban lawns, leaky motor vehicles, faulty septic systems, air deposition, and urban areas.

V

Vaiont dam Located in Italy, one of the highest dams in the world and was completed in 1961 for generating electricity. After heavy rains in 1963, landslides into the Vaiont reservoir caused the stored water to spill over the dam, sweeping away the village of Longarone and flooding nearby hamlets; some 2,000 people drowned.

Valence state A measure of the number of chemical bonds an atom can form based on the number of electrons in its outer shell.

Valley fill Excess spoil that cannot be placed back into the mine environment safely during reclamation is moved to nearby valleys, altering the landscape, negatively impacting wildlife and stream habitats, and decreases property values, thereby depressing economic recovery after a coal mine closes.

Vibrio cholerae Bacterium responsible for the cholera disease; where left untreated, rapid dehydration causes death.

Viscosity/API gravity The American Petroleum Institute's (API) inverted scale for denoting the "lightness" or "heaviness" of crude oils and other liquid hydrocarbons.

Viscosity of flows The degree of liquidity of pyroclastic material during volcanic activity (*see* pyroclastic material).

Volcanic Explosivity Index (VEI) A scale used to measure the intensity of a volcanic eruption; measures plume height, volume, classifies the type of volcano, and how often it erupts.

W

Waste heat Heat generated from a power plant that cannot be stored or used immediately.

Waste incineration Process of burning waste material to ash in an effort to reduce or eliminate pollution.

Water sector Refers to the scientific, economic, and political infrastructure by which water supplies are located, distributed, and used worldwide.

Watershed A water body (i.e., river, lake, or small stream) with the total land surface that drains into that water body.

Water table Marks the top of the zone of water saturation in soil and rock (*see* aquifer, saturated zone).

Wave amplification Occurs where earthquake waves of certain frequencies pass from stronger into weaker geologic materials, causing the wave velocity to decrease and wave amplitude to increase.

Welded tuff Glass-rich mixtures of volcanic rock, fragments, and ash.

Wet ponds Urban runoff storage ponds used to control stormwater runoff.

Withdrawal use of water Removes water from the water body where it may be consumed and not returned, or ultimately returned either in a degraded form, or relatively unchanged. Irrigation water is the biggest consumptive use, and while municipal users may eventually return the water after use, it is usually degraded from its former state, rendering it less fit for reuse by others.

Y

Yucca Mountain waste repository A storage site for nuclear waste materials located in Nevada.

Z

Zone of accumulation The layer of regolith in which new minerals precipitate out of water passing through, thus leaving behind a load of fine clay. Area of a glacier in which snowfall adds to the glacier.

Zoning Local regulations that specify how a piece of property can be used within a city or county zoning district, which are created to attract and contain specific types of development (e.g., low- or high-end residential, industrial, office/business, recreation, etc.; *see* urban growth boundaries).

INDEX

uses, 707–708
valence state of, 707
Artesian, 215
Aseismic slip, 149
Ash, 110, 110*t*, 464, 752
incinerator, 686
Assumptions, 10
corporations, 12
government's role, protecting
environment, 10–11
proper level of government, 11
Asthenosphere, 78
Asthma, 633
Atmosphere, 595
gases and concentration, 595*t*
residence time, 595
Atrazine, 400, 401*f*
Attenuation, 164

B

Banded iron formations (BIFs), 417, 417*f*
Bangladesh
coastal population, 355–356
floods in, 59
location of, 60*f*
river flooding, 356
Bar-built estuaries, 330, 332
Barrel, 489
Barrier complexes, 334–336
Barrier islands, 335
Basaltic, 743
Basaltic crust, 77
Basaltic rocks, Bowen's research on, 758*b*
Basalts, 744
metamorphism of, 764*b*
Base flow, 220
Base metals, 414
Basel Convention, 1988, 686, 692, 734, 735*b*
Base-load plants, 528
Batholith, 754
Battisti, David, 633
Bauxite, 23–24, 23*f*, 383, 383*f*, 418–419, 419*f*
BCMA. *See* British Columbia (Canada)
Medical Association
Beach renourishment, 361, 361*f*
Beach sand, 345, 348
Bedding planes, 303
Bentonites, 317
Best management practices (BMPs), 309,
431, 578, 658
Better Site Design, 658, 661–662, 661*f*
Billion, 460
Binary cycle plants, 526

Bioaccumulates, 715
Bioaccumulation, 572–573
Biochemical oxygen demand (BOD), 684*f*
Biodiesel, 550–551, 551*t*
Biodiversity, 29–30
Biofuels, 549
Biomagnification, 564
Biomass, 540
energy, 549
advantages, 551–552
drawbacks, 552
ethanol and biodiesel, 550–551
ethics, 552–553
impact of, 551*t*
types, 550*t*
Bioremediation, intrinsic, 235
Bioslurping, 235
Biosolids, 431, 678
Biota, 657
Birth rate, 53, 54*t*
Biscayne aquifer, 231–232
Bituminous coal, 464
Black lung disease, 462
Blind thrusts, 159, 161
Blue baby syndrome, 198
Bottle bills, 688
Boulder County Comprehensive Plan
(BCCP), 664
Bowen's Reaction Series, 758*b*
Brackish water, 201–202
Bradford pear trees, 658*f*
Braided rivers, 258*f*, 259
Breadth, 9
Breakwaters, 360, 361*f*
British Columbia (Canada) Medical
Association (BCMA), 424
British Thermal Units (Btu), 447*b*, 448*f*
Brownfields, 663
Bunker Hill, Idaho, 431, 432*f*

C

CAFOs. *See* concentrated animal feeding
operations
Calderas, 114–116, 116*f*–117*f*
California Coast Ranges, 433, 434*f*
California Department of Conservation
(CDC), 180
California, seismic risk map of, 667*b*
California State Air Resources Board
(CALARB), 704
Callendar, Stewart, 624
Capital-intensive water sector, 192
Carbon cycle, 474, 627–629, 628*f*, 750
greenhouse gas in, 629, 630*f*

Carbon dioxide, 609, 609*f*
levels, 627–629, 628*f*
reservoirs, 629*f*
Carbon sequestration, 384
methods, 641–647
Carbonates, 744, 750
Carbonic acid, 623
Carnotite, 422, 423*f*
Carrying capacity, 46
and ecological footprints, 64–65
human population, 46, 58*b*, 63
Carson River, Nevada Watershed, 432–433,
433*f*
Catalytic converters, 603–604, 604*f*
Cation exchange capacities, 402
Cations, 746
Cause-effect *vs.* correlation, 5
Cedars of Lebanon, 641*f*
Central America, volcanic region, 132
CERCLA. *See* Comprehensive
Environmental Response,
Compensation and Liability Act
Channelization, 284, 286
China, mass wasting in, 295
Chlorination, 566
Chlorite as metamorphic rocks, 763
Chlorofluorocarbons (CFCs), 593, 594*f*,
627, 736*b*
Cholera, 567
Chronic effects, 708
Cinder, 113
Cinder cones, 113, 115*f*
Clarity, 8–9
Classicism, 136
Clay minerals, 317, 428, 756
Clays, 749
in sedimentary rocks, 760
structure, 749*f*
Clean Air Act, 1990, 734
"Clean coal" technology, 474–475
fluidized bed combustion, 476*f*
Clean Water Act (CWA), 717
Cleaning, air
internal combustion engines, 605
motor vehicles, 603–604
power plant emissions, 605–609
tires, as air pollution, 605
Climate change, 29, 616
and coral reefs, 337*b*–338*b*
floods, 278
historical research on, 623–625
impacts of, 631–639, 632*f*
mitigating, 640–647
and rise in sea level, 631–632
scientific evidence, 630–631
and sea level changes, 349–354
and water supplies, 202

Erosion, 423–424
  control, 431
Eruption of Lake Nyos, 103–105
Escarpment, 303
Estuaries, 34, 330, 330*f*, 331*f*
  bar-built, 330, 330*f*, 331*f*, 332
  circulation and stratification, 332–334, 332*f*
  coastal plain, 330, 331*f*
  delta, 330*f*, 331*f*, 332
  fjords, 330*f*, 331*f*, 332
  tectonic, 330*f*, 331*f*, 332
Ethanol, 549
  and biodiesel, 550–551
Euclid v. Ambler Realty Co., 665
European union, agricultural pollution in, 402–403, 403*t*
Eustatic processes, 348
Eutrophication, 659
Evidence, 2, 10
Expansive clays, 317–319
Exponential growth, 19, 55–57
Ex-situ remediation *vs.* in-situ remediation, 234–236
External factors, mass waiting, 297
Externalities, 12, 25, 461, 519

F

Fault, 96, 149
FBR SO$_2$ Abatement Project, 718
Federal Emergency Management Agency (FEMA), 283, 344
Federal Insecticide, Fungicide and Rodenticide Act (FIFRA), 242–243
Federal Water Pollution Control Act (FWPCA), 243, 582
Feldspars, 746–747
Felsic lava, 112
Fertile crescent region, 390, 390*f*
Fertility rate, 53
Fine fraction (PM2.5), 703–704
  hazards from, 705
Fine particulates, 703
  coarse fraction, 704
  fine fraction (PM2.5), 703–704
  PM10, 703–704
Fires, 169
Fish kills, source, 400, 400*f*
Fisheries, sustainable society, 30
Fjords, 330*f*, 331*f*, 332
Flaring methane, 684
Flaring of landfill gas, 684
Flash floods, 253
Flash steam plants, 525

Flood insurance, 282–283
Flood mitigation
  land-use, 286
  zoning regulations, 286
Flood stage, 263
Flood-control structures, 283–286
Flood-frequency diagram, 267–268, 269*f*
Floods
  categories, 257–258
  in earthquake, 169, 171
  factors for damage, 269–275, 278
  and hydrologic cycle, 254–257
  and land use, 276–278
  Passaic River Basin, 279–280
  storm surges
    definition, 281
    protection against, 282
Florida
  coastal population, 356
  groundwater in, 228–232
Floridan aquifer, 228, 230*f*
Fluorosis, 460, 461*f*
Focus, earthquake, 153, 153*f*
Foliated metamorphic rocks, 762*b*
Food
  crops, 550*t*
  waste, 550*t*
Fool's gold, 421
Forecasting, earthquake, 175
Forest impacts, due to greenhouse effect, 635–636, 635*f*–636*f*
Forests, sustainable society, 30
Fossil fuels
  advantages, 443
  environmental impact, 444
  liquefied natural gas (LNG), 444
  oil supplies, 444
  potential for, 555*t*
  subsidies and externalities in, 519
  transportation, 443
  U.S. primary energy consumption, 443, 444*f*
  world coal reserves, 443
  world energy consumption, 443*f*
Fracking, 184
Fracturing, 235
Friction, 215
Fuel rod assemblies, 528
Fuel switching, 607
Fuels
  biofuels, 549
  fossil. *See* fossil fuels
Fukushima Daiichi Nuclear Power Plant, 530–531, 534, 534*f*
Furans, 686
FWPCA. *See* Federal Water Pollution Control Act

G

Gamma radiation, 423
Gas, energy content of, 502–503
GDP. *See* gross domestic product
Gemstone, 425, 426*t*
Genetically modified (GM) foods, 391
Geohazards, 32, 50, 363, 384–386
  global impact, 36
  impacts on humans, 33*f*
  in Nepal, 34–35, 34*f*–35*f*
  risk assessment, 35
    financial loss, earthquakes, 37, 40
    flood losses, 40, 40*t*
Geological model concept, 71–72
Geological resource, coal
  classification, 467, 467*t*
  coal bed methane, 468–471
  coal formation, geology, 464–466
  coal reserves, 467–468
  composition, 464
  consumption and price, US, 462–463, 463*f*
  global reliance, 463
  grades and characteristics, 465*f*
Geological time scale, 740
Geothermal energy
  disadvantages, 526–527
  geothermal plants, history, 524
  geothermal power in United States, 524–525
  heat pumps, 527
  types, 525–526
Geothermal heat pumps, 527, 527*f*
Geothermal plants, history, 524
Geothermal power in United States, 524–525
Geysers geothermal field in Northern California, 525*f*
Global, 387, 388*f*
Global air pollutants, 593, 593*t*
Global air pollution, 590
  death, 591*f*
Global energy demand
  Btu, 447*b*
  burning fossil fuels, 445
  CO$_2$ emissions, 445, 446*f*
  global population growth, 447*f*
  growth in energy consumption in China, 445, 445*f*
  kilowatt, 447*b*
  quads, 447*b*–448*b*
Global hectare, 65
Global population growth, 447*f*
Global reliance, coal, 463
Global warming, 616

People's Republic of China, 197–198
in rocks, 151
slope stability control, 299, 301
twenty-first century, 202
withdrawal use of, 194
worth, 192
Water pollutants, categories
disease-causing agents, 563
heat, 565
organic chemicals, 564–565
oxygen-demanding wastes, 563–564
plant nutrients, 564
radioactive substances, 565
sediments, 565
water-soluble inorganic chemicals, 564
Water pollution
and chlorination, 566
legislation, 582–583
NPDES, 583
trihalomethanes, 566
Water quality, 583–584, 584f
KBP, 395–396
Water sector, 192–193
Water supply, 192–193
and climate change, 202
in Gaza, 199–200
Water table, 213
Water use
consumptive, 194
factors, 193–194
instream, 194
in United States, 195f, 196
Water velocity, 261

Water waves, 165–169
Watersheds, 260
Waterways, 191
Wave amplification, 165
Wave energy generating system, 537f–538f
Wave frequency, 341
Wave height, 340
Wave period, 340
Wave velocity, 341
Wave-dominated barrier islands, 335
Wavelength, 340
Waves, 339
aspects, 341, 343
attributes, 340–341
longshore currents, 343
seasonal variation, 344
Weathering, 302, 758b
Weathering products, 749
of common silicates, 759t
Weather-related mortalities, 632
Wegener, Alfred, 81
Weirs, 284, 285f
Welded tuffs, 115
West Coast vernal pools, 660
West Indies, volcanic region, 134
West Virginia Department of
Environmental Protection (WVDEP), 478
West Virginia economy, coal, 477–478
Western Correlative Rights, 239
Wetlands, 286, 287f, 659–661, 659f
intertidal, 330

Wetponds, 658
Wilson Cycle, 85f
Wind energy, 29
Wind energy systems, 546
advantages, 548
future, 548
potential sites, 547–548
subsidies for development, 546–547
Windbreaks, 388
World coal reserves, 443
World Commission on Environment and
Development, 26
World energy consumption, 443f
World Resources Institute, 197
World Trade Organization (WTO),
734–735
Wyoming, coal pit, 457f

Y

Yellowcake, 422
Yucca Mountain Nuclear Waste
Repository, 532–533

Z

Zone of accumulation, 381f
Zoned olivines, 757f
Zoning, 663, 664
map of Falmouth, 666f
zoning districts, 664